THE NEXT GENERATION OF EMERGING

CHEMISTRY

THE NEXT GENERATION OF EMERGING

CHEMISTRY

THE NEXT GENERATION OF EMERGING CHEMISTRY

HARMINDER GILL

WORKBOOK PRESS LLC
187 E Warm Springs Rd,
Suite B285 Las Vegas NV 89119 USA

Website: https://workbookpress.com/
Hotline: 1-888-818-4856
Email: admin@workbookpress.com

Ordering Information:
Quantity sales. Special discounts are available on quantity purchases by corporations, associations, and others. For details, contact the publisher at the address above.

Library of Congress Control Number:

ISBN-13: 978-1-963718-41-6 Paperback Version
978-1-961845-13-8 Hardback Version

REV. DATE: 03/11/2026

Table of Contents

ACKNOWLEDGEMENTS

I am deeply grateful to WorkBook Press for their unwavering dedication, professionalism, and the considerable amount of time they invested in bringing this book to publication. From the earliest stages of preparation to the final refinements, your commitment to quality and your willingness to support this project at every step have been truly invaluable. Your efforts not only shaped the finished work but also provided the guidance and encouragement necessary to see it through to completion. This book would not have been possible without your steadfast involvement, and for that, I extend my sincere appreciation.

Preface for the General Reader

Chemistry is the story of transformation—the invisible thread connecting the universe, our planet, life, and technology. It is a discipline that reveals the hidden patterns of matter, the forces that govern change, and the principles that allow us to shape the world. *The Next Generation of Emerging Chemistry* invites you on a journey through both the foundations and frontiers of chemical science, blending classical knowledge with the innovations that define the cutting edge.

Our journey begins with the roots of chemistry. From the alchemists' quest to turn base metals into gold, to the precise measurements that defined modern science, you will see how curiosity and careful observation led to the first systematic understanding of matter. The philosophical origins of atomic ideas illuminate how thinkers imagined the building blocks of the universe, while laboratory techniques and quantitative methods provide the tools to test these ideas in the real world.

We then explore the blueprint of chemical change, revealing the behavior of gases, the flow of energy in thermochemistry, and the evolution of atomic theory. You will learn how atoms connect to form bonds, how molecular orbitals govern shape and reactivity, and how subtle intermolecular forces influence the properties of liquids, solutions, and reactions. From the adventure of chemical kinetics to the balancing acts of equilibrium and aqueous dynamics, chemistry reveals itself as a precise yet endlessly fascinating language of matter.

The journey continues into the applied world. Acids, bases, and buffers harmonize life's delicate balances, while thermodynamics and electrochemistry show how energy drives transformation. The architecture of complex coordination compounds unlocks new possibilities in materials and medicine, while organic chemistry uncovers the molecules that make up life itself. Chemistry touches every facet of our existence—from the composition of the Earth and the periodic beauty of s-, p-, d-, and f-block elements, to innovations in smart healthcare, dentistry, forensics, and firefighting that demonstrate its profound impact on human well being.

Yet chemistry is not limited to practicality—it is also creativity. Explore the chemical artistry behind water purification, paintings, and cosmetics, and discover the transformative science of molecular gastronomy. Witness molecular computing and chemical logic gates, where molecules become tools of computation and design. Step into extreme environments and quantum biology, where chemistry illuminates life's most mysterious processes, and reach out into space with astrochemistry, uncovering complex molecules drifting among the stars.

The next generation of chemistry pushes even further. Self-healing materials, molecular origami, synthetic ecosystems, and chemically activated energy storage reveal a world where molecules are architects and engineers. Bioluminescence and biofluorescence show nature's molecular magic, while the chemistry of consciousness, neurochemistry, and epigenetics explore the very essence of thought, behavior, and life's adaptability. From chemical metamaterials to non-aqueous life possibilities, this book emphasizes chemistry's role in shaping both the imaginable and the extraordinary.

This text is not a static reference—it is an invitation to think, imagine, and create. Each chapter connects fundamental principles with real-world applications, showing how chemistry shapes medicine, sustainability, technology, and art. You will see how matter moves, interacts, and transforms, how molecular patterns reveal themselves, and how innovation emerges from understanding these principles. From the smallest molecule to the vastness of the cosmos, chemistry provides both explanation and inspiration.

Whether you are a student encountering these ideas for the first time, a researcher seeking new perspectives, or a lifelong learner captivated by the wonders of the molecular world, this book opens the door to a living, evolving discipline. Step inside to witness the elegance of chemical transformations, the excitement of discovery, and the endless possibilities that chemistry offers. This is the next generation of chemistry—dynamic, transformative, and deeply intertwined with the story of life, matter, and the universe.

Chapter 1

From Alchemy to Modern Chemistry

The study of chemistry, at its core, is the study of matter and the changes it undergoes. However, this field did not emerge fully formed; it has deep historical roots stretching back thousands of years to the practice of alchemy. Alchemy, often shrouded in mysticism and philosophy, served as the precursor to modern chemistry. While alchemists are commonly remembered for their attempts to turn base metals into gold, their legacy encompasses much more: systematic experimentation, laboratory techniques, and the conceptual frameworks that gradually evolved into the science of chemistry as we know it today.

Understanding the historical development of chemistry is not merely a lesson in antiquity. It provides context for why chemical theory developed the way it did, why certain experimental methods became standard, and how the philosophical underpinnings of alchemy influenced the rigorous scientific approach that followed. This chapter traces the journey from the mystical pursuits of ancient alchemists to the establishment of chemistry as a modern, quantitative science.

Alchemy originated in multiple regions, including ancient Egypt, Greece, India, and China, with each culture contributing unique perspectives and methodologies. In Egypt, the earliest records suggest a combination of metallurgy, medicine, and religious practices, where chemical processes were intertwined with spiritual beliefs. The Greco-Egyptian alchemical tradition later codified ideas about the transmutation of metals and the search for the "philosopher's stone", a substance believed to grant eternal life and convert base metals into gold.

Chinese alchemy, emerging around the 2nd century BCE, focused largely on immortality elixirs and medicinal compounds, combining practical experimentation with philosophical concepts like Taoist balance. In India, alchemical practices were closely linked to Ayurveda, emphasizing the transformation of natural substances for healing and longevity.

Despite their mystical goals, alchemists laid the groundwork for practical chemistry. They developed basic laboratory apparatus, such as furnaces, crucibles, distillation devices, and glassware, which would become essential tools for later chemists. Importantly, alchemists recorded their observations systematically, even if couched in allegory or symbolism, introducing a nascent form of scientific methodology.

The period from the 8th to the 13th century marked a major turning point for alchemy, particularly in the Islamic world, where scholars such as Jabir ibn Hayyan (Geber) codified experimental methods. Jabir emphasized empirical observation and reproducible procedures, separating mystical speculation from laboratory practice. He described processes like distillation, sublimation, calcination, and crystallization, many of which remain fundamental in chemistry labs today.

Islamic alchemists also introduced the idea of chemical substances as entities with properties that could be manipulated, paving the way for the modern concept of elements and compounds. Their works were translated into Latin during the Middle Ages, profoundly influencing European alchemists. By the late Middle Ages, alchemy had spread throughout Europe, blending with Christian mysticism and the intellectual ferment of the Renaissance. Figures such as Paracelsus emphasized practical chemistry for medicine, challenging purely mystical approaches and promoting the use of chemical remedies. Paracelsus's work foreshadowed the modern field of pharmacology and emphasized that chemical substances had real, measurable effects on the body.

Despite this, much of European alchemy remained secretive, symbolic, and interwoven with philosophical speculation. Alchemists still pursued transmutation, but their meticulous record-keeping and experimental methods laid the foundation for a more systematic and empirical approach to studying matter.

The transition from alchemy to chemistry began in earnest in the 17th and 18th centuries, driven by the Scientific Revolution. Alchemy's mystical framework gradually gave way to empirical observation, controlled experimentation, and quantitative measurement. Key figures in this transformation included:

- Robert Boyle (1627–1691), often called the father of modern chemistry, who emphasized careful experimentation and rejected speculative alchemical theories. His work, particularly the formulation of Boyle's Law, highlighted the importance of reproducible results.
- Antoine Lavoisier (1743–1794), who introduced the Law of Conservation of Mass, systematically defined chemical elements, and helped overthrow the phlogiston theory of combustion. Lavoisier's precise quantitative methods set the standard for modern chemical experimentation.

During this period, alchemy's symbolic language was replaced by chemical nomenclature, measurement, and theory. The focus shifted from mystical transmutation to understanding the composition, properties, and reactions of substances. Laboratory techniques developed by alchemists—distillation, filtration, and crystallization—were refined and rigorously applied in the emerging field of chemistry.

While alchemy's goals may have been misguided by modern standards, its contributions to chemistry are undeniable:

1. Laboratory Techniques and Apparatus: The use of furnaces, retorts, crucibles, and early glassware allowed precise manipulation of substances.
2. Systematic Observation and Documentation: Alchemists were meticulous in recording procedures and results, fostering a culture of reproducibility.
3. Conceptual Foundations: Ideas of transformation, combination, and purification of matter anticipated later theories of chemical reactions.
4. Medicinal Chemistry: The preparation of mineral and herbal compounds laid the foundation for pharmaceuticals and chemical therapeutics.

By the 18th century, chemistry had evolved into a quantitative, experimental science, shedding its mystical and esoteric trappings. The legacy of alchemy persisted not in gold-making but in the rigor and creativity of chemical investigation. The journey from alchemy to modern chemistry is a story of curiosity, experimentation, and intellectual transformation. Alchemy's mystical beginnings may seem distant from today's rigorous, laboratory-based science, but the underlying drive—to understand and manipulate the material world—remains the same. Through centuries of trial, error, and careful observation, alchemists laid the groundwork for the principles, techniques, and methodologies that define modern chemistry.

This historical perspective reminds us that science is a cumulative endeavor. The laboratory practices, experimental rigor, and conceptual insights of alchemists continue to resonate in chemistry classrooms and research labs worldwide. Modern chemistry, with its quantitative laws and predictive power, owes much to the imaginative and diligent efforts of those early seekers of knowledge which leads us now to the chemistry of our modern world.

Chemistry is the study of matter and the physical and chemical changes it undergoes. A key approach in chemistry is connecting the macroscopic world we observe with the microscopic world of atoms, molecules, and ions. The processes at the microscopic level have long fascinated scientists—not just out of curiosity, but because understanding these mechanisms reveals the behavior of matter itself. Everyday phenomena, such as a lamp emitting light, a rainbow appearing after rain, an iceberg floating on water, or a machine in operation, all involve chemical principles.

Chemistry intersects with numerous scientific disciplines, including biology, physics, medicine, and material science. Chemical research drives innovation across industries. For instance, silicone polymer chemistry has enabled advanced drug-delivery systems. Pure chemicals and automated dispensing systems are critical in semiconductor manufacturing, which powers the electronics industry. Alloys are essential for automobiles and aircraft, while ceramic chemistry involves both aqueous and non-aqueous syntheses to produce powders used in electronic components. In material science, even a single-atom laser demonstrates both classical and quantum properties. Cosmetic chemistry combines emulsions, surfactants, moisturizers, active ingredients, sunscreens, and preservatives to create everyday products. Pharmacy bridges chemical sciences with healthcare, ensuring the safe and effective use of drugs, while radiological chemistry develops imaging agents for diagnostics and therapy.

Chemistry enriches our lives much like art enriches culture. Just as an artist combines colors to create paintings, a chemist combines chemicals to produce novel substances. Everything around us is made of chemicals, and the transformations they undergo are called chemical reactions. These reactions form new compounds with unique physical and chemical properties. Beyond simple combinations of elements, large organic molecules can be synthesized for practical use, as seen in cosmetics and other daily products. The development of such compounds has profoundly shaped human life.

Both inorganic and organic compounds are ubiquitous on Earth. The chemistry of our planet—from the atmosphere to the oceans—is constantly changing. Concerns such as global warming, rising sea levels, and shifting climates pose significant challenges. Solutions may require bold approaches, from floating cities to subterranean or mountainous settlements, and perhaps even space-based habitats. By exploring the intricate details of chemistry, we can better understand the natural world and address the scientific problems that impact civilization.

At its core, chemistry is the science of matter and the transformations it undergoes. Matter exists in various states—solid, liquid, gas, and plasma—and each state has distinct properties and behaviors. Understanding these states, and the transitions between them, requires exploring atomic and molecular interactions.

Atoms are the fundamental units of matter, composed of protons, neutrons, and electrons. The arrangement and behavior of these subatomic particles determine an element's chemical properties. When atoms bond, they form molecules, which can range from simple diatomic species like oxygen (O_2) to complex macromolecules such as proteins and polymers. The nature of these bonds—ionic, covalent, metallic, or hydrogen—dictates the physical and chemical properties of substances, from melting points to solubility and reactivity.

Chemical reactions are processes in which reactants transform into products through the breaking and formation of bonds. Energy changes accompany these transformations. Exothermic reactions release energy, often in the form of heat or light, while endothermic reactions absorb energy from the surroundings. Thermodynamics provides the framework to predict whether a reaction will occur spontaneously, considering factors such as enthalpy, entropy, and Gibbs free energy.

Kinetics, another critical aspect of chemistry, examines the rate at which reactions occur and the mechanisms underlying them. Catalysts, substances that speed up reactions without being consumed, are fundamental in industrial chemistry, environmental applications, and biological systems. Enzymes, the natural catalysts in living organisms, demonstrate the remarkable efficiency with which chemistry orchestrates life processes.

Biochemistry bridges chemistry and biology, revealing how molecules sustain life. Proteins, nucleic acids, carbohydrates, and lipids are the macromolecules essential for structure, energy storage, and information transfer. For instance, DNA, composed of nucleotide sequences, stores genetic information, while enzymes catalyze the biochemical reactions necessary for cellular function.

Metabolism, the network of chemical reactions in living organisms, transforms nutrients into energy and building blocks for growth. Understanding these processes is crucial in medicine, pharmacology, and nutrition. Drug design relies on chemical principles to target specific enzymes or receptors, minimizing side effects while maximizing therapeutic benefit.

Environmental chemistry examines the impact of human activity and natural processes on the Earth. Pollutants, such as heavy metals, greenhouse gases, and synthetic organic compounds, interact with air, water, and soil, influencing ecosystems and human health. Analytical chemistry provides the tools to detect, quantify, and monitor these substances, enabling informed decisions and regulatory policies.

For example, understanding the chemistry of carbon dioxide and methane is essential in modeling climate change. Similarly, water treatment involves chemical principles, from the coagulation of suspended solids to the disinfection of pathogens. Green chemistry focuses on minimizing the environmental footprint of chemical processes, emphasizing sustainability, efficiency, and the use of renewable resources.

Materials chemistry lies at the interface of chemistry, physics, and engineering, shaping modern technology. Semiconductors, superconductors, polymers, ceramics, and nanomaterials each rely on precise chemical design. The discovery of graphene, a single layer of carbon atoms arranged in a hexagonal lattice, exemplifies how chemical manipulation at the atomic scale can yield extraordinary mechanical, electrical, and thermal properties.

In medicine, materials chemistry enables targeted drug delivery, biocompatible implants, and diagnostic imaging agents. In electronics, precise chemical control allows the production of microchips and displays. Energy storage, including batteries and fuel cells, depends on materials that can efficiently store and release chemical energy.

Organic chemistry studies carbon-containing compounds, which form the backbone of fuels, pharmaceuticals, polymers, and natural products. The versatility of carbon arises from its ability to form stable chains, rings, and multiple bonds, leading to virtually limitless molecular diversity. Synthetic organic chemistry allows chemists to design molecules with specific properties, such as medicines that inhibit disease pathways or polymers with exceptional strength and flexibility.

Inorganic chemistry explores compounds not centered on carbon, including metals, salts, and minerals. Transition metals, in particular, exhibit unique electronic configurations that enable catalytic activity, magnetic properties, and electronic conductivity. Coordination chemistry, studying how metals bind to ligands, underpins fields from bioinorganic chemistry to industrial catalysis.

Analytical chemistry provides the tools to identify and quantify substances. Techniques such as spectroscopy, chromatography, and mass spectrometry allow scientists to examine matter at increasingly precise levels. For example, nuclear magnetic resonance (NMR) spectroscopy reveals the structure of complex organic molecules, while inductively coupled plasma mass spectrometry (ICP-MS) detects trace metals at parts-per-billion concentrations.

These tools are essential not only in research but also in quality control, environmental monitoring, and forensic investigations. Analytical chemistry transforms observations into actionable knowledge, bridging the microscopic world of atoms and molecules with real-world applications.

The influence of chemistry extends beyond the laboratory into societal development. Agricultural chemistry improves crop yield and resistance to disease, while food chemistry enhances safety, flavor, and nutritional content. Industrial chemistry fuels economic growth, from the production of fertilizers and plastics to energy generation and waste management.

Ethical considerations in chemistry are increasingly important. The development of sustainable processes, responsible disposal of chemicals, and minimizing the risks of synthetic compounds are central to modern chemical practice. Chemists must balance innovation with environmental stewardship and public safety, ensuring that chemical progress benefits society without causing harm.

As science advances, chemistry continues to evolve. Emerging fields such as computational chemistry, artificial intelligence in molecular design, and quantum chemistry are pushing the boundaries of what we can predict and create. Space exploration introduces new challenges, such as synthesizing materials under microgravity conditions or recycling waste products for life support. Chemistry is not just a collection of facts; it is a lens through which we understand the universe. By connecting the behavior of atoms and molecules to observable phenomena, chemists unlock insights that drive technology, medicine, and environmental solutions. The ongoing study of chemistry empowers humanity to shape a sustainable, innovative, and scientifically informed future.

Chapter 1: Word Problems

1. An alchemist from the 16th century claims that he can convert 10 kg of lead into gold using a secret combination of herbs, heating, and mystical procedures. Considering modern atomic theory, nuclear physics, and isotopic stability, critically analyze why this transformation is impossible through chemical reactions alone. Calculate the approximate energy required to transmute lead-208 into gold-197 via nuclear reaction, and discuss the practical challenges and safety concerns if such a nuclear process were attempted.

2. Paracelsus believed that chemical remedies could cure diseases and often recommended mineral-based elixirs for internal use. Imagine you are evaluating a modern pharmacy preparation derived from a similar mineral source. Discuss, using pharmacokinetics, solubility, toxicity, and receptor binding, how you would critically assess the effectiveness and safety of his remedies. Compare the rationale behind historical alchemical therapeutics to modern drug design principles.

3. An alchemist records that mixing two metals produces a vivid color change, but provides no quantitative details, measurements, or controls. Design a modern experiment to replicate his observation. Include selection of metals, ratio control, temperature regulation, and analytical methods such as spectroscopy or colorimetry to quantify changes. Discuss statistical methods to confirm reproducibility and distinguish between surface oxidation and true alloy formation.

4. A Chinese alchemist in the quest for immortality consumes herbal and mineral elixirs containing cinnabar, arsenic, and mercury. Using modern toxicology knowledge, identify the chemical hazards associated with such practices, including chronic exposure effects, bioaccumulation, and acute toxicity. Propose a modern experimental approach to measure heavy metal content and evaluate safe dosage limits, comparing it to historical anecdotal practices.

5. A European alchemist heating a mixture observes the evolution of a "spirit" that rises and condenses in a separate vessel. Analyze the chemical and physical processes he might have witnessed, such as volatilization, sublimation, or distillation of volatile components. Describe these phenomena using modern thermodynamics, including phase diagrams, vapor pressure, and latent heat concepts, and suggest how this observation could be quantified today.

6. Compare the experimental goals and methodologies of Jabir ibn Hayyan, who sought transformation and purification of substances, with those of Lavoisier, who emphasized mass balance and reaction stoichiometry. Identify which principles, such as systematic observation, controlled experiments, and quantitative measurement, are foundational in modern chemistry and which aspects remain purely philosophical or mystical.

7. A medieval alchemist performs calcination on a metal ore and observes a change in color and texture, keeping no quantitative record. Design a modern experimental

workflow to analyze the calcination residue using X-ray diffraction (XRD), Fourier-transform infrared spectroscopy (FTIR), and inductively coupled plasma mass spectrometry (ICP-MS). Discuss how each technique would contribute to understanding the chemical composition and oxidation states of the products.

8. Suppose you are tasked with evaluating the efficacy of a historical "elixir" said to dissolve base metals. Design a modern laboratory experiment, specifying which metals to test, solvent choice, pH adjustments, and safety measures to avoid toxic exposure. Include quantitative endpoints such as mass loss, spectroscopic measurements, or titration to determine dissolution efficiency.

9. Paracelsus promoted the use of chemicals as medicinal remedies. Identify a modern therapeutic compound (e.g., arsenic trioxide for leukemia) with historical roots in alchemical practice. Trace its conceptual origin from mystical alchemical preparation to evidence-based pharmacology, and discuss how modern dosing, formulation, and safety guidelines differ from historical recommendations.

10. A European alchemist observes a color change when heating a metal compound, attributing it to mystical forces. Using modern chemistry, explain how such a color change can be accounted for by oxidation state transitions, coordination chemistry, or thermal decomposition. Propose a set of experiments, including spectrophotometry, to verify your hypothesis.

11. An alchemist's manuscript contains symbolic instructions for distillation, referencing "spirits" and "sublime essences." Translate these symbols into a modern experimental procedure, specifying apparatus, heating methods, collection techniques, and expected chemical compounds. Include potential yield calculations and analytical methods to verify purity.

12. Considering that most alchemical experiments were qualitative, discuss how you would assess the reproducibility of a historical procedure. Propose methods to convert qualitative observations into quantitative measurements using modern instrumentation such as thermocouples, balances, and spectrophotometers.

13. A student hypothesizes that the legendary philosopher's stone could act as a catalyst to transmute base metals into gold. Critically analyze this claim using modern chemical principles, including reaction kinetics, activation energy, and the

impossibility of changing atomic number through chemical catalysis. Suggest an experiment to test this safely in principle.

14. An alchemist reports producing a substance that emits light when heated. Explore possible explanations using modern chemical phenomena such as incandescence, chemiluminescence, phosphorescence, or fluorescence. Design an experiment to measure emission spectra and determine the chemical identity of the luminescent material.

15. Compare the use of symbolic notation in alchemical texts with modern chemical notation. Discuss how standardization of symbols, formulae, and units improves reproducibility, data interpretation, and global communication among chemists. Illustrate with examples of ambiguous alchemical symbols versus modern molecular representation.

16. A medieval alchemist heats a mixture in a retort and collects a distillate in a separate vessel. Predict the likely chemical components based on the starting materials and describe the separation principles involved, such as volatility differences, azeotropes, and condensation behavior. Propose analytical methods to confirm the identity of the distillate.

17. Design a controlled modern experiment using sublimation to isolate iodine from a mixture containing potassium iodide and impurities. Discuss how this reflects the original alchemist's approach and highlight the improvements modern analytical techniques provide, such as purity assessment and yield quantification.

18. An alchemist repeatedly observes gas evolution from a heated mineral but cannot identify it. Design a modern laboratory experiment to collect, measure, and identify the gas using gas chromatography, mass spectrometry, or infrared spectroscopy. Include safety precautions and quantitative analysis.

19. A 17th-century European alchemist attempts to crystallize a substance but keeps no quantitative records. Using stoichiometry, calculate the theoretical maximum yield of crystals from 50 g of a known starting compound, assuming complete reaction. Discuss the factors that could cause deviations from this theoretical yield.

20. Compare the distillation techniques used by Islamic alchemists to modern fractional distillation. Identify at least two major improvements in apparatus,

temperature control, and purity outcomes, and explain the chemical rationale for each improvement.

21. An alchemist performs a calcination experiment on a metal ore and collects the residue. Describe how modern chemistry would determine whether the residue contains metal oxides, sulfates, or elemental metals using XRD, thermogravimetric analysis, and spectroscopic methods.

22. Design a modern laboratory procedure to quantify the purity of a metal after alchemical purification attempts. Include analytical methods such as atomic absorption spectroscopy, ICP-MS, and gravimetric analysis, and discuss sources of error and reproducibility concerns.

23. A student recreates a medieval elixir by combining mercury and sulfur. Predict the products, such as mercury sulfide, discuss their chemical properties, and assess potential hazards using occupational safety protocols. Include guidance for safe disposal.

24. Using modern analytical tools, describe how you would verify an alchemist's claim that a specific elixir could dissolve gold. Include solubility testing, spectroscopic verification, and quantitative mass balance calculations to confirm or refute the claim.

25. An alchemist observes that heating a substance produces water condensation in a cold receiver. Explain this observation using vapor pressure, phase equilibrium, and latent heat principles. Design an experiment to measure condensation rates under controlled conditions.

26. Design a modern experiment to measure mass change during a medieval calcination procedure. Compare your results to Lavoisier's Law of Conservation of Mass and discuss sources of discrepancy in historical observations.

27. A retort breaks while distilling an acidic solution in a medieval lab. Discuss modern laboratory safety measures, including the use of fume hoods, PPE, chemical-resistant glassware, and emergency protocols, to prevent similar accidents.

28. Translate a symbolic alchemical diagram describing sublimation into a step-by-step modern experimental procedure, including heating, collection, and product analysis. Include expected chemical identities and potential side reactions.

29. An alchemist claims that combining two powders produces a "universal solvent." Using solubility rules, thermodynamics, and modern chemical reasoning, explain why this claim violates established principles of chemistry and is physically impossible.

30. How would you modernize the alchemist's process of purifying salts via recrystallization? Discuss solvent choice, temperature gradients, nucleation control, filtration, and quantification to improve yield, purity, and reproducibility.

31. Analyze how the philosophical goals of alchemy, such as the pursuit of perfection and transformation, influenced the development of experimental design, hypothesis formation, and analytical methodology in modern chemistry.

32. Compare the symbolic approach of European alchemy with the systematic approach of Islamic alchemy, emphasizing methodology, documentation, and experimental rigor. Evaluate which aligns more closely with modern chemistry practices.

33. A medieval alchemist asserts that metals contain a hidden "essence" that determines their properties. Relate this idea to atomic theory, electron configuration, and modern understanding of elemental identity, bonding, and reactivity.

34. Explain how the discovery of the Law of Conservation of Mass disproved mystical alchemical concepts such as elemental transmutation through heat alone. Illustrate with a reaction example showing measured mass balance.

35. Identify a modern chemical element, such as antimony or arsenic, and trace its conceptual history from alchemical theories, mystical properties, and symbolic representation to its modern understanding, including physical properties, common compounds, and applications.

36. Analyze why quantitative measurement became essential in the transition from mystical alchemy to empirical chemistry. Discuss historical examples where lack of measurement led to flawed conclusions and how standardization improved reproducibility.

37. Compare the goals of immortality elixirs with modern pharmaceutical chemistry, focusing on extraction, isolation, dosage, and toxicity assessment. Highlight methodological similarities and key differences in evidence-based design.

38. Explain how recording experimental observations in alchemy foreshadowed the modern practice of lab notebooks, including reproducibility, traceability, and validation of chemical experiments. Provide an example of a modern experiment referencing historical notation.
39. Discuss the role of trial and error in alchemy and explain how modern chemistry formalizes this approach using the scientific method, hypothesis testing, controlled experiments, and statistical analysis.
40. Analyze how mystical symbolism in alchemical texts both hindered and promoted chemical discovery. Provide examples where symbolic representation obscured understanding but indirectly guided methodological innovations.
41. You are a modern chemist attempting to reproduce a medieval experiment where a metal turns black upon heating. Propose a hypothesis for the observed change (e.g., oxide formation, surface sulfidation) and design a controlled experiment to test it, including apparatus, temperature control, and analytical verification.
42. A student discovers a recipe for a “gold-making elixir” containing lead nitrate and vinegar. Using chemical reasoning, predict all plausible reactions, identify products, and explain why formation of gold is chemically impossible.
43. Design a modern experimental protocol to determine the solubility of a mineral compound used in alchemical elixirs. Include sample preparation, temperature control, analytical quantification, and discussion of solubility equilibria.
44. A historical manuscript describes a color change when heating copper compounds. Using modern chemical knowledge, identify possible oxidation or thermal decomposition reactions responsible and propose an experiment to confirm the chemical transformations.
45. Compare the experimental tools of Jabir ibn Hayyan with those of Robert Boyle, identifying three major improvements in methodology, precision, instrumentation, and reproducibility. Explain the chemical rationale behind each advancement.
46. Explain how Lavoisier’s experiments disproved the phlogiston theory using modern chemical language. Present a step-by-step reasoning, including mass balance measurements, gas identification, and stoichiometry, to show why combustion is oxidation, not phlogiston release.

47. A medieval alchemist observes gas evolution but cannot quantify it. Design a modern experiment to collect, measure the volume, and determine the composition of the gas using appropriate analytical instrumentation, while ensuring laboratory safety.

48. A historical alchemical procedure instructs "drying the substance in the sun." Propose a modern equivalent method, such as vacuum drying or oven heating, and explain the advantages in temperature control, reproducibility, and yield.

49. Critically evaluate the claim that alchemy contributed directly to modern pharmacology. Provide examples where alchemical practices informed chemical isolation, preparation, and therapeutic application, and contrast them with evidence-based modern pharmacology.

50. Imagine reconstructing a medieval alchemical laboratory for educational purposes. Propose three modern experiments—distillation, crystallization, sublimation—that teach fundamental chemical concepts while referencing historical alchemical techniques, including a discussion of historical context, apparatus, safety, and quantitative measurement.

Chapter 2

Scientific Measurements And Reasoning

Scientific investigation is grounded in the effort to understand natural phenomena through systematic observation, logical reasoning, and careful experimentation. At the university level, the scientific method is not merely a simple sequence of steps but rather a dynamic and iterative framework that helps researchers move from initial curiosity to robust, evidence-based conclusions. Scientists begin by observing the natural world, gathering both qualitative and quantitative information. Qualitative observations describe characteristics such as color, texture, odor, or physical state—details that rely on sensory perception and descriptive language rather than numbers. Quantitative observations, by contrast, provide measurable and numerical data that include both a number and a unit, such as mass in grams, time in seconds, or temperature in kelvin. These quantitative measurements form the backbone of scientific reasoning because they allow comparisons, mathematical treatment, and statistical analysis. Both types of observations are essential because qualitative details often inspire initial questions, while quantitative measurements make it possible to test explanations rigorously.

From these observations, scientists formulate hypotheses, which are tentative, testable, and falsifiable explanations for the phenomena they have observed. A hypothesis must be specific enough to generate clear predictions and must be structured so that evidence can either support or contradict it. Importantly, hypotheses are not equivalent to predictions; rather, the hypothesis proposes a mechanism or underlying reason, while the prediction describes what will happen if the hypothesis is correct. The strength of a hypothesis lies in its ability to be examined through well-designed experiments. Once a hypothesis is established, experiments are constructed to test it under controlled conditions. Experimental design requires careful management of variables: the independent variable is deliberately manipulated by the investigator, the dependent variable is measured to assess the effect of that manipulation, and all other factors are held constant to ensure that any observed changes can be attributed to the variable of interest. Scientific experiments must also be reproducible, meaning that the same procedures performed by independent researchers should lead to comparable results. Reproducibility is a foundational requirement for scientific credibility and ensures that findings are not merely the product of chance or hidden bias.

As experiments are conducted, results may support the hypothesis, contradict it, or suggest the need for modification. The scientific method is therefore cyclical and self-correcting; hypotheses may be revised repeatedly, and experimental approaches refined to gather clearer or more reliable evidence. Over time, when a hypothesis consistently withstands experimental testing and is supported by a substantial body of evidence, it may evolve into a scientific theory. In university-level science, a theory is not an untested idea but rather a comprehensive and thoroughly validated explanation of natural behavior. Scientific theories are conceptual or mathematical frameworks that integrate many observations and experiments into a coherent whole, enabling both explanation and prediction. Examples include atomic theory, plate tectonics, and quantum theory. Theories often serve as models—simplified representations of complex systems that are accurate within a well-defined domain. Importantly, scientific theories remain open to revision. As new techniques, new data, or new technologies emerge, a theory may be refined, expanded, or replaced by a more accurate model.

In contrast to theories, natural laws summarize consistent patterns observed in nature without explaining why those patterns occur. Natural laws often take mathematical form and describe relationships that hold under specific conditions, such as Boyle's law describing the inverse relationship between gas pressure and volume. Laws arise directly from empirical data and reflect reproducible behavior in the physical world. While laws describe the "what," theories explain the "why," and both are essential components of scientific understanding. Together, observations, hypotheses, experiments, theories, and natural laws form an interconnected structure that defines scientific inquiry. Understanding these foundations is critical before exploring the quantitative tools—scientific notation, SI units, significant figures, dimensional analysis, temperature scales, density, and specific gravity—that allow scientists to measure the world with precision and communicate results with clarity.

Scientific notation is an indispensable mathematical tool in the sciences because it enables researchers to express extremely large or extremely small quantities in a compact, standardized form. Many numerical values encountered in chemistry, physics, astronomy, and engineering fall far outside the range of ordinary counting numbers, making conventional decimal notation impractical. Scientific notation represents any numerical quantity as the product of a coefficient between 1 and 10 and an integer power of ten. This structure, written as $N \times 10^n$, provides both clarity and efficiency. For example, a number like 734.267 becomes 7.34267×10^2, and a very small value such as 0.00000881 becomes 8.81×10^{-6}. The exponent conveys how many places the decimal point has been shifted from the coefficient's position. When converting from standard decimal form to scientific notation, shifting the decimal to the left creates a positive exponent, while shifting it to the right creates a negative exponent. This mechanism provides a universal method for handling magnitudes that would otherwise be cumbersome to write and difficult to interpret.

The usefulness of scientific notation extends far beyond simplified writing. It plays a critical role in scientific calculation because its structure mirrors the laws of exponents, allowing large computations to be performed quickly and systematically. Addition and subtraction in scientific notation require the exponents of the numbers to be equal, meaning that one or both numbers may need to be rewritten so their powers of ten match. For instance, to add 3.2×10^5 and 8.1×10^4, the second term must be rewritten as 0.81×10^5, after which the coefficients can be added directly. Multiplication and division are even more streamlined because they take advantage of fundamental exponent rules: coefficients are multiplied or divided in the usual way, and exponents are added or subtracted. A multiplication such as $(6.0 \times 10^3)(2.5 \times 10^4)$ becomes 15.0×10^7, which is then rewritten as 1.50×10^8 to maintain proper scientific-notation form. Division follows the same principle but requires subtracting exponents. These operations not only simplify computation but also minimize rounding errors by allowing coefficients to remain within an easily manageable numeric range.

Understanding scientific notation also helps avoid common misconceptions. One frequent mistake occurs when students assume that a larger coefficient indicates a larger overall number, ignoring the exponent entirely. In reality, the exponent dictates order of magnitude, which dominates any variation in the coefficient. Another common error arises from incorrectly shifting decimal places or failing to rewrite terms so that coefficients remain in the required interval between 1 and 10. Because scientific notation is standardized across disciplines, adherence to its conventions is essential for clear scientific communication.

Exponent notation is deeply connected to logarithms, another mathematical tool widely used in science. Many scientific relationships, including pH in chemistry, sound intensity in physics, and orders of magnitude in astronomy, use logarithmic scales. In base-10 logarithms, only the digits in the mantissa—the decimal portion of the logarithm—reflect the significant figures of the original number. The characteristic, or integer part of the logarithm, indicates the number's order of magnitude but not its precision. This distinction ensures that measurements retain their appropriate level of certainty when expressed logarithmically. When converting from logarithms back to numbers through antilogarithms, the number of significant figures in the mantissa determines the number of significant figures that must appear in the resulting value.

Scientific notation is more than a mathematical convenience; it is an essential language of science. It provides the structure needed to compare the mass of an atom with the mass of a planet, the width of a cell membrane with the distance between galaxies, and the concentration of ions in solution with the amount of matter in a star. By compressing vast numerical ranges into a uniform representation, scientific notation allows scientists to reason about scales that would otherwise be unmanageable. Whether dealing with Avogadro's number, Planck's constant, astronomical distances, or microscopic dimensions, the scientific community depends on this notation to communicate quantitative information efficiently and with precision. Mastery of scientific notation is therefore foundational to any quantitative scientific discipline and forms a natural bridge to more advanced topics in measurement, units, and numerical analysis.

Scientific measurement depends on a standardized and coherent system of units that allows researchers across the world to communicate results without ambiguity. The International System of Units (SI) serves as the universal foundation for quantitative work in science because it provides a consistent set of base units, derived units, and prefixes that span every conceivable scale of measurement. The SI system is rooted in the metric system, which was originally designed during the Enlightenment to create a rational, decimal-based alternative to the jumble of traditional regional units. Today, SI units are used in laboratories, research institutions, engineering applications, and industries worldwide. Each measurement within SI expresses a physical quantity in terms of a numerical value and a standardized unit, a pairing that gives the measurement both magnitude and meaning. The seven fundamental SI base units—meter for length, kilogram for mass, second for time, ampere for electric current, kelvin for temperature, mole for amount of substance, and candela for luminous intensity—form the basis from which all other scientific units are derived. By combining these base units algebraically, scientists can express quantities such as speed (meters per second), force (kilogram meter per second squared), pressure (newtons per square meter), and concentration (moles per liter).

In scientific practice, it is rarely sufficient to measure quantities only at human scales. Many scientific phenomena involve objects far larger or far smaller than everyday experience—such as interstellar distances, molecular dimensions, and subatomic masses. For that reason, the SI system incorporates a prefix structure that modifies base units by powers of ten. Prefixes such as kilo- (10^3), mega- (10^6), and giga- (10^9) extend measurements to large scales, while milli- (10^{-3}), micro- (10^{-6}), nano- (10^{-9}), and even smaller prefixes allow precise expression of microscopic and submicroscopic quantities. A scientist working with chemical reactions may express a solute mass in milligrams or micrograms, while a physicist studying electromagnetic waves may calculate frequency in megahertz or gigahertz. These prefixes eliminate the need for unwieldy numbers and maintain consistency with the decimal nature of scientific notation. Converting between prefixed units and base units requires only the application of powers of ten, which reinforces the mathematical harmony between the SI system and the exponential notation used throughout science.

The process of measuring physical quantities relies on carefully designed instruments whose precision and reliability determine the quality of the resulting data. Length may be measured using meter sticks, calipers, or laser-based devices; volume with graduated cylinders, volumetric flasks, or burettes; mass with electronic balances calibrated to high accuracy; and temperature with digital thermometers or thermocouples. Each instrument carries its own limitations, including resolution—the smallest increment it can meaningfully measure—and calibration accuracy. Because no instrument is perfect, every measurement inherently contains uncertainty, and scientists must report measurements in ways that honestly reflect this limitation. For example, a graduated cylinder marked in 1-milliliter increments allows estimation to the nearest tenth of a milliliter, while a more refined instrument marked in 0.1-milliliter increments permits estimation to the nearest hundredth of a milliliter. This difference in instrument resolution directly affects the number of digits that can be reported with confidence. Understanding how instrument design constrains measurement is essential for interpreting data properly and for recognizing when two measurements are meaningfully different or statistically indistinguishable.

Unit relationships within the SI system are designed to be straightforward because they rely on simple powers of ten rather than on arbitrary conversion factors. This uniform structure allows scientists to convert between units by multiplying by ratios that represent equivalent quantities. For instance, because 1 meter equals 100 centimeters, the conversion factor 100 centimeters per 1 meter can be used to transform measurements in either direction, depending on which unit must be canceled. Similarly, volume is linked to length through the relationship that one cubic meter contains one million cubic centimeters, or equivalently, that one liter equals 1,000 cubic centimeters. These relationships are defined rather than measured, which means they carry no experimental uncertainty. The mole, a uniquely important SI unit in chemistry, connects macroscopic quantities to the number of particles present. Defined as exactly $6.02214076 \times 10^{23}$ entities, the mole links microscopic behavior to measurable laboratory quantities such as mass and volume. This definition provides a bridge between atomic-scale properties and the macroscopic quantities scientists can directly observe.

Having a unified system of units is not merely a matter of convenience; it is critical for accurate scientific communication. Before the adoption of SI, scientists often had to reconcile conflicting unit systems across countries and disciplines, introducing unnecessary complexity and increasing the likelihood of error. Even today, when mixing SI and non-SI units in calculations—such as using inches with centimeters, or calories with joules—misunderstandings can arise unless careful conversions are applied. The strict use of SI units in scientific literature reduces these risks, facilitates international collaboration, and ensures that experimental procedures can be precisely replicated anywhere in the world. In experimental sciences, clarity is as essential as accuracy, and the SI system provides both by maintaining a universally accepted structure for measurement.

Ultimately, mastery of the SI system allows scientists to work fluidly across scales, disciplines, and methodologies. Whether expressing the mass of a nanoparticle, the energy of a photon, the dosage of a medication, or the distance between planets, SI units anchor these quantities in a common framework. This coherent system, combined with scientific notation and the principles of significant figures, equips researchers with the tools needed to quantify, compare, and analyze the physical world with precision. Understanding how units function, how measurements are made, and how prefixes extend the practical range of numeric representation is foundational for all subsequent topics in scientific measurement and essential for any rigorous study of chemistry, physics, or related fields.

All scientific measurements contain some degree of uncertainty, a fact that reflects the inherent limitations of both instruments and human perception. Even the most advanced measuring devices—whether laser interferometers, mass spectrometers, or digital balances—are constrained by physical, electronic, and calibration-based boundaries. When a scientist records a measurement, the digits reported consist of those known with absolute certainty, derived directly from the instrument's scale, and one final digit that is estimated. This final digit embodies the uncertainty of the measurement and communicates the resolution or precision of the measuring device. For example, a ruler marked in millimeters allows a length to be read with certainty to the nearest millimeter and estimated to the nearest half millimeter, whereas a digital caliper displaying values to the nearest hundredth of a millimeter permits a much finer level of precision. Understanding uncertainty is not merely a formality; it determines how reliably measurements can be compared and how confidently conclusions can be drawn from experimental results.

Scientific discipline requires that uncertainty be reflected explicitly in the way measurements are recorded through the use of significant figures. Significant figures are the digits in a measurement that carry meaningful information about its precision, including all certain digits and the first uncertain one. Their purpose is to ensure that numerical results faithfully represent the reliability of the underlying measurements. The rules governing significant figures are designed to capture the structure of precision in a consistent manner. Nonzero digits are always significant because they unambiguously represent measured quantities. Zeros, however, require more careful interpretation. Zeros between nonzero digits, known as captive zeros, are always significant because they mark measured positions within the number. Trailing zeros—zeros at the end of a number—are significant only if a decimal point is present, indicating that those zeros were measured or estimated rather than merely serving as placeholders. Leading zeros, which precede the first nonzero digit in a decimal number, are not significant because they merely locate the decimal point and do not convey measurement precision. These conventions allow scientists to express numerical values that accurately reflect how much information a measurement truly contains.

Scientific notation plays a crucial role in clarifying significant figures by removing ambiguity surrounding trailing zeros. When a number is written in scientific notation, the digits in the coefficient explicitly represent all significant figures. For example, the number 3500 is ambiguous because it is unclear whether the zeros are measured or simply placeholders; however, if it is written as 3.500×10^3, the four significant figures become unmistakable. This precision is especially important in scientific communication, where the difference between three significant figures and five can determine the validity of a calculation or the interpretation of experimental data.

Handling significant figures in calculations requires careful adherence to standard rules. In addition and subtraction, the limiting factor is the position of the decimal place rather than the total number of digits. The result must be rounded to the same decimal position as the least precise measurement. For example, adding 12.56 g and 0.4 g yields 13.0 g, because the second measurement is precise only to the tenths place. In multiplication and division, the total number of significant figures determines the precision of the result. The calculated value must contain no more significant figures than the measurement with the fewest. These rules ensure that precision is not artificially inflated—an essential consideration, because producing an excessively precise result from imprecise data is misleading and scientifically invalid. Measurements obtained from counting or defined relationships, such as the definition of a meter or the number of atoms in a mole, are considered exact and carry infinitely many significant figures; such values do not limit the precision of a calculation.

Rounding plays a critical part in expressing final answers while maintaining correct precision. Standard rounding rules dictate that if the first digit to be discarded is less than five, the preceding digit remains unchanged; if greater than five, the preceding digit increases by one. When the discarded digit is exactly five and is followed only by zeros, scientists employ a special rule known as "round to the nearest even," or unbiased rounding. This method prevents systematic bias that could accumulate in large data sets. Importantly, rounding should be performed only when a calculation is complete. Rounding intermediate values, even slightly, can introduce compounding errors that lead to incorrect final results. High-level scientific work often requires maintaining extra guard digits throughout calculations and rounding only at the end to preserve accuracy.

Logarithmic calculations introduce additional considerations for significant figures. In base-10 logarithms, the characteristic, or integer part of the logarithm, expresses the order of magnitude but does not reflect the precision of the measurement; the mantissa, or decimal part, contains all significant digits. This distinction ensures that values such as pH, which depend on logarithms of concentration, convey the correct level of precision. Conversely, when taking antilogarithms, the number of digits in the mantissa dictates the number of significant figures in the final numerical result. These rules maintain consistency between linear and logarithmic representations of data and preserve the scientific meaning embedded in the original measurements.

Together, uncertainty, significant figures, and rounding form a framework that ensures scientific measurements are reported with honesty, consistency, and respect for the limits of empirical data. These principles prevent overstatement of precision and maintain the integrity of quantitative work. They ensure that conclusions drawn from measurements—whether about reaction rates, physical constants, or environmental conditions—are supported by data whose reliability is transparently communicated. Mastery of these concepts is essential for any scientific discipline, forming the bridge between raw measurement and meaningful numerical analysis.

Scientific investigation frequently requires converting measurements from one unit to another, and this process must be both precise and logically consistent. Dimensional analysis, often referred to as the factor-label method, is a systematic approach that allows scientists to manipulate units mathematically to ensure correct results. This method relies on conversion factors, which are ratios of equivalent quantities. For example, because one meter equals 100 centimeters, the ratio 1 m / 100 cm or its reciprocal 100 cm / 1 m can be used to transform lengths between meters and centimeters, depending on which units need to cancel. Correct application of dimensional analysis ensures that equations remain dimensionally consistent, preventing errors that could arise from unit mismatches. This method extends to more complex problems, including the conversion of derived units such as volume, energy, or pressure. For instance, converting liters to cubic meters or joules to calories requires the careful combination of base unit ratios and dimensional logic.

Dimensional analysis also plays a critical role in checking the plausibility of scientific equations. A derived physical equation must be dimensionally homogeneous: every term must have the same overall units. If units do not match on both sides of an equation, this discrepancy signals a fundamental error in the formulation of the problem. Such unit-based reasoning provides a powerful verification tool that can catch mistakes before extensive calculations are performed. Consequently, scientists routinely perform dimensional checks to validate experimental formulas, theoretical derivations, and computational results.

Temperature measurement is a fundamental aspect of experimental science, and several scales have been developed for different contexts. The Kelvin scale, used universally in physical sciences, is an absolute scale where zero represents the absence of thermal energy. Temperature in kelvins is directly related to the Celsius scale by the simple linear relationship K =°C + 273.15. The Celsius scale itself is widely used in laboratory and industrial settings due to its familiarity and convenient reference points, such as the freezing and boiling points of water. The Fahrenheit scale, primarily used in the United States, converts to Celsius via °F = (°C × 9/5) + 32. Other less common scales, including Rankine, Rømer, Newton, Delisle, and Réaumur, have historical or specialized uses in engineering, thermodynamics, and scientific research. Mastery of temperature conversions is essential, especially when integrating data from multiple sources, performing calculations that involve thermodynamic quantities, or analyzing results that depend on absolute temperature, such as gas laws or reaction kinetics.

Temperature, like any other measured quantity, must be expressed with appropriate significant figures and uncertainty. For instance, recording the temperature as 25 °C in a laboratory may imply uncertainty of ±1 °C depending on the thermometer's resolution, whereas a high-precision digital sensor might measure to 25.12 °C with uncertainty on the order of ±0.01 °C. Correctly acknowledging these limits preserves the reliability of derived calculations, such as heat capacity, reaction rate constants, or energy changes. When combined with dimensional analysis, careful temperature measurement ensures that physical quantities maintain internal consistency and accuracy across calculations, whether in chemistry, physics, or engineering contexts.

The principles of dimensional analysis and temperature conversion extend naturally into all areas of quantitative science. Without the ability to reliably transform units and maintain consistent measurements across experiments, results would lack reproducibility and comparability. In advanced applications, dimensional analysis also aids in deriving new relationships between physical quantities, such as the form of equations describing diffusion, fluid flow, or thermodynamic processes. Through these methods, scientists maintain a structured and rigorous framework for analyzing the world, ensuring that numerical results carry clear meaning and can be interpreted universally, regardless of the units in which they were originally measured.

Density is a fundamental physical property that describes the compactness of matter in a substance and is defined as the mass per unit volume. It serves as a key parameter in characterizing materials, understanding buoyancy, and predicting the behavior of substances under varying conditions. In practical terms, objects with a density greater than that of the surrounding fluid will sink, whereas those with lower density will float. Density is quantitatively expressed in units such as kilograms per cubic meter (kg/m^3) or grams per cubic centimeter (g/cm^3), depending on the scale of observation. Measuring density requires accurate determination of both mass and volume. Mass can be obtained using electronic balances or precision scales, while volume may be measured directly for regularly shaped objects or indirectly via displacement methods for irregularly shaped solids. In chemistry, the density of liquids and gases is critical for determining concentrations, designing processes, and interpreting experimental results.

Specific gravity is a related concept that compares the density of a substance to that of a reference substance, typically water for liquids and solids or air for gases.

Because specific gravity is a ratio of two densities, it is a dimensionless quantity and therefore has no units. The determination of specific gravity is particularly important in fields such as mineralogy, metallurgy, and gemology, where it provides a rapid and reliable method of identifying materials. One classical technique for measuring the specific gravity of a solid is based on Archimedes' principle, which relates the apparent loss of weight of an object when submerged in a fluid to the volume of fluid displaced. The specific gravity is calculated by dividing the weight of the object in air by the difference between its weight in air and its weight in water. This approach not only provides insight into the intrinsic density of the material but also serves as a practical means of distinguishing substances that are visually similar but compositionally distinct.

The concepts of density and specific gravity are intimately connected with other quantitative principles in science, including unit consistency, significant figures, and dimensional analysis. Accurate reporting of density requires careful attention to the precision of mass and volume measurements. If a mass is measured to three significant figures, and the volume to four, the resulting density should be expressed with the limiting precision of three significant figures. Furthermore, when density is used in calculations, such as determining buoyant force, fluid pressure, or concentration, dimensional analysis ensures that units are correctly handled, preserving the physical meaning of the results. Temperature also affects density, particularly for gases and liquids, because thermal expansion alters the volume without changing the mass. Therefore, density measurements are often reported at a specified reference temperature to maintain comparability.

Taken together, the scientific method, measurement principles, significant figures, unit systems, dimensional analysis, and the quantitative treatment of properties such as density and specific gravity constitute a coherent framework for modern science. Mastery of these concepts allows scientists to observe natural phenomena with rigor, quantify results with precision, and communicate findings unambiguously. From atomic-scale investigations to astronomical measurements, these principles enable the systematic exploration of the physical world, providing the foundation for theory development, experimental design, and technological application. The integrated understanding of measurement, notation, unit conversion, uncertainty, and physical properties is therefore essential for success in chemistry, physics, engineering, and all quantitative sciences.

In conclusion, the process of scientific inquiry rests upon a foundation of careful observation, hypothesis formulation, and rigorous experimentation. Measurements provide the bridge between empirical observation and theoretical understanding, while scientific notation and the SI system ensure clarity and consistency across scales and disciplines. Recognizing and reporting uncertainty through significant figures, applying dimensional analysis, converting between temperature scales, and analyzing material properties such as density and specific gravity are all critical for producing valid and reliable scientific results. Together, these principles exemplify the logical, systematic, and precise nature of modern science, equipping students and researchers with the tools necessary to explore, quantify, and ultimately understand the natural world with intellectual rigor.

Chapter 2: Word Problems

1. A researcher observes a newly synthesized nanomaterial and records qualitative properties including color change upon heating, variations in surface texture under a scanning electron microscope, and changes in optical reflectivity. Propose a hypothesis about the material's thermal stability based on these qualitative observations. Then, design a comprehensive set of quantitative experiments—including thermogravimetric analysis (TGA), differential scanning calorimetry (DSC), and X-ray diffraction (XRD) at elevated temperatures—to rigorously test the hypothesis, describing the type of data each technique will provide and the statistical methods to compare experimental replicates.

2. A chemist hypothesizes that increasing the concentration of a heterogeneous catalyst decreases reaction time through a specific reaction mechanism. Design an experiment specifying the independent variable (catalyst concentration), dependent variable (reaction rate or half-life), and controlled variables (temperature, solvent, reactant concentrations). Include a plan for replicates, randomization, and blinding, and describe how statistical analysis such as ANOVA could be used to ensure reproducibility and evaluate confidence intervals.

3. A physicist studies a gas at varying temperatures and pressures to validate Boyle's Law. How could the experiment be modified to detect systematic errors such as parallax in volume measurement or leakages in the system? Discuss how precision

instruments, error propagation, and reporting significant figures influence the reliability and interpretability of the results, including how to differentiate between random and systematic errors.

4. In a solution-phase experiment, two competing hypotheses are proposed to explain an observed color change under varying pH. Develop a statistical analysis plan using spectrophotometric absorbance data collected at multiple wavelengths and time points. Discuss methods such as linear regression, goodness-of-fit tests, and hypothesis testing to determine which hypothesis is better supported, including the role of confidence intervals and error bars.

5. Convert the astronomical distance from Earth to Proxima Centauri (4.246×10^{13} km) into meters using SI prefixes and scientific notation. Discuss the implications of rounding and significant figures in astronomical calculations, including how uncertainties in distance measurements from parallax or redshift could propagate through further calculations in astrophysical modeling.

6. A biochemist measures enzyme activity using spectrophotometric absorbance values recorded with known uncertainties. Explain in detail how to propagate these uncertainties when calculating reaction rates using Michaelis–Menten kinetics and how to correctly report the final rates with appropriate significant figures, including considerations when performing logarithmic transformations of data.

7. A materials scientist compares the density of a newly developed alloy to that of pure metals. Explain how Archimedes' principle can be applied to an irregularly shaped sample, including detailed steps for water displacement, corrections for buoyancy in air, and error propagation. Discuss how measurement uncertainties in mass and volume affect the calculated specific gravity and the comparison to theoretical predictions.

8. A physicist records a body temperature of 98.6 °F and must report it in kelvins with correct significant figures. Explain the Fahrenheit-to-Celsius conversion, the Celsius-to-Kelvin conversion, and how measurement uncertainty propagates through both steps. Discuss the relevance of significant figures in thermodynamic calculations where absolute temperatures are required.

9. A researcher measures the mass of a powder as 0.0423 g ± 0.0002 g and dissolves it in 50.0 mL of solution. Discuss how the mass uncertainty affects the calculated

molarity, including the propagation of errors, appropriate rounding rules, and implications for further kinetic or equilibrium calculations.

10. A graduate student measures the volume of a cylindrical flask using both a caliper and a micrometer. Compare how the resolution of each instrument impacts the calculated density of a liquid filled in the flask. Propose a methodology for reconciling discrepancies between instruments, including statistical averaging and error analysis.

11. During an electrical experiment, a student measures current in a circuit using an ammeter with ±0.02 A uncertainty and voltage with a multimeter ±0.05 V. Describe in detail how to combine these uncertainties to calculate resistance using Ohm's Law, and explain the treatment of significant figures in the resulting resistance value.

12. Two students independently perform the same experiment and report values of 2.31×10^5 and 2.3×10^5. Analyze whether these results are consistent, taking into account the meaning of significant figures, uncertainty, and rounding, and discuss how one might reconcile the differences for reporting in a publication.

13. A chemist uses dimensional analysis to convert 50.0 mL of solution to cubic meters. Explain the process in detail, including SI unit consistency, how to handle prefixes, and how these conversions impact the calculation of molar concentration in a multi-step preparation.

14. A physicist tests a novel theoretical equation relating pressure, volume, and temperature. Describe how dimensional homogeneity can be applied to validate the equation prior to performing experiments, including a systematic approach to checking units and identifying potential conceptual errors.

15. A student records temperatures in Celsius with a standard deviation of ±0.5 °C. Explain how the thermometer resolution affects the calculation of coefficients of thermal expansion and discuss the influence of measurement uncertainty on derived material properties.

16. A materials engineer measures the mass and volume of a metal rod at several temperatures. Describe how density versus temperature data can be used to model thermal expansion, including polynomial regression, error propagation, and reporting results with correct significant figures.

17. A biologist measures enzyme activity at different pH values and logs the results to base 10. Explain how to propagate uncertainties through logarithmic transformations and determine the appropriate number of significant figures in both log-transformed and antilogarithmic data sets.

18. A graduate student converts a speed of 5.0×10^4 cm/s into m/s using SI prefixes. Show all steps and discuss the effect of rounding on significant figures and the potential impact on subsequent calculations in fluid dynamics or kinetics experiments.

19. In a high-precision chemical experiment, the concentration of H^+ ions is 3.2×10^{-5} M. Calculate the pH, and discuss how the mantissa and exponent affect the number of significant figures in the reported pH and subsequent equilibrium calculations.

20. A physicist weighs a sample in air and then submerged in water, obtaining 23.456 g and 20.123 g. Calculate the specific gravity of the material, discuss how measurement precision affects the final result, and explain how uncertainties propagate from mass measurements to derived density values.

21. A student attempts to sum 2.56×10^3 and 3.4×10^2. Explain the detailed steps for addition in scientific notation, discuss the effect on significant figures, and provide guidance for reporting results in a way consistent with proper measurement standards.

22. A chemist studies the solubility of a salt at different temperatures. Analyze how temperature measurement uncertainties influence the derived solubility curve, and describe how to represent these uncertainties graphically, including error bars and confidence intervals.

23. In an experiment to determine density, a sample has a mass of 12.3 g and volume of 4.5 cm^3. Calculate the density with correct significant figures and discuss implications if instrument resolution changes, including the potential impact on material identification.

24. A physics lab reports a distance of 0.000045 m. Rewrite this in scientific notation, and discuss how the exponent conveys the order of magnitude and affects the interpretation of small-scale measurements in precision experiments.

25. A student calculates the gas constant R from multiple experiments with varying uncertainties. Describe a method to combine these results using weighted averages and propagate uncertainties to achieve a rigorous final value.

26. A graduate-level lab requires converting 75 °F to Kelvin for thermodynamics experiments. Perform the conversion, explain the significance of using absolute temperatures, and discuss how measurement uncertainty affects thermodynamic property calculations.

27. A chemical engineer calculates pressure from measured force and area, both with uncertainties. Explain how to propagate uncertainties through the derived pressure calculation, including partial derivatives for complex error propagation.

28. A physicist measures the volume of irregular rocks using water displacement. Discuss how meniscus reading, parallax errors, and instrument resolution influence the accuracy of density calculations and their reported significant figures.

29. A biologist measures reaction rates at varying temperatures resulting in an exponential relationship. Explain how logarithmic transformation aids linearization for kinetic analysis and how significant figures must be treated in both raw and transformed data.

30. A student calculates the volume of a cube from side length measurements with uncertainty. Show how uncertainty propagates into the calculated volume and affects the reported significant figures, including potential implications for material property analysis.

31. A chemist compares the densities of two liquids using a hydrometer. Discuss limitations of this method in terms of instrument resolution, meniscus reading, and significant figures, and suggest alternative methods to improve accuracy.

32. In an advanced optics lab, a student measures light intensity in lux. Explain how uncertainties in calibration, detector sensitivity, and ambient conditions affect quantitative comparisons of results, including the propagation of error in derived ratios.

33. A physicist observes deviations from a theoretical law due to experimental error. Discuss how dimensional analysis can determine whether the deviations are physically meaningful or arise from inconsistent units or scaling errors.

34. A student measures mass using a balance accurate to ±0.001 g. Determine the number of significant figures appropriate when calculating derived quantities such as moles, molarity, or energy content, including the effect of rounding errors.

35. A chemist reports Avogadro's number as 6.022×10^{23} molecules/mol. Explain why this is considered an exact value and does not limit significant figures in calculations, and discuss its role in stoichiometric computations in complex reactions.

36. A student calculates the product of $(3.45 \times 10^4) \times (2.1 \times 10^3)$. Show how to determine the correct scientific notation, propagate uncertainties, and assign significant figures correctly in the final answer.

37. A graduate student evaluates reproducibility by performing three independent reaction trials. Explain how to calculate the mean, standard deviation, and confidence intervals, and how significant figures should be reported in the final data set for publication.

38. A physicist calculates the gravitational force between two spheres with measured masses and distances, each with uncertainties. Describe the method for propagating uncertainty in multiplicative and power-law formulas to determine the range of possible gravitational forces.

39. A chemist converts 0.00032 L to milliliters and microliters. Demonstrate the calculation using SI prefixes and scientific notation, explaining how significant figures are preserved and how rounding might affect downstream calculations.

40. A student examines how the density of a liquid changes with temperature. Discuss how uncertainties in both temperature and density measurements are combined in regression analysis, including weighting data points and reporting uncertainties in the slope of a linear model.

41. In a quantum mechanics experiment, a particle's wavelength is measured as 4.56×10^{-7} m. Convert this to nanometers, and discuss how the exponent influences the representation of measurements on the nanoscale and in high-precision optical calculations.

42. A biochemist prepares a solution of precise molarity using a balance and volumetric flask. Analyze how uncertainties in mass and volume measurements propagate

to the final molarity, and discuss the implications for enzyme kinetics or titration experiments requiring high precision.

43. A student measures current at multiple points in a circuit using instruments with varying resolution. Explain how to reconcile these measurements, propagate uncertainties, and ensure proper significant figures when calculating derived quantities such as power or efficiency.

44. A physics experiment measures gravitational acceleration using a pendulum. Discuss how uncertainties in both length and period measurements propagate into the calculated value of g, and describe methods to minimize systematic errors.

45. A chemist converts gas volumes measured in liters at STP to moles. Discuss how SI units, Avogadro's number, and scientific notation are used to ensure accurate and consistent results, including the role of measurement uncertainty in subsequent stoichiometric calculations.

46. A student analyzes a titration curve to determine the endpoint volume. Describe how instrument resolution, reading uncertainty, and interpolation affect the determination of molarity and the number of significant figures to report, especially for weak acid–base systems.

47. A physicist records sound intensity in decibels (logarithmic scale). Explain how to convert the measurements back to linear intensity values, and discuss the proper treatment of significant figures during logarithmic transformations.

48. A materials scientist measures the mass of a powder with an analytical balance and its volume by displacement. Explain how the resulting specific gravity calculation can be used to identify the material, considering measurement uncertainties and the propagation of error.

49. A graduate student measures liquid density at multiple temperatures and observes nonlinearity. Discuss how regression analysis, careful treatment of significant figures, and uncertainty analysis are critical to interpreting thermal expansion and predicting behavior outside measured ranges.

50. A chemistry researcher performs multi-step calculations involving moles, mass, and volume. Discuss how significant figures, uncertainty propagation, and unit consistency must be applied at each stage to ensure scientifically valid and reproducible results, including strategies for minimizing rounding errors and documenting assumptions.

Chapter 3

Philosophical Roots of Atomic Ideas

The idea that matter is made of tiny, indivisible particles has roots stretching back to ancient civilizations, but John Dalton was the first to transform this philosophical idea into a scientific theory supported by experimental evidence. Before Dalton's work, the nature of matter was largely speculative. Early Greek philosophers such as Democritus proposed that matter consisted of minute, indestructible units called "atomos," yet they lacked the empirical tools to validate such claims. For centuries, this concept remained only a philosophical curiosity while alchemists and early chemists conducted practical experiments without a unifying theoretical framework. Dalton changed this landscape by grounding atomic theory in measurable chemical behavior, offering explanations for why substances combine in consistent proportions and why mass is conserved during reactions.

Dalton's Atomic Theory comprises several concise principles that fundamentally shifted chemistry into a quantitative science. He asserted that all matter is composed of atoms—extremely small particles that cannot be further subdivided by chemical means. According to Dalton, atoms of the same element are identical in mass and properties, whereas atoms of different elements exhibit distinct characteristics. His idea that atoms combine in fixed, whole-number ratios to form compounds explained long-standing observations such as definite composition and constant mass ratios. Dalton also emphasized that atoms are never created or destroyed in chemical reactions; instead, they simply rearrange, forming new substances without any change in total mass. Though later discoveries revealed that atoms themselves are divisible and can undergo nuclear transformations, Dalton's framework remains the conceptual backbone of modern chemistry. His ideas introduced the particulate model of matter, enabling chemists to interpret reactions in terms of atomic interactions rather than visible changes alone.

Although Dalton considered atoms indivisible, the late nineteenth and early twentieth centuries brought discoveries that revealed a rich internal structure. Experiments with cathode rays led to the discovery of the electron, showing that atoms contained negatively charged particles far smaller than the atom itself. Later, the gold-foil experiment demonstrated that atoms have a dense, positively charged nucleus containing most of their mass. As research progressed, scientists identified two nuclear particles: the proton, carrying a +1 charge, and the neutron, carrying no charge. This structural understanding refined Dalton's framework, revealing that the identity of an element is defined by the number of protons in its nucleus, while variations in neutron number produce isotopes with differing masses. Electrons, extremely light and negatively charged, occupy regions around the nucleus called electron clouds or orbitals, forming patterns that influence chemical bonding, reactivity, and the physical properties of matter.

The three primary subatomic particles—protons, neutrons, and electrons—play distinct roles within the atom. Protons and neutrons form the nucleus, a compact region at the center of the atom where nearly all mass is concentrated. Although both particles have similar mass, neutrons are slightly heavier and contribute to nuclear stability by offsetting proton-proton repulsion. Electrons, by contrast, have negligible mass but occupy most of the atom's volume. Their arrangement in various energy levels determines how atoms interact to form chemical bonds. The number of electrons equals the number of protons in a neutral atom, balancing the charges and maintaining electrical neutrality. Because electrons occupy quantized energy levels, they can absorb or release energy when transitioning between states—phenomena that underpin spectroscopy, atomic emission lines, and numerous analytical techniques used in modern chemistry.

Atoms of the same element can differ in the number of neutrons, creating isotopes with unique mass numbers. All isotopes of an element share the same atomic number (Z), representing the number of protons, but differ in mass number (A), the total number of protons and neutrons. This distinction is critical in nuclear chemistry, medicine, geology, and environmental science. For example, carbon-12 and carbon-14 are both carbon isotopes, but carbon-12 is stable while carbon-14 is radioactive and used for radiocarbon dating. Isotope notation clearly identifies each form by placing the mass number above the atomic number beside the element symbol. Understanding isotopes enables chemists to analyze elemental abundances, track chemical pathways in biological systems, and harness nuclear processes for energy generation.

The periodic table organizes all known elements according to increasing atomic number and recurring patterns in chemical behavior. Its structure allows scientists to predict properties, anticipate reactivity, and understand relationships among elements. Element symbols are standardized for clarity, typically consisting of one or two letters derived from either English or Latin names. The periodic table includes naturally occurring elements and synthesized ones created in laboratories through nuclear reactions. Many early-known elements—gold, silver, copper, iron, and sulfur—were identified long before formal chemical science existed, while modern research continues to expand the table with new, short-lived synthetic elements. The placement of elements in groups and periods reveals periodic trends in atomic size, ionization energy, electronegativity, and metallic character, providing a framework for understanding the behavior of atoms across the table.

The modern periodic table is more than a listing of the chemical elements—it is a powerful organizational tool that reveals the underlying structure of matter. Built around the principle that chemical properties repeat periodically when elements are arranged by increasing atomic number, the table is a map of atomic behavior. Each square represents a unique element and typically contains its atomic number, symbol, name, and atomic mass. The horizontal rows, called *periods*, reflect the filling of electron shells or principal energy levels, while the vertical columns, known as *groups* or *families*, consist of elements with similar chemical properties because they share the same number of valence electrons. This arrangement allows chemists to predict bonding tendencies, reactivity patterns, and even physical properties such as melting points and conductivity. The table's brilliance lies not just in its organization but in its predictive power: many elements were discovered precisely where the table predicted they must exist, confirming the periodic law and validating the structure still used today.

Each group on the periodic table represents a family of elements with shared characteristics. Group 1, the *alkali metals*, includes highly reactive metals such as lithium, sodium, and potassium. They react vigorously with water and readily form +1 cations due to their single valence electron. Group 2, the *alkaline earth metals*, also form stable cations but carry a +2 charge. The transition metals, occupying the wide central block of the table, exhibit multiple oxidation states, superior conductivity, and the ability to form complex ions—behavior tied to the involvement of d-electrons in bonding. On the right side lie the reactive *halogens* (Group 17), which form −1 anions and exist naturally as diatomic molecules. Group 18 contains the *noble gases*, which are exceptionally stable due to their complete valence electron shells. The periodic relationship among these families allows chemists to anticipate how unfamiliar elements will behave by comparing them with well-studied neighbors.

The physical and chemical distinctions between metals, nonmetals, and metalloids form another essential layer of classification. Metals, which dominate the left and central areas of the table, typically have high electrical and thermal conductivity, a shiny luster, and the ability to be hammered into sheets (malleability) or drawn into wires (ductility). They readily lose electrons to form cations and often exhibit high melting and boiling points. Nonmetals, concentrated on the right side, show much greater diversity. Some, like sulfur or phosphorus, are brittle solids, while others like oxygen and nitrogen are gases at room temperature. Nonmetals tend to gain electrons to form anions or share electrons in covalent bonds. Between these categories lie the metalloids, such as silicon, boron, and arsenic, which exhibit intermediate behavior.

Their semiconducting properties make them essential in electronics, and their dual nature often allows them to form both ionic and covalent compounds depending on reaction partners.

Several elements exist naturally as *diatomic molecules*, meaning they consist of two atoms bound together. These include hydrogen (H_2), nitrogen (N_2), oxygen (O_2), fluorine (F_2), chlorine (Cl_2), bromine (Br_2), and iodine (I_2). Diatomic formation arises because certain atoms achieve greater stability by pairing to complete their valence shells. For instance, oxygen atoms bond to share electrons and achieve a more stable electron configuration, resulting in O_2. Bromine and mercury are also unique in that they are liquids at room temperature; bromine remains diatomic in liquid form. Understanding the natural molecular form of elements is critical for writing correct equations, predicting reaction behavior, and interpreting laboratory and industrial processes that involve elemental substances.

Some elements occur in multiple structural forms known as allotropes, which differ not in composition but in the arrangement of atoms. These structural differences result in dramatically different properties. Carbon is the classic example: diamond's rigid three-dimensional network yields extraordinary hardness, while graphite's layered structure produces a slippery solid ideal for lubricants and pencil lead. More recently discovered allotropes—such as fullerenes, graphene, and carbon nanotubes—exhibit remarkable electrical and mechanical properties and have revolutionized materials science. Oxygen also has allotropes: O_2 is essential for respiration and combustion, while ozone (O_3) absorbs harmful ultraviolet radiation in the stratosphere. Phosphorus exists in white, red, black, and violet allotropes with vastly different reactivities, illustrating how structure alone can dramatically influence an element's behavior.

Among the metals, certain elements are classified as noble metals due to their resistance to corrosion and oxidation. Gold, silver, and platinum are the most well-known examples. Their reluctance to react with environmental agents such as oxygen, acids, or moisture has made them valuable for centuries, particularly in coinage, jewelry, and catalytic applications. In industrial chemistry, noble metals are prized for their ability to catalyze reactions without degrading, especially in processes involving hydrogenation, automobile exhaust treatment, and fuel cells. Their inertness stems from their electron configurations and relatively high ionization energies, which make the loss of electrons less favorable and chemical attack more difficult.

Atomic behavior is strongly governed by the gain or loss of electrons. Metals tend to lose electrons because their valence shells are sparsely populated, forming positively charged *cations*. For example, sodium (Na) loses one electron to become Na^{+}, and magnesium (Mg) loses two to form Mg^{2+}. Nonmetals, by contrast, tend to gain electrons to fill their valence shells, forming negatively charged *anions* such as Cl^{-} or O^{2-}. These processes reflect fundamental trends: elements on the left of the periodic table have low ionization energies, whereas elements on the right have high electron affinities. When cations and anions attract, they form ionic bonds, creating ionic compounds whose structures and properties reflect the electrostatic attraction between oppositely charged ions.

Most main-group elements form predictable ions based on their group number, but transition metals are more complex. Because they involve d-electrons in bonding, they often form multiple ions with different charges. Iron, for example, can form Fe^{2+} or Fe^{3+}; copper can form Cu^{+} or Cu^{2+}. These varying oxidation states allow transition metals to participate in diverse chemical reactions and make them indispensable in biological systems (such as iron in hemoglobin), industrial catalysts, and electronic materials. The Roman numeral system is used in compound names to specify which ion is present—for instance, iron(II) oxide versus iron(III) oxide—ensuring clarity in chemical communication.

Dalton's atomic framework helped solidify two major laws of chemical composition. The Law of Definite Proportions states that a chemical compound always contains the same elements in the same fixed mass ratio, regardless of its source. For example, water is always composed of 2 hydrogen atoms and 1 oxygen atom; any sample of pure water on Earth or Mars will have the same proportion of hydrogen to oxygen by mass. The Law of Multiple Proportions explains how the same two elements may combine in different ratios to form different compounds—such as carbon monoxide (CO) and carbon dioxide (CO_2). These laws provided strong evidence that matter is composed of discrete particles combining in whole-number ratios, supporting Dalton's theory long before atoms were directly observed

Ionic compounds are characterized by a rigid lattice structure composed of alternating cations and anions arranged in a repeating three-dimensional pattern. This arrangement leads to distinctive properties: high melting and boiling points, brittleness, and the ability to conduct electricity only when dissolved or molten.

In solid form, ions are locked in place and cannot move freely, so solid ionic compounds do not conduct electricity. When dissolved in water, however, the ions separate and disperse, allowing electrical charge to flow through the solution. This distinction forms the basis for understanding electrolytes, conductivity tests, and numerous biological and industrial processes that rely on ionic mobility.

As chemists continued exploring the nature of matter, Dalton's atomic theory served as a foundational framework that guided experimental design and interpretation. Although several of Dalton's postulates were later refined or modified through the discoveries of subatomic particles and quantum mechanics, his basic assumptions still form the conceptual backbone of modern chemistry. At its core, the theory emphasizes three fundamental ideas: the discrete nature of matter, the uniformity of atoms within a single element, and the predictable combination of atoms to form compounds. Each of these principles opened the door for further advancements and inspired generations of scientists to investigate the underlying structure of the physical world.

One of the most influential aspects of Dalton's theory was his insistence that atoms behave as indestructible units during chemical processes. Although we now know that atoms themselves can be split or transformed during nuclear reactions, Dalton correctly observed that ordinary chemical reactions do not alter the identity of the atom. When wood burns, for example, the original atoms simply rearrange to form carbon dioxide, water vapor, and ash; none of the atoms themselves disappear. This observation was vital, because it helped cement the Law of Conservation of Mass—a principle that remains fundamental in chemistry and physics.

Dalton also introduced the notion that atoms combine in simple whole-number ratios. This was an intuitive yet powerful insight. It explained why elements do not combine in arbitrary proportions but instead follow strict numerical patterns. For instance, carbon and oxygen can form carbon monoxide (CO), where the ratio is 1 carbon atom to 1 oxygen atom, or carbon dioxide (CO_2), where the ratio is 1 carbon atom to 2 oxygen atoms. The ratios never involve fractions such as half of an atom, because atoms themselves cannot be subdivided within a chemical reaction. This reasoning directly supports the Law of Multiple Proportions, another foundational law that Dalton helped articulate.

In addition, Dalton's early work on atomic weights, although crude by modern standards, marked one of the first attempts to quantify the masses of atoms relative to each other. He proposed that hydrogen be assigned a mass of 1 (as the lightest known element), and other elements should be measured relative to it. While his exact values were sometimes inaccurate due to limited experimental tools, the concept paved the way for the modern periodic table, where atomic masses and chemical properties are systematically organized.

As the nineteenth century progressed, improvements in experimental techniques—especially advancements in gas studies, electrochemistry, and spectroscopy—provided mounting evidence that Dalton's atomic assumptions were essentially correct. Later discoveries such as electrons, protons, neutrons, isotopes, and nuclear forces refined the picture, but the core logic of Dalton's theory remained intact. Atoms are still considered the smallest units that retain the chemical identity of an element, and chemical reactions are still understood as processes of atom rearrangement, not creation or destruction.

Dalton's emphasis on reasoning from experimental data was also groundbreaking. Instead of relying on philosophical speculation, he grounded his claims in observable patterns and measurable quantities. This shift helped move chemistry firmly into the realm of empirical science. His theory provided a structure—a conceptual map—that allowed chemists to not only explain known reactions but also predict new ones. In this sense, Dalton's atomic theory was not just a description of matter; it was a model, one that guided scientific inquiry for more than a century.

Even today, almost every topic in chemistry—from stoichiometry and bonding to thermodynamics and quantum mechanics—can be traced back to Dalton's original propositions. The notion that matter is composed of discrete units underlies theories of molecular geometry, reaction mechanisms, gas behavior, and the interactions between light and matter. Without Dalton's insights, the development of modern chemistry would have been dramatically delayed.

Building on Dalton's atomic theory, chemists soon discovered that atoms themselves are not indivisible but consist of smaller subatomic particles: electrons, protons, and neutrons. This revelation led to the development of the periodic table, where elements are organized by their atomic number and recurring chemical properties. Understanding atomic structure allowed scientists to explain chemical bonding in terms of electron sharing or transfer, giving rise to the concepts of covalent and ionic bonds. Furthermore, the study of electrons and energy levels became foundational to quantum chemistry, which describes how electrons occupy orbitals and predict chemical reactivity with remarkable precision.

Modern chemistry also relies heavily on the principles of thermodynamics and kinetics, disciplines that explain why reactions occur, how fast they proceed, and under what conditions they are most efficient. From designing pharmaceuticals to developing sustainable energy solutions, these principles guide chemists in manipulating matter at the molecular and atomic levels. Techniques such as spectroscopy, chromatography, and X-ray crystallography allow scientists to probe the structure and behavior of molecules with extraordinary detail, further connecting the macroscopic properties of substances to their microscopic components.

Ultimately, the legacy of Dalton's atomic theory endures not just as a historical milestone but as the conceptual framework upon which contemporary chemistry continues to expand. Every discovery, from novel materials to life-saving medicines, rests on the understanding that matter is composed of discrete, interacting units, each following predictable laws. The journey that began with the simple postulates of atoms has evolved into a sophisticated exploration of matter, energy, and the intricate rules that govern the physical universe.

Chapter 3: Word Problems

1. A research chemist studying deviations in Dalton's Law of Multiple Proportions analyzes two oxygen–fluorine compounds. Mass spectrometry shows that a 2.35-g sample of compound X contains 1.52 g of oxygen, while a 3.66-g sample of compound Y contains 2.44 g of oxygen. Assuming the fluorine mass in each compound arises from whole-number ratios of atoms, determine the simplest integer ratio of fluorine atoms between the two compounds and comment on whether this supports Dalton's postulate of whole-number atomic combinations.

2. A geochemist analyzing a carbon-bearing mineral finds that its natural carbon isotope distribution differs from global averages: the sample contains 98.3% ^{12}C, 1.5% ^{13}C, and 0.2% ^{14}C. Given that the mineral formed 70,000 years ago and assuming no nuclear reactions occurred within the sample over time, calculate

the expected radiocarbon content today based solely on radioactive decay and evaluate whether the measured 0.2% ^{14}C is consistent with Dalton's assumption that atoms retain identity in chemical processes.

3. During a precision elemental analysis, a compound containing nitrogen and an unknown halogen is decomposed, yielding 0.844 g of nitrogen and 3.36 g of the halogen. A related compound yields 0.844 g of nitrogen and 5.04 g of halogen. Determine the likely halogen involved based on periodic trends and mass ratios, then evaluate how the mass comparison illustrates the Law of Multiple Proportions.

4. A nuclear chemist compares stability among isotopes of an element with Z = 46. One isotope has 60 neutrons, another 62 neutrons. Using nuclear binding-energy trends and the neutron–proton ratio required for mid-mass nuclei, predict which isotope should exhibit greater stability. Then analyze how the behavior of these isotopes both affirms and challenges early atomic models.

5. An analytical chemist dissolves an ionic solid composed of a +2 metal cation and a −1 anion, producing an aqueous solution with a measured conductivity of 13.5 mS/cm at 25°C. If the molar conductivity of the cation at infinite dilution is 119 $S{\cdot}cm^2{\cdot}mol^{-1}$ and that of the anion is 76 $S{\cdot}cm^2{\cdot}mol^{-1}$, estimate the molarity of the solution and comment on how this experiment reinforces atomic-level explanations of ionic mobility.

6. A sample of gaseous element M exists naturally as a diatomic molecule. When 1.00 g of M_2 reacts completely with fluorine to form 3.16 g of MF_4, determine the atomic mass of M and identify the element. Discuss how the stoichiometry of the reaction supports Dalton's proposal about atoms combining in fixed integer ratios.

7. A metallurgist studying noble metals measures the reactivity of gold nanoparticles toward aqua regia under controlled conditions. When the surface area doubles, the dissolution rate increases by a factor of 2.8. Explain whether this nonlinear increase aligns with Dalton's atomic-level interpretation of surface-dependent reactions and propose a mechanism involving electron configuration and coordination environment.

8. Copper exists naturally as ^{63}Cu and ^{65}Cu. A scientist analyzing an ore finds that its ratio ^{63}Cu:^{65}Cu is 2.77:1. Calculate the average atomic mass of copper in the ore, then discuss how isotopic differences challenge Dalton's original assertion that all atoms of an element are identical.

9. A compound contains only carbon and oxygen. Combustion of 0.850 g of the compound yields 1.32 g CO_2 and 0.540 g H_2O (from trace hydrogen contamination). After correcting for hydrogen, determine the empirical formula and analyze how the result reflects Dalton's quantitative methodology for deducing atomic composition.

10. Manganese forms several stable ions. In an electrochemical experiment, MnO_4^- is reduced stepwise to Mn^{2+}. If the total electrons transferred per manganese atom is experimentally measured as 5.00 ± 0.03, evaluate whether this supports the classical oxidation-state model and discuss its significance in understanding variable valence in transition metals.

11. A chemist prepares a 0.250 M solution of KBr and measures its freezing-point depression as –1.38°C. Given the cryoscopic constant of water is 1.86 °C·kg/mol, calculate the van't Hoff factor (i) for KBr in solution, and discuss how this observation confirms the lattice structure of ionic compounds as predicted by the behavior of discrete cations and anions.

12. A carbon allotrope sample contains 0.150 mol of carbon atoms in diamond form and 0.200 mol in graphite form. If both are exposed to identical thermal conditions, predict the relative heat capacities and thermal conductivity differences. Then explain how atomic arrangement, rather than composition, dictates the macroscopic properties, illustrating Dalton's idea that atoms themselves are indivisible in ordinary reactions but structure matters.

13. Two gases, A_2 and B_2, both diatomic, are introduced into a rigid 5.00 L vessel at 298 K. The partial pressures are measured as 0.400 atm and 0.600 atm, respectively. Calculate the mole fraction of each gas and determine the average molecular weight of the gas mixture. Discuss how Dalton's law of partial pressures and atomic theory combine to explain these observations.

14. A sample of lithium metal reacts completely with excess water, generating hydrogen gas. If 0.345 g of lithium produces 0.110 L of H_2 at 25°C and 1 atm, calculate the empirical formula of the hydride formed and analyze how the stoichiometry aligns with Dalton's principle of whole-number ratios.

15. A chemist isolates a transition-metal complex with the formula $[Fe(CN)_6]^{3-}$. Spectroscopic analysis indicates that Fe is in the +3 oxidation state. If 0.125 g of the complex contains 0.0178 g of iron, determine the molar mass of the complex

and discuss how variable oxidation states illustrate the refinement of Dalton's assumption that atoms of the same element are identical.

16. A 1.50 g sample of a binary compound containing calcium and oxygen yields 0.935 g of calcium after thermal decomposition. Determine the empirical formula of the compound and compare it to theoretical expectations based on atomic weights, highlighting how quantitative analysis underpins atomic theory.

17. A chemist observes that two isotopes of chlorine, ^{35}Cl and ^{37}Cl, have natural abundances of 75.8% and 24.2%, respectively. Calculate the average atomic mass of chlorine and evaluate how isotopic variation challenges the original notion that all atoms of an element are identical in mass.

18. A noble gas sample at 300 K has a density of 3.52 g/L at 1.20 atm. Determine the molar mass of the gas, then discuss how this aligns with the placement of noble gases on the periodic table and their full valence shells predicted from electron configurations.

19. A halogen X_2 reacts with hydrogen to form HX. If 0.872 g of X_2 reacts with excess H_2 producing 1.356 g of HX, determine the molar mass of X and identify the halogen. Then explain how the reaction exemplifies Dalton's whole-number combination principle.

20. A sample containing 1.25 g of magnesium reacts with 0.95 g of oxygen. After forming the oxide, a second experiment with 2.50 g of magnesium yields 1.90 g of oxide. Determine the law of multiple proportions ratio and analyze its significance for validating Dalton's atomic postulates.

21. A chemist studies the electrolysis of molten NaCl. Passing a current sufficient to transfer 2.00×10^{20} electrons produces 0.064 g of sodium. Calculate the number of moles of sodium formed and discuss how Faraday's constant and atomic theory jointly explain the stoichiometry of the process.

22. In a study of metallic conductivity, silver, copper, and gold wires of identical length are measured. Their resistivities are 1.59×10^{-8}, 1.68×10^{-8}, and 2.44×10^{-8} $\Omega \cdot m$, respectively. Analyze how electron mobility, as predicted by atomic structure, governs these differences and how this relates to Dalton's emphasis on discrete particle behavior.

23. An unknown compound contains only nitrogen and oxygen. Combustion analysis of 0.500 g of the compound yields 0.350 g of nitrogen oxides. Determine the simplest whole-number ratio of N to O atoms and relate this to Dalton's assertion that atoms combine in fixed integer ratios.

24. A chemist synthesizes two copper salts: CuCl and $CuCl_2$. When 0.400 g of Cu reacts with chlorine, the mass ratios of Cl in each salt differ. Calculate the multiple proportion and comment on its historical significance for confirming Dalton's atomic theory.

25. A 0.250 g sample of H_2O_2 decomposes into O_2 and H_2O. If the volume of oxygen collected at 1 atm and 298 K is 0.056 L, determine the mole fraction of O_2 and verify whether the decomposition stoichiometry aligns with atomic-level predictions.

26. In a periodic table exercise, a student compares first ionization energies across Period 3. The sequence of Na → Mg → Al shows 496, 738, and 578 kJ/mol. Explain the anomaly in aluminum using electron configurations, then discuss how this reinforces modern refinements to Dalton's atomic ideas.

27. A sample of zinc contains two isotopes, ^{64}Zn (48.6%) and ^{66}Zn (27.9%). If 0.300 g of zinc is analyzed, determine the number of nuclei of each isotope and discuss how isotopic composition affects atomic mass calculations, challenging Dalton's original assumptions.

28. A compound containing 0.200 g of sulfur reacts completely with 0.800 g of oxygen to form a sulfate. Determine the empirical formula and explain how the fixed mass ratio supports Dalton's postulate of discrete, indivisible atoms.

29. A noble metal catalyzes hydrogenation of an alkene at a rate proportional to surface area. If the reaction rate doubles when particle size halves, explain how atomic arrangement on the surface, electron configuration, and particle geometry support Dalton's later refinements to atomic theory.

30. A chemist studies the formation of Fe_2O_3 and Fe_3O_4. If 2.80 g of iron reacts with oxygen to form Fe_2O_3 and 2.80 g reacts with oxygen to form Fe_3O_4, determine the multiple proportion of oxygen involved and analyze the result in light of Dalton's laws.

31. A gaseous element with molar mass 32.0 g/mol is confined in a 1.50 L container

at 0.980 atm and 300 K. Calculate the number of molecules and explain how Avogadro's hypothesis extends Dalton's idea of discrete atoms.

32. A chemist examines bromine's diatomic molecular form in liquid versus gas phase. If the vapor density is measured as 159 g/L at 1 atm, determine whether the molecular formula aligns with Br_2 and discuss Dalton's atomic rationale.

33. A 1.00 g sample of an unknown oxide containing a transition metal M is heated with carbon, producing 0.574 g of pure metal. Determine the oxidation state of M in the oxide and evaluate how this experiment reflects the predictability of atomic combination ratios.

34. Hydrogen chloride gas is synthesized by the reaction of 0.100 g of hydrogen with excess chlorine. If the theoretical yield is 3.65 times the observed yield, calculate the actual amount of HCl formed and discuss potential experimental errors within the context of Dalton's atomic framework.

35. A chemist examines the ionization energy trend of Group 1 elements. For Li, Na, and K, the first ionization energies are 520, 496, and 419 kJ/mol. Explain these trends in terms of atomic structure and electron shielding, extending Dalton's atomic postulates.

36. In studying carbon allotropes, 0.100 mol of diamond and 0.100 mol of graphite are exposed to identical pressures. Predict the relative compressibility and analyze how atomic bonding arrangement overrides elemental identity, a refinement beyond Dalton's original model.

37. A sample of oxygen is enriched in ^{18}O. If 0.050 mol of this sample weighs 1.03 g, calculate the isotopic ratio and discuss its significance for geochemical tracing, showing the limits of Dalton's assertion of identical atoms.

38. A chemist analyzes the conductivity of 0.050 M Na_2SO_4 at 25°C, measuring 9.50 mS/cm. Using limiting molar conductivities, determine the degree of dissociation and explain how ionic mobility reflects discrete atomic charges.

39. A compound contains 0.400 g nitrogen and 0.900 g oxygen. Combustion analysis confirms the compound is N_2O_5. Calculate the mass ratio and explain how this result validates Dalton's atomic combination rules.

40. A sample of chlorine gas reacts with excess potassium bromide, producing 0.150 g of bromine. If the reaction involves halogen displacement, determine the moles of Cl_2 used and relate the process to periodic reactivity trends.

41. A chemist studies the thermal decomposition of $KClO_3$ to KCl and O_2. If 2.45 g $KClO_3$ produces 0.73 g O_2, calculate the molar ratio of potassium to oxygen in the compound and confirm the integer relationship predicted by Dalton.

42. A solution of $Mg(NO_3)_2$ is electrolyzed, depositing 0.244 g Mg on the cathode after passing 0.250 F of charge. Calculate the Faraday constant implied and relate it to Dalton's atomic concept applied in electrochemistry.

43. A chemist investigates Fe^{2+} oxidation to Fe^{3+} by dichromate. If 0.500 g Fe^{2+} reacts with 0.312 g $Cr_2O_7^{2-}$, determine the stoichiometric coefficient of iron to dichromate and relate to whole-number atomic ratios.

44. A halogen X has a molar mass of 70 g/mol. Its diatomic molecule reacts with hydrogen to form HX. Calculate the stoichiometric coefficients for the reaction and identify X, then explain the atomic basis for diatomic formation.

45. A noble gas sample with density 2.90 g/L at 1.50 atm and 298 K is analyzed. Determine its molar mass and identify the element. Discuss the relationship between full valence shells, inertness, and atomic stability.

46. A chemist measures ionization energies for the second period: Li 520, Be 899, B 801 kJ/mol. Predict the chemical reactivity trends and explain anomalies using orbital configurations, refining Dalton's view of identical atoms.

47. A compound of Mg and O contains 0.243 g Mg and 0.324 g O. Determine the empirical formula, calculate the expected formula mass, and discuss how this experiment supports Dalton's quantitative approach to chemical composition.

48. A chemist studies the solubility of NaCl in water at 25°C. Dissolving 5.85 g NaCl in 100 g water yields a saturated solution. Calculate the molar concentration and explain how the dissolution process reflects discrete ion formation and mobility.

49. A sample of $CaCO_3$ is decomposed thermally, producing 0.441 g of CO_2. Determine the mass of $CaCO_3$ decomposed and the empirical formula of the oxide formed. Analyze how atomic-level rearrangement illustrates Dalton's idea that atoms themselves are not destroyed in chemical reactions.

50. In a study of isotopic enrichment, 0.200 g of uranium contains 0.198 g ^{238}U and 0.002 g ^{235}U. Calculate the isotopic fraction of ^{235}U, discuss its nuclear significance, and explain how isotopes refine, but do not invalidate, Dalton's original atomic theory.

Chapter 4

Lab Techniques and Quantitative Chemistry

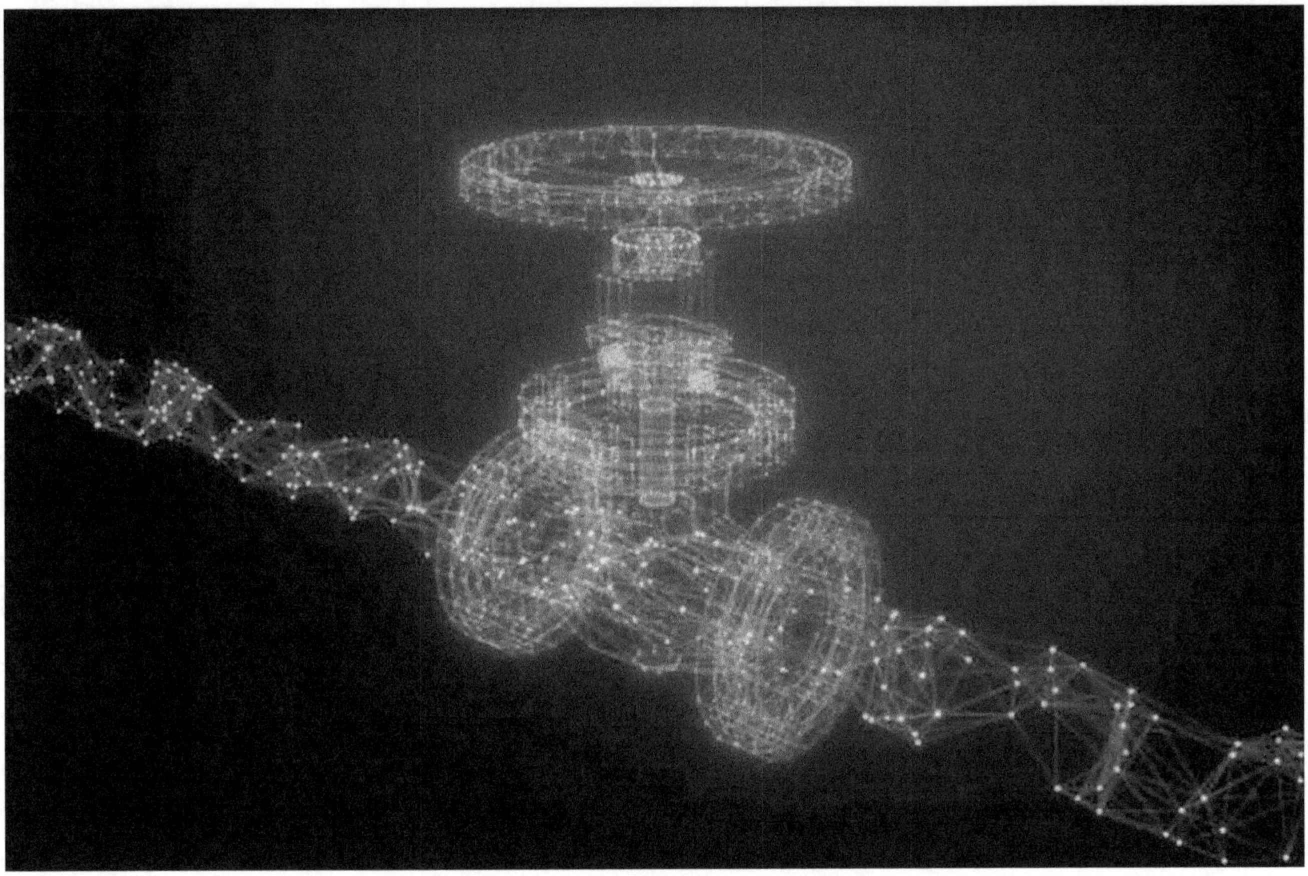

Chemistry is fundamentally an experimental science, and the practice of chemistry is inseparable from careful observation and precise measurement. Unlike purely theoretical sciences, chemistry relies on the ability to manipulate matter under controlled conditions, allowing chemists to uncover the underlying principles that govern the behavior of substances. The laboratory serves as both a testing ground and a window into the molecular world, where hypotheses are verified through experimentation, patterns are discerned, and predictive models are developed. At the heart of this process lies the importance of accurate measurement. A single error in weighing a substance or measuring a volume of liquid can propagate through an experiment, resulting in faulty conclusions or inconsistent results. Understanding how to use laboratory equipment effectively is therefore a critical skill for anyone working in the chemical sciences.

Balances are among the most fundamental tools in a chemistry laboratory, allowing chemists to determine the mass of substances with precision. Analytical balances, in particular, are designed to measure extremely small differences in mass, often to 0.01 grams or even less, and are indispensable when preparing solutions or conducting reactions that require exact quantities of reactants. Such precision is crucial, as many chemical reactions depend on strict stoichiometric relationships, where the ratio of reactants determines the efficiency and outcome of the reaction. Digital balances, with their rapid response and clear readout, provide convenience and accuracy, whereas traditional beam balances, though requiring more careful handling and calibration, offer a reliable alternative when digital instruments are unavailable. Regardless of type, all balances must be properly maintained, calibrated, and used according to standardized procedures to ensure consistency in experimental work.

Volumetric glassware, including pipettes, burettes, and volumetric flasks, is equally essential in chemistry laboratories, enabling accurate measurement of liquid volumes. Unlike solid masses, liquids can be challenging to measure precisely due to meniscus formation and the risk of parallax error, which occurs when measurements are read from an angle rather than at eye level. Proper technique in using volumetric glassware involves careful alignment with the eye, gentle handling to avoid splashes, and the selection of containers appropriate for the desired measurement. Additionally, calibration of glassware prior to use is critical for accuracy, as even minor deviations in volume can lead to significant errors in subsequent calculations. The combination of precise mass measurement and accurate volume measurement allows chemists to conduct experiments with reproducibility and confidence, providing reliable data that forms the basis for both qualitative and quantitative analysis.

When dealing with collections of objects, counting individual items directly is often impractical, particularly when the quantities involved are large. For example, if a sample of peanuts weighs fifty grams and a request is made for five hundred peanuts, manually counting each peanut would be tedious and inefficient. Instead, by weighing a representative sample of the peanuts, it is possible to determine the average mass of a single peanut and then calculate the total mass needed for five hundred peanuts through simple multiplication. Although individual peanuts vary slightly in mass, using the average provides an effective approximation that greatly simplifies the problem. This approach is not limited to peanuts; it can be applied to any collection of objects with variable masses, such as walnuts, almonds, or even manufactured components, by determining the mass-to-count ratio. This principle exemplifies a fundamental concept in chemistry: indirect counting. By knowing the average mass of individual units, chemists can infer the number of atoms, molecules, or ions in a given sample, even when direct counting is impossible.

Atoms and molecules are extraordinarily small, and their tiny size makes direct counting impractical, necessitating alternative methods for quantification. Chemists use the atomic mass unit, defined as approximately 1.66×10^{-24} grams, to express the mass of individual atoms. This unit provides a practical scale for discussing the masses of elements and compounds, which are otherwise exceedingly small. The periodic table, one of the most powerful tools in chemistry, provides the average atomic mass for each element, taking into account the natural abundance of isotopes. For instance, nitrogen has an average atomic mass of 14.01 amu, reflecting the weighted contribution of its isotopes. This means that a sample with a mass of 28.02 amu contains precisely two nitrogen atoms. The use of atomic masses expressed in decimal form captures these averages, allowing chemists to relate the mass of a sample to the number of constituent particles and to perform precise calculations in quantitative analysis.

The mole is a concept that bridges the gap between microscopic particles and macroscopic quantities that can be measured in the laboratory. Defined as 6.022×10^{23} entities—Avogadro's number—the mole allows chemists to connect the observable mass of a substance with the number of atoms, ions, or molecules it contains. For example, twelve grams of carbon corresponds to one mole of carbon atoms, containing 6.022×10^{23} individual atoms. This relationship enables direct calculations of particle numbers from measurable masses, forming the basis for stoichiometric calculations in chemical reactions. Mastery of mole-based conversions between grams, moles, and particles is essential for understanding chemical composition and predicting reaction outcomes.

Molar mass extends the mole concept to compounds, allowing chemists to calculate the mass of one mole of a given substance by summing the molar masses of its constituent elements. For instance, sodium chloride has a molar mass of 58.45 grams per mole, obtained by adding the molar masses of sodium (23.0 g/mol) and chlorine (35.45 g/mol). Water, with a molar mass of 18.016 grams per mole, illustrates the principle for a molecular compound: two hydrogen atoms (2.016 g/mol) combined with one oxygen atom (16.00 g/mol) yield the total molar mass. Knowledge of molar mass allows chemists to convert between the mass of a sample and the corresponding number of moles, facilitating accurate preparation of solutions, precise reaction planning, and effective scaling of experiments from the laboratory to industrial production.

Chemical formulas convey essential information about the composition of substances. The empirical formula represents the simplest whole-number ratio of elements in a compound, while the molecular formula reflects the actual number of atoms of each element present. Determining empirical formulas involves converting the mass of each element in a compound to moles, dividing by the smallest number of moles to find a ratio, and rounding to the nearest whole number. The molecular formula is then derived by comparing the molar mass of the compound to the mass of the empirical formula, adjusting subscripts accordingly. For example, a compound with a molar mass of 180 grams per mole and an empirical formula of CH_2O corresponds to a molecular formula of $C_6H_{12}O_6$. Understanding the relationships between empirical and molecular formulas allows chemists to predict chemical behavior, design reactions, and verify the composition of unknown substances.

Stoichiometry—the quantitative study of chemical reactions—relies on balanced chemical equations to provide the mole ratios necessary for calculating the amounts of reactants and products involved in reactions. To perform stoichiometric calculations, chemists convert the masses of reactants to moles, apply mole ratios derived from the balanced equation, and convert back to mass to predict the quantities of products formed. For example, in the combustion of methanol (CH_3OH) with oxygen to produce carbon dioxide and water, the balanced equation specifies the exact mole ratio between reactants and products, enabling precise determination of the amounts generated. A critical aspect of stoichiometry is the identification of the limiting reagent—the reactant that is fully consumed first—which dictates the maximum amount of product obtainable. Any remaining reactants are considered excess, and knowledge of these quantities is crucial for optimizing reaction efficiency and minimizing waste.

The theoretical yield of a reaction represents the maximum amount of product that could be formed based on the stoichiometric calculations involving the limiting reagent. In practice, however, the actual yield obtained in the laboratory is often lower due to incomplete reactions, side reactions, or practical limitations in handling and purification. The efficiency of a reaction is expressed through the percent yield, calculated as the ratio of actual yield to theoretical yield multiplied by one hundred. This metric allows chemists to assess reaction efficiency, troubleshoot experimental procedures, and refine laboratory techniques. A deep understanding of these concepts is essential not only for experimental work but also for industrial applications, where efficiency, cost-effectiveness, and scalability are critical considerations.

Quantitative chemistry extends far beyond laboratory exercises into numerous practical applications in industry, medicine, and environmental science. In pharmaceutical manufacturing, stoichiometric calculations ensure that active ingredients are combined in precise proportions, producing medications that are both safe and effective. Industrial chemical processes, such as the synthesis of ammonia or the controlled combustion of fuels, depend on accurate mole-based calculations to optimize yields, reduce waste, and minimize energy consumption. Environmental chemistry relies on quantitative methods to monitor pollutant formation, measure greenhouse gas emissions, and assess the impact of chemical reactions on ecosystems. By mastering laboratory techniques, understanding the mole, calculating atomic and molar masses, interpreting chemical formulas, and applying stoichiometry, chemists gain the ability to predict, control, and optimize chemical processes in a wide variety of contexts.

In conclusion, the study of chemistry integrates precise measurement, systematic observation, and rigorous quantitative analysis. Laboratory equipment, careful technique, and a thorough understanding of concepts such as atomic mass, the mole, molar mass, chemical formulas, stoichiometry, limiting reagents, and percent yield form the foundation of both theoretical and practical chemistry. By combining these principles, chemists can quantify chemical changes, design experiments that produce reliable results, and apply their knowledge to real-world problems ranging from pharmaceuticals to environmental protection. Through continuous practice and careful attention to detail, the chemist gains not only the ability to manipulate matter effectively but also the insight to understand the molecular and atomic structures that govern the natural world.

The accurate measurement of matter in a laboratory setting goes beyond simply reading numbers from an instrument; it involves a deep understanding of precision, accuracy, and the sources of error that can influence experimental outcomes. Precision refers to the consistency of repeated measurements, while accuracy denotes how close a measurement is to the true or accepted value. Achieving both requires careful technique, proper calibration of equipment, and attention to environmental conditions such as temperature, humidity, and vibrations. Even seemingly minor factors, such as the placement of a balance in a drafty area or touching the pan with ungloved hands, can introduce significant deviations in results. Recognizing and controlling these variables is a crucial skill for chemists, as small errors in measurement can propagate and affect subsequent calculations, leading to incorrect conclusions about the chemical processes under study.

Laboratory techniques for measuring solids, liquids, and gases vary depending on the substance and the required level of precision. For solids, analytical balances allow the measurement of powders, crystals, and other materials to fractions of a milligram. Careful transfer of samples using spatulas, weighing boats, or scoops prevents loss of material and contamination, both of which can compromise accuracy. Volumetric analysis for liquids demands mastery of pipetting techniques, proper rinsing of glassware to prevent dilution or concentration errors, and precise reading of menisci to avoid parallax error. Burettes, commonly used in titrations, require slow and controlled delivery of reagents, with readings recorded at the bottom of the meniscus to ensure reliability. Volumetric flasks, designed for preparing solutions of exact concentration, must be filled to the calibration mark using careful eye-level observation. Each of these practices contributes to the reproducibility of experimental results, which is essential for validating scientific hypotheses.

The concept of indirect measurement is particularly powerful in chemistry, where counting individual atoms or molecules is impossible due to their minute size. In the macroscopic world, indirect measurement can be illustrated through simple examples, such as estimating the number of seeds in a fruit or the number of grains of sand in a small container. By weighing a representative sample and calculating the average mass of a single unit, it is possible to extrapolate to much larger quantities with remarkable accuracy. This principle is directly applicable in chemistry, where chemists use mass and volume measurements to infer the number of particles present in a sample. The concept of the mole formalizes this approach, allowing the transformation of macroscopic quantities into microscopic counts and vice versa. By linking measurable quantities to the number of atoms or molecules, chemists gain a powerful tool for understanding and predicting chemical behavior.

The historical development of the atomic theory further underscores the importance of careful measurement and quantitative reasoning in chemistry. While the idea that matter is composed of indivisible units dates back to ancient Greek philosophers such as Democritus, it was John Dalton in the early nineteenth century who provided experimental evidence supporting the concept of atoms. Dalton's meticulous measurements of gas volumes and chemical reactions allowed him to propose that each element is composed of identical atoms, that atoms of different elements have characteristic masses, and that chemical reactions involve the rearrangement of these atoms in simple ratios. These insights laid the groundwork for modern chemical theory and established the critical link between experimental observation and scientific understanding.

Atoms themselves are extraordinarily small, with masses so minute that direct measurement in grams is impractical. The atomic mass unit (amu) provides a practical scale for expressing the mass of individual atoms, and the periodic table presents the weighted average of isotopic masses for each element. For example, the atomic mass of chlorine, 35.45 amu, reflects the combined natural abundance of its two stable isotopes, chlorine-35 and chlorine-37. Understanding these averages enables chemists to calculate the masses of molecules and compounds with precision, allowing the design of experiments that involve exact ratios of reactants. The ability to relate macroscopic mass measurements to microscopic particle counts is fundamental to quantitative chemistry and forms the basis for stoichiometric analysis.

Moles and molar mass provide chemists with a universal counting system that bridges the atomic and macroscopic scales. The definition of the mole as 6.022×10^{23} entities, known as Avogadro's number, allows chemists to relate the mass of a substance measured in grams to the number of atoms, ions, or molecules it contains. This conversion is vital in preparing chemical reactions, as the stoichiometric ratios derived from balanced chemical equations depend on the relative numbers of reactant particles. For example, in the reaction of hydrogen gas with oxygen to form water, two moles of hydrogen molecules react with one mole of oxygen molecules to produce two moles of water molecules. Understanding these relationships allows chemists to predict the quantities of products and optimize the amounts of reactants used, reducing waste and enhancing efficiency in both laboratory and industrial settings.

The calculation of molar mass extends this principle to compounds, providing a direct link between the chemical formula and measurable mass. For instance, calculating the molar mass of glucose, $C_6H_{12}O_6$, involves summing the molar masses of six carbon atoms, twelve hydrogen atoms, and six oxygen atoms to obtain 180.16 g/mol. This knowledge allows chemists to prepare solutions of known concentration, determine limiting reagents, and calculate theoretical yields with confidence. Accurate determination of molar mass is also critical in analytical chemistry, where it forms the basis for gravimetric analysis, volumetric analysis, and the identification of unknown compounds. In laboratory practice, a single miscalculation in molar mass can propagate through multiple steps of an experiment, highlighting the importance of precision in both measurement and computation.

Chemical formulas provide a concise way to represent the composition of compounds, and interpreting these formulas is essential for quantitative chemistry. The empirical formula represents the simplest ratio of elements, while the molecular formula specifies the actual number of each atom present. Determining these formulas requires the careful conversion of masses to moles, comparison of ratios, and, in some cases, the use of additional information such as molecular weight determined experimentally. For instance, a compound with a molar mass of 180 g/mol and an empirical formula of CH_2O has a molecular formula of $C_6H_{12}O_6$, revealing the true number of atoms in the molecule. Such calculations are foundational in organic chemistry, where understanding molecular structure is essential for predicting chemical reactivity and designing synthetic pathways.

Stoichiometry, the quantitative analysis of chemical reactions, relies on these fundamental concepts. Balanced chemical equations not only represent the identity of reactants and products but also provide the mole ratios necessary for predicting the amounts of substances involved. In practice, performing stoichiometric calculations involves several steps: converting the mass of a reactant to moles, applying mole ratios from the balanced equation, and converting the resulting moles back to mass or volume as required. Identifying the limiting reagent—the reactant that will be consumed first—ensures that calculations of maximum product yield are accurate. For example, in the combustion of ethanol, C_2H_5OH, with oxygen, determining which reactant is limiting allows chemists to predict the exact amount of carbon dioxide and water formed, an essential step in both laboratory and industrial applications.

The concepts of theoretical yield, actual yield, and percent yield provide a framework for assessing reaction efficiency. The theoretical yield represents the maximum quantity of product that could be formed based on stoichiometric calculations, while the actual yield is the amount obtained experimentally. Percent yield, expressed as the ratio of actual yield to theoretical yield multiplied by one hundred, provides a measure of the efficiency and effectiveness of the reaction. Differences between theoretical and actual yields may result from incomplete reactions, side reactions, or procedural losses during handling and purification. Understanding these differences is critical for evaluating experimental procedures, improving laboratory techniques, and optimizing chemical processes on an industrial scale.

Laboratory experiments often extend beyond simple calculations to include careful observation of chemical behavior, monitoring reaction rates, and understanding reaction mechanisms. Reaction kinetics, for example, examines how the speed of a reaction depends on factors such as concentration, temperature, and the presence of catalysts. By measuring changes in concentration over time, chemists can determine reaction rates and develop models to predict the behavior of similar reactions under different conditions. Such studies are essential in chemical engineering, environmental chemistry, and pharmaceuticals, where reaction efficiency and control are paramount.

Quantitative chemistry is also integral to solution preparation, titration, and analysis of unknown substances. Preparing a solution of known concentration requires precise weighing of solutes, accurate measurement of solvent volumes, and careful mixing to ensure homogeneity. In titrations, a solution of known concentration (the titrant) is gradually added to a solution of unknown concentration until a reaction is complete, as indicated by a color change or other observable endpoint. Accurate titration results depend on proper technique, correct selection of indicators, and careful calibration of glassware. These procedures exemplify how theoretical concepts, such as molarity, stoichiometry, and chemical reactivity, are applied directly in laboratory practice.

The mole concept and stoichiometric calculations also find application in real-world industrial processes. The Haber process for ammonia synthesis, for example, relies on precise stoichiometric ratios of nitrogen and hydrogen gases to optimize yield. Any deviation from the ideal ratio reduces efficiency and increases cost. In environmental chemistry, accurate quantification of pollutants such as carbon dioxide or nitrogen oxides depends on stoichiometric analysis, allowing for the design of strategies to reduce emissions and mitigate environmental impact. Similarly, in pharmaceutical manufacturing, stoichiometry ensures the correct formulation of drugs, preventing overdosing or underdosing and ensuring safety and efficacy.

Understanding chemical reactions and quantitative relationships further allows chemists to explore complex systems, such as equilibrium reactions, acid-base chemistry, and redox processes. In equilibrium, the relative concentrations of reactants and products remain constant over time, and the equilibrium constant provides a quantitative measure of the system's composition. Acid-base reactions involve the transfer of protons, and precise knowledge of molar concentrations and reaction stoichiometry is essential for calculating pH and preparing buffer solutions. Redox reactions, which involve the transfer of electrons, require careful attention to both mass and charge balance, emphasizing the interconnected nature of measurement, stoichiometry, and chemical theory.

In laboratory practice, attention to safety, precision, and procedure is as important as theoretical knowledge. Proper handling of chemicals, use of personal protective equipment, and adherence to standardized protocols prevent accidents and ensure reliable results. Accurate labeling of solutions, correct disposal of waste, and careful record-keeping are integral to good laboratory practice. The combination of careful observation, precise measurement, and quantitative reasoning allows chemists not only to understand chemical behavior but also to control it, leading to innovations in materials, energy, medicine, and environmental protection.

Chapter 4: Word Problems

1. A chemist calibrates an analytical balance using a 10.0000-g Class-S standard mass. Over a week of experiments involving preparation of a 0.0250 M primary-standard KHP solution, she notices that calculated concentrations drift upward by approximately 0.8%. She suspects balance drift caused by temperature

fluctuations that altered the buoyant force on the calibration weight. If the true mass is 10.0000 g but the balance systematically reads low by x%, determine x and calculate how the miscalibration propagates into the molarity determination of KHP when exactly 0.612 g is weighed for solution preparation.

2. A solution is prepared by dissolving an impure hydrated salt, $CoCl_2 \cdot xH_2O$, in a volumetric flask. A 1.214 g portion is dissolved and diluted to 250.0 mL. Titration with excess $AgNO_3$ shows that 0.01550 mol of chloride is present. Assuming all chloride originates from $CoCl_2 \cdot xH_2O$, determine the hydration number x and the percent purity of the sample. Explain how measurement uncertainties in the burette (±0.02 mL) and balance (±0.001 g) influence your final values.

3. A research group synthesizes a metal–organic complex requiring a 1:3 stoichiometric ratio of metal nitrate to ligand. Using a 50.00 mL burette, they add ligand solution (0.1025 M) to 0.248 g of metal nitrate (molar mass unknown). The reaction reaches its endpoint at 37.62 mL. Assuming complete reaction and that the complex contains no side products, determine the molar mass of the metal nitrate and propose which transition metal it may correspond to. Discuss how parallax error in burette readings could affect the identification.

4. A gas sample is collected over water at 23.0 °C in a 2.000 L flask. The total pressure measures 742.3 torr. After drying the gas, its mass is found to be 2.138 g. Given that the vapor pressure of water at 23.0 °C is 21.1 torr, identify the gas by calculating its molar mass. Evaluate how uncertainties in temperature (±0.5 °C) propagate to vapor pressure corrections and ultimately to molar mass identification.

5. A 250.00 mL volumetric flask is used to prepare a buffer containing acetic acid and sodium acetate. The student accidentally fills the flask above the calibration line and withdraws 2.50 mL using a volumetric pipette to compensate. After adjusting back to the mark, the final total acetic acid concentration is measured to be 0.212 M—0.9% lower than expected. Determine the original overfill volume and explain how volumetric error affects Henderson–Hasselbalch calculations for buffer pH.

6. A powdered catalyst is weighed into a tared porcelain crucible. Upon heating to 900 °C, its mass decreases from 7.382 g to 6.955 g due to decomposition. If the catalyst contains 13.2% physically adsorbed water and the remainder decomposes

according to $M_xO_y \rightarrow xM + y/2\ O_2(g)$, determine x and y given the observed mass loss and estimate the catalyst's empirical formula. Discuss how air drafts and buoyancy corrections in high-precision gravimetry alter your results.

7. A batch synthesis of an active pharmaceutical ingredient (API) requires reacting 0.250 mol of an amine with a stoichiometric amount of acyl chloride. The acyl chloride reagent decomposes slowly in moisture following first-order kinetics with $k = 0.0031\ min^{-1}$. If the reagent is exposed to ambient air for 25 minutes before use, determine the effective loss in available reactant and calculate how this changes the limiting reagent and theoretical yield. Assess the significance of reaction timing in industrial quality-control settings.

8. A sealed reaction vessel initially contains 4.50 g of $PCl_5(g)$. At 250 °C it reaches equilibrium according to $PCl_5(g) \rightleftharpoons PCl_3(g) + Cl_2(g)$,and the final pressure is 3.18 atm in a 1.000 L container. Using the ideal gas law, determine the equilibrium constant Kp. Evaluate how instrument precision in pressure (±0.01 atm) and volume (±0.005 L) affect the accuracy of the calculated equilibrium constant.

9. During a kinetic study of the iodination of acetone in acidic solution, the absorbance at 470 nm is recorded every 30 seconds. If a 1.00 cm cuvette is used and the molar absorptivity $\varepsilon = 855\ M^{-1}\ cm^{-1}$, determine the rate constant for the reaction assuming pseudo-first-order conditions and provide the integrated rate-law plot slope. Evaluate how stray light, cuvette fingerprints, and baseline drift influence kinetic accuracy.

10. A chemist must determine the empirical formula of a newly isolated organometallic compound. Combustion analysis of a 0.215 g sample yields 0.471 g of CO_2 and 0.193 g of H_2O. A separate elemental analysis reveals the presence of a transition metal whose mass fraction is 35.2%. Determine the empirical formula, then discuss how systematic analytical errors—such as incomplete combustion—would bias the formula.

11. A calorimetry experiment is conducted to determine the enthalpy of dissolution of an unknown ionic solid. A 1.842 g sample is added to 125.0 g of water contained in a coffee-cup calorimeter with a heat capacity of 42.3 J/K. The temperature rises from 21.4 °C to 24.9 °C. Determine the molar enthalpy of dissolution and identify whether the dissolution is exothermic or endothermic. Discuss how imperfect

insulation and temperature-probe lag influence the accuracy of the result.

12. A titration of a polyprotic acid H_3A is performed using standardized 0.2500 M NaOH. The equivalence volumes occur at 14.22 mL, 28.54 mL, and 42.77 mL. If 0.402 g of the acid was dissolved in a 100.00 mL volumetric flask before the titration, determine the molar mass of the acid and comment on how deviations from ideal polyprotic behavior could distort equivalence-point spacing.

13. A chemist attempts to determine the density of a volatile organic compound using a pycnometer. The pycnometer has a calibrated volume of 24.92 mL at 25 °C. When filled with distilled water, its mass is 45.332 g; when filled with the organic liquid, its mass is 39.118 g. Correct for buoyancy using air density 0.00119 g/mL and determine the true density of the organic compound. Discuss the significance of thermal expansion of both water and the pycnometer.

14. A mixture of three gases—CO_2, N_2, and an unknown inert gas—is trapped in a 3.000 L vessel at 298 K with total pressure 2.115 atm. Combustion analysis shows that removing all CO_2 decreases the pressure by 0.422 atm, and removal of N_2 decreases the remaining pressure by 1.118 atm. Determine the molar mass of the unknown inert gas and identify the most likely candidate. Comment on how nonideal behavior might shift partial pressures.

15. A chemist prepares a 0.1000 M Fe^{2+} solution but notices that after 48 hours the concentration has fallen due to slow oxidation to Fe^{3+} by dissolved oxygen. If the oxidation follows pseudo-first-order kinetics with $k = 2.6\times10^{-5}\ s^{-1}$, calculate the remaining Fe^{2+} concentration after 48 hours and explain how ignoring redox kinetics could significantly undermine quantitative redox titrations.

16. A metal alloy is analyzed using EDTA titration. A 0.3124 g sample is dissolved and treated with excess EDTA, requiring 22.14 mL of 0.05000 M EDTA to reach endpoint. The alloy contains two metals, $M1^{2+}$ and $M2^{3+}$, but only the divalent metal forms a 1:1 EDTA complex under the chosen pH. Determine the mass percent of the divalent metal and discuss how competitive complexation and pH stability of EDTA affect the measurement.

17. A sealed quartz cuvette containing 2.50×10^{-5} M dye solution is irradiated with a laser at 420 nm. The absorbance decreases from 0.812 to 0.517 in 160 seconds.

Assuming first-order photobleaching, determine the quantum yield if the photon flux is 3.2×10^{16} photons/s and the cuvette area is 1.0 cm^2. Discuss the influence of scattering and reflection losses on the calculated quantum yield.

18. A student prepares a solution of Na_2CO_3 as a primary standard for acid–base titrations. A 0.4562 g sample is dried at 110 °C for 2 hours, but incomplete dehydration may leave residual $Na_2CO_3\cdot H_2O$. After titration, calculated molarities of standardized HCl show a 1.7% positive bias. Estimate the fraction of monohydrate remaining and analyze the consequences for standardization accuracy.

19. A catalytic hydrogenation experiment consumes 14.7 mL of H_2 gas at 27 °C and 1.00 atm. The catalyst mass is 28.4 mg. Determine the turnover number (TON) and turnover frequency (TOF) if the reaction takes 12 minutes. Describe how gas-measurement uncertainties in syringe-based systems affect kinetic interpretation.

20. A buffer made from NH_3 and NH_4Cl is prepared to be pH 9.50. Using a 500.0 mL volumetric flask, the student dissolves 18.2 g NH_4Cl and adds concentrated NH_3(aq) (14.8 M) until the desired pH is reached. If the buffer capacity toward added acid is 0.123 mol/L·pH unit, determine the ratio $[NH_3]/[NH_4^+]$ and the volume of concentrated NH_3 required. Discuss how ionic-strength effects alter the Henderson–Hasselbalch prediction.

21. A polymerization reaction proceeds via free-radical chain growth with an initiator that decomposes with kd = 6.5×10^{-4} s^{-1} at 65 °C. If only 64% of the polymer has the desired degree of polymerization after 20 minutes, determine the initiator efficiency and discuss how temperature fluctuations alter propagating-radical concentrations and molecular-weight distributions.

22. A combustion calorimeter is used to determine ΔHc of an organic compound. A 0.682 g sample causes the temperature of the 2.45 kg water jacket to rise by 2.84 °C. If the calorimeter constant is 752 J/K, determine the molar heat of combustion. Evaluate how heat losses through the bomb head and incomplete combustion would skew the calculated enthalpy.

23. A 100.0 mL solution contains an unknown metal ion M^{n+}.

Spectrophotometric determination using a chelating dye gives an absorbance of 0.664 at 525 nm ($\varepsilon = 1.52\times10^4$ M^{-1} cm^{-1}, 1.00 cm path length). After adding excess complexing agent and diluting to 250.0 mL, the absorbance becomes 0.897. Determine the charge n assuming Beer's law linearity and complete complexation. Discuss deviations from Beer's law at high concentrations.

24. To determine chloride in seawater, Mohr titration is performed on a 15.00 mL sample requiring 24.88 mL of 0.1000 M $AgNO_3$ to reach the endpoint. Considering competing precipitation of carbonate and chromate indicator solubility, determine the chloride concentration and explain how endpoint visibility affects accuracy in high-salinity matrices.

25. An industrial reactor converts SO_2 to SO_3 via the contact process. At 450 °C, a 1.00 L vessel initially contains 0.800 mol SO_2 and 0.400 mol O_2. At equilibrium, 0.512 mol SO_3 is present. Determine Kc and evaluate how deviations from ideal gas behavior at elevated temperatures influence Kc determination.

26. A chemist determines the molecular weight distribution of a polymer by GPC. The detector response yields Mn = 27,500 g/mol and Mw = 45,200 g/mol. If trace cross-linking increases Mw by 8% but does not affect Mn, calculate the true polydispersity index and discuss how poor column calibration distorts molecular-weight averages.

27. A redox titration is performed where 0.01500 mol Fe^{2+} is oxidized by Ce^{4+}. The Ce^{4+} solution is standardized but contains 0.35% Ce^{3+} impurity. Determine the required volume of the "0.1000 M" Ce^{4+} solution to reach the endpoint and evaluate how impurities in oxidizing agents bias analytical chemistry results.

28. An electrogravimetric experiment deposits Cu metal from Cu^{2+} solution onto a platinum electrode. The current is held at 0.145 A for 18.5 minutes. Determine the mass of copper deposited and assess how current efficiency less than 100%—due to side reactions like hydrogen evolution—affects gravimetric precision.

29. A 0.2000 g impure sample containing $CaCO_3$ and inert silica is reacted with excess HCl(aq). The reaction releases 42.6 mL CO_2 at 22.0 °C and 0.987 atm. Determine the percent $CaCO_3$ by mass and discuss how dissolved CO_2 and water vapor corrections influence gas-evolution analyses.

30. A transition-metal complex exhibits ligand substitution kinetics with measured half-life 85 s. Temperature studies reveal an activation energy of 62 kJ/mol. Predict

the half-life at 45 °C and analyze how instrumental timing resolution influences kinetic-mechanism interpretation.

31. A chemist studies the solubility of $CaSO_4$ by measuring conductivity. A saturated solution at 25 °C has conductivity 2.12×10^{-4} S/cm with molar conductivity $\lambda = 156$ $S\cdot cm^2/mol$ at infinite dilution. Determine the molar solubility, and discuss how ion pairing in sparingly soluble salts can cause underestimation of solubility.

32. A sample of hydrated copper sulfate is heated strongly until completely dehydrated. If 3.482 g of $CuSO_4\cdot xH_2O$ becomes 2.223 g of anhydrous $CuSO_4$, determine x and analyze how residual moisture retained in the lattice affects gravimetric hydration numbers.

33. A gas-chromatographic analysis of a mixture of alkanes gives retention times sensitive to column temperature. If the temperature fluctuates ±1.5 °C around a 75 °C set point, causing retention-time shifts of 0.8%, determine how this affects quantitation when peak areas are normalized. Discuss the role of thermal stability in GC reproducibility.

34. A photochemical reaction converts compound A into B with quantum yield 0.45. If irradiated for 7 minutes with light power 18.5 mW at 365 nm, calculate the moles of B formed from an initial 1.50×10^{-3} mol A and discuss how internal filter effects influence actual photon absorption.

35. A titration curve for a weak diprotic acid shows pH values at half-equivalence volumes of 3.62 and 7.51. If 0.285 g of the acid is dissolved in 120.00 mL before titration with 0.1050 M NaOH, determine the molar mass and discuss how overlapping equilibria complicate pKa determination.

36. A membrane electrode used to measure fluoride ion activity has slope −54.2 mV/decade at 25 °C. If the measured potential for a solution is −182.4 mV, determine the fluoride concentration relative to a 1.00 mM standard. Discuss how ionic strength and liquid-junction potentials influence ion-selective electrode accuracy.

37. A gas-phase decomposition follows second-order kinetics with k = 0.0044 L/mol·s. If the initial pressure of reactant A is 415 torr in a 2.00 L vessel at 330 K, determine the time required for pressure to fall to 210 torr and discuss how using pressure instead of concentration affects rate-law interpretation.

38. A complexometric titration determines magnesium content in a mineral supplement. A 0.514 g sample requires 19.44 mL of 0.05000 M EDTA in buffered solution. However, Mn^{2+} impurities compete for EDTA with lower stability constant. Estimate the percent error in magnesium determination if 3.5% of EDTA is complexed by Mn^{2+}.

39. A laser-induced breakdown spectroscopy (LIBS) experiment analyzes trace metals in soil. A calibration curve for Pb gives slope 0.182 counts/ppm and intercept 1.55 counts. If the measured intensity is 11.4 counts, determine the Pb concentration and discuss how matrix effects and plasma temperature instability influence quantitation.

40. A 2.500 g sample of impure oxalic acid dihydrate is titrated with 0.09500 M $KMnO_4$. The endpoint occurs at 31.22 mL. Determine the purity of the sample and evaluate how manganese dioxide formation as a side reaction affects the observed equivalence volume.

41. A dynamic light scattering (DLS) experiment on a nanoparticle suspension yields an autocorrelation decay time $\tau = 1.82\times10^{-5}$ s. Using viscosity 0.890 cP at 25 °C, determine the hydrodynamic diameter and comment on how polydispersity and multiple scattering distort DLS sizing.

42. A UV–Vis spectrum shows two overlapping $\pi\rightarrow\pi^*$ bands at 260 nm and 278 nm. Deconvolution reveals absorbances 0.442 and 0.318 respectively at 1.00 cm path length. Given molar absorptivities $\varepsilon_1 = 1.1\times10^4$ and $\varepsilon_2 = 8.2\times10^3$ M^{-1} cm^{-1}, determine the concentrations of each chromophore and discuss limits of linear combination analysis.

43. A student measures the freezing-point depression of a nonelectrolyte solute. A 0.812 g sample lowers the freezing point of 25.00 g benzene by 0.426 °C. Given Kf = 5.12 °C·kg/mol, determine the molar mass and discuss how supercooling and solvent impurities alter freezing-point behavior.

44. A precipitation titration with Ba^{2+} is used to quantify sulfate. A 50.00 mL wastewater sample requires 17.88 mL of 0.1500 M $BaCl_2$. Determine sulfate concentration and comment on how colloidal $BaSO_4$ formation affects endpoint accuracy.

45. A mixed acid solution contains HCl and H_3PO_4. Titration with 0.2000 M NaOH

requires 12.44 mL to reach the first equivalence point and 28.33 mL to reach the second. Determine concentrations of both acids and discuss how weak-acid buffering affects equivalence-point detection.

46. A chemist uses a differential scanning calorimeter (DSC) to determine melting point and purity of an organic compound. The DSC trace shows an onset at 118.2 °C and peak maximum at 121.4 °C. If impurity depresses the melting point according to the van't Hoff equation, estimate impurity mole fraction and discuss baseline correction challenges in DSC.

47. A solid–liquid extraction isolates caffeine from tea leaves. A 5.12 g sample yields 42.8 mg caffeine after three sequential extractions. If individual extraction efficiency is E, determine E and analyze how solvent saturation and plant-matrix binding lower ideal extraction yields.

48. A redox reaction between dichromate and an organic alcohol proceeds with stoichiometry requiring 3 mol alcohol per mol dichromate. If 0.556 g dichromate yields 0.812 g oxidized product, determine the limiting reagent and percent yield. Discuss how incomplete oxidation pathways lead to side-product formation.

49. An acid–base titration determines the degree of polymerization (DP) of a polyamide. Hydrolysis generates monomers that titrate to 0.1570 M HCl with volume 12.22 mL from a 0.224 g polymer sample. Determine DP and comment on how incomplete hydrolysis biases polymer characterization.

50. A chemical reactor produces ester from carboxylic acid and alcohol with a 2:1 stoichiometric requirement of alcohol. Starting with 0.620 mol acid and 1.10 mol alcohol, the reaction reaches equilibrium with 0.224 mol ester formed. Determine the equilibrium constant and evaluate how temperature-dependent activity coefficients influence equilibrium calculations in liquid-phase systems.

Chapter 5

Blueprints of Chemical Changes

Chemical reactions form the very foundation of chemistry, representing the myriad ways in which atoms interact, rearrange, and combine to form new substances. At the most fundamental level, a chemical reaction occurs whenever bonds between atoms are broken and reformed, producing products that are chemically distinct from the original substances, known as reactants. To describe these processes with precision, chemists employ chemical equations, a symbolic language that represents the transformation of matter in a concise and systematic way. In a chemical equation, reactants are always placed on the left-hand side, while products appear on the right, separated by an arrow that indicates the direction of the reaction. This arrow is not merely symbolic; it signifies the transformation of matter under controlled conditions, whether in a laboratory, in nature, or in industrial applications. The ability to accurately write and balance chemical equations is central to understanding chemistry, because it directly reflects the Law of Conservation of Mass, which asserts that atoms can neither be created nor destroyed in a chemical reaction. This principle necessitates that every atom present in the reactants must also appear in the products, leading to the essential skill of balancing equations, which ensures that the number of each type of atom is identical on both sides of the reaction.

Balancing chemical equations begins with identifying the number of atoms of each element in the reactants and products. Consider the formation of water from hydrogen and oxygen. The unbalanced equation, $H_2 + O_2 \rightarrow H_2O$, initially shows two hydrogen atoms on both sides, which are already balanced. However, oxygen presents an imbalance, with two atoms on the reactant side but only one in the product. By placing a coefficient of two before H_2O, the equation becomes $H_2 + O_2 \rightarrow 2H_2O$, yet the hydrogen atoms are now unbalanced. Correcting this by placing a coefficient of two before H_2 results in the fully balanced equation, $2H_2 + O_2 \rightarrow 2H_2O$, reflecting equal numbers of both hydrogen and oxygen atoms on each side. This simple example illustrates the systematic approach required in all chemical reactions: first assessing the atom count, then applying coefficients to achieve balance, while ensuring the integrity of each molecular formula is maintained. Subscripts within chemical formulas must never be altered, as doing so would change the identity of the molecule itself, a fundamental violation of chemical principles. Coefficients, however, provide a flexible method to maintain stoichiometric balance without altering molecular identity.

Chemical reactions constitute the heart of chemistry, representing processes in which atoms are rearranged to form new substances. These rearrangements occur through the breaking and forming of chemical bonds, transforming the starting materials, called reactants, into new substances called products. Chemists use chemical equations to represent these reactions symbolically, providing a concise way to capture the shuffling of atoms. In such an equation, reactants are placed on the left side, products on the right, and an arrow separates them to indicate the direction of the reaction. A double-headed arrow can represent a reversible reaction, reflecting the dynamic equilibrium that may exist under certain conditions. The writing and balancing of chemical equations is governed by the Law of Conservation of Mass, which states that atoms cannot be created or destroyed in chemical reactions. Consequently, the number of each type of atom on the reactant side must equal that on the product side. Achieving this balance requires both logic and methodology, ensuring that chemical equations accurately reflect the stoichiometry of the reaction.

Balancing chemical equations often begins with simple inspection, especially for straightforward reactions. For example, consider the formation of water from hydrogen and oxygen: $H_2 + O_2 \rightarrow H_2O$. Counting atoms shows that hydrogen is already balanced, but oxygen is not. By placing a coefficient of two in front of H_2O, we temporarily obtain $H_2 + O_2 \rightarrow 2H_2O$, yet this unbalances hydrogen. Correcting this with a coefficient of two in front of H_2 gives $2H_2 + O_2 \rightarrow 2H_2O$, producing the final balanced equation. While inspection works for simple reactions, more complex systems often require a systematic method such as the Algebraic Method, which uses algebra to balance equations reliably, even when multiple reactants and products are involved.

The Algebraic Method involves assigning variables to each compound in the chemical equation and then writing equations based on the conservation of each element. For instance, consider the formation of iron(III) oxide from iron and oxygen: $a\ Fe + b\ O_2 \rightarrow c\ Fe_2O_3$. Assigning coefficients a, b, and c, we then construct equations for each element. For iron, the equation is $a = 2c$, since there are two iron atoms in each Fe_2O_3 molecule. For oxygen, the equation is $2b = 3c$, reflecting two oxygen atoms per O_2 molecule and three per Fe_2O_3 molecule. Solving this system by choosing the smallest integer values gives $c = 2$, $a = 4$, $b = 3$, resulting in the balanced equation: $4Fe + 3O_2 \rightarrow 2Fe_2O_3$. The Algebraic Method is particularly advantageous for reactions that are too complex for simple inspection, providing a systematic, error-free approach for balancing even the most complicated chemical equations.

A further example illustrates this approach: the reaction of aluminum with hydrochloric acid to produce aluminum chloride and hydrogen gas: $Al + HCl \rightarrow AlCl_3 + H_2$. Assign variables a, b, c, d to Al, HCl, $AlCl_3$, and H_2, respectively. Balancing aluminum gives $a = c$, chlorine gives $b = 3c$, and hydrogen gives $b = 2d$. Choosing $c = 2$ produces integer coefficients $a = 2$, $b = 6$, $d = 3$, yielding the balanced equation: $2Al + 6HCl \rightarrow 2AlCl_3 + 3H_2$. The Algebraic Method ensures precision and is indispensable in laboratory, industrial, and educational settings.

Physical states of reactants and products provide additional context in chemical equations. Solids, liquids, gases, and aqueous solutions are denoted by (s), (l), (g), and (aq), respectively. These notations are essential, especially for solution chemistry where ions interact in water. Strong electrolytes, such as sodium chloride, dissociate completely into ions, conducting electricity efficiently, whereas weak electrolytes, such as acetic acid, partially dissociate, producing a limited current. Non-electrolytes, such as sugar, do not dissociate and cannot conduct electricity. Solubility rules, which predict whether compounds will dissolve or precipitate, are essential for understanding reactions in solution. For example, nitrate salts are universally soluble, whereas halides and sulfates have important exceptions, such as silver chloride and barium sulfate. Insolubility leads to precipitate formation, which can be represented in net ionic equations that focus on only the reactive ions, omitting spectator ions that do not participate in the reaction.

Redox reactions, or oxidation-reduction reactions, are central to chemistry, involving the transfer of electrons between species. Oxidation refers to electron loss, while reduction refers to electron gain. For example, when magnesium reacts with oxygen to form magnesium oxide, Mg loses two electrons to become Mg^{2+}, and O gains two electrons to become O^{2-}, forming MgO. The half-reaction method allows systematic balancing of redox reactions by separating oxidation and reduction processes, balancing atoms and charges individually, then combining the half-reactions to produce a fully balanced overall equation. This method is particularly important in acidic or basic solutions, where H^+, OH^-, and H_2O may be used to achieve balance, ensuring that both mass and charge are conserved.

Chemical reactions are classified based on patterns of atom rearrangement. Acid-base reactions involve proton transfer: an acid donates a proton, and a base accepts it, producing water and a salt. Neutralization occurs when exact stoichiometric amounts of acid and base react completely. Precipitation reactions involve the formation of an insoluble solid from soluble reactants, such as the reaction of NaCl(aq) with $AgNO_3$(aq) to produce AgCl(s). Combustion reactions involve rapid oxidation, typically producing CO_2 and H_2O, and release significant energy, useful for heating, electricity, and transportation. Combination reactions involve the joining of elements or compounds to form a single product, while decomposition reactions break down compounds into simpler substances, often as a first step in chemical synthesis or energy production.

Nuclear chemistry extends these principles to the atomic nucleus, where stability is determined by the ratio of neutrons to protons. Nuclides are isotopes of elements with distinct neutron counts. Stable nuclides lie within the "zone of stability," whereas unstable nuclides undergo radioactive decay, emitting alpha particles, beta particles, positrons, or gamma rays until a stable configuration is reached. The energy changes in nuclear reactions, calculated via Einstein's equation $\Delta E = \Delta mc^2$, are immense compared to chemical reactions, with mass defects accounting for the energy released in fusion and fission. Nuclear fusion combines light nuclei to form heavier, stable nuclei, while fission splits heavy nuclei into smaller nuclei, releasing energy harnessed in reactors and weapons. Radioactive decay follows predictable kinetics, characterized by half-life, the time required for half of a sample to decay. The study of these processes is essential for energy production, medicine, and understanding fundamental atomic structure.

Balancing chemical equations accurately, whether by inspection or algebraic methods, underpins all of these concepts. Proper balancing ensures compliance with the Law of Conservation of Mass, provides the foundation for stoichiometric calculations, and allows chemists to predict yields, analyze reactions in solution, and study redox processes or nuclear transformations. The Algebraic Method, in particular, provides a systematic and reliable approach to equations that are too complex for trial-and-error methods, integrating seamlessly with the study of physical states, solubility, redox chemistry, acid-base reactions, and nuclear chemistry. Mastery of this method, along with a deep understanding of chemical principles, equips students, researchers, and professionals to analyze, predict, and apply chemical reactions across a wide spectrum of scientific, industrial, and technological contexts.

In addition to ensuring mass balance, chemists must consider the physical states of substances, as these states often influence the reaction pathway and outcome. Solids, liquids, gases, and aqueous solutions are indicated by the symbols (s), (l), (g), and (aq), respectively. Such notations allow chemists to visualize the reaction environment and understand how molecules interact, especially in solution chemistry where ions play a dominant role. For example, aqueous solutions of ionic compounds dissociate into their constituent ions, which can then participate in further reactions, whereas non-electrolyte solutions do not dissociate and therefore behave differently. When balancing equations and predicting reactions, knowledge of physical states is essential, as it affects solubility, reactivity, and the formation of precipitates, gases, or liquids.

Chemical reactions are accompanied by observable changes in matter, providing the practical means to detect and analyze them. Rusting of steel, bleaching of fabrics, mixing of culinary ingredients, and the spectacular combustion of fireworks all represent chemical changes. Key indicators of chemical reactions include the evolution of gas bubbles, formation of a solid precipitate, color changes, and energy transfer in the form of heat, light, or sound. These observable signs confirm that a chemical transformation has occurred, even if the molecular rearrangements are invisible to the naked eye. Each reaction, regardless of scale or context, involves the shuffling of atoms to produce new substances, and this rearrangement is faithfully captured through chemical equations, which describe both the reactants and products in a systematic manner. The arrow separating the reactants and products indicates not only the direction of the reaction but can also represent reversibility when a double-headed arrow is used, reflecting the dynamic equilibrium that exists in some chemical systems.

The types of chemical reactions are diverse, encompassing numerous pathways through which atoms and molecules interact. Acid-base reactions, for example, involve the transfer of protons (H^+) from an acid to a base, producing water and a salt in a process known as neutralization. A classic example is the reaction between hydrochloric acid and potassium hydroxide: $HCl(aq) + KOH(aq) \rightarrow H_2O(l) + KCl(aq)$. In this reaction, HCl serves as the proton donor, or acid, while KOH acts as the proton acceptor, or base, resulting in the formation of water and the ionic compound potassium chloride. Understanding the stoichiometry of such reactions is critical, as the molar ratios of reactants determine whether a solution is neutralized or remains acidic or basic. Similarly, precipitation reactions occur when two aqueous solutions of soluble salts are combined, resulting in the formation of an insoluble solid, or precipitate. For example, mixing aqueous solutions of sodium chloride and silver nitrate produces solid silver chloride and aqueous sodium nitrate: $NaCl(aq) + AgNO_3(aq) \rightarrow AgCl(s) + NaNO_3(aq)$. Spectator ions, which do not participate in the formation of the precipitate, are omitted in the net ionic equation, emphasizing the species directly involved in the reaction: $Ag^+(aq) + Cl^-(aq) \rightarrow AgCl(s)$.

Redox reactions, short for oxidation-reduction reactions, are another fundamental category of chemical transformations. These reactions involve the transfer of electrons between species, with one element undergoing oxidation (loss of electrons) and another undergoing reduction (gain of electrons). For instance, the reaction of magnesium metal with oxygen to form magnesium oxide can be written as $2Mg(s) + O_2(g) \rightarrow 2MgO(s)$. Analyzing this reaction at the atomic level, magnesium atoms each lose two electrons to form Mg^{2+} cations, while oxygen atoms gain electrons to form O^{2-} anions. This electron transfer explains why magnesium is oxidized and oxygen is reduced, exemplifying the core principle of redox chemistry. Redox reactions are ubiquitous in both industrial processes, such as metal refining and energy production, and biological systems, including cellular respiration and photosynthesis. The half-reaction method provides a systematic approach for balancing complex redox reactions, particularly in acidic or basic solutions, by separating the oxidation and reduction components, balancing atoms and charges individually, and then recombining them to form a complete balanced equation.

The solubility of compounds in water is intimately linked to their chemical behavior, influencing reaction pathways and the formation of electrolytes. Soluble salts, such as nitrates, as well as sodium and ammonium compounds, readily dissociate into their respective ions in aqueous solution, allowing them to conduct electricity as strong electrolytes. In contrast, weak electrolytes, such as acetic acid, partially dissociate, producing a limited flow of ions and weak conductivity. Non-electrolytes, like sugar, dissolve in water without generating ions and therefore do not conduct electricity. Predicting solubility involves understanding both the cation and anion properties, as well as exceptions to general rules, such as the insolubility of silver chloride or barium sulfate. Solubility rules are essential for determining whether a reaction in solution will produce a precipitate, a gas, or remain fully dissolved, which directly impacts both laboratory practices and industrial chemical processes.

Moving from chemical reactions to nuclear processes, the stability of an atomic nucleus is determined by the delicate balance of protons and neutrons, collectively called nucleons. The nucleus, despite its minuscule size, contains nearly all of an atom's mass and exhibits immense energy density. Protons carry a positive charge, neutrons are neutral, and both are composed of quarks, the fundamental constituents of matter. The atomic number, Z, represents the number of protons in a nucleus, while the mass number, A, is the total number of protons and neutrons. Isotopes share the same atomic number but differ in mass number, and a nuclide refers to a specific isotope of an element. Nuclear stability is influenced by the neutron-to-proton ratio, with lighter nuclei stable when the ratio approaches one, while heavier nuclei require a higher ratio to counterbalance the electrostatic repulsion between protons. Certain "magic numbers" of protons or neutrons, such as 2, 8, 20, 28, 50, 82, and 126, confer exceptional stability to nuclides, analogous to noble gases in chemical electron configurations.

Unstable nuclei undergo radioactive decay to achieve a more stable state, emitting alpha particles (helium nuclei), beta particles (electrons), positrons (antimatter electrons), or gamma rays (high-energy photons). Alpha decay decreases both the atomic and mass numbers, beta decay converts a neutron to a proton while increasing the atomic number, and positron emission converts a proton to a neutron. Gamma radiation allows an excited nucleus to release excess energy without changing its composition. The kinetics of radioactive decay follow a predictable pattern, characterized by half-life, the time required for half of the original nuclide to decay. Nuclear reactions, including fission and fusion, release energy orders of magnitude greater than chemical reactions due to the conversion of mass to energy, as described by Einstein's equation $\Delta E = \Delta mc^2$. In nuclear fusion, light nuclei combine to form a heavier, more stable nucleus, releasing tremendous energy as seen in stars, whereas nuclear fission involves splitting a heavy nucleus into lighter nuclei, a process harnessed in nuclear reactors and atomic weaponry.

The study of chemical and nuclear reactions highlights the extraordinary versatility of matter and energy transformations. In chemical systems, reactions are guided by the conservation of mass, the nature of bonding, and reaction kinetics, producing predictable patterns that allow chemists to design, control, and optimize reactions for industrial, environmental, and biological applications. In nuclear systems, the forces within the nucleus, the energy associated with binding nucleons, and the pathways of radioactive decay define the limits of stability and the potential for energy release. Both realms—chemical and nuclear—demonstrate the profound principle that matter and energy are intricately linked, governed by laws that dictate their behavior, and harnessed to improve technology, medicine, and our understanding of the universe.

Chemical reactions, whether involving ordinary molecules or the nuclei of atoms, are never random; they follow patterns dictated by stoichiometry, thermodynamics, and kinetics. Through careful study, chemists can predict which reactions will occur, under what conditions, and what products will result. Whether dissolving salts to produce electrolytes, combining acids and bases to form neutral solutions, generating energy through redox reactions, or manipulating nuclear processes to produce energy or medical isotopes, the principles outlined here serve as a roadmap for understanding the vast and complex landscape of chemical and nuclear phenomena. Mastery of these concepts, from balancing simple equations to analyzing radioactive decay series, equips students and scientists alike with the tools to explore, innovate, and apply the laws of nature to practical challenges.

Building on these foundational principles, modern chemistry extends beyond prediction and explanation into the realm of innovation and application. Analytical techniques such as spectroscopy, chromatography, and mass spectrometry allow chemists to probe the structure and composition of substances with unprecedented precision, revealing the subtle behaviors of molecules and ions. In parallel, computational chemistry leverages quantum mechanics and powerful algorithms to simulate reactions and molecular interactions before they are tested in the laboratory, accelerating discovery while minimizing risk. Materials chemistry harnesses these insights to design polymers, catalysts, and nanomaterials with tailored properties, enabling technologies ranging from renewable energy solutions to advanced electronics and medical devices. Biochemistry and chemical biology bridge chemistry with the life sciences, elucidating the molecular machinery of cells, the mechanisms of enzymes, and the pathways of metabolism, thereby opening doors to novel therapeutics and diagnostic tools. Ultimately, chemistry is not just a catalog of substances or reactions; it is a unifying framework that connects the microscopic and macroscopic worlds. By understanding the interplay of atoms, molecules, and energy, chemists transform theoretical knowledge into practical solutions, shaping the modern world and laying the groundwork for the scientific advances of tomorrow.

Chapter 5: Word Problems

1. A research team investigates the heterogeneous oxidation of methane in a catalytic reactor where methane and oxygen react on a platinum surface to form CO_2 and H_2O. The inlet stream contains methane and oxygen in a 1:2.5 molar ratio at 800 K, but catalytic microanalysis shows that side-reactions form trace CO and unreacted radicals. Assuming the global reaction pathway can still be represented

stoichiometrically as the complete combustion of methane, derive the balanced equation and compute the theoretical oxygen requirement per mole of methane. Then determine how many moles of O_2 remain unreacted per 100 moles of feed gas under the assumed inlet composition, ignoring all side products.

2. An industrial neutralization process feeds aqueous sulfuric acid of unknown concentration into a mixing tank where it reacts with excess aqueous potassium hydroxide. After mixing, calorimetric measurements reveal that exactly 1.54 kJ of heat is released per 100 g of the resulting solution. Given that the molar enthalpy of neutralization for strong acids and bases is –57.1 kJ/mol H^+, write the balanced chemical equation and determine the molarity of the original acid solution, assuming density effects and heat losses are negligible.

3. A geochemist studies the dissolution of calcium carbonate in acidic groundwater. The reaction $CaCO_3(s) + H^+(aq) \rightarrow Ca^{2+}(aq) + HCO_3^-(aq)$ proceeds until equilibrium is reached. If a sample of groundwater originally contains 4.0×10^{-4} M H^+ and is placed in contact with excess $CaCO_3$, determine the maximum amount (in moles per liter) of $CaCO_3$ that could dissolve, assuming the reaction proceeds until all available H^+ is consumed and no further equilibria (e.g., CO_2 evolution) occur.

4. A fuel cell uses hydrogen and oxygen to produce electricity and water vapor. Under operating conditions, the water vapor exits at 423 K and 1.5 atm. If 2.2 kg of H_2 is completely consumed and the reaction follows $2H_2 + O_2 \rightarrow 2H_2O(g)$, calculate the volume of water vapor produced at the stated conditions. Assume ideal-gas behavior at high temperature.

5. A materials scientist analyzes the corrosion of iron in seawater, where dissolved oxygen oxidizes Fe(s) to $Fe_2O_3{\cdot}xH_2O(s)$. Using the simplified redox reaction $4Fe + 3O_2 \rightarrow 2Fe_2O_3$, determine the mass of oxygen needed to fully oxidize a 3.75-kg steel panel that is 98.6% pure iron by mass. Then compute the mass percent of the oxide layer formed relative to the original iron mass.

6. A mineral processing operation burns pyrite (FeS_2) in air to produce SO_2 for downstream sulfuric acid production. The roasting reaction is $4FeS_2 + 11O_2 \rightarrow 2Fe_2O_3 + 8SO_2$. If the feed stream is 65.0% FeS_2 by mass and the remainder inert silica, calculate how many kilograms of SO_2 are produced per metric ton of feed. Assume complete conversion.

7. A pharmaceutical chemist observes that the hydration of an anhydrous metal salt, $MX_3(s)$, proceeds via $MX_3(s) + 6H_2O(l) \rightarrow MX_3{\cdot}6H_2O(s)$. If 38.4 g of the anhydrous salt absorbs water until fully hydrated and the molar masses of M and X are 55.8 g/mol and 35.5 g/mol, respectively, determine the mass of water absorbed and the percent mass increase resulting from hydration.

8. In an electrochemical cell intended to model biochemical redox pathways, NADH is oxidized to NAD^+ while O_2 is reduced to water. Using the half-reactions $NADH \rightarrow NAD^+ + H^+ + 2e^-$ and $O_2 + 4H^+ + 4e^- \rightarrow 2H_2O$, construct the balanced net ionic equation. Then compute how many moles of NADH are required to consume 0.250 mol of O_2, assuming idealized electron transfer.

9. A polymerization process requires chlorine gas to initiate free radical formation in styrene. The reaction forming radical chlorine is $Cl_2 \rightarrow 2Cl\cdot$. Although not a chemical equation to balance, production of Cl_2 occurs by the reaction $2NaCl + 2H_2O \rightarrow 2NaOH + H_2 + Cl_2$. Given a requirement of 0.450 mol of $Cl\cdot$ radicals, calculate the mass of NaCl required assuming 100% efficiency and that each Cl_2 molecule yields two radicals.

10. A ceramic manufacturing line uses the decomposition reaction of ammonium dichromate: $(NH_4)_2Cr_2O_7(s) \rightarrow Cr_2O_3(s) + N_2(g) + 4H_2O(g)$. If 5.00 g of the solid decomposes completely at 873 K and the produced gases expand to 15.0 L at that temperature under 1.0 atm, calculate the percent yield of gaseous products relative to theoretical ideal-gas volume.

11. A mixed-oxide catalyst converts CO and H_2 (syngas) to methanol over Cu/ZnO/Al_2O_3 in the overall reaction $CO + 2H_2 \rightarrow CH_3OH$. An experimental run feeds syngas with $CO{:}H_2 = 1{:}3$ (molar) and observes partial conversion with side-reactions producing CO_2 via the water-gas shift: $CO + H_2O \rightleftharpoons CO_2 + H_2$. Write balanced stoichiometric expressions for both reactions, then formulate (but do not solve) a system of algebraic equations that would determine the final mole fractions of CO, H_2, CO_2, H_2O, and CH_3OH given total feed, conversion of CO to both products, and one independent equilibrium constant for the WGS at reactor conditions.

12. A redox titration task requires determining the percent purity of iron metal in an ore sample. The ore is digested to produce Fe^{2+} in solution, then titrated with $KMnO_4$

in acidic medium, where MnO_4^- oxidizes Fe^{2+} to Fe^{3+}. Write the balanced net ionic redox equation for the titration in acidic solution and describe how you would use titrant volume and concentration to compute mass percent iron in the original ore, including stoichiometric factors.

13. During anodization of aluminum, the formation of an Al_2O_3 passivation layer consumes aluminum and oxygen in an electrochemical process. Propose balanced half-reactions and an overall reaction for oxide formation starting from Al(s) and O^{2-} (from water/oxide species), then calculate the theoretical mass of Al_2O_3 produced per coulomb of charge passed if three electrons per Al atom are involved (set up the calculation; numeric evaluation optional).
14. A buffer system composed of acetic acid and acetate keeps pH at 4.76. If 0.150 mol of acetic acid and 0.0250 mol of sodium acetate are mixed in 1.00 L, use the Henderson–Hasselbalch framework to estimate the pH, then discuss how the addition of 0.0100 mol strong base would change concentration ratios and pH (no numeric finishing required; show algebraic setup).
15. In an irreversible first-order decomposition A $\rightarrow$ products, a sample initially contains 0.500 mol of A and decays with rate constant $k = 0.0340\ s^{-1}$ at 350 K. Derive the expression for remaining moles of A as a function of time, then find the time required for the sample to drop to 10% of its initial amount (show steps algebraically).
16. A multi-step organic synthesis converts substrate S to product P by two consecutive reactions: S $\rightarrow$ I (k_1) and I $\rightarrow$ P (k_2). Under conditions where $k_1 \gg k_2$, describe qualitatively and quantitatively (rate expressions) how the concentration of intermediate I behaves in time, and propose which step determines overall yield and how you would modify conditions to minimize buildup of a reactive intermediate that decomposes sidewise.
17. A precipitation titration between Ba^{2+} and SO_4^{2-} forms $BaSO_4(s)$. Given a 25.00 mL aliquot of an unknown Ba^{2+} solution titrated with 0.01000 M Na_2SO_4 to the first detectable precipitate, write the balanced net ionic equation and describe how to calculate the molarity of Ba^{2+} from titrant volume, including stoichiometric coefficients and any assumptions about endpoint detection.

18. A radioisotope decays by successive β^- emissions: A → B → C (stable), with half-lives $t_{1/2,A}$ = 6.0 h and $t_{1/2,B}$ = 1.5 h. Starting with a pure 1.00-mol sample of A at t = 0, write the Bateman equations (or outline them) for the time-dependent activities and discuss qualitatively the time at which B reaches its maximum concentration relative to the half-lives.

19. An electroplating bath deposits copper from Cu^{2+} solution onto a cathode. If a current of 2.50 A is applied for 3.00 hours, calculate the theoretical mass of copper plated ($Cu \rightarrow Cu^{2+} + 2e^-$ reversed), and then discuss realistic factors (current efficiency, side reactions) that could reduce actual deposition yield and how to measure current efficiency experimentally.

20. A sealed high-pressure reactor carries out the gas-phase reversible reaction $N_2 + 3H_2 \rightleftharpoons 2NH_3$ at elevated pressure and 700 K. Using Le Chatelier's principle, predict how an increase in total pressure and an inert gas addition at constant volume affect the equilibrium position. Then write the balanced equation and express the equilibrium constant Kp in terms of partial pressures.

21. A mixed-valence manganese oxide catalyst undergoes redox cycling between Mn^{3+} and Mn^{4+} during selective oxidation. Propose balanced half-reactions (as ionic processes) for $Mn^{3+} \leftrightarrow Mn^{4+}$ interconversion in aqueous conditions, and explain how these redox couples can mediate electron transfer between substrate and molecular oxygen in catalytic turnover.

22. In a neutron irradiation experiment, stable ^{58}Fe captures a neutron to form ^{59}Fe which decays by β^- with a half-life of 44.5 days. Write the nuclear reaction for neutron capture and the decay equation, then describe how to compute the specific activity (Bq/g) of the produced ^{59}Fe after irradiation given flux, cross-section, irradiation time, and decay during and after irradiation (set up formulas).

23. A high-temperature solid oxide fuel cell (SOFC) uses the oxide ion conductor O^{2-} to transport charge. For the oxygen reduction at the cathode, write a balanced reaction converting $O_2(g)$ to O^{2-}(solid) and show how oxygen stoichiometry couples with electron flow—then connect this to the cell's current production per mole of O_2 consumed.

24. A synthetic chemist must balance a complex reaction in which chromium(VI) oxidizes an alcohol to a ketone in acidic medium, while Cr(VI) is reduced to Cr(III).

Propose the half-reactions for oxidation of a generic secondary alcohol $R_2CHOH \rightarrow R_2C{=}O + 2H^+ + 2e^-$ and for $Cr_2O_7^{2-} + 14H^+ + 6e^- \rightarrow 2Cr^{3+} + 7H_2O$, then combine them to give the balanced redox equation (include smallest integer coefficients) and discuss the role of acid concentration in the stoichiometry.

25. An environmental chemist analyzes nitrate reduction to nitrogen gas in anoxic sediments via multiple electron transfer steps ($NO_3^- \rightarrow NO_2^- \rightarrow NO \rightarrow N_2O \rightarrow N_2$). For the overall reduction $NO_3^- + 5e^- + 6H^+ \rightarrow 0.5\ N_2 + 3H_2O$, write the balanced equation (remove fractional coefficients if desired) and explain how to compute electron demand per mole of nitrate and the implications for organic carbon respiration requirements in sediments.

26. A graduate student must design a stoichiometric separation scheme for a metal complex: $FeCl_3(aq)$ reacts with excess KSCN to give the red $Fe(SCN)^{2+}$ complex in equilibrium with free ions. Write the net ionic equation for formation of $Fe(SCN)^{2+}$ from Fe^{3+} and SCN^-, discuss how Le Chatelier's principle can be used to shift the equilibrium toward complex formation, and propose an experimental protocol (brief) to quantify trace Fe^{3+} using this reaction.

27. A combustion engineer studies the stoichiometry of burning an industrial hydrocarbon C_xH_yS to form SO_2, CO_2, and H_2O. Given empirical composition $C_{3.2}H_{6.4}S_{0.1}$ (per mole basis), write the balanced combustion equation in O_2, include SO_2 formation, and determine the theoretical O_2 consumption per mole hydrocarbon (express algebraically).

28. In tracer experiments, a chemist replaces natural hydrogen with deuterium in a benzene ring to study kinetic isotope effects. Write the balanced hydrogen–deuterium exchange (conceptual) and explain how isotopic substitution alters zero-point energies and reaction rates, including which rate constants in a mechanism will show primary vs secondary kinetic isotope effects.

29. A pressure swing adsorption (PSA) unit separates O_2 from air by selectively adsorbing N_2 on a zeolite bed. If desorption is achieved by reducing pressure, explain the mass-balance and mole-fraction changes during adsorption and desorption steps, write stoichiometric balances for a simplified two-component system, and discuss how impurity levels in product O_2 are computed from cycle mass balances.

30. A classic fission energy calculation compares mass defects: ^{235}U undergoes fission to form ^{96}Kr and ^{137}Ba plus neutrons. Given atomic masses (symbolic variables), set up the mass-defect expression for energy released per fission using $\Delta E = \Delta mc^2$, and discuss how the distribution of kinetic energy among fragments and neutrons is determined experimentally (no numeric values required).

31. A proton transfer equilibrium between a weak acid HA and a base B ($HA + B \rightleftharpoons A^- + BH^+$) has measured equilibrium constant K. Derive expressions relating pK_a of HA, pK_b of B, and K; then show how the direction of the reaction depends on relative pK_a/pK_b values and how you would choose a suitable base to quantitatively deprotonate HA.

32. In an advanced titration, 0.200 g of an unknown dibasic acid H_2X is titrated to the first equivalence with 0.02500 M NaOH requiring 24.60 mL. Write the stoichiometric relation used to determine molar mass of H_2X, show the algebra to compute molar mass from titrant data, and discuss experimental sources of error that would bias the result (e.g., CO_2 absorption).

33. A heterogeneous catalytic hydrogenation uses Pd/C to reduce an alkene in the presence of acetic acid solvent. Propose a balanced equation for hydrogenation of a generic alkene RCH=CHR' to RCH_2CH_2R' using H_2, discuss how catalyst poisoning by sulfur compounds alters active site availability (Stoichiometric consequences), and suggest regeneration strategies.

34. A photochemical reaction converts cis- to trans-isomers via light absorption followed by isomerization. If the quantum yield is 0.35 and a 500-nm monochromatic light source delivers 2.00×10^{20} photons/s to the sample, calculate (set up expression) the rate of isomerization events per second, then discuss how back-isomerization and thermal relaxation may be included in a kinetic model.

35. In nuclear transmutation, ^{24}Mg captures an alpha to form ^{28}Si in an (α,γ) reaction. Write the nuclear reaction equation and discuss selection rules and Q-value considerations that determine whether radiative capture is energetically allowed, and outline how cross sections depend on projectile energy and nuclear structure.

36. A graduate lab investigates the kinetics of the iodide-iodate (Landolt) reaction in acidic conditions, which shows complex dependence on H^+. Present a balanced

net ionic equation for the reaction $IO_3^- + 5I^- + 6H^+ \rightarrow 3I_2 + 3H_2O$, then describe experimental strategies (mixed-flow, stopped-flow) to extract rate laws and determine reaction orders in $[IO_3^-]$, $[I^-]$, and $[H^+]$.

37. A battery material undergoes intercalation reaction $Li_xMO_2 \leftrightarrow Li_{x\text{-}\delta}MO_2 + \delta Li^+ + \delta e^-$ during charge. Write the stoichiometry of a small fractional deintercalation (δ per formula unit) and explain how capacity (mAh/g) is calculated from δ and molar mass, including the role of Faraday's constant and conversion factors.

38. A hazardous waste stream contains perchlorate (ClO_4^-) that must be reduced to chloride using catalytic hydrogenation on a metal catalyst. Propose plausible balanced redox pathways for stepwise reduction ($ClO_4^- \rightarrow ClO_3^- \rightarrow ClO_2^- \rightarrow Cl^-$) including electron and proton stoichiometry, and discuss potential formation of chlorinated byproducts and methods to suppress them.

39. A tracer study in metabolic chemistry uses ^{13}C-labeled glucose to follow carbon flow through glycolysis and the TCA cycle. Write a conceptual mass balance for the fate of one labeled carbon atom starting from glucose → pyruvate → acetyl-CoA → citrate, and explain how NMR or mass spec data can be used to deduce pathway branchpoints and scrambling.

40. In a continuous stirred-tank reactor (CSTR), the A + B → C reaction is second order overall (rate = k[A][B]). Given equal inlet concentrations CA0 of A and B and volumetric flow $F/V = \tau^{-1}$ (residence time inverse), derive the steady-state mass-balance equation for outlet concentration CA and show how multiple steady states could occur for autocatalytic modifications (describe qualitatively).

41. A graduate student examines ligand substitution in octahedral transition metal complexes where water is replaced by chloride: $[M(H_2O)_6]^{3+} + nCl^- \rightleftharpoons [MCl_n(H_2O)_{6-n}]$^(3⁻ⁿ) + nH_2O. For M = Cr^{3+}, write the net ionic equation for n = 3, discuss crystal-field effects on ligand preference, and explain how stepwise formation constants (β_1, β_2, β_3) relate to the observed distribution of species.

42. A novel radiopharmaceutical is produced by neutron activation of ^{97}Mo to ^{98}Mo followed by β decay to ^{98}Tc (fictional sequence for problem purposes). Outline the chain of nuclear reactions and decays, then propose how to separate chemically the technetium species from molybdenum using oxidation state differences and selective complexation, noting stoichiometric reagents.

43. In an advanced material synthesis, sol-gel hydrolysis of tetraethyl orthosilicate (TEOS) proceeds: $Si(OC_2H_5)_4 + 4H_2O \rightarrow Si(OH)_4 + 4C_2H_5OH$ (followed by condensation). Write the balanced hydrolysis and then the general condensation reaction between two silanols, and discuss how pH and water:TEOS ratio steer the network formation and final porosity in stoichiometric terms.

44. A high-precision isotope dilution mass spectrometry (IDMS) assay uses an enriched ^{15}N spike to quantify trace nitrate in water. Describe the stoichiometric basis of isotope dilution (mixing labeled spike with sample), write the mass-balance equations linking isotopic ratios to concentration, and outline steps to compute original nitrate concentration from measured isotope ratios.

45. An advanced oxidation process (AOP) generates hydroxyl radicals (·OH) by UV/H_2O_2 photolysis: $H_2O_2 + h\nu \rightarrow 2\cdot OH$. For a water treatment plant treating 1.00 m^3/min with $[H_2O_2]$ feed and known UV photon flux, set up the stoichiometric relation between H_2O_2 consumed and ·OH generated, discuss scavenging reactions with natural organic matter, and show how to estimate required H_2O_2 dose to achieve a target radical exposure (no final numeric answer required).

46. A reversible electron transfer between ferrocene (Fc) and a metal complex M^{3+} is studied: $Fc \rightleftharpoons Fc^+ + e^-$ and $M^{3+} + e^- \rightleftharpoons M^{2+}$. Combine these to write the chemical electron-transfer reaction and derive the Nernst equation expression for the cell potential E as a function of activities of Fc/Fc^+ and M^{3+}/M^{2+}, explaining how concentration changes shift redox equilibria.

47. A graduate course problem asks you to balance the combustion of a chlorinated hydrocarbon C_2HCl_3 (trichloroethene analog). Write the balanced combustion equation forming CO_2, H_2O, and HCl (assume complete combustion and chlorine ends up as HCl) and determine the stoichiometric oxygen requirement per mole fuel, discussing environmental implications of HCl formation.

48. In a transmutation reactor, successive neutron captures create long-lived actinide isotopes from ^{238}U. Describe the neutron capture and beta-decay chain that could lead from ^{238}U to plutonium isotopes, write generic nuclear reaction steps (n,γ) and β^-, and discuss how capture cross sections and neutron spectrum shape the isotopic inventory stoichiometrically.

49. A graduate calorimetry experiment measures the enthalpy change for the dissolution of NH_4NO_3 in water at infinite dilution. Propose the thermochemical equation for dissolution, describe how to design a calorimetric titration to obtain ΔH_dissolution per mole, and explain how solvation stoichiometry and heat capacity of the solution affect the measurement and derived molar enthalpy.

50. A heterogeneous catalytic cycle for olefin metathesis involves formation and breakdown of a metallacyclobutane intermediate. Write a conceptual balanced sequence (not full mechanistic detail) that consumes two olefin molecules to form two new olefins via the metal carbene catalyst, discuss the role of catalyst turnover number (TON) and turnover frequency (TOF) in stoichiometric productivity, and outline mass-balance considerations when scaling the reaction from lab to pilot plant.

Chapter 6

THE BEHAVIOR OF GASES UNVEILED

Matter exists in four fundamental physical states: solid, liquid, gas, and plasma, distinguished by particle arrangement, energy, and interactions. Solids have tightly packed particles that vibrate around fixed positions, giving solids a definite shape and volume. Liquids have particles that are closely packed but can slide past one another, giving liquids a definite volume but no fixed shape. Gases, however, consist of widely spaced particles moving at high velocities in all directions, which allows them to fill any container completely, regardless of shape, and makes them highly compressible. Plasma, often called the fourth state of matter, is an ionized gas with free electrons and ions, found naturally in stars, lightning, and fusion reactors. While all four states are important, in this chapter, we focus on gaseous states, their properties, and how they behave under different physical conditions.

Gases exhibit several fundamental behaviors that distinguish them from solids and liquids. They can expand indefinitely, are easily compressible, and mix completely with other gases without any mechanical intervention. This ability results from the very large distances between gas molecules relative to their size. The properties of gases, such as pressure, volume, and temperature, can be measured and related mathematically through various laws, which were developed over centuries of experimental observations. Gases are crucial in numerous natural and technological processes. For example, the Earth's atmosphere is composed mainly of nitrogen, oxygen, argon, and trace gases, and its behavior governs weather, climate, and life support systems. The greenhouse effect is another example, where gases like carbon dioxide, methane, and water vapor absorb infrared radiation and influence global temperature.

The pressure of a gas is one of its most fundamental properties. Pressure arises from the force of gas molecules colliding with the walls of their container. This is measured using instruments such as a barometer, which consists of a column of mercury in a glass tube. The height of the mercury column varies with atmospheric pressure: as pressure increases, mercury rises; as pressure decreases, mercury falls. This method led to the establishment of the unit millimeters of mercury (mmHg), also called a torr. Another commonly used unit is the standard atmosphere (atm), defined such that 1 atm = 760 mmHg = 760 torr. Pressure can also be expressed mechanically as force per area, using the equation $P = F/A$, where P is pressure, F is force in newtons (N), and A is area in square meters (m^2). This yields the SI unit pascal (Pa), where 1 Pa = 1 N/m^2. Converting between units, we have 1 atm = 101,325 Pa, which connects laboratory, atmospheric, and industrial measurements.

The first significant quantitative law for gases was discovered by Robert Boyle, an Irish chemist in the 17th century. Using a J-shaped glass tube filled with mercury, Boyle trapped a volume of air and varied the pressure by adding or removing mercury. He observed that increasing pressure caused the gas volume to decrease proportionally, and decreasing pressure allowed the volume to expand. This led to the law now known as Boyle's Law, mathematically expressed as: $PV = k$, where P is the pressure of the gas, V is its volume, and k is a constant at a given temperature. Boyle's Law demonstrates an inverse relationship: as one variable increases, the other decreases. Graphically, plotting V versus 1/P yields a straight line, confirming the hyperbolic nature of the relationship. For different conditions, the law can be applied as: $P_iV_i = P_fV_f$, where subscripts i and f refer to initial and final states of the gas. Boyle's Law is particularly accurate for low-pressure gases, where particle interactions are negligible, allowing gases to behave nearly ideally. This principle is applied in numerous fields, from diving and scuba physics to pneumatic systems and even the behavior of gases in lungs and syringes.

In the late 18th century, Jacques Charles, a French physicist, investigated the effect of temperature on gas volume. He observed that the volume of a gas increases linearly with its temperature, provided pressure is constant. This phenomenon occurs because heating increases the kinetic energy of gas particles, causing them to move faster and occupy more space. The mathematical representation of this relationship is Charles's Law, given by: $V = bT$, where V is volume, T is temperature in Kelvin, and b is a proportionality constant. To convert from Celsius to Kelvin, we use: $T_K = T_{°C} + 273$. Charles extrapolated the line to the temperature at which volume becomes zero, defining absolute zero (-273°C or 0 K). While absolute zero cannot be physically reached, laboratories have cooled gases to fractions of a Kelvin above this limit. For initial and final conditions, the relationship becomes: $V_i/T_i = V_f/T_f$. Charles's Law is foundational for hot-air balloon operation, thermal expansion calculations, and temperature-dependent gas volume predictions.

Italian chemist Amedeo Avogadro proposed in 1811 that equal volumes of gases at the same temperature and pressure contain the same number of molecules, regardless of chemical identity. This insight distinguishes between the mole of gas and its chemical identity. Avogadro's Law can be expressed mathematically as: $V = an$, where V is volume, n is the number of moles, and a is a constant. This relationship implies that doubling the number of moles at constant temperature and pressure doubles the volume. For comparing two states: $V_i/n_i = V_f/n_f$. Avogadro's Law is critical for calculating molar volumes, determining stoichiometry in reactions involving gases, and understanding gas mixtures in atmospheric science and industrial chemistry.

Building upon the individual gas laws of Boyle, Charles, and Avogadro, scientists recognized that pressure, volume, temperature, and the number of moles are interrelated, leading to the development of the Ideal Gas Law, which provides a comprehensive description of gas behavior under a wide range of conditions. The law is expressed mathematically as $PV = nRT$, where P represents the pressure of the gas, V is the volume, n is the number of moles, T is the temperature in Kelvin, and R is the universal gas constant. Depending on the units used, the value of R varies; in the common set of units where P is in atmospheres, V in liters, n in moles, and T in Kelvin, the value of R is 0.08206 L* atm/mol*K. This equation is sometimes referred to as the equation of state, because it defines the state of a gas when any three of its properties are known. Gases behave ideally when they obey this equation closely, which is generally true at high temperatures and low pressures, where particle interactions are minimal.

An important application of the Ideal Gas Law is in gas stoichiometry, which allows the prediction of gas volumes or moles in chemical reactions. For instance, consider a gas sample containing 1.00 mole at standard temperature and pressure (STP), defined as 0°C (273 K) and 1 atm. Using the Ideal Gas Law, the volume of this gas can be calculated as V = nRT/P = (1.00 mol)(0.08206 L*atm/mol*K)(273 K)/1.00 atm = 22.4 L. This result, 22.4 liters per mole, is referred to as the molar volume of an ideal gas at STP. This concept allows chemists to convert between moles and volume in chemical reactions, environmental analyses, and industrial gas measurements. For example, if 2 moles of oxygen are present at STP, they occupy 2 mol × 22.4 L/mol = 44.8 L, which can then be used to calculate reactant or product quantities in stoichiometric equations.

The Ideal Gas Law also provides a means to calculate the molar mass of a gas using its density. Since the number of moles n is related to mass m and molar mass M by n = m/M, substituting into the Ideal Gas Law gives P = nRT/V = mRT/MV. Rewriting in terms of density d = m/V, this becomes P = dRT/M , which can be rearranged to solve for molar mass: M = dRT/PM. This equation is particularly useful in identifying unknown gases by measuring their density at known pressure and temperature.

In mixtures of gases, each gas exerts a partial pressure, independent of the other components, as if it alone occupied the container. This principle, formulated by John Dalton in 1803, is known as Dalton's Law of Partial Pressures. The law states that the total pressure of a gas mixture is the sum of the partial pressures of its individual gases: $P_{Total} = P_1 + P_2 + P_3 + \ldots$. If the gases behave ideally, each partial pressure can be calculated using the Ideal Gas Law as $P_1=n_1RT/V$, $P_2=n_2RT/V$, $P_3= n_3RT/V$, and so on. Summing these pressures gives the total pressure $P_{Total} = (n_1 + n_2 + n_3 + \ldots)RT/V = n_{Total}RT/V$, illustrating that the total pressure depends only on the total number of moles in the system, rather than the chemical identity of the gases.

The mole fraction is a related concept, representing the fraction of the total number of moles contributed by a particular gas in a mixture. It is denoted by X and defined as $X_1 = n_1 n_{Total} = n_1 + n_2 + n_3 + \ldots$. Using the Ideal Gas Law, this can also be expressed in terms of pressure as $X_1 = P_1 P_{Total}$, $X_2 = P_2 P_{Total}$. $X_2 = P_{Tota} l\ P_2$, and so forth. Mole fractions are widely applied in chemical engineering, environmental science, and atmospheric chemistry, particularly when calculating partial pressures, predicting gas behavior in reactions, and analyzing compositions of gas mixtures.

While gas laws describe macroscopic relationships, the Kinetic Molecular Theory (KMT) provides a molecular-level explanation. KMT is based on several postulates: gas particles are considered to have negligible volume compared to the space between them, particles are in constant, random motion and collide elastically with one another and the container walls, there are no intermolecular forces acting between particles, and the average kinetic energy of the particles is directly proportional to the Kelvin temperature, given by $KE_{avg} = (3/2)kBT$ where kB is Boltzmann's constant. According to KMT, pressure arises from collisions of fast-moving particles with container walls, and temperature is a measure of the average kinetic energy of particles. The theory provides a molecular explanation for all gas laws: in Boyle's Law, increasing pressure compresses the gas because particles collide more frequently with walls; in Charles's Law, increasing temperature causes particles to move faster and expand the volume; in Avogadro's Law, increasing the number of moles increases volume at constant temperature and pressure; and in Dalton's Law, the total pressure is simply the sum of independent contributions from all particles.

Thus, the Ideal Gas Law, gas stoichiometry, molar volume, molar mass calculations, Dalton's Law, mole fraction, and the Kinetic Molecular Theory together form a comprehensive framework for understanding and predicting the behavior of gases. Each of these concepts is connected through equations that describe macroscopic measurements in terms of microscopic particle behavior, providing both theoretical insight and practical tools for scientific, industrial, and environmental applications.

In addition to macroscopic gas laws, the behavior of gases can be further understood through the concepts of diffusion and effusion, which describe the movement of gas molecules. Diffusion refers to the spontaneous mixing of gas particles due to their constant random motion. Graham's Law of Diffusion quantifies the rate at which one gas spreads into another, stating that the rates of diffusion of two gases are inversely proportional to the square roots of their molar masses. Mathematically, this is expressed as $Rate_1 Rate_2 = M_2 M_1$, where Rate 1 and Rate 2 are the diffusion rates of gases 1 and 2, and M_1 and M_2 are their respective molar masses. A lighter gas diffuses more rapidly than a heavier gas because its particles have higher average velocities at the same temperature. Effusion, on the other hand, is the process by which gas particles pass through a small hole or porous barrier without collisions, and it follows the same mathematical relationship, allowing predictions of the relative rates of effusion of different gases. These principles are not only important in laboratory experiments but also in industrial applications such as gas separation, leak detection, and even the design of airbags.

While the Ideal Gas Law provides a useful model, real gases deviate from ideal behavior under high pressures or low temperatures. These deviations occur because real gas particles occupy a finite volume and experience intermolecular attractions, unlike ideal gas assumptions where particle volume is negligible and no interactions exist. The observed pressure of a real gas, P_{obs}, can be expressed as $P_{obs} = nRTV - nb - a(n/V)^2$, where n is the number of moles, V is the container volume, T is the temperature, R is the gas constant, a is a constant correcting for intermolecular attractions, and b accounts for the finite volume occupied by gas molecules. The first term, $nRTV-nb$, represents the corrected pressure for volume exclusion, while the second term, $a(n/V)^2$, represents the pressure reduction due to attractions between particles.

The Van der Waals equation is a further refinement that provides a more accurate description of real gas behavior by combining these corrections into a single equation: $[P_{obs}+a(n/V)^2](V-nb)=nRT$. Here, the term $P_{obs} + a(n/V)^2$ is the corrected pressure, and $V-nb$ is the corrected volume, reflecting the true free space available for particle motion. By applying this equation, chemists and engineers can predict the behavior of gases under high pressures, low temperatures, or conditions approaching condensation, where ideal gas assumptions fail. For example, the Van der Waals equation accurately models the behavior of carbon dioxide in pressurized cylinders and explains the critical temperature and pressure conditions required for liquefaction.

Real gas behavior also explains the phenomena observed in compressed gases, refrigeration systems, and liquefied natural gas (LNG) storage, where intermolecular forces significantly influence pressure, volume, and temperature relationships.

By adjusting for volume exclusion and intermolecular attractions, the Van der Waals equation bridges the gap between theoretical models and experimental observations, allowing precise control and prediction in chemical engineering, meteorology, and physical chemistry.

Furthermore, the kinetic molecular theory provides a molecular-level understanding of why real gases deviate from ideal behavior. At high pressures, particles are forced closer together, increasing collisions and making their finite volume significant. At low temperatures, particle velocities decrease, allowing intermolecular forces, such as van der Waals attractions, to influence movement and reduce pressure relative to ideal predictions. These deviations highlight the importance of corrections when designing systems that involve high-pressure gas storage, gas liquefaction, or supercritical fluid applications.

In addition to theoretical implications, real gas behavior has practical applications in industries such as chemical manufacturing, petroleum processing, aerospace engineering, and environmental monitoring. For example, predicting the expansion of compressed natural gas during transport relies on the Van der Waals equation rather than the Ideal Gas Law, ensuring accurate safety calculations. Similarly, understanding effusion and diffusion is essential for separating isotopes in nuclear applications and purifying industrial gases such as oxygen, nitrogen, and helium.

In summary, Graham's Law of Diffusion and Effusion provides insight into the movement of gas particles, while the Van der Waals equation and real gas corrections explain deviations from ideal behavior under extreme conditions. Together with the Ideal Gas Law, Dalton's Law, and the Kinetic Molecular Theory, these principles form a complete framework for predicting and analyzing the behavior of gases in both theoretical and practical contexts. By integrating equations into the analysis, scientists and engineers can quantify gas behavior accurately, ensuring precise design and interpretation across a broad spectrum of chemical and physical applications.

Gases play a critical role in atmospheric chemistry, influencing climate, weather, and life on Earth. The Earth's atmosphere consists primarily of nitrogen (N_2) and oxygen (O_2), with smaller amounts of argon (Ar), carbon dioxide (CO_2), and trace gases. The behavior of these gases, including pressure, volume, and temperature relationships, can be analyzed using the Ideal Gas Law, $PV=nRT$, which allows calculation of air density, the number of molecules in a given volume, and pressure variations with altitude. At higher altitudes, atmospheric pressure decreases because fewer air molecules are present per unit volume, which can be quantified using the hydrostatic equation $P = \rho gh$, where ρ is the density of air, g is the acceleration due to gravity, and h is height above sea level. Understanding these relationships is essential for aviation, weather prediction, and modeling atmospheric phenomena.

The greenhouse effect is another phenomenon where gases play a central role. Certain gases, known as greenhouse gases—including CO_2, CH_4 (methane), N_2O (nitrous oxide), and water vapor—absorb and emit infrared radiation, trapping heat in the atmosphere. This effect can be analyzed using gas laws and kinetic theory. For instance, increasing the concentration of CO_2 increases the total number of gas molecules contributing to energy absorption. The Ideal Gas Law, $PV=nRT$, shows that for a given volume and temperature, an increase in n (number of moles) influences pressure and energy distribution. Moreover, the kinetic molecular theory explains that higher temperatures increase the average kinetic energy of gas molecules, leading to increased radiation absorption and emission rates. Quantifying these processes allows scientists to model global temperature changes and predict the impact of rising greenhouse gas concentrations on climate.

In industrial applications, the behavior of gases is exploited in a variety of chemical and engineering processes. Chemical reactors, for example, often operate under controlled temperature and pressure to optimize gas-phase reactions. The Ideal Gas Law, $PV=nRT$, is used to calculate the required volume of reactant gases, predict pressure changes during reaction, and ensure safety by avoiding over-pressurization. Compressed gases such as oxygen, nitrogen, and hydrogen are stored and transported in cylinders, where Van der Waals corrections $[P_{obs} + a(n/V)^2](V-nb)=nRT$, ensure that real gas behavior is accounted for, particularly at high pressures where particle volume and intermolecular forces significantly affect storage conditions. Similarly, industrial gas separation techniques, such as fractional distillation of air or isotope separation, rely on principles of diffusion and effusion. Using Graham's Law, Rate $_1$Rate$_2$ = M_2M_1, engineers can predict the relative rates at which gases pass through membranes, enabling selective separation of lighter and heavier molecules.

Gas laws also provide critical insight into environmental monitoring and pollution control. For instance, calculating the dispersion of pollutants in the atmosphere involves understanding the diffusion of gas molecules, as described by Graham's Law. The mole fraction, $X_i = n_i n_{Total}$ and $X_i = P_i P_{Total}$, allows scientists to quantify the contribution of specific pollutants to the overall atmospheric pressure. By monitoring mole fractions of gases such as CO_2 and CH_4, climate scientists can track emission levels and evaluate compliance with environmental regulations. Similarly, understanding partial pressures helps in modeling chemical equilibria in atmospheric reactions, such as ozone formation and depletion, where the partial pressure of oxygen and other reactive gases determines reaction rates.

Furthermore, gas laws are vital in thermodynamics and engineering processes, including heating, ventilation, and air conditioning (HVAC) systems. For example, predicting the expansion of air in response to heating is achieved using Charles's Law, $V=bT$, while understanding how pressure changes in compressed air systems follows Boyle's Law, $PV = k$. The Ideal Gas Law, $PV=nRT$, integrates these effects, providing a comprehensive tool for designing efficient systems. In rocket propulsion, calculating the behavior of exhaust gases relies on high-pressure, high-temperature gas laws, often using Van der Waals corrections for real gases, $[P_{obs} + a(n/V)^2](V-nb)=nRT$, to account for interactions between molecules at extreme conditions.

Gases are also essential in laboratory experiments where precise measurements of volumes, pressures, and temperatures are required. For example, determining the molar mass of an unknown gas involves measuring mass, volume, pressure, and temperature, then applying $M = dRT/P$, where $d = m/V$. Similarly, gas collection experiments often rely on water displacement or manometers to measure pressure, and corrections for real gas behavior using Van der Waals constants, a and b, are applied when high accuracy is needed. Understanding diffusion and effusion through Graham's Law, $Rate_1 Rate_2 = M_2 M_1$, allows chemists to separate gases, analyze isotopic compositions, and design sensitive analytical techniques such as mass spectrometry and gas chromatography.

In summary, gases are fundamental to Earth systems, environmental processes, and industrial applications, with their behavior accurately described by the combination of the Ideal Gas Law, gas stoichiometry, Dalton's Law, Graham's Law, Van der Waals equation, and kinetic molecular theory. Each equation provides a quantitative framework for predicting pressures, volumes, temperatures, rates of diffusion, and real gas deviations. Whether calculating the volume of a gas at STP, modeling the greenhouse effect, designing high-pressure reactors, or separating industrial gases, these principles provide the foundation for scientific understanding and technological advancement, linking microscopic particle behavior to macroscopic observations through precise mathematical relationships.

Chapter 6: Word Problems

1. A deep-sea engineering team traps a sample of atmospheric air inside a rigid cylinder at the ocean surface where the pressure is 1.00 atm and the internal air volume is 2.75 L. As the submersible descends to 350 meters, the ambient pressure on the chamber increases to 35.5 atm. Assuming temperature remains constant, determine the final volume of the trapped air using Boyle's Law and evaluate why, at such high pressures, ideal behavior becomes questionable and which intermolecular effects begin to dominate as gas molecules are driven much closer together.

2. A weather research balloon is filled at sea level with 5.00 m^3 of helium at 22°C and 1 atm. As it rises to 15 km altitude, its internal pressure is regulated to remain at 1 atm, but the surrounding temperature drops to −52°C. Using Charles's Law, calculate the balloon's new volume and discuss the mechanical and material risks introduced by such large thermal expansions, as well as why helium's behavior only approximates ideality in these atmospheric conditions.

3. An industrial ammonia reactor initially contains 500 mol of a nitrogen–hydrogen mixture at 350°C within a 150-L chamber. Engineers inject an additional 150 mol of hydrogen at the same temperature without changing the pressure, assuming ideal behavior. Calculate the new volume required to maintain constant pressure and explain how real-gas deviations at elevated temperature and pressure complicate the use of Avogadro's Law for precise reactor control.

4. A cryogenic storage cylinder contains neon at −185°C in a 120-L tank with an internal pressure of 2.20 atm. Using the Ideal Gas Law, compute the number of moles present. Then analyze whether neon remains nearly ideal under such cryogenic conditions and determine whether the attraction term or volume-exclusion term of the Van der Waals equation would dominate if deviations were accounted for.

5. A combustion analysis lab traps 0.725 g of an unidentified exhaust gas inside a 375-mL flask at 95°C and 1.15 atm. Using the Ideal Gas Law, determine the molar mass and assess whether the gas is more likely to be CO, CO_2, NO, or SO_2. Then discuss how real-gas behavior and vibrational modes could shift the apparent molar mass when measurements are taken at elevated temperatures.

6. A sealed climatic simulation chamber contains a mixture of 0.40 mol of CO_2, 1.25 mol of O_2, and 3.10 mol of N_2 at 27°C in a 50-L vessel. Calculate the partial pressure of each gas using Dalton's Law and determine the overall system pressure. Then discuss how increasing the mole fraction of CO_2, even slightly, alters radiative heat absorption and internal energy balance in the chamber.

7. A polluted-city air sample shows a mole composition of 78% nitrogen, 20% oxygen, 1.0% argon, 0.9% CO_2, and 0.1% pollutant gases. If total atmospheric pressure at the measurement site is 0.94 atm, compute the partial pressure of each component. Evaluate how these values help environmental scientists regulate emissions and distinguish between background gases and harmful atmospheric contaminants.

8. A gas separation membrane is engineered to preferentially effuse hydrogen (2.0 g/mol) over nitrogen (28.0 g/mol). If nitrogen effuses through the membrane at 0.80 L/min under identical conditions, use Graham's Law to determine the expected hydrogen effusion rate. Explain why real separation systems deviate from Graham's ideal assumptions due to molecular collisions, non-uniform pore geometry, and boundary-layer diffusion effects.

9. A spacecraft leak-detection system uses helium as a tracer because of its high effusion rate. If helium is observed escaping through a micro-perforation at 2.5×10^{-3} mol/min, estimate the corresponding effusion rate of xenon (131.3 g/mol) under identical conditions. Discuss how pressure variations, temperature gradients, and non-ideal wall interactions make effusion-based detection increasingly unreliable in low-gravity environments.

10. A fire-suppression cylinder contains CO_2 at 25°C with an internal pressure predicted by the Ideal Gas Law to be 70 atm. Using the Van der Waals constants for CO_2 and assuming a fixed number of moles, compute the corrected pressure. Explain why ignoring real-gas behavior under such high pressures can lead to

safety failures in industrial suppression systems where compressed CO_2 may approach supercritical conditions.

11. A natural-gas LNG tank stores 125 mol of methane at −160°C and 50 atm. Use the Van der Waals equation to determine the corrected gas volume and discuss why compressibility factors become essential to ensure the methane does not prematurely condense, potentially stressing tank walls or causing hazardous pressure oscillations.
12. At an altitude of 9,000 meters, hikers experience severely reduced atmospheric pressure. Using the hydrostatic relation $P = \rho gh$ with air density approximated at 0.467 kg/m^3, estimate the atmospheric pressure and compare it to the Ideal Gas Law prediction using typical stratospheric temperatures. Describe why the two predictions differ significantly due to variable temperature gradients and changing molecular composition with altitude.
13. A jet engine combusts 1.5 mol of air at 750°C and 15 atm, producing combustion gases that reach 2200°C under constant pressure. Calculate the expansion factor using temperature ratios and discuss why real-gas effects and dissociation phenomena become significant at temperatures exceeding 2000°C, altering engine efficiency predictions.
14. A desert air parcel begins at 1.00 atm, 40°C, and 2.5 L. As it rises, it expands to a region where pressure is 0.65 atm and temperature is 15°C. Using the combined gas law, compute the new volume and evaluate why atmospheric humidity and latent heat exchange introduce substantial deviations from ideal, dry-air assumptions.
15. A medical anesthesia machine compresses a gas mixture from 7.0 L to 2.0 L while its internal temperature rises from 20°C to 38°C. If initial pressure was 1.15 atm, calculate the final pressure. Discuss the clinical importance of accounting for such gas-law-driven pressure changes to avoid unsafe delivery rates during surgery.
16. A chemical-spill forensics team analyzes a 1.80-L sealed cylinder found at an accident site. The internal pressure is 2.25 atm at 45°C, and the gas mass inside is 3.15 g. Determine the molar mass and propose likely industrial gases that match

this value. Explain how Van der Waals corrections could refine the identification when molecular polarizability is high.

17. A uranium enrichment centrifuge must separate UF_6 containing U-235 from UF_6 containing U-238. Using Graham's Law, compute the relative diffusion rates of the two isotopic molecules. Then explain why the extremely small difference in rates necessitates thousands of cascade stages to achieve weapons-grade enrichment.

18. A supercritical CO_2 extractor operates at 88 atm and 32°C. Using Van der Waals constants, compute the corrected pressure for 35 mol of CO_2. Assess how real-gas deviations enable CO_2 to acquire liquid-like density and gas-like diffusivity simultaneously, making it an effective solvent for food and pharmaceutical extraction.

19. A hyperbaric therapy chamber contains 75% O_2 and 25% N_2 at a total pressure of 2.8 atm. Calculate the partial pressures and analyze how elevated oxygen partial pressure accelerates decompression recovery yet also raises the risk of oxygen toxicity through excessive oxidative stress in tissues.

20. A 12-L sample of nitrogen at 500°C and 1 atm is heated to 1150°C while the volume remains constant. Calculate the new pressure using the Ideal Gas Law and evaluate the limitations imposed by furnace wall materials due to thermal expansion and pressure-induced mechanical strain.

21. An industrial pipeline transports natural gas at 350 K and 8.0 atm through a 600-m segment. A sudden drop in temperature to 270 K occurs due to an external coolant leak while the volume remains fixed. Calculate the resulting pressure and discuss how thermal shocks create dangerous oscillations and stress fractures in metal pipelines.

22. A meteorology drone collects a 1.20-L air sample at −18°C and 0.72 atm. Upon returning to the base at 24°C, the sample warms at constant volume. Compute the new pressure and explain how rapid warming alters the measured concentrations of trace gases by changing partial pressures.

23. A submarine life-support system maintains a breathable mixture by adding oxygen to compressed air. The system contains 8.0 mol of air at 4.0 atm in a rigid 10-L

tank. If 0.50 mol of pure O_2 is added without changing temperature or volume, compute the new total pressure and discuss why oxygen enrichment must be tightly regulated to avoid fire hazards.

24. A sealed 3.5-L cylinder contains sulfur hexafluoride at 1.20 atm and 35°C. After cooling to −10°C at constant volume, determine the new pressure. Discuss why SF_6's large molecular size makes it deviate slightly from ideality and how this affects HVAC leak-diagnostic calibrations.

25. A power plant turbine receives 2500 mol of steam superheated to 600°C at 15 atm. As the steam enters an expansion chamber where pressure drops to 3 atm at constant temperature, compute the new volume and discuss why steam deviates strongly from ideal gas behavior due to hydrogen-bonding remnants.

26. A spacecraft air-recycling module stores CO_2 scrubbing cartridges with a pressurized internal reservoir of 1.5 mol of pure CO_2 at 18°C in a 3.00-L tank. When exposed to solar heating, the reservoir temperature rises to 66°C. Calculate the final pressure and discuss why CO_2's near-critical behavior could cause unexpected surges.

27. A mining ventilation shaft pumps air at 2.1 atm and 30°C through a rigid conduit. If a fire incident heats the air to 290°C before it escapes, determine the final pressure and analyze why sudden over-pressurization poses severe structural risks.

28. A 4.50-L scuba tank contains 1850 g of compressed air at 22°C and 215 atm. Using average molar mass of air, compute the number of moles and determine the expected tank volume at 1.00 atm and 22°C. Discuss why such extrapolations are only approximate due to real-gas effects at 215 atm.

29. A steel cylinder holds ammonia gas at 5.0 atm and 15°C. After being transported to a desert site, its temperature rises to 61°C while volume remains constant. Calculate the final pressure and discuss how NH_3's polar nature leads to real-gas deviations affecting storage risk assessments.

30. A 2.20-L tank of neon at 1.10 atm and −50°C warms to 25°C at constant volume. Compute the final pressure and analyze why neon, being monatomic and nonpolar, remains close to ideal across wide temperature ranges.

31. A manufacturing facility uses argon shielding gas stored in 25-L cylinders at 8.0 atm and 20°C. When left in direct sunlight, temperature rises to 67°C. Determine the new pressure and examine how cylinder aging and metal fatigue modify safe-pressure margins.

32. A lab stores 0.80 mol of krypton in a 2.00-L flask at 0°C. If the flask is heated to 85°C, calculate the final pressure. Discuss why heavier noble gases deviate slightly from ideality due to stronger dispersion forces.

33. A 500-mL reaction vessel contains a gaseous reagent at 745 torr and 28°C. If the vessel is heated to 96°C at constant pressure, compute the new volume and discuss how reaction kinetics could be altered by the expansion of the gas phase.

34. A 7.00-L cylinder of hydrogen at 1.50 atm and 22°C is cooled to −40°C at constant pressure. Determine the final volume and explain how the ideal assumption may break down at extremely low temperatures as gases approach liquefaction thresholds.

35. A forensic lab examines an explosion residue by heating trapped gas from 12°C to 155°C. Initial conditions were 1.8 L at 1.05 atm. Compute the final volume assuming constant pressure and discuss why real-gas corrections may be needed if the gas contains heavy combustion byproducts.

36. A chemical reactor contains 2.00 mol of NO_2 and 1.00 mol of N_2O_4 in equilibrium at 40°C within a 10-L vessel. Compute partial pressures using mole fractions and total pressure. Then explain how Le Châtelier's principle interacts with gas-law calculations when equilibrium shifts with temperature.

37. An aircraft cabin pressurization system keeps cabin air at 0.78 atm. If the oxygen mole fraction is 0.21, calculate the partial pressure of oxygen and discuss why maintaining adequate O_2 is critical for preventing hypoxia in passengers.

38. A gas-powered refrigeration system uses isobutane vapor compressed from 1.2 atm to 4.8 atm at constant temperature. Determine the new volume using Boyle's Law and explain why isobutane's significant real-gas deviations require the use of compressibility charts rather than ideal approximations.

39. A 10-L tank holds 25 mol of nitrogen at 320 K. Use the Van der Waals equation

to compute the corrected pressure and discuss why nitrogen's small but finite attractions affect high-pressure predictions.

40. A 3.30-L vessel contains 4.5 mol of CO at 350 K. Calculate the ideal pressure, then apply Van der Waals corrections. Discuss how CO's moderate polarity alters the corrected pressure and why this matters for toxic-gas containment systems.

41. A laboratory stores 1.2 mol of air at 670 torr and 20°C in a sealed 2-L flask. After heating to 140°C, compute the new pressure and assess how water vapor contamination could shift partial pressures.

42. A hospital ventilator compresses air to 1.6 atm at 40°C before releasing it through a tubing system. If this compressed air cools to 21°C at constant volume, calculate the final pressure and discuss how temperature changes influence delivered oxygen concentration.

43. A 15-L tank contains 18 mol of helium at 27°C. Using the Ideal Gas Law, compute the pressure, then assess whether Van der Waals corrections meaningfully change the answer given helium's extremely weak intermolecular forces.

44. A vehicle's airbag inflator rapidly produces 0.75 mol of hot nitrogen gas at 65°C inside a 23-L bag. Calculate the internal pressure and discuss how real-gas behavior and cooling during deployment influence passenger safety.

45. A 50-L industrial container holds 62 mol of CO_2 at 350 K. Calculate the ideal pressure and then the Van der Waals corrected pressure. Discuss why CO_2's substantial intermolecular attractions make ideal calculations unreliable near moderate pressures.

46. A laser lab stores neon-argon mixtures used for plasma generation. A 4.0-L tube contains 1.2 mol of neon and 0.8 mol of argon at 300 K. Calculate partial and total pressures and discuss how different ionization energies influence the mixture's plasma behavior even though gas laws treat them ideally.

47. A climate-control system cools a 6.0-L sample of indoor air from 32°C to 10°C at constant pressure. Determine the new volume and discuss how moisture condensation complicates ideal-gas predictions for HVAC calculations.

48. A fuel-cell research team analyzes hydrogen diffusion by measuring effusion

through a porous ceramic disk. If hydrogen effuses at 1.25×10^{-2} mol/min, compute the expected effusion rate of deuterium gas (D_2, 4.0 g/mol). Evaluate how quantum tunneling introduces deviations from Graham's Law at very small pore sizes.

49. A sealed meteorite-analysis chamber contains 0.055 mol of trapped Martian atmospheric gas at 245 K in a 2.2-L chamber. Calculate the internal pressure and explain how the CO_2-heavy composition of Martian air influences real-gas behavior compared to Earth's primarily nitrogen-oxygen blend.

50. A geothermal power plant injects superheated steam into a re-expansion chamber at 210°C and 1.3 atm. If the steam cools to 120°C at constant pressure, determine the new volume and discuss why steam's proximity to its saturation curve limits the usefulness of the Ideal Gas Law without additional thermodynamic corrections.

Chapter 7

Thermochemistry

Ignited Insight

Energy is the invisible force that powers life, society, and the machinery of modern civilization. It is present in the air we breathe, the food we eat, the movement of water, and the fuels that drive industries. At the simplest level, humans consume food to obtain chemical energy stored in carbohydrates, proteins, and fats. This energy fuels our bodies, enabling movement, cognitive activity, and the complex orchestration of physiological processes that sustain life. A single bite of food contains an intricate array of chemical bonds, each representing stored potential energy. Through metabolism, the body releases this energy gradually, powering muscles, neurons, and organs. From the beating of the heart to the firing of neurons in the brain, every action is a transformation of chemical energy into mechanical work and heat.

Yet, humans are only one part of a much larger societal picture in which energy fuels entire systems. Modern civilization depends heavily on external sources of energy, predominantly derived from fossil fuels. Coal, petroleum, and natural gas have served as the engines of industrial development, providing the heat, power, and transportation fuels that define modern life. Petroleum, in particular, has revolutionized the world, offering energy-dense fuels such as gasoline, diesel, and jet fuel. These fuels drive vehicles, airplanes, and ships, facilitating global trade and communication. At the same time, petroleum serves as a feedstock for countless chemical products, including plastics, fertilizers, and synthetic materials. This dependence on finite resources highlights the critical importance of understanding and managing energy systems, as the balance between supply, demand, and environmental sustainability becomes increasingly precarious.

Fossil fuels are essentially chemical batteries, storing energy in the bonds of hydrocarbons. These organic compounds, formed from decayed plants and marine organisms over millions of years, release energy when they react with oxygen in the atmosphere. The combustion of hydrocarbons produces carbon dioxide and water, a simple reaction in chemical terms but one with profound environmental implications. While plants naturally absorb CO_2 during photosynthesis and oceans act as carbon sinks, the rapid burning of fossil fuels has overwhelmed these systems. Atmospheric CO_2 levels have risen sharply over the past century, contributing to global warming, climate change, and disruptions to ecosystems worldwide.

Beyond CO_2, fossil fuel combustion releases other pollutants. Sulfur-containing fuels generate sulfur dioxide, which reacts with water to form acid rain, damaging forests, soils, and aquatic ecosystems. Nitrogen oxides contribute to smog, impairing air quality and human health. The chemistry of energy, therefore, is inseparable from the chemistry of environmental impact. Understanding the molecular structure of hydrocarbons, their combustion pathways, and the byproducts they produce is essential for developing cleaner fuels and mitigating environmental harm.

At a microscopic level, energy governs life itself. Cells can be seen as miniature chemical factories, extracting energy from biomolecules like glucose through processes such as glycolysis and cellular respiration. This cellular energy powers biosynthesis, transport, and mechanical work, analogous to how engines convert fuel into motion and heat. The principles are universal: energy is conserved, transformed, and ultimately dissipated, whether in the microscopic machinery of a cell or the macroscopic machinery of an industrial plant.

Defining energy precisely is deceptively difficult, yet it remains a central concept in physics, chemistry, and engineering. Broadly, energy is the capacity to do work or produce heat. The Law of Conservation of Energy asserts that energy cannot be created or destroyed but only transformed from one form to another. This principle governs everything from a falling ball to a nuclear reactor.

Energy exists in two primary forms: potential and kinetic. Potential energy arises from position or configuration, such as a stretched spring, a raised weight, or an electrically charged capacitor. Kinetic energy is associated with motion and is mathematically described by the equation K.E. = (1/2)mv2, where m is mass and v is velocity. These forms of energy are interconvertible: a ball rolling down a hill converts gravitational potential energy into kinetic energy, which can then be transferred to other objects or dissipated as heat.

The First Law of Thermodynamics extends this concept to thermodynamic systems. It states that the change in internal energy (ΔU) is equal to the heat (q) added to a system plus the work (w) done on the system: $\Delta U = q + w$. Heat is energy transfer due to temperature differences, while work is energy transfer due to forces acting over distances. Understanding these transfers is critical for designing engines, evaluating chemical reactions, and managing energy in biological systems.

Thermodynamics provides the tools to quantify and understand energy transformations. For gases confined in cylinders with movable pistons, pressure (P) is force per unit area (F/A), and work done by an expanding gas is $w = -P\Delta V$. The negative sign indicates energy leaving the system, as when gas pushes a piston outward. Enthalpy (H), a state function defined as $H = U + PV$, allows us to track heat changes at constant pressure. Exothermic reactions release heat ($\Delta H < 0$), while endothermic reactions absorb heat ($\Delta H > 0$). Consider the combustion of methane: $CH_4(g) + 2O_2(g) \rightarrow CO_2(g) + 2H_2O(g)$ + energy. Here, the products' chemical bonds are stronger than the reactants', releasing energy as heat. Conversely, photosynthesis absorbs energy from sunlight to form glucose, illustrating the energy storage capabilities of endothermic reactions.

Calorimetry measures these energy changes. Constant-pressure calorimetry relates heat transfer to mass, specific heat, and temperature change ($Q = ms\Delta T$), while bomb calorimetry measures combustion energy at constant volume ($w = 0$, $\Delta U = q_v$). Hess's Law further simplifies energy calculations by asserting that the total enthalpy change is independent of the reaction pathway. Standard enthalpies of formation (ΔH_f°) provide reference points for predicting reaction energetics, crucial for fuel analysis, environmental chemistry, and metabolic studies.

Earth's energy resources are fundamentally solar in origin. Plants capture sunlight through photosynthesis, storing energy in chemical bonds. Over geological timescales, decayed plant and marine organisms transformed into fossil fuels under heat and pressure. Petroleum, a mixture of hydrocarbons, is fractionated into gasoline, kerosene, diesel, lubricating oils, and asphalt. Pyrolytic cracking enhances fuel yield by breaking large molecules into smaller, more useful hydrocarbons. Coal, derived from terrestrial plant matter, progresses from lignite to anthracite as carbon content and energy density increase.

Fossil fuels remain dominant due to high energy density and existing infrastructure. However, environmental consequences are severe: CO_2 emissions contribute to climate change, while sulfur compounds generate acid rain. Reducing reliance on fossil fuels is essential for environmental sustainability and energy security.

To mitigate environmental impact and ensure future energy supply, alternative energy sources are being actively developed. Coal gasification produces syngas, a mixture of carbon monoxide and hydrogen, which can be burned or converted into methanol and gasoline. Hydrogen, stored in metal hydrides or high-pressure tanks, offers a clean fuel source. Biofuels such as ethanol and biodiesel are renewable alternatives derived from fermentation or plant oils. Gasohol and methanol blends are already in use, demonstrating practical applications. Solar energy and nuclear power offer low-carbon options with long-term sustainability. Each alternative comes with technical, economic, and societal challenges, but the pursuit of diverse energy sources is critical to reducing global dependence on fossil fuels.

Energy flows on Earth are intimately connected to climate. The planet receives solar radiation, roughly 30% of which is reflected back into space. The remainder is absorbed by soil, water, and vegetation and eventually emitted as infrared radiation. Greenhouse gases—CO_2, H_2O, and methane—trap some of this energy, warming the surface. Methane, produced in large quantities by livestock, is particularly potent. Even small temperature increases can disrupt agriculture, ecosystems, and water resources, highlighting the urgent need for sustainable energy management.

Energy chemistry bridges the microscopic and macroscopic worlds, from the metabolism of a cell to the combustion of fuels in a power plant. By understanding chemical bonds, reaction energetics, thermodynamics, and energy transfer, scientists and engineers can innovate alternative fuels, improve efficiency, and reduce environmental impact. The future of energy relies on a deep understanding of these principles, coupled with technological innovation, policy frameworks, and global cooperation. Sustainable energy systems, renewable technologies, and smarter consumption patterns will define the next era of civilization, ensuring that energy remains abundant, accessible, and environmentally responsible.

Energy is the foundation of life and civilization, shaping everything from cellular processes to global economies. Fossil fuels have driven progress but at significant environmental cost. Alternative energy sources—hydrogen, biofuels, solar, and nuclear power—offer paths to sustainable energy use. The chemistry of energy provides the tools to understand, manage, and optimize these resources. As the world faces climate change, population growth, and finite fossil fuel reserves, mastering energy chemistry will be critical for maintaining modern standards of living while protecting the planet for future generations.

The story of energy is also a story of human civilization. From the discovery of fire to the development of industrial machinery, the pursuit and mastery of energy have shaped societies and defined progress. Early humans first harnessed energy through fire, learning to cook food, keep warm, and ward off predators. Fire was more than survival; it became a social and technological catalyst, allowing humans to settle, form communities, and expand their cognitive capacities.

With the advent of agriculture, humans began to rely on biological energy stored in crops and livestock. Animal labor became a critical source of mechanical energy, powering plows, mills, and transportation. Wind and water were harnessed in rudimentary ways, driving sails and waterwheels to grind grain or irrigate fields. These early energy systems, though modest compared to today's technologies, laid the foundation for more sophisticated energy conversion methods.

The Industrial Revolution marked a pivotal era in the history of energy. Coal-fired steam engines transformed manufacturing, transportation, and urban life. Factories could produce goods at unprecedented scales, while railways and steamships revolutionized global trade. Energy became closely linked to economic power and national development, with industrialized nations expanding their influence through control of coal, oil, and other energy resources. Petroleum, initially discovered in the mid-19th century, quickly became a transformative fuel. It powered internal combustion engines, giving rise to automobiles, airplanes, and mechanized agriculture, reshaping both the economy and society.

As civilizations have advanced, the environmental cost of energy use has grown. The burning of fossil fuels, while central to modern life, has led to profound changes in the atmosphere, land, and water. Carbon dioxide, a byproduct of combustion, accumulates in the atmosphere, trapping heat and altering the Earth's climate. This process has contributed to rising global temperatures, melting ice caps, shifting weather patterns, and more frequent extreme weather events.

Pollutants such as sulfur dioxide, nitrogen oxides, and particulate matter have created additional challenges. Acid rain damages forests, soils, and freshwater systems, while smog impairs air quality in urban areas. Oil spills, whether accidental or deliberate, have devastated marine ecosystems, threatening fish, birds, and other wildlife. Even the extraction of fossil fuels has environmental impacts: coal mining scars landscapes, natural gas extraction can cause soil and water contamination, and petroleum drilling alters fragile ecosystems.

The awareness of these environmental consequences has prompted a global conversation about sustainable energy. Scientists, policymakers, and communities have begun exploring ways to meet energy demands without compromising the health of the planet. The challenge lies in balancing energy access, economic growth, and environmental stewardship—a task that requires both innovation and cooperation.

Alternative energy sources are no longer theoretical possibilities; they are practical solutions already transforming the world. Solar power captures sunlight and converts it into electricity using photovoltaic panels or concentrated solar thermal systems. Entire regions, from deserts to rooftops in cities, now harness this abundant energy, reducing reliance on fossil fuels. Wind energy, generated by turbines that capture the kinetic energy of moving air, has become a major source of renewable electricity in many countries. Offshore wind farms, in particular, demonstrate the potential of harnessing vast, untapped energy resources.

Hydropower has been used for centuries, initially in simple watermills and more recently in large-scale dams that generate electricity. By controlling the flow of water, these systems provide reliable, renewable energy, though they must be managed carefully to minimize ecological disruption. Geothermal energy, derived from heat within the Earth, is another remarkable source, supplying electricity and heating in regions with active geothermal activity.

Biofuels offer yet another pathway. Ethanol from fermented crops, biodiesel from vegetable oils, and other bio-based fuels provide renewable alternatives to petroleum. When blended with conventional fuels, they can reduce carbon emissions and enhance energy security. Each of these alternative sources reflects a growing understanding of energy as not just a commodity but as a responsibility, requiring consideration of environmental impact, sustainability, and long-term availability.

Among alternative fuels, hydrogen holds exceptional promise. When used as a fuel, hydrogen produces energy and water, making it a clean and efficient source. Its potential extends across transportation, industrial applications, and even residential power. Hydrogen fuel cells, which convert hydrogen directly into electricity, are being incorporated into vehicles, buses, and even entire buildings. While challenges remain—such as safe storage, production efficiency, and infrastructure development—the ongoing research in hydrogen technology reflects humanity's commitment to a sustainable energy future.

Metal hydrides and other storage methods are being explored to make hydrogen practical for everyday use. In combination with renewable energy sources, hydrogen could serve as an energy carrier, storing excess energy produced from intermittent sources like solar and wind for use when demand is high. This flexibility is essential for creating resilient, low-carbon energy systems that can meet global needs without harming the environment.

Energy is deeply intertwined with social, economic, and political systems. Access to reliable energy shapes economic growth, education, healthcare, and quality of life. Societies with abundant energy resources have historically experienced rapid industrialization and urbanization, while regions with limited access often struggle with poverty and limited opportunities.

The global energy landscape is also influenced by geopolitics. Control of oil, natural gas, and coal reserves can confer significant strategic advantages, affecting international relations, trade policies, and even conflicts. Renewable energy technologies, by contrast, offer the potential for decentralized energy production, empowering communities and reducing dependence on imported fuels. Distributed energy systems, such as rooftop solar panels or small wind turbines, can transform energy from a centralized commodity into a local asset, providing resilience and independence.

Humans are not merely consumers of energy; they are creators, innovators, and stewards. From inventors of the steam engine to pioneers of solar technology, human ingenuity has shaped the ways energy is produced and used. Education and awareness are critical: societies that understand energy, its impacts, and its potential are better equipped to make informed decisions.

Behavioral choices, such as energy conservation, efficiency improvements, and the adoption of renewable technologies, collectively contribute to sustainable energy use. Small actions—like reducing household energy waste, using public transportation, or supporting green technologies—can scale up to significant impacts when embraced by communities, cities, and nations. Energy is not only a technical challenge but a social one, requiring collaboration, communication, and shared responsibility.

Looking forward, the future of energy is intertwined with technological innovation and environmental stewardship. Advances in battery storage, smart grids, and artificial intelligence are enabling more efficient energy use and better integration of renewable sources. Energy storage technologies allow for the capture of intermittent energy from solar and wind, ensuring reliable supply even when natural conditions fluctuate.

Emerging concepts, such as artificial photosynthesis, fusion energy, and advanced biofuels, hold the potential to revolutionize the global energy landscape. Fusion energy, for instance, promises almost limitless power with minimal environmental impact, replicating the processes that fuel the Sun. While still in the experimental stage, these technologies demonstrate the creativity and determination that humanity brings to solving complex energy challenges.

Addressing climate change is inseparable from managing energy. Rising global temperatures, melting ice sheets, and shifting weather patterns are direct consequences of how energy has been produced and consumed. Reducing carbon emissions, transitioning to renewables, and improving efficiency are essential strategies. International agreements, national policies, and corporate commitments all play a role in guiding the transition toward sustainable energy systems.

Adaptation strategies are also necessary. Urban planning, resilient infrastructure, and disaster preparedness help communities withstand the impacts of climate change. Energy systems themselves must be resilient, capable of responding to extreme weather, natural disasters, and fluctuations in supply and demand. By integrating energy planning with climate strategy, humanity can build a sustainable future that meets the needs of both people and the planet.

Energy is a shared global resource and responsibility. The choices made in one country can affect the climate, economy, and security of others. International cooperation, knowledge sharing, and investment in global energy solutions are crucial for addressing the challenges of the 21st century. Collaborative efforts, from global climate accords to joint research initiatives, demonstrate that energy issues transcend borders and require collective action. Equity is also central. Ensuring that all communities have access to reliable, sustainable energy is a moral imperative. Millions of people worldwide still lack electricity or clean cooking fuels. Expanding access while minimizing environmental harm represents one of the most pressing challenges and opportunities for the coming decades.

The future of energy depends on understanding. Energy literacy—knowledge of where energy comes from, how it is used, and its impact on the environment—is essential for informed decision-making. Public engagement, education, and outreach empower individuals to participate in energy transitions. Schools, media, and community programs play vital roles in raising awareness, shaping habits, and fostering innovation. Energy is not an abstract concept; it is a lived reality that affects health, prosperity, and the natural world. By cultivating an informed and engaged society, the potential of alternative energy, conservation efforts, and technological breakthroughs can be fully realized.

Energy links together science, technology, society, and the environment. Biological systems, industrial processes, and climate dynamics all rely on the principles of energy flow. Understanding these connections allows for more holistic energy management, integrating technical efficiency with environmental protection and social equity. The study of energy is not confined to laboratories or power plants. It encompasses policy decisions, economic planning, community action, and cultural attitudes. By appreciating energy as a central, unifying concept, humanity can navigate the challenges of the future while maintaining the delicate balance between growth and sustainability.

As the 21st century unfolds, energy will continue to shape every aspect of life. The challenge is monumental: balancing economic growth, environmental protection, technological advancement, and social equity. Yet the opportunities are equally vast. Renewable energy, advanced technologies, and collective action provide pathways to a sustainable, equitable, and resilient future. Energy is more than a physical quantity; it is a lens through which we understand the world, a tool to shape societies, and a responsibility we bear for future generations. By learning from history, embracing innovation, and acting with foresight, humanity can ensure that energy remains abundant, clean, and accessible for all, forging a path toward a sustainable and prosperous future.

Chapter 7: Word Problems

1. A coastal city relies on a hybrid energy grid composed of coal-fired power plants, offshore wind turbines, and solar farms. Over a one-month period, the coal plants operate at 70% capacity, while solar and wind provide intermittent energy depending on weather. If the coal plants emit 0.95 kg of CO_2 per kWh and the city consumes 5×10^6 kWh in total, determine the total CO_2 emissions for the month. Additionally, discuss the potential effects of integrating energy storage systems on reducing coal dependency and the environmental implications of seasonal variability in renewable generation.

2. In human metabolism, the oxidation of one mole of glucose provides approximately –2870 kJ of energy, part of which is stored in ATP molecules. Assuming that one mole of glucose produces 30 moles of ATP, each with an energy content of 30 kJ, calculate the fraction of glucose energy captured as ATP. Discuss the thermodynamic significance of the remaining energy released as heat and its role in maintaining human body temperature, including considerations for basal metabolic rate fluctuations under varying environmental conditions.

3. A gasoline-powered vehicle consumes 40 liters of fuel to travel 400 km, with gasoline having an energy density of 34 MJ per liter. Assuming ideal stoichiometric combustion, calculate the total chemical energy available and the effective energy converted to mechanical work if the car's efficiency is 25%. Evaluate the contribution of energy losses due to engine heat, friction, and aerodynamic drag, and propose design modifications or alternative fuels that could improve overall vehicle efficiency in real-world driving conditions.

4. A 60-meter radius offshore wind turbine experiences fluctuating wind speeds averaging 10 m/s over a year, with periods of calm and storms. Using the Betz limit, estimate the maximum theoretical power output and compare it with practical power generation, considering blade aerodynamics, electrical conversion efficiency, and wake effects. Additionally, discuss how intermittent wind patterns affect grid stability and the potential role of battery storage or pumped hydro in smoothing energy supply for coastal urban centers.

5. Methane combustion releases 890 kJ per mole of CH_4. A gas-fired power plant consumes 2×10^5 moles of methane per hour. Calculate the total energy released per hour and the electrical energy output assuming 35% thermal-to-electric conversion efficiency. Further, analyze the environmental consequences of the associated CO_2 emissions over one year of operation, and propose mitigation strategies such as carbon capture, fuel blending, or transitioning to lower-emission alternatives.

6. A hydroelectric dam channels 200 m^3/s of water through turbines from a height of 150 m. Determine the gravitational potential energy available per second and the expected mechanical energy output, accounting for realistic turbine efficiency and hydraulic losses due to friction and turbulence. Evaluate the long-term implications of sedimentation, fluctuating river flow, and climate change on energy generation capacity and discuss how adaptive management strategies could ensure sustainable hydropower production.

7. A photovoltaic solar array with an area of 20 m^2 receives an average solar irradiance of 800 W/m^2 for six hours daily. Assuming a conversion efficiency of 15%, calculate the total energy output in one day. Discuss the effects of seasonal angle changes, dust accumulation, and temperature-induced efficiency drops

on energy generation, and propose optimization strategies including tracking systems, cleaning schedules, and thermal management.

8. In photosynthesis, six molecules of CO_2 react with six molecules of water to produce one molecule of glucose and six molecules of O_2. If 1 kg of glucose is synthesized, calculate the solar energy captured based on the chemical energy stored in glucose (15.6 MJ per 100 g). Discuss the implications for large-scale bioenergy production, considering land use, water availability, seasonal light variation, and the efficiency limitations of natural photosynthetic processes compared to artificial systems.

9. A nuclear fission reactor produces 1000 MW of power continuously. If each uranium-235 fission releases 200 MeV, determine the mass of uranium-235 consumed per day, accounting for energy conversion efficiency to electricity. Discuss the challenges associated with fuel enrichment, spent fuel management, and long-term radioactive waste storage, including the thermodynamic and safety considerations that affect reactor operation and lifecycle energy assessment.

10. An industrial chemical process generates 50 kJ of waste heat per mole of reactant processed. If 1000 moles are reacted per day, propose a method to recover a significant portion of this heat for preheating incoming reactants or generating steam, and calculate the potential energy savings. Further, analyze how heat recovery integration affects overall process efficiency, carbon footprint, and economic viability, considering trade-offs between capital cost and energy savings.

11. Combustion of sulfur-containing coal releases sulfur dioxide, contributing to acid rain. For 1 ton of coal containing 2% sulfur, estimate the mass of SO_2 emitted. Discuss the broader environmental consequences, including soil acidification, water body pH changes, and forest ecosystem degradation, and evaluate methods such as flue gas desulfurization, cleaner coal technologies, and policy interventions to mitigate these impacts.

12. Hydrogen fuel cells generate electricity through the reaction of H_2 with O_2. If a fuel cell provides 10 kW of continuous power, calculate the mass of hydrogen consumed per hour assuming ideal efficiency. Discuss challenges in storage, transport, and infrastructure development for hydrogen as a scalable alternative fuel, including safety considerations and energy losses during compression or liquefaction.

13. A lithium-ion battery stores 100 Ah at 3.7 V. Estimate the total chemical energy available and compare it to the energy required to move a 1500 kg electric vehicle 100 km at an average speed of 60 km/h, considering rolling resistance and air drag. Discuss strategies for enhancing battery energy density, thermal management, and lifecycle sustainability in the context of large-scale adoption of electric mobility.

14. Coal gasification converts solid coal into syngas ($CO + H_2$). If 1 ton of coal yields 500 m^3 of CO and 300 m^3 of H_2, calculate the total chemical energy in the syngas and compare its efficiency with direct combustion. Evaluate the environmental implications, including CO_2 emissions, particulate formation, and potential integration with carbon capture or Fischer-Tropsch fuel synthesis.

15. A jet engine consumes 5 kg of kerosene per second while cruising. With an energy density of 43 MJ/kg and 40% engine efficiency, determine the mechanical power delivered and the fraction lost as heat. Discuss how aerodynamic design, altitude, temperature, and fuel composition influence overall propulsion efficiency, and propose strategies to improve energy utilization in aviation.

16. In a thermodynamic system, a gas absorbs 1000 kJ of heat at constant pressure and performs 400 kJ of work on the surroundings. Calculate the change in internal energy and discuss the interplay between work, heat, and internal energy in chemical and biological systems. Evaluate implications for designing engines, refrigerators, and metabolic energy budgets under varying environmental conditions.

17. A residential building consumes 10 GJ of natural gas monthly, but 60% of the heat is lost due to poor insulation and air leakage. Propose a combination of passive and active retrofitting measures, estimate potential energy savings, and discuss trade-offs between upfront capital investment, payback period, and environmental benefits. Include analysis of seasonal variations in energy consumption.

18. The standard enthalpy of formation for CO_2 is –393.5 kJ/mol and for H_2O is –241.8 kJ/mol. Calculate the enthalpy change for methane combustion and explain how bond energies and reaction pathways determine fuel energy density and combustion efficiency. Discuss how this knowledge informs fuel selection for residential, industrial, and transportation applications.

19. A solar thermal power plant employs a 1 km^2 field of parabolic mirrors concentrating sunlight at 800 W/m^2. With 30% conversion efficiency, estimate the daily electrical energy output given six peak sun hours. Analyze the impact of dust deposition, optical misalignment, and seasonal solar angle variation on energy yield, and propose monitoring and maintenance strategies to maintain high efficiency.

20. Consider a city of 1 million inhabitants, each consuming 2000 kcal/day from food. Calculate the total daily chemical energy intake and compare it with daily per capita electricity consumption of 3 kWh. Discuss implications for urban energy planning, including the potential for bioenergy recovery from food waste and integration with renewable energy systems.

21. A geothermal plant extracts water at 200°C at a flow rate of 50 kg/s, while the surrounding environment is at 25°C. Calculate the rate of thermal energy extraction and analyze the efficiency of converting this energy into electricity. Discuss geological constraints, sustainability of geothermal reservoirs, and long-term environmental impacts on local ecosystems.

22. Particulate pollution from coal-fired plants can affect human health and climate. If 0.05% of 1 million tons of coal burned annually becomes airborne particulates, estimate the total mass released. Evaluate the consequences for urban air quality, public health, and climate forcing, and propose engineering and regulatory strategies to mitigate particulate emissions.

23. A hybrid electric vehicle operates with 60% of energy from gasoline and 40% from a battery. Analyze the net CO_2 reduction potential compared to a conventional gasoline vehicle for a 100 km journey. Discuss the influence of driving patterns, battery recharge sources, and life cycle emissions on overall environmental impact.

24. Bioethanol derived from sugarcane has an energy density of 27 MJ/kg. If 100,000 kg are produced, calculate total energy content and compare with an equivalent quantity of gasoline. Evaluate the sustainability of large-scale biofuel production, considering land-use change, water consumption, and food security implications.

25. Hydrogen can be stored in metal hydrides at a 3% weight ratio. Determine the mass of hydride required to store 10 kg of hydrogen and analyze the energy density relative to conventional fuels. Discuss the trade-offs between storage safety, weight, and energy efficiency for vehicular or stationary applications.

26. A solar water heater is used to raise 500 liters of water from 20°C to 60°C. Calculate the energy required and propose improvements through insulation, selective coatings, and heat exchanger optimization. Evaluate the cost-benefit of these modifications in residential versus commercial applications.

27. Nuclear fusion of deuterium and tritium releases 17.6 MeV per reaction. Estimate the number of reactions required to produce 1 MJ of energy, and discuss the technological challenges in achieving and maintaining plasma confinement, fuel availability, and net energy gain for practical power generation.

28. A 10 MW wind turbine operates at a 25% capacity factor due to variable wind speeds. Calculate its annual energy output and discuss the implications for grid integration. Evaluate the need for energy storage, demand-side management, and complementary renewable sources to ensure reliable power supply.

29. An internal combustion engine converts 30% of fuel chemical energy into mechanical work. For 1 liter of gasoline, calculate the mechanical energy output and the energy lost as heat. Discuss strategies to reduce losses, including advanced combustion techniques, hybridization, and waste heat recovery.

30. A regional power grid integrates solar (20%), wind (15%), and coal-fired plants for the remainder. Given coal emissions of 0.9 kg CO_2/kWh, calculate overall grid emissions per MWh and propose optimization strategies, including energy storage, demand response, and renewable expansion, to reduce the carbon footprint.

31. In a fusion reaction producing helium and neutrons, 0.3% of the mass is converted to energy. For a 1 g mass of fuel, calculate the kinetic energy of neutrons and discuss containment and radiation shielding challenges in a commercial fusion reactor.

32. Biomass combustion generates 18 MJ/kg of energy. If a community uses 5000 kg/day, calculate total energy output and CO_2 emissions, assuming 50% carbon content. Analyze the potential for biomass to replace fossil fuels, considering land use, collection logistics, and carbon neutrality assumptions.

33. A dam stores 1×10^9 kg of water at a height of 100 m. Calculate the total gravitational potential energy and the energy delivered to turbines considering 85% efficiency. Discuss seasonal water flow variation, sedimentation, and environmental impacts on local ecosystems and fish populations.

34. A photovoltaic array experiences a 20% efficiency drop due to elevated temperature. If initial output is 1 MW, calculate energy lost and propose mitigation strategies such as active cooling, selective surface coatings, or optimal panel orientation. Evaluate the cost and practicality of each method.

35. During acid rain formation, emitted SO_2 reacts with water to form H_2SO_4. If 1000 mol SO_2 are released, calculate the potential mass of sulfuric acid produced. Discuss ecosystem vulnerability, mitigation strategies, and the role of regulatory standards in limiting sulfur emissions.

36. A household uses 30 kWh/day in summer and 20 kWh/day in winter. Estimate annual energy consumption and evaluate strategies to reduce peak demand through renewable integration, smart appliances, and energy-efficient building design.

37. A coal seam contains 50% carbon. For a 10,000-ton seam, calculate the total chemical energy potential and compare with the energy contained in an equivalent volume of crude oil. Discuss the environmental trade-offs and feasibility of exploiting such a resource.

38. A wind turbine blade experiences fatigue and aerodynamic efficiency drops by 10% over 20 years. Calculate cumulative energy loss and propose predictive maintenance strategies to prolong operational lifespan and maximize energy output.

39. Methanol has an energy density of 20 MJ/kg. A vehicle consumes 50 kg per week. Estimate weekly energy usage and compare it with gasoline for the same distance. Evaluate environmental benefits, fuel economy, and sustainability of using methanol as a transportation fuel.

40. Industrial heat exchangers recover 500 kJ/min of waste energy at 60% efficiency. Calculate total energy recovered per day and discuss how integrating these systems affects process efficiency, emissions, and operating costs in large chemical plants.

41. A solar-powered desalination plant requires 1000 kWh/day. Calculate the required panel area at 15% efficiency with 5 sun hours/day and discuss the feasibility of scaling the system for arid regions while accounting for storage and water distribution infrastructure.

42. Combined cycle natural gas plants operate at 60% efficiency. If 10,000 m^3 of gas per day is consumed, with 38 MJ/m^3 energy content, calculate total electricity produced and CO_2 emissions. Compare with equivalent coal-fired generation and discuss trade-offs in emissions and energy security.

43. A building with passive solar design saves 15% of annual energy use. For a baseline consumption of 50,000 kWh/year, calculate energy savings and discuss the combination of building orientation, insulation, window glazing, and thermal mass that contributes to this reduction.

44. Flywheel energy storage stores 500 kJ at 3000 rpm. Determine the required moment of inertia and discuss practical limits imposed by material strength, friction, and system integration for grid-scale energy storage.

45. A tidal power station extracts 1×10^6 m^3 of water per tidal cycle across a 5 m height difference. Calculate the energy per cycle and estimate annual potential energy, considering tidal variability. Discuss ecological impacts on estuarine environments.

46. Hydrogen production via electrolysis requires 237 kJ/mol. For a plant producing 1000 mol/hour, calculate electrical energy input and discuss efficiency, cost competitiveness, and integration with renewable sources for sustainable hydrogen generation.

47. Methane leaks during extraction, with 0.5% of 1 million m^3 escaping. Calculate CO_2-equivalent emissions, assuming a GWP of 28 for methane, and evaluate mitigation strategies including capture, flaring, and leak detection technologies.

48. An urban fleet of electric buses is charged from a grid with 40% renewable energy. Compare net CO_2 emissions for a 100 km trip to conventional diesel buses with 1 kg CO_2/km. Discuss the implications for urban air quality, policy incentives, and energy planning.

49. Biofuel production consumes 40% of its energy output in cultivation, harvesting, and processing. For a 10,000 MJ output, calculate net energy gain and discuss the sustainability and scalability of biofuels in comparison to fossil fuels and advanced renewable alternatives.

50. A fusion reactor aims to deliver 500 MW of electricity. If containment losses account for 10% and thermal-to-electric conversion efficiency is 45%, calculate the required thermal power input. Discuss the technological barriers to achieving net energy gain, including plasma stability, material constraints, and fuel availability.

Chapter 8

A Journey into Atomic Theory

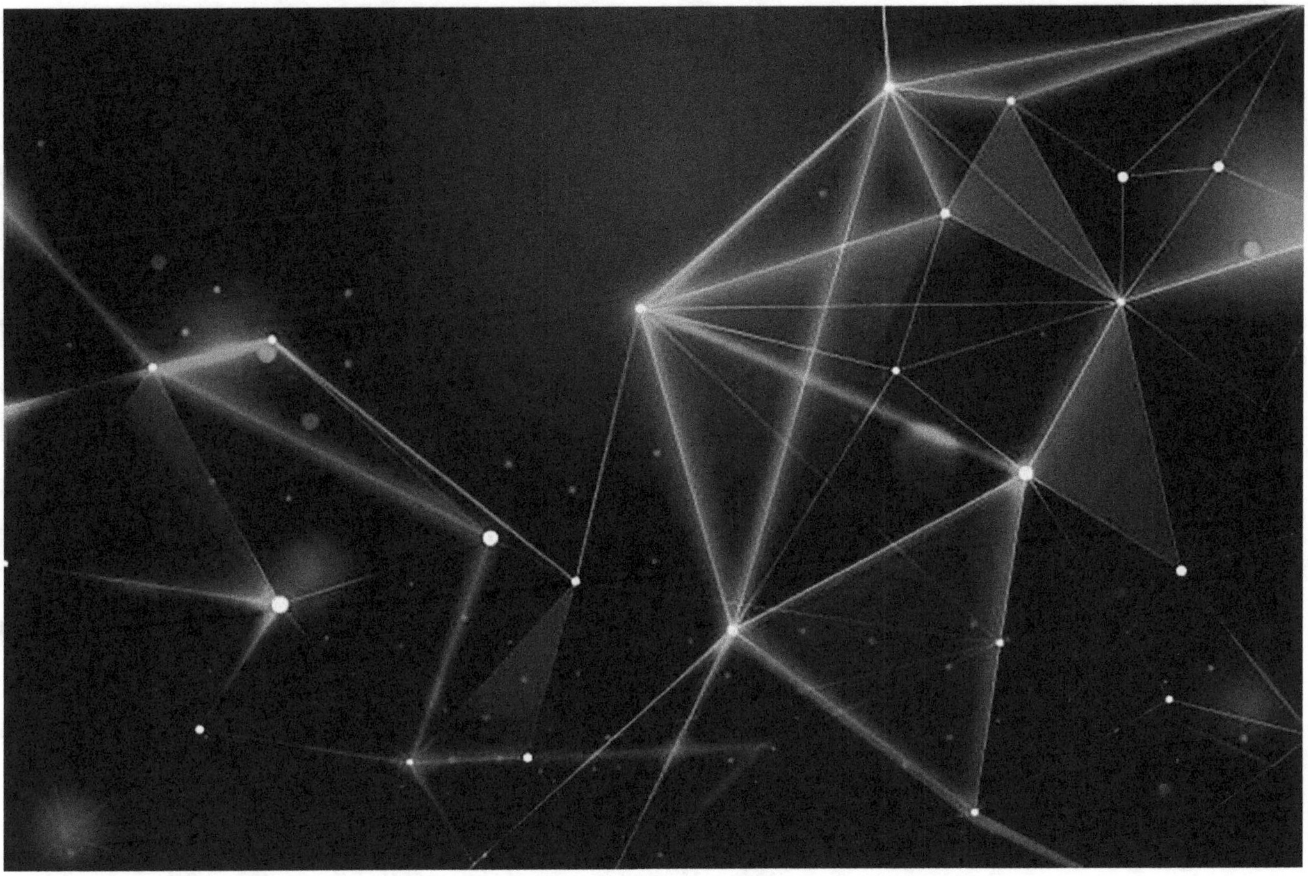

The concept of the atom traces back to a time when human reasoning itself was beginning to stretch toward abstractions no longer tethered to direct sensory experience. In ancient Greek philosophy, long before the arrival of any instruments capable of probing nature at the microscopic scale, thinkers such as Leucippus and Democritus speculated that matter must ultimately be composed of discrete, indivisible particles whose movements and interactions give rise to all observable change. They lacked empirical data, yet their vision of atoms as tiny building blocks, differing in shape, size, and motion, offered a framework for understanding why substances behaved differently under the same external conditions. This early philosophical atomism was remarkable because it implied an underlying order to the chaos of natural phenomena, suggesting that the complexities of the macroscopic world might arise from the dynamics of entities too small to see. Though their work remained speculative for centuries, the essential idea—matter built from hidden, indivisible units—continued to exert intellectual influence, lingering in the margins of scientific thought until experimental methods matured sufficiently to take the question seriously.

As scientific inquiry transitioned from speculative reasoning into experimental discipline during the Enlightenment, the foundational work of Antoine Lavoisier established the principle that chemical reactions obey precise quantitative laws. Lavoisier's meticulous measurements of mass before and after combustion overturned the phlogiston theory and demonstrated that mass is conserved in chemical transformations. This shift in perspective was transformative: matter behaved systematically, not mysteriously, and this regularity was measurable. Joseph Proust soon followed with his law of definite proportions, asserting that chemical compounds always contain elements in fixed ratios by mass. These laws offered empirical scaffolding strong enough for John Dalton to articulate the first scientific atomic theory. Dalton proposed that each element consisted of identical atoms, distinct from those of other elements, and that compounds were formed when atoms combined in simple whole-number ratios. His model resurrected the ancient idea but framed it in empirical terms, lending atoms scientific legitimacy rather than philosophical speculation. Although Dalton had no conception of internal atomic structure, his theory built the intellectual foundation upon which the entire edifice of modern chemistry would rise.

Further developments in chemical analysis and the discovery of dozens of new elements in the nineteenth century revealed an intriguing regularity: properties of the elements recurred periodically when the elements were arranged by increasing mass. Dmitri Mendeleev perceived in this pattern a deeper order and constructed what would become the Periodic Table. His table not only organized known elements but boldly predicted the existence and properties of elements yet undiscovered. The accuracy of these predictions, confirmed as new elements were isolated, provided powerful evidence that the periodicity of chemical behavior reflected an underlying atomic order. At the time, the mechanism behind periodicity was unknown, but the structure in the data suggested that atoms themselves must contain internal patterns—something governing their behavior beyond what Dalton's indivisible spheres could explain. Eventually, the development of atomic spectroscopy, electron discovery, and the early 20th-century revolution in physics would unravel those hidden patterns, revealing electrons arranged in quantized energy levels as the origin of periodic trends.

The study of electromagnetic radiation proved essential for probing the internal structure of atoms. Electromagnetic waves, traveling at the speed of light, carry energy across space in oscillating electric and magnetic fields. These waves exhibit a spectrum ranging from extremely energetic gamma rays, with incredibly short wavelengths, to the long radio waves used in communication technologies. The visible spectrum represents only a tiny fraction of this continuum, yet it played a crucial role in early investigations, as the distinct colors emitted or absorbed by elements provided clues about the internal workings of atoms. Key to understanding this radiation was the relationship $\lambda v = c$, linking the wavelength and frequency of any electromagnetic wave to the universal constant c, the speed of light. The higher the frequency, the shorter the wavelength, and the more energetic the radiation. But as scientists examined thermal emission more closely, particularly the ultraviolet catastrophe predicted by classical physics, they discovered that classical wave theory alone could not account for the observed behavior.

Max Planck's revolutionary insight was that electromagnetic energy is not continuous but quantized. To explain thermal radiation, he hypothesized that atoms emit or absorb energy only in discrete packets, or quanta, with energies proportional to the frequency of radiation: $\Delta E = nhv$. This idea, which seemed at first a mathematical trick, ultimately proved fundamental to the nature of reality. Albert Einstein extended this notion by proposing that light itself consisted of particles—photons—each carrying energy $E = hv$. His explanation of the photoelectric effect, in which electrons are ejected from a metal surface when exposed to light of sufficient frequency, demonstrated that light sometimes behaved as a particle rather than a wave. Einstein's equation $E = mc^2$ further unified energy and mass, allowing the mass-equivalent of photon energy to be expressed as $m = h/(\lambda c)$. The consequence of these ideas was profound: the distinction between waves and particles could no longer be cleanly drawn, and matter itself, as Louis de Broglie later suggested, must exhibit dual behavior. De Broglie's wavelength $\lambda = h/(mv)$ predicted that even massive particles like electrons possess wave-like properties, a prediction soon confirmed by diffraction experiments involving beams of electrons scattered by crystals. Such observations revealed that particles interfere and diffract like waves, firmly establishing wave–particle duality as a universal property of nature.

The hydrogen atom, with its single electron, became the ideal system for uncovering the quantized structure of atoms. When hydrogen gas is energized by high-voltage sparks, electrons absorb energy, transition to higher energy levels, and emit photons of specific energies as they return to lower states. Passing this emitted light through a prism reveals a line spectrum—discrete wavelengths corresponding to quantized energy differences between electron levels. Johann Balmer's empirical formula for the visible spectrum further hinted at a mathematical order within the atom. Niels Bohr integrated this spectral data with Planck's quantization to propose a model in which the electron moves in stable circular orbits around the nucleus, each with a quantized energy given by $E = -2.178 \times 10^{-18}$ J (Z^2/n^2). In Bohr's model, electrons occupy only certain orbits, and transitions between them produce photons with energies equal to the difference between these quantized states. Although the Bohr model reproduced the hydrogen spectrum with precision, it could not account for multi-electron atoms, nor could it explain fine structure or spectral line splitting in magnetic fields. Still, it marked a crucial step toward the quantum mechanical model that would replace it.

Werner Heisenberg, Erwin Schrödinger, and Louis de Broglie advanced the understanding of atomic structure by shifting from particle trajectories to wave functions. Schrödinger's equation, $\hat{H}\psi = E\psi$, describes electrons as standing waves bound to the nucleus, with ψ representing a probability amplitude whose square gives the probability density of finding an electron in a particular region of space. This probabilistic interpretation emerged partly from Heisenberg's Uncertainty Principle, which states that the position and momentum of an electron cannot both be precisely known: $\Delta x \cdot \Delta(mv) \geq h/4\pi$. The electron does not travel in a well-defined orbit but occupies a probabilistic cloud shaped by quantum numbers arising from the solutions to the Schrödinger equation. Each electron in an atom is characterized by a unique set of four quantum numbers: the principal quantum number n, which determines orbital size and energy; the angular momentum quantum number l, which defines orbital shape; the magnetic quantum number m_l, specifying orientation in space; and the spin quantum number m_s, describing intrinsic angular momentum.

Pauli's Exclusion Principle ensures that no two electrons can share the same set of quantum numbers, giving rise to structured electronic shells and subshells that dictate chemical behavior.

In multi-electron atoms, electron-electron repulsions complicate the picture. The effective nuclear charge experienced by outer electrons is reduced by shielding from inner electrons, and the extent of penetration—how close an orbital's electron density approaches the nucleus—affects orbital energy. An electron in a 2s orbital penetrates closer to the nucleus and experiences a higher effective nuclear charge than an electron in a 2p orbital, making the 2s orbital lower in energy. Similar patterns arise across higher shells, leading to the ordering of orbital energies within each principal level. When constructing electron configurations, the Aufbau principle dictates filling orbitals in order of increasing energy, though this order deviates slightly in transition metals and lanthanides where subshell energies are closely spaced. Hund's Rule explains the distribution of electrons in degenerate orbitals: electrons fill each equivalent orbital singly before pairing, minimizing electron-electron repulsion by maintaining parallel spins. Such rules, though simple in appearance, reflect deep quantum mechanical principles governing electron correlation and exchange energy.

Periodic trends in the elements arise naturally from these quantum principles. Ionization energy increases across a period as effective nuclear charge strengthens, drawing electrons closer and making removal more difficult. Down a group, ionization energy decreases due to larger orbitals and increased shielding. Electron affinity generally becomes more negative across a period as atoms more effectively attract additional electrons, while atomic radius decreases as electrons are pulled inward by increasing nuclear charge. Conversely, atomic radius grows down a group as electrons occupy orbitals of higher principal quantum number. These trends form the basis for predicting chemical reactivity, metallic character, and bonding behavior. The periodic table itself becomes a map of electron configurations, where families of elements share characteristic properties due to similar valence electron arrangements.

The propagation of energy through electromagnetic radiation, the quantization of light and matter, and the probabilistic behavior of electrons converge into a unified picture of the atom: a quantum mechanical system governed by wave functions rather than classical trajectories, by discrete energy levels rather than continuous ranges, and by probabilistic distributions rather than deterministic paths. The hydrogen emission spectrum, once a puzzling pattern of colored lines, becomes a direct fingerprint of quantized electron energies. Bohr's orbital model provides a conceptual stepping stone, but it is the Schrödinger equation that fully describes atomic structure. The interplay of quantum numbers explains orbital shapes and energy ordering, while shielding, penetration, and electron correlation determine the behavior of multi-electron atoms. From these principles emerge the periodic trends that dictate the structure of molecules, the reactivity of elements, and the properties of materials. Quantum mechanics thus provides not only a description of the atom itself but a comprehensive framework connecting electromagnetic radiation, atomic structure, and the behavior of matter on all scales.

The gradual refinement of atomic theory increasingly revealed that the internal architecture of atoms was more intricate than early models could ever have predicted. As experimental physicists explored the behavior of charged particles in electric and magnetic fields, the atom emerged as a composite system rather than an indivisible unit. J. J. Thomson's discovery of the electron in 1897 demonstrated conclusively that atoms contained subatomic particles, and his measurements of charge-to-mass ratio suggested electrons were far smaller and lighter than atoms themselves. This discovery undermined Dalton's indivisibility postulate and forced scientists to confront the unsettling idea that atoms possessed internal structure. Thomson's "plum pudding" model attempted to reconcile electrons with atomic neutrality by positing them embedded within a diffuse positive matrix. Although ultimately incorrect, the model introduced the concept that atoms could be partitioned into smaller charged components whose arrangement determined macroscopic properties.

The real breakthrough came with Ernest Rutherford's gold foil experiment, in which alpha particles were directed at thin metal films. Classical expectations predicted only minor deflections if Thomson's model were correct, but instead some alpha particles rebounded almost directly backward. Rutherford interpreted this as evidence of a small, dense nucleus containing most of the atom's mass and positive charge. This revolutionary finding shifted the atom's conceptual center inward, replacing the diffuse positive cloud with a concentrated nucleus orbited by electrons. The planetary model that followed was elegant in its simplicity, yet it carried inherent contradictions, particularly the classical prediction that accelerating electrons should radiate energy continuously, spiraling inward and collapsing into the nucleus. That atoms were stable suggested the inadequacy of classical mechanics at these small scales.

Quantum theory resolved these contradictions by replacing continuous trajectories with quantized energy states. It became clear that electrons in atoms do not radiate energy arbitrarily but only during transitions between discrete energy levels. Experimental evidence accumulated rapidly through the analysis of spectral lines not only in hydrogen but in other elements. Fine structure, caused by spin–orbit coupling, revealed that electron spin interacts with orbital angular momentum, splitting energy levels into slightly different sublevels. Hyperfine structure, resulting from interactions between the electron cloud and nuclear magnetic moments, introduced additional splitting. These increasingly subtle features provided new constraints on theoretical models, confirming that electrons respond to complex internal forces that classical theories could not capture. The atom became a microcosm of quantum mechanics, where the behavior of particles is governed by probabilistic distributions, quantized angular momentum, and wave interference patterns.

The wave mechanical model further advanced the understanding of electron distribution, revealing that orbitals are not simple paths but regions of space defined by mathematical solutions to differential equations. These orbitals possess distinct shapes, from the spherical symmetry of s orbitals to the dumbbell-like p orbitals and the more elaborate geometries of d and f orbitals. Their nodal structures, regions of zero probability, reflect interference between components of the electron wave function. The number of radial and angular nodes increases with principal and angular momentum quantum numbers, shaping electron density in ways that influence chemical bonding and reactivity. Electrons in higher-energy orbitals exhibit more nodes and more diffuse distributions, contributing to the increasing size of atoms down a group. The mathematical nature of orbitals, once an abstract concept, became concrete as computational methods allowed visualization of these electron probability clouds, offering a window into the invisible dynamics governing atomic behavior.

As quantum mechanics matured, it became evident that electron correlation—the way electrons avoid one another due to their mutual repulsion—plays a significant role in determining atomic and molecular energies. The Hartree–Fock method provided an early approximation by treating electron interactions in an averaged way, yet it could not fully capture instantaneous correlations. More sophisticated methods such as configuration interaction and coupled-cluster theory were developed to address these limitations, each providing more accurate descriptions of multi-electron systems.

These computational techniques demonstrated that the electronic structure of even simple atoms is far more nuanced than the independent-electron approximation suggests. Correlation energy, though small relative to total energy, significantly affects spectral properties, ionization potentials, and chemical reactivity.

The behavior of electrons under extreme conditions offered further insight into the atomic fabric of matter. High-pressure environments, such as those found within planetary interiors, compress electron clouds to the point where atomic orbitals overlap and conventional bonding models no longer apply. Electrons become delocalized, forming electron seas that give rise to metallic properties. At even higher pressures, electrons can be forced into higher-energy states or into paths constrained by magnetic fields, leading to exotic states of matter such as degenerate electron gases or even superconducting phases. These extreme behaviors underscore the flexibility and robustness of quantum theory, which remains applicable across vast ranges of pressure, temperature, and electromagnetic field strength.

Another layer of complexity emerged from the study of nuclear structure and its influence on atomic behavior. Although electrons determine chemical properties, the nucleus anchors the atom and contributes to isotopic variation. Variations in neutron number affect nuclear size, stability, and magnetic moment, which in turn influence hyperfine splitting and certain spectroscopic signatures. Nuclear stability reflects the balance between strong nuclear forces and electrostatic repulsion, governed by shell structures reminiscent of electronic shells but arising from different quantum mechanical principles. Some nuclei exhibit magic numbers of protons or neutrons that confer exceptional stability, similar to noble gases in electronic systems. Nuclear decay processes—alpha, beta, and gamma emission—transform one element into another, demonstrating the fluidity of atomic identity under certain circumstances. These transformations, though occurring within the nucleus, can indirectly affect electron configurations and chemical behavior, particularly in short-lived isotopes where electronic readjustment must occur rapidly following nuclear change.

Quantum electrodynamics (QED), the quantum field theory of electromagnetic interactions, added yet greater precision to the understanding of atomic structure. QED predicts subtle corrections to electron energy levels, such as the Lamb shift, which results from interactions between electrons and transient vacuum fluctuations. These phenomena indicate that even empty space is not truly empty but populated by virtual particles that momentarily emerge and vanish, influencing measurable energies within atoms. Such corrections, though minuscule, validate the extraordinary predictive power of modern physics and reveal that atoms exist within a dynamic quantum vacuum, not a static void. The discovery of these small but significant effects deepened the conceptual complexity of atomic theory, integrating atoms into the broader framework of quantum fields.

Magnetic interactions further illuminate atomic behavior. Electrons possess intrinsic magnetic moments arising from both orbital and spin angular momentum, causing them to interact with external magnetic fields in ways that split energy levels into distinct components. This phenomenon, known as the Zeeman effect, demonstrates that magnetic fields break degeneracy among orbitals that would otherwise share the same energy. In strong magnetic fields, electron motion becomes quantized into discrete Landau levels, profoundly altering the structure of atomic states. Such conditions occur naturally in astronomical environments, such as the surfaces of neutron stars, where magnetic fields millions of times stronger than those achievable on Earth compress atoms and distort their electron clouds into elongated shapes unknown under normal circumstances.

The interplay between atomic structure and electromagnetic radiation forms the basis for nearly every spectroscopic technique used in modern science. Infrared spectroscopy probes molecular vibrations determined by the arrangement of atoms and the nature of chemical bonds. Ultraviolet-visible spectroscopy examines electronic transitions, revealing details of conjugated systems and transition-metal complexes. X-ray spectroscopy and crystallography provide information on electron density distribution within solids, enabling the determination of atomic positions in molecules and materials to astonishing precision. Nuclear magnetic resonance spectroscopy, founded on the interaction between nuclear spin states and magnetic fields, allows the probing of atomic environments within molecules, yielding structural and dynamic information. Each technique translates quantum-level interactions into measurable signals, effectively using the atom's internal structure as an analytical tool.

The evolution of atomic theory continues as new methods allow scientists to probe ever-smaller scales and faster timescales. Attosecond laser pulses capture electron dynamics in real time, revealing how electrons respond to light and rearrange during chemical reactions. Advances in scanning tunneling microscopy permit direct imaging of electron density on surfaces, offering unprecedented views of atomic-scale processes. Meanwhile, research into ultra-cold atoms manipulated by laser cooling techniques explores regimes where quantum behavior dominates entirely, allowing the creation of Bose–Einstein condensates whose atoms share a collective wave function. These developments reveal that atomic behavior is not merely a subject for theoretical analysis but a domain that can now be probed, manipulated, and observed with extraordinary control.

Through all these perspectives—historical, experimental, theoretical, and computational—the atom emerges not as a simple particle but as a multilayered structure governed by quantum laws that interconnect matter and energy at the most fundamental levels. The discrete spectral lines that once hinted at quantized energies now serve as gateways to an ever-deepening understanding of the quantum structure of reality. Atoms, once purely conceptual constructs, have become dynamic entities described with remarkable precision by mathematical frameworks that unify classical observations and quantum behavior. The progression from speculative Greek philosophy to the rigorous quantum mechanical model illustrates the power of scientific inquiry to reveal nature's underlying order. And yet, despite the immense progress, the atom remains an open frontier, continually offering new mysteries as technologies push the boundaries of resolution, precision, and theoretical exploration.

The atom, though often conceptualized as an isolated entity, exists in nature almost exclusively as part of larger assemblies, and its behavior in these contexts exposes additional subtleties of quantum mechanics. While single-atom models provided early insights into quantization and electron arrangement, the interactions between atoms reveal how the underlying principles scale into the complexity of molecules, solids, and extended materials. Chemical bonding, for instance, emerges not as a rigid linkage but as a quantum mechanical phenomenon arising from the sharing, transfer, or delocalization of electrons in ways that minimize the total energy of the system. The molecular orbital (MO) theory offered a profound reinterpretation of bonding by proposing that atomic orbitals combine to form molecular orbitals that extend over the entire molecule rather than remain localized on individual atoms. These molecular orbitals, classified as bonding, antibonding, or nonbonding depending on their symmetry and energy, illustrate how the constructive or destructive interference of electron wave functions determines whether atoms attract or repel one another.

The development of MO theory was grounded in the same principles that govern atomic structure: the Schrödinger equation, boundary conditions imposed by nuclear geometry, and the Pauli Exclusion Principle's requirement that electrons occupy distinct quantum states. The theory revealed that even the simplest molecule, hydrogen, possesses a bonding orbital lower in energy than the separate atomic orbitals, stabilizing the molecule when occupied by two electrons with opposite spins. As molecules increased in complexity, hybridization concepts emerged to account for shapes and geometries not readily explained by simple combinations of s and p orbitals. The mixing of atomic orbitals to form hybrid orbitals—such as sp, sp^2, and sp^3 hybrids—provided a quantitative basis for molecular geometry, explaining why carbon forms tetrahedral methane, trigonal planar ethene, and linear acetylene depending on the extent of s–p mixing. These hybrid orbitals, though mathematical constructs, represent solutions that better reflect the symmetrical distribution of electron density around molecules, illustrating how quantum mechanics reconciles molecular structure with energetic stability.

Electron delocalization in conjugated systems offered another window into the quantum mechanical nature of bonding. In systems with alternating single and double bonds, such as benzene or polyenes, electrons are not confined to individual bonds but occupy extended π orbitals spanning multiple atoms. This delocalization lowers the total energy of the system by distributing electron density more broadly, a concept captured elegantly by the Hückel method and subsequent quantum chemical models. Resonance structures, once introduced as heuristic tools to depict electron distribution, gained rigorous theoretical grounding as manifestations of the superposition of multiple contributing wave functions. Aromaticity, characterized by cyclic electron delocalization following the $4n + 2$ rule, represents a uniquely stable electronic configuration arising from the symmetry and energetics of conjugated π orbitals. Such systems highlight the interplay between electron wave behavior, molecular geometry, and stability, demonstrating how atomic principles extend seamlessly into more complex frameworks.

As the understanding of molecular bonding deepened, it became necessary to explore phenomena that could not be described solely by non-relativistic quantum mechanics. Heavy elements, with their high nuclear charges, accelerate inner-shell electrons to velocities approaching a significant fraction of the speed of light. Under these conditions, relativistic effects become essential for accurate descriptions.

According to Einstein's relativity, the increased velocity of electrons leads to a greater effective mass, contracting s orbitals and stabilizing them relative to p, d, and f orbitals. These relativistic contractions explain numerous empirical observations, such as the unusual liquid state of mercury at room temperature, the high reactivity of gold, and the distinct color of gold resulting from relativistic shifts in electronic transitions. Without incorporating relativity, calculations fail to reproduce such properties, underscoring the inseparability of relativistic and quantum mechanical frameworks for describing heavy atoms.

The periodic trends themselves also acquire deeper meaning in light of relativistic effects. As atomic numbers increase, the effective nuclear charge experienced by electrons grows, pulling orbitals closer and enhancing penetration effects. At the same time, electron shielding within inner shells moderates this attraction for outer electrons. In heavy elements, relativistic contraction of s orbitals and expansion of p, d, and f orbitals alter chemical behavior, contributing to phenomena such as the inert-pair effect in post-transition metals, where ns^2 electrons become relatively inert due to their stabilized, contracted orbitals. Such effects differentiate the chemistry of heavy elements from lighter analogues, showing that periodic behavior arises not only from straightforward electron filling order but from a complex interplay of quantum, relativistic, and electron-correlation effects.

The transition from isolated atoms to extended materials reveals further layers of quantum behavior. In solids, the proximity of atoms causes their orbitals to overlap and split into bands of energy levels so closely spaced that they form quasi-continuous ranges known as energy bands. The valence band, filled with electrons, and the conduction band, accessible at higher energy, define the electronic properties of materials. Metals possess partially filled bands or overlapping valence and conduction bands, allowing electrons to move freely and conduct electricity. Insulators, by contrast, have large band gaps preventing electron promotion under normal conditions. Semiconductors, with intermediate band gaps, exhibit tunable conductivity that forms the basis for modern electronics. The emergence of band theory represents a natural extension of molecular orbital concepts to macroscopic scales, illustrating how the same quantum principles govern behavior across orders of magnitude.

Electron behavior in solids becomes even more intricate when interactions with lattice vibrations, or phonons, are considered. These quantized vibrational modes influence thermal conductivity, electrical resistance, and phase transitions. The coupling between electrons and phonons underlies phenomena such as superconductivity, where electrons form Cooper pairs that move through the lattice without resistance. This pairing mechanism, described by the BCS theory, demonstrates how attractive interactions mediated by phonons overcome electron–electron repulsion under certain conditions. Superconductivity exemplifies the emergence of collective quantum states in macroscopic systems—states in which the quantum mechanical nature of electrons manifests not at the atomic scale but across entire materials.

Magnetism likewise arises from quantum mechanical behavior of electrons. Ferromagnetism, antiferromagnetism, and related phenomena depend on the alignment or opposition of electron spins and orbital contributions within materials. Exchange interactions, rooted in the symmetry properties of electron wave functions and the Pauli Exclusion Principle, dictate whether electron spins align parallel or antiparallel. In ferromagnetic materials, aligned electron spins produce large-scale magnetic ordering, while in antiferromagnetic substances, alternating spin alignments cancel out macroscopic magnetization. These magnetic behaviors illustrate the richness of quantum many-body interactions, where the collective properties of enormous ensembles of electrons emerge from simple, fundamental principles governing individual atoms.

Atomic behavior also extends into systems where electrons interact not only with each other and the nucleus but with the vacuum itself. Quantum field theory predicts that the vacuum is filled with fleeting pairs of virtual particles, and electrons within atoms interact with these fluctuations, producing measurable effects such as vacuum polarization. Such interactions shift energy levels slightly and contribute to phenomena that require exceptional precision to observe, such as the anomalous magnetic moment of the electron. These minute deviations between experimental measurements and theoretical predictions have become sensitive tests for physics beyond the Standard Model, demonstrating how atomic-scale experiments can probe the deepest structures of physical law.

At the outermost frontier of atomic behavior lies the study of coherent quantum systems, where atoms or electrons share correlated states across large distances. In ultracold gases cooled to temperatures near absolute zero, atoms lose their individual identity and condense into collective quantum states such as Bose–Einstein condensates. These condensates behave as single macroscopic wave functions, exhibiting phenomena like matter-wave interference, quantized vortices, and the ability to flow without viscosity. Such states confirm that quantum mechanics governs not only isolated atoms but entire ensembles when thermal agitation is minimized. Similarly, advanced control of trapped ions and neutral atoms enables precise manipulation of quantum states for emerging technologies such as quantum computation, where coherent superposition and entanglement allow for computational operations far beyond classical capabilities.

Across these increasingly diverse contexts—molecules, solids, magnetic materials, superconductors, relativistic heavy atoms, ultracold gases—the atom reveals itself not as a static entity but as a dynamic quantum system whose behavior depends sensitively on environment, interactions, and external conditions. The principles derived from the simplest atoms scale upward into elaborate and sometimes unexpected emergent phenomena, showing the unity of physical law from microscopic to macroscopic realms. The atom remains the fundamental nexus at which matter, energy, and quantum behavior converge, providing both the building block of all visible matter and the conceptual portal through which deeper layers of reality continue to be explored.

Chapter 8: Word Problems

1. A researcher studying atomic emission spectra observes a series of hydrogen spectral lines and notes that one prominent line corresponds to a photon of wavelength 486 nm, which lies within the Balmer series. Using Bohr's quantized energy-level expression for hydrogen, determine which pair of principal quantum numbers this transition most likely occurred between, and then compare the experimentally observed photon energy to the theoretical energy difference predicted by the Bohr model. Discuss possible sources of small discrepancies, including reduced-mass corrections arising from treating the proton as a finite-mass particle rather than an immovable nucleus.

2. A heated solid radiates energy with a spectrum that significantly deviates from classical predictions at short wavelengths. A student initially attempts to model this behavior using classical equipartition arguments but finds that the predictions

diverge at high frequency. By applying Planck's quantization assumption for oscillators in thermal equilibrium and using a measured peak emission temperature of 1400 K, calculate the approximate peak wavelength with Wien's displacement law. Explain how Planck's introduction of energy quantization resolves the ultraviolet catastrophe that plagued classical physics.

3. In a photoelectric-effect experiment, monochromatic light is shined on a clean metal surface. Electrons are observed to be ejected only when the incident radiation frequency exceeds 7.2×10^{14} s^{-1}. When light of frequency 8.0×10^{14} s^{-1} is used, the emitted electrons have kinetic energy of 1.1 eV. Determine the metal's work function from these data, and explain how the threshold-frequency behavior and the linear relationship between electron kinetic energy and photon frequency support Einstein's photon model rather than classical wave energy predictions.

4. An electron in an atom is observed to have a de Broglie wavelength of 2.5×10^{-10} m while moving at its most probable speed as predicted by quantum mechanical probability distributions. Calculate the electron's momentum corresponding to this wavelength, and discuss how wave–particle duality leads to standing-wave solutions around the nucleus. Explain how these standing waves give rise to quantized orbitals and discrete allowed energy levels.

5. A high-resolution spectroscopy instrument measures a hyperfine splitting of 5.9 μeV in the ground state of atomic hydrogen. This splitting arises from the magnetic interaction between the electron's intrinsic spin magnetic moment and the proton's magnetic moment. Explain qualitatively how this hyperfine structure emerges from quantum electrodynamics and magnetic dipole interactions, and calculate the electromagnetic frequency corresponding to this hyperfine transition energy.

6. During an electron-diffraction experiment, a beam of electrons accelerated to a kinetic energy of 500 eV is directed through a thin crystalline nickel foil. Compute the corresponding de Broglie wavelength of the electrons and explain how this wavelength determines the diffraction angle observed using Bragg's law. Discuss how the periodic spacing of atomic planes in the crystal produces constructive and destructive interference of the electron wavefronts.

7. A student analyzing hydrogenic orbitals models the radial probability distribution of a 2s electron and identifies a radial node at a specific distance from the nucleus.

Explain mathematically, using radial wavefunctions, why the 2s orbital contains one radial node whereas the 2p orbital contains none. Discuss how differences in penetration and shielding allow 2s electrons to approach the nucleus more closely, leading to lower energy than 2p electrons in multi-electron atoms.

8. A multi-electron atom is observed to have electrons occupying both 3d and 4s subshells. Ionization experiments reveal that upon removing electrons, the 4s electrons are preferentially ionized before the 3d electrons, even though the 4s subshell is filled first in the neutral atom. Using concepts of shielding, penetration, and electron–electron repulsion, explain why the energy ordering of 3d and 4s orbitals reverses between the neutral and ionized states.

9. A chemist examining periodic ionization-energy data notices an irregularity between elements in groups 2 and 13. Using Slater's rules, compute approximate effective nuclear charges (Z_{eff}) for a representative group 2 element and a representative group 13 element. Explain why, despite the higher nuclear charge of group 13 elements, their first ionization energies are lower because of shielding effects and the occupancy of a higher-energy p orbital.

10. A hydrogen atom absorbs a photon that excites its electron from the n=2 level to the n=5 level. Calculate the energy absorbed during this upward transition. Then enumerate every allowed radiative transition pathway the electron may follow as it returns to n=2, and compute the wavelengths of all photons that may be emitted in the process.

11. A scanning tunneling microscope (STM) probes a metallic surface at atomic resolution by measuring tunneling current between the tip and surface. Since the tunneling probability decays exponentially with tip-sample separation, use a given decay constant to calculate the ratio of currents at two different tip distances. Explain how the electronic density of states at the Fermi level contributes to tunneling behavior and ultimately determines the spatial resolution of STM images.

12. A superconducting material exhibits zero electrical resistance when cooled below a critical temperature of 7 K. Given a characteristic phonon frequency and an electron–phonon coupling constant, estimate the critical temperature using a simplified form of BCS theory. Discuss how Cooper pairs form through an effective attractive interaction mediated by lattice vibrations, and how these paired electrons condense into a collective quantum state.

13. Heavy elements such as gold display distinctive colors—gold appears yellow—due in part to relativistic effects. Using relativistic corrections to orbital energies, calculate the approximate amount of stabilization of the 6s orbital compared to a non-relativistic model. Explain how relativistic orbital contraction shifts the 5d→6s transition energy into the visible spectrum, producing the characteristic color of metallic gold.

14. An attosecond laser pulse is employed to observe ultrafast electron dynamics during atomic ionization. Given the pulse duration and intensity, estimate the minimum uncertainty in energy using the time–energy uncertainty principle. Explain how such brief pulses allow experimentalists to resolve electron rearrangements and continuum wave-packet motion on timescales previously inaccessible.

15. In a magnetic spectroscopy experiment, the Zeeman splitting of a p orbital is measured by applying an external magnetic field of 2 T. Calculate the energy separation between adjacent magnetic sublevels using the Bohr magneton. Explain how the degeneracy of magnetic quantum numbers (m_l values) is lifted and why the splitting magnitude depends on both orbital angular momentum and field strength.

16. A nuclear magnetic resonance (NMR) experiment examines a carbon-13 nucleus in three distinct molecular environments. Using provided shielding tensor components, compute the effective magnetic field experienced by the nucleus in each environment. Explain how differences in electron distribution produce varying degrees of shielding, thereby leading to the observed chemical shifts.

17. An electron gas confined to two dimensions in a strong magnetic field displays quantized energy levels called Landau levels. Given the applied magnetic-field strength, calculate the spacing between these levels and discuss how quantized cyclotron motion differs from the classical picture of continuous circular orbits. Explain how Landau quantization affects transport properties in high-field regimes.

18. In a semiconductor, laser light is used to excite electrons from the valence band to the conduction band. Given the band-gap energy, determine the minimum photon energy required for excitation and discuss how the density of states—particularly near the band edges—affects the intensity and sharpness of absorption features in the optical spectrum.

19. A metal is known to have a work function of 4.2 eV. Calculate the threshold frequency required for the photoelectric effect to occur. Explain why, in accordance with the photon model of light, increasing the intensity of sub-threshold radiation produces no electrons despite increasing energy flux in the classical sense.

20. Molecular orbital theory predicts paramagnetism for the O_2 molecule due to unpaired electrons. Using the MO diagram, calculate the bond order of O_2 and explain how the filling of degenerate π^* antibonding orbitals leads to the presence of unpaired electrons, accounting for the experimentally observed magnetic behavior.

21. A conjugated polyene contains eight π electrons arranged in an extended chain of overlapping p orbitals. Using Hückel molecular-orbital theory, determine whether the system qualifies as aromatic and calculate the π molecular orbital energies. Discuss how delocalization of the π electrons lowers overall molecular energy relative to isolated C=C double bonds.

22. In ethene (C_2H_4), each carbon atom forms three sp^2 hybrid orbitals. Calculate the angle between these hybrid orbitals and explain how the remaining unhybridized p orbitals overlap to form a π bond. Discuss why this π bond restricts free rotation around the carbon–carbon double bond.

23. A computational chemist performs a Hartree–Fock calculation for a neon atom and obtains a total electronic energy that lacks correlation energy. Using experimental ground-state energy values, estimate the missing correlation energy and explain why Hartree–Fock theory fails to fully capture electron–electron correlation effects.

24. A relativistic quantum-chemistry calculation predicts that the 1s orbital in uranium is contracted by several percent relative to its size in a non-relativistic model. Explain how high electron velocities near massive nuclei lead to an effective relativistic mass increase, altering orbital shapes and necessitating a relativistic treatment for heavy atoms.

25. A neutron star exhibits an extremely strong magnetic field on the order of 10^8 T. Predict qualitatively how such intense magnetic fields distort the electron cloud of a hydrogen atom, and explain how the atom becomes elongated along the magnetic field direction due to the strong magnetic confinement of charged particles.

26. In an X-ray diffraction study, a crystal with a known interplanar spacing is irradiated by X-rays of wavelength 0.15 nm. Using Bragg's law, calculate the reflection angle for the first-order diffraction peak. Explain how the diffraction patterns provide a map of atomic positions within the crystal lattice.

27. A Raman spectrum of an organic molecule displays a prominent vibrational peak associated with C=C stretching. Using the observed vibrational frequency and the reduced mass of the carbon atoms, calculate the corresponding force constant. Discuss how differences in molecular geometry and hybridization influence vibrational energy levels.

28. A nuclear isotope emits gamma radiation with an energy of 500 keV. Calculate the wavelength of this photon using the energy–wavelength relation. Explain why nuclear transitions typically involve energies many orders of magnitude larger than electronic transitions in atoms.

29. A degenerate electron gas in the interior of a white dwarf experiences enormous Fermi pressure. Given the electron density, calculate the Fermi energy and explain how degeneracy pressure—arising from the Pauli exclusion principle—prevents gravitational collapse in compact stellar remnants.

30. A Bose–Einstein condensate (BEC) is formed when a dilute atomic gas is cooled to temperatures near 50 nK. Using the de Broglie wavelength formula, show why the thermal wavelengths of atoms overlap at this temperature, causing the atoms to occupy a single quantum state. Discuss how long-range coherence emerges across the condensate.

31. A quantum computer employs trapped ions as qubits, where quantum transitions are controlled using laser beams. Given the Lamb–Dicke parameter and the trap frequency, calculate the transition probability for a sideband excitation. Explain the role of laser cooling in reducing motional excitations and preserving coherence in the qubit system.

32. In a one-dimensional particle-in-a-box model of length 1.0 nm, an electron undergoes an electronic transition from n=1 to n=3. Calculate the energy absorbed during this transition, and discuss how confinement in a small spatial region increases energy spacing relative to atomic orbitals.

33. A precise spectroscopy experiment measures the Lamb shift in hydrogen—a small correction to energy levels. Using quantum electrodynamics concepts such as vacuum polarization and self-energy, explain why this shift arises and estimate the equivalent frequency difference corresponding to the observed energy correction.

34. A paramagnetic solid obeys Curie's law, relating magnetic susceptibility to temperature. Using experimental susceptibility data, calculate the effective magnetic moment of the magnetic ions. Discuss how unpaired electron spins contribute to macroscopic magnetic behavior in such materials.

35. A transition-metal complex exhibits a d–d electronic transition at a wavelength of 650 nm. Using crystal field theory, calculate the octahedral splitting energy Δ_o. Explain how differences in ligand field strength influence the color of transition-metal complexes.

36. A superconducting loop sustains a persistent current indefinitely due to the absence of electrical resistance. Using the magnetic flux quantization condition, calculate the allowed values of quantized flux in the loop. Explain how the superconducting state enforces quantized circulation of Cooper pairs.

37. A solid material shows a temperature-dependent band gap that decreases as the lattice expands. Using data for thermal expansion and orbital overlap, explain how increased atomic spacing reduces band dispersion and modifies the electronic band structure with temperature.

38. A heavy atom forms an ion with a filled 4f subshell. Using shielding rules, explain why 4f electrons provide poor shielding for outer electrons and calculate an approximate effective nuclear charge (Z_{eff}) for a valence s electron. Discuss how this weak shielding influences chemical behavior.

39. An ultrafast pump–probe experiment measures electron relaxation in a semiconductor. From the decay curve of the probe signal, calculate the relaxation time constant and discuss how this timescale reflects electron–phonon coupling and carrier scattering rates.

40. A molecule exhibits resonance involving two major contributing structures. Using resonance energy data, calculate the stabilization relative to a hypothetical localized structure. Explain how electron delocalization influences molecular stability and reactivity.

41. A crystalline material undergoes a metal–insulator transition at low temperatures. Using given electron bandwidths and on-site Coulomb repulsion values, analyze the conditions that produce a Mott transition. Explain how electron–electron interactions outweigh band dispersion to localize electrons.

42. A hydrogen molecule is analyzed using molecular orbital theory. Calculate the bond order and determine whether the molecule remains stable if one electron is promoted to the σ^* antibonding orbital. Explain how this promotion alters bond strength.

43. Spectral lines from helium exhibit fine structure that the Bohr model cannot explain. Using electron spin–orbit coupling concepts, explain why fine structure arises and calculate the approximate splitting energy for given quantum numbers.

44. A nuclear fission event releases 200 MeV of energy. Using $E=mc^2$, calculate the mass defect associated with this release. Explain how trends in nuclear binding energy make heavy nuclei susceptible to fission.

45. A semiconductor nanoparticle shows a size-dependent band gap due to quantum confinement. Using the particle-in-a-sphere approximation, calculate the expected band-gap shift for a nanoparticle of given radius and explain why decreasing size causes a blue shift in emission.

46. A polyatomic molecule undergoes normal-mode vibrations that include both IR-active and IR-inactive motions. Using symmetry operations from the molecule's point group, classify these modes and explain why only vibrations that change the dipole moment lead to infrared absorption.

47. A radioactive isotope undergoes beta decay, altering its neutron-to-proton ratio. Using the Q-value of the reaction, calculate the maximum kinetic energy of the emitted beta particle. Discuss how nuclear shell structure and magic numbers influence nuclear stability.

48. A crystal containing a single point defect exhibits localized electronic states within the band gap. Using tight-binding principles, explain how the absence of an atom perturbs orbital overlap and calculate approximate defect-state energies relative to the conduction and valence bands.

49. An electron spin resonance (ESR) experiment measures spin precession in a magnetic field of 0.35 T. Calculate the Larmor frequency for the electron and explain how deviations of the electron g-factor from 2.000 arise due to relativistic and quantum electrodynamical effects.

50. A gas-phase pericyclic reaction requires precise information about molecular orbital energies. Using ab initio calculations, compare the HOMO–LUMO gaps of two reactants and explain how orbital symmetry rules (Woodward–Hoffmann) determine whether the reaction proceeds under thermal or photochemical conditions.

Chapter 9

Chemical Connections of Bonding

The exploration of chemical bonding has continually reshaped humanity's understanding of matter, evolving from speculative philosophical musings to the mathematically rigorous quantum discipline that lies at the heart of modern chemistry. Although the earliest conceptualizations of the atom were little more than intuitive abstractions, they provided the necessary imaginative scaffold upon which centuries of scientific thought gradually erected a comprehensive theory of bonding capable of explaining not only the stability of compounds but also the full spectrum of observable physical and chemical behavior. The ancient Greek conjectures of Democritus and

Leucippus, though scientifically ungrounded by contemporary standards, demonstrated a remarkable attempt to reduce matter to fundamental entities. Their "atomos" were conceived as the indivisible endpoints of subdivision, the smallest units constituting the myriad forms of matter. Despite their lack of empirical basis, these notions contained an essential germ of truth: the notion that matter possesses discrete underlying structure. Yet these ideas remained largely dormant, as the Western world lacked the experimental apparatus, philosophical inclination, and mathematical sophistication required to scrutinize matter at a fundamental level.

The scientific stagnation that enveloped atomic theory for nearly two millennia was not solely the result of technological limitation. It also reflected a broader conceptual environment dominated by Aristotelian thought, in which matter was composed of continuous, mutable elements—earth, air, fire, and water—whose transformations were governed by qualities rather than quantitative parameters. This worldview, resistant to atomistic rationalization, dominated natural philosophy until the intellectual upheaval accompanying the Scientific Revolution. During this transformative period, measurement, instrumentation, and systematic experimentation emerged as the defining features of scientific inquiry. Chemistry itself began shedding the mysticism of alchemy and adopting a methodological rigor that would culminate in the work of Antoine Lavoisier.

Lavoisier's contribution to chemical science was nothing short of revolutionary. By insisting on precise mass measurements and carefully controlled experimental conditions, he established that mass remains conserved during chemical transformations. This simple yet profound observation directly contradicted the notion that reacting substances dissolved into or emerged from the classical elements. Instead, it suggested that during a reaction, matter merely changed form; it did not vanish or appear spontaneously. Lavoisier's systematic analyses of combustion, calcination, and gaseous reactions demonstrated that oxygen, not phlogiston as previously believed, was involved in these processes. His work catalyzed a shift toward the quantitative, predictive, and empirical framework that allowed chemistry to evolve into a mature science.

Building upon Lavoisier's foundation, John Dalton proposed in the early nineteenth century a set of principles that became known collectively as Dalton's atomic theory. Recognizing patterns in empirical laws—such as constant composition, which states that a given chemical compound always contains the same proportion of elements by mass, and multiple proportions, which governs how elements combine to form different compounds—Dalton argued that only a particulate model of matter could account for such regularities. His assertion that elements are composed of unique, indivisible atoms capable of combining in fixed whole-number ratios offered a unifying explanation for a vast array of experimental data. Dalton's theory represented the first coherent scientific model of atomic behavior, and though it lacked the structural detail and nuance of later developments, it provided the conceptual scaffolding that allowed subsequent generations to probe matter at ever deeper levels.

The unraveling of atomic structure began in earnest at the end of the nineteenth century, driven in part by advances in experimental physics. J.J. Thomson's cathode ray experiments revealed the presence of the electron, a negatively charged subatomic particle whose properties contradicted the longstanding belief that atoms were indivisible. Thomson's demonstration that cathode rays bent in electric and magnetic fields established that they carried charge and possessed mass far smaller than that of the lightest atom. This discovery prompted Thomson to propose a new atomic model in which electrons were embedded within a positively charged sphere, forming the so-called "plum pudding" configuration. Though short-lived, this model marked a fundamental shift away from Dalton's monolithic atomic spheres and toward a more complex internal architecture.

The inadequacy of the Thomson model was dramatically exposed through the gold foil experiments conducted under Ernest Rutherford's supervision. Alpha particles—positively charged helium nuclei—were directed at a thin sheet of gold foil, and their scattering patterns were analyzed. While most particles passed through with little deviation, a small fraction were deflected at large angles, and some even rebounded directly backward. Rutherford famously remarked that this outcome was as surprising as firing a shell at tissue paper and having it bounce back. The only plausible explanation was that the atom's positive charge and mass were concentrated in an extraordinarily small nucleus, surrounded by mostly empty space in which electrons resided. This nuclear model redefined the atom's internal structure and set the stage for further refinements.

Yet Rutherford's model, which envisioned electrons orbiting the nucleus like planets around a star, was fundamentally unstable according to classical electrodynamics.

Orbiting electrons should radiate energy continuously, spiraling into the nucleus. The resolution to this paradox arrived through the advent of quantum theory. Niels Bohr introduced a quantized model of the hydrogen atom in 1913, proposing that electrons occupy discrete energy levels and emit or absorb energy only when transitioning between them. Although the Bohr model explained the hydrogen spectrum with impressive accuracy, it lacked generality, failing for atoms with more than one electron and offering no insight into chemical bonding.

The deeper revolution arrived with the development of quantum mechanics by Schrödinger, Heisenberg, Dirac, and others. Schrödinger's wave equation transformed atomic theory, representing electrons not as particles with precise trajectories but as wavefunctions whose squared magnitude described the probability distribution of finding an electron in a given region of space. Quantum numbers emerged naturally from the solutions to the Schrödinger equation, defining the size (n), shape (l), orientation (m_l), and spin (m_s) of the orbitals that electrons can occupy. These orbitals—spherical s orbitals, dumbbell-shaped p orbitals, and the more complex d and f orbitals—describe spatial distributions that dictate both the energy and directional characteristics of electrons. Chemical bonding, once viewed through the lens of mechanical linkage or electrostatic pairing, could now be understood as the quantum mechanical interaction between electron wavefunctions.

At the heart of bonding lies Coulomb's law, which states that the electrostatic energy between two charged particles varies inversely with the distance separating them. This fundamental relationship explains why the attraction between electrons and nuclei stabilizes atoms, while electron–electron and nucleus–nucleus repulsions counteract stabilization. Chemical bonding thus emerges as a delicate balance of attractive and repulsive interactions mediated by quantum mechanical constraints. The wave nature of electrons allows them to occupy regions of space between nuclei, lowering the system's energy and forming covalent bonds. Conversely, when electron density is drawn disproportionately toward one atom, ionic character emerges. Though classical electrostatics provides useful intuition, the full richness of chemical bonding arises only when electron delocalization, orbital overlap, exchange interactions, and symmetry considerations are incorporated into the quantum mechanical framework.

The interplay between ionic and covalent character becomes especially evident when examining bonding across the periodic table. In ionic compounds, electrons are transferred from electropositive metals to electronegative nonmetals, producing oppositely charged ions whose electrostatic attraction forms extended crystalline lattices. The stability of ionic solids is governed by their lattice energy, the energy released when gaseous ions assemble into a solid lattice. The Born–Landé equation captures this relationship by combining Coulombic attraction with a short-range repulsive term, characterized by the Born exponent, which reflects the ions' resistance to compression. Ionic lattices adopt structures that maximize attraction while minimizing repulsion, giving rise to repeating geometrical patterns such as rock-salt, fluorite, cesium chloride, and zinc blende structures. These arrangements dictate not only stability but also melting point, hardness, solubility, electrical conductivity in molten or dissolved states, and thermal properties.

Ionic bonding, however, is rarely purely ionic. Even the most electropositive metals induce some degree of polarization in their partner anions. Fajans' rules articulate how small, highly charged cations polarize electron clouds, imparting covalent character. As such, aluminum nitride, magnesium chloride, and lithium iodide all exhibit notable deviations from ideal ionic behavior. These subtleties reflect the continuum between ionic and covalent bonding and highlight the importance of considering electron cloud distortion in predicting properties.

Covalent bonding, in contrast, arises from the mutual sharing of electron pairs. Gilbert Lewis's insight that stable bonds form through shared electron pairs laid the foundation for valence bond theory. Lewis structures simplify bonding into diagrams that obey the octet rule, though this rule occasionally breaks down in cases of electron-deficient or hypervalent species. The capacity of atoms to share electrons stems from the quantum mechanical overlap of orbitals. Linus Pauling advanced this framework by introducing hybridization, positing that atomic orbitals mix into hybrid forms—sp^3, sp^2, sp, dsp^3, and d^2sp^3—each with defined geometry and directionality. These hybridized orbitals explain molecular geometries predicted by VSEPR theory but provide a deeper rationale rooted in orbital mechanics and energetics.

The consequences of hybridization manifest across countless molecular systems. In methane, sp^3 hybridization yields a tetrahedral symmetry that maximizes spatial separation among electron pairs. In ethene, sp^2 hybridization leads to a trigonal planar arrangement with an orthogonal unhybridized p orbital forming a π bond responsible for restricted rotation and reactivity associated with electrophilic addition.

In alkynes, the linear geometry of sp hybridization results in cylindrical electron density around the internuclear axis, strengthening the bond and enhancing its acidity relative to alkenes and alkanes. Hybridization also plays a central role in organic reactivity, influencing electrophilicity, nucleophilicity, bond strength, and steric accessibility.

Electronegativity differences provide a powerful predictor of bond polarity and chemical behavior. Elements with high electronegativity, such as fluorine and oxygen, attract electron density, leading to polarized bonds with partial ionic character. The degree of polarity influences reaction mechanisms, intermolecular forces, and phase behavior. Bonds with moderate polarity may engage in dipole–dipole interactions, whereas molecules capable of hydrogen bonding exhibit strikingly high boiling points relative to molecular mass. Water exemplifies this phenomenon: its bent geometry and large electronegativity difference produce a strong molecular dipole, enabling extensive hydrogen-bonding networks that underlie its anomalously high heat capacity, surface tension, and solvent capabilities.

Bond polarity also plays an essential role in determining molecular shape and symmetry. Even in molecules with polar bonds, symmetry can lead to dipole cancellation. Carbon dioxide's linear geometry results in zero net dipole moment despite the polarity of each C=O bond, whereas sulfur dioxide's bent geometry produces a measurable dipole. These distinctions influence physical behavior such as infrared absorption patterns, reaction pathways, and solubility profiles.

The limitations of classical Lewis structures become apparent when attempting to depict molecules with delocalized electrons. Resonance theory provides a conceptual tool for representing such species, though it must be understood as a heuristic approximation of the true quantum mechanical distribution. Benzene is the canonical example, with its equal bond lengths arising from delocalized π electrons spanning the entire ring. Resonance stabilization lowers the energy of such systems relative to hypothetical localized alternatives. The concept extends to numerous functional groups—nitrate, carbonate, carboxylate, amide, and aromatic heterocycles—each exhibiting properties dependent on delocalized electron density.

As molecular complexity increases, valence bond theory becomes insufficient to account for magnetic properties, spectral transitions, and electron delocalization extending beyond localized pairs. Molecular orbital theory addresses these limitations by treating electrons as occupying orbitals formed through the linear combination of atomic orbitals. These molecular orbitals can be bonding, antibonding, or nonbonding. Their energy ordering determines bond order, stability, magnetism, and reactivity. MO theory elegantly explains why O_2 is paramagnetic, a fact that baffled chemists relying solely on Lewis structures. It also provides insight into conjugation, aromaticity, and the photophysical behavior of chromophores.

The gradual fusion of quantum mechanics, spectroscopy, and computational modeling has brought modern bonding theory to unprecedented levels of precision. Advanced techniques such as density functional theory, Hartree–Fock methods, coupled-cluster theory, and multiconfigurational wavefunction approaches allow the prediction of bond energies, reaction pathways, charge distributions, and electronic transitions with remarkable accuracy. Bonding environments in solids, surfaces, and nanostructures can now be examined with tools such as X-ray diffraction, neutron scattering, nuclear magnetic resonance, electron paramagnetic resonance, and ultrafast laser spectroscopy. These methods reveal the dynamic, time-dependent aspects of bonding, including electron relaxation, excited-state behavior, and photochemical transformations.

As the field advances, chemical bonding is increasingly understood as a manifestation of quantum mechanical interactions shaped by symmetry, electron distribution, and the delicate balance between kinetic and potential energy. The continuous development of theoretical tools and experimental techniques ensures that the study of chemical bonding remains a vibrant, evolving frontier, connecting the microscale behavior of electrons and nuclei to the macroscale properties of materials, biological systems, and emerging technologies.

The centuries that followed Dalton's revival of atomism marked a turning point in the relationship between philosophical speculation and empirical science. No longer was the atom a purely conceptual entity posited to explain macroscopic consistency; it gradually became an experimentally interrogated structure, its properties inferred from the behavior of matter under extreme conditions. The nineteenth century introduced experimental methodologies that fractured the façade of indivisibility, exposing the atom as a composite entity woven from electrical and mass-bearing subcomponents. Each discovery in this era reshaped the scientific imagination, revealing layers of structure previously obscured by the limits of instrumentation.

One of the first cracks in the classical view of the atom came from the study of electricity and its interaction with gases at low pressure. Cathode rays—enigmatic luminous streams observed within evacuated discharge tubes—had long puzzled scientists. While some argued they were immaterial disturbances of the ether, others speculated they might be streams of charged particles. The instrumental breakthrough by J. J. Thomson in 1897 clarified the issue decisively. By subjecting cathode rays to simultaneous electric and magnetic fields, Thomson demonstrated that the deflection pattern was consistent with a particle carrying a negative charge far smaller in mass than any known atom. This was the electron, the first subatomic particle ever discovered, an entity nearly two thousand times lighter than a hydrogen atom yet possessing a universal presence in all forms of matter. The discovery shattered the ancient notion of indivisible atoms, revealing instead a composite architecture embedded with lighter constituents.

The implication was radical: if atoms contained negatively charged particles, they must also contain counterbalancing positive charges to maintain electrical neutrality. Thomson himself proposed the "plum pudding" model, in which electrons resided within a diffuse sphere of positive charge, much like plums embedded in a pudding. Although short-lived, the model represented an essential step toward an experimentally grounded theory of atomic structure, bridging speculative philosophy and quantitative physics. It envisioned atoms not merely as particulate units but as dynamic electrical systems capable of responding to external fields, radiating energy, and undergoing ionization.

The next major leap came from a line of inquiry that seemed initially unrelated: the study of radioactivity. In 1896, Henri Becquerel observed that uranium salts emitted spontaneous radiation capable of fogging photographic plates even in darkness. Following Becquerel's discovery, Marie and Pierre Curie embarked on a systematic study of radioactive substances, identifying new elements such as polonium and radium that displayed extraordinary emission intensities. The existence of spontaneous atomic radiation suggested that atoms could undergo internal transformations, releasing energy without external stimulus. This observation dealt another blow to the idea of atomic permanence and suggested that atomic structure must include internal stresses, instabilities, or constituent components capable of rearrangement.

Ernest Rutherford, working initially with Becquerel rays, later classified radioactive emissions into three types—alpha, beta, and gamma radiation—each with distinct penetrating powers and charge characteristics. His work culminated in one of the most famous experiments in the history of science: the 1909 gold-foil experiment conducted by Hans Geiger and Ernest Marsden. By bombarding thin metal foils with alpha particles, they observed that while most particles passed straight through, a small fraction were deflected at large angles and some even reversed direction entirely. This behavior was inexplicable under Thomson's diffuse-charge atom, which lacked localized concentrations of mass or charge capable of deflecting such energetic particles.

Rutherford interpreted these deflections as evidence of a dense, positively charged nucleus located at the atom's center, containing nearly all the mass of the atom and surrounded by a largely empty volume inhabited by orbiting electrons. The nuclear model redefined the atom as a miniature solar system, with electrons occupying a peripheral region much larger in scale than the nucleus itself. This revelation overturned nearly two millennia of thought and opened the door to a more precise quantitative treatment of atomic behavior, particularly through emerging quantum theory.

Yet even this nuclear model immediately presented new puzzles. Classical electromagnetism predicted that electrons orbiting a positively charged nucleus would continuously radiate energy, spiraling inward and collapsing within fractions of a second. But atoms were evidently stable across cosmic timescales. This contradiction signaled the inadequacy of classical physics to describe atomic systems and highlighted the need for a new theoretical framework that could incorporate stability, quantization, and wave-particle duality. Enter Niels Bohr, whose 1913 model proposed that electrons occupy discrete energy levels, transitioning between them only by emitting or absorbing quantized amounts of energy. Bohr's formulation successfully explained the spectral lines of hydrogen, marking the first successful quantitative theory of atomic structure.

Bohr's atom, however, was only a stepping stone toward the more complete quantum-mechanical picture that would emerge over the next two decades. Louis de Broglie's hypothesis of matter waves, Werner Heisenberg's matrix mechanics, and Erwin Schrödinger's wave equation all converged to reveal that electrons do not travel in classical orbits but instead occupy probabilistic distributions described by wavefunctions. The resulting quantum mechanical model replaced fixed trajectories with regions of high probability density—atomic orbitals—whose shapes and energies arise from solutions to the Schrödinger equation under Coulombic potentials. These mathematical structures brought unprecedented predictive power to chemistry, explaining not only atomic spectra but also the periodic table, chemical bonding, molecular geometry, and reactivity trends.

The discovery of the neutron in 1932 by James Chadwick completed the foundational picture of nuclear composition. Neutrons, electrically neutral yet nearly as massive as protons, explained the anomalous mass patterns of nuclei and provided the mechanism for isotopic variation. They also introduced nuclear forces—short-range attractions between nucleons—that govern stability, fission, fusion, and radioactive decay. With protons and neutrons established as the nucleus's structural units and electrons as the carriers of chemical behavior, the core architecture of the atom approached modern form. Yet from this foundation emerged even more intricate questions about the forces binding nucleons, the quark composition of hadrons, and the quantum fields underlying particle interactions.

By the mid-twentieth century, technological innovations such as particle accelerators and bubble chambers extended the experimental reach of physicists into realms far smaller than atomic nuclei. High-energy collisions revealed an array of previously unknown particles—pions, kaons, muons, baryons, and more—demanding an even deeper classification system. These discoveries gave rise to the Standard Model of particle physics, which describes matter as composed of quarks and leptons interacting via gauge bosons corresponding to fundamental forces. In this modern view, atoms are merely one scale within a hierarchy of structure stretching from macroscopic matter down to quantum fields. Yet even this elegant theoretical framework, powerful as it is, represents only one chapter in an unfolding scientific narrative rather than a finished story. As soon as the Standard Model began to reveal its predictive success, it also exposed profound mysteries: the absence of gravity from its formalism, the enigmatic behavior of neutrinos, the prevalence of matter over antimatter in the universe, and the shadowy existence of dark matter that appears to outweigh visible matter by a substantial factor.

As particle accelerators grew in scale, from early tabletop devices to monumental facilities spanning kilometers, humanity's quest to probe the fabric of nature began to acquire the character of a global engineering enterprise. The construction of high-luminosity colliders required not only deep theoretical insight, but also the collaboration of thousands of engineers, technicians, and computing experts whose combined expertise made it possible to orchestrate conditions resembling those of the early universe. Particle physics became a crossroads where scientific curiosity intersected with political will, economic investment, and international diplomacy. Laboratories such as CERN, Fermi National Accelerator Laboratory, and the later facilities in Asia and North America became emblematic of a new era—places where nations pooled resources to pursue knowledge that no single country could obtain alone. Despite their complexity, these massive machines served a purpose rooted in ancient philosophical questions: What is matter? What is motion? What are the ultimate constituents of reality? In answering these questions, physicists unearthed a vision of the universe grounded not merely in particles, but in symmetries and interactions, and in the peculiar behavior of quantum fields whose fluctuations shape everything from the formation of stars to the chemistry of life.

The rise of quantum field theory both clarified and complicated the understanding of matter. It clarified reality by demonstrating that particles could be understood as excitations of underlying fields, thereby unifying a diverse zoo of observed phenomena under a single conceptual umbrella. Yet it complicated matters by revealing that the vacuum itself is not empty but teems with transient virtual fluctuations. In this new worldview, even the simplest event—such as two electrons interacting—encapsulates an astonishingly rich dance of intermediate states that appear and disappear in femtoseconds. These theoretical structures, though mathematically demanding, gradually became the scaffolding of modern physics, shaping how researchers understand not only particle interactions but the evolution of the cosmos itself. The success of quantum electrodynamics, the electroweak theory, and quantum chromodynamics confirmed that the universe possesses a deep internal order encoded within gauge symmetries. The breaking of these symmetries during the early moments of the universe set the stage for the formation of the fundamental particles we now observe. Because of this, physicists began to view the cosmos not as a static backdrop but as a dynamic participant in the unfolding history of matter.

As particle physics matured, scientists also turned outward to cosmology, realizing that the structure of the universe on the largest scales could reveal hints about its smallest constituents. Observations of the cosmic microwave background, the distribution of galaxies, and the expansion rate of the universe provided windows into conditions that no accelerator could replicate. Paradoxically, the distant past left signatures that modern technology can detect in the faint radiation permeating space. These cosmological signals indicated that ordinary atoms—protons, neutrons, and electrons—constitute only a fraction of all matter. Most of the mass in the universe appears to be dark matter: invisible, non-luminous material detectable only through its gravitational influence. While the Standard Model offers no explanation for this silent majority, its existence suggests that the universe contains additional particles or fields waiting to be discovered. Dark energy, an even more mysterious entity driving the accelerated expansion of the cosmos, deepens the enigma. Together, dark matter and dark energy imply that the known scientific framework captures only a sliver of cosmic reality. This recognition has propelled the search for physics beyond the Standard Model, including theories of supersymmetry, extra dimensions, and new families of particles or interactions that might account for the unseen mass and energy permeating the universe.

Meanwhile, advances in experimental techniques have allowed scientists to probe increasingly subtle aspects of matter. The development of neutrino detectors—massive underground caverns filled with ultra-pure liquids—revealed that these ghostly particles, once thought to be massless, actually oscillate between different identities as they travel. This discovery, though small in numerical terms, shook the foundations of particle physics, proving that the Standard Model was incomplete. Similarly, the precise measurement of the muon's magnetic moment continues to raise questions about whether undiscovered particles are influencing observable quantities. Experiments involving ultra-cold atoms, high-precision spectroscopy, and rare decays also contribute to the search for new physics, demonstrating that the frontier of knowledge is not limited to high-energy collisions but includes subtle laboratory measurements where minute deviations from theoretical predictions may reveal profound truths. These diverse approaches underscore the versatility of scientific inquiry and the fact that answers to fundamental questions may emerge from unexpected directions.

At the same time, the practical implications of particle physics extend far beyond its theoretical ambitions. The technologies developed for accelerators and detectors have transformed fields ranging from medical imaging to materials science. Magnetic resonance imaging, positron emission tomography, superconducting magnets, and advanced radiation therapies owe much of their existence to the underlying physics originally pursued for purely scientific reasons. Even the World Wide Web began as a tool at CERN to share data among researchers scattered across continents. Thus, the study of fundamental particles has contributed not only to philosophical understanding but also to tangible improvements in human health, communication, and industry. Through these technological derivatives, the quest to understand the universe's smallest constituents enriches everyday life, reinforcing the interconnectedness of basic science and societal development.

As research continued into the late twentieth and early twenty-first centuries, scientists increasingly recognized that matter cannot be understood in isolation from energy. In the subatomic realm, particles transform into one another, annihilate into pure energy, and materialize from energetic interactions. The classical distinction between matter and energy blurred into a unified concept governed by conservation laws that operate at quantum scales. Moreover, emerging fields such as condensed matter physics revealed that collective behavior among many particles can produce phenomena as rich and fascinating as those seen in particle accelerators. Superconductivity, superfluidity, topological materials, and quantum phase transitions showcased how the interactions among countless electrons and atoms could give rise to exotic states of matter with properties unanticipated by early theories. These systems behave as if new particles—quasiparticles—emerge to describe collective excitations. In this sense, nature displays a creative versatility: structures and patterns arise not only from fundamental particles but also from the interplay between energy, order, and symmetry in complex environments.

This new appreciation for emergence reframed how scientists grasp the hierarchy of matter. Instead of viewing atoms, quarks, and electrons as the sole building blocks, researchers came to see each level of organization—from molecules to cells, from ecosystems to galaxies—as governed by distinct principles that cannot always be reduced to lower-level components. Chemistry, biology, and astrophysics each investigate phenomena that, while ultimately rooted in particle interactions, require their own frameworks to explain behavior at larger scales. The emergence of life, for instance, cannot be fully captured by listing quarks or leptons; it reflects a tapestry of interactions shaped by evolutionary pressures, environmental conditions, and self-organizing processes. This broadened perspective emphasizes that matter is not just a list of particles but a dynamic continuum of structures whose complexity grows with scale. From this viewpoint, the Standard Model stands not at the center of all sciences, but as one essential layer in a much larger hierarchical architecture.

As scientists explored these layers, the notion of fields took on increasing significance. Fields provide a unifying principle linking electromagnetism, the weak and strong forces, and potentially gravity as well. The idea that particles are excitations in fields raises profound questions about what these fields originate from, how they acquire mass, and whether they might be manifestations of deeper structures not yet discovered. The Higgs field, introduced to explain how particles attain mass, exemplifies this shift in thinking. The discovery of the Higgs boson was celebrated as a triumph, yet it also opened further questions: Why does the Higgs field have the value it does? Why is it stable? Does it hint at new symmetries that might unify the known forces? These inquiries drive ongoing research, suggesting that the map of fundamental physics, though extensively charted, still contains vast regions waiting to be explored.

In tandem with theoretical advances, improvements in computational power have allowed scientists to model physical systems with unprecedented fidelity. Lattice quantum chromodynamics, for example, enables researchers to simulate the behavior of quarks and gluons confined inside protons and neutrons. Though computationally demanding, such simulations offer insights impossible to obtain through direct measurement. Likewise, computational astrophysics models galaxy formation, stellar evolution, and the large-scale structure of the universe, revealing how small-scale physics influences cosmic evolution across billions of years. These digital explorations complement laboratory experiments, forming a triad—experimental, theoretical, computational—that drives contemporary science. With artificial intelligence now assisting in data analysis, anomaly detection, and experimental design, the pace of discovery continues to accelerate, allowing researchers to sift through colossal datasets in search of subtle patterns or rare events that might signal the presence of new physics.

Yet despite these successes, each new insight seems to enlarge the horizon rather than bring it closer. Physicists are now grappling with questions that span scales from the Planck length to the cosmic horizon. What is the nature of space-time itself? Is gravity a fundamental force or an emergent phenomenon arising from more primitive ingredients? If space-time is quantized, what are the fundamental units from which it is built? These questions remain unanswered, but they guide the search for quantum gravity theories, including string theory, loop quantum gravity, and other frameworks that attempt to reconcile the principles of quantum mechanics with the geometry of general relativity. The resolution of these issues promises to reshape our understanding of matter and energy, perhaps revealing that the foundations of the universe are more intricate—and more surprising—than anything imagined by earlier generations of scientists.

In this evolving landscape, the concept of matter has undergone a profound transformation. It no longer refers simply to solid, liquid, and gas, nor even to atoms or subatomic particles. Matter has become a multi-layered phenomenon shaped by interactions, symmetries, fields, and the interplay between quantum fluctuations and cosmic evolution. Each layer of understanding enriches the others, weaving a comprehensive picture in which humans themselves are participants—constructed from the same fields and particles that animate stars and galaxies. Through scientific inquiry, humanity continually deepens its relationship with the universe, revealing that beneath the complexity of life and the vastness of the cosmos lies a subtle fabric of energy and structure. Even as mysteries persist, these explorations reflect an unending drive to understand existence at its most fundamental level, echoing the ancient quest to comprehend the nature of reality and our place within it.

Chapter 9: Word Problems

1. In a newly synthesized diatomic molecule formed between a heavy p-block element and a mid-row transition metal, spectroscopy reveals a bond length significantly shorter than expected from covalent radii, while photoelectron spectra indicate partial charge transfer from the metal to the p-block element. Assuming hybridization on the p-block atom leads to strong directional bonding, and the metal contributes a d orbital capable of π back-donation, determine the bond order and approximate degree of ionic vs covalent character required to reconcile the anomalously short bond distance with the observed partial negative charge on the p-block element.

2. A researcher studying lattice vibrations in an ionic crystal notes that the measured phonon dispersion differs markedly from predictions based on a simple

harmonic lattice model with nearest-neighbor Coulomb interactions. Suspecting that polarization of the large, easily deformable anions introduces significant anharmonic corrections to the short-range repulsive term of the Born–Landé equation, calculate how much the Born exponent must shift to explain a 12% decrease in the predicted longitudinal optical phonon frequency at the Brillouin zone edge.

3. In a surface chemistry experiment, CO binds to a metal nanoparticle, shifting the stretching frequency by 180 cm^{-1} compared to gas-phase CO. Given that σ donation from CO to the metal weakens the C–O bond, while π back-donation from metal d orbitals strengthens it, determine the relative contributions of σ and π interactions required to produce the observed spectral shift, and estimate the effective change in bond order of the C–O bond upon adsorption.
4. A synthetic chemist observes that the reaction rate of an oxidative addition step in a catalytic cycle increases tenfold when replacing a phosphine ligand with a more π-accepting N-heterocyclic carbene. Using frontier orbital theory, determine how the stronger σ donation and altered π back-bonding of the carbene modify the metal's d-electron occupancy, lowering the activation energy sufficiently to account for the rate increase.
5. In a laser-cooling experiment with alkaline-earth atoms, the cooling limit is found to be significantly above the theoretical Doppler limit predicted from the natural linewidth of the cooling transition. Assuming that fine-structure splitting leads to off-resonant excitations and introduces additional heating channels, calculate the required detuning and laser intensity to minimize these effects and reach at least 80% of the Doppler limit.
6. A newly discovered solid-state material exhibits superconductivity at 38 K, even though the phonon-mediated pairing mechanism predicts a maximum Tc of only 20 K for its known lattice structure. Propose how strong electron correlation effects or unconventional pairing symmetries modify the density of states and electron-phonon coupling constants enough to justify the observed Tc, and estimate the necessary enhancement.
7. A quantum chemist models a molecule with a three-center, four-electron bond and finds that simple MO theory underestimates its stability by ~25 kJ/mol compared

to experiment. Considering hyperconjugation and resonance-like electron delocalization into an adjacent π system, calculate how much electron density must flow into the bridging orbital to reproduce the experimental stabilization.

8. In a rare-gas excimer formation study, an excited xenon dimer emits at 172 nm. Using the known repulsive ground-state potential and attractive excited-state potential, estimate the Franck–Condon factors required to explain why only the 172 nm emission band dominates the spectrum despite multiple vibrational states being energetically accessible.

9. A researcher observes that a sulfide mineral displays an unexpectedly low bandgap of 0.9 eV. Considering the orbital mixing between metal d orbitals and chalcogen p orbitals, determine how strong the hybridization must be to push the conduction band edge downward by at least 0.3 eV relative to predictions from simple ligand-field models.

10. In an experiment involving Rydberg atoms, electrons excited to n = 60 show lifetimes significantly shorter than hydrogenic scaling predicts. Assuming blackbody radiation from the laboratory environment induces transitions between nearby Rydberg states, calculate the ambient photon occupation numbers and transition probabilities needed to account for the measured lifetimes.

11. A molecular beam is directed through an inhomogeneous electric field, causing deflection proportional to the dipole moments of the molecules. If a molecule with two nearly isoenergetic conformers differs in dipole moment by 1.8 D, determine the relative populations and barrier height between conformers necessary for the measured deflection pattern at 400 K.

12. An organic semiconductor exhibits high charge-carrier mobility despite possessing no obvious π-stacked geometry. Propose how transient localization, dynamic disorder, and reorganization energy considerations lead to bandlike transport, and calculate the effective electronic coupling required to achieve the observed mobility of 12 $cm^2\ V^{-1}\ s^{-1}$.

13. A nuclear magnetic resonance experiment on a metal complex reveals a ^2J coupling constant much larger than anticipated. Considering that relativistic effects can enhance s-character in heavy-element bonding, estimate the required shift in orbital hybridization and relativistic contraction necessary to produce the coupling enhancement.

14. A thin film of a perovskite material exhibits a shift in its optical absorption edge depending on strain applied by the substrate. Using deformation potential theory, calculate the magnitude of strain needed to account for a 100 meV redshift in the bandgap, given the known dependence of electronic energy levels on lattice parameter.

15. In catalytic ammonia synthesis, a new iron alloy demonstrates unexpectedly high turnover frequencies despite weaker N_2 adsorption energies predicted by DFT. Considering kinetic compensation effects and entropy contributions to the free-energy landscape, determine how much entropic stabilization of the transition state is necessary to overcome weakened adsorption.

16. A cryogenic matrix isolation experiment reveals that a highly reactive radical displays vibrational frequencies inconsistent with predictions from simple valence bond arguments. Using a two-configuration mixing model, estimate the degree of configuration interaction needed to shift the predicted values to match the experimental spectrum.

17. During a neutron scattering experiment on a magnetic material, the measured magnon dispersion curve shows asymmetry between positive and negative wavevectors. Propose how Dzyaloshinskii–Moriya interactions arising from lack of inversion symmetry introduce this asymmetry, and calculate the associated interaction strength.

18. A molecule with a nominally forbidden $n \rightarrow \pi^*$ transition displays unexpectedly high intensity in UV–vis spectroscopy. Using vibronic coupling theory, determine how much distortion from equilibrium geometry is needed in the excited state to relax selection rules and enhance the transition moment.

19. A lanthanide complex displays a surprisingly large magnetic susceptibility despite possessing a nearly filled 4f shell. Assuming strong spin-orbit coupling and crystal-field splitting contribute to significant orbital angular momentum, calculate the effective magnetic moment and its deviation from the spin-only value.

20. A transition-metal oxide undergoes a metal–insulator transition at 200 K. Using the Mott–Hubbard framework, determine the ratio of on-site Coulomb repulsion U to bandwidth W that would drive such a transition, and estimate how lattice distortions modify W across the transition.

21. A researcher is examining a transition metal complex whose ligand field splitting appears unusually small compared to predicted values from both crystal field theory and ligand field theory. The complex exhibits an unexpected high-spin configuration even though its ligands are typically strong-field donors. Propose a detailed explanation accounting for covalency effects, metal–ligand orbital mismatch, and potential relativistic contributions that may be reducing the effective splitting. Formulate how one could experimentally distinguish between these competing effects using spectroscopic or computational tools.

22. Consider a molecule with extensive π-electron delocalization that displays anomalously high thermodynamic stability relative to its localized analogues. Experimental bond lengths show near equivalence across conjugated regions. Explain, using resonance theory and molecular orbital analysis, how the energetic contribution of delocalization stabilizes the system. Then design a method to quantify the delocalization energy experimentally, considering spectroscopic, calorimetric, and computational approaches.

23. A chemist synthesizes a new crystalline solid where cations and anions are similar in size, yet the compound unexpectedly crystallizes into a highly asymmetric structure rather than a simple body-centered lattice. Discuss how polarization effects, ion softness, and covalent character distort predictions based on purely ionic radii. Describe how manipulating temperature or pressure might shift the structure toward a more symmetric arrangement.

24. In a multi-electron atom, an anomaly is observed in the ordering of orbital energies, where a typically higher subshell becomes preferentially occupied. Analyze how electron–electron repulsion, penetration, shielding, and relativistic contraction influence orbital energies. Develop a theoretical framework predicting under what conditions this anomalous reversal becomes likely for heavy elements.

25. A molecule exhibiting sp^2 hybridization unexpectedly displays a bent geometry rather than trigonal planar. Provide an explanation that incorporates lone-pair repulsion, vibronic coupling, dynamic Jahn–Teller distortions, and potential multi-reference electronic character. Propose how vibrational spectroscopy could confirm or refute the presence of these distortions.

26. A researcher measures the dipole moment of a symmetric molecule and finds it to be nonzero, contradicting basic symmetry considerations.

Explore how dynamic motion, hindered rotation, or out-of-plane vibrations may break instantaneous symmetry. Devise an experiment to determine whether the dipole moment arises from static asymmetry, dynamic behavior, or an external perturbation.

27. A catalytic reaction involving a metal surface proceeds at rates far exceeding predictions from classical kinetic models. Explain how electron delocalization across the surface, formation of transient chemisorption states, and d-band center alignment facilitate the reaction. Then outline a computational method to evaluate how altering the metal's composition could tune catalytic effectiveness.

28. A diatomic molecule exhibits an observed bond length significantly shorter than predicted by simple covalent radii summation. As a graduate-level problem, analyze how partial multiple-bond character, orbital hybridization, and exchange interactions contribute to bond contraction. Describe how one could use high-resolution spectroscopy to determine the correct bond order.

29. A synthesized aromatic heterocycle shows unusual magnetic anisotropy and deviates from classical aromaticity rules. Evaluate the molecule's electronic structure using both Hückel and advanced MO perspectives, taking into account heteroatom contributions, orbital energies, and spin–orbit effects. Propose how NMR chemical shifts could be used to infer the true aromatic character.

30. A nanoparticle displays optical absorption peaks that do not align with bulk material properties. Analyze how quantum confinement, surface states, and excitonic effects alter electronic transitions. Then propose how changing particle size or ligand environment might tunably shift these optical features.

31. A researcher measures vibrational frequencies of a polar diatomic molecule and discovers strong deviations from harmonic predictions. Provide an explanation involving anharmonicity, bond polarity, and long-range electrostatic interactions. Suggest experimental techniques capable of capturing higher-order vibrational effects.

32. In a metallic solid, electrical conductivity drops sharply at temperatures where phonon scattering should still be minimal. Explain how impurities, electron correlation, band splitting, or magnetic ordering may induce this deviation. Design an experiment capable of differentiating between electronic and structural origins of the conductivity drop.

33. A reaction predicted to proceed via a simple SN1 mechanism instead exhibits stereoselectivity inconsistent with complete carbocation rearrangement. Evaluate how neighboring group participation, ion-pair retention, and solvent dynamics could influence the reaction pathway. Propose computational methods to model the transition state landscape.

34. A coordination complex demonstrates strong color changes under applied pressure. Assess how pressure influences metal–ligand orbital overlap, d–d transition energies, and ligand field strength. Propose how high-pressure optical spectroscopy could quantify these effects.

35. An organometallic compound undergoes oxidative addition under conditions where classical electron counting predicts it should be inert. Explain how ligand constraints, steric strain, and metal–ligand backbonding lower activation barriers. Suggest how computational energy decomposition could clarify the governing interactions.

36. A superconducting material shows a critical temperature far higher than predicted by conventional electron–phonon coupling models. Discuss how unconventional mechanisms—such as spin fluctuations, strong correlations, or emergent quasi-particles—might raise Tc. Propose how neutron scattering could test these hypotheses.

37. A laser-driven excitation in a molecular crystal produces transient states with lifetimes unexpectedly long for the energy involved. Explore how exciton delocalization, coherent vibrational coupling, and protective lattice dynamics may stabilize these states. Design a pump–probe experiment to measure the relaxation pathway.

38. A quantum dot emits photons of variable wavelength even under monochromatic excitation. Discuss the roles of charge trapping, surface recombination, and multi-exciton processes in producing this variability. Propose how surface passivation could reduce spectral wandering.

39. A heavy-element compound exhibits bonding patterns that appear to violate classical valence rules. Analyze how relativistic contraction, spin–orbit coupling, and expanded orbital availability produce unexpected bonding modes. Suggest spectroscopic techniques to detect these relativistic effects.

40. During polymerization, a monomer unexpectedly forms a cross-linked network rather than a linear chain. Examine how steric factors, radical stability, and propagating chain dynamics influence the reaction outcome. Devise a method to kinetically control chain growth to favor linear formation.

41. A high-temperature gas-phase reaction exhibits unexpected selectivity favoring a high-energy product. Analyze how entropic contributions, transition-state degeneracy, and statistical (RRKM-like) considerations dominate product distribution. Propose a method to experimentally confirm the presence of multiple competing transition states.

42. A liquid crystal phase displays emergent behavior leading to spontaneous pattern formation under weak electric fields. Explain how dipole alignment, quadrupolar interactions, and collective molecular motion contribute to the instability. Describe how modifying molecular side groups might alter the resulting textures.

43. A biochemical enzyme shows unexpectedly low activity despite favorable thermodynamic parameters for substrate binding. Provide an explanation incorporating dynamic disorder, conformational gating, and quantum tunneling in the active site. Propose experiments to separate dynamic effects from electronic contributions.

44. A particle trapped in an optical lattice demonstrates tunneling rates inconsistent with the potential depth inferred from laser calibration. Explore how zero-point energy, band structure effects, and inter-site coherence modify classical expectations. Propose how adjusting lattice geometry could tune tunneling behavior.

45. A vibrationally excited molecule undergoes a photochemical reaction that bypasses the lowest excited electronic state. Explain the role of conical intersections, non-adiabatic coupling, and ultrafast nuclear motion in determining this pathway. Suggest how femtosecond spectroscopy could capture the transition dynamics.

46. A gas composed of polar molecules exhibits anomalous compressibility at moderate pressures. Analyze how orientational ordering, dipole–dipole attraction, and local clustering influence bulk properties. Suggest how molecular dynamics simulations could probe the microscopic structure of the gas.

47. A semiconductor heterojunction displays charge separation efficiencies higher

than theoretical predictions. Discuss how interface states, band bending, and exciton dissociation contribute to enhanced performance. Propose how altering doping levels could modulate the effect.

48. A plasma of partially ionized atoms exhibits collective oscillations with frequencies deviating from predictions of simple charge density models. Evaluate how electron temperature, screening effects, and collisional dynamics influence the oscillations. Suggest experiments using laser probing to characterize plasma waves.

49. A colloidal suspension undergoes spontaneous self-assembly into ordered structures despite weak intermolecular forces. Analyze how entropy-driven ordering, depletion interactions, and solvent-mediated forces contribute. Propose how particle size or shape modifications could tune the assembly.

50. An astrophysical observation reveals a molecular cloud exhibiting spectral lines inconsistent with expected rotational or vibrational transitions. Assess how extreme temperatures, strong magnetic fields, or isotopic fractionation could generate the anomalies. Propose how laboratory astrochemistry experiments might recreate and analyze these unusual conditions.

Chapter 10

Atomic Orbitals to Functional Materials

The revelations of nuclear structure and subatomic particles set the stage for a profound theoretical revolution: the formulation of quantum mechanics. Unlike classical physics, which governed macroscopic phenomena with deterministic laws, quantum theory revealed that at the atomic and subatomic level, nature behaves according to principles that are inherently probabilistic, non-intuitive, and governed by discrete quantization. This paradigm shift not only resolved the paradoxes surrounding atomic stability but also provided the conceptual and mathematical framework necessary to explain the formation of chemical bonds in terms of electron distribution, orbital overlap, and energy minimization.

Central to this new understanding is the Schrödinger equation, which treats the electron as a wavefunction $\psi(r, t)$ whose squared magnitude represents the probability density of finding the electron in a given region of space. Solutions to this equation in the context of a Coulombic potential yield atomic orbitals, each characterized by quantum numbers (n, l, m_l, m_s) that define the electron's energy, angular momentum, spatial orientation, and intrinsic spin. The resulting orbitals—spherically symmetric s orbitals, dumbbell-shaped p orbitals, cloverleaf d orbitals, and complex f orbitals—provide a rigorous mathematical description of where electrons are likely to be found around a nucleus. Beyond mere geometric forms, these orbitals govern the directional properties of chemical bonds, the multiplicity of bonding interactions, and the capacity for electron delocalization in polyatomic systems.

Valence electrons, occupying the outermost orbitals, are the primary architects of chemical bonding. Covalent bonds form through the quantum mechanical overlap of atomic orbitals, producing molecular orbitals that lower the system's total energy. Sigma (σ) bonds arise from end-to-end overlap, creating cylindrical electron density along the internuclear axis. Pi (π) bonds form from lateral overlap of p orbitals, generating regions of electron density above and below the bond axis and restricting rotational freedom. Multiple bonds—double and triple bonds—incorporate combinations of sigma and pi interactions, resulting in enhanced bond strength, shorter bond length, and altered reactivity. The quantum mechanical framework predicts that bond energies, lengths, and angles emerge naturally from the solutions to the Schrödinger equation, eliminating the need for purely empirical descriptions.

The concept of orbital hybridization, introduced by Linus Pauling, further refines our understanding of molecular geometry. Atomic orbitals mix to form hybrid orbitals with specific directional properties that explain the observed symmetry of molecules. In methane (CH_4), carbon's one 2s and three 2p orbitals combine to produce four sp^3 hybrid orbitals arranged tetrahedrally with bond angles of 109.5°, perfectly matching experimental measurements. In ethene (C_2H_4), sp^2 hybridization generates three coplanar orbitals forming sigma bonds, while unhybridized p orbitals on each carbon participate in π bonding, producing planar geometry with 120° angles. Ethyne (C_2H_2) exhibits sp hybridization, resulting in linear geometry and the formation of two orthogonal π bonds. These hybridization schemes are not merely heuristic; they correspond to the energy-minimized configuration predicted by quantum mechanics, and are supported by spectroscopic and diffraction data.

Electronegativity, a property quantifying an atom's ability to attract shared electrons, introduces another layer of sophistication to bonding theory. Pauling's empirical scale allows for the calculation of bond polarity and, indirectly, partial ionic character in covalent bonds. The degree of bond polarity influences molecular dipole moments, reactivity, solubility, and intermolecular interactions. Water exemplifies the impact of high electronegativity differences and bent geometry: its substantial dipole moment enables extensive hydrogen bonding networks, which in turn govern thermodynamic properties such as heat capacity, surface tension, and solvent ability. In contrast, molecules such as carbon dioxide, despite containing polar bonds, exhibit net zero dipole due to symmetry, demonstrating that molecular geometry is as critical as atomic electronegativity in determining macroscopic properties.

Resonance, another consequence of quantum mechanical delocalization, arises when a single Lewis structure is insufficient to describe electron distribution. Benzene, with six conjugated π electrons delocalized over a planar ring, exemplifies aromatic stabilization. All C–C bond lengths in benzene are equivalent at 1.39 Å, intermediate between single and double bonds, reflecting the distribution of electron density. Similarly, the nitrate ion (NO_3^-) and carbonate ion (CO_3^{2-}) exhibit equivalent resonance forms, resulting in trigonal planar geometries with uniform bond lengths. Resonance is not a matter of electrons oscillating between fixed structures; it represents a quantum superposition, an energy-lowering delocalization that stabilizes the molecule relative to any hypothetical localized arrangement.

Molecular orbital theory provides a more comprehensive description, particularly for conjugated systems and homonuclear diatomic molecules. Molecular orbitals form through the linear combination of atomic orbitals (LCAO), producing bonding, antibonding, and nonbonding orbitals that dictate bond order, magnetism, and chemical stability. Oxygen, for example, is paramagnetic due to the occupation of degenerate π^* antibonding orbitals, a property inaccessible to classical Lewis structures. Molecular orbital theory further explains electron delocalization in extended π systems, conjugated polymers, and aromatic species, providing predictive power for spectroscopic transitions, reactivity patterns, and photophysical behavior.

Beyond traditional covalent and ionic interactions, metallic bonding introduces the concept of delocalized electrons in extended lattices of metal cations. The “electron sea” model, formalized in quantum terms through band theory, explains electrical conductivity, malleability, ductility, and optical properties of metals. Valence electrons occupy delocalized bands rather than discrete molecular orbitals, with partially filled conduction bands enabling current flow. The presence of d orbitals in transition metals introduces additional complexity: crystal field splitting, ligand field effects, and variable oxidation states contribute to magnetic, spectroscopic, and catalytic behavior. Organometallic complexes, where covalent and metallic character coexist, illustrate the versatility and subtlety of electron distribution in advanced chemical systems.

Intermolecular forces, though weaker than intramolecular bonds, govern the aggregation, phase behavior, and biological function of matter. London dispersion forces arise from instantaneous induced dipoles in all atoms and molecules, scaling with polarizability and molecular size. Dipole–dipole interactions stabilize polar species, while hydrogen bonding—directional interactions involving H atoms bonded to electronegative atoms—imparts remarkable cohesion and specificity in liquids, solids, and biological macromolecules. In proteins, nucleic acids, and carbohydrates, hydrogen bonding defines secondary and tertiary structures, influences folding kinetics, and mediates enzyme-substrate specificity. Collectively, these weak interactions exemplify the continuum between covalent, ionic, and van der Waals forces in determining chemical behavior.

The intersection of quantum mechanics with experimental methods allows unprecedented insight into bonding. X-ray crystallography, neutron diffraction, and electron diffraction provide atomic-scale structural data, confirming bond lengths, angles, and symmetry predicted by theory. Nuclear magnetic resonance (NMR) spectroscopy reveals local electronic environments and connectivity. Ultrafast laser spectroscopy probes electron dynamics, vibrational modes, and excited-state behavior on femtosecond timescales, linking theoretical predictions with temporal molecular behavior. Computational chemistry—employing methods ranging from Hartree–Fock and post-Hartree–Fock techniques to density functional theory—quantitatively predicts geometries, energies, charge distributions, and reaction mechanisms with remarkable fidelity. These tools enable the exploration of systems from simple diatomics to large biomolecules, crystalline solids, and functional materials, providing a unified framework in which bonding theory, quantum mechanics, and experimental validation converge.

In extending bonding theory into applied domains, one observes the interplay between microscopic principles and macroscopic phenomena. Materials chemistry leverages hybridization, orbital symmetry, and delocalization to design semiconductors, superconductors, catalysts, and energy-storage compounds. Organic synthesis exploits electronic effects, steric hindrance, and resonance stabilization to construct molecules with targeted functionality. In biochemistry, the principles of hydrogen bonding, electrostatics, and molecular orbital interactions underlie enzymatic catalysis, DNA replication, protein folding, and ligand recognition. Even environmental chemistry and nanotechnology draw upon the understanding of atomic-level interactions, highlighting the universal applicability of bonding theory across scales.

Through the lens of quantum mechanics, the chemical bond ceases to be an abstract notion and becomes a tangible manifestation of underlying physical laws: electrons distributed in orbitals, interacting through electrostatics, exchange, and delocalization; nuclei stabilized through strong and weak forces; and energy landscapes defined by the interplay of kinetic and potential terms. This perspective unifies disparate phenomena—bond energies, reactivity trends, phase behavior, and spectroscopic signatures—within a single conceptual framework, enabling predictions, rational design, and the ongoing discovery of novel materials and molecules.

As the conceptual framework of chemical bonding matured, it became increasingly evident that simple ionic and covalent models were insufficient to describe the full diversity of chemical structures observed in nature. The discovery of hypervalent molecules, electron-deficient species, coordination complexes, and extended solids revealed the need for a more sophisticated understanding that integrates orbital theory, quantum mechanics, and symmetry principles. Each of these systems highlights the profound influence of electron distribution, delocalization, and orbital interactions on molecular and material properties, bridging the gap between fundamental theory and real-world applications.

Hypervalent molecules, in which a central atom appears to exceed the octet rule, exemplify the limits of classical bonding models. Phosphorus pentachloride (PCl_5), sulfur hexafluoride (SF_6), and xenon tetrafluoride (XeF_4) all feature central atoms surrounded by more than eight valence electrons. The resolution of this apparent paradox lies in the utilization of d orbitals and three-center four-electron bonding models. For example, in SF_6, the sulfur atom employs sp^3d^2 hybridization to generate six equivalent bonding orbitals, each accommodating a pair of electrons. These configurations yield octahedral geometries, with bond lengths and angles consistent with both experimental crystallography and computational predictions. Hypervalency demonstrates that the octet rule is a convenient heuristic rather than a fundamental restriction and that the expansion of valence shells is permissible when energy minimization and orbital availability are satisfied.

Electron-deficient compounds present another challenge. Boranes, such as diborane (B_2H_6), contain fewer valence electrons than required for classical two-center two-electron bonds. The solution emerges in the concept of three-center two-electron (3c–2e) bonds, wherein a single pair of electrons is delocalized over three atoms, stabilizing the molecule while accommodating the electron deficiency. In diborane, two bridging hydrogen atoms form such bonds, each connecting two boron centers with a shared electron pair. Molecular orbital analysis of these species confirms the delocalization of electron density across the bond framework, illustrating the predictive power of quantum mechanics in nonclassical bonding scenarios.

Coordination complexes, particularly those of transition metals, introduce additional layers of complexity. The central metal atom or ion, often a transition element with partially filled d orbitals, interacts with ligands through coordinate covalent bonds. Ligands, electron-pair donors, stabilize the metal center while modulating its electronic structure, geometry, and reactivity. Crystal field theory (CFT) and ligand field theory (LFT) formalize these interactions, predicting electronic splitting of d orbitals in various geometries—octahedral, tetrahedral, square planar, and beyond. Octahedral complexes, for instance, experience splitting of the five d orbitals into t_2g and e_g subsets, with the energy gap Δ_oct governing spectroscopic transitions, magnetic properties, and reactivity. This splitting explains the colors of transition metal complexes, the distinction between high-spin and low-spin configurations, and the thermodynamic preference for specific geometries. Ligand field stabilization energy (LFSE) quantifies the energetic advantage conferred by orbital interactions, providing a unifying principle that connects electronic structure with molecular stability and chemical behavior.

The bonding in these complexes often exhibits partial covalency, especially when ligands possess π-donor or π-acceptor capabilities. Carbonyl ligands (CO), for example, participate in synergic bonding: donation of electron density from the ligand's filled σ orbital to the metal center, coupled with back-donation from filled metal d orbitals into the ligand's π* antibonding orbitals. This interplay strengthens the metal-ligand interaction while modulating bond orders within both the ligand and the metal coordination sphere. Organometallic complexes, featuring direct metal–carbon bonds, further illustrate the hybrid nature of bonding: covalent, ionic, and metallic character coexist, contributing to catalytic activity, electronic tunability, and structural diversity.

Extended solids present yet another dimension in the exploration of chemical bonding. Ionic lattices, metallic crystals, covalent network solids, and molecular crystals each exemplify unique bonding patterns at the mesoscale. Sodium chloride, as an archetypal ionic solid, crystallizes into a face-centered cubic lattice stabilized by long-range Coulombic interactions. Its lattice energy, determined by the charges and interionic distances, governs macroscopic properties such as melting point, hardness, and solubility. In contrast, covalent network solids, such as diamond and silicon carbide, feature directional, strong covalent bonds extending throughout the crystal. Diamond's tetrahedral sp^3 hybridized carbon atoms form an incompressible lattice with extremely high bond dissociation energy, accounting for its remarkable hardness and thermal conductivity. Graphite, by contrast, demonstrates the effect of anisotropic bonding: strong in-plane σ bonds within hexagonal layers coexisting with weak van der Waals forces between planes, yielding lubricating properties, electrical conductivity along layers, and facile intercalation chemistry.

Metallic solids exhibit yet another form of bonding. The delocalized "sea of electrons" model, formalized quantitatively through band theory, explains conductivity, ductility, and malleability. In metals, the overlap of atomic orbitals forms conduction bands populated by delocalized electrons, while valence electrons occupy filled bands. The occupancy, width, and energy gaps of these bands govern electronic behavior: metals exhibit partially filled bands allowing electron flow, semiconductors possess small band gaps permitting controlled conduction, and insulators display wide gaps precluding free electron motion. The incorporation of transition metals with partially filled d bands introduces further nuance: localized d electrons contribute to magnetic phenomena, coloration, and catalytic properties, revealing the intricate relationship between electronic structure and bulk material properties.

The concept of aromaticity, extending from classical organic chemistry, bridges molecular bonding with delocalized electron systems in solids and polymers. Benzene's planar, conjugated π system exemplifies σ–π separation, delocalized bonding, and resonance stabilization. Its energy, structure, and reactivity can be rationalized through Hückel's rule (4n+2 π electrons) and molecular orbital theory. Larger polycyclic aromatic hydrocarbons, fullerenes, and graphene sheets extend this principle to macroscopic scales, exhibiting delocalized electronic conduction, mechanical strength, and chemical tunability. In graphene, for example, the extended conjugation across sp^2-hybridized carbon atoms produces Dirac cones in the electronic band structure, enabling high electron mobility and quantum Hall effects. The emergence of such materials underscores the continuity between molecular bonding principles and condensed matter phenomena.

Advances in computational chemistry allow precise modeling of these complex bonding environments. Density functional theory (DFT) provides approximate solutions to the many-body Schrödinger equation for systems containing hundreds of electrons, balancing computational tractability with accuracy. Post-Hartree–Fock methods—Møller–Plesset perturbation theory, coupled-cluster theory, and multireference configuration interaction—capture electron correlation effects essential for describing weak interactions, transition states, and excited states. Natural bond orbital (NBO) analysis, quantum theory of atoms in molecules (QTAIM), and electron localization function (ELF) mapping offer interpretive tools linking electronic structure with chemical intuition. These methods illuminate the delicate balance between covalency, ionicity, metallicity, and delocalization in molecules, complexes, and solids, enabling predictive design of novel functional materials, catalysts, and supramolecular assemblies.

Nonclassical bonding further expands the scope of chemical interaction. Three-center two-electron bonds, agostic interactions, halogen bonding, and hydrogen–

halogen cooperativity illustrate the continuum between localized and delocalized electron distribution. In supramolecular chemistry, weak interactions—including π–π stacking, cation–π interactions, and van der Waals forces—directly dictate self-assembly, molecular recognition, and host–guest complexation. Such phenomena, while individually weak, collectively produce highly stable architectures capable of selective catalysis, molecular encapsulation, and nanoscale device fabrication. These systems exemplify the idea that bonding is not merely an intrinsic property of isolated molecules, but a context-dependent feature emerging from electron distribution, orbital overlap, and environmental interactions.

Even in the most exotic cases, bonding principles are preserved. Hypercoordinate species, organometallic clusters, quantum dots, and low-dimensional materials obey the same fundamental balance of electrostatics, orbital interactions, and energy minimization. What differs is the scale, symmetry, and delocalization domain over which these principles operate. Quantum confinement in nanomaterials, for example, alters band gaps and optical properties, revealing that electronic behavior is intimately linked to size, geometry, and electron delocalization. Similarly, extended covalent-organic frameworks (COFs) and metal-organic frameworks (MOFs) illustrate how classical bonding motifs—metal-ligand coordination, covalent linkages, π-stacking—can be assembled into crystalline networks with tunable porosity, reactivity, and electronic properties. Such materials bridge molecular chemistry with solid-state physics, emphasizing that bonding theory remains a unifying language across scales.

The integration of experimental observations, theoretical formalism, and computational prediction has thus elevated the study of chemical bonding from a descriptive endeavor to a predictive science. Bonding is no longer perceived solely as a static feature of molecules but as a dynamic interplay of forces, orbitals, and electrons. Each atom, ligand, or molecular fragment participates in a web of interactions that collectively define the structure, stability, reactivity, and function of matter. Understanding these interactions enables rational design, whether in the creation of pharmaceuticals with precise target affinity, catalysts that accelerate reactions with high selectivity, or materials exhibiting specific electronic, optical, or mechanical properties. The depth and breadth of bonding theory highlight its centrality to all domains of chemistry, physics, and materials science, demonstrating that the principles first glimpsed in the philosophical musings of the ancient Greeks have evolved into a comprehensive framework encompassing the full diversity of chemical phenomena.

As the theoretical and experimental foundations of chemical bonding matured, the discipline entered a new era defined by the confluence of computational power, high-resolution spectroscopy, and the exploration of exotic molecular and materials systems. No longer limited to classical ionic, covalent, and metallic frameworks, chemists now probe the subtleties of electron correlation, relativistic effects, and non-classical interactions across scales ranging from isolated atoms to complex supramolecular architectures. Computational chemistry, in particular, has become an indispensable tool, allowing the prediction and rationalization of chemical behavior in systems that are experimentally challenging, transient, or otherwise inaccessible.

Quantum chemical calculations, grounded in solutions to the Schrödinger equation or its approximations, provide detailed insights into electronic structure, reaction pathways, and bonding energetics. Methods such as Hartree–Fock theory offer a first-order description of electron distributions, while post-Hartree–Fock approaches—including Møller–Plesset perturbation theory and coupled-cluster methods—account for electron correlation, crucial for accurately describing dispersion interactions, transition states, and excited states. Density functional theory (DFT), by incorporating electron density as the fundamental variable, enables practical treatment of large molecular systems and solid-state lattices, balancing computational efficiency with accuracy. These approaches quantify bond orders, charge delocalization, orbital overlap, and potential energy surfaces, providing predictive power that complements experimental observations and guides synthetic design.

The application of computational methods has revealed subtle effects in bonding that extend beyond classical models. Hyperconjugation, an interaction in which σ electrons stabilize adjacent empty or partially filled orbitals, influences conformational preferences, acidity, and reactivity in organic systems. Agostic interactions, characterized by weak coordination between C–H bonds and electron-deficient metal centers, exemplify how transient electron density can stabilize unusual geometries in organometallic chemistry. Similarly, halogen bonding, chalcogen bonding, and pnictogen bonding illustrate that atoms traditionally considered inert or peripheral can engage in directional, energetically significant non-covalent interactions, analogous in many ways to hydrogen bonding yet governed by subtle electrostatic and polarization effects.

Supramolecular chemistry leverages these weak interactions to construct highly ordered assemblies, from host–guest complexes to self-assembling polymers and nanostructures. In these systems, bonding is not confined to localized electron pairs but emerges collectively from multiple cooperative interactions. The predictability and tunability of supramolecular architectures rely on an intimate understanding of orbital interactions, dipole alignment, π–π stacking, van der Waals forces, and hydrogen bonding networks. These principles have enabled the design of molecular machines, stimuli-responsive materials, and selective catalysts, demonstrating that the concept of the chemical bond extends well beyond traditional two-center frameworks.

In the realm of materials chemistry, the principles of chemical bonding underlie the development of functional solids. Covalent-organic frameworks (COFs), metal-organic frameworks (MOFs), and perovskites exemplify how hybrid bonding motifs—combining coordination bonds, covalent linkages, and non-covalent stabilization—can yield materials with tunable porosity, electronic structure, and catalytic activity. In MOFs, for example, the metal nodes act as coordination centers while organic linkers define the dimensionality and rigidity of the lattice. Quantum chemical modeling allows precise prediction of adsorption sites, electronic band structures, and ligand-field effects, guiding the rational design of materials for gas storage, catalysis, and energy conversion. Similarly, in covalent semiconductors, orbital hybridization, electron delocalization, and band structure dictate conductivity, optical properties, and defect tolerance, demonstrating the direct connection between bonding theory and functional performance.

Exotic bonding motifs also challenge traditional chemical paradigms. Superatoms, clusters of atoms that collectively exhibit properties reminiscent of single elements, demonstrate delocalized electron shells extending over multiple nuclei. These clusters obey "magic number" rules for electron shell filling and often exhibit unexpected stability, reactivity, and electronic properties. Similarly, cluster compounds, including polyhedral boranes, metal clusters, and heteronuclear cages, rely on delocalized multi-center bonding that cannot be described by simple Lewis structures. The understanding of these systems draws heavily on molecular orbital theory, electron counting rules, and quantum mechanical simulations, highlighting the adaptability of bonding principles to non-classical contexts.

Topological materials, including graphene, transition metal dichalcogenides, and topological insulators, illustrate how bonding and electron delocalization can lead to emergent quantum phenomena. In graphene, sp^2 hybridization produces a planar lattice with delocalized π electrons that enable high electron mobility, fractional quantum Hall effects, and Dirac cone formation in the electronic band structure. In topological insulators, strong spin–orbit coupling, relativistic effects, and orbital hybridization produce materials that are insulating in the bulk yet conductive along surfaces or edges, a behavior fundamentally tied to the topology of bonding and orbital interactions. These examples underscore that bonding is not merely a local interaction but can have macroscopic consequences dictated by symmetry, electron delocalization, and quantum coherence.

Non-covalent interactions remain central to biological systems, where subtle differences in electron density and orbital alignment govern macromolecular structure and function. Protein folding, DNA base pairing, enzyme catalysis, and ligand recognition all rely on hydrogen bonding, π–π stacking, cation–π interactions, and van der Waals forces. Quantum chemical methods, combined with molecular dynamics simulations, enable detailed mapping of these interactions, revealing transition states, energetic landscapes, and cooperative effects that drive biological activity. Computational predictions often match or surpass experimental resolution, providing mechanistic insight into processes such as proton transfer, allosteric regulation, and enzymatic turnover.

Photochemistry and excited-state chemistry further extend the conceptual domain of bonding. Upon absorption of light, electrons are promoted to antibonding or Rydberg orbitals, altering bond orders, geometries, and reactivity. Time-resolved spectroscopy and nonadiabatic molecular dynamics capture the fleeting electronic configurations that drive photochemical reactions, enabling the design of light-responsive materials, photoredox catalysts, and photovoltaic devices. In such systems, the distinction between bonding and antibonding orbitals becomes dynamic, dependent on time, environment, and electronic configuration, illustrating that chemical bonds are not static entities but evolving manifestations of electronic structure.

At the most fundamental level, bonding principles extend to relativistic and high-pressure regimes. Heavy elements, where relativistic effects significantly alter orbital energies and contraction, exhibit unusual oxidation states, bond lengths, and chemical reactivity. For example, gold's relativistic contraction of 6s orbitals contributes to its exceptional stability and inertness, while lead and bismuth display bonding patterns that deviate from expectations based on lighter congeners. Under extreme pressures, new bonding motifs emerge in solids, including polymerized nitrogen, metallic hydrogen, and unusual coordination geometries in transition metal oxides, revealing that electron density, orbital overlap, and electrostatics are modulated by environmental extremes, yet still obey fundamental quantum mechanical principles.

The synthesis of these insights underscores a unifying theme: chemical bonding, once conceived as a simple pairing of electrons or electrostatic attraction, is now recognized as a multi-scale phenomenon governed by quantum mechanics, orbital interactions, electron correlation, relativistic effects, and environmental context. Whether in molecules, solids, biological macromolecules, or engineered nanostructures, the principles of bonding dictate structure, stability, reactivity, and function. Advanced computational methods, coupled with high-resolution spectroscopy and diffraction, allow predictive modeling of bonding in unprecedented detail, enabling rational design of materials, catalysts, and functional molecules. In this sense, bonding theory has evolved from an abstract philosophical notion into a comprehensive, predictive, and universally applicable framework, connecting the microscopic behavior of electrons to the macroscopic properties of matter.

As research continues, new frontiers emerge. Quantum information materials, molecular electronics, supramolecular machines, and topologically complex compounds all depend critically on precise control and understanding of bonding interactions. The integration of experimental, theoretical, and computational perspectives ensures that the study of chemical bonding remains not only central to chemistry but also to physics, materials science, biology, and engineering. The once-simple concept of the atom and its bonds has expanded into a profound intellectual edifice, encompassing nearly every domain of modern science and continually inspiring new discoveries at the interface of theory and experiment.

Chapter 10: Word Problems

1. A chemist is investigating the bonding in ozone (O_3) using quantum chemical calculations. Considering that O_3 exhibits resonance and delocalized π electrons, explain how molecular orbital theory accounts for the experimentally observed equivalent bond lengths. Further, discuss how the inclusion of electron correlation effects in post-Hartree–Fock methods might refine bond length predictions and

energy calculations. Finally, predict how perturbing the molecule with an external electric field might alter the electron distribution and whether this could influence reactivity toward nucleophilic attack.

2. In methane (CH_4), carbon adopts sp^3 hybridization, producing a tetrahedral geometry with bond angles of 109.5°. If a computational chemist artificially constrains one of the H–C–H bond angles to 120° in a geometry optimization, discuss the likely energetic consequences. Explain how density functional theory (DFT) or post-Hartree–Fock methods could quantify destabilization, including contributions from electron repulsion and orbital overlap. Predict which vibrational modes would be most affected and how the IR spectrum might change.

3. Consider diborane (B_2H_6), which exhibits three-center two-electron (3c–2e) bonds. Using molecular orbital analysis, describe the distribution of electrons across the bridging hydrogen atoms and how it stabilizes the molecule. Predict how replacing boron with aluminum would affect orbital overlap, bond lengths, and molecular stability. Discuss the potential differences in reactivity toward nucleophiles, considering the electron deficiency in these clusters.

4. A transition metal octahedral complex contains a d^5 metal ion in a high-spin configuration. Using crystal field theory, calculate the ligand field stabilization energy (LFSE) for the system. Predict how changing ligands from weak-field to strong-field affects the spin state, d orbital occupancy, and magnetic properties. Additionally, discuss how these changes might influence UV–vis absorption spectra and the color of the complex.

5. Benzene (C_6H_6) has delocalized π electrons over a planar ring. Using Hückel molecular orbital theory, determine the relative energies of bonding and antibonding π orbitals. Predict how substituting electron-donating groups at the 1 and 4 positions alters electron density, resonance stabilization, and reactivity toward electrophilic aromatic substitution. Finally, discuss the influence of these substitutions on aromaticity and NICS (nucleus-independent chemical shift) values.

6. The nitrate ion (NO_3^-) exhibits resonance with equivalent bond lengths. Design a computational experiment to compare bond lengths and electron density predicted by Hartree–Fock, MP2, and DFT methods. Discuss the strengths and limitations

of each method in capturing electron delocalization. Predict how vibrational frequencies of the N–O stretching modes differ among computational approaches and with experimental IR measurements.

7. In ethene (C_2H_4), the carbon atoms are sp^2 hybridized, forming a planar geometry with σ and π bonds. Explain how these bonds arise from orbital overlap. Predict the effect on bond lengths, π-bond strength, and rotational barriers if one carbon is substituted with silicon (Si), considering its larger atomic radius and altered orbital energies. Discuss the impact on chemical reactivity in addition reactions compared to ethene.

8. Molecular oxygen (O_2) is paramagnetic, possessing two unpaired electrons. Using molecular orbital theory, identify the orbitals responsible for these unpaired electrons and explain why classical Lewis structures fail to predict paramagnetism. Predict how replacing oxygen with sulfur in S_2 would affect paramagnetic behavior, bond length, and bond order. Discuss the influence of relativistic effects on heavier chalcogens.

9. Water exhibits extensive hydrogen bonding networks, critical for its unique physical properties. Calculate how replacing hydrogen with deuterium affects vibrational frequencies, bond strength, and heat capacity. Discuss how isotopic substitution could be modeled computationally using DFT or MP2 methods, including zero-point energy corrections, and predict changes in predicted hydrogen bond geometries and thermodynamic properties.

10. Phosphorus pentachloride (PCl_5) exhibits trigonal bipyramidal geometry due to sp^3d hybridization. Using molecular orbital theory, predict how the energies of axial versus equatorial bonds differ and how these differences influence chemical reactivity toward nucleophiles or halide exchange. Additionally, discuss how hypervalent bonding in PCl_5 can be analyzed using three-center four-electron bonding models and validated computationally.

11. In graphite, sp^2-hybridized carbon atoms form planar sheets stabilized by delocalized π electrons. Analyze the effect of intercalating lithium ions between layers on electronic band structure, conductivity, and mechanical properties. Predict changes in interlayer spacing and vibrational frequencies. Discuss how DFT calculations with van der Waals corrections could model these effects and compare to experimental X-ray diffraction data.

12. Sulfur hexafluoride (SF_6) is octahedral and highly symmetric. Using both VSEPR and molecular orbital theory, predict how relativistic effects for sulfur influence bond lengths, electron density distribution, and vibrational frequencies. Discuss implications for chemical inertness, reactivity with nucleophiles, and potential deviations from ideal octahedral geometry under extreme conditions.

13. In acetylene (C_2H_2), carbon atoms are sp hybridized, forming a linear geometry with orthogonal π bonds. Discuss how the π bond orthogonality impacts molecular orbital formation and bond order. Predict how bending the molecule by external forces affects orbital overlap, electron density, and energy. Suggest computational approaches to simulate these distortions and their consequences for reactivity in addition reactions.

14. A computational chemist investigates the band structure of graphene using DFT. Explain how sp^2 hybridization and π delocalization contribute to Dirac cone formation and high electron mobility. Predict how nitrogen doping or vacancy defects modify electronic properties, local charge distribution, and chemical reactivity. Discuss how these changes could be detected experimentally through scanning tunneling microscopy or angle-resolved photoemission spectroscopy.

15. A metal carbonyl complex, $Fe(CO)_5$, is analyzed for synergic bonding between metal and ligand. Using MO theory, predict how back-donation from filled metal d orbitals into CO π* orbitals alters CO bond lengths compared to free CO. Discuss how substitution of CO with phosphine ligands would change metal–ligand bond strength, bond order, and electronic spectra, providing a detailed rationale based on orbital interactions.

16. A Cr^{3+} octahedral complex with weak-field ligands is studied using crystal field theory. Calculate d orbital splitting, LFSE, and predict the spin state. Predict how replacing ligands with cyanide alters orbital splitting, favors low-spin configuration, and changes absorption spectra. Discuss the impact on magnetic susceptibility and reactivity trends in redox reactions.

17. Borane clusters exhibit delocalized 3c–2e bonding. Using molecular orbital theory, predict how electron deficiency affects stability, bond order, and reactivity. Propose a hypothetical substitution that increases electron density in the cluster and rationalize how it influences bond lengths, geometry, and potential for nucleophilic attack. Compare predictions to experimental data if available.

18. In covalent-organic frameworks (COFs), π–π stacking interactions stabilize the lattice. Predict how altering linker geometry or substituent polarity influences porosity, lattice energy, and electronic conductivity. Suggest computational approaches, including periodic DFT or molecular dynamics simulations, to quantify these changes and predict structural stability under thermal or mechanical stress.

19. A supramolecular host–guest system relies on hydrogen bonding and cation–π interactions. Calculate the binding energy using DFT or MP2 methods and predict how solvent polarity, temperature, and ion strength influence complex formation, selectivity, and thermodynamics. Discuss strategies for enhancing host–guest affinity through orbital engineering and non-covalent interactions.

20. Consider a diatomic transition metal molecule, Cr_2. Using molecular orbital theory, determine bond order, expected magnetism, and bond length. Predict how including relativistic corrections and electron correlation would modify these predictions. Discuss the implications for spectroscopic transitions and reactivity in metal cluster formation.

21. A conjugated polymer exhibits extended π delocalization along the backbone. Predict how the electronic band gap changes as the number of repeating units increases. Suggest computational approaches to model electronic structure, optical absorption spectra, and charge transport properties. Discuss how substituents or chain defects could modulate conductivity.

22. In metallic silver (Ag), the conduction band is partially filled. Using band theory, explain the relationship between electron delocalization, conductivity, and mechanical properties such as malleability and ductility. Predict how alloying with copper (Cu) modifies band structure, electron density, and electrical conductivity, and discuss experimental methods to verify these predictions.

23. Hyperconjugation in the tert-butyl carbocation stabilizes the intermediate through delocalization of σ electrons. Using molecular orbital analysis, predict which σ orbitals participate and how stabilization energy compares with resonance methods. Discuss implications for reactivity in SN1 reactions and how computational methods could quantify energy differences.

24. A quantum dot exhibits size-dependent optical properties due to quantum

confinement. Predict how decreasing the radius affects the band gap, exciton energy, and electron delocalization. Suggest computational methods, such as time-dependent DFT, to simulate absorption spectra and correlate theoretical predictions with experimental photoluminescence measurements.

25. Ethene interacts with a transition metal catalyst via π coordination during hydrogenation. Using frontier orbital theory, predict the preferred orientation of binding, which molecular orbitals participate, and how orbital overlap affects activation energy. Discuss how this understanding could guide catalyst design to improve selectivity and turnover frequency.

26. In transition metal oxides under high pressure, unusual coordination geometries can emerge due to compression of lattice parameters. Predict how pressure affects orbital overlap, electronic band structure, magnetic properties, and reactivity. Discuss computational strategies to model these effects, including periodic DFT with pressure constraints, and explain how experimental techniques like high-pressure X-ray diffraction could validate predictions.

27. Superatoms, clusters of atoms exhibiting closed electronic shells, often mimic elemental properties. Predict stability trends based on total valence electron count, symmetry, and delocalized bonding. Compare predicted chemical reactivity to analogous monatomic species and discuss how ligand adsorption might modify electronic structure and cluster stability. Consider computational approaches for predicting HOMO–LUMO gaps and ionization potentials.

28. Consider an organometallic cluster containing multiple metal–metal bonds. Using molecular orbital theory, predict electron distribution, bond order, and magnetic properties. Suggest possible substitutions or ligand modifications to tune catalytic activity, electronic conductivity, or reactivity. Discuss how cluster geometry affects orbital overlap and energy splitting.

29. A π-conjugated molecule absorbs UV light, promoting electrons to antibonding orbitals. Predict the transient changes in bond lengths, bond orders, and molecular geometry. Discuss how time-resolved spectroscopy and nonadiabatic molecular dynamics simulations could validate theoretical predictions, and suggest factors that influence photochemical reaction pathways.

30. Hydrogen fluoride (HF) exhibits strong hydrogen bonding networks affecting boiling point and viscosity. Predict how replacing fluorine with chlorine or bromine alters hydrogen bond strength, vibrational frequencies, and macroscopic thermodynamic properties. Suggest computational methods to quantify these effects, including DFT with dispersion corrections and explicit solvent models.

31. A fullerene (C_{60}) molecule has delocalized π electrons over a spherical lattice. Predict electronic properties, frontier orbital energies, and likely reactive sites using MO theory. Compare to graphene and discuss the influence of curvature on orbital overlap, strain, and reactivity in cycloaddition reactions. Include possible computational methods to map electron density and predict electrostatic potentials.

32. In a metal–organic framework (MOF), electron-rich organic linkers coordinate to transition metal nodes. Predict how substituting linkers with electron-withdrawing or electron-donating groups affects band structure, adsorption properties, and catalytic activity. Discuss how periodic DFT and molecular dynamics simulations can quantify changes in pore size, electron density distribution, and stability.

33. A benzene derivative undergoes electrophilic aromatic substitution. Using frontier molecular orbitals, predict regioselectivity based on electron density distribution. Consider substituent effects, resonance stabilization, and steric hindrance. Compare theoretical predictions to experimental product distributions and suggest how computational modeling can refine regioselectivity predictions.

34. In an aqueous ionic lattice such as NaCl, solvation stabilizes the crystal. Using electrostatics and molecular simulations, predict the effect of replacing Na^+ with K^+ or Li^+ on lattice energy, solubility, and hydration shell structure. Discuss how DFT or classical MD simulations can quantify energetic differences and structural changes in solution.

35. Relativistic effects in gold (Au) lead to contraction of 6s orbitals. Predict how this contraction influences Au–Au bond length, reactivity in organogold complexes, and color compared to lighter congeners such as Ag or Cu. Discuss computational methods capable of incorporating relativistic effects and how experimental spectroscopy could validate predictions.

36. In protein folding, hydrogen bonding networks define secondary structure. Using quantum chemical calculations and MD simulations, predict the effect of

substituting alanine residues with serine on hydrogen bond distribution, backbone conformation, and folding kinetics. Discuss implications for stability, solvent interactions, and enzymatic function.

37. A small molecule exhibits halogen bonding. Predict how electron-withdrawing substituents on the halogen influence bond strength, directionality, and intermolecular geometry. Suggest computational approaches such as DFT with dispersion corrections to quantify interaction energies and analyze electron density maps.

38. In a topological insulator, strong spin–orbit coupling and orbital hybridization produce surface conductivity. Predict how substituting atoms with heavier congeners affects band inversion, Dirac cone properties, and electronic transport. Discuss the role of bonding and orbital interactions in emergent macroscopic quantum phenomena.

39. A photochemical reaction promotes electrons to Rydberg or antibonding orbitals. Predict how these electronic excitations alter bond orders, molecular geometry, and reaction pathways. Suggest time-dependent DFT or nonadiabatic molecular dynamics approaches to model excited-state behavior, and explain how ultrafast spectroscopy could validate predictions.

40. In a supramolecular assembly stabilized by cooperative π–π stacking, predict how steric bulk, linker rigidity, or solvent polarity affects self-assembly, binding constants, and thermodynamic stability. Discuss how computational modeling of interaction energies and conformational dynamics can inform design of stable architectures.

41. Consider an excited-state intramolecular proton transfer (ESIPT) reaction. Using molecular orbital theory, predict how electron rearrangements facilitate proton transfer, influence excited-state lifetimes, and modulate fluorescence emission. Suggest computational methods for mapping potential energy surfaces and predicting photophysical properties.

42. In metal hydrides, agostic interactions involve weak coordination between C–H bonds and electron-deficient metal centers. Predict how changing the metal center or ligand environment affects interaction strength, geometry, and catalytic activity.

Suggest computational tools such as NBO or QTAIM analysis to visualize electron density distribution in these interactions.

43. A conjugated polymer exhibits alternating single and double bonds. Predict how bond length alternation affects electronic delocalization, band gap, optical absorption, and charge mobility. Suggest spectroscopic techniques (UV–vis, Raman) to validate computational predictions and discuss the influence of defects or chain length on electronic properties.

44. In organophosphorus chemistry, lone pair–π interactions stabilize molecular conformations. Predict how substituting phosphorus with arsenic or antimony alters orbital overlap, energy stabilization, and reactivity patterns. Discuss computational strategies to quantify these effects and correlate with NMR or crystallographic data.

45. Layered transition metal dichalcogenides exhibit interlayer van der Waals interactions. Predict how mechanical exfoliation or strain affects electronic band structure, electron mobility, and optical properties. Suggest computational methods for simulating few-layer systems and experimental techniques such as ARPES or Raman spectroscopy for validation.

46. Halogen–halogen interactions contribute to crystal packing in solids. Predict how molecular polarizability, orientation, and substituents influence lattice energy, bond geometry, and macroscopic stability. Discuss computational methods for resolving subtle energetic differences, including DFT with dispersion corrections and periodic boundary conditions.

47. Perovskite materials exhibit hybrid covalent–ionic bonding. Predict how substitution of cations or anions affects band gap, light absorption, thermal stability, and electronic conductivity. Suggest computational approaches for modeling electronic structure, lattice dynamics, and defect formation, and discuss implications for photovoltaic or optoelectronic applications.

48. Aromatic polymers rely on π-electron delocalization for conductivity. Predict how chain length, defects, and interruptions in conjugation influence charge transport and band structure. Suggest computational methods to model electron mobility, and discuss how spectroscopic and transport measurements could validate theoretical predictions.

49. Hypervalent iodine compounds, such as IF_5 or ICl_3, feature bonding beyond the octet. Predict the bonding scheme using three-center four-electron models, orbital hybridization, and electron density analysis. Compare predicted geometries, bond lengths, and bond strengths to X-ray crystallography data and discuss implications for reactivity and oxidative properties.

50. In nanoclusters, quantum confinement and surface effects dominate electronic structure and catalytic properties. Predict how cluster size, composition, and degree of electron delocalization influence chemical reactivity, optical absorption, and stability. Suggest computational strategies for designing clusters with targeted electronic and catalytic properties, including DFT, TD-DFT, and ab initio molecular dynamics.

Chapter 11

Hidden Strength of Intermolecular Forces

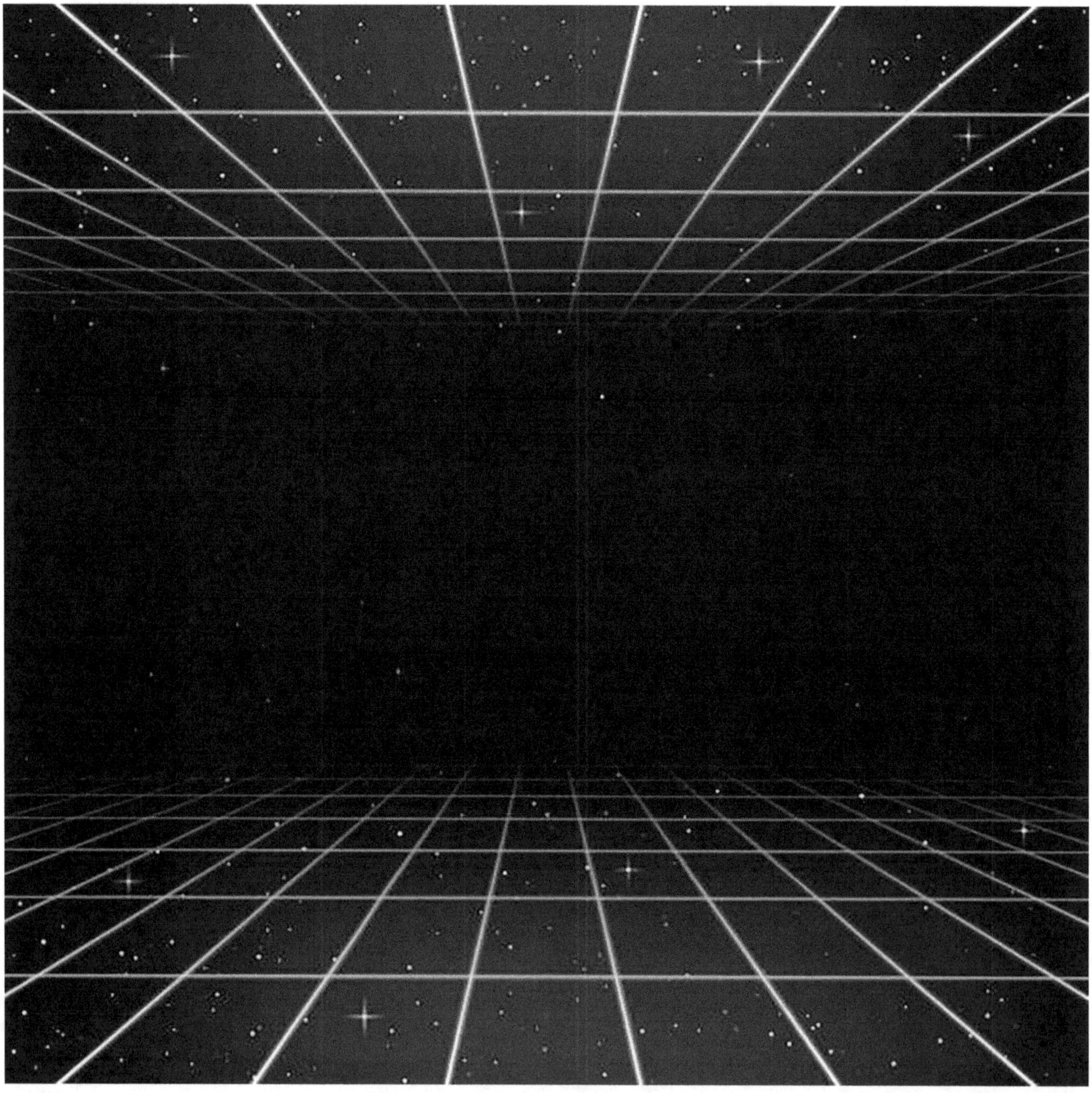

Matter exists in various states—solid, liquid, gas, and plasma—and each state exhibits distinct physical and chemical properties that arise from the arrangement and interactions of atoms and molecules. Solids are characterized by closely packed particles arranged in either long-range ordered structures, known as crystalline solids, or disordered arrangements, referred to as amorphous solids. Crystalline solids include ionic, covalent, metallic, and molecular solids, each with unique bonding and physical properties. For instance, ionic solids such as sodium chloride form a lattice structure in which positive and negative ions alternate in a highly organized three-dimensional array. The stability of this lattice is quantified by the lattice energy, given by $U = kQ_1Q_2/r$, where Q_1 and Q_2 are the charges of the ions, r is the interionic distance, and k is a proportionality constant depending on the dielectric constant of the medium. The high lattice energy contributes to the high melting and boiling points of ionic solids and their poor electrical conductivity in the solid state, as ions are fixed in the lattice but excellent conductivity when molten due to mobile ions. Covalent network solids, such as diamond and quartz, consist of atoms connected through extended networks of covalent bonds, giving rise to extreme hardness, high melting points, and low electrical conductivity. Diamond's tetrahedral sp^3 bonding forms a rigid three-dimensional network that is responsible for its remarkable mechanical strength. In contrast, graphite, composed of sp^2 bonded carbon atoms in planar hexagonal sheets, exhibits high electrical conductivity along the planes and lubrication properties due to weak interlayer van der Waals interactions. Molecular solids, such as ice, dry ice, and iodine, are held together primarily by intermolecular forces, including hydrogen bonding, dipole-dipole interactions, and London dispersion forces. Hydrogen bonding, exemplified by water, significantly influences melting and boiling points, specific heat capacity, and the density anomaly observed in ice, where solid water is less dense than its liquid form due to the open hydrogen-bonded lattice.

Liquids possess intermediate properties between gases and solids, characterized by higher densities than gases but lower than solids, low compressibility, and the ability to flow. The behavior of liquids is influenced by the balance of cohesive forces between liquid molecules and adhesive forces with surrounding surfaces. Surface tension arises from the unbalanced intermolecular forces at the surface, creating a net inward force, and can be quantified experimentally through methods such as the capillary rise method. Capillary action, the rise of liquid in narrow tubes, occurs when adhesive forces with the tube walls overcome the cohesive forces within the liquid.

The viscosity of a liquid, or its resistance to flow, increases with stronger intermolecular interactions and molecular size. For example, glycerol exhibits a viscosity several orders of magnitude higher than that of water due to extensive hydrogen bonding, whereas nonpolar liquids such as hexane flow readily due to weak dispersion forces. Liquids also exhibit vapor pressure, a measure of the tendency of molecules to escape into the gas phase. The Clausius-Clapeyron equation, $\ln(P_2/P_1) = \Delta H_{vap}R(1/T_1 - 1/T_2)$, describes the relationship between vapor pressure and temperature, where ΔH_{vap} is the enthalpy of vaporization, R is the universal gas constant, and T_1 and T_2 are the initial and final temperatures. For water, with ΔH_{vap} = 40.7 kJ/mol, the vapor pressure rises exponentially with temperature, explaining the rapid evaporation of water at elevated temperatures.

Gases, in contrast, exhibit neither fixed shape nor fixed volume, expanding to occupy the container in which they are confined. Their behavior can be described by the ideal gas law, PV = nRT, where P is pressure, V is volume, n is the number of moles, R is the gas constant, and T is temperature. Real gases deviate from ideal behavior at high pressures and low temperatures due to intermolecular attractions and finite molecular volume, and these deviations are accounted for by the Van der Waals equation, $(P + an^2/V^2)(V-nb)=nRT$, where a corrects for intermolecular attractions and b for molecular volume. The kinetic molecular theory provides a microscopic explanation for gas laws, stating that gas particles move in random motion, exerting pressure on container walls through elastic collisions. Temperature is directly proportional to the average kinetic energy of the gas particles, expressed as $\langle E_k \rangle = (3/2)kBT$, where kB is Boltzmann's constant. Diffusion and effusion of gases, described by Graham's law, r_1/r_2 = Square Root(M_2/M_1), depend on molecular masses, with lighter molecules diffusing faster.

Energy plays a central role in the behavior of matter. Bond formation and bond breaking are associated with energy changes, where bond enthalpies provide a quantitative measure. For a reaction $A-B+C-D \rightarrow A-D+C-B$, the enthalpy change, ΔH, can be approximated as the sum of bond energies of bonds broken minus bonds formed, $\Delta H \approx \sum D(\text{broken}) - \sum D(\text{formed})$. Thermodynamic principles further govern phase transitions and chemical reactions. The first law of thermodynamics states that the change in internal energy, ΔU, of a system equals the heat added to the system, q, minus the work done by the system, w, or $\Delta U = q - w$. Enthalpy, H =

U + PV, is particularly useful at constant pressure, where $\Delta H = qp$, the heat absorbed or released at constant pressure. The second law introduces the concept of entropy, S, which quantifies the dispersal of energy and the degree of disorder in a system. Spontaneous processes occur when the Gibbs free energy, $G = H - TS$, decreases, or $\Delta G < 0$. The entropic contributions of phase transitions, such as melting, vaporization, and sublimation, illustrate how energy distribution and disorder govern the feasibility of processes.

Chemical bonding can be further classified into ionic, covalent, metallic, and coordinate covalent bonds. Ionic bonds result from electrostatic attraction between oppositely charged ions, forming crystalline lattices with high lattice energies. Covalent bonds involve sharing of electron pairs between atoms, with bond strength dependent on bond order, bond length, and electronegativity differences. Polar covalent bonds arise when electrons are unequally shared, generating partial charges and a molecular dipole moment, $\mu = \delta \cdot r$, where δ is the partial charge and r is the distance between charges. Metallic bonding, featuring delocalized electrons across a lattice of cations, accounts for electrical conductivity, malleability, and ductility in metals. Transition metals often exhibit variable oxidation states and complex coordination chemistry, leading to applications in catalysis, electronic devices, and materials design.

Semiconductors, such as silicon and germanium, possess energy band structures with a valence band filled with electrons and an empty conduction band separated by a band gap, E_g. Doping introduces extrinsic carriers that enhance conductivity: n-type semiconductors introduce electrons through donor atoms, while p-type semiconductors create holes through acceptor atoms. P-n junctions, formed by combining n-type and p-type regions, enable rectification and are fundamental to diodes, transistors, and photovoltaic cells. Advanced semiconductor devices exploit quantum mechanical tunneling, carrier recombination, and mobility effects to achieve high-speed and energy-efficient operation.

Experimental techniques play a pivotal role in elucidating atomic and molecular structure. X-ray diffraction, based on Bragg's law $n\lambda = 2d\sin\theta$, provides information about interplanar spacings in crystals. Spectroscopic methods, including infrared spectroscopy, nuclear magnetic resonance, and UV-visible absorption, probe molecular vibrations, chemical environments, and electronic transitions, allowing the determination of bond strengths, functional groups, and molecular geometries.

Thermochemical measurements, such as calorimetry, reveal enthalpy changes in chemical reactions and phase transitions. Computational chemistry, employing quantum mechanical and molecular mechanics approaches, enables prediction of molecular structures, reaction pathways, and material properties.

Phase diagrams summarize the equilibrium relationships between solid, liquid, and gas phases under varying temperatures and pressures. The triple point defines the unique conditions where all three phases coexist in equilibrium, while the critical point marks the end of the liquid-gas coexistence curve. Supercritical fluids, obtained above the critical temperature and pressure, combine liquid-like density with gas-like diffusivity, enabling applications in extraction, chromatography, and green chemistry. Pressure-induced phase transitions, such as graphite-to-diamond transformation, illustrate how thermodynamic variables influence crystal structure and material properties.

Intermolecular interactions dictate macroscopic phenomena such as solubility, boiling and melting points, viscosity, and surface tension. Polar molecules dissolve in polar solvents due to dipole-dipole and hydrogen-bonding interactions, whereas nonpolar molecules dissolve in nonpolar solvents, reflecting the principle “like dissolves like.” Colligative properties, including vapor pressure lowering, boiling point elevation, freezing point depression, and osmotic pressure, depend on solute concentration rather than chemical identity, and are quantitatively described by formulas such as $\Delta T_b = K_b m$ and $\pi = MRT$.

The behavior of matter results from an intricate interplay of atomic structure, bonding, energy, and intermolecular forces. Experimental observations, including calorimetry, spectroscopy, diffraction, and electrical measurements, provide quantitative data to validate theoretical models. The study of these principles underpins the design of materials, industrial processes, chemical reactions, and emerging technologies in electronics, energy storage, and materials science. Mastery of these concepts enables chemists and engineers to predict the properties of novel substances, manipulate reaction conditions, and develop innovative solutions to real-world problems.

At the heart of chemistry lies the atom, the fundamental unit of matter. Each atom consists of a dense nucleus, containing positively charged protons and electrically neutral neutrons, surrounded by electrons that occupy discrete energy levels or orbitals. The discovery of the electron by J.J. Thomson, the nuclear model proposed by Rutherford, and the quantized energy levels introduced by Niels Bohr set the stage for modern quantum chemistry.

Quantum mechanics dictates that electrons occupy orbitals defined by wave functions, which describe the probability distribution of finding an electron in space. These orbitals—s, p, d, and f—possess characteristic shapes and energies, which in turn govern how atoms interact. Electron configurations, governed by the Pauli exclusion principle and Hund's rule, determine the chemical properties of elements and their propensity to form specific types of bonds. The periodic table itself is a direct reflection of these electronic structures, with periodic trends in electronegativity, ionization energy, and atomic radius emerging from quantum considerations.

Atoms rarely exist in isolation; they interact through chemical bonds to form molecules and extended structures. Covalent bonds arise when atoms share electrons, with bond polarity determined by differences in electronegativity. Ionic bonds form through the complete transfer of electrons from one atom to another, resulting in electrostatic attraction between oppositely charged ions. Metallic bonding, characteristic of metals, involves a delocalized "sea" of electrons that confers conductivity, malleability, and luster.

Molecular geometry, as described by Valence Shell Electron Pair Repulsion (VSEPR) theory, predicts the spatial arrangement of atoms around a central atom based on electron-pair repulsion. Hybridization theory further explains deviations from simple geometries, such as the tetrahedral bonding in methane (sp^3) or the trigonal planar bonding in boron trifluoride (sp^2). These structural principles influence not only physical properties but also chemical reactivity, solubility, and interaction with electromagnetic radiation.

Beyond the covalent or ionic bonds holding molecules together, weaker intermolecular forces play a crucial role in determining the bulk properties of substances. These include hydrogen bonding, dipole–dipole interactions, and London dispersion forces. For example, water's high boiling point and surface tension result from extensive hydrogen bonding networks, while nonpolar gases like helium exhibit minimal intermolecular attraction.

The understanding of these forces enables chemists to rationalize phenomena such as solubility, volatility, viscosity, and phase transitions. Thermodynamic principles—enthalpy, entropy, and Gibbs free energy—govern these processes, providing predictive power for chemical behavior. By controlling intermolecular interactions, scientists design novel solvents, polymers, and supramolecular assemblies with tailored properties.

Chemical reactions are governed by both thermodynamics, which determines whether a reaction is energetically favorable, and kinetics, which dictates how rapidly the reaction proceeds. The first and second laws of thermodynamics underpin the concepts of energy conservation and entropy change, while the Gibbs free energy equation, $\Delta G = \Delta H - T\Delta S$, predicts the spontaneity of reactions. Exothermic reactions release energy, often as heat or light, while endothermic reactions absorb energy from their surroundings.

Reaction kinetics, on the other hand, depend on the frequency and energy of molecular collisions. The Arrhenius equation links the rate constant to temperature and activation energy, providing insights into how conditions influence reaction speed. Catalysts, whether enzymatic or synthetic, lower activation barriers without altering overall thermodynamics, enabling industrial and biological reactions to occur efficiently under mild conditions. Mastery of these principles allows chemists to optimize reactions for yield, selectivity, and sustainability.

Experimental observation is the cornerstone of chemistry, bridging theory and practical application. Spectroscopic techniques, such as infrared (IR), nuclear magnetic resonance (NMR), ultraviolet-visible (UV-Vis), and mass spectrometry (MS), allow precise identification of molecular structure, functional groups, and isotopic composition. Diffraction methods, including X-ray crystallography and electron diffraction, reveal atomic arrangements in solids, while electrochemical measurements probe redox behavior and ionic transport.

These methods have enabled remarkable technological advances. In materials science, chemists design semiconductors, superconductors, and polymers with specific electrical, optical, and mechanical properties. In energy storage, electrochemical principles guide the development of high-capacity batteries and fuel cells. In medicine, molecular analysis informs drug design, diagnostics, and imaging. Across every domain, the deep understanding of atomic interactions, energy transformations, and molecular behavior drives innovation and improves the human experience.

The influence of chemistry on society extends far beyond laboratory walls. Modern environmental science, for instance, relies on chemical principles to monitor pollutants, develop sustainable materials, and mitigate the effects of climate change. Atmospheric chemistry explains the formation and depletion of the ozone layer, the behavior of greenhouse gases, and the reactions that govern air quality. By understanding these molecular and elemental processes, chemists contribute to regulatory policies, cleaner technologies, and strategies for conservation. Similarly, water chemistry provides essential insights into treatment processes, contamination detection, and the purification of drinking water. These practical applications demonstrate that chemistry is not an abstract science but a foundational tool for maintaining ecosystems and protecting public health.

Industrial chemistry illustrates another dimension of applied chemical knowledge. From petrochemicals to pharmaceuticals, chemical engineering translates fundamental principles into large-scale production while balancing efficiency, safety, and sustainability. Catalysis, a core concept in both chemical and industrial processes, allows reactions to proceed faster or more selectively, reducing waste and energy consumption. Advanced materials, such as biodegradable plastics, smart polymers, and nanostructured composites, are engineered through careful control of chemical composition and structure. Each innovation, while rooted in atomic and molecular theory, reflects an interplay between experimental observation and theoretical insight, highlighting the integrative nature of modern chemical research.

On a molecular level, the study of chemical reactivity continues to reveal profound patterns. The understanding of reaction mechanisms, transition states, and energy landscapes enables chemists to predict how molecules will interact, transform, or decompose. Quantum chemistry offers a window into electron distribution, orbital hybridization, and bond strengths, providing a predictive framework for designing molecules with desired properties. Thermodynamics and kinetics remain indispensable in evaluating reaction feasibility and rate, ensuring that theoretical predictions align with experimental outcomes. This rigorous methodology underpins the rational development of new drugs, catalysts, and materials, illustrating chemistry's capacity to anticipate and shape the outcomes of complex molecular interactions.

In medicine and biotechnology, chemistry serves as the bridge between molecular structure and biological function. The design of pharmaceuticals depends on a precise understanding of molecular interactions with biological targets. Techniques such as high-throughput screening, structure-based drug design, and computational modeling accelerate the identification of therapeutic candidates while minimizing side effects. Analytical chemistry provides critical tools for diagnosing disease, monitoring treatment efficacy, and ensuring the safety of food, water, and pharmaceuticals. Moreover, the integration of chemical knowledge with biology has given rise to fields such as chemical biology and medicinal chemistry, where the manipulation of biomolecules leads to innovations in gene therapy, enzyme engineering, and personalized medicine.

Energy research represents another arena where chemistry drives societal transformation. Renewable energy technologies—solar cells, hydrogen fuel, and energy-efficient batteries—rely on a detailed understanding of electron flow, redox reactions, and material stability. Photochemistry and electrochemistry are central to converting sunlight into usable electricity, storing energy in chemical bonds, and reducing the reliance on fossil fuels. The design of catalysts that operate efficiently under mild conditions exemplifies the synergy between theoretical chemistry and practical engineering. Through these advances, chemistry not only addresses pressing global challenges but also shapes the future of energy production, storage, and utilization.

Finally, chemistry's role in expanding human knowledge and technological potential cannot be overstated. From the nanoscale design of molecular machines to the creation of artificial enzymes, chemical innovation pushes the boundaries of what is scientifically achievable. The iterative cycle of hypothesis, experimentation, and theory enables continuous refinement and discovery, whether in understanding fundamental interactions, developing life-saving technologies, or creating materials that transform industry. Ultimately, chemistry embodies both a deep intellectual pursuit and a pragmatic force, connecting atomic-level phenomena to large-scale societal outcomes. Its contributions are evident in every facet of modern life, from healthcare and energy to the environment and beyond, affirming that the study of matter is inseparable from the advancement of human civilization.

In conclusion, chemistry stands as a central pillar of modern science and technology. Its principles illuminate the workings of the natural world, guide the creation of innovative solutions, and enhance the quality of life across the globe. By linking theory with experiment, fundamental insights with practical applications, and molecular understanding with societal needs, chemistry continues to drive progress in virtually every domain of human endeavor. The ongoing exploration of matter at all scales—from subatomic particles to complex materials—promises further breakthroughs, ensuring that chemistry will remain at the forefront of discovery, innovation, and sustainable development for generations to come.

Chapter 11: Word Problems

1. A computational chemist is modeling the vibrational spectra of a hypervalent sulfur compound resembling SF_6 but containing substituted ligands that reduce symmetry from octahedral O_h to D_{4h}. She observes that the degeneracy of several IR-active modes predicted for SF_6 splits into multiple distinct peaks. Explain, using group theory and molecular orbital considerations, why this degeneracy lifting occurs and how the symmetry reduction affects the predicted normal modes, their energies, and selection rules. Propose a method (DFT vs. MP2 vs. CCSD(T)) most appropriate for accurately capturing these subtle effects and justify your choice.

2. You are studying a transition-metal complex that can exist in either high-spin or low-spin octahedral configurations depending on ligand field strength. When coordinated with a mixed ligand array (three strong-field ligands and three weak-field ligands in facial geometry), the complex exhibits anomalous magnetic behavior that varies with temperature. Construct a theoretical explanation reconciling the observed magnetic moment with intermediate spin states, spin crossover phenomena, and possible distortions from ideal octahedral symmetry. Identify which spectroscopic measurements could confirm your hypothesis.

3. A researcher investigates the photochemical reaction of a conjugated organic molecule whose UV-Vis absorption shows an unusually intense $\pi \rightarrow \pi^*$ transition and a weak $n \rightarrow \pi^*$ band. Upon substituting an electron-withdrawing group, the $\pi \rightarrow \pi^*$ band red-shifts while the $n \rightarrow \pi^*$ band intensifies. Analyze this behavior using MO diagrams, orbital mixing, and excited-state stabilization arguments. How would solvent polarity and hydrogen bonding influence the spectral trends further?

4. In a study of diborane analogues containing different bridging atoms (H, F, Cl), experimental bond lengths differ significantly from predictions of simple 3-center–2-electron bonding theory. Develop a quantitative explanation based on orbital energy alignment, electronegativity differences, and the relative ability of different bridging atoms to delocalize electron density. How might NBO analysis help distinguish among the proposed explanations?

5. A graduate student synthesizes a series of metal–carbonyl complexes and observes that the C–O stretching frequency decreases systematically as the metal center is changed from Fe(II) → Ru(II) → Os(II). Provide a molecular orbital explanation grounded in synergic bonding, π-backbonding efficiency, and relativistic effects. Predict how isotopic substitution (^{13}CO) and solvent polarity would influence the observed IR frequencies.

6. An organic semiconductor displays a conductivity change of four orders of magnitude when exposed to compressive strain. Using concepts of orbital overlap, band dispersion, and π-stacking interactions, explain how mechanical distortion modifies the material's band structure and charge-transport properties. Propose an experiment using time-resolved spectroscopy or scanning probe microscopy to validate your model.

7. A topological insulator exhibits conductive surface states despite having a bulk band gap of ~0.3 eV. A theorist claims that strong spin-orbit coupling and orbital hybridization create an inverted band structure. Using bonding theory and relativistic quantum mechanics, construct a detailed rationale for how orbital symmetry, band inversion, and SOC interact to produce the emergent topological behavior. Suggest how ARPES could experimentally confirm the surface state dispersion.

8. A diatomic molecule with heteronuclear atoms shows a dipole moment significantly smaller than predicted from electronegativity differences. Using MO theory, polarization functions, and electron density mapping, explain how orbital mixing, partial covalency, or back-donation could reduce the effective charge separation. Predict how applying a strong electric field would affect the observed dipole.

9. Researchers studying a metal–organic framework (MOF) find that replacing one ligand with an electron-rich linker increases gas adsorption capacity while

reducing framework stability. Analyze this using orbital overlap, π-backbonding, and ligand-field effects, and discuss how electron donation from linkers alters the metal node's geometry, charge distribution, and interaction with guest molecules.

10. A cluster compound containing 12 metal atoms displays superatom behavior consistent with a closed electron shell of 58 electrons. Using the jellium model and MO theory, explain why certain electron counts yield increased stability. Predict how substituting one metal atom with another of different valence would influence the stability, magnetic moment, and optical properties.
11. A chemical reaction proceeds through an asynchronous concerted mechanism in which bond formation is partially advanced before bond breaking begins. Using potential energy surfaces, Hammond's postulate, and transition-state orbital interactions, analyze how asynchronous progress affects activation energy, stereoselectivity, and substituent sensitivity. Explain how computational methods could identify the synchronicity of the transition state.
12. A metal hydride complex shows an unusual agostic interaction between the metal center and a β-C–H bond. Propose an explanation based on electron donation, orbital symmetry, and steric accessibility. Predict how the agostic interaction would affect the NMR chemical shifts, IR stretching frequencies, and reactivity toward oxidative addition.
13. A graphene nanoribbon with varying edge termination (zigzag vs. armchair) displays different magnetic and conductive properties. Using bonding theory, Dirac cone physics, and π-electron localization, explain the origin of these edge effects. Predict how hydrogenation or doping would modify the band gap and spin density distribution.
14. A catalyst containing a transition metal undergoes a ligand-induced spin-state switch that dramatically enhances turnover frequency. Explain how ligand field strength, orbital splitting, and electron pairing energies combine to produce the observed reactivity enhancement. Suggest kinetic experiments that could differentiate between spin-accelerated and ligand-accelerated mechanisms.
15. A heavy-element compound (involving Bi, Pb, or Au) exhibits bond lengths inconsistent with non-relativistic predictions. Using relativistic contraction, spin-

orbit coupling, and changes in orbital energies, explain how relativistic effects alter bonding patterns in heavy elements. Predict how these modifications would be reflected in spectroscopic data.

16. A molecule capable of intramolecular hydrogen bonding shows dramatically different IR spectra depending on whether it is in gas phase or solution. Provide an explanation using dielectric effects, vibrational coupling, and the competition between intra- and intermolecular hydrogen bonding. How might temperature variations modulate this behavior?

17. A supramolecular capsule binds guest molecules with extraordinary selectivity for shape over polarity. Explain this selectivity using concepts of induced fit, weak interactions (cation–π, π–π, CH–π), and entropic contributions from solvent reorganization. Predict how altering capsule rigidity would change binding thermodynamics.

18. A molecular dynamics researcher is studying a dense ionic liquid whose local structuring is dominated by competing Coulombic attractions and steric repulsions. She observes that introducing a small concentration of multivalent cations causes a dramatic shift in the radial distribution function, producing unexpectedly deep minima and sharper coordination shells. Formulate a graduate-level problem requiring analysis of how multivalency perturbs both the potential of mean force and the liquid's emergent mesoscale ordering, and ask the student to propose a method for distinguishing genuine ion-bridging from artifacts arising from finite-size simulation effects.

19. In a plasma processing reactor used for thin-film fabrication, the electron temperature unexpectedly becomes decoupled from the heavy-particle temperature when a trace impurity gas is introduced. The engineering team suspects that vibrational excitation pathways in the impurity species are siphoning off electron kinetic energy through inelastic collisions. Construct a problem that challenges students to quantify how the impurity concentration alters the electron energy distribution function and to determine under what operating conditions the plasma transitions from a quasi-thermal to a distinctly non-Maxwellian regime.

20. A surface chemist investigates catalytic oxidation on a polycrystalline metal substrate and discovers that certain grain boundaries induce localized oscillatory

reaction fronts. These spatiotemporal oscillations appear only when the feed ratio of reactants lies within a narrow window, suggesting a delicate interplay among adsorption kinetics, lateral diffusion, and heat release. Write a problem requiring students to analyze how structural heterogeneity modulates reaction–diffusion coupling and to predict conditions under which self-organized surface patterns either stabilize or collapse.

21. A materials scientist studies a ceramic lattice doped with rare-earth ions that exhibit anomalous thermal expansion—shrinking instead of expanding as temperature increases within a specific range. She hypothesizes that electronically driven changes in coordination symmetry lead to competing lattice relaxation pathways. Frame a problem asking students to evaluate how electronic degeneracy, crystal-field splitting, and phonon coupling could jointly produce negative thermal expansion and to design an experiment capable of distinguishing electronic effects from purely vibrational ones.

22. An atmospheric chemist monitoring upper-tropospheric aerosol formation finds that certain organic nitrates trigger unexpectedly rapid particle nucleation events. These compounds appear to undergo photolysis and heterogeneous reactions on existing aerosols, producing highly oxygenated molecules with low vapor pressures. Formulate a problem requiring quantitative analysis of the competing timescales governing photochemistry, gas-particle partitioning, and cluster stabilization to determine when runaway nucleation becomes thermodynamically plausible.

23. A quantum chemist simulating a transition-metal cluster observes that its ground-state wavefunction exhibits strong multireference character, making single-determinant approximations inadequate. The cluster shows dynamic Jahn–Teller distortions that reshuffle the relative energies of several near-degenerate electronic states. Construct a problem challenging the student to articulate why multireference approaches are essential, how symmetry breaking manifests in potential-energy surfaces, and what diagnostics can reliably quantify nondynamic versus dynamic correlation.

24. In an electrochemical device designed for CO_2 reduction, intermittent fluctuations in electrode potential are detected even under nominally steady-state conditions.

The fluctuations correlate with the sporadic formation of nanobubbles that alter mass-transport pathways and modify the local electric double layer. Develop a problem requiring analysis of bubble nucleation energetics, surface tension effects, double-layer restructuring, and resulting deviations in Faradaic efficiency. Ask the student to propose a monitoring strategy that would reveal whether fluctuations arise from bubble dynamics or from intrinsic kinetic oscillations.

25. A geochemist analyzing mineral dissolution in hydrothermal vents detects unexpected isotopic fractionation patterns in silicon and magnesium. These patterns do not match equilibrium predictions and instead suggest rapid, nonequilibrium dissolution fronts coupled with fluid mixing gradients. Frame a problem that investigates how temperature gradients, fluid residence times, and surface roughness influence kinetic isotope effects and how these signatures could be used to reconstruct subsurface fluid transport histories.

26. A polymer physicist observes that a block copolymer thin film undergoes a spontaneous reorientation transition when the substrate is coated with a weakly interactive monolayer. The shift in orientation—from perpendicular to parallel lamellae—suggests that subtle changes in interfacial energy dominate the free-energy landscape. Create a problem challenging the student to analyze enthalpy–entropy competition in microphase ordering, evaluate film-thickness effects, and predict how altering the monolayer's chemical functionality would shift the reorientation threshold.

27. A photochemist studying ultrafast charge separation in organic solar-cell blends finds that increasing molecular ordering paradoxically slows charge generation. She suspects that excessive crystallinity restricts energetic disorder, thereby reducing the driving force for charge delocalization. Write a problem requiring students to evaluate the role of vibronic coupling, exciton diffusion length, and energetic landscape topography in determining the efficiency of charge-transfer states, and to model conditions under which increased ordering becomes detrimental.

28. A nuclear chemist exploring molten-salt reactor chemistry identifies transient redox oscillations arising from parasitic reactions between fission products and salt constituents. These oscillations change the speciation of key actinides, altering their solubility and transport. Formulate a problem requiring the student to analyze

the redox buffering capacity of the melt, to predict how specific fission products destabilize redox equilibrium, and to design a control strategy that maintains stable actinide oxidation states during reactor operation.

29. A computational physicist studying strongly correlated electron systems finds that increasing electron density in a layered material causes a sudden collapse of the Mott insulating state. The transition appears driven by interlayer tunneling that quenches local magnetic moments. Write a problem asking students to compare Hubbard-model predictions with more realistic band-structure calculations and to determine how dimensional crossover affects the critical parameters governing the metal–insulator transition.

30. A biophysical chemist analyzing protein folding dynamics discovers that certain residues undergo correlated fluctuations that precede the major folding transition. These correlations suggest the presence of hidden collective variables not captured by traditional reaction-coordinate definitions. Craft a problem requiring the student to use principal-component analysis or manifold-learning techniques to identify emergent coordinates and to evaluate how these coordinates predict folding intermediates or transition-state ensembles.

31. An astrochemical model predicts that in cold molecular clouds, barrierless ion–molecule reactions should dominate the production of certain prebiotic molecules. Yet observations show abundances inconsistent with purely gas-phase chemistry, implying unaccounted surface-mediated reactions on icy grains. Design a problem prompting students to quantify how adsorption, diffusion on amorphous ice, and cosmic-ray-driven desorption reshape the chemical network and to estimate the relative contributions of gas-phase versus grain-surface pathways.

32. A spectroscopist studying vibrational strong coupling in an optical cavity notices that modifying the cavity length alters not only Rabi splitting but also reaction selectivity in a tautomerization reaction. Construct a problem that forces students to dissect how cavity-modified potential-energy surfaces, polariton lifetimes, and mode-specific coupling strengths cooperate to generate altered product distributions.

33. A battery researcher finds that lithium plating on a fast-charging anode initiates beneath the solid–electrolyte interphase rather than on its outer surface.

This suggests that subsurface defects or local ionic-transport anisotropies govern the earliest stages of dendrite nucleation. Create a problem asking students to evaluate defect-mediated transport, interface mechanical stresses, and SEI elasticity to determine conditions that trigger internal plating.

34. An environmental chemist monitoring groundwater contamination from a landfill discovers that heavy-metal chelates undergo unexpected photochemical transformations even at modest depths. These transformations produce reactive intermediates that mobilize additional contaminants. Frame a problem asking the student to analyze photon penetration, ligand-to-metal charge transfer dynamics, and radical scavenging pathways to predict when such photochemistry becomes environmentally significant.

35. A physicist studying dusty plasmas finds that micron-scale grains organize into hexagonal lattices whose rigidity fluctuates with neutral-gas pressure. The fluctuations do not follow simple kinetic-theory expectations and appear to involve collective charge rearrangements. Write a problem requiring students to quantify grain charging, plasma shielding, and coupling parameters, then to predict the onset of melting in the plasma crystal.

36. A chemical engineer designing a membrane-separation process discovers that trace surfactants in the feed spontaneously reorganize into dynamic mesoscale structures within the pores, drastically reducing permeance. Construct a problem requiring the student to model adsorption–desorption equilibria, confinement-driven micelle deformation, and the subsequent impact on flow resistance.

37. A quantum-information theorist working with trapped ions observes that motional heating rates strongly depend on electrode surface contamination. Hypothesizing that fluctuating patch potentials induce decoherence, she seeks a predictive model. Craft a problem that challenges students to analyze spectral noise densities, ion–surface distance scaling, and mitigation strategies involving in-situ cleaning or cryogenic operation.

38. A metallurgist studying high-entropy alloys finds that adding a small concentration of a seemingly inert element dramatically alters dislocation mobility. The behavior suggests the formation of short-range chemical order that perturbs the local energy landscape. Write a problem asking students to use thermodynamic models of

configurational entropy and atomistic simulations to quantify how such ordering modifies mechanical properties.

39. A kineticist examining enzyme–substrate encounters in crowded cellular environments observes that reaction rates exceed predictions from classical diffusion models. Transient confinement and anomalous diffusion appear to enhance encounter probability. Construct a problem requiring students to incorporate fractal diffusion, molecular crowding, and transient binding sites into a predictive kinetic scheme.

40. A combustion researcher finds that adding small quantities of bio-derived oxygenates to traditional jet fuel suppresses soot formation in ways not fully explained by known radical pathways. Create a problem prompting students to analyze fuel-molecule decomposition pathways, oxygen-containing radical chemistry, and how these shifts suppress aromatic ring growth.

41. A crystallographer discovers that applying modest uniaxial stress to an organic semiconductor induces polymorphic interconversion accompanied by dramatic shifts in charge-carrier mobility. Construct a problem requiring analysis of how lattice slip planes, π-stacking geometry, and stress-coupled free-energy differences govern polymorph stability.

42. A physical chemist studying solvation dynamics in supercritical fluids finds that local density enhancements persist for unexpectedly long times, creating "solvation cages" that impede reaction progress. Frame a problem that asks students to compute how local compressibility, density fluctuations, and critical slowing-down influence reaction kinetics.

43. A planetary scientist analyzing cryovolcanic plumes on an icy moon detects complex organics that seem incompatible with the observed surface temperatures. She proposes that subsurface reservoirs undergo episodic heating from tidal stresses. Write a problem requiring estimation of mechanical-to-thermal energy conversion, solute–solvent equilibria at high pressure, and resulting chemical signatures observable in the plume.

44. A soft-matter theorist finds that colloidal particles functionalized with DNA strands self-assemble into topologically complex structures that defy simple pairwise-

interaction models. Create a problem asking students to evaluate multivalent binding, cooperative hybridization, and emergent constraints that lead to nontrivial topology.

45. A photophysics researcher studying triplet–triplet annihilation upconversion notes that in certain host matrices, triplet diffusion becomes anomalously rapid. This suggests transient lattice fluctuations mediating long-range hopping. Develop a problem requiring students to analyze coupling between excitonic transport and dynamic disorder.

46. A soil chemist examining mineral–root interactions finds that root exudates create microscale pH gradients that selectively mobilize micronutrients while immobilizing toxic metals. Frame a problem involving diffusion–reaction coupling, microbial mediation, and mineral dissolution equilibria to predict nutrient profiles around root surfaces.

47. A semiconductor physicist measures unexpectedly long spin lifetimes in a 2D material subjected to modest magnetic fields. She suspects that spin–valley coupling creates protected states. Craft a problem asking students to analyze spin–orbit effects, Berry curvature, and valley-selective scattering mechanisms.

48. A climatochemist models ocean–atmosphere gas exchange and discovers that certain halogenated compounds show flux behaviors inconsistent with Henry's-law predictions. She hypothesizes microbubble-mediated supersaturation events. Write a problem concerning bubble dynamics, surfactant effects, and nonequilibrium partitioning.

49. A nanotechnologist finds that electron-beam lithography on hybrid organic–inorganic films induces unexpected cross-linking patterns, forming hierarchical nanostructures. Create a problem asking students to analyze radical pathways, electron-scattering profiles, and proximity effects that determine the observed structures.

50. A theoretical chemist modeling coupled proton–electron transfer in a biomimetic complex discovers that introducing asymmetry into the ligand scaffold produces a strong directionality in the reaction coordinate. Construct a problem requiring evaluation of vibronic coupling, reorganization energies, and the role of solvent polarity in biasing coupled transfer pathways.

Chapter 12

Mystery of Solution Chemistry

Most solutions encountered in everyday life are mixtures composed of two or more substances combined in varying proportions. Common examples—wood, milk, seawater, air, gasoline, blood, and even alloys—illustrate how widespread mixtures are in nature and technology. When a mixture is uniform in composition throughout, with no visible boundaries separating its components, it is classified as a homogeneous mixture, also known as a *solution*. Solutions can exist in any physical state: gases (such as air, a mixture of nitrogen, oxygen, carbon dioxide, and trace gases), liquids (such as vinegar or saltwater), or solids (such as brass, a copper–zinc alloy). Among these, liquid solutions, especially those with water as the solvent, are of exceptional significance. The predominance of water-based solutions in nature and industry arises from water's unique molecular structure, polarity, extensive hydrogen-bonding ability, and its capacity to dissolve an extraordinary range of ionic and polar molecular substances. Consequently, the vast majority of biological, environmental, and industrial chemical reactions take place in aqueous solutions, making the study of solution chemistry central to all scientific disciplines.

Because mixtures consist of variable proportions of components, it becomes essential to specify their composition quantitatively, especially when performing calculations involving reaction stoichiometry or physical properties. When only a small amount of solute is present relative to the solvent, the solution is considered dilute; conversely, when the solute proportion is large, the solution is concentrated. However, these qualitative descriptions are too vague for scientific or industrial use, so chemists employ several more precise methods of expressing solution concentration.

One of the most common measures is molarity (M), defined as the number of *moles of solute per liter of solution*. Molarity is especially useful because chemical equations naturally link quantities in moles, enabling direct stoichiometric calculations. The substance being dissolved is the solute, while the medium that performs the dissolving is the solvent. Another measure, mass percent, sometimes called weight percent, represents the ratio of the solute's mass to the total mass of the solution, multiplied by 100%. Mass percent is particularly important in industries such as food processing, metallurgy, and pharmaceuticals, where masses are easier to measure than volumes.

A more thermodynamically fundamental measure is the mole fraction, symbolized by the Greek letter χ. Mole fraction expresses the proportion of moles of a single component relative to the total number of moles in the mixture. It is dimensionless, independent of temperature, and useful in Gibbs energy calculations and vapor-pressure relationships. When two liquids are mixed, the component present in the larger amount is typically considered the solvent. Molality *(m)*, defined as moles of solute per kilogram of solvent, is another important concentration unit. Unlike molarity, molality depends only on the mass of the solvent and thus does *not* vary with temperature, making it ideal for quantitative studies of colligative properties such as boiling-point elevation and freezing-point depression.

Another concentration unit, less common but still used in specific contexts, is normality (N). Normality refers to the number of *equivalents* per liter of solution. The equivalent depends on the nature of the reaction. For acid–base reactions, one equivalent corresponds to the amount of acid capable of donating one mole of H^+ ions or the amount of base capable of accepting one mole of H^+. For example, sulfuric acid (H_2SO_4) donates two protons per molecule; thus, its equivalent mass is half its molar mass. In oxidation–reduction reactions, an equivalent corresponds to the amount of species that gains or loses one mole of electrons. Normality is particularly useful in titration chemistry because it directly reflects reactive capacity.

Dissolution is an energetic process. Familiar examples include dissolving salt in water during cooking, dissolving carbon dioxide in water to make carbonated beverages, or dissolving ethanol in gasoline to produce gasohol. Whether a substance will dissolve, and to what extent, depends on energetic factors and the molecular nature of the solute and solvent. A useful heuristic is "like dissolves like": polar or ionic solutes tend to dissolve in polar solvents, and nonpolar substances dissolve more readily in nonpolar media.

Solution formation typically occurs in three conceptual steps. First, solute particles must separate from one another, requiring energy to overcome intermolecular forces or ionic attractions (ΔH_1). Second, some solvent molecules must move apart to create space for the solute, requiring additional energy to disrupt solvent–solvent interactions (ΔH_2). Finally, solute and solvent particles interact to form new solute–solvent attractions (ΔH_3), which generally *releases* energy. The enthalpy of solution (ΔH_{soln}) represents the sum of these contributions:

$$\Delta H_{soln} = \Delta H_1 + \Delta H_2 + \Delta H_3$$

Depending on the relative magnitudes of these terms, dissolution may be endothermic or exothermic. A related quantity is the enthalpy of hydration (ΔH_{hyd}), which applies specifically to ionic solutes dissolving in water. It combines the energy required to expand the solvent (ΔH_2) with the energy released when ions become surrounded by water molecules (ΔH_3). Hydration enthalpies are typically quite large and negative because water molecules strongly stabilize ions through ion–dipole interactions.

While energetic factors are important, entropy also plays a major role. Nature tends toward states of higher disorder and greater probability. Mixing—whether of gases, liquids, or particles—creates more possible arrangements for the components and thus increases entropy, often driving solution formation even when dissolution is slightly endothermic.

The solubility of a substance is determined by its molecular structure, polarity, hydrogen-bonding ability, and intermolecular forces. A striking example lies in the differing solubilities of vitamins. Fat-soluble vitamins (A, D, E, and K) have long hydrophobic hydrocarbon chains or ring structures and dissolve easily in nonpolar lipids. Water-soluble vitamins (such as vitamin C and the B-complex vitamins) possess multiple polar or ionic functional groups, enabling strong interactions with water. These solubility patterns explain why fat-soluble vitamins accumulate in body tissues, whereas water-soluble vitamins must be consumed regularly because excess amounts are excreted in urine. Understanding the relationship between structure and solubility is critical not only in biochemistry but also in drug design, chromatography, and environmental chemistry.

Pressure has minimal effect on the solubility of solids and liquids but significantly influences the solubility of gases in liquids. Henry's Law states that the concentration of a dissolved gas (C) is directly proportional to the partial pressure of that gas (P) above the solution:

$$C = kP$$

where k is Henry's constant, which depends on both the gas and the solvent. This relationship explains why carbonated beverages lose carbonation when opened—the pressure above the liquid suddenly decreases, reducing gas solubility and causing CO_2 bubbles to escape. Henry's Law applies best when the gas does not chemically react with the solvent. Temperature influences solubility differently for solids and gases. Generally, solubility of most solids increases with temperature, but exceptions exist—for example, sodium sulfate and cerium sulfate decrease in solubility as temperature rises. Because of these complexities, experimental measurement is the only reliable way to determine a substance's temperature–solubility relationship.

Gas solubility, on the other hand, nearly always decreases with increasing temperature. This principle has important environmental implications. Industrial facilities often use lakes or rivers for cooling. The warmed water returned to the environment contains less dissolved oxygen—a phenomenon known as thermal pollution—which threatens aquatic life by reducing oxygen availability in deeper, cooler regions.

Temperature also affects the formation of boiler scale in heated water systems. As water warms, dissolved CO_2 escapes, shifting equilibrium among carbonate species and leading to the precipitation of calcium carbonate, which accumulates in pipes and appliances, reducing efficiency.

Adding a nonvolatile solute to a solvent reduces the solvent's vapor pressure because fewer solvent molecules occupy the surface and escape into the gas phase. Raoult's Law quantifies this reduction:

$$P_{solution} = X_{solvent} \times P^{\circ}_{solvent}$$

where X_solvent is the solvent's mole fraction and P°_solvent is its pure vapor pressure. This principle permits the determination of solute molar masses by measuring vapor-pressure lowering. If the solute dissociates, as in the case of ionic compounds like KCl, vapor-pressure changes reveal the number of particles present in solution. When both components of a liquid mixture are volatile, each contributes to the total vapor pressure according to a modified form of Raoult's Law:

$$P_{Total} = X_A P^{\circ}_A + X_B P_{^{\circ}B}$$

Solutions that obey Raoult's Law exactly are termed ideal solutions, a condition approached when intermolecular forces between unlike molecules are similar to those between like molecules. Deviations—positive or negative—occur when solute–solvent interactions are significantly weaker or stronger than solvent–solvent interactions. Changes in vapor pressure lead to changes in phase-transition temperatures. The presence of a nonvolatile solute raises a solvent's boiling point and lowers its freezing point—phenomena collectively known as colligative properties, because they depend only on the number of dissolved particles, not their identity.

Boiling-point elevation is expressed as:

$$\Delta T_b = K_b m$$

Freezing-point depression follows:

$$\Delta T_f = K_f m$$

These principles explain why adding ethylene glycol antifreeze prevents water in car radiators from freezing in winter and boiling in summer.

When a solution is separated from a pure solvent by a semipermeable membrane, only solvent molecules can pass through. Solvent naturally flows into the solution, causing the solution level to rise until hydrostatic pressure balances the osmotic tendency. The pressure needed to stop this flow is the osmotic pressure (π):

$$\pi = MRT$$

Osmotic pressure measurements are extremely sensitive and allow accurate determination of molar masses, especially for large biomolecules such as proteins or polymers. Osmosis is fundamental in biological systems. Cells rely on osmotic balance to maintain volume and function. If red blood cells are placed in hypertonic solutions, water leaves the cells, causing them to shrink (crenation). In hypotonic solutions, water enters cells, potentially causing them to burst (hemolysis). Dialysis, a process similar to osmosis but allowing small solutes to pass through a semipermeable membrane, is used medically in artificial kidney machines to remove waste molecules from the blood. Reverse osmosis, achieved by applying pressure greater than the osmotic pressure, forces solvent to flow from the solution to the pure solvent side. This method is widely used in desalination plants to produce fresh water from seawater.

For electrolyte solutions, colligative properties must account for the number of ions produced upon dissociation. The van't Hoff factor (i) expresses the ratio of particles actually present to the number of formula units dissolved. Ideal dissociation gives integer values (e.g., $i = 2$ for NaCl), but real solutions show slightly smaller values due to ion pairing, where oppositely charged ions temporarily associate. As solutions become more dilute, ion pairing decreases and i approaches the expected value.

Colloids occupy an unusual and fascinating niche within chemistry—larger in structure than molecular solutions but far too small to be classified as suspensions. Although their particle sizes and heterogeneity make them appear deceptively simple, colloids embody an entire field of interfacial phenomena, surface chemistry, and physicochemical interactions that stretch across classical and modern chemical theory. Their stability hinges on delicate balances of intermolecular forces, electrolyte concentrations, surface potentials, and molecular adsorption events. A typical colloidal particle carries a shell of counter-ions that assemble spontaneously around its charged surface, forming the electrical double layer. The repulsion created by these like-charged surfaces prevents neighboring particles from fusing into larger aggregates, thus allowing the colloid to remain dispersed for exceptionally long periods. When the ionic environment changes—whether by heating, adding electrolytes, altering pH, or exposing the system to mechanical agitation—this protective double layer can collapse, triggering coagulation. Such sensitivity to environmental conditions is one reason colloid chemistry is fundamental in environmental engineering, materials design, pharmaceutical formulation, and biological function.

The Tyndall effect, the optical scattering phenomenon characteristic of colloids, provides a visually compelling confirmation of particle size and dispersion. Because colloidal particles are large enough to intercept light yet small enough to remain suspended, they scatter light beams into observable cones or shafts. This optical signature is essential in identifying colloidal systems in both laboratory settings and industrial quality control processes. It has been instrumental historically as well: early studies of colloidal light scattering contributed to the development of theories about molecular size, particle charge, and Brownian motion, eventually intersecting with Einstein's interpretation of molecular fluctuations and diffusion. Thus, colloids sit at the intersection of optics, statistical mechanics, and interfacial science, linking seemingly disparate fields through the behavior of particles invisible to the naked eye yet profoundly influential in chemical systems.

In industrial applications, colloidal stability has enormous practical value. Many everyday products—paints, inks, cosmetics, milk, fog, and even blood—owe their macroscopic properties to the stability of their dispersed phases. Paints, for example, require pigments to remain evenly dispersed in a solvent or polymer matrix for color uniformity and shelf-life longevity. Engineers must carefully tailor dispersing agents, surfactant concentrations, and polymeric stabilizers so pigments do not interact too strongly, avoiding clumping that would ruin paint quality. Similarly, detergents rely on micelles—colloidal aggregates of surfactant molecules—to trap oils and dirt within their hydrophobic cores, allowing water to wash them away. Without the self-assembling behavior of surfactant colloids, modern cleaning, formulation science, and even wastewater remediation would be vastly less effective.

Biological colloids introduce a realm of exquisite complexity. Proteins in blood plasma, lipid bilayers forming cell membranes, and intracellular organelles all function as colloidal assemblies governed by finely tuned electrostatics and hydration dynamics. The balance of protein structure and colloidal stability in blood, for instance, affects osmotic pressure, nutrient transport, clotting mechanisms, and immune responses. Protein denaturation—whether triggered by heat, pH shifts, or chemical agents—can destabilize biological colloids and disrupt cellular homeostasis. Even the cytoplasm behaves as a crowded colloidal gel, its viscosity and mechanical properties essential for molecular diffusion, enzymatic activity, and organelle movement. Biological colloids do not merely exist as passive structural materials; they form dynamic networks whose stability directly influences metabolism, signaling, and physiological resilience.

Electrostatic precipitators exemplify the practical harnessing of colloid destabilization. By imparting electrical charges to particulate matter in industrial exhaust streams, these systems cause small airborne particles to coagulate or be drawn to oppositely charged collection plates. This principle reduces particulate pollution dramatically, preventing harmful aerosols from entering the atmosphere. The success of electrostatic precipitators highlights how a concept originally studied in seemingly benign laboratory colloids can scale to industrial and environmental engineering, demonstrating the real-world value of mastering colloidal behavior. In a similar way, water purification technologies—including flocculation tanks, sedimentation basins, and membrane filters—rely on controlled colloidal destabilization. Coagulants such as alum or iron salts neutralize the surface charges of suspended particles, promoting aggregation into flocs large enough to be filtered or settled out. These principles underlie modern municipal water treatment, helping deliver clean water to millions of people.

The roles of colloids extend deeper into materials science, where nanoparticle dispersions, polymer emulsions, and gels form the basis of advanced technologies. Nanoparticles, though often smaller than classical colloids, behave colloidally in suspension, showcasing phenomena such as self-assembly, quantum confinement, and enhanced surface reactivity. Engineers exploit these properties to create conductive inks for flexible electronics, catalytic nanoparticles for fuel cells, and targeted drug-delivery vehicles capable of entering specific cells. Stabilizing such nanoparticles requires mastering surface functionalization—attaching ligands, polymers, or small molecules that prevent aggregation while supporting desired chemical reactions. This interface between colloid science and nanotechnology has become a frontier of modern chemistry, where manipulating interparticle forces becomes a tool for constructing new materials with emergent properties.

Another dimension of colloidal science appears in food chemistry, where emulsions, foams, and gels define the textures, flavors, and stability of culinary products. Milk is a natural colloid composed of fat droplets dispersed in aqueous solution, stabilized by proteins and phospholipids. Ice cream, whipped cream, mayonnaise, and salad dressings likewise owe their textures to colloidal architecture. The consistency of yogurt results from bacterial fermentation altering protein–protein interactions, thereby forming a colloidal network. Food scientists manipulate pH, ionic strength, temperature, and mechanical mixing to control colloidal structure, tailoring the sensory qualities of foods while extending shelf life and safety. Even chocolate production utilizes colloidal physics: the tempering process arranges cocoa butter crystals into specific polymorphs, producing desirable melting behavior, gloss, and snap. Thus, colloid chemistry plays a surprisingly intimate role in gastronomy.

The stability of colloids often depends on the dynamic tug-of-war between attractive van der Waals forces and various repulsive interactions. Van der Waals attractions encourage particles to approach and stick together, but repulsive interactions—including electrostatic repulsion, hydration forces, and steric hindrance—push particles apart. When repulsive interactions dominate, colloids remain stable; when attractions win, coagulation or flocculation occurs. The fine balance between these forces is captured conceptually in classical theories such as DLVO theory, which examines the net interaction potential between colloidal particles. Although the theory simplifies real systems, it has served as a foundation for understanding stability across industries. In more complex colloids containing polymers or biomolecules, steric repulsion becomes important. When polymer chains extend into a solution from the particle surface, they form an entangled, brush-like layer that mechanically prevents particles from approaching closely enough to aggregate. This steric protection is vital in formulations such as paints and pharmaceuticals, where long-term stability is essential.

Polymer colloids, including latexes and hydrogels, illustrate how colloidal stabilization can be engineered to create sophisticated materials. Latex paints rely on dispersed polymer particles that coalesce upon drying, forming continuous films. Hydrogels—three-dimensional polymer networks swollen with water—are colloidal at the microscopic level yet macroscopic in appearance. These materials exhibit remarkable elasticity, tunable porosity, and biocompatibility, making them ideal for wound dressings, contact lenses, drug delivery matrices, and tissue engineering scaffolds. The polymer chains within hydrogels interact with water through hydrogen bonding and hydrophilic functional groups, creating materials that respond to temperature, pH, or ionic strength. Such "smart hydrogels" are opening pathways toward responsive materials that can adjust their properties autonomously.

Micelles and emulsions form another foundational class of colloids. Surfactants assemble into micelles when their amphiphilic molecules arrange with hydrophobic tails inward and hydrophilic heads outward. This structural organization captures hydrophobic substances—such as oils, fragrances, or drugs—within the micellar core. Pharmaceutical scientists exploit micelles to increase the solubility and bioavailability of hydrophobic drugs. Cosmetics use emulsions to blend oil-soluble ingredients with water-soluble ones, producing creams, lotions, and ointments. Emulsion stability depends on interfacial tension, droplet size, and the nature of emulsifying agents, which can be proteins, small surfactants, or polymeric stabilizers. Fine-tuning these variables allows manufacturers to control viscosity, spreadability, absorption rate, and overall product performance.

Foams, a less intuitive form of colloid, consist of gas bubbles dispersed in a liquid or solid matrix. They are inherently unstable without stabilizers because gas bubbles tend to coalesce or escape from the liquid. Yet through structural additives or protein films, foams can be remarkably robust. Beer foam, for instance, depends on proteins and hop resins that stabilize gas bubbles, while fire-suppressing foams are engineered to cover surfaces with thermally insulating layers that smother flames. Polymer foams, including polystyrene and polyurethane foams, are solid colloidal structures formed through gas incorporation during polymerization. Their cellular architecture gives them exceptional insulation properties and mechanical versatility, playing critical roles in packaging, construction, and impact absorption.

Gel formation—another hallmark of colloidal behavior—occurs when dispersed particles or polymers form continuous three-dimensional networks. In biological systems, gels provide structural integrity to tissues while facilitating diffusion of nutrients and metabolites. Agar, pectin, and gelatin are natural gel-forming colloids used widely in microbiology cultures, confectionery, and biomedical applications. Synthetic gels have progressed into the realm of ionic gels, organogels, and hybrid gels combining inorganic nanoparticles with polymer matrices. These advanced gels exhibit thermal responsiveness, selective binding, or unusually high mechanical strength, making them candidates for next-generation sensors, separation membranes, and artificial organs.

Colloidal interactions also play pivotal roles in geochemistry and environmental science. Clay particles in soil, which behave colloidally due to their extreme surface area and charge density, determine how nutrients, heavy metals, and organic pollutants bind and move through terrestrial environments. The ability of clays to adsorb ions makes them natural filters, but when destabilized, they can contribute to erosion or turbidity in waterways. Understanding colloidal forces in soil helps agricultural scientists optimize fertilizer delivery, conservation strategies, and soil conditioning treatments. In marine environments, colloidal silica, iron oxides, and organic macromolecules influence nutrient cycles and water clarity. Even the formation of some minerals—such as opal—occurs through colloidal deposition and solidification.

In the pharmaceutical world, colloids serve as delivery vehicles, stabilizers, and controlled-release matrices. Liposomes—tiny vesicles composed of phospholipid bilayers—encapsulate active compounds, protecting them from degradation while enabling targeted delivery to specific tissues. Nanocolloids enhance the solubility of poorly soluble drugs, improving their therapeutic effectiveness. Injectable suspensions of protein colloids or polymer nanoparticles can be tuned for slow release, allowing medications to act over days or weeks. The interplay of colloidal stability, biodegradability, and surface functionalization is transforming drug development and personalized medicine.

Even atmospheric chemistry is shaped by colloidal processes. Aerosols, consisting of tiny liquid or solid particles suspended in air, influence climate patterns through light scattering, cloud formation, and chemical reactions occurring on particle surfaces. Aerosols arise from natural sources—such as sea spray, volcanic eruptions, and forest emissions—as well as human activities, including combustion, industrial emissions, and agriculture. Their colloidal nature allows them to remain airborne for extended periods, with global implications for climate science and human health. The ability of aerosol particles to serve as nucleation sites for cloud droplets is central to understanding precipitation patterns and the energy balance of Earth's atmosphere.

In the electronics industry, colloidal materials are enabling technologies once considered science fiction. Quantum dots—semiconductor nanoparticles with size-dependent optical properties—form colloidal suspensions that can be incorporated into displays, solar cells, and sensors. Their ability to emit light in narrow, tunable wavelengths makes them invaluable for high-definition screens and color-stable LEDs. The stability of these nanoparticle colloids is crucial, requiring ligands that prevent aggregation while allowing electronic coupling and charge transport. Similarly, colloidal inks containing conductive nanoparticles are used in printed electronics, where they form flexible circuits on plastic substrates. These advances rely on managing colloidal forces at the nanoscale to yield macroscopic technological breakthroughs.

Soft matter physics, which studies materials that deform easily under external forces, encompasses colloids, polymers, gels, and liquid crystals. In this realm, colloidal particles can exhibit collective behaviors such as self-assembly into crystal-like arrays, formation of glassy states, or emergence of non-Newtonian rheology. Colloidal crystals, formed when monodisperse particles arrange into regular lattices, have remarkable optical properties that depend on their periodicity. These materials can create photonic band gaps—regions where light cannot propagate—enabling the design of optical filters, sensors, and even components for future quantum communication systems. Controlling colloidal self-assembly requires exquisite command over particle interactions, solvent conditions, and processing kinetics.

The study of interfacial chemistry within colloids has also progressed into the molecular realm. Advances in spectroscopy, microscopy, and scattering techniques allow chemists to visualize particle surfaces, measure zeta potentials, and resolve hydration structures in unprecedented detail. Techniques such as atomic force microscopy, cryogenic electron microscopy, and neutron scattering provide windows into structure and dynamics previously inaccessible. These experimental innovations are complemented by computational modeling, where molecular simulations predict how ions, polymers, or surfactants interact with nanoparticle surfaces. Such modeling guides the development of new colloidal systems for catalysis, imaging, and materials assembly.

Catalysis itself has embraced colloidal nanoparticles as powerful tools for accelerating chemical transformations. Metallic nanoparticles—such as platinum, gold, palladium, and copper—exhibit unique catalytic properties due to their high surface-area-to-volume ratios and quantum-scale effects. In solution, colloidal nanoparticles can catalyze reactions ranging from hydrogenation and oxidation to carbon-carbon coupling. Their surfaces can be tuned with ligands to control selectivity, activity, and resistance to poisoning. Industrial catalysts often begin as colloidal dispersions before being immobilized on supports, ensuring uniform distribution and high reactivity.

The field of colloid chemistry has also expanded into sustainability and green technology. Bio-based surfactants derived from renewable resources reduce toxicity and environmental impact in detergents, cosmetics, and cleaners. Nanocolloids made from biodegradable polymers offer biodegradable packaging, environmentally safe agricultural treatments, and sustainable drug-delivery matrices. Water purification technologies increasingly rely on magnetically responsive colloidal particles, which can adsorb contaminants and then be removed easily with magnetic fields. Such innovations demonstrate how colloid science contributes directly to environmental responsibility and resource conservation.

Advanced energy technologies benefit +as well. Colloidal quantum dots enhance the light-absorbing properties of solar cells, improving efficiency and enabling flexible photovoltaic materials. Colloidal catalysts in fuel cells and electrolyzers accelerate energy conversion reactions with lower metal content, reducing costs and improving sustainability. Battery researchers explore colloidal dispersions of conductive particles to create lightweight, scalable electrode materials with improved charge capacity and stability. At the frontier of energy storage, colloidal gels known as "aerogels" and "cryogels" provide ultralight, highly porous structures for supercapacitors and thermal insulation.

Interdisciplinary research continues to reveal new phenomena arising from colloidal behavior. In medicine, colloidal gold and magnetic nanoparticles enhance imaging techniques such as MRI and optical contrast imaging. In environmental monitoring, colloidal sensors detect pollutants, pathogens, and trace chemicals with high sensitivity. In cultural heritage preservation, colloidal dispersions are used to clean aging artworks and restore delicate materials without destructive chemicals. In soft robotics, colloidal hydrogels and elastomers enable flexible, adaptive devices inspired by biological tissue.

As we understand colloids more deeply, the boundary between classical colloid chemistry and nanoscience becomes increasingly fluid. Interdisciplinary collaborations among chemists, physicists, biologists, and engineers are revealing that colloids are not merely intermediate states of matter but foundational building blocks of complex systems. Whether stabilizing emulsions, constructing nanoscale electronics, powering future energy devices, or maintaining the flow of life within living organisms, colloids continue to demonstrate an extraordinary capacity for shaping technological and natural worlds.

In this sense, colloid chemistry is not a narrow subfield but a vast conceptual atlas that connects phenomena ranging from the microscopic structure of cells to the macroscopic behavior of materials and environments. It exemplifies the richness of chemical science: the interplay of structure, dynamics, stability, and transformation across multiple scales of organization. By probing the subtle forces that govern colloidal interactions—electrostatic, steric, entropic, and van der Waals—scientists unlock a deeper understanding of matter itself, revealing how complexity emerges from fundamental principles. As research progresses, colloids will continue to inspire new technologies, refine global processes, and illuminate the hidden architecture of both biological life and engineered systems.

Chapter 12: Word Problems

1. A pharmaceutical company is attempting to formulate an intravenous drug whose active ingredient is sparingly soluble in water but highly soluble in ethanol. However, ethanol concentration in blood must remain extremely low to avoid toxicity. The formulation chemists consider using micelles to solubilize the hydrophobic drug but worry that serum proteins may destabilize the micelles upon injection. Describe the molecular mechanisms by which micelles can enhance solubility of hydrophobic compounds, analyze how competitive binding or adsorption by serum proteins could lead to micelle collapse or drug precipitation, and propose a chemically reasonable strategy—including potential polymeric or lipid-based carriers—to ensure the drug remains solubilized and bioavailable after systemic administration.

2. A lake adjacent to an industrial facility exhibits declining fish populations over several summers. Measurements indicate that dissolved oxygen concentrations decrease significantly near the facility's warm-water discharge points.

Using thermodynamic principles of gas solubility, kinetic diffusion, and the entropy-driven behavior of water–gas systems, explain why elevated water temperature lowers oxygen content, how seasonal thermal stratification exacerbates oxygen depletion in the hypolimnion, and evaluate both physical (mixing, aeration) and chemical (oxidizing agents, oxygen carriers) interventions that could be implemented to restore ecological balance without causing further environmental disruption.

3. A research team studying nanomedicine synthesizes citrate-stabilized gold nanoparticles that remain stable in ultrapure water but aggregate rapidly in physiological saline. Explain how increased ionic strength compresses the electrical double layer, modifies DLVO (Derjaguin–Landau–Verwey–Overbeek) interactions, and favors van der Waals-driven aggregation. Discuss how steric stabilization, PEGylation, or zwitterionic surface functionalization could maintain nanoparticle stability in high-salt environments, and evaluate the trade-offs between surface passivation and biological activity.

4. A chemist attempts to dissolve a large quantity of ammonium nitrate in water for an endothermic cooling application. As dissolution progresses, the solution temperature drops sufficiently to form ice crystals on the surface. Analyze the enthalpic and entropic contributions driving dissolution, explain why the system spontaneously cools, and discuss how the decreasing temperature itself creates a thermodynamic feedback limiting further solubility. Consider how solvent choice, stirring, and initial temperature affect maximum achievable solute concentration.

5. An industrial plant employs reverse osmosis (RO) membranes for seawater desalination but observes gradual permeate flow decline due to fouling by colloidal silica and organic macromolecules. Explain the physicochemical mechanisms by which colloids adhere to membrane surfaces, how osmotic pressure and shear forces influence fouling rates, and propose a combination of chemical pretreatment, membrane surface modification, and operational adjustments that could mitigate fouling while maintaining high desalination efficiency.

6. A winery finds that wine stored in oak barrels develops a haze over time, although initially the product was crystal clear. Chemical analysis reveals that polyphenols and proteins form colloidal aggregates. Explain how changes in pH, ionic strength, oxygen exposure, and temperature during barrel aging can destabilize initially

soluble molecules, promoting aggregation, and propose interventions such as fining agents, filtration, or careful pH control to prevent haze formation without adversely affecting flavor or aroma compounds.

7. A materials scientist prepares a polymer latex intended for film-forming coatings. During storage, the latex loses stability, with polymer particles coalescing into clumps. Analyze how surfactant desorption, ionic impurities, divalent cations, and pH fluctuations could collapse electrostatic or steric stabilization, leading to coagulation. Discuss formulation strategies—such as stabilizer optimization, ionic chelators, or controlled ionic environment—to maintain colloidal stability over extended storage.

8. A biochemist studying protein purification using dialysis observes unexpected aggregation when protein solutions are placed in tubing immersed in buffer. Discuss how osmotic pressure differences, changes in ionic strength, local pH gradients, and interactions with membrane surfaces could partially unfold proteins and induce aggregation. Propose adjustments such as buffer composition optimization, protein stabilizers, or controlled ionic environments to prevent unwanted aggregation while maintaining protein activity.

9. Farmers using geothermal water for irrigation observe mineral scaling in pipes and equipment. The water contains high bicarbonate concentrations, and warming during sunlight exposure drives CO_2 escape, leading to precipitation of calcium carbonate. Explain the chemical equilibria and Le Chatelier's principle governing carbonate speciation, why warming shifts equilibria toward solid carbonate formation, and suggest practical approaches such as pH adjustment, ion exchange, or periodic acidification to reduce scaling.

10. A chemical engineer designs a process to dissolve hydrogen gas into organic liquids under high pressure. Explain how Henry's Law quantitatively predicts gas solubility, how temperature variations influence the Henry's constant, and why solvent polarity, hydrogen bonding capability, and molecular size lead to differing solubility behaviors. Propose criteria for selecting an optimal solvent that maximizes gas dissolution while minimizing energy costs and safety risks.

11. Heavy metal ions remain suspended in a river for unusually long distances, associating with clay colloids. Describe how the negatively charged surfaces of

clay particles adsorb cations, how changes in ionic strength or the presence of competing ions affect sedimentation, and how environmental conditions such as pH, dissolved organic matter, and redox potential influence metal mobility and bioavailability in freshwater ecosystems.

12. A team designing next-generation battery electrolytes discovers that adding small amounts of polymeric additives dramatically alters ion mobility. Using solvation, polymer–ion interactions, and entropy-based constraints, explain how polymer chains modify ionic transport, alter the solvation shell, and change viscosity. Discuss why excessive polymer concentration can hinder conductivity and propose optimal formulations for balancing stability, conductivity, and electrochemical performance.

13. Food scientists investigate why some salad dressings remain stable while others separate rapidly. Explain how emulsifiers reduce interfacial tension and form protective layers around oil droplets, how droplet size distributions influence creaming and coalescence rates, and why minor changes in pH, ionic strength, or temperature can destabilize emulsions. Suggest formulation strategies to maximize shelf-life without altering taste or texture.

14. A clinical team observes that red blood cells in hypertonic solutions shrink and lose functionality. Explain the molecular mechanism of osmotic water flow, the resulting effects on membrane tension and cytoskeletal structure, and discuss how careful control of solution osmolarity in intravenous fluids prevents hemolysis while maintaining physiological compatibility.

15. A graduate student studies nanoparticle-based inks for printable electronics. Increasing nanoparticle loading results in non-linear viscosity increases and shear-thinning behavior. Using colloidal rheology, describe how interparticle forces, clustering, and microstructure influence macroscopic flow, and discuss design considerations for optimizing printability, stability, and electrical performance of high-concentration suspensions.

16. A catalyst manufacturer produces colloidal platinum nanoparticles for hydrogenation. Over time, activity decreases due to particle aggregation. Explain how ligand desorption, Ostwald ripening, and van der Waals interactions contribute to loss of surface area, and propose strategies including surface functionalization,

polymeric stabilizers, or controlled ionic environments to preserve catalytic performance over extended use.

17. A lake undergoing algal bloom treatment is dosed with aluminum sulfate, causing suspended organic colloids to flocculate and settle. Explain the mechanism by which alum neutralizes surface charges, promotes floc formation, and reduces turbidity. Discuss potential pitfalls, such as pH dependence and aluminum speciation, that may limit treatment efficacy or introduce environmental risks.
18. A polymer hydrogel designed for controlled drug release swells excessively in physiological fluid, accelerating drug diffusion. Using concepts of osmotic swelling pressure, polymer–solvent interactions, and network elasticity, explain the causes of swelling, the role of crosslink density, and how controlled polymer design could tailor release kinetics without compromising mechanical integrity.
19. A student studying the Tyndall effect observes that colloids scatter light differently depending on particle size. Explain how Rayleigh and Mie scattering mechanisms depend on particle size relative to wavelength, why smaller particles preferentially scatter shorter wavelengths, and how this principle can be used to characterize colloidal suspensions quantitatively.
20. A beverage company struggles to retain CO_2 in sparkling drinks shipped to high-altitude regions. Explain how decreased atmospheric pressure reduces gas solubility, why this leads to rapid degassing, and suggest packaging design, bottle geometry, or pressurization strategies to maintain carbonation during transport.
21. A cryobiologist studies the effects of water-soluble cryoprotectants on cellular freezing. Explain how colligative properties, hydration shell formation, and reduced molecular mobility depress freezing points, prevent intracellular ice formation, and promote vitrification. Compare solute effectiveness and propose rational criteria for cryoprotectant selection in biological preservation.
22. Environmental chemists find that acid rain increases the solubility of certain metals in groundwater. Discuss how protonation of mineral surfaces, metal–ligand complexation, and shifts in equilibrium solubility influence mobility, bioavailability, and potential ecological impact. Consider implications for remediation strategies.
23. A cosmetic chemist notices that lotion formulations become gritty at low temperatures due to wax crystallization within the emulsion. Explain how temperature-dependent

solubility, polymorphism, and nucleation kinetics contribute to crystal formation, and why surfactant layers at interfaces may modulate crystal growth. Propose formulation modifications to prevent undesirable texture changes.

24. A graduate student designing colloidal photonic crystals finds that particle polydispersity disrupts ordered assembly. Explain why uniform particle size is critical for entropy-driven self-organization, how interparticle potentials govern ordering, and suggest synthesis and purification strategies to reduce size variability and enhance photonic properties.

25. An industrial evaporator concentrates brine solutions and observes non-linear boiling-point elevation at high solute concentrations. Using colligative principles and ionic strength effects, explain deviations from ideal solution behavior, how ion pairing or activity coefficients modify vapor pressure, and discuss implications for evaporator design and energy efficiency.

26. A student attempting to recrystallize an organic compound observes oiling-out rather than crystal formation. Discuss how supersaturation, solubility, interfacial tension, and nucleation kinetics contribute to oil-phase formation, and propose solvent selection, temperature control, and seeding strategies to favor proper crystallization.

27. A biomedical researcher studies liposomes loaded with anticancer drugs and observes slow leakage over time. Explain how membrane fluidity, phase transitions, and osmotic gradients drive leakage, and discuss how lipid composition, cholesterol incorporation, and surface modifications could optimize retention and targeted release.

28. A fish hatchery experiences sudden mortality after heavy rainfall, coinciding with increased colloidal clay runoff. Explain how suspended colloids interfere with gas exchange, reduce light penetration, and affect photosynthetic oxygen production, and propose mitigation strategies to protect aquatic life during sedimentation events.

29. A ceramics manufacturer struggles to achieve uniform particle dispersion in clay slips, resulting in casting defects. Explain the influence of zeta potential, ionic strength, and organic dispersants on colloid stability, and suggest formulation or

processing adjustments to achieve homogeneity and improve final ceramic quality.

30. Protein folding studies reveal that some salts stabilize proteins while others destabilize them. Using hydration shell theory, Hofmeister series concepts, and solvation thermodynamics, explain how different ions influence protein stability, solubility, and folding pathways. Discuss implications for buffer selection and protein formulation.

31. Industrial inkjet printers clog because pigment colloids aggregate during drying at the printhead. Explain how solvent evaporation concentrates electrolytes, destabilizes electrostatic repulsion, and promotes aggregation, and propose strategies such as surfactant optimization or humidity control to maintain stable ink flow.

32. A researcher observes that nanoparticle suspensions remain visually stable but slowly sediment over months. Explain how gravitational settling, Brownian motion, and particle size distributions determine sedimentation rates, and how steric stabilization or viscosity modifiers may slow, but not fully prevent, settling.

33. A desalination plant experiences silica scaling even with low calcium concentrations. Explain how soluble silica polymerizes to colloidal gels, how pH and temperature accelerate polymerization, and why common antiscalants may fail. Discuss alternative scaling control strategies.

34. A student observes Ostwald ripening in emulsions over time. Explain how differences in Laplace pressure drive mass transfer from small to large droplets, why volatile oils exacerbate the effect, and how emulsifier type and concentration influence ripening rates.

35. A metallurgist studies brass dissolution in acidic baths for cleaning purposes. Explain how surface potentials, complexation reactions, and solute–solvent interactions influence dissolution kinetics, and why passivation layers can inhibit uniform etching.

36. A synthetic chemist dissolves a salt in water and observes an exothermic temperature rise. Explain why some dissolution processes release heat despite requiring solute–solute bond disruption, considering solvation enthalpies and lattice energies.

37. An atmospheric scientist studies hygroscopic aerosols that grow under humid conditions. Explain how water adsorption layers form on particle surfaces, how growth depends on relative humidity, and why this process influences cloud condensation nuclei and atmospheric processes.

38. A packaging scientist observes that hydrogels used for moisture control dry out in low-humidity warehouses. Explain how osmotic gradients drive water loss, how polymer composition affects water retention, and suggest material design strategies to maintain functional hydration under varying storage conditions.

39. A chemical engineer must select between aluminum-based and polymer-based coagulants for water treatment. Compare their charge-neutralization mechanisms, floc formation kinetics, effectiveness under different pH and ionic strength conditions, and potential limitations or environmental concerns.

40. A researcher observes depletion flocculation when nonadsorbing polymers are added to a colloid. Explain the entropic origin of depletion attraction, how polymer size relative to colloids governs strength, and discuss practical implications for formulation design in foods, pharmaceuticals, or coatings.

41. Developers of quantum-dot LEDs find that emission intensity decreases over time due to ligand desorption. Explain how surface chemistry affects electronic states, aggregation tendencies, and non-radiative pathways, and propose stabilization strategies such as ligand engineering or shell passivation.

42. A wastewater treatment plant experiences slower oil separation in winter. Explain how lower temperature increases viscosity, reduces Brownian motion, and alters interfacial tension, slowing coalescence. Propose operational adjustments or formulation modifications to improve separation efficiency.

43. Food scientists note that repeated freeze–thaw cycles in ice cream result in sandy texture. Explain how ice crystal growth outpaces stabilizer protection, disrupts colloidal fat structures, and suggest formulation or storage strategies to maintain smooth texture.

44. A pharmaceutical suspension rapidly forms a dense cake that resists redispersion. Explain the balance between gravitational, Brownian, van der Waals, and electrostatic forces that control suspension stability, and why some formulations

exhibit irreversible aggregation. Suggest methods for achieving long-term redispersible suspensions.

45. A soft-matter physicist observes shear-induced ordering in colloidal dispersions. Explain how applied flow aligns particles, reduces free energy, and creates anisotropic structures. Discuss implications for processing and material properties of colloidal gels or liquid crystals.

46. A geochemist studying opal formation observes that silica colloids slowly assemble into hydrated networks. Explain how polymerization, dehydration, and colloidal interactions govern the transition from sol to solid opal, and propose experimental approaches to control optical properties through particle assembly.

47. A nanofabrication researcher observes crack patterns in dried colloidal films. Explain how capillary forces, solvent evaporation, and differential mechanical stresses produce cracks, and discuss strategies—such as controlled drying rates or substrate modification—to minimize defects.

48. A biophysicist describes cytoplasm as a viscoelastic colloidal gel. Explain how macromolecular crowding, transient binding interactions, and hydration layers contribute to mechanical properties, and discuss implications for intracellular transport and cellular mechanics.

49. Environmental scientists investigate why microplastics remain suspended in seawater for decades. Explain the roles of particle density, biofilm formation, surface charge, and aggregation dynamics in long-term suspension, and suggest environmental factors that could accelerate sedimentation or removal.

50. A chemical engineer finds that adding surfactants to a foam sometimes destabilizes it instead of enhancing stability. Explain how surfactant molecular structure influences film elasticity, drainage, and bubble coalescence, and why higher surfactant concentration does not necessarily improve foam stability due to Marangoni effects or micelle formation.

Chapter 13

Adventure of Kinetic Reactions

Chemical reactions unfold across an extraordinary landscape of molecular possibilities, each process governed by thermodynamic favorability and kinetic accessibility. Spontaneity sets the direction in which nature tends to move, but kinetics determines whether motion along that path is meaningfully observable on human timescales. Thus, any rigorous study of chemical change demands a dual focus: on the energetic destination of a transformation and on the dynamic journey by which reactants traverse the energy landscape to become products. This interplay becomes especially vivid when one considers reactions such as diamond transforming into graphite—spontaneous yet imperceptibly slow—or the photochemical splitting of water, a nonspontaneous reaction that can be explosively fast when supplied with sufficient energy. The subtlety of this distinction drives chemists to develop quantitative frameworks that go well beyond simple thermodynamic predictions, probing instead the molecular choreography underlying reaction rates. Kinetics therefore becomes the bridge between the theoretical potential for change and the practical realization of chemical transformation, enabling fields as diverse as atmospheric chemistry, metabolic engineering, and materials synthesis to predict, manipulate, and optimize chemical pathways with ever-increasing precision.

At the most foundational level, kinetics is anchored in the measurement of reaction rate—the velocity at which a chemical system evolves. Concentration serves as the natural coordinate for tracking progress, reflecting the relative abundance of species in solution or gas phase. The time-dependent evolution of concentration allows the rate to be expressed as $\Delta[A]/\Delta t$, capturing either the disappearance of reactants or the appearance of products. Yet the deceptively simple algebra masks a wealth of complexity. As reactions proceed, the molecular environment shifts; concentrations fall, intermediates accumulate, and competing pathways may emerge. Consequently, the instantaneous rate, defined as the slope of the tangent to the concentration–time curve at a given moment, often provides the most accurate snapshot of system behavior. Because early times minimize complications from reverse reactions and product inhibition, initial-rate methods serve as powerful tools for determining how rate depends on reactant concentration. Such experiments reveal critical information encoded in the rate law, typically written as $\text{Rate} = k[A]^n[B]^m\ldots$, where the exponents reflect reaction orders determined experimentally rather than derived from stoichiometry. These orders provide glimpses into the sequence of molecular encounters guiding the reaction forward, enabling inference of mechanistic features that would remain hidden without kinetic evidence.

Differential rate laws describe how concentration changes with time based on the instantaneous rate, but integrated rate laws reveal the trajectory of concentrations as a reaction unfolds. Their forms differ markedly across reaction orders, reflecting fundamental differences in molecular behavior. First-order processes exhibit exponential decay, ubiquitous in natural systems because the rate is directly proportional to the amount of material present. Radioactive isotopes, pharmacokinetic drug clearance, atmospheric pollutant degradation, and neuronal neurotransmitter breakdown all follow first-order kinetics, with half-lives that remain constant regardless of initial amounts. This built-in self-similarity of exponential decay underpins many biological and geological dating methods, lending remarkable utility to first-order kinetic models. In contrast, second-order reactions amplify the effect of concentration changes: as reactants become scarce, the likelihood of productive collisions diminishes dramatically. The mathematical consequence is a hyperbolic decay, with half-lives extending progressively as the reaction proceeds. This behavior becomes pronounced in bimolecular reactions such as radical recombination, dimerization, or nucleophilic substitution under certain conditions. Zero-order kinetics, on the other hand, emerge in saturated systems—ones in which reaction sites or catalytic surfaces become fully occupied. Here, the rate becomes independent of reactant concentration entirely, a phenomenon frequently observed in enzymatic catalysis under high substrate load, photochemical reactions under constant irradiation, and heterogeneous catalytic systems in which all active sites are engaged. Such distinctions in kinetic form are not mere mathematical curiosities; they encode essential information about molecular interactions and constraints governing the reaction environment.

The concept of reaction mechanism brings the kinetic picture into sharper resolution. While balanced chemical equations succinctly indicate the net transformation occurring, they veil the intricate parade of fleeting species and elementary processes that connect the starting materials to the final products. Reaction mechanisms dissect these transformations into elementary steps whose molecularity—unimolecular, bimolecular, or occasionally termolecular—describes the number of molecules involved in each step. Unimolecular steps often reflect processes such as internal rearrangements, bond dissociation, or isomerization, where a single molecule undergoes transformation. Bimolecular steps involve direct encounters between two entities, making them sensitive to concentration, diffusion, and orientation. Termolecular steps, though theoretically possible, are exceedingly rare in the gas phase and virtually nonexistent in the condensed phase due to the improbability of three-body collisions with proper spatial and energetic alignment. Mechanistic hypotheses must satisfy two stringent criteria: they must reproduce the overall stoichiometry, and they must predict rate laws consistent with experimental data. These conditions allow rate laws to act as molecular signatures, helping chemists discriminate between plausible mechanistic pathways. Moreover, the identification of the rate-determining step—a uniquely slow elementary process—offers crucial explanatory power by linking the bottleneck of the mechanism to the observed kinetic behavior. Sophisticated techniques such as spectroscopy, mass spectrometry, isotopic labeling, transient absorption, and computational quantum chemistry enhance mechanistic understanding, revealing intermediates and transition states that exist for only femtoseconds yet govern macroscopic measurables.

Underlying mechanistic behavior is the fundamental question of why reactions have the rates they do, which leads directly to collision theory and transition-state theory. Collision theory provides an intuitive picture: reactions occur when molecules collide with sufficient energy and in the proper orientation. It emphasizes three factors: collision frequency z, the steric factor p describing orientation probability, and the fraction of collisions possessing energy exceeding the activation barrier. While simplified, this theory captures important qualitative trends: increasing temperature accelerates molecular motion, increasing the number of energetic collisions; raising concentrations enhances collision frequency; and molecular geometry influences the likelihood of productive encounters. However, it soon becomes evident that collision theory oversimplifies molecular behavior by treating molecules as hard spheres. Transition-state theory (TST) extends the picture by invoking a high-energy activated complex—also known as the transition state—formed transiently when reactants reach the crest of the potential energy barrier. This transition state sits at a saddle point on the multidimensional potential energy surface describing the system, representing a configuration in which old bonds are partially broken and new bonds partially formed. The rate constant k is then expressed through the Eyring equation, $k = (k^B T/h)e^\wedge(-\Delta G^{\ddagger}/RT)$, where $\Delta G^{\ddagger}$ is the Gibbs free energy of activation. TST integrates thermodynamics and kinetics by highlighting contributions from enthalpy and entropy to the activation barrier. A large positive $\Delta H^{\ddagger}$ slows reactions because more energy is needed to reach the activated complex, while $\Delta S^{\ddagger}$ characterizes ordering or disordering in the transition state: a negative activation entropy suggests a more ordered transition configuration, typical of bimolecular association, whereas a positive activation entropy indicates an expanded or dissociative transition state. This nuanced decomposition gives chemists powerful levers for interpreting reaction behavior, particularly in complex environments such as enzymes, atmospheric aerosols, and solid catalytic surfaces.

The temperature dependence of reaction rates is elegantly described by the Arrhenius equation, which expresses the rate constant as $k = Ae^\wedge(-E_a/RT)$. Here, E_a embodies the activation energy barrier, while A captures steric and collision factors that reflect molecular geometry, frequency of encounters, and reaction pathway probabilities. When the natural logarithm is taken, $\ln(k) = -E_a/R\ (1/T) + \ln(A)$, enabling experimental determination of E_a by plotting ln(k) versus 1/T. The slope yields $-E_a/R$, while the intercept furnishes ln(A). Such linear Arrhenius plots are characteristic of elementary reactions; deviations from linearity suggest changes in mechanism, phase transitions, or complex temperature-dependent effects such as tunneling or diffusion limitation. Quantum tunneling, particularly relevant for reactions involving light atoms like hydrogen, allows particles to penetrate energy barriers rather than surmounting them, leading to rates that exceed classical predictions, especially at low temperatures. This phenomenon is critical in biological proton transfer, low-temperature combustion processes, cryogenic chemistry, and astrochemical reactions occurring on interstellar dust grains. These subtleties demonstrate that while the Arrhenius equation provides a compelling first approximation, the deeper reality of reaction kinetics is governed by the interplay of quantum mechanical constraints and thermal activation.

Catalysis introduces a profound dimension to kinetic control by offering alternative reaction pathways with lower activation energies. A catalyst operates by stabilizing the transition state or by providing a sequence of energetically favorable steps, thereby increasing the fraction of molecules able to react at a given temperature. Importantly, catalysts affect kinetics but leave thermodynamic equilibria unchanged, as they lower both forward and reverse activation energies equally. Catalysts can be classified broadly into homogeneous and heterogeneous types. Homogeneous catalysts share the same phase as the reactants, enabling intimate molecular interaction and often allowing fine control over reaction selectivity. Acid-base catalysis, organometallic complexes such as Wilkinson's catalyst, and enzymatic catalysis fall into this category. Heterogeneous catalysis, in contrast, operates at interfaces—typically solid surfaces upon which gas or liquid reactants adsorb, react, and desorb. Metals such as platinum, palladium, nickel, and rhodium play pivotal roles in surface-mediated hydrogenation, dehydrogenation, cracking, and pollution abatement processes. These catalysts provide adsorption sites that weaken specific bonds, align reactants in favorable geometries, and facilitate stepwise transformations that would be improbable in solution. Reaction rates in heterogeneous catalysis often depend on surface coverage, diffusion rates, and the availability of active sites, leading to complex rate laws described by Langmuir–Hinshelwood or Eley–Rideal models. Such models incorporate adsorption equilibria and surface reaction kinetics, revealing how saturation, competitive adsorption, and site heterogeneity influence overall rates.

Enzymes deserve special attention as biological catalysts of extraordinary specificity and efficiency. They operate under physiological conditions, enabling life to maintain chemical order in the face of thermodynamic tendencies toward disorder. Enzymes achieve rate enhancements of up to 10^{17} relative to uncatalyzed reactions, an astonishing capacity resulting from precise active-site architecture, transition-state stabilization, cofactor assistance, and dynamic conformational changes. Michaelis–Menten kinetics, one of the earliest quantitative models of enzyme behavior, portrays the catalytic cycle as a sequence in which the enzyme binds substrate to form an enzyme–substrate complex, converts it to product, and regenerates free enzyme. The resulting rate law, $v = V_{max}[S]/(K_m + [S])$, captures the characteristic saturation behavior of enzymes, with Km representing the substrate concentration at which the reaction proceeds at half-maximal velocity. More sophisticated kinetic descriptions incorporate cooperativity, allosteric regulation, product inhibition, and enzyme activation, reflecting the complexity of metabolic networks. Enzyme kinetics is inseparable from thermodynamic and mechanistic considerations: the rate depends on activation free energy barriers, conformational landscapes, and the subtle interplay between entropy and enthalpy in the transition state. These principles extend far beyond biology, informing fields such as medicinal chemistry, protein engineering, synthetic biology, and biomimetic catalyst design.

To contextualize the theories of kinetics within the broader framework of chemical reactivity, it is essential to examine the role of energy landscapes—multidimensional surfaces describing potential energy as a function of molecular geometry. Reactants navigate this landscape through a combination of thermal motion, collisions, and internal vibrations. Valleys represent stable intermediates, while ridges and saddle points correspond to transition states. Reaction pathways trace minimum-energy routes across this terrain, shifting with changes in temperature, pressure, solvent polarity, or catalytic environment. Saddle points on these surfaces define activation barriers whose heights control the rate and whose locations dictate the sensitivity to external perturbations. Computational chemistry allows visualization and quantitative evaluation of these landscapes through methods such as density functional theory (DFT), ab initio calculations, and hybrid quantum mechanics/molecular mechanics (QM/MM) techniques. These approaches reveal the microscopic features governing macroscopic rate behavior, illuminating why certain pathways dominate, how substituents influence reactivity, and how environmental factors reshape mechanisms.

In condensed phases, the solvent becomes an integral part of the energy landscape, providing stabilization through dielectric interactions, hydrogen bonding, and dynamic fluctuations that can either facilitate or hinder reaction progress. Solvent reorganizational energy contributes to the activation barrier, as highlighted in Marcus theory, which revolutionized understanding of electron-transfer kinetics. Marcus theory predicts that electron transfer rates depend on the free-energy driving force and the reorganization energy required to rearrange both reactants and solvent. The resulting parabolic dependence leads to an "inverted region" where increasing thermodynamic driving force paradoxically slows the reaction—a striking reminder that kinetic behavior often defies simple intuition.

Diffusion introduces another layer of complexity, especially in solution-phase and biological systems. When reactants must migrate through a medium to encounter each other, the rate of diffusion can limit the overall reaction rate. In such diffusion-controlled reactions, rate constants approach the upper limits allowed by Brownian motion, as described by the Smoluchowski equation. These reactions display near-instantaneous kinetics once molecules encounter each other, making the diffusion process itself the bottleneck. Enzyme catalysis sometimes operates near this limit, leading to the concept of "catalytic perfection," in which enzymes accelerate reactions to the threshold at which molecular encounters, rather than chemical barriers, dominate the timescale. Conversely, reactions taking place within crowded cellular environments often experience diffusion hinderance, as molecular motion is restricted by macromolecular crowding, cytoskeletal structures, and compartmentalization. These constraints lead to fractal diffusion, anomalous kinetics, and spatial heterogeneity that challenge classical kinetic models. To account for these effects, stochastic simulations, molecular dynamics, and spatially resolved kinetic frameworks are increasingly employed, highlighting the growing importance of computational approaches in modern chemical kinetics.

Kinetics also exerts a profound influence on reaction selectivity and product distribution. In many systems, thermodynamics predicts the most stable product, but kinetics determines the actual outcome when multiple pathways compete. This distinction becomes evident in scenarios exhibiting kinetic vs. thermodynamic control. Under kinetic control, the product forming fastest dominates, even if it is not the most stable. Under thermodynamic control, the product with the lowest free energy prevails, provided the system is allowed to equilibrate. Temperature, solvent, catalysts, and reaction time all modulate which regime governs the outcome. Organic pericyclic reactions, isomerizations, and carbocation rearrangements exemplify systems in which kinetic and thermodynamic products differ significantly. Similarly, polymerization reactions require careful kinetic management to control molecular weight distribution, branching, and stereoregularity. Techniques such as living polymerization rely on precisely balanced propagation and termination kinetics to yield polymers with narrow molecular weight distribution and well-defined architectures. Industrial chemical processes likewise depend critically on kinetic optimization to maximize yield, minimize byproducts, and ensure environmental and economic sustainability.

In heterogeneous catalysis, kinetics intersect with surface science, materials chemistry, and fluid mechanics. Catalytic behavior depends not only on the intrinsic kinetics of surface reactions but also on mass transport to and from catalytic sites, heat transfer, pore diffusion, and surface reconstruction. Real catalysts often exhibit complex surface structures with terraces, edges, defects, and vacancies, each offering unique reactivity. Nanoparticles introduce quantum size effects, altered electronic densities, and enhanced surface-to-volume ratios that reshape catalytic activity. Reaction rates in porous catalysts can experience internal diffusion limitations, leading to effectiveness factors that quantify the fraction of catalytic potential actually utilized. Reaction engineering integrates these kinetic phenomena into reactor design, accounting for temperature gradients, concentration profiles, flow patterns, and catalyst deactivation. Microkinetic modeling, which constructs detailed mechanistic networks incorporating adsorption equilibria, elementary rate constants, and surface coverage, provides a bridge between molecular-scale understanding and reactor-scale optimization. Such models guide the design of catalytic converters, ammonia synthesis reactors, petrochemical processes, and fuel cells, underscoring the centrality of kinetics to industrial innovation.

Thermodynamics and kinetics converge in complex chemical systems that display oscillatory behavior, autocatalysis, or feedback regulation. Chemical oscillators such as the Belousov–Zhabotinsky reaction reveal that far-from-equilibrium dynamics can produce rhythmic concentration changes, traveling chemical waves, and pattern formation reminiscent of biological morphogenesis. Autocatalytic systems, in which products catalyze their own formation, exhibit sigmoidal kinetic profiles, burst behavior, and bistability. These phenomena emerge from nonlinear kinetic laws and feedback loops that amplify or dampen fluctuations. In biological systems, metabolic pathways harness such nonlinear kinetics to maintain homeostasis, respond to stimuli, and coordinate cellular processes. Gene regulatory networks rely on feedback-driven kinetics to control enzyme expression, metabolic flux, and developmental patterning. Systems chemistry extends these concepts to synthetic networks designed to mimic biological complexity, demonstrating how intricate behaviors arise from simple kinetic rules governed by energy flow and molecular recognition.

Electrochemical kinetics represents yet another domain in which reaction rate theories merge with electron transfer, ion migration, and interfacial phenomena. At electrode surfaces, reaction rates depend on overpotential, surface adsorption, electric double-layer structure, and charge-transfer resistance. The Butler–Volmer equation relates the rate of electrochemical reactions to these parameters, providing fundamental insight into batteries, fuel cells, corrosion, and electrodeposition. Tafel analysis enables extraction of kinetic parameters such as exchange current densities and transfer coefficients, shedding light on mechanisms of charge transfer. The coupling between mass transport and electron transfer leads to mixed kinetic-diffusion control, captured by the Nernst–Planck equations and the concept of limiting current. Modern electrochemical systems increasingly rely on nanoscale electrode modifications, redox mediators, and electrocatalysts that lower activation barriers, enhance selectivity, and resist degradation under operational conditions.

Across all these systems, chemical kinetics emerges not merely as a descriptive tool but as a predictive and transformative science. It connects microscopic molecular behavior to macroscopic observables, bridging quantum mechanics, statistical thermodynamics, and continuum modeling. It offers a lens through which to interpret natural phenomena—from atmospheric ozone depletion and combustion dynamics to enzymatic regulation and geochemical cycles. It also empowers technological advancement, guiding the development of energy storage systems, pharmaceuticals, sustainable manufacturing processes, polymeric materials, and environmental remediation strategies. The breadth of kinetic phenomena reflects the richness of chemical reactivity itself: dynamic, multidimensional, and profoundly sensitive to both molecular structure and environmental context.

In conclusion, spontaneity outlines whether a reaction *can* occur, but kinetics determines whether it *does* occur on any meaningful timescale. By integrating rate laws, mechanistic analysis, transition-state theory, thermodynamic principles, and catalytic strategies, chemical kinetics provides the quantitative and conceptual frameworks essential for understanding the tempo of chemical change. Zero-, first-, and second-order behaviors each embody distinct mechanistic principles; activation energies and temperature dependencies reveal the energy landscape that molecules must traverse; and catalysts offer pathways for controlling rates without altering thermodynamic constraints. Whether examining biological enzymes, industrial reactors, atmospheric processes, or complex reaction networks, kinetics remains indispensable to predicting and controlling chemical behavior. Through its insights, chemists translate possibility into practice, harnessing the laws of molecular motion to shape the natural and technological world.

A deeper appreciation of chemical kinetics emerges when one frames reaction rates within the broader statistical landscape of molecular motion. Although rate laws and transition-state theory provide phenomenological and mechanistic descriptions, the true origin of kinetic behavior lies in the statistical distribution of molecular energies described by the Maxwell–Boltzmann distribution. This distribution reflects the probability that molecules possess certain translational energies at a given temperature, and thus determines the fraction of collisions that can surmount the activation barrier. At higher temperatures, the tail of the distribution lengthens, allowing a greater proportion of molecules to exceed E_a—even if the average energy increases only modestly. This statistical perspective clarifies why small temperature increases can produce disproportionately large increases in reaction rate, and why low-temperature reactions often depend on quantum mechanical phenomena such as tunneling to proceed at appreciable speeds. In dense phases such as liquids and solids, additional complexity arises from vibrational and rotational motions, solvent-solute interactions, and the energy redistribution that follows molecular collisions. These effects necessitate treating reactions not as isolated encounters but as collective phenomena shaped by the ensemble behavior of billions of molecules, each sampling the potential energy landscape according to thermodynamic and kinetic rules.

Extending these ideas to reactions in complex solvents underscores the role of solvation dynamics in shaping kinetic pathways. Solvent molecules surround reactants and transition states, forming solvation shells that reorganize as the reaction progresses. This reorganization carries energetic cost reflected in the solvent's reorganization energy, a key parameter in Marcus theory. When solvent dynamics become slow relative to chemical timescales, solvation can no longer reach equilibrium instantaneously, leading to non-equilibrium solvation effects. These effects can accelerate or decelerate reactions depending on how well the solvent stabilizes the transition state compared to reactants. Polar solvents often enhance the rate of reactions involving charge separation by stabilizing polar transition states, whereas nonpolar environments may favor pathways involving radical intermediates or concerted rearrangements. Hydrogen-bonding networks introduce yet another layer of complexity, enabling proton-coupled electron transfer, cooperative activation of substrates, and solvent-mediated catalysis. Ultrashort laser spectroscopy has revealed that solvent reorganization occurs on femtosecond to picosecond timescales, implicating solvation dynamics as an active participant in the kinetic process rather than a passive background feature. In biological systems, structured water molecules and confined hydration layers can dramatically modify reaction rates within enzymes, ion channels, and cellular membranes, demonstrating that the environment is inseparable from the kinetic behavior it shapes.

Non-equilibrium thermodynamics offers a powerful framework for understanding reactions that operate far from equilibrium—conditions common in living systems, atmospheric reactions, and driven chemical networks. In these regimes, reaction rates are influenced not only by thermodynamic driving forces but also by continuous energy input, gradients, or dissipative processes. Systems maintained far from equilibrium can exhibit phenomena such as chemical oscillations, bistability, and autocatalytic cycles, each arising from feedback loops embedded in the kinetic structure of the reaction network. Dissipative self-organization, in which ordered structures emerge through the continuous consumption of energy, allows chemical systems to display behaviors reminiscent of biological function, including signal amplification, adaptation, and homeostasis. For instance, autocatalytic reaction networks can amplify small fluctuations into macroscopic chemical waves, while feedback inhibition stabilizes oscillatory cycles. In prebiotic chemistry, such far-from-equilibrium dynamics may have played a crucial role in the emergence of primitive metabolic cycles, templating processes, and early chemical information systems. Insights from non-equilibrium kinetics thus connect fundamental physical principles to the origins of life and the behavior of complex adaptive systems.

Mechanochemistry—reactions driven or accelerated by mechanical force—provides another frontier in modern kinetics. Traditionally, chemical reactions are driven by thermal energy, photons, electrons, or catalysts, but mechanical forces can dramatically alter potential energy surfaces by distorting bonds and lowering activation barriers. In bulk mechanochemical systems such as ball milling, grinding induces localized heating, high pressures, and shear forces that can break covalent bonds, initiate polymerizations, or drive solid-state transformations without solvent. At the molecular scale, single-molecule force spectroscopy reveals that applying tension to a bond modifies its energy landscape, effectively "tilting" the potential energy surface and enabling selective activation of specific reaction pathways. This principle has practical implications for polymer mechanochemistry, where mechanophores—molecular motifs that undergo predictable reactions when stretched—allow stress-responsive materials to change color, self-heal, or release cargo upon mechanical deformation. In biological systems, mechanochemical kinetics is central to processes such as protein unfolding, cellular adhesion, mechanotransduction, and motor protein dynamics. These examples illustrate how mechanical energy provides an orthogonal means of controlling reaction rates, expanding the scope of kinetic manipulation beyond traditional thermal and catalytic routes.

Advances in computational chemistry have revolutionized the study of kinetics by enabling prediction of reaction mechanisms, transition states, and rate constants with unprecedented accuracy. Density functional theory (DFT), molecular dynamics (MD), and hybrid QM/MM methods allow researchers to map potential energy surfaces, estimate activation barriers, and simulate reaction trajectories under realistic conditions. Machine learning models trained on high-level quantum mechanical data can now predict kinetic parameters for vast chemical spaces, accelerating catalyst design, materials discovery, and synthetic route optimization. Transition path sampling, a computational technique that focuses on rare but critical events such as barrier crossing, provides insight into the statistical nature of reactions that occur infrequently yet dominate system behavior. These methods reveal that reaction pathways often involve multiple competing trajectories rather than a single well-defined route, reflecting the inherent stochasticity of motion on high-dimensional energy landscapes. Enhanced sampling techniques, including metadynamics and accelerated MD, further enable exploration of kinetics in complex environments such as solvated clusters, biomolecular assemblies, and heterogeneous catalytic surfaces. As simulations become increasingly predictive, they complement experimental kinetics, allowing hypotheses to be tested in silico before being validated in the laboratory.

Together, these extended perspectives demonstrate that chemical kinetics is not merely the study of how fast reactions occur, but a multidisciplinary science connecting molecular dynamics, statistical mechanics, quantum theory, and complex systems behavior. Whether examining ultrafast reactions in photochemistry, stress-induced bond rupture in materials, charge-transfer kinetics in electrochemical devices, or nonlinear reaction networks in biology, kinetics provides the language and conceptual tools to describe the temporal evolution of chemical systems. By embracing the full richness of energy landscapes, environmental influences, and non-equilibrium processes, chemists gain the ability not only to understand existing reactions but also to design new pathways, construct adaptive chemical systems, and manipulate matter across scales. In this way, kinetics stands as a cornerstone of modern chemical science, transcending its origins in rate measurement to become a powerful lens for interpreting and controlling the dynamic behavior of the molecular world.

Chapter 13: Word Problems

1. A photochemical reaction converts A → B under constant-intensity irradiation. Experiments reveal that at high [A], the reaction rate becomes independent of [A], while at low [A], it exhibits first-order kinetics. Design a mechanistic scheme consistent with these observations, considering excited-state formation, intersystem crossing, and quenching by solvent or impurities. Derive the rate law using steady-state approximation for the excited species, and predict how the reaction rate would change if the photon flux were doubled. Discuss how factors such as light absorption cross-section, quantum yield, and the presence of triplet quenchers could modify observed kinetics. Suggest an experimental setup using a laser flash photolysis system to measure the lifetime of the excited state and its correlation with observed rates.

2. A unimolecular decomposition X → Y occurs in the gas phase, and Arrhenius plots at three different temperatures show downward curvature at low temperatures, indicating deviations from classical behavior. Explain how quantum tunneling can account for this curvature and propose a modified Arrhenius rate expression that

incorporates a tunneling correction factor. Using hypothetical activation energies and pre-exponential factors, calculate the tunneling contribution at 200 K and 300 K. Discuss how the width and height of the energy barrier influence tunneling probability, and explain why classical transition-state theory fails to predict low-temperature rates in this system. Suggest experimental strategies, such as isotopic substitution or cryogenic measurements, to validate the presence of tunneling effects.

3. During radical chain polymerization of monomer M via a chain-growth mechanism, the propagating radical concentration remains nearly constant, but the molecular weight distribution broadens unexpectedly at high pressures. Analyze how diffusion-limited termination could produce this effect, taking into account increased solution viscosity and reduced radical mobility. Quantitatively predict how the rate of termination changes with viscosity using the Smoluchowski equation for diffusion-controlled reactions. Consider how chain length and radical size influence diffusion rates, and propose an experiment to measure radical mobility using pulsed-laser ESR spectroscopy. Discuss the implications for controlling polymer molecular weight distribution in industrial polymerization processes.

4. An enzymatic reaction follows Michaelis–Menten kinetics, but product inhibition becomes significant at moderate conversion. Experiments show that the presence of product P doubles the apparent K_m without altering V_{max}. Identify the type of inhibition (competitive, noncompetitive, uncompetitive), derive the modified rate law incorporating product concentration, and illustrate how Lineweaver–Burk or Eadie–Hofstee plots would reveal this mechanism. Propose an experiment to determine the inhibition constant K_i using initial rate measurements and simulate hypothetical data. Discuss how the inhibition mechanism affects enzyme efficiency in metabolic pathways and how it could be modulated pharmacologically.

5. A reaction between two ions in aqueous solution exhibits negative activation entropy and slightly negative activation enthalpy. Using transition-state theory and considering solvent reorganization, interpret these results in terms of ordered transition-state complexes and solvent cage effects. Predict how the reaction rate would change in a less polar solvent, and justify your reasoning using both enthalpic and entropic arguments. Design a series of experiments where the dielectric constant and viscosity of the solvent are systematically varied, and

propose data analysis methods to extract activation parameters. Discuss how ion pairing and hydrogen-bonding networks could further influence observed kinetics.

6. In a catalytic hydrogenation, H_2 dissociatively adsorbs on a metal surface before reacting with an adsorbed alkene. It is observed that increasing H_2 pressure beyond a certain point decreases the overall rate of hydrogenation. Propose a Langmuir–Hinshelwood mechanism that accounts for this surface inhibition by H_2, derive the rate law, and predict surface coverages of H_2 and alkene at various pressures. Suggest an experimental protocol using temperature-programmed desorption (TPD) and steady-state kinetic measurements to validate the proposed model. Discuss how the metal type, surface defects, and particle size could modify observed inhibitory effects.

7. A reaction network contains an autocatalytic step $A + B \rightarrow 2B$ and a slower consumption step $B \rightarrow C$. Show how these two reactions can produce sigmoidal kinetics in B. Derive conditions under which bistability occurs, and explain how the initial concentrations of A and B determine which steady state is reached. Discuss potential applications of such autocatalytic networks in synthetic biology or oscillatory chemical systems. Suggest experimental methods to measure transient concentrations of B using fluorescence or NMR spectroscopy and simulate hypothetical kinetic curves under different initial conditions.

8. A gas-phase isomerization proceeds via a high-energy intermediate I*. Using RRKM theory, derive the microcanonical rate constant k(E) and discuss how the density of states influences the overall reaction rate. Explain how collisional energy transfer affects falloff behavior at intermediate pressures and predict the transition from high-pressure limit to low-pressure limit rates. Propose a master-equation approach to model pressure-dependent kinetics and suggest experimental setups to measure energy-resolved rate constants, such as time-resolved mass spectrometry.

9. A bimolecular reaction between two neutral molecules shows second-order behavior at high dilution but first-order behavior at high concentration. Propose a mechanistic explanation involving reversible dimer formation or solvent cage effects, and derive a composite rate law that accounts for both regimes. Discuss how temperature and solvent viscosity could shift the crossover between regimes.

Design an experiment using UV–Vis spectroscopy or NMR to monitor monomer and dimer populations, and simulate kinetic curves for different initial concentrations and temperatures.

10. A solution-phase redox reaction follows Marcus kinetics. Given observed rate constants at varying thermodynamic driving forces $\Delta G°$, estimate the reorganization energy λ and predict the presence of an inverted region where the rate decreases with increasing driving force. Explain why classical transition-state theory alone cannot capture this behavior. Suggest an experimental design using cyclic voltammetry with varying electrode potentials and solvent polarity to validate Marcus predictions, and discuss the implications for electron-transfer reactions in biological and electrochemical systems.

11. A protein-folding reaction monitored by fluorescence shows two distinct kinetic phases: an initial rapid collapse followed by a slower structural rearrangement. Model this as a sequential kinetic process involving an intermediate state I, derive time-dependent expressions for [I] and [folded protein], and explain how stopped-flow or temperature-jump experiments could validate this mechanism. Discuss how environmental factors such as pH, ionic strength, and crowding agents might alter the relative rates of collapse and rearrangement. Propose a series of hypothetical kinetic traces under varying temperatures to extract activation parameters for each phase, and suggest how single-molecule FRET could complement bulk measurements.

12. A ligand substitution reaction in a transition-metal complex exhibits rate saturation at high incoming ligand concentrations. Compare associative (A) and dissociative (D) mechanisms, derive the corresponding rate laws, and use hypothetical kinetic data to determine which pathway predominates. Include discussion of how the metal's oxidation state, ligand field strength, and steric effects influence the mechanism. Design an experiment using stopped-flow UV–Vis spectroscopy to measure transient intermediates, and explain how temperature variation and isotopic labeling could further distinguish between associative and dissociative pathways.

13. A heterogeneous catalyst deactivates over time due to irreversible site poisoning by impurity Q. Derive a time-dependent rate law for the catalytic reaction,

considering the fraction of active sites blocked by Q, and predict how the apparent lifetime of the catalyst depends on [Q]. Propose regeneration strategies such as thermal treatment or chemical washing, and discuss how surface characterization techniques (XPS, TEM) could be used to monitor site occupancy. Simulate kinetic curves under different impurity concentrations, and analyze how turnover frequency and selectivity degrade over time.

14. A bimolecular reaction exhibits a kinetic isotope effect (KIE) of $k_H/k_D = 8$ at room temperature. Explain why such a large KIE suggests tunneling contributions, and show why classical transition-state theory underestimates this effect. Predict how the KIE changes at lower temperatures and for isotopes with greater mass. Suggest experimental designs using isotopically labeled reactants and temperature-dependent rate measurements to quantify tunneling contributions. Discuss how multidimensional tunneling models and semiclassical corrections could improve agreement between theory and experiment.

15. A reaction occurring in micelles shows a change in apparent order with respect to substrate as surfactant concentration increases. Using pseudo-phase ion-exchange (PPIE) models, derive expressions relating observed rate constants to micellar binding constants and micelle aggregation number. Explain how substrate partitioning between aqueous and micellar phases alters kinetics, and propose experiments using fluorometric probes or NMR to measure binding constants. Simulate kinetic curves for low versus high surfactant concentrations and discuss implications for designing micellar catalysts or drug-delivery systems.

16. A reaction in supercritical CO_2 occurs significantly faster than in liquid CO_2 under otherwise identical conditions. Analyze how compressibility, density fluctuations, and solvent-cage effects in the supercritical phase alter both activation entropy and diffusional encounter rates. Derive expressions for diffusion-limited rates incorporating supercritical density fluctuations, and suggest hypothetical kinetic data comparing liquid, supercritical, and near-critical CO_2. Discuss how solvent polarity, pressure, and temperature could be tuned to optimize reaction rates in supercritical fluids.

17. In atmospheric chemistry, the reaction of NO with O_3 produces oscillatory behavior when coupled with NO_2 photolysis under sunlight. Formulate a minimal kinetic model

incorporating both the photolytic and bimolecular reactions, determine conditions for limit-cycle oscillations, and describe how photon flux, rate constants, and initial concentrations affect the oscillatory period. Propose laboratory experiments using flow reactors and actinometry to monitor oscillatory concentrations, and discuss implications for atmospheric ozone depletion and smog formation.

18. A solid-state reaction between two powders follows shrinking-core kinetics. Derive integrated rate expressions for diffusion-limited, reaction-interface-limited, and ash-layer diffusion-limited regimes. Using hypothetical time vs. conversion data, determine the rate-controlling step and analyze deviations caused by particle size distribution or porosity. Propose experimental approaches using in situ X-ray diffraction or Raman spectroscopy to monitor reaction progress, and discuss how heat and mass transfer limitations influence observed kinetics.

19. A catalytic cycle involves a spin-forbidden step requiring intersystem crossing (ISC). Explain how ISC influences the observed rate constant, derive a composite rate expression including both spin-allowed and spin-forbidden pathways, and predict how varying ligand-field strength or heavy-atom substitution could modulate the rate. Suggest experimental approaches using photoluminescence, transient absorption, or phosphorescence lifetimes to probe ISC contributions. Discuss how spin-orbit coupling affects selectivity and efficiency in catalytic transformations.

20. Radical chain chlorination of methane exhibits an induction period followed by sudden acceleration. Using the radical chain mechanism, derive the conditions under which radical concentration increases explosively and explain how chain branching influences the onset of runaway kinetics. Simulate hypothetical concentration-time profiles under varying light intensity, temperature, and initiator concentration. Suggest experimental techniques using laser flash photolysis or EPR spectroscopy to monitor radical formation in real time. Discuss implications for industrial safety and reaction control.

21. A temperature-jump experiment shows that the relaxation time τ varies nonlinearly with temperature. Model this system with coupled pre-equilibria preceding the rate-determining step, derive expressions for $\tau(T)$, and explain deviations from simple Arrhenius behavior. Propose hypothetical T-jump data and demonstrate how to extract rate constants and equilibrium constants. Discuss the impact of

solvent viscosity, ionic strength, and pressure on relaxation times, and suggest methods to distinguish coupled equilibria from sequential reaction intermediates.

22. A hydrolysis reaction exhibits general acid catalysis. Using Bronsted relationships, determine the catalytic coefficients (β) for a series of acids, interpret the slope α, and explain how it relates to charge distribution in the transition state. Propose a mechanistic model that rationalizes the observed α value, and design experiments with substituted acids or buffers to validate the model. Discuss how kinetic isotope effects or temperature-dependent measurements could provide additional insight into proton-transfer contributions.

23. A conformational isomerization in a biomolecule exhibits non-exponential kinetics. Show how a heterogeneous distribution of activation energies produces stretched-exponential behavior, and propose experimental observables to distinguish this from sequential intermediates. Suggest approaches using temperature-jump, pressure-jump, or single-molecule spectroscopy to probe the distribution of relaxation times. Discuss the implications of heterogeneity for protein folding landscapes, enzymatic dynamics, and molecular recognition.

24. An electron-transfer reaction at an electrode follows the Butler–Volmer equation. Using measured current–overpotential data, extract the transfer coefficient α and exchange current density i_0. Discuss how the double-layer structure, surface roughness, and electrode material could distort apparent kinetic parameters. Propose hypothetical voltammograms at varying temperatures and electrolyte compositions, and simulate the effect of changing α on the current response. Suggest experimental strategies to isolate charge-transfer kinetics from mass-transport limitations.

25. A two-step catalytic reaction A → I → P exhibits steady-state behavior for intermediate I only under specific conditions. Derive these conditions explicitly and determine how deviations from steady state manifest in concentration-time profiles. Propose hypothetical kinetic data showing pre-steady-state accumulation and decay of I. Discuss experimental approaches using rapid-mixing or stopped-flow techniques to observe intermediate concentrations, and analyze how changes in substrate concentration, catalyst loading, or temperature affect the steady-state approximation.

26. A mechanochemical reaction is accelerated by tension applied to a polymer chain. Using the Bell equation, model how the applied force modifies the activation free energy and predict the change in rate upon doubling the force. Discuss the conditions under which force could switch the reaction pathway or favor alternative mechanochemical products. Propose an experimental setup using atomic force microscopy or optical tweezers to measure force-dependent reaction rates. Simulate hypothetical kinetic traces under varying tension and discuss how polymer chain length, flexibility, and solvent viscosity influence mechanochemical kinetics.

27. A reaction exhibits pressure-dependent kinetics with a falloff curve characteristic of Lindemann–Hinshelwood behavior. Derive the general rate expression, determine high- and low-pressure limits, and evaluate how collisional efficiencies and energy transfer probabilities influence observed curvature. Propose a set of hypothetical kinetic data at multiple pressures, and demonstrate how to extract the effective rate constants and collisional efficiency factor. Discuss experimental approaches such as shock-tube or flow reactor studies for verifying pressure-dependent behavior.

28. Catalytic CO oxidation proceeds via competing pathways: an associative Langmuir–Hinshelwood route and an Eley–Rideal pathway. Design kinetic experiments to distinguish these mechanisms, including surface coverage variation, temperature-programmed reaction rates, and isotope labeling. Derive the expected rate laws for both pathways and simulate hypothetical data showing divergence between associative- and Eley–Rideal-dominated regimes. Discuss the impact of catalyst surface structure, particle size, and promoter species on pathway preference.

29. In a crowded cellular environment, molecular diffusion becomes anomalous, with mean-square displacement scaling as MSD $\propto$ t^α, α < 1. Explain how subdiffusive behavior alters diffusion-controlled rate constants and propose a modified Smoluchowski expression accounting for anomalous transport. Suggest experimental approaches, such as single-particle tracking or fluorescence correlation spectroscopy, to measure α and effective diffusion rates. Simulate kinetic curves for diffusion-limited reactions in crowded versus dilute media, and discuss implications for intracellular signaling and enzymatic cascades.

30. A reaction A → B competes with A → C. Although A → B has a higher activation energy, its pre-exponential factor is larger than that of A → C. Determine conditions under which kinetic versus thermodynamic control dominates product distribution using explicit Arrhenius expressions. Simulate concentration-time profiles for varying temperatures, and discuss how solvent polarity, catalysts, and reaction time influence selectivity. Propose experimental methods to probe product distribution, including NMR, GC-MS, or HPLC analysis.

31. A reversible enzyme–inhibitor binding equilibrium exhibits slow on/off kinetics. Derive the time-dependent inhibition curve, considering both association and dissociation rate constants, and explain how surface plasmon resonance (SPR) experiments can extract individual kinetic parameters. Simulate hypothetical sensorgrams for inhibitors with varying affinity and residence times. Discuss how slow-binding inhibitors impact pharmacodynamics and enzymatic regulation in vivo, and suggest complementary techniques such as stopped-flow or isothermal titration calorimetry for verification.

32. A molecular motor converts ATP hydrolysis into directional motion via a series of irreversible stochastic steps. Model the kinetics as a biased random walk, derive an expression for net velocity, and explain how detailed balance must be violated to achieve directional movement. Suggest experimental approaches using single-molecule fluorescence or optical tweezers to measure step sizes and velocities. Simulate hypothetical traces showing motor processivity under varying ATP concentrations and load forces, and discuss how ATP hydrolysis rate and mechanical strain affect efficiency.

33. A photoredox catalyst undergoes intersystem crossing, forward and back electron transfer, and product formation. Construct a kinetic scheme capturing all processes, apply steady-state approximations to intermediate species, and derive an expression for the quantum yield of product formation. Propose hypothetical experimental data using transient absorption spectroscopy and steady-state photolysis to validate the model. Discuss how solvent polarity, redox potential, and light intensity affect branching between productive and unproductive pathways.

34. An SN1 reaction in polar solvents exhibits dramatically higher rates than in nonpolar solvents. Using continuum solvation models, derive how the activation

Gibbs free energy depends on solvent dielectric constant. Predict the rate enhancement quantitatively and simulate hypothetical kinetic data for solvents ranging from nonpolar to highly polar. Discuss the role of carbocation stabilization, ion pairing, and nucleophile strength, and propose experiments with isotopically labeled substrates or varying leaving groups to probe solvent effects.

35. A reaction inside a nanopore shows enhanced selectivity for one product. Explain how confinement alters activation entropy, restricts transition-state geometries, and may lead to entropic trapping. Propose a kinetic model that accounts for confinement, calculate expected rate enhancements, and suggest experimental approaches using nanoporous silica or carbon nanotubes. Simulate concentration-time curves for reactions in bulk versus confined environments, and discuss how pore size, surface functionalization, and solvent viscosity influence kinetics and selectivity.

36. A catalytic oxidation requires electron transfer between the catalyst and substrate via a bridging ligand. Using superexchange theory, derive a distance-dependent rate constant and analyze how electronic coupling between donor and acceptor influences overall kinetics. Propose hypothetical data varying donor–acceptor distance, ligand identity, or bridging orbital energy, and predict effects on turnover frequency. Suggest experimental strategies using ultrafast spectroscopy or redox titration to validate superexchange-mediated electron transfer rates.

37. A three-step reaction mechanism involves a pre-equilibrium followed by a slow irreversible step. Determine how the apparent rate law depends on the magnitude of the pre-equilibrium constant and the rate of the slow step. Propose experimental tests, including varying reactant concentrations or monitoring intermediates, to distinguish pre-equilibrium from steady-state conditions. Simulate hypothetical concentration-time curves under both limiting cases and discuss implications for reaction optimization.

38. A proton-coupled electron transfer (PCET) reaction shows distinct kinetic isotope effects for proton and electron analogs. Develop a model describing coupled transitions, derive expressions for observed rates, and explain why classical KIE analysis fails. Propose experimental designs using isotopically labeled proton donors and electron donors to separate contributions. Discuss implications for

enzymatic PCET systems and synthetic catalysts, including the role of solvent reorganization and hydrogen-bond networks.

39. A catalytic cycle contains two parallel pathways with comparable activation energies. Using microkinetic modeling, derive how small changes in activation barriers or entropy differences influence turnover frequency (TOF) and branching ratios. Simulate hypothetical scenarios where slight changes in temperature or catalyst ligand structure shift product distribution. Propose experimental kinetic studies and isotopic labeling to quantify pathway contributions and validate model predictions.

40. A reaction exhibits oscillatory kinetics when continuously fed with reactants. Using nonlinear dynamics, identify necessary positive and negative feedback loops and derive conditions for Hopf bifurcation leading to stable oscillations. Propose experimental setups using continuously stirred tank reactors (CSTR) and in situ spectroscopic monitoring. Simulate concentration-time profiles under varying feed rates and initial concentrations, and discuss how oscillatory behavior could inform chemical clock reactions or synthetic biochemical circuits.

41. A combustion reaction involves radical branching and chain termination. Model induction time as a function of temperature, radical initiator concentration, and fuel composition. Analyze how small perturbations can lead to runaway reactions, and propose hypothetical kinetic curves showing induction periods followed by explosive acceleration. Suggest experimental approaches using shock tubes or rapid compression machines, and discuss how these models inform safety protocols and combustion engine design.

42. A solvent reorganization barrier controls an electron-transfer step embedded in a catalytic cycle. Show how changing solvent polarity or viscosity can shift the rate-determining step, producing turnover-frequency maxima. Derive rate expressions accounting for solvent relaxation times and reorganization energy, and propose experiments using solvents with systematically varied dielectric constants or viscosities. Simulate hypothetical kinetic traces and discuss how solvent optimization can enhance catalytic efficiency.

43. A bi-enzyme cascade converts A $\rightarrow$ B $\rightarrow$ C. If the first enzyme saturates but the second does not, derive the overall rate of appearance of C and predict how

intermediate accumulation depends on relative V_{max} values. Propose hypothetical kinetic data for enzyme concentrations, substrate saturation, and enzyme affinities, and simulate concentration-time profiles. Discuss implications for metabolic pathway control, synthetic enzyme cascades, and possible bottlenecks in intermediate accumulation.

44. A nucleophilic substitution reaction in ionic liquids exhibits different activation entropy than in water. Explain how ionic-liquid structuring affects solvation dynamics, ion pairing, and transition-state stabilization. Propose kinetic experiments comparing rates in ionic liquids of varying cation/anion combinations, and simulate hypothetical data for activation parameters. Discuss how these effects could be exploited to enhance selectivity or design green solvent systems.

45. A diffusion-coupled reaction in a polymer matrix depends on the diffusion rates of penetrant molecules. Develop a coupled kinetic–diffusion model and predict how reaction fronts propagate through the polymer. Simulate hypothetical concentration profiles for different polymer viscosities, temperatures, and penetrant sizes. Suggest experimental approaches such as confocal microscopy or FTIR imaging to measure reaction fronts, and discuss implications for polymer functionalization and controlled-release systems.

46. A reaction under high electric fields experiences Stark tuning of transition-state energies. Model field-dependent rate constants, explain how alignment of the field with the reaction coordinate can enhance or suppress reactivity, and simulate hypothetical rate enhancements. Propose experiments using parallel-plate cells or electrochemical setups to apply uniform fields, and discuss implications for electrocatalysis, field-assisted organic reactions, and mechanistic control of polar reactions.

47. A photochemical isomerization shows ultrafast excited-state decay followed by slower ground-state rearrangement. Construct a kinetic model incorporating femtosecond and nanosecond timescales, derive time-dependent populations of intermediates, and explain how transient absorption spectroscopy can reveal excited-state lifetimes and branching ratios. Simulate hypothetical decay curves for multiple wavelengths, and discuss how solvent polarity, temperature, and light intensity influence isomerization efficiency.

48. A supramolecular catalyst accelerates a Diels–Alder reaction by pre-organizing reactants inside a host cavity. Determine how encapsulation changes activation entropy and derive a rate enhancement factor based on statistical probability of reactive orientation. Simulate hypothetical kinetic data for host–guest complexes of varying binding strengths, and propose experimental strategies using NMR, calorimetry, or stopped-flow spectroscopy to validate pre-organization effects.

49. A reaction inside a vesicle exhibits different rates on interior vs exterior leaflets. Explain how curvature, local polarity, lipid packing, and membrane asymmetry influence reaction kinetics. Propose a kinetic model incorporating leaflet-specific rate constants, simulate hypothetical time-course data, and suggest experimental approaches using fluorescent probes or membrane-mimetic systems. Discuss implications for intracellular compartmentalization and localized catalysis in biological membranes.

50. A complex network involves reversible dimerization, catalytic turnover, and product inhibition, producing nonlinear concentration–time curves. Formulate differential equations for the full network, identify steady states, and analyze stability to determine whether multiple kinetic regimes are supported. Simulate hypothetical data showing bistability, oscillations, or plateau behavior, and discuss strategies for parameter estimation. Suggest experimental approaches using time-resolved spectroscopy, stopped-flow kinetics, or continuous-flow reactors to validate model predictions, and discuss relevance to systems chemistry, synthetic biology, and metabolic networks.

Chapter 14

Equilibrium as a Balancing Act

Chemical stoichiometry is often introduced with the simplifying notion that chemical reactions proceed fully from reactants to products, consuming at least one reagent entirely while sometimes leaving others in excess. This assumption greatly simplifies introductory calculations, allowing chemists to determine theoretical yields, limiting reagents, and reaction efficiencies with straightforward proportional reasoning. Yet as chemical understanding deepens, this premise reveals itself to be a convenient approximation rather than a universal truth. Many real reactions, especially those taking place in closed systems or involving reversible processes, do not proceed to completion but instead approach a state of balance known as chemical equilibrium. Equilibrium represents a condition in which the macroscopic concentrations of reactants and products remain constant over time. This steadiness does not indicate stasis at the molecular level; rather, it conceals a dynamic behavior where forward and reverse reactions continue simultaneously at equal rates. Molecules are constantly transformed, exchanged, and reconfigured, yet the overall composition of the system stays the same because the rates of production and consumption of each species are precisely matched.

Understanding equilibrium requires a shift in conceptual perspective from the deterministic stoichiometric endpoint to a probabilistic and energetically governed balance. In some reactions, such as strong acid–base neutralizations or precipitation reactions with highly insoluble products, equilibrium lies overwhelmingly toward the products, and the assumption of completion holds reasonably well. In other cases, particularly those involving weak acids, weak bases, gas-phase reactions, or reversible molecular rearrangements, equilibrium lies closer to the reactants, resulting in only partial transformation. A classic example is the dissociation of acetic acid in water. Although the reaction is spontaneous in terms of free energy, only a small fraction of acetic acid molecules actually ionize at equilibrium, because the energetic cost of producing additional hydronium ions eventually offsets the thermodynamic drive to produce them. The equilibrium position therefore reflects a delicate balance between the energetics of bond breaking and bond formation and the statistical likelihood of molecular collisions with appropriate orientation and energy.

To articulate this balance quantitatively, chemists rely on the law of mass action, a foundational principle of chemical thermodynamics. For a general reversible reaction of the form $aA + bB \rightleftharpoons cC + dD$, the law of mass action states that the equilibrium constant, K, is defined by the ratio $K = [C]^{\wedge c}[D]^{\wedge d} / [A]^{\wedge a}[B]^{\wedge b}$, where square brackets denote equilibrium concentrations. This deceptively simple expression embodies a profound insight: no matter how a system begins, whether it starts with pure reactants, pure products, or some mixture thereof, it will always settle into a state where this ratio takes on a fixed value for a given temperature. The constancy of K reveals the underlying thermodynamic constraints dictating molecular behavior. It shows that while kinetics governs the path a system takes toward equilibrium and how quickly it arrives there, thermodynamics governs the final position of equilibrium itself.

The equilibrium constant is intimately related to the difference in free energy between reactants and products. A large K indicates that products are thermodynamically favored, meaning the formation of products lowers the system's free energy. A small K suggests that reactants are more stable. Yet even reactions with small equilibrium constants can proceed significantly if conditions such as concentrations or pressures are altered to favor product formation. The interplay between these factors becomes especially important in systems where external conditions—pressure, temperature, volume, solvent composition, ionic strength—can be manipulated to adjust equilibrium behavior.

In gas-phase systems, equilibrium can also be expressed in terms of partial pressures, leading to the equilibrium constant K_p. The relationship between Kc, the concentration-based constant, and K_p is given by $K_p = K_c(RT)^{\wedge\Delta n}$, where Δn represents the difference between the moles of gaseous products and gaseous reactants. This expression arises from the ideal gas law, which relates concentration to pressure for gases under standard conditions. Gas-phase equilibria are particularly sensitive to changes in volume and pressure because gaseous species occupy variable amounts of space depending on system constraints. A compression of the reaction vessel increases gas concentrations, shifts collision frequencies, and can dramatically alter equilibrium positions, especially for reactions in which the number of gas molecules changes during the process.

Heterogeneous equilibria introduce additional considerations. In systems where different phases coexist—for instance, solids reacting with gases or liquids—the activities of pure solids and liquids are considered constant and thus do not appear explicitly in the equilibrium expression. Their contribution is effectively incorporated into the value of K. This principle explains why equilibria such as $CaCO_3(s) \rightleftharpoons CaO(s) + CO_2(g)$ depend only on the partial pressure of carbon dioxide. No matter how much solid calcium carbonate or calcium oxide is present, the equilibrium CO_2 pressure at a given temperature remains the same, provided both solids are present to maintain phase balance. This property underlies numerous geological and industrial processes, including limestone decomposition in cement manufacturing and mineral stabilization in Earth's crust.

While the equilibrium constant specifies the state toward which a system evolves, it says nothing about how far from or close to equilibrium the system is at any given moment. To assess instantaneous conditions, chemists use the reaction quotient, Q, which has the same mathematical form as K but is calculated using initial or non-equilibrium concentrations. The comparison of Q to K provides a powerful diagnostic tool. If $Q < K$, the system has insufficient products relative to reactants, so the reaction proceeds in the forward direction to generate more products. If $Q > K$, the system has excess products, and the reverse reaction is favored. These relationships guide predictions about spontaneous shifts and enable controlled manipulation of chemical processes in laboratory and industrial settings. For example, in the Haber–Bosch process for synthesizing ammonia, engineers continually adjust reactant pressures to drive Q below K, thereby maximizing ammonia yield under equilibrium constraints.

Le Chatelier's principle provides a qualitative rule for predicting how a system at equilibrium responds to disturbances. When concentration, pressure, or temperature changes are imposed, the system adjusts to counteract the perturbation, seeking to re-establish equilibrium. This principle captures the essential self-regulatory nature of equilibrium systems. If a reactant is added, the system absorbs the perturbation by consuming some of the added reactant in the forward reaction. If a product is removed, the system shifts forward to replenish it. These shifts do not violate thermodynamic laws; instead, they reflect the system's tendency to move toward the equilibrium composition dictated by K.

For gaseous equilibria, changes in volume or pressure alter partial pressures and thereby influence equilibrium. Decreasing the container volume increases pressure and shifts equilibrium toward the side of the reaction with fewer gas molecules, reducing the total pressure. Conversely, increasing the volume favors the side with more gas molecules. This behavior can be interpreted in terms of the law of mass action or in terms of entropy: gases naturally expand to fill available space, and equilibrium will shift to maximize the number of microstates under the imposed constraints. The addition of inert gases at constant volume does not change partial pressures of reacting gases and thus does not shift equilibrium directly; however, if added at constant pressure, inert gases increase total volume and can influence equilibria by effectively altering concentrations.

Temperature changes affect equilibrium in fundamentally different ways because temperature directly modifies the value of the equilibrium constant itself. For endothermic reactions, heat is a reactant, so increasing temperature shifts equilibrium toward the products. For exothermic reactions, heat is a product, so raising temperature shifts equilibrium toward reactants. These changes are quantifiable through the van't Hoff equation, which relates the temperature dependence of K to the enthalpy change of the reaction. The van 't Hoff equation reveals that even modest temperature changes can significantly alter equilibrium positions, especially for reactions with large enthalpy changes.

A classic application of these principles is the synthesis of ammonia from nitrogen and hydrogen gases: $N_2(g) + 3H_2(g) \rightleftharpoons 2NH_3(g)$. This reaction is exothermic and involves a reduction in the number of gaseous molecules; therefore, high pressures and low temperatures favor ammonia formation. However, low temperatures also slow reaction kinetics, presenting a conflict between thermodynamic favorability and reaction rate. Industrial processes therefore operate at compromise conditions—moderate temperatures where kinetics are acceptable and high pressures where thermodynamics drive equilibrium toward ammonia production. Catalysts composed of iron with various promoters are used to increase reaction rates without altering equilibrium positions, illustrating the distinct roles of kinetics and thermodynamics.

Catalysts, in fact, play no direct role in shifting equilibrium; their influence lies solely in providing alternate pathways with lower activation energy, thereby increasing both forward and reverse reaction rates equally. A catalyst enables the system to reach equilibrium more quickly but does not affect the equilibrium composition. This distinction is crucial: misinterpreting catalysts as agents that change equilibrium positions leads to misconceptions about their function. Instead, catalysts lower the energetic barrier for the reaction, enabling the system to achieve the thermodynamically determined equilibrium composition in a shorter time.

The concept of equilibrium extends beyond simple molecular reactions to encompass acid–base chemistry, solubility equilibria, complex ion formation, redox systems, and even biochemical processes such as enzyme catalysis. Acid–base equilibria illustrate the interplay between proton donors and acceptors and highlight how equilibrium constants relate to pH and buffer capacity. Weak acids and bases dissociate only partially, and their equilibrium constants (K_a and K_b) quantify their strength. The Henderson–Hasselbalch equation provides a practical tool for calculating pH in buffer systems by relating pH to the ratio of conjugate base to acid concentrations. Buffer systems maintain relatively stable pH values even when small amounts of acid or base are added, exemplifying how equilibrium principles contribute to the stability of chemical environments, including those within living organisms.

Solubility equilibria govern the dissolution and precipitation of ionic compounds in aqueous solution. The solubility product constant, K_{sp}, quantifies the equilibrium between a sparingly soluble solid and its dissolved ions. When ion concentrations in solution exceed the solubility product, precipitation occurs; when they fall below it, dissolution is favored. These principles are central to environmental chemistry, geochemistry, and analytical techniques such as gravimetric analysis. For instance, the formation of scale in pipes involves precipitation of calcium carbonate when temperature or pressure changes shift equilibrium conditions. In biological systems, the solubility equilibrium of calcium phosphate influences bone mineralization and dissolution.

Complex ion equilibria demonstrate how ligands interact with metal ions to form coordination complexes with distinct stability constants. These equilibria underlie numerous industrial applications, such as metal extraction, water softening, and analytical titrations involving EDTA. The formation constants of complex ions quantify the tendency of ligands to coordinate metal centers, and the overall stability of a complex ion depends on both enthalpic and entropic contributions. Chelation effects, in which multidentate ligands form more stable complexes than monodentate ligands, illustrate how molecular organization influences equilibrium. Redox equilibria introduce electron transfer considerations and are governed by the Nernst equation, which relates cell potential to the reaction quotient. The equilibrium position of redox reactions determines whether spontaneous electron flow can occur and under what conditions. These principles underpin battery technology, corrosion processes, and electrochemical synthesis.

Even biochemistry relies heavily on equilibrium concepts. Enzymatic reactions approach steady-state conditions rather than true equilibrium, but the same thermodynamic principles apply. Enzymes accelerate reaction rates but cannot alter equilibrium positions. Biological systems often exploit coupled reactions, using energy from favorable processes such as ATP hydrolysis to drive unfavorable reactions. This coupling relies on the additive nature of free energy changes and illustrates how complex networks of equilibria sustain life. Ultimately, the study of chemical equilibrium reveals the inherent stability and self-regulating nature of chemical systems.

While stoichiometric assumptions provide valuable introductory tools, they are merely approximations of a deeper reality. Real systems continually adjust in response to internal and external perturbations, seeking positions determined by thermodynamics but accessible only through kinetics. By mastering equilibrium concepts—laws of mass action, reaction quotients, Le Chatelier's principle, solubility and acid–base behavior, gas-phase equilibria, and the temperature dependence of K—chemists gain the ability to predict and manipulate the composition of chemical systems with precision. Whether optimizing industrial reactions, designing pharmaceuticals, modeling environmental processes, or exploring the molecular basis of life, understanding equilibria is fundamental to interpreting and controlling the behavior of matter.

The role of chemical equilibrium in shaping our understanding of molecular transformation cannot be overstated. Although typically introduced in foundational chemistry courses, the concept continues to influence advanced theoretical frameworks and applied chemical technologies. A deeper exploration reveals equilibrium as not merely a mathematical condition characterized by specific concentrations, but as a profound manifestation of competing microscopic forces. At the molecular scale, equilibrium arises from a delicate interplay between energy distribution, molecular motion, interaction potentials, and probabilistic transitions among available quantum states. This perspective places equilibrium within the larger narrative of statistical thermodynamics, where macroscopic observables emerge from the behavior of countless particles operating under thermal agitation. Every molecule in a reacting system experiences a fluctuating environment shaped by collisions, short-range repulsions, long-range attractions, rotational and vibrational motions, and transient alignments that determine the probability of reaction. The equilibrium constant thus reflects, in condensed form, the totality of these competing influences and the extent to which the system's accessible microstates favor one distribution of species over another.

Examining equilibrium through the lens of molecular energy surfaces further enriches this understanding. Chemical reactions are often visualized in terms of potential energy surfaces, where valleys represent stable species and peaks correspond to transition states. At equilibrium, the depth of the energy wells for reactants and products dictates relative stability, while the height of the barrier influences kinetics. A system at equilibrium does not remain confined to a single energy well; instead, thermal energy allows molecules to traverse the surface, occasionally acquiring sufficient energy to surmount the transition state barrier. In a multi-dimensional energy landscape—where molecular geometry, intermolecular distance, bond angles, and electronic distribution all contribute to potential energy—the ensemble of molecules continuously explores nearby coordinates. Equilibrium does not imply stasis on this surface but reflects a balance in the relative population of energy wells based on their statistical favorability. The global minimum of Gibbs free energy corresponds not to a static point on the surface but to the most probable distribution of molecules across the available energy configurations.

Another dimension of equilibrium emerges in multi-step or coupled reactions, where the overall reaction is composed of sequential elementary processes. Many biological pathways, catalytic cycles, and environmental systems involve chains of reactions where intermediates form and decay continuously. In such systems, individual steps may reach their own micro-equilibria long before the overall system stabilizes. These transient equilibria greatly influence reaction rates and product distributions, giving rise to steady-state approximations in kinetics. The steady state occurs when the concentration of an intermediate remains approximately constant because its rate of formation equals its rate of consumption, a concept reminiscent of but distinct from thermodynamic equilibrium. The presence of coupled reactions also allows systems to exhibit behavior such as equilibrium shifting through energy input, as occurs in biochemical cycles driven by ATP hydrolysis. Here, thermodynamically unfavorable steps become feasible because they are linked to highly exergonic transformations that pull the reaction network forward. Such couplings illustrate how equilibrium thermodynamics, while powerful, does not prohibit biological systems from sustaining non-equilibrium states essential for life.

Beyond coupled reactions, equilibrium in open systems introduces further complexity. Most theoretical treatments assume closed systems—those in which matter does not enter or leave. Yet many real systems exchange matter or energy with their surroundings. In open systems, equilibrium can only be achieved when external flows are balanced by internal processes, a condition known as dynamic steady state. For example, rivers receiving dissolved minerals, biological tissues releasing gases to the bloodstream, and industrial reactors with continuous feed streams all operate far from traditional equilibrium. Nevertheless, they maintain constant concentrations due to continuous flux. This broadens the conceptual relevance of equilibrium, showing that the principle of balance applies even in systems where classical equilibrium cannot be reached. Chemical engineers exploit this fact in continuous reactors, where equilibrium constants determine conversion efficiency within the reactor, while designers adjust flow rates, pressures, and temperatures to maintain optimal overall outputs. Understanding how equilibrium and flow interact enables precise control over complex reaction systems.

At the opposite extreme, equilibrium considerations apply to extremely slow processes in geological and planetary chemistry. Minerals forming deep within the Earth's crust, gases released from magma chambers, and equilibria among atmospheric gases on other planets evolve over time periods spanning millions or billions of years. High pressures, extreme temperatures, and long timescales allow reactions far too sluggish under laboratory conditions to advance appreciably in geological environments. Equilibrium constants help geochemists infer historical temperatures, pressures, and compositional changes by analyzing mineral assemblages that reflect equilibrium distributions established long ago. The presence of carbonate minerals, silicate equilibria, and isotopic ratios often serves as evidence of ancient equilibria that have remained effectively frozen by changes in environmental conditions. This perspective reinforces the idea that equilibrium is a universal concept, bridging contexts from microscopic timescales of femtoseconds to planetary evolutions spanning eons.

Thermodynamic equilibrium also integrates elegantly with quantum chemistry, especially when discussing energy quantization in molecular vibrations and rotations. At the molecular scale, equilibrium reflects populations distributed among quantized energy levels according to the Boltzmann distribution. For reactions involving changes in molecular electronic structure, the equilibrium constant can be expressed in terms of contributions from partition functions, which summarize accessible energy states. This results in expressions for K that incorporate translational, rotational, vibrational, and electronic energies of the participating molecules. Vibrational frequencies, dipole moments, and electronic degeneracies all contribute to the temperature dependence of equilibrium. Researchers can calculate equilibrium constants from first principles using quantum mechanical data, enabling predictions for systems too reactive or hazardous for empirical study. Such treatments highlight the unifying power of equilibrium theory across disciplines, from macroscopic reaction vessels to quantum-scale descriptions of atoms and molecules.

In many chemical systems, entropy changes drive equilibrium behavior more strongly than enthalpy changes. Processes in which disorder increases—such as dissolution of solids, formation of gases, or isotropic mixing—often exhibit equilibrium constants that reflect entropy's dominant role. For example, when a solid dissolves into ions, the entropy change is typically positive because dispersed ions possess far more accessible microstates than a structured crystalline lattice. Even if the enthalpy change is slightly endothermic, the increase in entropy may drive the dissolution forward, resulting in substantial solubility. Conversely, reactions that lead to ordering—such as crystallization or condensation—experience decreases in entropy that oppose product formation. These reactions often require external stimuli such as cooling or compression in order to reach equilibrium favoring the condensed phase. The degrees to which enthalpy and entropy contribute are embedded in the Gibbs free energy equation $\Delta G = \Delta H - T\Delta S$, linking equilibrium behavior to the competition between energy minimization and disorder maximization.

A further extension of equilibrium theory involves electrochemical systems, in which chemical equilibria intersect with electron transfer. Redox reactions occurring at electrodes exhibit equilibrium potentials predicted by the Nernst equation, which directly incorporates the equilibrium constant through its relationship to Gibbs free energy. The Nernst equation shows that cell potential decreases as product concentrations increase, implying that electrochemical driving force diminishes as the reaction approaches equilibrium. In reversible electrochemical cells, equilibrium is achieved when the cell potential drops to zero, and no net current flows. This equilibrium represents a balance between oxidation and reduction processes across electrodes. Applications such as batteries, corrosion protection, sensors, and biochemical assays all rely on this interplay between equilibrium and electron transfer. When oxygen concentration in blood is measured using electrode-based sensors, the equilibrium between oxygen molecules and the electrode surface dictates the measured potential, allowing precise quantitation of physiological conditions.

Returning to equilibrium in solutions, the behavior of electrolytes highlights the importance of ionic interactions. In concentrated ionic solutions, ions are not independent but interact through long-range Coulombic forces. These interactions reduce the effective concentration, or activity, of ions relative to their nominal molarity. As a result, equilibrium constants expressed in terms of activities differ from those expressed in terms of raw concentrations. The Debye–Hückel theory provides a framework for estimating activity coefficients, enabling equilibrium calculations to remain accurate even in moderately concentrated solutions. As ionic strength increases, activity coefficients deviate further from unity, and thermodynamic calculations must incorporate these corrections. Without them, predictions of solubility, acid–base behavior, and complex ion formation would become increasingly inaccurate as ionic strength rises. In biochemical systems, high concentrations of ions in cellular environments necessitate equilibrium treatments that account for non-ideal behavior, enabling accurate modeling of processes such as enzyme binding, protein folding, and ligand association.

Equilibrium principles also extend to polymer chemistry, where chain-growth and step-growth polymerizations exhibit distinct equilibrium characteristics. In some polymerization reactions, equilibrium limits the maximum attainable molecular weight. Reversible condensation polymerizations, for example, generate small molecules such as water or methanol as byproducts that must be removed to drive polymer formation toward higher degrees of polymerization. Failure to remove these products results in an equilibrium distribution that restricts polymer chain length. This constraint manifests in the equilibrium constant for the condensation reaction, which indicates the maximum achievable conversion under given conditions. Designing industrial polymerization processes requires careful control of temperature, pressure, catalysts, and removal of byproducts to push equilibrium toward long-chain polymer formation. This interplay between reaction design and equilibrium manipulation underscores the importance of thermodynamics in materials science.

Equilibrium further manifests in colloidal systems, where the distribution of dispersed particles between aggregated and dispersed states reflects a balance of intermolecular forces. Colloid stability depends on equilibrium between attractive van der Waals forces and repulsive electrostatic interactions. These interactions govern whether particles remain suspended or aggregate into larger clusters that precipitate. Equilibrium models such as DLVO theory describe how ionic strength, pH, and surface potential influence colloid stability. For example, increasing ionic strength compresses the electrical double layer around particles, reducing repulsion and shifting equilibrium toward aggregation. Conversely, reducing ionic strength expands the double layer, stabilizing the suspension. These ideas have direct consequences in environmental chemistry, wastewater treatment, pharmaceuticals, and food science, where controlling particle dispersion is essential.

In atmospheric chemistry, equilibrium governs processes such as the formation and dissociation of ozone, partitioning of pollutants between air and water, and aerosol formation. The equilibrium between nitrogen dioxide and dinitrogen tetroxide, for instance, depends strongly on temperature, with lower temperatures favoring the dimer. Seasonal variations in equilibrium therefore influence pollutant distribution and visibility. Similarly, Henry's law describes the equilibrium dissolution of gases into liquids, essential for predicting how atmospheric pollutants enter bodies of water or how carbon dioxide exchanges between the atmosphere and oceans. Temperature, pressure, salinity, and chemical composition of water all influence gas solubility equilibria, shaping global climate patterns and ecological health.

Equilibrium principles also provide insight into the microscopic structure of liquids and solids. In liquid–vapor systems, the equilibrium vapor pressure reflects the balance between molecules escaping from the liquid surface and those returning from the vapor phase. Increasing temperature increases molecular kinetic energy, raising vapor pressure and shifting equilibrium toward the vapor phase. This equilibrium explains boiling, condensation, and distillation, processes crucial to chemical purification. In crystalline solids, lattice defects—vacancies, interstitial atoms, and dislocations—exist in equilibrium concentrations determined by thermodynamics. At higher temperatures, entropy favors the formation of more defects, altering material properties such as conductivity, strength, and diffusion rates. These examples show that equilibrium affects not only reactions but also physical properties of materials.

Finally, a comprehensive understanding of chemical equilibrium illuminates the boundary between reversible and irreversible processes. While many reactions are reversible under appropriate conditions, some appear irreversible because equilibrium lies overwhelmingly toward products or because reaction pathways lead to states from which return is kinetically inaccessible. Combustion reactions, for example, produce stable molecules such as CO_2 and H_2O with extremely deep energy wells, making reverse reactions effectively impossible without enormous energy input. Polymorphic transformations of solids may also exhibit kinetic barriers so high that once formed, a polymorph persists indefinitely despite not being the thermodynamically favored phase at current conditions. These limitations remind us that equilibrium describes the natural end point of a system unconstrained by kinetic barriers, but real systems often become trapped in metastable states. Understanding the interplay between thermodynamics and kinetics therefore remains crucial for interpreting experimental observations and designing chemical processes.

Through these broad perspectives—spanning quantum mechanics, statistical thermodynamics, geological processes, environmental systems, electrochemistry, polymer science, colloid stability, and materials behavior—the concept of chemical equilibrium emerges as a unifying principle governing transformations across the chemical sciences. It offers a powerful framework for predicting the fate of molecules under a wide range of conditions, explaining not only how reactions proceed but how the universe organizes matter at every scale. By integrating stoichiometric reasoning with equilibrium theory, chemists gain a comprehensive understanding of both the idealized and the real behaviors of chemical systems, enabling accurate predictions, optimized reaction design, and deeper insight into the molecular foundations of the physical world.

Chapter 14: Word Problems

1. A sealed reaction vessel contains a mixture of $SO_2(g)$, $O_2(g)$, and $SO_3(g)$ at high temperature, undergoing the equilibrium $2SO_2(g) + O_2(g) \rightleftharpoons 2SO_3(g)$. An engineer discovers that after adding a porous solid catalyst, the system returns to the same equilibrium composition much faster than before, but the catalyst also adsorbs a small fraction of SO_2 onto its surface through a reversible, non-reactive interaction. She notices that the apparent equilibrium constant calculated from measured gas-phase concentrations is slightly different from the theoretical K_p. She suspects adsorption perturbs measured partial pressures. She must determine whether the discrepancy arises from altered thermodynamic equilibrium or from experimentally distorted gas-phase concentrations, and how to correct measured values to extract the true equilibrium constant.

2. A weak organic base B is dissolved into an aqueous buffer already containing a weak acid HA and its conjugate base A^-. The base B undergoes protonation to form BH^+ with equilibrium constant K_b, but the presence of a competing acid–base pair introduces coupled equilibria. As temperature increases, the buffer's

pKa shifts, changing the proton availability for B. A researcher must determine whether a measured change in the total protonated fraction of B arises from shifts in the intrinsic acidity of HA, changes in water autoprotolysis, or coupled ionic-strength effects that alter activity coefficients.

3. A sparingly soluble metal carbonate $MCO_3(s)$ is exposed to an aqueous solution containing both H^+ and a chelating ligand L^{2-}. The mineral dissolves through two simultaneous equilibria: acid-driven dissolution to produce CO_2, and ligand-driven complexation of M^{2+}. At intermediate pH, both processes contribute comparably. Given field measurements of dissolved inorganic carbon and metal–ligand complexes, a geochemist must disentangle the contributions of acid-promoted dissolution and ligand-enhanced solubility to determine whether dissolution is limited by equilibrium constraints or by surface-reaction kinetics.

4. A catalytic cycle for CO oxidation on a metal surface proceeds through several intermediates, including adsorbed CO*, O*, and CO_2*. Experiments reveal that two steps—O_2 dissociation and CO_2 desorption—achieve quasi-equilibria under reaction conditions. Because the overall rate depends on the slight deviation from these quasi-equilibria, a kinetic modeler must determine equilibrium constants for the surface species using measured coverages that are distorted by lateral interactions, complicating the extraction of thermodynamic parameters.

5. A sealed 2-phase system contains liquid water and water vapor with added volatile solutes, including methanol and acetone. Each solute has a temperature-dependent Henry's law constant, and the total pressure of the vapor phase differs from ideal behavior because both solutes significantly modify the water activity. When the temperature is raised by 12 K, the vapor composition becomes enriched in one solute and depleted in the other, contrary to naive expectations based solely on vapor pressures. The problem is to reconcile the observed changes with coupled equilibria involving activity coefficients, non-ideal Raoult behavior, and solute–solute interactions.

6. A reaction network involves the reversible hydration of CO_2, the acid–base equilibria of carbonic acid, and the complexation of bicarbonate with a metal ion M^{3+}. When titrating the system with a strong base, the apparent pKa values shift significantly because metal complexation perturbs the distribution of carbonate

species. A chemist must determine the intrinsic pKa values of carbonic acid and the formation constants of the M^{3+}–carbonate complexes using only titration data in which species cannot be isolated directly.

7. A continuously stirred tank reactor (CSTR) carries out the exothermic equilibrium reaction $A \rightleftharpoons B + C$. Because fresh reactants are fed continuously and products removed, the system never reaches true equilibrium but instead reaches a steady state where the reaction quotient Q differs from K. At high inflow rates, the reactor exhibits multiple steady states due to thermal feedback. The problem requires determining under what combinations of feed composition, temperature, and residence time the reactor transitions from a single stable steady state to multiple steady states and whether thermal runaway is possible even though the reaction is reversible.

8. An electrochemical concentration cell contains identical electrodes immersed in solutions of the same redox couple but with different ionic strengths due to added inert salts. Even though the formal concentrations of the redox species are equal, the measured cell potential is nonzero. The challenge is to derive the effective equilibrium constant of the redox reaction under the differing activity conditions and determine how much of the observed potential originates from ion-activity differences rather than concentration differences.

9. An atmospheric chemist studies the $NO_2(g) \rightleftharpoons N_2O_4(g)$ equilibrium inside a high-altitude haze layer where temperature fluctuates rapidly. Because light absorption by NO_2 induces photodissociation, the reaction is no longer purely thermal. The problem is to determine under what solar-irradiance and temperature ranges the photostationary steady state deviates from the true thermal equilibrium constant, and whether the haze composition can ever be interpreted as being in equilibrium despite constant photolytic perturbation.

10. A natural water sample contains Fe^{2+}, Fe^{3+}, dissolved O_2, organic ligands, and colloidal iron oxide particles. Fe^{2+} oxidizes to Fe^{3+}, Fe^{3+} hydrolyzes to form $Fe(OH)_x$ species, and these species aggregate into colloids; meanwhile, organic ligands stabilize dissolved iron through coordination. Because each equilibrium is strongly pH-dependent, small environmental pH fluctuations dramatically shift total soluble iron. A geochemist must determine whether an observed drop in dissolved iron

results from oxidation kinetics, precipitation equilibria, or competition between ligand binding and hydrolysis.

11. A polymerization reaction proceeds through reversible step-growth between monomers A and B, producing oligomers and releasing a volatile byproduct that slowly escapes the reactor. Because the byproduct removal is incomplete, the reaction mixture approaches but never reaches the ideal equilibrium conversion predicted by K. The challenge is to quantify the maximum attainable number-average molecular weight under partial byproduct removal and to determine whether diffusion limitations in the viscous polymer phase shift the effective equilibrium.

12. A biological enzyme catalyzes the interconversion of two metabolites via $A \rightleftharpoons B$. The enzyme accelerates both directions and reaches a rapid internal equilibrium among its enzyme-bound intermediates, but the cellular environment continuously drains B into another pathway. The intracellular ratio [B]/[A] therefore deviates from the thermodynamic ratio K. The problem is to determine the true equilibrium constant from observed nonequilibrium steady-state data while accounting for enzyme binding, substrate channeling, and rapid exchange between bound and free states.

13. A molten-salt reactor contains the equilibrium $U^{4+} + e^- \rightleftharpoons U^{3+}$, with redox potential that depends on interactions between uranium species and the ionic melt. Because the melt contains high concentrations of other multivalent ions, activity coefficients deviate strongly from unity. A nuclear chemist must determine whether the redox potential predicted by the Nernst equation should incorporate activity coefficients derived from Debye–Hückel–type models or whether specific ion–ion interactions dominate such that empirical calibration is required.

14. A gas-phase industrial reaction $A(g) + 2B(g) \rightleftharpoons C(g)$ is carried out at extremely high pressure inside a reactor where non-ideal gas behavior is substantial. Fugacity coefficients for all species vary significantly with pressure. Even so, operators traditionally use Kp based on ideal-gas partial pressures. The puzzle is to determine how far the actual equilibrium composition deviates from the ideal-gas prediction and whether using fugacities alters the optimal pressure recommended for maximizing C(g) production.

15. A pharmaceutical formulation contains a weak acid drug HA, its sodium salt NaA, and several excipients that alter microenvironmental pH within the solid tablet. Upon dissolution, the tablet creates local pockets of varying pH as excipients dissolve at different rates. Because the drug's solubility depends on its ionization state, non-uniform dissolution leads to a time-dependent and spatially heterogeneous equilibrium between HA and A^-. The problem is to determine whether the experimentally observed dissolution curve is limited by acid–base equilibrium or by mass-transfer barriers within the dissolving matrix.

16. A photochemical reaction converts a reactant R into an excited state R*, which then undergoes reversible isomerization to I. The $R \rightleftharpoons I$ thermal equilibrium lies strongly toward R, but under continuous illumination, the photo-steady state favors I. However, because I absorbs light differently than R, increasing the light intensity eventually decreases the I fraction. The challenge is to model how the dynamic photostationary equilibrium depends simultaneously on the quantum yields, extinction coefficients, and thermal equilibrium constant.

17. A crystallization process involves a salt that forms multiple hydrated and anhydrous solid phases. Depending on temperature and humidity, the equilibrium among these solids changes, and the aqueous solubility varies by orders of magnitude. A materials chemist must determine which hydrate is thermodynamically stable at a given temperature, how the presence of metastable hydrates shifts the apparent Ksp, and whether rapid dehydration during filtration causes the solid to fall out of equilibrium with the mother liquor.

18. A metalloprotein binds a metal ion M^{2+} through a reversible equilibrium with dissociation constant Kd. However, the protein exists in two conformational states, folded and partially unfolded, each with different affinities for M^{2+}. Temperature and ionic strength shift the equilibrium between protein conformers. A biophysical chemist must determine the intrinsic metal-binding constants for each conformer using only overall binding data that averages over the conformational ensemble.

19. In a semiconductor synthesis, gaseous $SiCl_4$ and H_2 react to form solid silicon and gaseous HCl. The equilibrium constant is temperature-dependent and strongly favors silicon production at high temperatures. Yet experimental yields fall short of equilibrium predictions due to formation of volatile intermediate species such

as $SiHCl_3$. The problem is to determine whether the reduced yield is due to side equilibria forming volatile silicon chlorides or due to transport limitations that prevent gaseous reactants from reaching the growing surface.

20. A solubility experiment on a protein reveals that its apparent solubility increases with added salt up to a point ("salting-in") and then decreases ("salting-out"). The system involves equilibrium between protein monomers in solution and aggregated forms, while ionic strength modulates both electrostatic repulsion and water structure. A biochemist must distinguish whether the solubility maximum arises from equilibrium shifts in protein–protein interactions or from non-equilibrium aggregation kinetics that are slowed temporarily by moderate salt concentrations.

21. A solid catalyst promotes the reversible dehydration of an alcohol to an alkene. Under high conversion, the product alkene also adsorbs onto the catalyst surface and undergoes reversible oligomerization. This creates a network of equilibria between gas-phase alcohol, adsorbed species, and polymeric surface oligomers. Catalytic activity drops over time, and the question is whether the deactivation reflects a shift in thermodynamic equilibrium toward long-lived surface oligomers or kinetic trapping in metastable polymeric states.

22. A deep-sea hydrothermal vent contains dissolved metals, sulfide ions, and organic ligands. Pressure significantly affects the solubility of gases and the equilibrium constants of aqueous complexes. A geochemist must determine whether an observed metal-sulfide precipitation event at depth can be attributed to equilibrium shifts caused by depressurization during fluid ascent or whether microbial redox activity consumes key species, pulling the system away from equilibrium.

23. An aqueous system containing the weak acid H_2A exhibits two dissociation steps, both influenced by ionic strength. When titrated with strong base in a background electrolyte, the apparent pKa_2 value shifts more strongly with ionic strength than pKa_1. The problem is to determine why the second dissociation is more sensitive to activity corrections and whether the result implies stronger interactions between A^{2-} and spectator ions.

24. A gas-phase photochemical smog system involves dozens of reversible reactions among ozone, nitrogen oxides, and organic radicals. Although many reactions have well-defined equilibrium constants, continuous sunlight and pollutant

emissions keep the system far from equilibrium. A researcher must determine whether a subset of fast reversible reactions can be considered in equilibrium at any given moment and whether imposing pseudo-equilibrium assumptions in the model improves or degrades predictive accuracy.

25. A supramolecular host molecule H binds a guest molecule G through hydrogen bonding, but water competes strongly for the same binding sites. Binding follows the equilibrium $H + G \rightleftharpoons HG$ with constant K, but the effective equilibrium in aqueous solution depends on activity of water and on hydrophobic interactions. A chemist must determine the intrinsic gas-phase binding enthalpy of the host–guest pair from experimentally measured aqueous binding constants distorted by solvation effects.

26. A nanoparticle suspension undergoes reversible aggregation governed by DLVO forces. As electrolyte concentration slowly increases due to solvent evaporation, the system crosses a threshold at which aggregation becomes favored. However, optical measurements suggest the system begins aggregating before the predicted equilibrium transition point. The challenge is to determine whether early aggregation reflects non-equilibrium kinetics or whether short-range hydration forces shift the true equilibrium beyond classical DLVO predictions.

27. A battery system using a redox-active organic molecule exhibits apparent self-discharge during open-circuit rest. The molecule exists in multiple oxidized and reduced forms that interconvert via disproportionation equilibria. Because one disproportionation product is volatile, its slow evaporation alters equilibrium concentrations. The task is to determine whether the voltage decay corresponds to relaxation toward thermodynamic equilibrium or to irreversible loss of material through gas-phase escape.

28. A molten-metal alloy undergoes a reversible order–disorder transition between a random atomic mixture and an ordered intermetallic compound. The equilibrium depends strongly on temperature, but slow diffusion at low temperatures traps the alloy in metastable states. A metallurgist must determine whether a partially ordered sample represents a true equilibrium mixture of solid phases or a kinetically frozen structure that would transform given sufficient time.

29. A biological receptor binds two different ligands, L_1 and L_2, each with its own affinity and each influencing the receptor's conformational equilibrium.

Because binding of either ligand shifts the receptor to an active conformation, the apparent binding constants depend on ligand concentrations. The challenge is to derive intrinsic equilibrium constants for each ligand and conformational transition using data from competitive binding experiments.

30. A catalytic hydrogenation reaction in supercritical CO_2 exhibits unusual pressure dependence. As pressure increases, the supercritical solvent density rises, altering solubilities of both reactants and catalyst ligands. The observed equilibrium constant for hydrogen addition increases steeply with density, much more than predicted by simple activity models. The problem is to explain whether the enhanced equilibrium conversion arises from solvent-induced changes in reaction enthalpy or from preferential solvation of transition states.

31. An industrial reactor converts ethylene to polyethylene through a coordination-polymerization mechanism involving a metal catalyst. Trace impurities such as water and oxygen reversibly bind to the metal center, inhibiting polymerization. Because binding is reversible and temperature-dependent, the system reaches an equilibrium among active sites, reversibly poisoned sites, and dormant sites. The problem is to determine how much catalyst activity can be recovered by temperature adjustments that shift these equilibria.

32. A geological sample contains coexisting mineral phases whose compositions indicate the equilibrium temperature of crystallization. However, subsequent cooling allowed solid-state diffusion to partially alter compositional gradients. A geochemist must determine whether the mineral assemblage reflects true equilibrium at high temperature or re-equilibration at lower temperatures, and whether isotope ratios can constrain the time at which equilibrium was last achieved.

33. An enzyme that catalyzes a reversible redox reaction contains a metal center that can adopt multiple oxidation states. Under turnover conditions, the distribution of oxidation states deviates from thermodynamic equilibrium because steady-state flux continuously consumes one state. A biochemist must deduce the true redox equilibrium potentials from transient relaxation experiments performed immediately after the reaction is quenched.

34. A phase-transfer reaction involves an organic cation exchanging between an aqueous phase and an organic solvent. The distribution ratio depends on the

cation's hydration equilibrium, the ion pairing with its counter-ion, and the ligand environment in the organic phase. Because ion pairs have lower polarity, the apparent partition coefficient changes with ionic strength. The problem is to derive the intrinsic hydration free energy of the cation from partition data complicated by competing equilibria.

35. A polymer gel swells in water according to osmotic equilibrium between fixed charges in the gel and mobile ions in solution. When multivalent ions are added, they bind to the gel's charged groups and collapse the gel. Because binding is reversible but cooperative, the swelling curve shows discontinuous transitions. The task is to determine whether the collapse reflects a thermodynamic phase transition or a kinetic barrier between swollen and collapsed states.

36. A catalytic converter operates via reversible adsorption and reaction of CO, NO, and O_2 on metal surfaces. At high exhaust temperatures, some surface sites sinter, changing the equilibrium site distribution. The challenge is to determine whether reduced CO conversion at high mileage reflects a shift in adsorption equilibrium due to altered site energies or irreversible loss of active surface area.

37. A gas-liquid reactor dissolves CO_2 into an alkaline solution, forming CO_3^{2-} and HCO_3^-. As the reactor operates, bubbles coalesce and change gas–liquid interfacial area, altering mass transfer rates. Even though the acid–base equilibria are fast, the reactor does not reach the equilibrium distribution of carbonate species predicted by pH. The problem is to determine how far the system may deviate from equilibrium due to mass-transfer limitations alone.

38. A solid oxide fuel cell operates with oxygen ions diffusing through a ceramic membrane. On each side, the equilibrium O_2 partial pressure determines the oxide ion chemical potential. Because the electrode reactions are reversible, the overall cell potential reflects the equilibrium constant for O^{2-} formation. The problem is to determine whether the measured voltage drop during operation arises from deviation from equilibrium at the electrodes or from slow ionic transport inside the membrane.

39. A peptide undergoes reversible folding governed by equilibrium constant Kf. A denaturant gradually increases its concentration during an experiment because of slow evaporation of solvent. Surprisingly, the peptide appears to refold slightly

during late stages of evaporation. The challenge is to determine whether this reflects a subtle temperature drift that shifts equilibrium or an artifact arising from changes in activity coefficients at high denaturant concentration.

40. A metal-organic framework (MOF) adsorbs CO_2 via equilibrium binding to open metal sites. However, at higher CO_2 pressures, the framework undergoes a subtle structural distortion that changes the binding enthalpy. The adsorption isotherm therefore deviates from Langmuir behavior. The problem is to determine whether the material exhibits true cooperative binding or two distinct equilibrium states corresponding to different framework conformations.

41. A zeolite-catalyzed isomerization reaction exhibits high selectivity toward one product at low temperature but shifts toward a different product at high temperature. Both products form through reversible reactions, but the transition-state entropies differ, making the temperature dependence of the apparent equilibrium unusual. The problem is to determine whether entropic contributions alone explain the temperature-dependent selectivity or if pore-diffusion limitations perturb the equilibrium.

42. A high-pressure reactor performing the methane steam-reforming equilibrium $CH_4 + H_2O \rightleftharpoons CO + 3H_2$ shows unexpectedly low hydrogen yields at very high pressure. Engineers suspect that methane adsorption on the catalyst shifts the effective equilibrium toward reactants by reducing free concentrations. The problem is to quantify the effect of adsorption equilibria on the overall thermodynamic equilibrium constant.

43. A liquid mixture of acetic acid and an alcohol forms an ester via equilibrium esterification, producing water. Because all species strongly hydrogen-bond, activity coefficients deviate significantly from unity. A chemical engineer must determine the true equilibrium constant from VLE data while accounting for excessive non-ideality and the presence of azeotropes that complicate product separation.

44. A lake receives intermittent acidic rainfall, temporarily lowering pH. Because carbonate equilibria buffer the water, pH recovers slowly after each event. A limnologist must determine whether the slow recovery arises from equilibrium shifts constrained by dissolved CO_2 exchange with the atmosphere or from slow dissolution of carbonate minerals in the sediment.

45. A protein–ligand binding study using fluorescence spectroscopy suggests two binding sites with different affinities. However, the protein forms dimers at concentrations used in the experiment, and dimerization itself is a concentration-dependent equilibrium. The observed binding isotherm is therefore a convolution of ligand binding and protein dimerization equilibria. The problem is to separate the two and determine intrinsic binding constants.

46. An organic reaction involves the reversible formation of an intermediate carbocation, which then reacts irreversibly with solvent. Under highly ionizing conditions, the carbocation concentration reaches a small but finite equilibrium, significantly influencing reaction rate. The problem is to determine whether variations in solvent polarity shift the equilibrium concentration of the carbocation or simply alter the rate of solvent capture.

47. A clay mineral surface exchanges cations with surrounding solution through equilibrium ion-exchange processes. However, some sites are buried within layers that expand only slowly when hydrated. A soil chemist must determine whether time-dependent cation release reflects kinetic limitations of layer swelling or equilibrium partitioning between external and internal sites.

48. A multi-step atmospheric oxidation of a hydrocarbon forms an equilibrium between a peroxy radical RO_2 and its dimer $(RO_2)_2$. Because the dimer has low vapor pressure, it contributes to aerosol formation. The problem is to determine whether aerosol growth reflects equilibrium partitioning of RO_2 species or whether irreversible oligomerization traps species away from the gas-phase equilibrium.

49. A hydrogen-bonded molecular crystal exhibits reversible proton transfer between two tautomeric forms. As pressure increases, the relative stability of the tautomers changes. Raman spectroscopy shows multiple peaks attributed to both forms. The challenge is to determine whether the observed spectral intensities correspond to equilibrium populations or to kinetically trapped domains unable to interconvert under high pressure.

50. A synthetic reaction network in a microfluidic chip couples an endergonic reaction to an exergonic one through mass transport between channels. Because diffusion distances are short, concentration gradients are steep, and local equilibria differ from global ones. The problem is to determine whether the overall product distribution reflects the true equilibrium constant of the coupled system or merely a spatially averaged nonequilibrium state created by flow geometry.

Chapter 15

Harmony of Acids, Bases, and Buffers

Acids and bases have been woven into human experience since long before the rise of formal chemical science, their presence felt in the tang of fermented foods, the sting of plant resins, the soothing qualities of alkaline ashes, and the practical uses of lye-rich soaps. Early civilizations lacked the conceptual framework to understand why certain substances burned the tongue with sourness or why others countered that sensation with a chalky bitterness, yet they learned—often through trial, error, and careful observation—to make use of these materials in cooking, cleaning, preservation, tanning, metallurgy, and medicine. Vinegar served not only as a culinary ingredient but also as a cleaning agent and preservative; citrus juices were prized for their ability to alter the taste and texture of foods; and ash-derived alkaline solutions found use in soap making and textile treatments. Long before the development of modern theories, artisans recognized that acidic mixtures could dissolve mineral deposits, etch metals, or aid in dyeing fabrics, while alkaline mixtures could soften plant fibers or neutralize harsh acidic components in natural materials. These early encounters laid the groundwork for later scientific inquiry by highlighting the distinct, reproducible behaviors associated with what we now classify as acids and bases.

With the emergence of experimental science in the Renaissance and the Enlightenment, thinkers such as Robert Boyle began to challenge older mystical theories about matter. Boyle observed that acids often produced effervescence when mixed with certain minerals, and he noted consistent properties—sharp taste, ability to corrode metals, and capacity to change the color of plant extracts—that hinted at an underlying principle. Yet without modern atomic theory or an understanding of ions, these observations remained descriptive rather than explanatory. As chemical knowledge deepened in the eighteenth and nineteenth centuries, aided by Lavoisier's work on oxygen and the development of early theories of chemical bonding, it became possible to categorize acidic and basic behavior more precisely. Still, definitions were tied to macroscopic characteristics and did not reveal the microscopic nature of the reactions.

It was not until the late nineteenth century that chemists such as Arrhenius began to connect acidity with the presence of specific particles in solution. Arrhenius, studying the electrical conductivity of water and salt solutions, recognized that certain substances dissociated into ions, and he proposed that acids were compounds releasing hydrogen ions (H^+) in water, whereas bases released hydroxide ions (OH^-). This theory was revolutionary because it tied macroscopic phenomena—corrosion, taste, reactivity—to microscopic events: the separation of molecules into charged species. Yet despite its insights, the Arrhenius model remained limited, applicable chiefly to aqueous systems and unable to account for substances like ammonia, which behaved as bases despite containing no hydroxide ions. The inadequacy of the Arrhenius definition for describing nonaqueous systems, gas-phase reactions, and many organic processes motivated the development of more flexible theories.

The Brønsted–Lowry concept, introduced in the 1920s by Johannes Brønsted and Thomas Lowry independently, transformed the study of acid–base reactions by focusing on proton transfer. In this theory, an acid is a proton donor and a base is a proton acceptor. This simple yet powerful redefinition extended acid–base chemistry beyond water, allowing chemists to analyze proton-transfer reactions in solvents such as ammonia, acetic acid, or even the gas phase. This framework also enabled the systematic description of conjugate acid–base pairs, illuminating how the loss or gain of a proton altered the chemical nature of a species. What emerged was a dynamic view in which acids and bases existed in equilibrium, constantly engaging in reversible proton exchange, with the direction of the reaction depending on the relative strengths of the acid and base involved.

Water, in this expanded framework, assumed a special position as an amphoteric substance capable of acting as either an acid or a base. Its molecular structure, featuring a highly electronegative oxygen atom bonded to two hydrogen atoms, endowed it with both the ability to donate a proton—under the right conditions—and the capacity to accept one via its lone electron pairs. This dual behavior became especially evident in reactions involving strong acids such as hydrobromic acid, where water accepts a proton to form the hydronium ion, H_3O^+. Conversely, when in the presence of species with strong proton affinity, water can donate a proton, demonstrating its acidic character. The equilibrium between an acid HA and water, $HA + H_2O \rightleftharpoons H_3O^+ + A^-$, illustrates the interplay between acid strength and base strength. The extent to which the reaction proceeds depends on the tendency of the proton to reside either with water or with the conjugate base A^-. When water is a stronger proton acceptor, the equilibrium shifts toward hydronium formation, while a stronger conjugate base pulls the proton back, limiting dissociation.

The acid dissociation constant, Ka, quantifies this balance. Defined as $K_a = [H^+][A^-] / [HA]$, it reflects the degree to which an acid donates protons in solution. Since water's concentration remains effectively constant in dilute solutions, it is omitted from the expression, focusing analysis on the species whose concentrations change during dissociation. Large K_a values signify strong acids that dissociate extensively; small K_a values denote weak acids in which most molecules remain undissociated. Although proton-transfer reactions can occur in any medium, the quantitative treatment of acid strength is especially meaningful in aqueous environments, where equilibrium expressions can be measured with precision. Even in the gas phase, where reactions such as $NH_3 + HCl \rightarrow NH_4Cl$ take place, the conceptual framework of proton transfer remains relevant, though equilibrium behavior differs from that observed in solution.

As chemists explored acid strength in greater depth, they recognized clear patterns. Strong acids such as hydrochloric, nitric, sulfuric, and perchloric acid dissociate nearly completely in water, driving their equilibria heavily toward hydronium formation. Their conjugate bases are exceedingly weak, possessing minimal ability to reaccept protons. Sulfuric acid presents an interesting case because it is diprotic: while its first proton dissociates essentially completely, the second does not, leaving measurable contributions from both dissociation steps in determining the acidity of its solutions. Weak acids like acetic acid or hydrofluoric acid dissociate only slightly, yielding solutions in which the majority of molecules remain intact. Their conjugate bases, being relatively stronger, compete effectively with water for protons. Hydrofluoric acid, despite possessing a highly polar H–F bond, is unexpectedly weak due to the exceptional strength of that bond, which resists dissociation. By contrast, hydrogen halides such as HCl, HBr, and HI grow stronger down the group, their bonds becoming weaker and thus more readily broken.

Further nuance arises when considering oxyacids—acids containing hydrogen bonded to oxygen atoms attached to a central atom. Their strengths correlate with the electronegativity and oxidation state of the central atom. Nitric acid (HNO_3), with a more oxidized nitrogen and an additional oxygen relative to nitrous acid (HNO_2), is significantly stronger because the additional oxygen withdraws electron density, stabilizing the conjugate base and easing proton release. In organic chemistry, carboxylic acids display moderate acidity due to resonance stabilization of their conjugate bases, while alcohols and hydrocarbons remain very weak acids due to strong, nonpolar C–H or O–H bonds lacking such stabilization.

The autoionization of water further enriches the picture. Water molecules undergo a spontaneous, though rare, proton exchange according to $H_2O + H_2O \rightleftharpoons H_3O^+ + OH^-$, producing equal amounts of hydronium and hydroxide ions. The equilibrium constant for this reaction, $K_w = [H^+][OH^-]$, equals 1.0×10^{-14} at 25 °C, leading to hydronium and hydroxide concentrations of 1.0×10^{-7} M each in pure water. This autoionization provides the basis for defining neutral, acidic, and basic solutions: when hydronium exceeds hydroxide, the solution is acidic; when hydroxide exceeds hydronium, it is basic. Although the hydronium contribution from water is negligible in solutions containing strong acids or bases, it becomes significant in very dilute or highly controlled environments, and it establishes the foundation of the pH scale.

Because hydronium concentrations span many orders of magnitude, the logarithmic pH scale offers a convenient means of expressing acidity. Defined as pH = $-\log[H^+]$, it converts small concentrations into manageable numbers. The pOH scale, similarly defined as pOH = $-\log[OH^-]$, complements pH, and at 25 °C, their sum equals 14 due to the value of Kw. Modern pH measurement relies on glass electrodes that develop a voltage proportional to hydronium ion activity. These devices permit rapid, accurate, and reproducible readings that surpass the capabilities of dyes, litmus paper, or other colorimetric indicators, making them essential in laboratory work, industrial processes, water treatment, and clinical diagnostics.

Bases, in parallel to acids, can be interpreted through both Arrhenius and Brønsted–Lowry models. Strong bases such as sodium hydroxide, potassium hydroxide, and other Group 1A hydroxides dissociate almost entirely in solution, producing hydroxide ions directly. Group 2A hydroxides like calcium hydroxide are less soluble but nevertheless strongly basic in the proportion that does dissolve. Bicarbonate, carbonate, and phosphate ions also display basic behavior, reacting with water to produce hydroxide through proton acceptance. Even without containing hydroxide ions, molecules such as ammonia function as bases by accepting protons from water, forming ammonium ions and generating hydroxide indirectly. Their behavior is quantified by the base dissociation constant $K_b = [BH^+][OH^-]/[B]$, which expresses the strength of the proton-accepting species.

Polyprotic acids add another layer of complexity. Sulfuric, carbonic, and phosphoric acids can each lose multiple protons in a stepwise fashion, each dissociation having its own equilibrium constant. Because successive protons are removed from increasingly negatively charged species, each step becomes more difficult, and Ka values decrease sharply with each stage. In most practical cases, only the first dissociation significantly affects solution pH, while later steps matter mainly in buffering systems, very dilute solutions, or specialized industrial and biological contexts. Carbonic acid, for example, plays a vital role in the blood bicarbonate buffering system, where the interplay between CO_2, H_2CO_3, HCO_3^-, and CO_3^{2-} helps maintain physiological pH.

Salt solutions highlight the interplay between acid and base properties in ionic compounds. A salt formed from a strong acid and strong base, such as sodium chloride, yields a neutral solution because neither ion reacts significantly with water. By contrast, salts of weak acids like sodium acetate produce basic solutions because the conjugate base reacts with water to generate hydroxide. Salts derived from weak bases, such as ammonium chloride, yield acidic solutions because the conjugate acid donates protons to water. Some metal ions, especially those with high charge density like Al^{3+} or Fe^{3+}, strongly polarize the water molecules surrounding them, weakening O–H bonds and facilitating proton release, thus creating acidic environments. When both the cation and anion of a salt can undergo hydrolysis, the dominant Ka or Kb determines the resulting pH.

Structural features exert profound influence on acidity. The presence of a removable proton is only a starting point; the strength of the bond to that proton, the polarity of the bond, the stability of the conjugate base, and the molecular environment all play crucial roles. Hydrocarbons, with strong nonpolar C–H bonds, remain extremely weak acids, while molecules with highly polarized bonds or resonance-stabilized conjugate bases exhibit stronger acidity. Electron-withdrawing substituents enhance acid strength by stabilizing the negative charge on the conjugate base, whereas electron-donating groups weaken acids. In inorganic chemistry, oxides of nonmetals tend to form acids in water due to their covalent structures and strong electron-withdrawing tendencies, while ionic oxides of metals form hydroxides, acting as bases. Transition-metal ions, depending on their oxidation state and ligand environment, may acidify water by polarizing its bonds. All these structural factors reveal how molecular architecture shapes macroscopic acid–base behavior.

The Lewis definition, introduced by Gilbert N. Lewis in 1923, broadened the concept even further by focusing on electron-pair interactions. A Lewis acid is an electron-pair acceptor, while a Lewis base is an electron-pair donor. This model expanded acid–base chemistry beyond proton transfer entirely, encompassing complex formation, electrophile–nucleophile interactions, and catalytic processes that did not fit the Brønsted–Lowry framework. Boron trifluoride exemplifies a Lewis acid due to its electron-deficient boron center with an empty p orbital capable of accepting a lone pair. Ammonia, with its nitrogen lone pair, readily acts as a Lewis base, forming adducts such as $BF_3 \cdot NH_3$. Transition-metal ions, often possessing vacant d orbitals, function as powerful Lewis acids that bind ligands through coordinate covalent bonds, producing a spectrum of colorful and reactive complex ions. This perspective encompasses a vast range of reactions, including many that occur in organic and inorganic synthesis, enzyme active sites, and industrial catalysis.

The relevance of acid–base chemistry extends far beyond theory, permeating biological, environmental, and industrial domains. In living organisms, acid–base balance is essential for survival. Human blood, for instance, must maintain a narrow pH range of 7.35–7.45. Even slight departures from this interval can disrupt protein folding, reduce oxygen-carrying capacity by altering hemoglobin conformation, impair enzyme activity, and compromise membrane integrity. The body employs a multilayered regulatory system to maintain pH: chemical buffers such as bicarbonate, phosphate, and plasma proteins respond almost instantly; respiratory adjustments modulate carbon dioxide levels; and renal mechanisms provide long-term control by reabsorbing or excreting hydrogen and bicarbonate ions. Similar homeostatic strategies appear throughout the biological world, from plant sap regulation to oceanic buffering systems that support marine life.

Industrial processes rely heavily on both acids and bases. Sulfuric acid remains one of the most widely produced industrial chemicals, supporting fertilizer manufacture, petroleum refining, mineral processing, steel pickling, battery production, and polymer synthesis. Hydrochloric, phosphoric, and nitric acids are indispensable in metallurgy, food processing, and the production of explosives and dyes. Bases such as sodium hydroxide and potassium hydroxide are central to paper manufacturing, textile preparation, water purification, soap production, and pharmaceutical synthesis. Despite their utility, acids and bases can exert destructive effects when introduced into the environment in uncontrolled ways. Acid rain, formed by the atmospheric oxidation of sulfur and nitrogen oxides, accelerates the degradation of buildings and monuments, leaches essential nutrients from soils, acidifies lake waters, and disrupts delicate ecological balances. Understanding the acid–base chemistry driving these phenomena is crucial for mitigating their impact and developing more sustainable technological practices.

Across all these contexts—biological, environmental, industrial, and theoretical—acid–base chemistry forms a unifying theme that links molecular behavior to macroscopic outcomes. From the earliest sensory encounters of ancient peoples to the sophisticated physical models of modern chemistry, our understanding has evolved from qualitative impressions to quantitative precision. The Arrhenius model explains the formation of hydronium and hydroxide ions in aqueous solution. The Brønsted–Lowry model reveals the dynamic equilibrium of proton transfer and conjugate pairs.

The Lewis model expands this knowledge into the domain of electron-pair interactions, encompassing countless processes unrelated to classical proton exchange. Together, these models weave a comprehensive, multi-layered understanding of acidic and basic behavior. They illuminate the roles played by electronegativity, bond strength, resonance stabilization, solvation effects, and charge density in determining acid and base strength. They clarify the behavior of salts, complexes, polyprotic species, and amphoteric molecules. And they link microscopic structural features to vast natural cycles, intricate biochemical networks, industrial production systems, and global environmental changes.

In the end, the story of acids and bases is not merely a chapter within chemistry; it is a lens through which we can view the interplay between structure and reactivity, between micro-scale dynamics and macro-scale consequences. It reveals how the transfer or sharing of something as small as a proton or an electron pair can shape processes as grand as atmospheric chemistry or as intimate as cellular metabolism. The evolution of acid–base theory reflects the broader progression of science—from empirical observation to mechanistic explanation and, ultimately, to integrative frameworks that unite diverse phenomena under coherent principles. Understanding acids and bases thus opens a window into the logic of chemical change itself, enriching our appreciation of the natural world and enhancing our capacity to harness chemical processes for human benefit.

The evolution of acid–base chemistry over the centuries also reflects the broader relationship between human curiosity, technological needs, and theoretical advancement. As new tools emerged—more precise balances, improved thermometers, advances in spectroscopy, and eventually powerful computational models—the ability to measure acidity and basicity grew increasingly refined. What once could only be detected through taste or crude indicators became measurable through quantitative titration methods, standardized reagents, and colorimetric scales. Early titration work by Gay-Lussac and Mohr brought acid–base analysis into the realm of careful stoichiometry, laying the foundation for modern analytical chemistry. Later, the development of indicators such as phenolphthalein, methyl orange, and bromothymol blue enabled the detection of tiny shifts in pH through visible color changes, bridging the macroscopic and microscopic realms by allowing scientists to visualize proton-transfer reactions in real time. These tools were revolutionary not only for laboratory research but also for industries dependent on consistent acidity control, such as brewing, textile dyeing, and leather tanning.

As the twentieth century progressed, the study of acids and bases expanded beyond classical aqueous systems into wholly new environments. Superacids—those stronger than pure sulfuric acid—captured the attention of chemists interested in carbocations and high-energy intermediates. Compounds such as fluoroantimonic acid, created by combining hydrogen fluoride with antimony pentafluoride, demonstrated acidity thousands of times stronger than traditional mineral acids, capable of protonating molecules normally considered inert. These superacids opened new frontiers in synthetic chemistry, enabling the stabilization of delicate species such as carbenium ions and allowing researchers to probe the limits of molecular stability. Their study required specialized equipment due to their extreme corrosiveness and reactivity, but their existence pushed acid–base theory into regions unimaginable a century earlier.

At the opposite end of the spectrum, superbases—extremely strong proton acceptors—also broadened chemical horizons. Organometallic compounds such as butyllithium or the alkali metal amides demonstrated powerful basicity in organic solvents, capable of deprotonating molecules far too weakly acidic to interact with conventional bases. The emergence of non-nucleophilic superbases like lithium diisopropylamide (LDA) transformed organic synthesis, allowing precise control over proton abstraction at specific molecular sites. These reagents made it possible to form carbanions, initiate polymerizations, and create new carbon–carbon bonds with remarkable selectivity. In these systems, the classical notion of hydroxide as the primary base fades into the background, replaced by a vast landscape of proton-accepting species whose behavior depends on solvent effects, solvation energy, steric hindrance, and electronic structure.

Solvents themselves became key actors in acid–base chemistry once the limitations of aqueous models became apparent. Chemists learned that the strength of an acid or base depends heavily on the solvent environment. A substance considered a strong acid in water, such as hydrochloric acid, may be weak in acetic acid due to the leveling effect, where all acids stronger than the conjugate acid of the solvent appear equally strong because they all fully donate protons to the solvent. Conversely, the differentiation effect of solvents like liquid ammonia or sulfuric acid enables the ranking of acid strengths that would otherwise be indistinguishable in water.

Solvent leveling and differentiation have important implications across synthetic chemistry, biochemistry, and industrial processing, underscoring that acidity is not an intrinsic property of a substance alone but a function of the entire chemical environment.

These insights became even more pronounced with the rise of physical chemistry and thermodynamics. By examining Gibbs free energy changes associated with proton-transfer reactions, chemists could quantify the driving forces behind acidity and basicity. Entropy, enthalpy, solvation energies, and hydrogen bonding all emerged as crucial determinants of acid strength. For instance, the acidity of oxoacids correlates strongly with the stability of their conjugate bases, which in turn depends on resonance stabilization, inductive effects, and electron delocalization. Thermodynamic cycles allowed the calculation of acid dissociation constants even in environments where direct measurement was difficult or impossible. These advances provided a theoretical backbone to observational chemistry, allowing predictions of acidity in novel compounds, guiding synthetic strategies, and improving understanding of biological proton-transfer processes.

Biochemistry, perhaps more than any other field, showcases the delicate interplay of acid–base interactions in sustaining life. Inside living cells, enzymes rely on carefully tuned microenvironments where specific amino acid side chains act as acids or bases. Histidine, with its imidazole ring capable of accepting or donating a proton near physiological pH, plays a pivotal role in catalytic mechanisms, stabilizing transition states or facilitating substrate transformations. The functioning of ATP synthase, the operation of proton pumps, the folding and stability of proteins, and the structure of DNA all rely on protonation states regulated by acid–base chemistry. Even small deviations from optimal pH can drastically alter enzymatic rates, binding affinities, and structural conformations. Many biological processes exploit subtle pKa differences to control when and where proton transfer occurs, allowing molecular machinery to operate with extraordinary precision.

The pH gradients maintained across biological membranes, such as those in mitochondria or chloroplasts, exemplify the profound importance of proton concentration differences. The chemiosmotic theory proposed by Peter Mitchell revolutionized understanding of cellular energy conversion by revealing that proton gradients drive ATP production. The movement of protons down their electrochemical gradient through ATP synthase couples acid–base chemistry directly to the energy economy of the cell. The maintenance of these gradients requires tightly regulated transport proteins, buffer systems, and metabolic pathways, illustrating how acid–base chemistry underlies even the most fundamental processes of life.

In the broader environment, acid–base reactions shape ecosystems and global cycles. Natural bodies of water rely on carbonate buffering systems to maintain relatively stable pH despite fluctuations in dissolved carbon dioxide, organic acids from decaying vegetation, or acidic inputs from rainfall. In lakes and oceans, the interplay between carbonic acid, bicarbonate, and carbonate ions regulates not only water chemistry but also the ability of organisms to form calcium carbonate shells and skeletons. Ocean acidification, driven by rising atmospheric CO_2, alters this balance by increasing carbonic acid formation, reducing carbonate ion concentrations, and threatening marine life such as corals, mollusks, and plankton. Understanding the acid–base equilibria in these systems is crucial for predicting ecological impacts and developing strategies for conservation and restoration.

In soils, acid–base chemistry governs nutrient availability, microbial activity, and plant growth. Soil pH influences the solubility of essential minerals such as iron, phosphorus, and magnesium, determining whether plants can absorb them. Acidic soils can mobilize toxic metals like aluminum, inhibiting root growth, while alkaline soils can cause nutrient deficiencies. Farmers routinely adjust soil pH with lime or sulfur amendments to optimize growing conditions, relying on a practical understanding of acid–base reactions even without formal chemical training. Environmental scientists monitor soil and water pH as key indicators of ecosystem health, recognizing that shifts in acidity can signal pollution, acid rain effects, or changes in land use.

Air pollution provides another striking example of large-scale acid–base interactions. Sulfur dioxide and nitrogen oxides released from industrial processes and combustion engines undergo complex atmospheric chemistry, ultimately producing sulfuric and nitric acids that fall as acid rain. This acidification can damage forests, accelerate the decay of stone buildings and monuments, and disrupt aquatic life. International agreements such as the Clean Air Act in the United States and transboundary air pollution treaties in Europe were informed by decades of research into the atmospheric acid–base chemistry responsible for these effects. These policies illustrate how scientific understanding of proton-transfer processes can shape public policy and environmental protection efforts.

In industrial contexts, the control of acidity and basicity is indispensable to manufacturing, materials processing, and quality assurance. The production of semiconductors requires ultra-pure acids and bases to clean silicon wafers without introducing contaminants. Petroleum refining relies on acid-catalyzed cracking processes to break long hydrocarbons into useful fractions. The synthesis of pharmaceuticals often depends on acid–base extraction techniques, salt formation to adjust drug solubility and stability, and pH-controlled crystallization steps. The paper industry uses sodium hydroxide to break down lignin, while water treatment plants employ lime and alum to adjust pH and remove impurities. Every one of these processes hinges on predictable acid–base behavior, and small variations in pH can have significant economic consequences.

Advances in computational chemistry have opened new avenues for understanding acidity at an even deeper level. Quantum mechanical calculations allow chemists to model proton-transfer pathways, predict pKa values in complex molecules, and simulate solvent effects with remarkable accuracy. Molecular dynamics simulations reveal how solvation shells reorganize during acid–base reactions, while density functional theory sheds light on the electronic structures that drive acidity trends. These methods enable the exploration of hypothetical molecules, guiding experimental chemists toward compounds with desired acidity or basicity properties before laboratory synthesis begins. In drug design, computational predictions of pKa help ensure that pharmaceutical molecules exist in the correct protonation state at physiological pH, influencing absorption, distribution, and receptor binding.

In the field of materials science, acid–base interactions play central roles in the synthesis of catalysts, polymers, and nanomaterials. Zeolites, for example, are porous aluminosilicate materials whose catalytic activity depends on acidic sites within their framework. These sites facilitate cracking, isomerization, and other reactions essential to petrochemical industries. Metal–organic frameworks (MOFs) often incorporate Lewis acidic metal centers that bind guest molecules or activate substrates. Polymers can be engineered with acidic or basic functional groups that respond to environmental pH, enabling smart materials that swell, contract, or release cargo under specific conditions. These innovations demonstrate how acid–base chemistry underlies the design of advanced technologies.

Even in everyday life, acid–base chemistry quietly influences countless activities. The fizz of carbonated beverages comes from carbonic acid. Baking relies on the reaction between acidic ingredients and basic baking soda to release carbon dioxide, leavening dough. Swimming pool maintenance requires constant pH monitoring to ensure safety and comfort. Household cleaners often contain strong bases to dissolve organic residues or strong acids to remove mineral deposits. Human senses have evolved to detect acidity as sourness, likely as a protective mechanism against spoiled or unripe foods, while bitterness associated with some bases warns of potentially harmful substances. These mundane examples illustrate how deeply ingrained acid–base interactions are in daily human experience, even when unrecognized.

Philosophically, the study of acids and bases represents a shift from qualitative natural philosophy toward quantitative modern science. Early thinkers marveled at the corrosive or cleansing powers of certain substances without understanding their essence. With Arrhenius came the realization that invisible particles—ions—governed these behaviors. Brønsted, Lowry, and Lewis further abstracted the concept, demonstrating that acidity and basicity were not inherent traits but relational properties defined by interactions. This progression from sensory description to theoretical abstraction mirrors the broader trajectory of chemistry from alchemy to a rigorous physical science. It illustrates how simple observations can inspire profound frameworks that overturn previous assumptions and reveal deeper unity among seemingly disparate phenomena.

As research continues, acid–base chemistry remains a dynamic field with new discoveries emerging in supramolecular chemistry, biophysics, green energy, and environmental science. Proton-coupled electron transfer, a process linking acidity to redox chemistry, plays a vital role in photosynthesis, fuel cell design, and artificial photosynthesis systems. Ionic liquids with tunable acidity offer environmentally friendly alternatives to volatile organic solvents. Solid acids and bases integrated into catalytic membranes are being developed for sustainable chemical manufacturing. Even extraterrestrial chemistry raises questions about acidity in environments radically different from Earth's oceans and atmosphere, inviting fresh perspectives on the universality of acid–base principles.

Through all these developments, one central theme persists: acidity and basicity are not isolated concepts but interconnected threads running through the fabric of chemical reactivity. Whether in the rusting of iron, the functioning of enzymes, the buffering of oceans, the synthesis of medicines, or the crafting of everyday materials, acid–base reactions underpin transformations both grand and subtle. The journey from ancient practices to sophisticated theories illustrates humanity's enduring effort to understand and manipulate the natural world through careful observation, creative theorizing, and technological innovation. Acid–base chemistry, in all its complexity and elegance, stands as a testament to the power of scientific inquiry to illuminate the hidden workings of matter and to harness those insights for the benefit of society.

Chapter 15: Word Problems

1. A chemist prepares a 0.0500 M solution of a monoprotic weak acid HA (pK_a = 5.30) and measures the pH using a high-precision electrode. After preparation, she notices that the pH is slightly lower than expected due to measurable autoionization of water at this low concentration. Considering the temperature is 25 °C, where water contributes 1.0×10^{-7} M hydronium, determine the actual hydronium concentration. Calculate the percent error in hydronium concentration if the autoionization of water is ignored, and discuss how this correction might influence pH measurements in very dilute weak acid solutions in analytical chemistry.

2. In a laboratory studying atmospheric acid rain formation, researchers bubble a gas mixture of SO_2 and O_2 through deionized water containing trace amounts of a transition metal catalyst that accelerates the oxidation of sulfur dioxide to sulfuric acid. After 30 minutes, the pH drops from 6.2 to 2.8. Assuming no buffering species are present, estimate the concentration of sulfuric acid formed. Evaluate whether the second dissociation of H_2SO_4 significantly contributes to the final pH and discuss implications for acid rain modeling in environmental chemistry.

3. An archaeologist analyzing residues on ancient pottery discovers evidence of early soapmaking using ash-derived alkaline solutions. She recreates the process by dissolving 15 g of potassium carbonate in 1.0 L of water and notes its alkalinity. Given that carbonate undergoes stepwise protonation with pKa_2 = 10.33 for $HCO_3^- \rightarrow H_2CO_3$, estimate the pH of the resulting solution. Discuss whether early artisans could have produced a strongly basic solution capable of saponifying fats, and how the chemistry of carbonate contributes to this process.

4. A solution initially contains 0.100 M acetic acid and 0.200 M sodium acetate at 25 °C. After accidentally adding 15 mL of 0.500 M HCl to 1.00 L of buffer, the laboratory technician wants to know if the solution remains within one pH unit of the pK_a (4.76). Determine the new pH, evaluate the buffer's effectiveness, and discuss how buffer capacity is influenced by both concentration and volume changes during titration experiments.

5. A pharmaceutical chemist needs to prepare a buffer at physiological pH 7.40 using the $H_2PO_4^-/HPO_4^{2-}$ pair (pKa_2 = 7.21). She has stock solutions of 0.250 M NaH_2PO_4 and 0.250 M Na_2HPO_4. If she requires 500 mL of buffer while keeping ionic strength below 0.10 M, determine the appropriate volumes of each stock solution, justify your choice, and explain the importance of controlling ionic strength in drug formulation for parenteral administration.

6. A 1.00-g sample of an impure diprotic acid is titrated with 0.100 M NaOH. The first equivalence point occurs at 22.50 mL, and the second at 44.80 mL. Determine the molar mass of the pure acid, calculate the percent purity of the sample, and discuss potential sources of error in titration, including the influence of coexisting impurities on the titration curve shape.

7. A biochemist studying an enzyme that requires histidine in its protonated form notes that the local environment of the active site shifts the pK_a of histidine's imidazole side chain from 6.0 to 7.2. If the cytosolic pH is 7.4, calculate the fraction of histidine residues that remain protonated. Discuss whether this fraction is sufficient for enzymatic activity and explore how microenvironmental effects modulate amino acid protonation in proteins.

8. A 2.0-L container holds a gaseous mixture of HCl and NH_3 at 1.00 atm total pressure and 298 K. When the gases react completely to form solid NH_4Cl, the pressure drops to 0.72 atm. Determine the initial partial pressures of HCl and NH_3 and identify which gas was in excess. Discuss how stoichiometry and gas laws inform the interpretation of partial pressures in reactive gas mixtures.

9. A strong base is dissolved in liquid ammonia, a solvent with a conjugate acid pK_a of 9.25. A researcher adds sodium amide (a superbase) to ammonia containing a weak acid with pK_a = 14.0 (in NH_3). Determine whether the acid is fully deprotonated, explain the concept of the leveling effect in nonaqueous solvents, and discuss how solvent properties alter acid–base equilibria compared to aqueous solutions.

10. A coastal scientist analyzes a seawater sample containing 2.30×10^{-3} M dissolved CO_2, 1.80×10^{-3} M bicarbonate, and 2.10×10^{-4} M carbonate. Using the carbonate system equilibria ($pKa_1 = 6.35$, $pKa_2 = 10.33$), estimate the seawater pH and assess whether the observed pH indicates the onset of acidification relative to the historical average of 8.20. Discuss the ecological implications of small pH changes on marine calcifying organisms.

11. A student titrates an unknown triprotic acid and observes a curve with three roughly equally spaced equivalence points. The midpoints of the buffer regions occur at pH 2.1, 7.4, and 12.6. She wants to determine the acid's identity and understand its buffering capacity over the pH range. Estimate the three pKa values and discuss the acid's classification as strong, weak, or polyprotic. Evaluate how overlapping pKa values could affect resolution of equivalence points, and suggest experimental approaches to accurately identify polyprotic acids with closely spaced dissociation constants.

12. A student dissolves 0.015 mol of the weak base pyridine ($K_b = 1.7 \times 10^{-9}$) in 500 mL of water. She then adds 0.010 mol of HCl and measures the pH. Analyze the resulting pH, determine whether the solution behaves primarily as a buffer or as a salt solution, and calculate the fraction of pyridine converted to pyridinium ions. Discuss how relative concentrations of base and strong acid influence buffer region, and consider practical implications for preparing pH-adjusted solutions in pharmaceutical formulations.

13. A winery monitors the acidity of grape must containing tartaric acid (diprotic; $pKa_1 = 3.02$, $pKa_2 = 4.37$) at a concentration of 6.00 g/L. The viticulturist is interested in understanding how acidity influences microbial fermentation. Estimate the pH of the must, analyze whether spontaneous fermentation microorganisms would thrive, and discuss the influence of weak acid equilibria on fermentation kinetics. Consider how minor pH fluctuations could affect wine quality and safety.

14. A municipal water system adds lime ($Ca(OH)_2$) to raise the pH of a soft water supply. The water initially contains 1.5×10^{-5} M dissolved CO_2 and has pH 5.75. Calculate the mass of $Ca(OH)_2$ required per 10,000 L to raise pH to 7.0, assuming complete reaction with carbonic acid. Discuss how the carbonate equilibria, water hardness, and buffer capacity interact, and analyze potential side effects on alkalinity and downstream infrastructure.

15. A chemist studying Lewis acid–base interactions dissolves 0.0200 mol of BF_3 gas into a solvent containing 0.0150 mol of NH_3. Assuming adduct formation occurs until one reactant is consumed, determine the equilibrium composition of BF_3, NH_3, and $BF_3 \cdot NH_3$. Discuss how stoichiometry, binding constants, and solvent effects influence equilibrium. Consider how this knowledge informs industrial reactions involving electron-deficient boron species and amine ligands.

16. During semiconductor processing, a silicon wafer is cleaned using a mixture of HF and NH_4F, which forms the complex ion $[HF_2]^-$. The worker prepares a solution containing 0.800 M HF and 1.20 M NH_4F. Estimate the concentration of free fluoride considering HF dissociation ($Ka = 6.6 \times 10^{-4}$) and complex formation with $[HF_2]^-$. Discuss how fluoride speciation affects etching efficiency, wafer surface chemistry, and safety considerations in semiconductor labs.

17. An industrial reactor uses 98% sulfuric acid as a solvent. A chemist adds 0.500 mol acetic acid and notices that it behaves as a very strong acid in this medium. Using the concept of solvent leveling, explain quantitatively which species donate protons, estimate relative acidity compared to aqueous solutions, and discuss implications for reactions carried out in concentrated acid media. Evaluate how the solvent influences equilibrium positions and the reactivity of weak acids.

18. A 250-mL buffer containing 0.100 M NH_3 and 0.100 M NH_4Cl is heated from 25 °C to 37 °C. The K_b of ammonia increases slightly with temperature (from 1.8×10^{-5} to 1.908×10^{-5}). Calculate the pH change due to heating, predict whether the solution becomes more acidic or basic, and discuss the importance of temperature effects in designing biological buffers. Consider implications for enzyme kinetics and protein stability in biochemical experiments.

19. A soil sample releases 12.0 mmol H^+ per kg when extracted with dilute salt solution. A farmer aims to neutralize soil acidity by applying powdered calcium carbonate. Calculate how many kilograms of $CaCO_3$ are needed per hectare for a 20-cm soil depth with bulk density 1.3 g/cm³. Discuss the role of soil buffering capacity, carbonate equilibria, and how maintaining optimal pH affects crop productivity and nutrient availability.

20. A marine biologist measures the pH of coral reef water at 7.92, with total alkalinity of 2.40×10^{-3} M primarily composed of bicarbonate and carbonate.

Estimate the carbonate concentration, evaluate whether the water supports optimal calcium carbonate deposition, and discuss ecological consequences of slight pH decreases on reef calcification and marine biodiversity. Consider both natural and anthropogenic contributions to seawater acidification.

21. A chemist prepares a 0.200 M solution of a weak monoprotic acid and finds that it is 3.5% dissociated. Determine its K_a and then predict the percent dissociation if the solution is diluted to 0.0200 M. Discuss quantitatively how dilution affects weak acid dissociation, consider ionic strength effects, and evaluate the relevance of this concept in designing low-concentration buffer solutions for analytical applications.

22. A 10.0-mL sample of an unknown polyprotic acid requires 35.0 mL of 0.150 M NaOH to reach the second equivalence point. Given that the molar mass of the acid is estimated to be between 80 and 100 g/mol, determine whether the acid is more likely diprotic or triprotic. Discuss how overlapping equivalence points could complicate experimental titration analysis, and consider strategies for resolving closely spaced titration steps in a graduate-level lab.

23. In a synthetic organic experiment, a researcher uses lithium diisopropylamide (LDA) at –78 °C to deprotonate a carbonyl compound with pK_a = 19.5. If the conjugate acid of LDA has pKa = 36, calculate the favorability of deprotonation and estimate the equilibrium constant. Discuss how low temperature influences kinetic versus thermodynamic control in enolate formation, and analyze potential side reactions that could occur if the superbase were added in excess.

24. A metal ion M^{3+} forms a hydrated complex $[M(H_2O)_6]^{3+}$ that acidifies water with K_a = 1.0×10^{-3}. If 0.0100 mol of the salt is dissolved in 1.00 L of water, calculate the pH and compare its acidity to that of carbonic acid. Discuss how hydrolysis of highly charged metal ions influences water chemistry, metal toxicity, and coordination chemistry equilibria in environmental and industrial contexts.

25. A 1.0-L buffer at pH 9.25 contains boric acid and borate (pK_a = 9.24). When 0.010 mol HCl is added, calculate the new pH and evaluate the buffer's capacity at high pH. Discuss the role of borate as a Lewis acid, its interaction with hydroxide, and how high-pH buffering is applied in chemical synthesis, analytical chemistry, and industrial processes.

26. A researcher dissolves 0.250 g of benzoic acid in an unknown organic solvent and measures the hydronium concentration as 1.2×10^{-6} M. Using the known K_a in water (6.5×10^{-5}), determine whether the solvent levels or differentiates acid strengths. Discuss how solvent polarity, hydrogen-bonding ability, and dielectric constant influence acid dissociation, and how this affects nonaqueous titrations and synthetic organic procedures.

27. A food scientist measures the titratable acidity of yogurt containing lactic acid (pK_a = 3.86). A 10.00-g sample requires 22.5 mL of 0.100 M NaOH to reach pH 8.3. Determine the lactic acid concentration and compare it to the concentration implied by the pH measurement. Discuss how titratable acidity and pH inform food quality, fermentation monitoring, and flavor development in dairy products.

28. In a biological experiment, CO_2 is bubbled into water initially at pH 7.0. After equilibrating with atmospheric CO_2 at 400 ppm, the pH drops to 5.61. Estimate the dissolved CO_2 concentration and relate it to the first dissociation equilibrium. Discuss the implications for oceanic carbon uptake, acidification, and the maintenance of physiological pH in aquatic organisms.

29. A buffer containing 0.30 M Tris base and 0.20 M Tris-HCl (pK_a = 8.06) is adjusted to pH 7.40 at 25 °C. When the temperature rises to 35 °C, the pKa decreases to 7.80. Calculate the new pH, quantify the change in protonation fraction of Tris, and discuss the importance of temperature compensation in biochemical assays, enzyme kinetics, and cell culture applications.

30. An industrial neutralization tank contains effluent with 0.0200 M HNO_3, which is mixed with an equal volume of 0.0150 M $Ca(OH)_2$. Determine the final pH, assess whether additional base is needed to meet regulatory pH limits of 6–9, and discuss the role of stoichiometry, excess base, and the buffer capacity in industrial wastewater treatment systems. Consider how process control and monitoring affect compliance with environmental regulations.

31. A lake receives runoff containing weak organic acids with pK_a values around 4.5. The lake is initially at pH 7.8 with alkalinity dominated by bicarbonate. Estimate whether the pH will change significantly, quantify the buffer capacity, and discuss ecological implications for aquatic life and the resilience of freshwater ecosystems to episodic acidification events.

32. A researcher mixes 0.100 mol of a weak base B ($K_b = 3.0 \times 10^{-4}$) with 0.0500 mol of HCl in 1.00 L. Determine the final pH, identify whether the solution behaves primarily as a buffer or as a salted weak base, and discuss how the ratio of base to acid affects buffer effectiveness, protonation state, and practical considerations for solution preparation.

33. A hydrofluoric acid etching bath contains 2.0% HF by mass. When 50.0 g of silica is added, it reacts according to $SiO_2 + 6\ HF \rightarrow H_2SiF_6 + 2\ H_2O$. Determine whether the bath becomes more or less acidic, calculate the remaining HF concentration, and discuss how silica consumption influences reaction kinetics, industrial etching efficiency, and safety precautions for handling HF.

34. A researcher studying ocean acidification finds that doubling atmospheric CO_2 decreases seawater pH from 8.20 to 8.04. Estimate the percent increase in $[H^+]$, analyze the impact on marine calcifying organisms, and discuss how the carbonate buffer system mitigates pH changes. Consider long-term ecological effects on coral reefs and shell-forming species.

35. A titration of the amino acid glycine shows buffering regions at pH 2.4 and 9.6. If 25.0 mL of glycine solution requires 18.0 mL of 0.200 M NaOH to reach the first equivalence point, calculate glycine concentration. Discuss the zwitterionic nature of glycine, how it influences titration curves, and implications for biochemical buffer design.

36. An autoclave sterilization system must operate at a pH that prevents corrosion of steel components. If steam condensate contains 0.020 M carbonic acid, determine whether raising the pH to 9.0 using ammonia is sufficient to prevent acid-driven corrosion. Discuss the chemical principles underlying corrosion prevention, buffering strategies, and industrial process control.

37. A chemical plant uses $AlCl_3$ as a Lewis-acid catalyst. Traces of water (1.0×10^{-3} mol) enter a vessel containing 0.050 mol $AlCl_3$. Calculate how much HCl gas is formed via hydrolysis, and assess whether the resulting acidity significantly alters reaction conditions. Discuss implications for catalyst stability, reaction selectivity, and pH control in industrial chemistry.

38. A solution of oxalic acid (diprotic; $pK_{a_1} = 1.25$, $pK_{a_2} = 4.28$) is titrated with NaOH. If 50.0 mL of solution requires 30.0 mL of 0.250 M NaOH to reach the second

equivalence point, determine the molarity of oxalic acid and discuss how diprotic acid titrations differ from monoprotic titrations, including buffering at intermediate pH values.

39. A triprotic acid H_3A is partially neutralized, neutralizing only the first proton. With $pK_{a_1} = 2.5$ and $pK_{a_2} = 7.8$, estimate the pH of the resulting 0.050 M H_2A^- solution. Discuss the stepwise protonation behavior of polyprotic acids and its importance for multi-step titrations, buffer design, and biochemical systems.

40. A weak acid with $pK_a = 8.50$ is dissolved in a buffer at pH 7.50. Determine the ratio of protonated to deprotonated forms and discuss whether the acid exists primarily in neutral or ionized form at physiological pH. Explain implications for drug formulation, solubility, and biological activity.

41. A 0.0100 M solution of a metal ion produces pH 3.62 due to hydrolysis. Determine Ka for the hydrolysis reaction, identify whether the ion is likely a highly charged transition metal, and discuss the effect of metal ion charge density on hydrolysis, water acidity, and coordination chemistry.

42. A basic industrial scrubber uses NaOH to neutralize SO_2 from exhaust gases. If the incoming gas contains 150 ppm SO_2 at 1 atm total pressure with a flow rate of 2.0 m^3/min, estimate the moles of NaOH required per hour for complete removal. Discuss stoichiometry, gas flow, and the engineering considerations necessary to maintain scrubbing efficiency.

43. A student performs a back-titration: a solid weak acid is treated with excess 0.200 M NaOH, then back-titrated with 0.150 M HCl. If 0.500 g of the acid requires 30.0 mL NaOH and 12.0 mL HCl, determine its molar mass. Discuss advantages and limitations of back-titration for poorly soluble acids, including practical considerations in laboratory experiments.

44. A protein contains surface residues whose pK_a values shift significantly during folding. If a glutamate residue's pK_a increases from 4.3 to 6.0 in a hydrophobic core, determine its protonation state at pH 7.0 and discuss how this affects protein stability, folding energetics, and electrostatic interactions within the tertiary structure.

45. Acid rain with pH 4.3 falls on a 5.0×10^8 L lake containing bicarbonate at 5.0 ×

10^{-4} M. If 1.0×10^7 L of rain mixes uniformly, calculate the resulting pH. Discuss how bicarbonate buffers episodic acidification, and the environmental implications for aquatic ecosystems.

46. A 0.0200 M solution of a weak base B gives pH 10.62. Determine K_b and predict the pH if the solution is diluted to 0.00500 M. Discuss how dilution affects weak base equilibria, buffer capacity, and analytical measurements in laboratory practice.

47. A catalytic cycle involves a Lewis-acid metal center binding a ligand L. If equilibrium constants for ML^{2+} formation and protonation of L are $K_1 = 1.0 \times 10^6$ and $K_a = 2.0 \times 10^{-7}$ respectively, determine which pathway dominates in a mildly acidic solution. Discuss the implications for reaction kinetics, selectivity, and catalyst design in coordination chemistry.

48. A 1.00-L mixture contains 0.0500 M HCN ($pK_a = 9.31$) and 0.100 M NaCN. After adding 0.0200 mol HCl, compute the new pH and evaluate whether the solution remains an effective cyanide buffer. Discuss safety concerns, buffer maintenance, and the chemistry of cyanide handling in laboratory environments.

49. A sample of an unknown weak acid exhibits a half-equivalence pH of 3.18. If 25.0 mL requires 20.0 mL of 0.100 M NaOH to reach equivalence, determine Ka and molar mass. Discuss how half-equivalence points are used in identifying weak acids and evaluating experimental accuracy in graduate-level titration experiments.

50. A researcher studies the behavior of a weak acid HA in liquid ammonia. Its pKa in water is 5.0, but in NH_3 it is 19.0. Explain quantitatively why the acid behaves nearly neutral in ammonia, determine the fraction deprotonated in a 0.10 M solution, and discuss the implications of solvent-dependent acidity for nonaqueous synthesis, reactivity, and industrial applications in superbase chemistry.

Chapter 16

Dynamics of Aqueous Equilibria

When a weak acid such as HA is combined with the salt of its conjugate base, such as NaA, the resulting solution exhibits behaviors that are subtle yet critically important in chemistry and in the natural world. Hydrofluoric acid, HF, paired with sodium fluoride, NaF, provides a particularly illustrative example. In water, sodium fluoride dissociates completely, producing sodium ions and fluoride ions. Hydrofluoric acid, in contrast, is only partially dissociated, so most of the molecules remain intact as HF, while a minor fraction releases hydrogen ions and fluoride ions. The fluoride ions are therefore present from two sources: the dissociation of NaF and the limited dissociation of HF itself. This dual origin creates a "common ion" situation, which directly impacts the acid's equilibrium and lowers the concentration of hydrogen ions relative to pure HF.

Historically, the observation of such equilibrium shifts predates modern chemistry. Early alchemists and natural philosophers noticed that the sourness of some solutions could be diminished by adding certain salts, although they lacked the conceptual tools to describe ions and dissociation. Over centuries, chemists such as Svante Arrhenius and Walther Nernst developed the understanding of ions in solution and the principles of chemical equilibrium, providing a framework for phenomena like the common ion effect. The common ion effect also plays a crucial role in polyprotic acids, such as phosphoric acid, which can release multiple protons sequentially. The hydrogen ions released in the first dissociation step increase the concentration of ions in solution, which suppresses subsequent proton dissociations. This self-limiting behavior illustrates the delicate balance inherent in acid-base chemistry, and it underpins the design of buffers for biological and industrial use.

Buffers are solutions designed to resist changes in pH when small amounts of acids or bases are added. They are indispensable in biological systems. Human blood, for instance, is buffered primarily by the bicarbonate system. The bicarbonate ions interact with hydrogen ions to form carbonic acid, which can then equilibrate with dissolved carbon dioxide. This interplay maintains blood pH within a narrow range of approximately 7.35 to 7.45, a range critical for the function of enzymes and the stability of cellular membranes. Small deviations from this pH range can impair oxygen transport, disrupt metabolic reactions, and cause severe physiological effects.

In the laboratory, buffers are often made using a weak acid and its conjugate base, sometimes supplied as a salt, or a weak base and its conjugate acid. The effectiveness of a buffer depends both on the ratio of the weak base to acid and on their absolute concentrations. When hydroxide ions are introduced into a weak acid buffer, the weak acid reacts preferentially with them to form water and additional conjugate base, minimizing any rise in pH. Conversely, when protons are added, they react with the conjugate base to regenerate the weak acid, preventing a significant drop in pH. In practice, a buffer is most effective when the concentrations of the acid and its conjugate base are equal, because then the ratio of conjugate base to acid remains most stable even as small amounts of acid or base are added.

This principle underlies the common laboratory practice of preparing buffers by mixing nearly equal amounts of a weak acid and its salt.

The design of buffers extends into industrial applications as well. In pharmaceuticals, for example, drug formulations often require strict pH control to ensure chemical stability and optimal absorption. Intravenous solutions, mouthwashes, and even contact lens solutions rely on buffered formulations to prevent irritation and maintain compatibility with biological tissues. In food chemistry, buffers stabilize flavors, preserve color, and prevent spoilage by inhibiting reactions that are pH-sensitive. Vinegar, citric acid, and phosphate salts are examples of natural and synthetic buffering agents commonly used in food systems.

Titration curves provide an accessible visual representation of buffer behavior. When a solution is gradually neutralized with a titrant of known concentration, the pH changes in a predictable manner. Strong acid–strong base titrations are straightforward, as the neutralization produces water, and the pH changes rapidly near the equivalence point. However, when a weak acid is titrated with a strong base, the solution initially behaves as a buffer, resisting significant changes in pH. The mid-point of the titration, where equal amounts of the weak acid and its conjugate base are present, corresponds to the pKa of the acid and represents the point of maximum buffering capacity. Weak base–strong acid titrations follow similar principles, with the weak base producing a conjugate acid that contributes to the equilibrium and influences the pH curve.

Indicators are themselves examples of weak acids. Phenolphthalein, for instance, is colorless in its protonated form but turns pink when deprotonated. The color change occurs when the ratio of deprotonated to protonated form reaches approximately ten to one or one to ten, depending on the titration direction. This behavior allows indicators to signal the approach to the equivalence point in a titration, guiding chemists in quantitative analysis. The choice of an indicator depends on its color change range relative to the expected equivalence point of the titration, ensuring accurate detection of the endpoint.

Solubility equilibria describe the behavior of sparingly soluble salts in water. Magnesium fluoride, for example, dissolves into magnesium ions and fluoride ions. Initially, the dissolution reaction dominates, but as ion concentrations increase, precipitation becomes more likely until equilibrium is established. The solubility product constant provides a measure of the extent to which the salt dissolves under saturated conditions. Notably, increasing the amount of solid salt does not change the equilibrium concentrations of ions; it only affects the rate at which equilibrium is reached.

The solubility of salts is influenced by the common ion effect and by pH. For salts containing basic anions, lowering the pH can increase solubility by protonating the anions and shifting the equilibrium toward dissolution. This principle has practical applications in environmental chemistry, where acidification can enhance the solubility of metal-containing pollutants, affecting their mobility and toxicity. Complex ion formation further modifies solubility behavior. A metal ion can bind to one or more ligands, which are electron-rich molecules capable of donating lone pairs. The sequential binding of ligands generates a series of equilibria characterized by formation constants. By forming complex ions, chemists can increase the solubility of otherwise insoluble salts, manipulate the distribution of ions in solution, and selectively isolate metal ions. This strategy is widely used in analytical chemistry, water treatment, and industrial processes. For example, chelating agents such as EDTA bind metal ions tightly, allowing removal of heavy metals from wastewater or stabilization of metal ions in pharmaceutical formulations.

Biological systems exploit these same principles. Hemoglobin, the oxygen-carrying protein in red blood cells, relies on reversible binding of oxygen to iron ions within the heme group. The binding equilibrium is sensitive to pH and carbon dioxide concentration, demonstrating a real-world instance of complex ion and acid-base equilibria working together to sustain life. Similarly, enzymatic catalysis often depends on the local pH, which is regulated by buffering systems within cellular compartments. In industrial applications, solubility and complexation principles are used to recover valuable metals, purify chemicals, and control scaling in pipes and reactors. For instance, selective precipitation can separate calcium from magnesium in hard water, while complexation can solubilize rare earth metals for extraction and refinement. Acid addition, ligand introduction, and temperature control all serve as levers to adjust solubility and optimize chemical processes.

In essence, the study of weak acids, weak bases, buffers, titration behavior, solubility equilibria, and complex ion formation reveals a cohesive framework that underpins both theoretical and practical chemistry. From the maintenance of blood pH to the design of industrial chemical processes, these principles govern the behavior of ions and molecules in solution. Understanding them allows chemists to predict outcomes, manipulate reaction conditions, and develop solutions with desired properties. The interplay of equilibrium, common ion effects, buffering, titration, solubility, and complexation forms a rich tapestry that is central to both the science and application of chemistry.

Imagine a chemist in the early 1800s, peering into a glass of weak acid and pondering why adding a pinch of salt sometimes made the liquid taste less sour. Long before the formal concept of ions or chemical equilibrium, these observations hinted at a subtle balance in nature. Today, we understand that this phenomenon is a result of what we call the common ion effect. When a weak acid, like hydrofluoric acid, is mixed with the salt of its conjugate base, such as sodium fluoride, the fluoride ions from both the salt and the minor dissociation of HF combine to shift the equilibrium toward the undissociated acid. In practical terms, this means that the solution produces fewer hydrogen ions and is less acidic than the pure weak acid alone. In the laboratory, this effect is a tool for controlling acidity and designing buffers, but in nature, it explains why certain waters taste milder after flowing through mineral-rich soils that supply common ions.

Polyprotic acids, which can donate multiple protons, offer another layer of fascination. Phosphoric acid, for example, releases protons one after another, but the first release increases the concentration of ions in solution, which in turn suppresses subsequent proton release. This self-regulation is a beautiful example of chemical balance at work. Ancient cultures may not have measured pH, but they exploited similar principles in food preservation and fermentation. Vinegar, pickled vegetables, and sourdough starters all rely on weak acids and their salts to create stable, predictable chemical environments, protecting food from spoilage and enabling controlled microbial growth.

Buffers take this concept even further. Consider the human body, where the blood's pH must remain around 7.4 to sustain life. The bicarbonate buffer system, consisting of carbonic acid and bicarbonate ions, absorbs small shocks of acid or base. When extra hydrogen ions enter the blood, they react with bicarbonate ions to form carbonic acid. When hydroxide ions appear, carbonic acid releases protons to neutralize them. This invisible balancing act is crucial for enzyme function, oxygen transport, and even neurological activity. A tiny shift in pH can affect how hemoglobin binds oxygen, illustrating that buffers are not just chemical curiosities—they are life-sustaining mechanisms.

Laboratory buffers mirror these natural systems. Chemists mix weak acids with their salts or weak bases with their conjugate acids to create solutions that resist pH change. The key to their effectiveness lies in both the ratio of conjugate base to acid and their absolute concentrations. A buffer with equal amounts of acid and base has the maximum capacity to absorb added hydrogen or hydroxide ions without a noticeable change in pH. In practical terms, if a chemist adds hydroxide ions to such a buffer, the weak acid reacts to form water and more conjugate base, leaving the pH almost unchanged. Similarly, added protons react with the conjugate base to regenerate the acid. This dynamic is not abstract—it is used in pharmaceuticals to stabilize drugs, in cosmetics to prevent irritation, and in industrial processes to maintain optimal chemical conditions.

To understand buffer behavior more vividly, imagine performing a titration. A student slowly adds a strong base to a flask containing a weak acid. At first, the pH barely changes because the acid neutralizes the base. This part of the titration curve is flat, representing the buffer region. Midway through, when equal amounts of acid and base are present, the pH reaches a value corresponding to the acid's intrinsic dissociation tendency, known as its pKa. Beyond this point, adding more base causes the pH to rise sharply, illustrating the limits of buffering. In real-world laboratories, indicators like phenolphthalein signal these transitions by changing color. The color shift occurs when the ratio of the deprotonated form of the indicator to its protonated form reaches a threshold, effectively translating a molecular equilibrium into a visible signal.

Solubility equilibria offer another stage for chemical intrigue. Consider magnesium fluoride dissolving in water. Initially, the solid dissociates into ions, but as concentrations rise, precipitation becomes more favorable. Equilibrium is reached when the rates of dissolution and re-precipitation balance. Here, the concept of the solubility product emerges, providing a measure of how much solid can dissolve. If one adds more solid to the solution, the equilibrium concentrations do not change; only the rate at which equilibrium is achieved is affected. The interplay of solubility and the common ion effect also manifests in nature. In rivers flowing through limestone or volcanic regions, mineral solubility is modulated by pH and ion composition, influencing water hardness, aquatic life, and the formation of deposits like stalactites.

Complex ions deepen the story. Imagine a central metal ion surrounded by a circle of ligands, each donating electron pairs. Step by step, these ligands attach, each interaction governed by its own stability. By forming such complexes, chemists can dissolve otherwise insoluble salts, control the availability of metals, or selectively isolate elements. This principle is central to the extraction of precious metals, the remediation of heavy metals in polluted water, and even the transport of oxygen by hemoglobin. The iron in hemoglobin binds oxygen reversibly, responding to the local chemical environment. This is chemistry in action at a scale that sustains life.

Chemistry is not just laboratory manipulation; it is a lens through which we can view natural processes and technological innovation. Early chemists could taste, smell, and observe changes in acidity and solubility without knowing about ions. Today, we understand the molecular underpinnings of these observations and can exploit them with precision. Buffered solutions prevent abrupt pH shifts in industrial fermentations, titration curves guide the formulation of medicines, and solubility equilibria dictate how metals are purified or removed from wastewater. Even in art and culinary chemistry, pH control affects pigment stability, flavor intensity, and texture.

Thought experiments further illuminate these ideas. Imagine adding a spoonful of NaF to a glass of HF. The solution becomes less acidic—not because HF has changed, but because the added fluoride shifts the balance of equilibrium. Or consider an aqueous solution of calcium hydroxide into which carbon dioxide is bubbled. As carbonic acid forms, it reacts with the hydroxide ions, gradually adjusting the pH. Each example reinforces that chemical equilibria are dynamic, self-correcting, and profoundly interconnected with the environment. Historical anecdotes enrich our understanding. In the 19th century, chemists carefully observed precipitation reactions to determine solubility rules, laying the groundwork for modern quantitative analysis. They noted that salts like silver chloride would not dissolve beyond a certain point, leading to the concept of the solubility product. Meanwhile, pioneers of acid-base theory, such as Arrhenius and Brønsted-Lowry, formalized the language of protons and bases, turning qualitative observations into predictive science. Today, these principles guide everything from industrial process design to pharmaceutical formulation.

Ultimately, the story of weak acids, weak bases, buffers, solubility equilibria, and complex ions is a story of balance, responsiveness, and control. It illustrates how subtle changes at the molecular level have profound macroscopic consequences. From the taste of mineral water to the survival of cells, from titration in the laboratory to the extraction of metals, chemistry demonstrates the elegance of equilibrium and the power of well-chosen conditions. By exploring these systems through narrative, historical insight, real-world application, and thought experiments, we see chemistry not just as a collection of equations, but as a living science that connects the microscopic to the macroscopic, theory to practice, and curiosity to discovery.

Imagine walking along the banks of a limestone river, observing the clear water flowing over pale rocks. The water carries dissolved calcium and magnesium ions, the result of natural solubility equilibria between minerals and acidic rainwater. These ions are present in just the right concentrations to maintain aquatic life, yet shifts in pH from acid rain or industrial runoff can disrupt the balance. When rainwater containing dissolved carbon dioxide meets the calcium carbonate in limestone, a delicate equilibrium forms. Calcium ions and bicarbonate ions dissolve in the water, creating a dynamic system in which dissolution and re-precipitation occur continuously. If additional carbonate ions are introduced, perhaps from upstream effluent, the system adjusts, and precipitation may occur. This natural demonstration of the common ion effect and solubility equilibria mirrors what chemists observe in laboratories with more controlled conditions.

The interplay between weak acids and salts extends to food chemistry, a field as old as civilization itself. Ancient cultures discovered, through trial and error, that adding salt to fermented foods could prevent sourness from overwhelming flavors. Modern chemistry explains this as a shift in equilibrium: the common ion effect suppresses excessive acid formation, stabilizing the food's taste and preventing microbial overgrowth. In pickling, vinegar supplies acetic acid, while added salts provide acetate ions. The resulting buffer system maintains a stable pH, allowing lactic acid bacteria to thrive without producing excess acidity that would spoil the food. Similarly, in winemaking, the balance of tartaric acid, its salts, and other organic acids determines the final taste profile and stability of the wine.

Biology provides even more compelling examples. Consider intracellular pH regulation. Cells produce acids through metabolic processes, including lactic acid during muscle activity and carbonic acid from carbon dioxide metabolism. Buffers within the cytoplasm, often involving phosphate and bicarbonate systems, absorb these protons to prevent cellular damage. The Henderson–Hasselbalch principle, though derived from simple acid-base chemistry, explains how the relative concentrations of conjugate acid-base pairs dictate intracellular pH. Even slight perturbations in these concentrations can affect enzyme activity, membrane potential, and nutrient transport. In the bloodstream, the same principles apply on a larger scale. Hemoglobin's oxygen-binding affinity is directly influenced by pH and carbon dioxide concentration—a phenomenon known as the Bohr effect. The dynamic equilibrium between oxygenated and deoxygenated hemoglobin depends on local chemical environments, demonstrating that acid-base chemistry is not confined to the laboratory but is woven into the very fabric of life.

Industrial processes also exploit these principles on a massive scale. In wastewater treatment, for example, the solubility of heavy metals must be controlled to prevent environmental contamination. By adjusting pH or introducing ligands that form complex ions with the metals, engineers can selectively precipitate or solubilize ions. This mirrors the laboratory technique of selective precipitation but applies it to rivers, treatment plants, and industrial effluents. Similarly, in metallurgy, complex ion formation allows extraction of precious and rare earth metals from ores. By binding metal ions with carefully chosen ligands, chemists increase their solubility and mobility, enabling efficient separation and purification.

Thought experiments deepen our understanding of these processes. Picture a solution of weak acid HA, to which a small amount of NaA is added. The resulting shift in equilibrium suppresses hydrogen ion release, decreasing acidity. Now imagine a gradual increase of temperature: kinetic energy allows molecules to move faster, potentially favoring either dissolution or dissociation depending on the system. The same principle underlies why heating water can dissolve more sugar or why industrial reactions often require controlled temperature regimes. By mentally manipulating concentrations, temperature, and pH, chemists can predict outcomes, optimize reactions, and design processes that take advantage of equilibrium behavior.

The study of weak acids and their salts also sheds light on pharmacology. Many drugs are weak acids or bases, and their absorption in the body depends on pH and the relative concentrations of ionized and un-ionized forms.

Aspirin, a weak acid, is absorbed primarily in the stomach, where the pH favors the uncharged form. In the intestines, where pH is higher, more of the drug exists in its ionized form, slowing absorption. Buffering agents are sometimes included in formulations to maintain optimal pH, ensuring that the active ingredient is delivered efficiently. Even antacids operate on simple principles of acid-base chemistry: hydroxide ions neutralize excess stomach acid, and the resulting salts may interact with other ions in solution, demonstrating the practical application of chemical equilibria.

Historical anecdotes provide context to these discoveries. In the 19th century, chemists like Justus von Liebig observed that small additions of certain salts could dramatically affect reaction outcomes. They did not have the benefit of modern analytical instruments, relying instead on taste, color change, and precipitation to draw conclusions. The systematic study of solubility, common ion effects, and buffering emerged gradually, transforming chemistry from observation to predictive science. Later, pioneers like Svante Arrhenius proposed the concept of ions and acid dissociation, providing a theoretical framework for these empirical observations. The development of pH meters and modern indicators further refined the ability to measure and manipulate chemical systems precisely.

Even in seemingly unrelated fields, the principles of weak acids, bases, and solubility equilibria play a role. In cosmetics, buffers stabilize formulations against changes in pH caused by skin contact or air exposure. In agriculture, soil pH affects nutrient availability; limestone is often added to acidic soils to raise pH, influencing the solubility of essential ions like calcium, magnesium, and phosphate. In geology, the formation of stalactites and stalagmites in caves arises from the slow precipitation of calcium carbonate as water equilibrates with carbon dioxide and minerals, a natural illustration of solubility equilibria and buffering in action over millennia.

Consider the classroom experiment in which students mix silver nitrate and sodium chloride solutions. Initially, the solution is clear, but a white precipitate forms as silver chloride reaches its solubility limit. By gradually changing concentrations or adding complexing agents, students can observe the principles of ion product versus solubility product, the common ion effect, and the influence of complexation on solubility. These simple demonstrations mirror processes occurring in natural waters, industrial reactors, and physiological fluids, illustrating the universality of chemical equilibria.

Even artistic endeavors can connect to chemistry. The brilliant colors in stained glass arise from metal ions embedded in the glass matrix. The final color depends on the oxidation state of the metal, which is influenced by the pH and the presence of other ions during fabrication. Glassmakers of the Middle Ages may not have had formal chemical theory, but through observation and experimentation, they effectively exploited the same principles modern chemists describe quantitatively. To imagine the scale of chemical influence, one can picture the ocean, a massive buffered system. Carbonate, bicarbonate, and dissolved carbon dioxide maintain ocean pH near neutrality, allowing marine life to thrive. When excess carbon dioxide from the atmosphere dissolves into the ocean, equilibria shift, and carbonate ions are consumed to form bicarbonate. This subtle chemical change can lead to ocean acidification, affecting shell formation in mollusks and coral reefs, demonstrating again that equilibrium chemistry connects directly to ecosystems and planetary health.

In conclusion, extending our narrative reveals that the study of weak acids, weak bases, buffers, solubility equilibria, common ion effects, and complex ions is not just abstract theory—it is deeply embedded in the natural world, human history, industrial innovation, and daily life. The principles learned in a chemistry classroom echo in the rivers we walk along, the foods we eat, the medicines we take, and the cells within our bodies. By exploring these systems through thought experiments, historical anecdotes, biological examples, and industrial applications, we see chemistry as a living, breathing discipline that connects molecular behavior to observable outcomes. Equilibrium, balance, and dynamic response are not merely concepts—they are a window into the elegance of the chemical world, revealing a harmony that has guided nature, human endeavor, and technological progress for centuries.

Chapter 16: Word Problems

1. A chemical engineer is tasked with studying the effect of adding salts to a weak acid solution in an industrial setting. She prepares a solution containing 0.10 M hydrofluoric acid and 0.10 M sodium fluoride, knowing that fluoride ions are contributed by both HF and NaF. She wants to calculate the hydrogen ion concentration, the pH of the solution, and compare it to the pH of a 0.10 M HF solution without NaF. Discuss the suppression of HF dissociation by the common

ion effect, calculate the expected pH using Ka, and explain how this principle can be applied to control acidity in industrial HF processes where corrosion must be minimized.

2. In a biochemical experiment, a graduate student prepares a mixture of phosphoric acid and its sodium salts to simulate the intracellular phosphate buffer system. She wants to determine the fractions of H_3PO_4, $H_2PO_4^-$, HPO_4^{2-}, and PO_4^{3-} at pH 7.0 and understand how the release of the first proton influences the release of subsequent protons. Calculate the species distributions using the appropriate Ka values, and discuss how the self-limiting nature of polyprotic acid dissociation contributes to intracellular pH stability.

3. A clinician is designing a protocol for intravenous administration of sodium bicarbonate to a patient experiencing mild acidosis. With a normal plasma bicarbonate concentration of 24 mM and a controlled infusion of 0.05 M HCl, the clinician wants to predict the resulting pH change in the blood. Using the Henderson–Hasselbalch equation, calculate the expected pH change and explain how the bicarbonate buffer interacts with carbonic acid to resist sudden acidification, considering implications for oxygen transport and enzyme activity.

4. A researcher prepares a laboratory buffer by mixing 0.2 M acetic acid with 0.1 M sodium acetate. She then introduces 0.01 M NaOH to the solution and observes a minimal change in pH. Calculate the precise pH before and after the addition, and discuss why the weak acid preferentially neutralizes the added hydroxide ions. Explain how the ratio of conjugate base to acid determines the buffer capacity and why such buffers are critical in pharmaceutical and biochemical applications.

5. During a titration of a weak acid with a strong base, a graduate student observes a flat region in the titration curve and identifies the midpoint where equal amounts of weak acid and conjugate base are present. Explain why this midpoint corresponds to the acid's pKa, and calculate the expected pH for an acid with $K_a = 1.8 \times 10^{-5}$. Discuss the significance of this buffer region in analytical chemistry and how it can be exploited to measure unknown concentrations accurately.

6. Phenolphthalein is used as an indicator in a titration of a weak acid. At the endpoint, the solution changes from colorless to pink at pH 8.3. Calculate the ratio of deprotonated to protonated indicator species at this pH and discuss how

this ratio relates to the indicator's equilibrium constant. Explain why selecting an appropriate indicator is critical for precise titration endpoints.

7. Magnesium fluoride is sparingly soluble in water, and its solubility decreases further when 0.05 M NaF is present. Derive an expression for MgF_2 solubility in the presence of the common ion, calculate the numerical solubility using $K_{sp} = 6.4 \times 10^{-9}$, and discuss the significance of the common ion effect in controlling fluoride levels in drinking water and industrial processes.
8. Bubbling carbon dioxide into an aqueous solution of calcium hydroxide leads to formation of calcium carbonate and bicarbonate ions. Predict the equilibrium concentrations of Ca^{2+}, CO_3^{2-}, and HCO_3^- after CO_2 addition, and explain how additional CO_2 will shift the equilibria. Relate this laboratory observation to natural limestone-water interactions and environmental buffering of river systems.
9. A coordination chemist investigates the stepwise complexation of a metal ion M with a ligand L, forming ML^+, ML_2, and ML_3, each with a known formation constant. Given total concentrations of M and L, calculate the distribution of species at equilibrium. Discuss how complexation can increase the solubility of otherwise insoluble salts, and describe applications in selective metal extraction and wastewater treatment.
10. A pharmaceutical chemist must design a weak acid drug formulation to remain stable at pH 4.5–5.5 using an acetate buffer. Determine the concentrations of acetic acid and sodium acetate necessary to maintain the pH, considering buffer capacity and drug concentration. Explain how maintaining the correct ratio of conjugate base to acid prevents degradation and ensures optimal drug absorption.
11. In industrial glass etching, uncontrolled hydrofluoric acid leads to excessive corrosion. A chemical engineer adds sodium fluoride to moderate acidity. Quantify how adding 0.05 M NaF suppresses HF dissociation and reduces hydrogen ion concentration. Calculate the expected pH change and explain the relevance of the common ion effect in safely controlling etching rates.
12. During traditional vegetable pickling, vinegar (acetic acid) and sodium salts create a stable acidic environment. For a mixture containing 0.1 M acetic acid and 0.05 M sodium acetate, calculate the pH and buffer capacity. Discuss how the common

ion effect stabilizes acidity, prevents spoilage, and allows controlled growth of lactic acid bacteria.

13. A student adds 0.02 M HCl to a solution of bicarbonate-carbonic acid buffer. Predict the resulting pH change and explain shifts among $CO_2(aq)$, H_2CO_3, HCO_3^-, and CO_3^{2-}. Quantify how the buffer resists acidification and discuss its relevance to physiological and laboratory systems.

14. In a titration of a weak base with a strong acid, the half-equivalence point occurs when half the base has been neutralized. Calculate the pH at this point and explain why it corresponds to the pK_a of the conjugate acid. Discuss the buffering behavior prior to equivalence and the importance of understanding this region in analytical work.

15. Adding NaCl to a saturated AgCl solution reduces the solubility of AgCl due to the common ion effect. Using $K_{sp} = 1.8 \times 10^{-10}$, calculate the new solubility of AgCl in a solution already containing 0.05 M NaCl. Explain quantitatively how the chloride ion from NaCl influences the solubility equilibrium, and discuss why understanding this effect is crucial in gravimetric analysis for precise mass determination of silver salts and in industrial processes such as photographic film development, wastewater treatment, and metal recovery from brines.

16. EDTA is widely used to chelate Ca^{2+} ions in wastewater treatment and environmental remediation. Determine the minimum concentration of EDTA required to complex all calcium ions in a 0.05 M solution. Explain the mechanism by which formation of the EDTA–Ca complex increases solubility of otherwise sparingly soluble salts and enables selective removal of metal ions. Discuss how these principles underpin water softening, removal of scale-forming ions, and heavy metal detoxification in industrial effluents.

17. Ocean acidification is modeled by dissolving 0.02 M CO_2 in seawater. Predict the resulting decrease in carbonate ion concentration and calculate the pH shift using appropriate equilibria for CO_2, HCO_3^-, and CO_3^{2-}. Discuss how these chemical changes affect calcifying organisms, such as corals, mollusks, and planktonic species, and illustrate the intersection of acid–base equilibria, environmental chemistry, and global ecological consequences in marine systems.

18. During lactic acid fermentation, intracellular phosphate buffers absorb protons generated by metabolic activity. Given intracellular concentrations of 10 mM $H_2PO_4^-$ and 10 mM HPO_4^{2-}, calculate the expected intracellular pH range using the Henderson–Hasselbalch equation. Explain how phosphate buffers protect enzymes from acid-induced denaturation, maintain metabolic flux, and support energy production in cells. Discuss how the dynamic balance of protonation states underpins cellular resilience during anaerobic metabolism.

19. A 0.10 M HF solution containing 0.10 M NaF has a measured pH of 3.17 at 25°C. When the temperature rises by 10°C, predict the direction of the pH change, considering the temperature dependence of the acid dissociation constant Ka and the solubility of HF. Discuss the practical significance of this effect in chemical manufacturing, laboratory storage, and industrial applications where precise pH control is critical for reactions involving weak acids.

20. Diluting 0.1 M acetic acid to 0.01 M results in a smaller pH change compared with diluting 0.1 M HCl to the same concentration. Calculate the pH of the diluted acetic acid solution, explain why weak acids display smaller pH changes due to partial dissociation, and discuss how the common ion effect further modulates pH. Relate these observations to titration practices, buffer design, and industrial processes where controlled acid strength is necessary.

21. A solution contains 0.05 M HF and 0.05 M HCl. Determine the fraction of undissociated HF and the resulting pH. Discuss how the presence of a strong acid suppresses weak acid dissociation, highlighting the chemical principles involved. Explain the implications for laboratory and industrial handling of HF, including safety considerations, controlled etching, and fluoride-based chemical processes.

22. Design a buffer at pH 7.0 using a polyprotic acid such as H2PO4− or citric acid. Choose appropriate conjugate base species and concentrations, calculate the ratio of each protonation state required to maximize buffer capacity, and discuss the underlying principles. Explain how polyprotic buffering systems stabilize pH in biological systems, pharmaceutical formulations, and analytical procedures requiring precise control of proton activity.

23. In winemaking, tartaric acid equilibrates with potassium bitartrate salts. Predict how precipitation of potassium bitartrate affects both the pH of the wine and the

concentration of free tartaric acid in solution. Discuss the role of solubility equilibria, the common ion effect, and supersaturation phenomena in maintaining wine stability, flavor profile, and preventing unwanted crystalline deposits in storage and bottling.

24. Titrate 0.05 M carbonic acid (H_2CO_3) with 0.10 M NaOH. Calculate the pH at the first equivalence point and describe how polyprotic acid behavior shapes the titration curve. Discuss the buffer regions preceding equivalence, the underlying acid–base equilibria, and their relevance to carbonate systems in environmental chemistry, such as freshwater buffering and oceanic pH regulation.

25. MgF_2 solubility is strongly influenced by the presence of fluoride ions. Derive the solubility expression in 0.05 M NaF and calculate the numerical solubility. Discuss applications in water treatment, fluoride regulation, and industrial processes where control of fluoride concentration is critical, including aluminum production, dental care products, and corrosion mitigation.

26. Phosphate salts stabilize intravenous solutions by buffering pH. Determine the necessary phosphate concentrations to maintain pH 7.4 under physiological conditions, calculate the buffer capacity, and explain how the common ion effect contributes to minimizing pH fluctuations during storage and administration. Discuss the implications for patient safety and pharmaceutical formulation design.

27. In a 0.1 M NH_3/0.1 M NH_4Cl buffer, 0.01 M HCl is added. Calculate the resulting pH and explain why NH_3 reacts preferentially with added protons. Discuss buffer dynamics, the importance of conjugate acid–base pairs, and the design of systems that resist sudden pH changes in laboratory, industrial, and biological contexts.

28. Industrial HF etching can be moderated by adding NaF. Calculate the suppression of HF dissociation and the resulting pH. Discuss the practical benefits for glass processing, metal etching, and corrosion control, emphasizing the chemical principles of common ion suppression and safety considerations in industrial chemical handling.

29. Addition of a ligand to $Cu(OH)_2$ increases solubility via complex formation. Calculate the new solubility using known formation constants and explain how complexation interplays with solubility equilibria. Discuss applications in selective metal

dissolution, hydrometallurgy, and analytical chemistry, including environmental recovery of trace metals from contaminated streams.

30. A buffer is required at pH 5. Calculate the optimal ratio of conjugate base to acid to achieve maximum buffering capacity, and explain why equal concentrations of acid and base provide the highest resistance to pH change. Discuss the practical applications of such a buffer in biochemical assays, pharmaceutical formulations, and industrial processes requiring precise pH control.

31. Seawater carbonate equilibria resist pH swings under CO_2 influx. Quantify the buffering action of HCO_3^- and CO_3^{2-} species, calculate expected pH changes upon CO_2 addition, and discuss the impact on oceanic pH stability. Relate this to environmental chemistry, marine biology, and climate change mitigation strategies.

32. A metal M forms sequential complexes ML^+, ML_2, and ML_3. Using known formation constants, calculate the fraction of each species at equilibrium. Discuss applications in selective metal extraction, purification techniques, and industrial recovery of valuable or toxic metals.

33. Predict changes in $Al(OH)_3$ solubility under acidic and basic conditions, quantify solubility at pH 4 and pH 10, and discuss the role of hydroxide equilibria and the common ion effect. Explain applications in water purification, industrial precipitation processes, and aluminum hydroxide-based antacids.

34. In a bicarbonate buffer containing equal concentrations of HCO_3^- and H_2CO_3, atmospheric CO_2 increases. Calculate the new pH and explain the buffering mechanism in terms of dynamic equilibria and protonation states. Discuss implications for physiological systems, oceanic carbon chemistry, and environmental monitoring.

35. Titration of a weak polyprotic acid shows a second equivalence point at higher pH. Explain the influence of ionic strength, protonation states, and the common ion effect on the titration curve. Quantify the expected pH values and relate these observations to analytical chemistry and environmental applications.

36. In a weak acid–strong base titration where 75% of the acid is neutralized, derive the pH expression using the Henderson–Hasselbalch equation and calculate the pH numerically givenKa and initial concentration. Discuss the practical relevance for buffer design and controlled neutralization reactions in chemical synthesis.

37. Mixing solutions of Ca^{2+} and CO_3^{2-} leads to $CaCO_3$ precipitation. Calculate the maximum soluble concentrations of ions using K_{sp} and discuss the common ion effect. Relate the problem to water hardness, scaling in industrial pipes, and calcium carbonate deposition in natural waters.

38. EDTA is used to remove Pb^{2+} from wastewater. Calculate the concentration required to complex 90% of Pb^{2+} in a 0.01 M solution and discuss the chemical basis for heavy metal remediation. Explain how selective chelation enables safe disposal, recycling, or recovery of toxic metals from industrial effluents.

39. Design a buffer for a pharmaceutical cream with a target pH range of 6.5–7.5. Select a weak acid/conjugate base pair, calculate necessary concentrations for effective buffering, and explain how buffer capacity ensures product stability and efficacy. Discuss the considerations of skin compatibility, shelf-life, and interaction with active ingredients.

40. In titrating phosphoric acid with NaOH, the first buffer region is broader than subsequent ones. Quantify species concentrations, explain the pH range of each buffer region, and discuss how polyprotic acids shape titration curves. Relate these observations to fertilizer chemistry, food processing, and biological buffering.

41. MgF_2 solubility in water is 0.001 M. Adding 0.05 M NaF decreases solubility via the common ion effect. Calculate the new solubility and explain the importance for fluoride control in water treatment, toothpaste production, and industrial applications where fluoride concentration must be tightly regulated.

42. Rapid glycolysis produces lactic acid, which lowers intracellular pH. Intracellular phosphate buffers mitigate the pH drop. Calculate the expected pH change and the $H_2PO_4^-/HPO_4^{2-}$ ratio, explaining how enzyme activity is protected and how metabolic homeostasis is maintained in highly active tissues such as muscles.

43. A solution contains HF, NaF, and HCl. Determine the pH and fraction of undissociated HF, illustrating the combined effects of a strong acid and a common ion. Discuss applications in analytical chemistry, controlled etching, and safety considerations for handling fluoride-containing solutions.

44. Adding 0.01 moles OH^- to 1 L of a 0.1 M acetic acid/acetate buffer changes the pH. Calculate the new pH and discuss the buffer's resistance to large changes

in proton concentration. Relate this to chemical process control, environmental chemistry, and formulation of laboratory reagents.

45. Hemoglobin's oxygen binding is pH-sensitive. Quantitatively show how a blood pH drop from 7.4 to 7.2 alters the fraction of oxygenated hemoglobin using the Bohr effect. Discuss the physiological significance for oxygen delivery, tissue metabolism, and clinical considerations in acid-base disturbances.

46. To separate Ca^{2+} from Mg^{2+} in hard water via selective precipitation, calculate the required ion concentrations for efficient separation. Explain the solubility equilibria, common ion effects, and how these principles are applied in industrial water softening, desalination, and mineral recovery processes.

47. In wine stabilization, tartaric acid equilibria precipitate potassium bitartrate. Calculate the resulting pH shift and remaining free tartaric acid in solution. Discuss the chemical principles that control wine stability, prevent unwanted crystallization, and maintain flavor integrity during storage.

48. In a 0.2 M/0.1 M NH_3/NH_4Cl buffer, addition of 0.01 M HCl changes pH. Predict the new pH, calculate changes in conjugate base and acid concentrations, and explain the dynamic equilibrium between NH_3, NH_4^+, and added protons. Discuss implications for industrial and biological buffer systems.

49. Temperature affects metal hydroxide solubility. Using solubility products and formation constants, predict trends in solubility for increasing temperature and quantify the effect for a specific metal hydroxide. Explain the implications for industrial processes such as scale prevention, wastewater treatment, and chemical synthesis.

50. Design an intravenous solution to maintain pH 7.4 ± 0.05 using a bicarbonate–phosphate buffer system. Calculate the necessary concentrations of each buffer component and explain how the two buffers interact to stabilize pH. Discuss physiological relevance, stability during storage, and considerations for patient safety and effective drug delivery.

Chapter 17

Thermodynamics as Power Play

The First Law of Thermodynamics, one of the foundational pillars of physical science, represents a formalized articulation of the ancient principle of energy conservation, a concept that intrigued natural philosophers long before the establishment of modern physics and chemistry. This law asserts that energy can neither be created nor destroyed, only transformed from one form to another, implying that the total energy of the universe remains invariant, even as individual systems undergo complex exchanges of energy with their surroundings in the form of heat or work. By providing this unshakable framework, the First Law enables scientists and engineers to "track" energy through myriad processes, from the microscopic motions of molecules to the macroscopic behavior of engines and batteries. For instance, when a battery illuminates a light, the chemical potential energy stored within its reactants is converted sequentially into electrical energy and then into light and thermal energy. The First Law guarantees that every joule of energy is accounted for, yet it offers no guidance on the spontaneous direction in which energy will flow. In other words, while it quantifies the energy involved in processes, it does not explain why some processes occur naturally while others require external intervention.

Spontaneous processes, which occur without external influence, manifest across a vast spectrum of timescales and magnitudes. Rapid, dramatic examples include the combustion of hydrogen in oxygen, where reactants instantaneously transform into water, releasing substantial heat and light. Conversely, the gradual rusting of iron over months or years represents a slow, yet thermodynamically favored, process governed by the same fundamental principles. Thermodynamics predicts whether a process is possible and which direction it will proceed, but it does not provide information about the rate at which the process occurs. The kinetics of a reaction, which considers activation energy, reaction pathways, concentration, temperature, and catalysts, governs the speed of transformation. Thus, thermodynamics and kinetics together offer a complete understanding: thermodynamics determines the possibility and spontaneity of change, while kinetics explains the timescale over which change occurs.

Central to the understanding of spontaneous processes is the concept of entropy, denoted as S, which quantifies the degree of disorder or randomness in a system. Entropy measures the number of microscopic configurations, or microstates, that correspond to a given macroscopic state. Processes that increase the number of accessible microstates are more likely to occur because nature favors states of higher probability. A classic example involves a gas initially confined to one compartment of a container. When the partition is removed, the gas expands spontaneously to fill the available volume. The number of possible microstates for the expanded gas is astronomically larger than for the confined gas, making spontaneous expansion overwhelmingly probable. The reverse process, in which all gas molecules spontaneously return to one compartment, corresponds to only a single microstate and is effectively impossible. This statistical perspective provides a molecular-level explanation for the directional behavior observed in macroscopic systems.

Entropy also illuminates the behavior of phase transitions and mixing phenomena. In solids, atoms or molecules occupy fixed positions within an ordered lattice, offering few degrees of freedom and hence low entropy. Upon melting, particles gain translational mobility while remaining in close contact, increasing positional disorder and entropy. In the gaseous state, molecules move freely and randomly, achieving maximal disorder. This establishes the general trend for a pure substance: $S_{solid} < S_{liquid} < S_{gas}$. Similarly, when two liquids are mixed, or when a solute dissolves in a solvent, the total number of possible arrangements increases. Molecules occupy larger volumes and more configurations become accessible, contributing to an increase in entropy. This effect explains why dissolution and mixing are often spontaneous processes, even in the absence of heat release from chemical reactions. The driving force of spontaneity, therefore, is not solely energy minimization, but the statistical favorability of higher-entropy states.

The Second Law of Thermodynamics codifies the directional tendency inherent in spontaneous processes, asserting that the entropy of the universe increases for any spontaneous change. Mathematically, the total entropy change is expressed as $\Delta S_{univ} = \Delta S_{sys} + \Delta S_{surr}$, where ΔS_{sys} is the entropy change of the system, and ΔS_{surr} is that of the surroundings. A positive ΔS_{univ} indicates spontaneity, a negative value indicates a process favors the reverse direction, and zero corresponds to thermodynamic equilibrium. In processes occurring at constant temperature, the surroundings' entropy change can be expressed in terms of heat exchange: $\Delta S_{surr} = -\Delta H_{sys}/T$, with ΔH_{sys} representing the enthalpy change of the system and T the absolute temperature. Exothermic processes, which release heat to the surroundings, generally increase the entropy of the environment, promoting spontaneity. Endothermic processes, which absorb heat, can decrease the surroundings' entropy, requiring a sufficiently large increase in the system's entropy to drive overall spontaneity. Thus, the Second Law elegantly links energy transformations to probabilistic molecular behavior, providing insight into why processes occur in the directions they do.

The Third Law of Thermodynamics complements these principles by establishing an absolute reference for entropy. It states that the entropy of a perfect crystal at absolute zero (0 K) is zero. At this temperature, atoms occupy a perfectly ordered lattice, corresponding to a unique microstate with no configurational disorder. As temperature increases, molecular vibrations and rotations introduce disorder, resulting in increasing entropy. The Third Law allows the calculation of absolute entropy values and makes meaningful comparisons between different systems possible. Moreover, because entropy is a state function, its value depends solely on the system's current condition, independent of the pathway by which the system reached that state. This property is crucial in thermodynamic calculations, allowing for the precise determination of entropy changes in chemical reactions, phase transitions, and other processes.

To predict the spontaneity of reactions under conditions of constant temperature and pressure, chemists utilize the concept of Gibbs free energy, defined as $G = H - TS$. Here, H is the enthalpy, T is the temperature, and S is the entropy. The change in free energy, ΔG, determines spontaneity: a negative ΔG indicates a spontaneous process, a positive ΔG a non-spontaneous one, and $\Delta G = 0$ denotes equilibrium. Standard free energy changes, $\Delta G°$, apply when reactants and products are in their standard states, permitting calculations from tabulated free energies of formation. For non-standard conditions, the reaction quotient Q allows computation of free energy changes via $\Delta G = \Delta G° + RT\ln Q$, linking thermodynamic potentials to measurable concentrations or pressures. This framework explains why certain endothermic reactions can occur spontaneously: the entropy contribution, $T\Delta S$, can outweigh the enthalpy cost, yielding a negative ΔG and driving the process.

Equilibrium, defined by a system's state of minimum free energy, occurs when the forward and reverse reaction rates are equal. For a simple reaction $A \rightleftharpoons B$, the equilibrium constant K relates to the standard free energy change: $\Delta G° = -RT\ln K$. Reactions with $\Delta G° < 0$ favor products ($K > 1$), while those with $\Delta G° > 0$ favor reactants ($K < 1$), and $\Delta G° = 0$ indicates equal favorability. The ability to predict equilibrium positions from thermodynamic principles is essential in chemical engineering, biochemistry, and materials science, where precise control over reaction outcomes is required. Furthermore, the maximum useful work obtainable from a process at constant temperature and pressure is given by wmax = ΔG. In reality, irreversibilities such as friction and heat loss reduce the work extractable, reflecting the Second Law's implications: while energy is conserved, its capacity to perform work diminishes as entropy increases.

Thermodynamics provides explanatory power across diverse domains. In chemical systems, it predicts the feasibility of reactions, the solubility of compounds, and phase behavior. Combustion reactions, acid-base neutralizations, redox processes, and synthetic transformations can all be understood in terms of enthalpy, entropy, and free energy changes. Phase transitions—melting, boiling, sublimation—are explained through the interplay of enthalpy and entropy, providing insight into temperature-dependent equilibria. Solution formation, including solvation and complexation, is driven by entropic considerations as well as enthalpic interactions, explaining why some salts dissolve readily while others resist dissolution. In biological contexts, thermodynamics underpins metabolic pathways, enzyme activity, protein folding, and cellular energetics. Proteins fold spontaneously because the overall Gibbs free energy decreases, balancing enthalpic stabilization from intramolecular interactions against entropic changes in the protein and surrounding water molecules. Similarly, metabolic reactions, including ATP hydrolysis and oxidative phosphorylation, exploit free energy changes to perform work and maintain cellular organization despite the entropy increase in the universe.

Engineering applications of thermodynamics are equally profound. Heat engines, refrigerators, and air conditioners operate on principles of energy conservation, work extraction, and entropy management. The efficiency of engines is bounded by the Second Law, which dictates the impossibility of complete conversion of heat into work without compensating increases in entropy. Refrigeration cycles employ controlled entropy manipulation to remove heat from cold reservoirs and expel it to hotter surroundings, illustrating practical exploitation of thermodynamic principles. Electrochemical cells, including batteries and fuel cells, convert chemical energy into electrical energy, with Gibbs free energy providing the theoretical limit for work extraction. Real-world deviations from ideal behavior highlight the significance of irreversibilities, emphasizing the necessity of integrating both thermodynamic and kinetic considerations in engineering design.

Thermodynamics also offers a unifying framework connecting microscopic molecular behavior with macroscopic observables. Statistical mechanics, for instance, provides the bridge between molecular microstates and thermodynamic quantities such as temperature, pressure, and entropy. The Boltzmann relation, $S = k \ln\Omega$, where Ω is the number of microstates and k is the Boltzmann constant, formalizes the connection between entropy and probability. This statistical view explains why gases expand spontaneously, why mixing increases disorder, and why certain configurations are overwhelmingly favored. It reconciles deterministic energy conservation with probabilistic entropy trends, illustrating the elegance and depth of thermodynamic reasoning. Moreover, it provides predictive power for systems of enormous complexity, from gases and liquids to polymers and biological macromolecules.

Entropy changes and free energy considerations also elucidate the subtleties of non-ideal behavior. Real gases deviate from ideal gas law predictions due to intermolecular interactions and finite molecular volumes, affecting ΔS and ΔG. Solutions exhibit non-ideal mixing, with deviations in entropy and enthalpy arising from specific interactions, such as hydrogen bonding or van der Waals forces. Thermodynamic models, including activity coefficients and fugacity corrections, incorporate these effects, allowing accurate predictions of phase equilibria, solubility, and reaction spontaneity under practical conditions. Such refinements demonstrate the adaptability of thermodynamic principles to complex, real-world systems.

Throughout chemistry and physics, thermodynamics guides experimentation and design. Calorimetry measures heat exchange to determine enthalpy changes, while spectroscopy provides insight into molecular vibrations and energies. Computational methods predict ΔH, ΔS, and ΔG for reactions that are challenging to measure experimentally, enabling the design of novel materials, catalysts, and pharmaceuticals. Thermodynamic consistency ensures that theoretical models align with fundamental laws, while the Second Law imposes constraints on the directionality and feasibility of processes. Engineers leverage these principles to optimize fuel efficiency, minimize energy loss, and design systems that respect both conservation laws and entropy limitations.

In biological systems, the principles of thermodynamics reveal the delicate balance between order and disorder. Living organisms maintain local order—highly structured proteins, membranes, and nucleic acids—by increasing the entropy of their surroundings, ensuring compliance with the Second Law. Metabolic networks are driven by free energy gradients, enabling cells to extract work from chemical potential differences. Enzyme catalysis, molecular motors, and transport processes are all governed by Gibbs free energy changes, which dictate whether a process occurs spontaneously and how efficiently it can perform work. Thermodynamics, therefore, provides a quantitative language for understanding life at the molecular level, linking energy, entropy, and function in a coherent framework.

Across chemistry, physics, engineering, and biology, thermodynamics serves as a universal lens through which the flow of energy and matter can be quantified, predicted, and harnessed. The First Law emphasizes the permanence of energy, the Second Law introduces directionality and the inexorable increase of disorder, the Third Law establishes a reference for absolute entropy, and Gibbs free energy integrates these insights to predict spontaneity and equilibrium under realistic conditions. By combining macroscopic measurements with molecular understanding, thermodynamics enables the rational design of reactions, materials, devices, and biological systems. Its principles extend from the simplest expansion of a gas to the operation of a nuclear power plant, from the dissolution of a salt in water to the folding of a protein within a living cell, demonstrating its universal applicability and profound explanatory power.

The study of thermodynamics is not merely an academic exercise but a practical necessity for understanding and controlling the natural and engineered world. By rigorously accounting for energy, entropy, and free energy, it provides the tools to predict whether a reaction will occur, how much work can be extracted, and how systems will respond to environmental changes. Its laws, which emerge from both empirical observation and statistical reasoning, unify the microscopic and macroscopic realms, offering insight into the spontaneous direction of processes, the constraints on energy transformation, and the ultimate limits imposed by the Second and Third Laws. Whether in designing a high-efficiency engine, developing a new drug, understanding metabolic pathways, or exploring the fundamental limits of molecular order, thermodynamics provides the foundational framework that ensures both predictive power and conceptual clarity, affirming its central role in the scientific understanding of the physical universe.

The historical development of thermodynamics is as rich as its modern applications are diverse. From the early philosophical musings of ancient Greek thinkers on the nature of energy and matter to the meticulous experiments of 18th and 19th-century scientists, the evolution of thermodynamic thought illustrates the convergence of empirical observation and theoretical abstraction. Figures such as Sadi Carnot, whose studies of steam engines revealed the intrinsic limitations imposed by heat flow, laid the groundwork for the Second Law, even before the concept of entropy had been formalized. Carnot's insights demonstrated that no heat engine could operate with perfect efficiency, foreshadowing the role of irreversibility and entropy in natural processes. Later, Rudolf Clausius formalized the notion of entropy and codified the Second Law mathematically, while William Thomson, later Lord Kelvin, introduced the concept of absolute temperature. The development of the First Law was intertwined with the work of James Joule, whose experiments established the equivalence of mechanical work and heat, providing direct experimental verification of energy conservation. These historical milestones underscore the empirical roots of thermodynamic principles, showing that the abstract formulations employed in modern science are deeply grounded in observation and measurement.

A deeper exploration of the First Law illustrates its mathematical and practical nuances. Expressed in differential form, $dU = \delta Q - \delta W$, where dU is the infinitesimal change in internal energy, δQ is the heat added to the system, and δW is the work done by the system, this formulation captures the balance of energy at every infinitesimal step. In processes involving gases, δW can often be expressed as $P\,dV$, the product of pressure and infinitesimal volume change, which integrates over reversible paths to yield the work done. In chemical systems, δQ may include enthalpy changes associated with bond formation and breaking, while dU encompasses translational, rotational, vibrational, and electronic contributions. These detailed considerations illustrate that while the First Law enforces energy bookkeeping, it does not predict the spontaneity of energy flow; for that, one must consider entropy and the Second Law.

Entropy, with its statistical underpinnings, provides a probabilistic framework for understanding macroscopic phenomena. The Boltzmann relation, $S = k \ln\Omega$, where Ω represents the number of microstates compatible with a system's macrostate, quantifies the statistical tendency toward disorder. This formulation explains phenomena ranging from the expansion of gases to the dissolution of salts and the mixing of liquids. In gases, the translational degrees of freedom dominate entropy, while in liquids, rotational and vibrational motions contribute significantly. In solids, vibrational entropy is limited but still increases with temperature. Entropy is also central to understanding phase equilibria: the balance between enthalpic stabilization and entropic favorability determines melting points, boiling points, and solubility limits. The intricate interplay between energy and disorder drives the directionality of natural processes, revealing that while energy is conserved, its distribution among accessible microstates dictates what changes will occur spontaneously.

The Gibbs free energy formalism, $G = H - TS$, offers a practical tool for predicting the spontaneity of reactions under constant temperature and pressure. Its applications extend across chemical, biological, and engineering contexts. For instance, in biochemical systems, ATP hydrolysis is exergonic, with a large negative ΔG, enabling cells to perform work such as active transport, muscle contraction, and biosynthetic reactions. Proteins fold into their native structures because folding decreases free energy, balancing enthalpic contributions from hydrogen bonding, van der Waals interactions, and electrostatic stabilization against the entropic cost of ordering the polypeptide chain. Enzyme-catalyzed reactions, while not altering ΔG, accelerate the attainment of equilibrium by lowering activation barriers, illustrating the distinction between thermodynamic favorability and kinetic accessibility. In materials science, the formation of alloys, the dissolution of salts, and phase transitions in polymers are all interpreted through the lens of free energy changes, demonstrating the universality of the concept.

Irreversibility is a key feature of real-world processes. While idealized thermodynamic analyses often consider reversible paths, actual processes invariably generate additional entropy due to friction, turbulence, heat transfer across finite gradients, and dissipative losses. This generation of entropy implies that the actual work extracted from a process is always less than the theoretical maximum predicted by ΔG. For example, in engines, some energy is inevitably lost as waste heat to the surroundings; in chemical reactions, some energy disperses into vibrational modes or radiates away as photons. Non-equilibrium thermodynamics further extends these concepts to systems far from equilibrium, where gradients in temperature, pressure, or chemical potential drive continuous fluxes of matter and energy. The formalism of Onsager reciprocal relations and linear transport theory allows the prediction of fluxes and the quantification of entropy production, bridging classical thermodynamics with dynamic, real-world processes.

Statistical mechanics provides an elegant framework connecting microscopic and macroscopic perspectives. By enumerating microstates and applying probability theory, it explains macroscopic thermodynamic quantities, including temperature, pressure, and entropy. For example, the Maxwell-Boltzmann distribution describes the spread of molecular velocities in a gas, while the partition function formalism allows calculation of internal energy, heat capacities, and free energy for systems ranging from ideal gases to interacting lattices. The profound insight of statistical mechanics is that macroscopic laws, including the Second Law, emerge from the collective behavior of large numbers of particles, rather than from deterministic rules for individual molecules. This perspective illuminates why spontaneous processes overwhelmingly proceed toward higher-entropy states, and why deviations from this behavior, while theoretically possible, are astronomically improbable.

The interplay between energy, entropy, and molecular structure governs solution chemistry. When a solute dissolves in a solvent, the process is driven by a balance of enthalpic interactions and entropic effects. Ion-dipole interactions, hydrogen bonding, and van der Waals forces contribute to enthalpy changes, while the increase in disorder associated with distributing solute particles throughout the solvent enhances entropy. In many cases, entropy gain outweighs modest enthalpy penalties, resulting in a negative ΔG and spontaneous dissolution. This balance explains phenomena such as the solubility of salts, the mixing of miscible liquids, and the formation of complex solutions like micelles or colloids. In biological systems, entropy-driven processes are pervasive: the hydrophobic effect, central to protein folding and membrane formation, arises from the tendency of water molecules to maximize their own entropy, indirectly driving hydrophobic molecules together.

Phase transitions illustrate the nuanced interplay of enthalpy and entropy. Melting occurs when thermal energy overcomes lattice forces, increasing positional freedom and entropy. Boiling introduces additional translational freedom in the gaseous phase, leading to a substantial entropy increase. Sublimation bypasses the liquid state entirely but similarly reflects a dramatic gain in disorder. In multicomponent systems, phase diagrams are constructed based on Gibbs free energy minimization, revealing regions of stability, coexistence, and metastability. Thermodynamic analysis of these diagrams allows prediction of melting points, eutectic compositions, azeotropes, and critical points, essential for materials design, chemical separations, and industrial processing.

Thermodynamics also underpins energy conversion and storage technologies. In electrochemical systems, the Nernst equation, derived from Gibbs free energy considerations, predicts cell potentials under non-standard conditions, accounting for ion concentrations and pressure. Batteries, fuel cells, and electrolyzers all exploit the conversion of chemical energy to electrical energy or vice versa, constrained by thermodynamic limits and efficiency losses. Heat engines, operating on Carnot cycles, illustrate the theoretical maximum efficiency dictated by temperature differences between reservoirs. Refrigeration cycles exploit controlled entropy changes to remove heat from low-temperature regions, emphasizing the practical importance of the Second Law in engineering design.

Biological systems provide compelling examples of thermodynamic principles at work. Metabolic networks are orchestrated to maximize energy extraction while maintaining homeostasis. Catabolic reactions release energy, increasing the entropy of the surroundings, while anabolic reactions consume energy to maintain cellular organization, resulting in local decreases in entropy offset by greater environmental entropy increases. Molecular motors, such as kinesin and ATP synthase, convert chemical potential energy into mechanical work with remarkable efficiency, constrained by thermodynamic laws. Thermodynamics also informs drug design, predicting binding affinities, solubility, and conformational preferences based on free energy landscapes.

Non-equilibrium thermodynamics extends these concepts to living and engineered systems that operate far from equilibrium. Biological systems maintain steady states by continuously exchanging energy and matter with their environments, producing entropy in the process. Industrial reactors, heat exchangers, and ecological systems similarly operate in non-equilibrium conditions, where gradients drive fluxes and sustain dynamic behavior. The formalism of irreversible thermodynamics quantifies these fluxes and their associated entropy production, providing predictive power for complex, real-world systems where equilibrium assumptions do not hold.

Finally, thermodynamics integrates with modern computational approaches. Molecular dynamics simulations, Monte Carlo methods, and quantum chemistry calculations allow the estimation of thermodynamic quantities for complex systems, from protein folding pathways to catalysis and materials design. Free energy perturbation methods and thermodynamic integration enable the calculation of ΔG for reactions that are experimentally challenging, facilitating rational design of pharmaceuticals, catalysts, and functional materials. The combination of classical thermodynamics, statistical mechanics, and computational tools represents a powerful framework for understanding, predicting, and optimizing energy transformations across disciplines.

In summary, thermodynamics provides a comprehensive, multi-scale framework that unites principles of energy conservation, entropy, and free energy to explain the behavior of physical, chemical, and biological systems. The First Law enforces energy conservation, the Second Law dictates directional tendencies, the Third Law establishes absolute entropy references, and Gibbs free energy quantifies spontaneity and equilibrium. By integrating historical insights, molecular-level interpretations, statistical mechanics, and practical applications, thermodynamics enables the prediction and manipulation of processes ranging from chemical reactions and phase transitions to energy conversion, biological metabolism, and advanced materials design. Its principles illuminate the hidden order within disorder, guiding both scientific understanding and technological innovation, and offering a profound and unifying lens through which the physical universe may be comprehensively interpreted.

Chapter 17: Word Problems

1. A chemist is studying an ideal monatomic gas confined in a rigid, insulated container of volume 5.00 L initially at 300 K. The gas is slowly heated by placing the container in a controlled thermal bath, and during the heating process, its internal energy increases by 1.50×10^3 J. Considering the First Law of Thermodynamics, calculate both the work done by the gas and the heat absorbed. Discuss how the use of a rigid container constrains work exchange, how the internal energy is distributed among translational degrees of freedom, and how the First Law governs energy bookkeeping in this scenario.

2. A laboratory experiment involves the isothermal expansion of an ideal gas from 2.0 L to 10.0 L at 350 K, with the gas in thermal equilibrium with the surroundings. Calculate the entropy change of the gas during this process and explain why, despite the temperature remaining constant, the entropy increases. Provide a molecular-level explanation relating the change in accessible microstates to the probability of configurations, and discuss how this demonstrates the statistical underpinnings of the Second Law of Thermodynamics.

3. A reversible heat engine operates between two thermal reservoirs maintained at 600 K and 300 K. The engine absorbs 5.00 kJ of heat from the high-temperature reservoir per cycle. Calculate the maximum work output per cycle and the

corresponding entropy change of the surroundings. Additionally, discuss why no actual engine can reach this maximum due to irreversibilities, friction, and non-idealities, and explain how this illustrates the practical limitations imposed by the Second Law.

4. A 1.0 mol sample of a crystalline solid melts at 400 K, absorbing 6.00 kJ of heat from the surroundings. Determine the entropy change of the system and of the surroundings during melting. Discuss the balance between enthalpy and entropy contributions to spontaneity, how increased molecular mobility in the liquid phase increases disorder, and why melting exemplifies a thermodynamically favored process even if the enthalpy change is significant.

5. Two miscible liquids are combined in a calorimeter, causing a measurable temperature rise of 2 K in the surroundings. The enthalpy change of mixing is −800 J, and the solution is maintained at 298 K. Calculate the entropy changes of the system and surroundings, determine ΔSuniv, and explain why the mixture spontaneously forms even if the enthalpy change were zero or slightly positive. Discuss the role of molecular disorder and the statistical favorability of increased microstates in driving the process.

6. The combustion of one mole of glucose releases 2.80×10^3 kJ of heat at 298 K under constant pressure. Calculate the entropy change of the surroundings resulting from this reaction. Using ΔS_{univ}, analyze why the reaction is spontaneous, emphasizing the interplay between energy release and the tendency of nature to favor states with higher entropy. Include a discussion of how exothermic reactions contribute to the directionality of spontaneous processes.

7. In a laboratory experiment, 0.50 mol of potassium chloride is dissolved in 1.0 L of water at 298 K. The dissolution process is slightly endothermic, with ΔH = +17 kJ/mol. Using Gibbs free energy concepts, determine whether the dissolution is spontaneous, and explain in detail how entropy compensates for the positive enthalpy. Discuss the microscopic mechanisms, including ion–dipole interactions and increased configurational freedom, that contribute to the entropy increase.

8. A system consisting of 2 mol of a diatomic ideal gas undergoes an adiabatic expansion from 2.0 L to 10.0 L. Calculate the final temperature and the change in internal energy. Explain how the First Law constrains the energy distribution in

the absence of heat exchange, and discuss how work is done at the expense of internal energy, considering both translational and rotational degrees of freedom.

9. A 1.0 mol ideal gas is isothermally compressed from 10.0 L to 2.0 L at 300 K. Compute the work done on the gas, the heat released to the surroundings, and the entropy change of the system. Provide a detailed explanation of why the gas's entropy decreases while the surroundings' entropy increases, and discuss the implications of these changes for the overall spontaneity of the process.

10. During the heating of 1 mol of water from 273 K to 373 K at constant pressure, calculate the change in entropy of the water. Then, assuming reversible heat transfer, determine the minimal entropy change of the surroundings. Discuss how the temperature-dependent distribution of molecular energies contributes to entropy, and explain how these calculations illustrate the relationship between heat flow, energy distribution, and the Second Law.

11. A Carnot engine operates between reservoirs at 500 K and 300 K, absorbing 1000 J from the hot reservoir per cycle. Determine the heat expelled to the cold reservoir, the net work output, and the total entropy change of the universe. Discuss how the Carnot cycle exemplifies the theoretical maximum efficiency dictated by the Second Law, and why real engines inevitably perform less work due to irreversibilities and dissipative processes.

12. A metal rod is heated at one end while the other end is in thermal contact with ice at 273 K. Using the formalism of irreversible thermodynamics, model the energy transfer along the rod and calculate the entropy production if 500 J flows in 10 s. Discuss why real processes generate additional entropy, how temperature gradients drive irreversible heat flow, and the practical implications for thermal engineering.

13. A protein in aqueous solution folds spontaneously at 310 K. The enthalpy change of folding is −40 kJ/mol, while the entropy change is −100 J/mol·K. Calculate ΔG and explain why folding occurs spontaneously despite the decrease in system entropy. Discuss the balance between enthalpic stabilization through hydrogen bonds, van der Waals interactions, and electrostatics, and entropic contributions from solvent reorganization.

14. A battery converts chemical energy into electrical energy, delivering 5.00×10^3 J of useful work at room temperature while releasing additional heat into the surroundings. If 1.20×10^4 J of chemical energy is consumed, calculate the increase in entropy of the surroundings and discuss how irreversibilities reduce the maximum achievable work. Relate this to the practical efficiency of energy conversion devices.

15. Two ideal gases, A and B, are initially separated by a partition in a container of total volume 10 L. When the partition is removed, the gases mix freely. Calculate the entropy change of the system and discuss why mixing occurs spontaneously even in the absence of heat or work exchange. Include a discussion of microstates, molecular probability, and the statistical interpretation of entropy.

16. A solution of NaOH is prepared by dissolving 1.0 mol in 500 mL of water, releasing 44 kJ of heat at 298 K. Determine the entropy changes of the system and surroundings, calculate ΔS_{univ}, and explain how exothermic dissolution contributes to spontaneity. Discuss the molecular basis for heat release and the accompanying increase in disorder of the water molecules.

17. An enzyme catalyzes a reaction with $\Delta G° = -25$ kJ/mol at 310 K. The substrate concentration is only half the standard state. Calculate ΔG under these conditions and predict the spontaneity of the reaction in the cellular environment. Discuss how free energy changes govern biochemical processes, the effect of substrate concentrations, and the importance of ΔG versus $\Delta G°$ in living systems.

18. In an industrial reactor, 5 mol of NO_2 decompose according to $2NO_2 \rightleftharpoons 2NO + O_2$ at 298 K. Given $\Delta G° = +10$ kJ/mol, calculate the equilibrium constant K and predict whether decomposition occurs spontaneously under standard conditions. Discuss how temperature and concentration affect the position of equilibrium and the practical implications for chemical manufacturing.

19. A heat pump removes 2.0×10^3 J of heat from a cold reservoir at 273 K and delivers it to a hot reservoir at 298 K. Calculate the entropy changes of the cold and hot reservoirs, the net change of the universe, and the minimum work required. Explain why this process cannot occur spontaneously without external work and how entropy considerations determine refrigeration efficiency.

20. A gas expands freely into a vacuum from 1.0 L to 5.0 L. Determine the entropy change of the gas and discuss why ΔS_{univ} is positive despite no heat transfer or work done. Explain the statistical reasoning behind the spontaneous expansion and why the reverse process is effectively impossible.

21. One mole of CO_2 undergoes vaporization at 194.7 K, absorbing 25 kJ of heat from a thermal reservoir. Calculate the entropy change associated with vaporization. Discuss how the gain in translational freedom of gas molecules relative to the solid phase contributes to this increase in entropy, and explain how phase transitions are driven by a balance between enthalpic input and entropy increase.

22. Two miscible liquids, one with a molar entropy of 75 J/mol·K and the other 85 J/mol·K, are mixed at constant temperature and pressure. Calculate the total entropy change of the system upon mixing. Discuss how spontaneous mixing occurs even in the absence of enthalpy change, highlighting the statistical interpretation of increased microstates and why nature favors disorder.

23. A 2-mol sample of a monatomic ideal gas undergoes a reversible adiabatic expansion that doubles its volume. Calculate the final temperature, work done by the gas, and the change in internal energy. Discuss how internal energy is redistributed among translational degrees of freedom and how the First Law constrains energy conservation during an adiabatic, reversible process.

24. A student performs an isothermal compression of air in a cylinder, doing 500 J of work on the system. Determine the heat released to the surroundings and the corresponding entropy change of the system. Discuss how compression reduces molecular positional freedom, why the surroundings' entropy increases, and the role of the Second Law in dictating the directionality of spontaneous energy dispersal.

25. In a muscle contraction experiment, 0.01 mol of ATP undergoes hydrolysis, releasing $\Delta G = -30.5$ kJ/mol at 310 K. Calculate the maximum work that can be extracted and explain how this energy is harnessed to perform mechanical work at the cellular level. Discuss the balance between enthalpy, entropy, and Gibbs free energy that allows ATP to function as a molecular energy currency.

26. A reversible heat engine absorbs 1.0 kJ of heat from a reservoir at 600 K and

expels heat to a reservoir at 300 K. Calculate the maximum work per cycle and discuss why no real engine can achieve this ideal output. Include an analysis of entropy generation and the practical implications for thermodynamic efficiency in engineering devices.

27. The dissolution of ammonium chloride in water is slightly endothermic, with $\Delta H = +14.8$ kJ/mol at 298 K. Calculate the entropy change of the system, the surroundings, and ΔG to determine whether the process is spontaneous. Discuss the microscopic mechanisms—hydration, ion dispersion, and configurational freedom—that allow dissolution to occur despite the enthalpy cost.

28. A polymer melts over a temperature range from 400 K to 420 K, absorbing 30 kJ/mol of heat. Calculate the entropy change associated with melting. Discuss how increased positional and conformational freedom in the molten polymer contributes to spontaneity, and explain why entropy can drive phase transitions even when enthalpy penalties are substantial.

29. Two ideal gases, A and B, initially occupy separate compartments in a 10-L container. Upon removal of the partition, the gases mix freely. Calculate the total entropy change of the system and discuss why the reverse process is virtually impossible. Include an explanation of microstates, molecular probability, and the statistical interpretation of entropy that underlies the Second Law.

30. In a chemical battery, 2 mol of zinc react with 2 mol of HCl to produce 2 mol of H_2. If 200 kJ of heat is generated and 120 kJ of electrical work is extracted, calculate ΔS_{univ}. Discuss how the First Law accounts for energy distribution and how the Second Law governs the maximum usable work, highlighting the role of irreversibilities.

31. A 0.1 mol ideal gas undergoes a reversible, isothermal expansion into a vacuum, doubling its volume. Calculate the entropy change and discuss why this process is effectively irreversible even though no work is performed. Include a discussion of molecular disorder, the statistical nature of entropy, and the improbability of spontaneous reversal.

32. A heat engine operating between 800 K and 300 K produces 500 J of work per cycle. Determine the heat absorbed from the hot reservoir, the entropy changes of

the reservoirs, and discuss how the Second Law limits efficiency. Explain why real engines yield less work due to irreversibilities and entropy production.

33. A gas mixture undergoes a chemical reaction at constant temperature, generating 2 mol of gas from 1 mol of reactants. Calculate ΔS for the reaction and discuss how the increase in molecular number contributes to spontaneity. Explain the thermodynamic reasoning behind why reactions producing more gaseous molecules are often favored.

34. A refrigerator removes 400 J of heat from the interior at 273 K and expels it to a room at 298 K. Calculate the entropy changes of the cold and hot reservoirs and the total entropy change of the universe. Discuss why external work is required to drive this non-spontaneous heat transfer and how entropy considerations dictate refrigeration efficiency.

35. The folding of a 50-residue protein has ΔH = −60 kJ/mol and ΔS = −150 J/mol·K at 310 K. Calculate ΔG and predict spontaneity. Discuss how enthalpic stabilization (hydrogen bonding, van der Waals forces) competes with entropic costs (loss of conformational freedom), and explain the enthalpy–entropy compensation that allows proteins to fold.

36. During electrolysis, 1 mol of electrons is transferred, performing 96.5 kC of work. If ΔG = −220 kJ/mol, calculate the maximum voltage that can be applied to achieve this reaction. Discuss the relationship between Gibbs free energy, work, and electrical energy, and explain the role of thermodynamic limits in electrochemical processes.

37. The mixing of 1 mol ethanol and 1 mol water is exothermic with ΔH = −4.0 kJ. Calculate the entropy change of the system and surroundings, determine ΔG, and explain why the mixture forms spontaneously. Discuss how intermolecular interactions and increased configurational freedom contribute to the overall thermodynamic favorability.

38. A diatomic ideal gas undergoes an adiabatic compression from 5.0 L to 1.0 L. Calculate the final temperature and discuss how energy from work done on the gas is partitioned into translational and rotational degrees of freedom. Explain how the First Law ensures energy conservation in the absence of heat exchange.

39. A chemical reaction at 298 K has $\Delta H = +50$ kJ/mol and $\Delta S = +180$ J/mol·K. Determine ΔG and discuss how the entropic contribution drives spontaneity despite the reaction being endothermic. Include a discussion of the molecular mechanisms that contribute to increased disorder.

40. In an industrial process, 3 mol of NaOH are dissolved in water, absorbing 12 kJ of heat. Calculate ΔS_{sys}, ΔS_{surr}, and ΔS_{univ}. Explain why endothermic processes can be spontaneous when the system entropy gain outweighs the enthalpic input, and discuss real-world industrial implications.

41. One mole of an ideal gas is heated at constant pressure from 300 K to 600 K. Determine the change in entropy and discuss how energy distribution among molecular translational, rotational, and vibrational modes contributes to increasing disorder. Relate this to the statistical interpretation of entropy.

42. In ATP synthesis, $\Delta G^\circ = +30.5$ kJ/mol at 298 K. If the proton-motive force provides 50 kJ of energy, determine whether ATP synthesis occurs spontaneously. Discuss the concept of energy coupling in biology and how Gibbs free energy predicts the feasibility of cellular processes.

43. A reversible engine absorbs 2 kJ from a hot reservoir at 500 K and rejects heat at 300 K. Calculate ΔS_{univ} and the maximum work output, explaining why real engines deliver less work due to entropy production. Discuss the practical importance of the Second Law in engineering design.

44. Two moles of water are vaporized at 373 K, absorbing 88 kJ of heat. Calculate ΔS_{vap} and explain how the dramatic increase in molecular freedom in the gas phase drives the phase transition. Discuss the balance of enthalpy and entropy in determining spontaneity.

45. Two ideal gases, A and B, expand into a combined volume of 10 L from initially separate 2-L compartments. Calculate the total entropy change and explain why diffusion occurs spontaneously even without heat transfer. Relate this to the statistical mechanics perspective on microstates.

46. A battery delivers 50 J of electrical work while releasing 100 J of heat into the surroundings. Calculate ΔS_{univ} and explain why energy conversion efficiency is limited by the Second Law. Discuss the implications for real electrochemical devices.

47. A reaction has $\Delta H = -20$ kJ/mol and $\Delta S = -50$ J/mol·K at 298 K. Determine ΔG and discuss the temperature conditions under which the reaction becomes spontaneous. Include a discussion of enthalpy–entropy compensation and its importance in predicting reaction behavior.

48. During a Carnot refrigeration cycle, 200 J of heat is removed from a cold reservoir at 270 K and expelled at 300 K. Calculate the entropy changes of the reservoirs and determine the minimum work required. Discuss why external work is necessary and the connection between entropy, reversibility, and efficiency.

49. The dissolution of sugar in water at 298 K has $\Delta H \approx 0$ and $\Delta S = +110$ J/mol·K. Calculate ΔG and explain why the process is spontaneous purely due to entropy increase. Discuss the molecular basis for solute dispersal and solvent reorganization contributing to entropy gain.

50. A gas undergoes a quasi-static isothermal expansion against an external pressure, increasing in volume from 1.0 L to 5.0 L. Derive the work done, calculate the entropy change, determine ΔG, and discuss how energy dispersal and molecular disorder govern the spontaneity of isothermal processes.

CHAPTER 18

ELECTROCHEMISTRY SPARKS CHANGE

Electrochemical reactions are ubiquitous in both natural phenomena and technological applications, forming the foundation for the generation and utilization of electrical energy. From the vehicles we drive to the radios that entertain us, from the digital watches that mark time to the calculators that simplify complex computations, electrochemistry is integral to everyday life. The development of compact batteries, in combination with advances in silicon-chip technology, has enabled the proliferation of portable devices that would have been inconceivable just a few decades ago. Batteries, in their simplest form, are devices that convert chemical energy directly into electrical energy, and their evolution has paralleled humanity's increasing need for mobile power sources. Yet, beyond these applications, electrochemistry governs processes that often escape immediate notice, such as corrosion.

The rusting of iron is not merely a surface blemish; it is an electrochemical phenomenon, representing the gradual oxidation of iron to its natural ore form in the presence of moisture and oxygen. Industrially, electrochemical processes underpin the production of critical materials such as chlorine, aluminum, and sodium hydroxide, demonstrating the immense versatility and utility of redox reactions. Electrodes, which are carefully chosen based on their affinity for specific ions such as hydronium, fluoride, or chloride, serve as the conduits for electron flow, enabling these reactions to be harnessed in both practical and analytical contexts. In analytical chemistry, electrochemical techniques are employed to detect trace pollutants in water or small concentrations of biomolecules in human blood, providing critical information that can predict disease development or monitor environmental quality. At its core, electrochemistry is the study of the relationship between chemical changes and the production or consumption of electrical energy, encompassing both the spontaneous generation of current from chemical reactions and the use of applied electrical energy to drive reactions that would not otherwise occur.

The foundational principle of electrochemistry lies in redox reactions, which involve the transfer of electrons between chemical species. Oxidation, defined as the loss of electrons or an increase in oxidation number, and reduction, the gain of electrons or decrease in oxidation number, are complementary processes that occur simultaneously. A classic example is the reaction between permanganate ions and ferrous ions in acidic solution: $8H^+ + MnO_4^- + 5Fe^{2+} \rightarrow Mn^{2+} + 5Fe^{3+} + 4H_2O$. In this reaction, Fe^{2+} ions lose electrons and are oxidized to Fe^{3+}, while MnO_4^- ions gain electrons and are reduced to Mn^{2+}. By physically separating the oxidizing and reducing agents and connecting them via a conductive wire, the electrons released during oxidation flow through the wire and can perform work, such as powering a light bulb or operating a motor. This arrangement is embodied in galvanic cells, where the compartment containing the oxidized species is the anode, and the compartment containing the reduced species is the cathode. To maintain electrical neutrality while allowing the transfer of ions, the compartments are connected by a salt bridge or porous disk, which permits the flow of ions without allowing the solutions to mix directly. The careful design of these cells, including the selection of electrodes and electrolytes, allows for the controlled conversion of chemical energy into electrical energy.

Standard reduction potentials provide a quantitative measure of a species' tendency to gain electrons and are essential for predicting the behavior of electrochemical cells. In a zinc-hydrogen cell, for instance, zinc metal at the anode undergoes oxidation to produce Zn^{2+} ions while releasing electrons, and hydrogen ions at the cathode are reduced to hydrogen gas. The standard hydrogen electrode (SHE) is assigned a potential of zero volts, serving as a reference against which the potentials of other half-reactions can be measured. These potentials are intensive properties and do not depend on the quantities of reactants used to balance the electron transfer. Chemists use line notation to represent electrochemical cells, placing the anode on the left, the cathode on the right, and indicating the salt bridge with double lines. By analyzing the standard reduction potentials of the half-reactions, one can predict the direction of electron flow, determine the maximum possible voltage, and assess the feasibility of the overall reaction.

The relationship between electrochemistry and thermodynamics is elegantly captured in the concept of cell potential, which reflects the maximum work that can be performed per unit of charge. The Faraday constant, approximately 96,485 coulombs per mole of electrons, provides a direct link between the number of electrons transferred and the electrical work performed. The Gibbs free energy change for a reaction, ΔG, is related to the cell potential through the equation $\Delta G = -nFE\_cell$, where n is the number of electrons involved. A positive cell potential corresponds to a negative Gibbs free energy, indicating a spontaneous reaction. While the maximum work is theoretically achieved when no current flows, real systems invariably experience energy losses, typically as heat, due to resistive elements in the circuit and inefficiencies in electron transfer. In addition to intrinsic properties, the potential of a cell depends on the concentrations of the reactants and products. Concentration cells, in which identical species exist at different concentrations in separate compartments, generate a potential that drives electrons to equalize the concentrations. The Nernst equation, $E = E^o - (0.0591/n) \log Q$ at 25°C, quantifies the relationship between cell potential and concentration, providing a powerful tool for predicting reaction behavior and determining equilibrium constants for redox processes.

Batteries represent one of the most tangible applications of galvanic cells, providing portable sources of electrical energy. Lead-acid batteries, which are widely used in automobiles, consist of lead anodes and lead dioxide cathodes immersed in sulfuric acid. Multiple cells are connected in series to achieve the desired voltage, typically 12 volts for standard car batteries. Dry cells, commonly found in household electronics, utilize zinc anodes and carbon cathodes with reactive pastes, while alkaline variants offer extended lifespans due to slower corrosion of the zinc anode. Specialized batteries, such as silver oxide and mercury cells, deliver stable voltages for devices that require precision and reliability, including calculators, watches, and hearing aids. Fuel cells differ in that they continuously receive reactants and convert chemical energy into electricity through reactions such as the combination of hydrogen and oxygen to produce water. These systems can achieve high efficiency and low environmental impact, making them promising candidates for sustainable energy technologies. The development of lithium-ion batteries, which rely on the reversible intercalation of lithium ions between electrodes, has further revolutionized portable electronics, enabling long-lasting power in smartphones, laptops, and electric vehicles.

Corrosion, the electrochemical degradation of metals, highlights the natural tendency of metals to revert to their more stable, oxidized forms. Iron, for example, rusts when exposed to moisture and oxygen, forming $Fe_2O_3{\cdot}xH_2O$. The process involves the migration of Fe^{2+} ions from anodic regions, where oxidation occurs, to cathodic regions, where reduction of oxygen or hydrogen ions takes place. Aluminum, in contrast, forms a thin, adherent oxide layer that protects it from further oxidation, demonstrating how different metals vary in susceptibility to corrosion. Preventive measures include protective coatings, such as paint or metal plating with chromium, tin, or zinc, as well as alloy formation, such as stainless steel, and cathodic protection, where a more reactive metal sacrifices itself to protect the primary metal. Understanding the mechanisms of corrosion and developing strategies to mitigate it are critical for the longevity of infrastructure, vehicles, and machinery, emphasizing the practical significance of electrochemistry in daily life and industry.

Electrolysis, the application of electrical energy to drive nonspontaneous reactions, is employed in numerous industrial processes. The Hall-Héroult process, for instance, produces aluminum by electrolyzing molten alumina dissolved in cryolite, while the chlor-alkali process generates chlorine and sodium hydroxide from the electrolysis of brine. Electrolysis also enables electroplating, where a thin layer of metal is deposited onto a surface for protection or decorative purposes. The efficiency of these processes depends on careful control of current density, voltage, temperature, and electrode composition, as well as an understanding of overpotential effects, which influence the energy required to drive the reactions. Advances in electrolysis technology continue to improve energy efficiency and reduce environmental impact, making it a vital tool in modern industry and sustainable chemical production.

In addition to industrial applications, electrochemistry plays a central role in analytical chemistry. Techniques such as potentiometry, voltammetry, and amperometry allow for the detection of trace amounts of pollutants in water, soil, and air, as well as minute concentrations of biomolecules in biological samples. Electrochemical sensors, including biosensors, provide high sensitivity and specificity, often detecting analytes at parts-per-billion concentrations. These capabilities are essential in environmental monitoring, medical diagnostics, and food safety, where early detection of harmful substances can prevent disease outbreaks and environmental damage. The ongoing development of electrochemical methods in nanotechnology, renewable energy, and green chemistry further illustrates the versatility and continuing relevance of the field, enabling innovations in energy storage, environmental protection, and medical diagnostics.

The study of electrochemistry illuminates the profound influence of electron transfer on both natural processes and engineered systems. Galvanic cells exemplify the conversion of chemical energy into electrical energy, while electrolytic cells demonstrate the reverse, using electrical energy to induce chemical change. Concepts such as standard reduction potentials, the Nernst equation, cell potential, and electrode behavior provide the framework for understanding batteries, fuel cells, corrosion, and industrial electrolysis. The applications of these principles are extensive, affecting everything from daily conveniences to large-scale industrial operations, from medical diagnostics to environmental monitoring. By elucidating the fundamental mechanisms that govern redox reactions and electron flow, electrochemistry bridges the gap between chemical energy and electrical work, offering both practical solutions and theoretical insights that continue to drive technological innovation, improve sustainability, and enhance human well-being.

Electrochemical processes are governed not only by qualitative understanding but also by precise quantitative relationships that define the behavior of cells under varying conditions. One of the most significant relationships is encapsulated in the Nernst equation, which provides a bridge between thermodynamic principles and observable cell potentials. For any redox reaction, the cell potential under nonstandard conditions can be calculated by $E = E^o - (0.0591/n) \log Q$ at 25°C, where Q represents the reaction quotient, reflecting the concentrations of reactants and products. This equation arises from the relationship between Gibbs free energy and electromotive force: $\Delta G = -nFE$, and $\Delta G = \Delta G^o + RT \ln Q$. By equating these two expressions and solving for E, the equation provides a predictive tool that enables chemists to determine the voltage output of a cell when the concentrations of the species involved deviate from standard conditions. In practice, this equation is critical for understanding the behavior of concentration cells, where the same species exists in compartments of differing concentration, driving electron flow to restore equilibrium. Such cells illustrate a profound principle of electrochemistry: the movement of electrons and ions is always governed by the tendency of systems to minimize free energy, achieving a state of thermodynamic equilibrium.

Batteries, as applied galvanic cells, exhibit a remarkable diversity of chemistries, each tailored to specific applications. Lead-acid batteries, ubiquitous in automotive systems, rely on the reversible reactions of lead, lead dioxide, and sulfuric acid. During discharge, the lead anode is oxidized to lead sulfate while the lead dioxide cathode is reduced to lead sulfate as well, generating an electric current. Upon recharging, these reactions are reversed, allowing the battery to be used repeatedly. Lithium-ion batteries, in contrast, utilize the intercalation of lithium ions between layered electrodes, typically graphite for the anode and lithium metal oxide for the cathode. The movement of lithium ions during charge and discharge is accompanied by electron flow through the external circuit, producing a stable voltage over extended cycles. Nickel-metal hydride and zinc-carbon cells offer alternative chemistries with distinct advantages in energy density, cycle life, and cost, illustrating the way electrochemistry has enabled tailored solutions for diverse technological needs. Modern research continues to explore innovations in solid-state batteries, sodium-ion cells, and hybrid chemistries, with the dual goals of increasing energy density and ensuring safety, sustainability, and efficiency.

The phenomena of corrosion illustrate the practical importance of electrochemistry in everyday materials and engineering. Iron, when exposed to oxygen and moisture, undergoes a series of anodic and cathodic reactions that convert it into hydrated iron oxide, commonly known as rust. The migration of electrons from anodic regions of the metal to cathodic regions facilitates the reduction of oxygen or hydrogen ions, accelerating the degradation process. Aluminum, by forming a stable oxide layer, exemplifies how natural passivation can prevent extensive corrosion. Engineers exploit this principle, applying controlled oxide layers to other metals and using sacrificial anodes in cathodic protection to divert corrosion to more reactive metals, such as magnesium or zinc, thereby preserving critical infrastructure. Real-world examples abound, from the protection of steel in bridges and pipelines to the preservation of naval vessels, demonstrating the critical role of electrochemistry in maintaining structural integrity and reducing maintenance costs. The study of corrosion also extends to high-temperature and marine environments, where chloride ions or elevated temperatures accelerate deterioration, requiring sophisticated alloys and protective coatings.

Electrolysis serves as the foundation for a wide range of industrial chemical processes, illustrating the controlled application of electrical energy to drive nonspontaneous reactions. In aluminum production via the Hall-Héroult process, alumina is dissolved in molten cryolite and subjected to a direct current. Aluminum ions migrate to the cathode, where they are reduced to aluminum metal, while oxygen ions migrate to the carbon anode, producing carbon dioxide. The process requires careful regulation of temperature and voltage to minimize energy losses and prevent unwanted side reactions. The chlor-alkali process, employed to produce chlorine gas and sodium hydroxide from brine, similarly relies on electrolysis, with specialized diaphragm or membrane cells preventing the recombination of products and ensuring high purity. These processes highlight how electrochemistry enables the large-scale production of essential chemicals, providing materials for water treatment, paper manufacturing, and chemical synthesis. Advances in electrolytic technology, including the use of renewable electricity and improved membrane materials, are increasingly important in reducing the carbon footprint of chemical industries.

Electrochemical sensors exemplify the application of electrochemistry in analytical science, allowing the detection of trace amounts of chemical species with exceptional sensitivity. Potentiometric sensors, such as pH electrodes, measure the potential difference between two electrodes in contact with a solution, providing rapid and accurate information about the hydrogen ion concentration.

Voltammetric techniques, including cyclic voltammetry and square wave voltammetry, involve the application of a varying potential to an electrode and measurement of the resulting current, revealing both the identity and concentration of electroactive species. Amperometric sensors, by maintaining a constant potential and measuring current proportional to analyte concentration, enable real-time monitoring of gases, ions, and biomolecules. These technologies are critical in environmental monitoring, medical diagnostics, and industrial process control. For example, electrochemical biosensors can detect glucose in blood, heavy metals in rivers, or pesticide residues on food, providing essential information for human health and environmental protection. The combination of high sensitivity, low cost, and portability has made electrochemical sensors indispensable in modern analytical chemistry.

Fuel cells offer another frontier in the practical application of electrochemistry, providing a highly efficient means of converting chemical energy directly into electricity. Hydrogen-oxygen fuel cells combine hydrogen fuel at the anode with oxygen at the cathode, producing water and electrical energy in a reaction that is both highly efficient and environmentally benign. Proton exchange membrane (PEM) fuel cells, solid oxide fuel cells, and alkaline fuel cells represent different technologies optimized for specific applications, from portable electronics to power generation in vehicles and stationary plants. The continuous supply of reactants ensures sustained electricity generation, with the efficiency of fuel-to-electricity conversion often exceeding that of conventional combustion engines. Research in fuel cell technology focuses on improving catalyst performance, reducing costs, and developing durable membranes capable of withstanding prolonged operation. The integration of fuel cells with renewable energy sources, such as electrolytically generated hydrogen from water splitting, points toward a future of sustainable, low-emission power systems that can complement or replace traditional fossil fuel technologies.

In addition to practical applications, electrochemistry offers profound insight into fundamental scientific concepts, including thermodynamics, kinetics, and transport phenomena. The kinetics of electron transfer at electrodes, described by the Butler-Volmer equation, governs the rate at which redox reactions proceed, while overpotential describes the additional voltage required to overcome kinetic barriers and drive a reaction at a practical rate. Mass transport, including diffusion, migration, and convection of ions in solution, further influences the efficiency and effectiveness of electrochemical reactions. Understanding these principles allows chemists and engineers to design electrodes, electrolytes, and cell geometries that optimize reaction rates, minimize energy losses, and achieve desired chemical transformations. The integration of theoretical and experimental insights into electrochemical systems has enabled advances ranging from micro-scale sensors to large-scale energy storage solutions, illustrating the versatility and scope of the field.

The impact of electrochemistry extends into cutting-edge research areas, including nanotechnology and green energy solutions. Nanostructured electrodes, with their high surface area and tunable properties, enable enhanced electron transfer rates and improved catalytic activity, allowing for faster and more selective electrochemical reactions. Flow batteries, in which reactants are stored externally and pumped through electrochemical cells, offer scalable energy storage solutions suitable for grid-level applications, addressing the intermittent nature of renewable energy sources like wind and solar. Electrochemical water splitting, involving the generation of hydrogen and oxygen from water using electrical energy, provides a promising pathway for producing clean hydrogen fuel. These advancements underscore the ongoing relevance of electrochemistry in addressing global energy challenges, environmental sustainability, and technological innovation.

Electrochemical principles are also integral to understanding biological processes. Cellular respiration and photosynthesis, for example, involve electron transfer reactions that generate energy for living organisms. The electron transport chain in mitochondria relies on redox reactions to create a proton gradient that drives the synthesis of ATP, the universal energy currency of the cell. Similarly, photosystem complexes in chloroplasts use light energy to initiate electron transfer, ultimately producing chemical energy stored in glucose molecules. Electrochemical gradients across membranes, ion fluxes, and redox-active cofactors highlight the universality of electrochemical principles in both synthetic and natural systems. Understanding these processes has led to the development of bioelectrochemical devices, including microbial fuel cells, which harness the metabolic activity of microorganisms to generate electricity, further bridging the gap between biology and electrochemistry.

The theoretical foundation of electrochemistry, grounded in the concepts of Gibbs free energy, Faraday's laws, and the relationship between thermodynamics and kinetics, enables precise predictions of reaction feasibility, cell potential, and energy output. Faraday's first law of electrolysis states that the amount of substance altered at an electrode is proportional to the quantity of electricity passed, while the second law relates this quantity to the chemical equivalent weight of the substance. These principles allow chemists to calculate the amounts of products generated in electrolytic processes, optimize reaction conditions, and design industrial-scale operations with predictable yields. By integrating thermodynamic calculations with experimental measurements, electrochemists can develop systems that balance energy efficiency with practical constraints, achieving sustainable chemical production and energy conversion.

Electrochemistry, therefore, is not merely a subject of academic interest but a discipline with profound implications for technology, industry, medicine, and environmental stewardship. Its applications range from powering everyday devices to generating industrial chemicals, protecting materials from corrosion, monitoring environmental pollutants, and enabling advanced energy storage and conversion systems. As society faces increasing energy demands and environmental challenges, electrochemistry provides both the theoretical framework and practical tools to develop innovative, efficient, and sustainable solutions. The continuous interplay between chemical energy and electrical energy, mediated by redox reactions, electrode behavior, and ionic transport, forms the basis for a vast array of technologies that define modern life. By exploring the mechanisms, applications, and emerging trends in electrochemistry, we gain not only a deeper understanding of the natural and engineered world but also the ability to shape it for the betterment of human society and the planet.

The quantitative foundations of electrochemistry are as important as the qualitative principles, allowing chemists to predict, control, and optimize reactions. Central to this understanding is the derivation of the Nernst equation from first principles. For a general redox reaction, $aA + bB \rightleftharpoons cC + dD$, the Gibbs free energy change under nonstandard conditions is expressed as $\Delta G = \Delta G^o + RT \ln Q$, where Q is the reaction quotient, R is the gas constant, and T is the absolute temperature. Simultaneously, the relationship between Gibbs free energy and electrical work is $\Delta G = -nFE$, where n is the number of electrons transferred and F is the Faraday constant. Equating these expressions yields $-nFE = -nFE^o + RT \ln Q$, which, upon rearrangement and conversion to base-10 logarithms, gives the familiar Nernst equation: $E = E^o - (0.0591/n) \log Q$ at 25°C.

This derivation highlights the intrinsic link between thermodynamics and electrochemical potential, emphasizing that the driving force for electron transfer depends not only on the inherent tendencies of chemical species but also on their relative concentrations. In practical applications, the Nernst equation allows chemists to calculate the potential of concentration cells, evaluate shifts in cell potential during discharge of batteries, and predict equilibrium positions for redox reactions. For instance, in a copper-copper(II) sulfate concentration cell, electrons flow from the compartment with lower Cu^{2+} concentration to that with higher concentration, generating a potential difference precisely predicted by the equation.

Batteries, as engineered galvanic cells, illustrate the principles of electrochemistry in tangible, practical devices. Lead-acid batteries, first developed in the mid-19th century, remain widely used in automotive applications due to their reliability and high surge currents. During discharge, the lead anode is oxidized to lead sulfate, while the lead dioxide cathode is reduced, generating a voltage of approximately 2 volts per cell. Multiple cells connected in series provide the 12 volts required to start most vehicles. Rechargeable lithium-ion batteries, a revolutionary 20th-century development, rely on the intercalation of lithium ions between a graphite anode and a lithium metal oxide cathode. During discharge, lithium ions move from the anode to the cathode, releasing electrons into the external circuit, while during charging, the process reverses. Lithium-ion batteries offer high energy density, low self-discharge, and long cycle life, powering everything from smartphones to electric vehicles. Other battery chemistries, such as nickel-cadmium (NiCd), nickel-metal hydride (NiMH), zinc-carbon, and silver-oxide cells, provide varying trade-offs in terms of energy density, cost, environmental impact, and voltage stability. Modern research focuses on developing solid-state lithium batteries, sodium-ion alternatives, and hybrid chemistries, aiming to improve safety, capacity, and sustainability while reducing reliance on scarce or toxic elements.

Corrosion is another domain in which electrochemistry demonstrates its practical significance. Metals naturally tend toward their most stable, oxidized states, with electron transfer processes driving this transformation. Iron, for example, corrodes when exposed to water and oxygen, forming rust through a series of electrochemical reactions. Anodic regions of the metal release electrons as $Fe \rightarrow Fe^{2+} + 2e^-$, while cathodic regions reduce oxygen in aqueous solution, producing hydroxide ions. The migration of ions and electrons results in the gradual deterioration of the metal.

Protective strategies exploit electrochemical principles to mitigate this process: sacrificial anodes, composed of more reactive metals like magnesium or zinc, corrode preferentially, protecting the primary metal; passivation layers, such as the oxide on aluminum, prevent further electron transfer; and coatings, including paints or metal platings, physically isolate the metal from corrosive agents. Industrial applications of these principles are vast, ranging from protecting steel infrastructure in bridges and pipelines to maintaining naval vessels and offshore platforms. High-chloride environments, elevated temperatures, and mechanical stress accelerate corrosion, necessitating alloys engineered for specific conditions, demonstrating how electrochemistry informs materials science and engineering solutions.

Industrial electrolysis represents a cornerstone of modern chemical manufacturing, demonstrating the controlled application of electrical energy to drive nonspontaneous reactions. In the Hall-Héroult process, alumina is dissolved in molten cryolite and electrolyzed using carbon electrodes. Aluminum ions are reduced at the cathode to produce metallic aluminum, while oxygen ions react at the anode to release carbon dioxide. Precise control of temperature, voltage, and current density is essential to optimize yield and minimize energy losses. The chlor-alkali process, essential for producing chlorine and sodium hydroxide, employs electrolytic cells with membranes or diaphragms to prevent product recombination and ensure purity. In the diaphragm cell, brine passes through a porous barrier; at the anode, chloride ions are oxidized to chlorine gas, while at the cathode, water is reduced to produce hydroxide ions and hydrogen gas. Membrane cells further enhance efficiency and environmental safety by selectively transporting sodium ions, reducing energy consumption and chemical contamination. These industrial processes underscore the scalability of electrochemical principles, translating fundamental electron transfer concepts into the production of millions of tons of materials annually, from metals to alkalis and industrial chemicals.

Fuel cells extend the principles of galvanic cells to systems with continuous reactant flow, offering highly efficient energy conversion. Hydrogen-oxygen fuel cells, in particular, convert chemical energy into electricity and water with efficiencies exceeding 60% under ideal conditions. In a proton exchange membrane (PEM) fuel cell, hydrogen gas at the anode is oxidized to protons, releasing electrons that flow through an external circuit to produce current. Oxygen at the cathode combines with the protons and incoming electrons to form water. Variations in fuel cell design, including solid oxide fuel cells (SOFCs), alkaline fuel cells, and direct methanol fuel cells, are optimized for applications ranging from portable electronics to stationary power generation and automotive propulsion. Research in this field focuses on improving catalysts to reduce reliance on expensive platinum, enhancing membrane durability, and integrating fuel cells with renewable energy sources such as electrolytically generated hydrogen. These systems exemplify how electrochemistry can contribute to sustainable energy solutions, addressing global concerns about fossil fuel dependence and greenhouse gas emissions.

Beyond engineered systems, electrochemistry illuminates fundamental biological processes. Cellular respiration, photosynthesis, and nerve signal transmission all involve electron transfer and electrochemical gradients. In mitochondria, the electron transport chain mediates the transfer of electrons through a series of redox-active proteins, pumping protons across the inner mitochondrial membrane to generate a proton-motive force that drives ATP synthesis. Similarly, photosystem complexes in chloroplasts convert light energy into chemical energy via electron transfer reactions, ultimately producing glucose and oxygen. Electrochemical principles also apply to bioelectrochemical devices, such as microbial fuel cells, which exploit the metabolism of bacteria to generate electrical energy from organic matter. These applications illustrate the universality of redox chemistry, bridging the gap between chemical engineering, biology, and energy technology, and demonstrating that the principles governing batteries, corrosion, and industrial electrolysis also underpin the fundamental processes of life.

In addition to these applications, electrochemistry provides essential tools for analytical and environmental chemistry. Electrochemical sensors, potentiometric and amperometric electrodes, and voltammetric techniques enable the detection of ions, gases, and biomolecules at extraordinarily low concentrations. Potentiometric sensors, such as pH electrodes, measure the voltage difference resulting from ion activity in solution, providing immediate information about acidity or alkalinity. Voltammetric methods involve sweeping potential at an electrode while measuring current, revealing the identity and concentration of electroactive species. Amperometric sensors, which maintain a constant potential and measure current proportional to analyte concentration, allow continuous monitoring of species like oxygen, glucose, and heavy metals. These technologies have broad implications, from monitoring water quality to detecting trace contaminants in blood, food, or environmental samples. The sensitivity, speed, and versatility of electrochemical sensors make them indispensable in modern analytical chemistry, enabling high-precision measurements that guide public health, environmental protection, and industrial quality control.

The kinetics of electrochemical reactions are as critical as their thermodynamic foundations. Electron transfer rates at electrode surfaces, described by the Butler-Volmer equation, depend on both the overpotential and the intrinsic reactivity of the species involved. Overpotential represents the additional potential beyond the equilibrium value required to drive a reaction at a measurable rate, influenced by factors such as electrode surface area, material, and the presence of catalysts. Mass transport phenomena, including diffusion, migration, and convection, further affect reaction efficiency, determining how quickly reactants reach the electrode surface and products are removed. Understanding these kinetic and transport considerations allows for the design of electrodes, electrolytes, and cell geometries that optimize performance, minimize energy loss, and achieve desired chemical transformations. Industrial electrolysis cells, batteries, and fuel cells all rely on careful management of these variables to maximize efficiency and longevity.

Historically, the development of electrochemistry involved a series of pioneering experiments and discoveries that laid the foundation for modern applications. Luigi Galvani's 18th-century experiments on "animal electricity" demonstrated that electricity could induce muscle contractions, hinting at the physiological relevance of electron flow. Alessandro Volta's invention of the voltaic pile provided the first practical source of continuous electrical current, enabling systematic studies of electrochemical reactions. Michael Faraday's rigorous experiments in the 19th century established the quantitative laws of electrolysis, relating charge passed to the amount of substance transformed. John Daniell's invention of the Daniell cell offered a more stable galvanic cell, solving problems of hydrogen bubble formation in early designs. These foundational studies transformed electrochemistry from a curiosity into a precise scientific discipline, providing the theoretical and experimental basis for the batteries, fuel cells, and industrial electrolysis processes that shape modern life.

Modern research in electrochemistry continues to push the boundaries of energy storage, sustainability, and sensor technology. Nanostructured electrodes, with large surface areas and tailored catalytic properties, enhance electron transfer rates and selectivity in reactions, improving efficiency in batteries, fuel cells, and sensors. Flow batteries, in which electrolytes are stored externally and pumped through electrochemical cells, offer scalable solutions for grid-level energy storage, capable of handling the intermittent supply of renewable energy sources like wind and solar. Electrochemical water splitting, involving the conversion of water into hydrogen and oxygen, offers a pathway to green hydrogen production, providing a carbon-free fuel for transportation and industry. Advances in bioelectrochemical systems, including microbial fuel cells and enzyme-based biofuel cells, exploit biological processes for electricity generation, highlighting the interdisciplinary nature of modern electrochemistry. These developments demonstrate that the field remains vibrant, with ongoing research driving innovation across multiple domains, from energy and environment to medicine and nanotechnology.

Chapter 18: Word Problems

1. A researcher is designing a galvanic cell using the Fe^{2+}/Fe^{3+} redox couple in one compartment and the MnO_4^-/Mn^{2+} redox couple in acidic solution in the other compartment. The Fe^{2+} concentration in the anode compartment is 0.10 M, and the MnO_4^- concentration in the cathode compartment is 0.050 M. The researcher is interested not only in calculating the cell potential at 25°C using the Nernst equation but also in understanding how changes in concentration over time affect the cell performance. Determine the initial cell potential, and then predict how the potential evolves as MnO_4^- is gradually reduced during operation. Discuss whether the reaction remains spontaneous throughout discharge, and analyze how factors such as activity coefficients, solution acidity, and ionic strength might alter the observed potential in real systems. Additionally, propose how you could experimentally monitor the concentration changes in each compartment to validate your theoretical predictions.

2. Consider a zinc-hydrogen galvanic cell in which the hydrogen ion concentration at the cathode is initially 1.0 M. The cell is connected to a resistive load, and the reaction proceeds until the hydrogen ion concentration decreases to 0.10 M. Using Faraday's laws, calculate the total charge transferred during this process,

and estimate the mass of zinc oxidized at the anode. Beyond these calculations, discuss the assumptions made, such as complete dissociation of acids, negligible overpotentials, and uniform current distribution. Consider how real-world factors like electrode polarization, ohmic losses, and hydrogen evolution side reactions could alter the total charge delivered. Explore the implications for designing practical zinc-hydrogen cells for long-duration operation.

3. A lithium-ion battery uses graphite as the anode and $LiCoO_2$ as the cathode. During discharge, 2 moles of Li^+ ions intercalate into graphite. Given an average cell voltage of 3.7 V, calculate the total electrical work done by the battery. Discuss how the electrode surface area, particle morphology, and porosity influence the battery's internal resistance, rate capability, and overall energy efficiency. Analyze how changes in temperature or electrolyte composition could affect lithium-ion mobility and electrode kinetics. Finally, propose experimental approaches to measure energy losses due to side reactions and SEI (solid electrolyte interphase) formation.

4. Consider a concentration cell made of two Cu/Cu^{2+} electrodes immersed in solutions of different concentrations, $[Cu^{2+}]_1$ = 0.010 M and $[Cu^{2+}]_2$ = 1.0 M. Determine the cell potential at 25°C using the Nernst equation and describe the direction of electron flow. Explain the thermodynamic driving force that causes electrons to move from the dilute solution to the concentrated solution and relate this to entropy changes. Discuss how non-idealities such as activity coefficients, temperature variation, and electrode surface effects could modify the observed potential. Propose an experimental procedure to measure the voltage as a function of concentration over time, and interpret how this data could be used to determine diffusion coefficients in the electrolyte.

5. In a lead-acid battery, the reaction $Pb + PbO_2 + 2H_2SO_4 \rightarrow 2PbSO_4 + 2H_2O$ occurs. Calculate the Gibbs free energy change per mole of reaction using the standard cell potential of 2.05 V. Discuss how repeated charge-discharge cycles affect the morphology of the Pb and PbO_2 electrodes, including dendrite formation, surface passivation, and pore clogging, and relate these changes to battery efficiency. Explore the influence of electrolyte concentration, temperature, and sulfation on cell performance. Finally, propose experimental techniques to monitor electrode degradation and predict battery lifetime under various operating conditions.

6. An aluminum rod is immersed in seawater, which contains high concentrations of chloride ions. Predict the anodic and cathodic sites, and identify the electrochemical reactions occurring at each site. Estimate the amount of aluminum lost after one month if a current density of 0.50 mA/cm^2 flows due to corrosion. Discuss how chloride ions accelerate pitting corrosion and how natural oxide layer formation may protect the surface. Analyze how temperature, dissolved oxygen, and flow conditions influence the corrosion rate. Suggest methods to mitigate corrosion, including coatings, sacrificial anodes, and alloying strategies.

7. During the electrolysis of molten NaCl, 96,500 C of charge is passed. Calculate the mass of sodium produced at the cathode and chlorine at the anode. Discuss how overpotential at both electrodes affects the efficiency of the process and the energy required for industrial-scale production. Consider safety challenges, including handling of reactive sodium metal and chlorine gas. Explore how cell design, electrode material, and temperature control influence current efficiency, gas collection, and product purity.

8. In a proton exchange membrane (PEM) fuel cell, hydrogen is oxidized at the anode, and oxygen is reduced at the cathode to produce water. The cell operates at 80°C, delivering 2 A at 0.85 V. Calculate the power output and analyze the effect of proton conductivity, membrane thickness, and membrane hydration on overall efficiency. Discuss the impact of gas crossover, catalyst degradation, and temperature fluctuations on cell performance. Propose methods to experimentally measure internal resistance, mass transport limitations, and electrode utilization in operating fuel cells.

9. A voltammetric sensor is used to detect 10^{-6} M Pb^{2+} ions in water, with a diffusion coefficient of 1×10^{-5} cm^2/s and a microelectrode of radius 10 μm. Estimate the steady-state current using appropriate electrochemical equations. Discuss how mass transport, electrode geometry, and surface roughness influence sensor sensitivity and detection limits. Analyze potential interferences from other metal ions, pH variations, and complexing agents. Suggest ways to improve reproducibility and calibration accuracy in environmental monitoring applications.

10. A microbial fuel cell uses acetate as a substrate for electricity generation via microbial oxidation. If 0.5 moles of acetate are completely oxidized according

to $CH_3COO^- + 2H_2O \rightarrow 2CO_2 + 7H^+ + 8e^-$, estimate the total charge generated. Discuss how electrode design, biofilm thickness, and substrate concentration affect current density and coulombic efficiency. Explore the influence of pH, temperature, and oxygen diffusion on microbial metabolism. Propose strategies to optimize long-term performance and stability of the microbial fuel cell.

11. In a silver-silver chloride electrode immersed in 0.10 M KCl, calculate the electrode potential relative to the standard hydrogen electrode. Discuss how changes in chloride ion concentration influence the electrode potential and Nernstian behavior. Analyze the effect of temperature and ionic strength on potential stability. Consider applications in electrochemical measurements, including reference electrodes for potentiometric titrations and bioelectrochemical sensors.

12. A zinc-carbon cell has an internal resistance of 2 Ω. If the cell delivers 0.5 A to a 10 Ω resistive load, calculate the terminal voltage. Discuss how internal resistance evolves with cell aging, electrolyte depletion, and temperature changes. Analyze how internal resistance impacts maximum power output, energy efficiency, and the design of portable electronic devices. Propose methods to experimentally measure internal resistance under load and interpret the results.

13. A nickel-metal hydride (NiMH) battery exhibits a nominal voltage of 1.2 V. If 1 mole of electrons flows during discharge, calculate the total energy delivered. Discuss how electrode surface passivation, hydrogen absorption kinetics, and electrolyte composition affect cycle life. Analyze the mechanisms of self-discharge, capacity fading, and thermal effects. Propose strategies to extend battery life in consumer electronics or hybrid electric vehicles.

14. During the electrolysis of water using inert electrodes, a voltage of 1.80 V is applied. Calculate the theoretical amount of oxygen produced after passing 96,500 C of charge. Discuss how overpotential at the anode and cathode affects energy efficiency, and how solution pH influences decomposition kinetics. Explore experimental approaches to minimize gas bubble formation, improve electrode surface area, and enhance overall efficiency of water splitting systems.

15. A galvanic cell is constructed using Ag^+/Ag and Zn^{2+}/Zn couples, with a measured cell potential of 1.56 V at 25°C. Calculate the Gibbs free energy change per mole of electrons and determine whether the reaction remains spontaneous at elevated

temperatures. Discuss the effects of concentration changes, ionic strength, and temperature on cell potential. Consider practical applications in primary batteries and explore potential limitations in high-temperature operation.

16. In a cyclic voltammetry experiment with the Fe^{3+}/Fe^{2+} couple, the peak current is measured at 10 μA for a 1 mm^2 electrode with a sweep rate of 0.1 V/s. Calculate the diffusion coefficient and discuss how peak separation and scan rate indicate the reversibility of the redox reaction. Explore the effects of electrode surface roughness, solution viscosity, and supporting electrolyte concentration on voltammetric response. Propose how experimental parameters can be adjusted to optimize sensitivity and resolution.

17. A copper pipe is protected from corrosion using a sacrificial zinc anode. The pipe has a surface area of 1000 cm^2, and a current density of 0.2 mA/cm^2 flows through the anode. Calculate the mass of zinc consumed over six months. Discuss environmental and practical considerations of using sacrificial anodes in marine environments, including zinc availability, maintenance schedules, and potential ecological impact. Analyze alternative materials and coating strategies for corrosion prevention.

18. Consider a lithium-ion battery with a $LiFePO_4$ cathode and a graphite anode, delivering 3.3 V with a capacity of 1 Ah. Calculate the number of moles of lithium intercalated during discharge. Discuss how temperature fluctuations affect lithium mobility, SEI formation, and battery safety. Analyze the effects of charging/discharging rates, electrolyte composition, and electrode microstructure on overall performance and lifespan.

19. During the chlor-alkali process, 5 M NaCl brine is electrolyzed at 100 A in a diaphragm cell. Calculate the amounts of Cl_2, H_2, and NaOH produced per hour. Discuss the role of the diaphragm in preventing recombination, maintaining product purity, and ensuring cell efficiency. Explore energy requirements, electrode material selection, and safety considerations for large-scale industrial operation.

20. During the corrosion of iron, the anodic reaction produces Fe^{2+} and electrons. If the rate of corrosion is 0.01 mm/year, calculate the average current density. Discuss how corrosion inhibitors alter anodic and cathodic reactions to reduce corrosion rates. Analyze the impact of environmental factors such as dissolved oxygen,

pH, temperature, and salinity. Propose monitoring and mitigation strategies for infrastructure exposed to marine environments.

21. A concentration cell is constructed using Ni^{2+}/Ni electrodes, with $[Ni^{2+}]_1 = 0.001$ M and $[Ni^{2+}]_2 = 1.0$ M. Calculate the maximum work obtainable from this cell at 25°C, using the Nernst equation. Discuss how the cell potential varies with temperature and concentration. Analyze the thermodynamic driving force, including entropy contributions, and predict the direction of electron flow. Consider non-ideal effects such as activity coefficients and ionic strength. Finally, propose an experiment to measure the energy output and evaluate deviations from theoretical predictions in real systems.

22. During the Hall-Héroult process, 1 mole of alumina (Al_2O_3) is electrolyzed to produce aluminum metal. Calculate the theoretical energy required. Discuss the role of the carbon anode in the process, including its consumption and CO_2 emission. Analyze how current density, electrolyte composition, and temperature affect efficiency. Explore industrial strategies to reduce energy use, minimize anode degradation, and manage environmental impact.

23. A silver oxide battery powers a calculator and draws a current of 0.5 mA over 200 hours. Calculate the total charge consumed and the mass of silver oxidized. Discuss advantages of silver oxide batteries, including high energy density and stable voltage output. Analyze the effects of temperature, self-discharge, and electrode morphology on performance. Explore applications in precision devices such as medical instruments and timing devices, considering reliability requirements.

24. A stack consists of 10 PEM fuel cells connected in series. Each cell has an active area of 100 cm^2 and delivers 0.7 V at 2 A. Calculate the total power output of the stack. Discuss how membrane hydration, proton conductivity, gas diffusion layers, and catalyst layer structure affect overall performance. Analyze thermal management issues, pressure drop across the stack, and strategies to minimize voltage losses due to crossover or mass transport limitations.

25. In an alkaline battery, Zn is oxidized while MnO_2 is reduced. If 0.2 moles of Zn react, calculate the amount of MnO_2 consumed and the total charge transferred. Discuss how electrode porosity, particle size, and electrolyte concentration affect reaction kinetics and battery lifespan. Analyze potential failure modes such as

electrolyte depletion, dendrite formation, or passivation, and propose strategies to extend operational life under high-drain conditions.

26. In a voltammetric experiment, dopamine is detected at a carbon electrode with a peak current of 5 μA. The electrode area is 0.1 cm^2. Calculate the surface coverage and discuss how the scan rate, supporting electrolyte, and electrode surface treatment influence signal resolution and sensitivity. Consider the effects of adsorption, diffusion, and kinetic limitations. Suggest methods to calibrate and improve reproducibility in neurotransmitter sensing applications.

27. During the electrolysis of molten KCl, 0.5 moles of potassium are deposited at the cathode. Calculate the corresponding volume of Cl_2 gas produced at STP. Discuss energy requirements, electrode material selection, and safety precautions for industrial-scale operations. Analyze how current density, temperature, and cell design influence current efficiency, gas purity, and overpotential losses. Explore environmental and economic considerations for large-scale alkali metal production.

28. A microbial fuel cell oxidizes 0.1 M glucose solution over 5 hours, generating a current of 0.5 A. Calculate the total number of electrons transferred. Discuss how pH, temperature, electrode spacing, and microbial biofilm thickness affect performance. Analyze the effects of substrate concentration and mass transport limitations on coulombic efficiency. Propose strategies for scaling up microbial fuel cells for sustainable energy applications.

29. A Li-S battery exhibits a discharge voltage of 2.1 V. If 0.05 moles of S_8 react, calculate the total energy output. Discuss the role of polysulfide shuttling in capacity fade, the effects of electrode porosity and sulfur loading, and methods to mitigate performance loss. Analyze challenges related to electrolyte selection, lithium dendrite formation, and cycle stability for practical battery applications.

30. An iron bridge is exposed to a marine environment. Using electrochemical data, estimate the potential difference between anodic and cathodic sites. Discuss how alloying elements such as chromium and nickel influence passivation and corrosion resistance. Analyze the effects of temperature, oxygen concentration, and chloride ion exposure. Propose monitoring strategies and maintenance protocols to prolong the structure's lifespan.

31. During cyclic voltammetry of a ferricyanide/ferrocyanide system, the peak current ratio deviates from unity. Analyze possible causes including electron transfer kinetics, electrode surface contamination, or solution composition. Discuss how scan rate, supporting electrolyte, and temperature affect reversibility. Propose methods to quantify the standard rate constant and diffusion coefficient using experimental data.

32. A zinc electrode in a $ZnSO_4$ solution is used in a concentration cell. If the Zn^{2+} concentration varies from 0.01 M to 1.0 M, calculate the change in cell potential. Explain the direction of electron flow and the underlying thermodynamic driving force. Discuss the effects of ionic strength, electrode surface area, and non-ideal solution behavior on measured potentials. Explore practical applications of concentration cells in electrochemical sensing or battery diagnostics.

33. Electrolysis of an aqueous $CuSO_4$ solution is performed with a platinum electrode at 2 A for 1 hour. Calculate the mass of copper deposited at the cathode. Discuss competing hydrogen evolution and its effect on current efficiency. Analyze the influence of electrode surface area, agitation, and temperature on deposition rate. Propose methods to optimize plating uniformity and minimize energy consumption.

34. In a Ni-Cd battery, cadmium is oxidized during discharge. Calculate the Gibbs free energy change for one mole of electrons transferred, given a standard cell potential of 1.25 V. Discuss implications for battery storage, cycling, and memory effect. Explore degradation mechanisms including electrolyte decomposition, dendrite formation, and electrode passivation. Propose strategies to maximize cycle life in practical applications.

35. A lithium-air battery reacts 1 mole of O_2 with Li to form Li_2O_2. Calculate the theoretical energy output at 2.96 V. Discuss challenges in controlling oxygen solubility, reaction kinetics, and electrode degradation. Analyze the effects of electrolyte choice, electrode porosity, and oxygen diffusion on overall efficiency. Propose experimental approaches to improve cycle stability and minimize side reactions.

36. During corrosion monitoring, the potential of iron in seawater is −0.45 V vs SHE. Determine whether the corrosion reaction is spontaneous and estimate the electron flow rate based on current density data. Discuss environmental factors,

such as chloride concentration, dissolved oxygen, and temperature, that influence corrosion. Analyze the effects of inhibitors, coatings, and cathodic protection on corrosion mitigation strategies.

37. A voltammetric sensor detects Pb^{2+} in soil at a concentration of 1 μM using a 0.01 cm^2 electrode. Calculate the expected current response using the Cottrell equation. Discuss factors affecting detection limits, including electrode surface roughness, diffusion layer thickness, and supporting electrolyte. Explore reproducibility issues, interference from other ions, and strategies to improve signal-to-noise ratio in environmental sensing.

38. During electrolysis of brine in a membrane cell, NaOH, Cl_2, and H_2 are produced. If 90% of chloride ions produce Cl_2 and 10% undergo side reactions, calculate the current efficiency. Discuss methods to improve efficiency, including electrode material selection, membrane design, and operational parameters. Analyze energy consumption, safety, and environmental considerations in industrial electrolysis.

39. A vanadium redox flow battery passes 1.5 moles of electrons during discharge with a cell voltage of 1.26 V. Calculate the total energy produced. Discuss how electrolyte flow rate, electrode porosity, and electrode surface area affect power density. Analyze limitations due to ion transport, viscosity, and membrane selectivity. Propose methods to optimize flow cell performance for grid-scale energy storage.

40. A lithium-ion capacitor undergoes 10^4 cycles, each delivering 50 Wh. Estimate the cumulative energy throughput. Discuss degradation mechanisms, including SEI formation, electrode structural changes, and electrolyte decomposition. Analyze how cycling rate, temperature, and voltage window affect long-term performance. Propose strategies to extend cycle life and maintain high energy efficiency.

41. During microbial fuel cell operation, acetate concentration decreases from 0.1 M to 0.01 M. Estimate the total electrons transferred and predict changes in current over time. Discuss how biofilm morphology, electrode spacing, and substrate diffusion influence performance. Analyze effects of pH, temperature, and competing microbial species. Propose strategies to improve long-term stability and coulombic efficiency.

42. A galvanic cell uses Fe^{2+}/Fe^{3+} and Cr^{3+}/Cr^{2+} redox couples. Using standard potentials, calculate the equilibrium constant for the reaction and determine

whether electron flow is spontaneous under standard conditions. Discuss how concentrations, temperature, and ionic strength influence spontaneity. Explore experimental methods to measure cell potential and verify thermodynamic predictions.

43. Nickel is electroplated onto a steel object using a current of 0.5 A for 2 hours. Calculate the thickness of nickel deposited, assuming 100% current efficiency. Discuss how solution pH, temperature, agitation, and additives affect plating quality. Analyze factors causing roughness, porosity, or uneven deposition. Propose strategies to optimize plating uniformity in industrial applications.

44. Alithium-ion battery experiences a 0.1 V voltage drop at 5 A due to internal resistance. Calculate the resistive power loss. Discuss design strategies to minimize internal resistance, including electrode formulation, electrolyte conductivity, and separator properties. Analyze how aging, temperature, and cycling history influence internal resistance and overall energy efficiency.

45. During water electrolysis at pH 1, calculate the theoretical potential required to evolve oxygen and hydrogen at 25°C. Discuss how overpotential, electrode material, and surface area affect decomposition efficiency. Analyze the effects of ionic strength and pH on energy requirements and practical considerations for acid electrolysis.

46. A copper electrode is immersed in a solution containing 0.01 M Cu^{2+} and 0.001 M Ag^{+}. Determine which metal deposits preferentially and explain the role of standard reduction potentials. Discuss factors influencing deposition order, including concentration, overpotential, complex formation, and electrode surface properties. Explore practical applications in electroplating and metal recovery.

47. During PEM fuel cell operation, 5% of hydrogen gas bypasses the membrane. Estimate the associated energy loss. Discuss engineering approaches to minimize gas crossover, including membrane design, operating pressure, and humidity control. Analyze the impact of crossover on fuel efficiency, cell lifetime, and safety.

48. A zinc-air battery is discharged at 50 mA/cm². Estimate the total zinc consumed per hour per cm². Discuss challenges associated with oxygen reduction kinetics, electrolyte management, and cathode design. Analyze strategies to improve energy density, power output, and cycle stability in practical applications.

49. Electrochemical water splitting using a nanostructured electrode shows enhanced current density. Quantitatively explain how surface area, catalyst activity, and electrode morphology influence reaction rate and energy efficiency. Discuss the role of mass transport, bubble detachment, and overpotential reduction in performance improvement. Propose experimental methods to characterize nanostructured electrodes.

50. In a lithium-sulfur battery, polysulfide formation leads to capacity fade. Using Nernstian considerations, calculate how the potential of the sulfur cathode changes as polysulfide concentration increases. Discuss methods to stabilize electrode performance, minimize the shuttle effect, and improve cycle life. Explore electrolyte design, separator modification, and cathode architecture as strategies for mitigation.

Chapter 19

Architecture of Complex Coordination

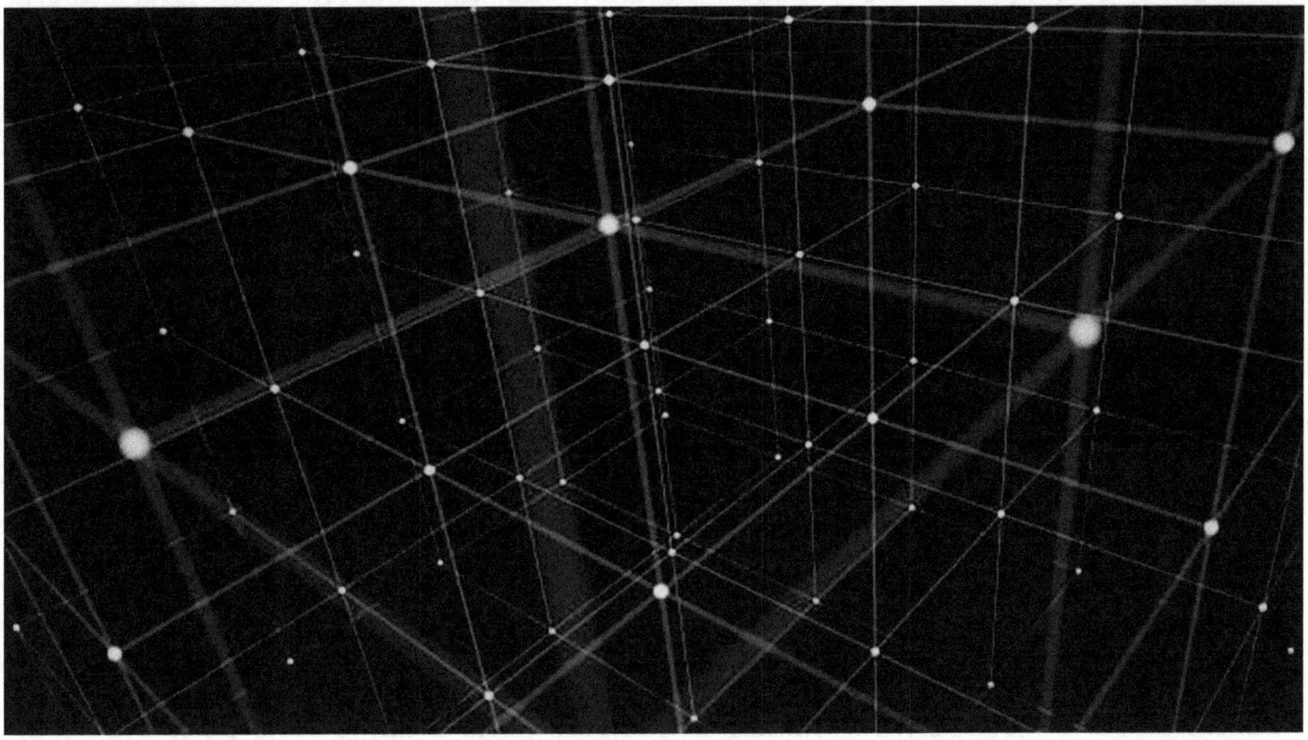

Transition metals, by virtue of their intricate electron configurations, display a chemical richness unparalleled by most other elements. Their partially filled d orbitals not only enable multiple oxidation states but also provide a fertile ground for subtle electronic interactions that manifest in magnetic, optical, and catalytic phenomena. For example, the magnetism of iron, cobalt, and nickel is directly tied to the presence of unpaired d electrons. These metals exhibit ferromagnetism at room temperature, a property arising from the alignment of electron spins within domains, which can be influenced by temperature, pressure, and alloying. Other transition metals, such as manganese or chromium, may display antiferromagnetic or ferrimagnetic ordering due to the opposing arrangement of spins in their crystal lattices. This interplay between electronic structure and magnetism is central not only to physical chemistry but also to materials science, forming the basis for data storage, permanent magnets, and spintronic devices.

The role of transition metals in catalysis is equally profound. Catalysts function by providing an alternative reaction pathway with lower activation energy, often via transient metal-ligand intermediates. In homogeneous catalysis, transition metal complexes dissolved in the reaction medium facilitate selective transformations. Wilkinson's catalyst, a rhodium complex, efficiently hydrogenates alkenes under mild conditions, while palladium complexes are central to carbon-carbon bond-forming reactions such as the Suzuki, Heck, and Stille couplings. Heterogeneous catalysis relies on metals dispersed on solid supports, where surface atoms provide sites for adsorption, bond activation, and product formation. Platinum and palladium catalysts, supported on alumina or carbon, are indispensable in automobile catalytic converters, reducing harmful nitrogen oxides, carbon monoxide, and hydrocarbons into benign gases.

The interplay between electronic structure and chemical reactivity is exemplified in the varied oxidation states of manganese. From Mn^{2+} to Mn^{7+}, manganese participates in redox chemistry with remarkable versatility. Permanganate ions (MnO_4^-) are powerful oxidizing agents, capable of accepting electrons in aqueous media to produce Mn^{2+} or MnO_2, depending on conditions. Such reactions are widely employed in analytical chemistry, environmental remediation, and organic synthesis. Chromium similarly demonstrates a duality: Cr^{3+} is relatively stable and inert, forming octahedral complexes, whereas Cr^{6+}, as in chromate or dichromate ions, is a potent oxidant with significant industrial applications but also notable toxicity. Understanding the delicate balance between stability and reactivity in transition metal ions is critical for designing safe and efficient chemical processes.

The coloration of transition metal compounds is not merely aesthetic but a direct manifestation of electronic transitions. When light interacts with a metal complex, electrons can be excited from lower-energy d orbitals to higher-energy orbitals split by the ligand field. The resulting absorption of specific wavelengths produces the vivid colors observed. Copper sulfate, with its characteristic blue hue, owes its color to transitions between d orbitals in the presence of water ligands. Similarly, the deep purple of potassium permanganate reflects electronic transitions in the MnO_4^- ion. This phenomenon is intricately linked to ligand identity, coordination geometry, and oxidation state. Spectroscopic methods, such as UV-Vis and infrared spectroscopy, exploit these properties to probe the electronic and structural features of transition metal complexes with high precision.

Beyond simple compounds, the chemistry of organometallic transition metal complexes further demonstrates their versatility. Organometallics, characterized by metal-carbon bonds, bridge the realms of inorganic and organic chemistry. Grubbs and Schrock catalysts, which contain ruthenium and molybdenum centers respectively, revolutionized olefin metathesis—a reaction fundamental to polymer synthesis, pharmaceuticals, and petrochemical transformations. Ferrocene, the archetypal metallocene, exhibits remarkable stability due to the delocalized π-electron interactions between the iron center and cyclopentadienyl ligands, exemplifying the synergy between electronic structure and bonding in transition metals. Such compounds have inspired new classes of catalysts, sensors, and materials that harness the unique properties of d-block elements.

Transition metals also manifest fascinating structural chemistry in clusters and extended solids. Metal clusters, small aggregates of metal atoms, often exhibit bonding reminiscent of both molecules and metallic lattices. These clusters can display unusual electronic, magnetic, and catalytic behavior, with potential applications in nanotechnology, photovoltaics, and catalysis. Extended solids, such as the transition metal oxides, demonstrate a rich interplay between structure and properties. Vanadium pentoxide (V_2O_5) is used in catalysis and batteries due to its layered structure and ability to accommodate reversible oxidation states. Titanium dioxide (TiO_2) serves as a photocatalyst, pigment, and semiconductor, where subtle changes in crystal phase and doping significantly affect electronic and optical behavior.

The environmental and sustainable aspects of transition metals are increasingly relevant in modern chemistry. Metals such as copper, nickel, and iron are integral to renewable energy technologies. Nickel and cobalt are essential in lithium-ion battery electrodes, where redox cycling allows energy storage and release. Platinum-group metals facilitate hydrogen evolution and oxygen reduction reactions in fuel cells, enabling clean energy conversion. Recycling and recovery of these metals from electronic waste, catalysts, and industrial residues are critical to sustainable technology development, underscoring the need for coordination chemistry, selective extraction techniques, and metallurgical innovations.

Biological systems exploit the chemical versatility of transition metals in intricate ways. Metalloenzymes employ metal centers to catalyze reactions with remarkable selectivity and efficiency. Zinc fingers, structural motifs in proteins, stabilize folds through Zn^{2+} coordination, regulating gene expression and DNA binding. Copper-containing enzymes, such as cytochrome c oxidase, mediate electron transfer in cellular respiration. Iron-sulfur clusters, ubiquitous in redox-active proteins, serve as electron shuttles in metabolic pathways. The interplay of geometry, ligand identity, and redox potential in these biological systems mirrors principles observed in inorganic chemistry, highlighting the universal relevance of transition metal chemistry.

Transition metals are also central to medical and diagnostic applications. Cisplatin, a platinum-based chemotherapeutic, binds selectively to DNA, inducing apoptosis in cancer cells. Gadolinium complexes are employed in magnetic resonance imaging (MRI) as contrast agents, exploiting paramagnetic properties of unpaired electrons to enhance imaging resolution. Radiometals, including technetium and rhenium isotopes, are used in nuclear medicine for both diagnostic and therapeutic purposes. The design of metal-based drugs relies on a nuanced understanding of ligand field effects, coordination stability, and biological interactions, demonstrating the interface of chemistry, biology, and medicine.

From a theoretical perspective, the study of transition metals informs fundamental concepts of chemical bonding, electronic structure, and reactivity. Molecular orbital theory, ligand field theory, and density functional theory are essential tools for predicting properties and guiding the synthesis of new compounds. For instance, computational studies can predict the preferred spin state of a metal complex, the relative stability of isomers, or the energy barriers for catalytic transformations. The integration of theory and experiment accelerates innovation, enabling chemists to design tailored materials, efficient catalysts, and selective therapeutics.

The chemistry of alloys further illustrates the profound impact of transition metals. Steel, an alloy of iron with carbon and other metals, displays properties tunable through composition and heat treatment. Stainless steel, incorporating chromium and nickel, resists corrosion due to the formation of passive oxide layers. Superalloys, containing nickel, cobalt, and refractory metals, withstand extreme temperatures in turbines and aerospace engines. Shape-memory alloys, such as nickel-titanium (Nitinol), exploit reversible phase transitions to recover their original shape, finding applications in medical devices, actuators, and robotics. The principles governing alloy formation—atomic size, electron density, and lattice energy—are grounded in the physics and chemistry of transition metals.

Electrochemical behavior of transition metals underscores their versatility in energy conversion and storage. The ability of a metal to donate or accept electrons underlies both their redox chemistry and practical applications in batteries, fuel cells, and electroplating. For instance, the Fe^{2+}/Fe^{3+} redox couple in aqueous solution serves as a fundamental model for electron transfer reactions, while in nonaqueous media, transition metal complexes provide controlled, multi-electron redox processes vital for synthetic applications. Electroplating, which deposits a thin metal layer onto a substrate, exploits the controlled reduction of transition metal ions from solution. Copper, nickel, and chromium are commonly used, improving corrosion resistance, electrical conductivity, and aesthetic appeal. Transition metals' varied oxidation states allow precise control over the electrochemical potential, enabling tailored deposition processes essential in electronics, aerospace, and decorative industries.

Photochemistry is another dimension where transition metals shine. Their d orbitals facilitate absorption of visible and ultraviolet light, inducing electronic transitions that can drive chemical reactions. Ruthenium and iridium complexes, for example, are central to photoredox catalysis, where light energy promotes electron transfer events, enabling transformations that are otherwise difficult under thermal conditions. This approach has revolutionized organic synthesis, allowing reactions such as C–H activation, cross-coupling, and radical cyclizations to occur under mild, energy-efficient conditions. Additionally, transition metal complexes in dye-sensitized solar cells (DSSCs) harness light absorption to generate electricity. The archetypal ruthenium bipyridyl complexes efficiently inject electrons into semiconductor materials, converting solar energy into electrical energy with impressive efficiency, illustrating the intersection of coordination chemistry, photophysics, and renewable energy technology.

Advanced catalysis often relies on transition metals' capacity to stabilize high-energy intermediates and facilitate multi-step transformations.

In homogeneous catalysis, metals provide defined coordination environments that enhance selectivity and reaction rates. The hydroformylation of alkenes, catalyzed by cobalt or rhodium complexes, adds formyl groups to produce aldehydes, a key step in producing fine chemicals and fragrances. In heterogeneous catalysis, transition metals dispersed on supports such as silica, alumina, or zeolites provide active surfaces for industrial reactions. Nickel catalyzes hydrogenation of unsaturated hydrocarbons in fuel processing, while palladium on carbon enables selective hydrogenation in fine chemical synthesis. The interplay of electronic structure, surface geometry, and adsorption energies governs catalytic efficiency, making transition metals indispensable for chemical industry sustainability.

Solid-state chemistry of transition metals opens a window into materials with extraordinary electronic, magnetic, and optical properties. Transition metal oxides, sulfides, and halides form lattices where d-electron interactions lead to phenomena such as superconductivity, metal-insulator transitions, and colossal magnetoresistance. Copper oxides, for instance, exhibit high-temperature superconductivity, a property arising from the cooperative behavior of electrons in layered lattice structures. Vanadium oxides display reversible changes in electrical conductivity upon temperature variation, forming the basis for smart windows and switching devices. Spinel and perovskite structures, incorporating transition metals, exhibit unique magnetic and electronic behavior, finding applications in batteries, sensors, and catalysis. Understanding these materials requires integrating crystallography, electronic structure theory, and solid-state thermodynamics, reflecting the sophistication of transition metal chemistry.

Supramolecular assemblies leverage the directional bonding tendencies of transition metals to construct well-defined, nanoscale architectures. Metal-organic frameworks (MOFs), coordination polymers, and polyhedral clusters exemplify the ability of transition metals to organize ligands into highly ordered, porous structures. MOFs, composed of metal nodes linked by organic ligands, exhibit tunable pore sizes, exceptional surface areas, and chemical modularity. They are used for gas storage, separation, catalysis, and sensing. For example, copper- and zirconium-based MOFs efficiently adsorb carbon dioxide, providing strategies for carbon capture and environmental mitigation. The precise control of geometry and connectivity in these systems illustrates how fundamental principles of coordination chemistry translate into functional, macroscopic materials with societal relevance.

Nanotechnology further extends the impact of transition metals. Metal nanoparticles, typically in the 1–100 nm range, display size-dependent electronic, optical, and catalytic properties. Gold and silver nanoparticles exhibit localized surface plasmon resonance, producing vivid colors and enhancing spectroscopic techniques like surface-enhanced Raman scattering (SERS). Platinum, palladium, and ruthenium nanoparticles catalyze reactions more efficiently than bulk metals due to increased surface area and under-coordinated atoms at edges and corners. Bimetallic nanoparticles combine properties of two metals, tuning electronic structure and improving catalytic selectivity. Quantum confinement in transition metal nanoclusters modifies energy levels, influencing photophysical behavior and opening avenues for next-generation electronics, photonics, and medical diagnostics. The ability to manipulate metals at the nanoscale illustrates a confluence of coordination chemistry, solid-state physics, and materials science.

Transition metals are central to environmental chemistry, providing solutions to pollution, energy, and sustainability challenges. Iron and manganese oxides serve as natural oxidants for water treatment, degrading organic pollutants and immobilizing toxic metals. Transition metal complexes, designed with specific ligands, catalyze the breakdown of persistent organic compounds in wastewater or contaminated soil. Photocatalytic titanium dioxide nanoparticles harness sunlight to oxidize pollutants, while copper- and iron-based catalysts promote sustainable chemical syntheses using oxygen as a green oxidant. These applications demonstrate the integration of electronic structure, redox chemistry, and environmental engineering, highlighting the practical importance of transition metal science.

Electrochemical energy storage relies heavily on transition metals. Lithium-ion batteries, ubiquitous in portable electronics and electric vehicles, utilize transition metal oxides such as $LiCoO_2$, $LiNiMnCoO_2$ (NMC), and $LiFePO_4$ as cathode materials. These metals undergo reversible redox cycling, allowing lithium ions to shuttle between electrodes during charging and discharging. The stability, voltage, and capacity of these materials are intimately connected to the oxidation states, electronic configuration, and crystal lattice structure of the transition metals involved. Emerging battery technologies, including sodium-ion, magnesium-ion, and solid-state systems, continue to explore novel transition metal compounds for enhanced performance, safety, and sustainability.

Transition metals also enable precision in chemical sensing and imaging. Metal complexes can respond to specific analytes with changes in color, fluorescence, or electrochemical signal. For instance, ruthenium and iridium complexes serve as oxygen sensors due to their phosphorescence quenching in the presence of O_2. Iron and copper complexes can selectively bind biomolecules, providing tools for monitoring biological processes in real time. Magnetic nanoparticles containing iron or cobalt enhance contrast in magnetic resonance imaging, enabling non-invasive diagnostics. These examples illustrate how the fundamental chemical properties of transition metals—coordination geometry, electronic transitions, redox activity—translate into practical technologies for health, safety, and industry.

The field of organometallic polymer chemistry exemplifies the creative use of transition metals in functional materials. By incorporating metal centers into polymer backbones or side chains, chemists create materials with tunable conductivity, magnetism, or photoreactivity. Polymers containing ferrocene units can act as redox-active materials for batteries and sensors. Palladium- or nickel-mediated polymerizations produce conductive polymers with applications in flexible electronics and photovoltaics. These systems highlight the fusion of synthetic methodology, coordination chemistry, and materials science to generate advanced, multifunctional materials.

Transition metals remain at the forefront of chemical innovation. Their ability to mediate multi-electron reactions, stabilize unusual oxidation states, and form intricate coordination architectures ensures their relevance in energy, medicine, environment, and materials. New frontiers, such as single-atom catalysis, where isolated transition metal atoms on supports provide unprecedented selectivity, and bioinspired catalysis, which mimics enzymatic metal centers, push the boundaries of efficiency and sustainability. Additionally, computational design and machine learning are accelerating the discovery of novel transition metal compounds with tailored properties, bridging the gap between theory and practical application.

In conclusion, transition metals represent a unique nexus of chemical versatility, physical robustness, and functional diversity. Their partially filled d orbitals endow them with rich electronic landscapes, enabling magnetism, catalysis, color, and multiple oxidation states. Across industrial, technological, environmental, and biological domains, they are indispensable, shaping modern civilization and enabling sustainable innovation. From ancient metallurgy to cutting-edge nanotechnology, the story of transition metals is one of continual discovery, illustrating how fundamental chemistry underpins the complex tapestry of human progress and the natural world. The depth, adaptability, and enduring relevance of transition metal chemistry ensure that it will remain a central pillar of scientific inquiry and technological advancement for generations to come.

The industrial significance of transition metals is illustrated by their central role in catalysis, both homogeneous and heterogeneous. In the mid-20th century, the development of Ziegler-Natta catalysts revolutionized polymer chemistry. Titanium tetrachloride, combined with organoaluminum compounds, enabled the stereospecific polymerization of ethylene and propylene into high-density polyethylene and isotactic polypropylene, respectively. These materials underpin modern plastics, fibers, and packaging industries. The precise control of stereochemistry in polymer chains exemplifies how subtle differences in metal-ligand interactions govern macroscopic material properties. Similarly, the Monsanto acetic acid process, relying on a rhodium-carbonyl catalyst, converts methanol and carbon monoxide into acetic acid with remarkable efficiency, illustrating the power of transition metals in carbonylation reactions and industrial synthesis.

Comparisons between transition metals and lanthanides illuminate the unique features of d-block chemistry. While both families exhibit multiple oxidation states and coordination tendencies, lanthanides primarily use their 4f electrons, which are largely shielded from the chemical environment, resulting in predominantly ionic interactions and weak ligand field effects. Transition metals, in contrast, employ the more exposed d orbitals, leading to covalent character in bonds, pronounced ligand field splitting, and versatile redox chemistry. This distinction explains the vivid colors, rich magnetic phenomena, and catalytic versatility unique to transition metals, as opposed to the more uniform chemistry of most lanthanides. The lanthanide contraction, however, causes certain 4d and 5d metals to closely mimic the size and chemistry of their 3d counterparts, enabling predictable periodic trends and the design of analogous compounds for advanced materials and catalysts.

Magnetic phenomena in transition metal solids reveal the interplay between electronic configuration and cooperative effects. The ferromagnetism of iron, cobalt, and nickel, as mentioned, arises from exchange interactions aligning spins across domains. Antiferromagnetic ordering occurs in materials like MnO, where neighboring spins orient oppositely, canceling net magnetization. Ferrimagnetism, observed in magnetite (Fe_3O_4), involves unequal antiparallel spins, producing partial magnetization. Spin-crossover compounds, often iron(II) complexes with octahedral geometry, can reversibly switch between high-spin and low-spin states in response to temperature, pressure, or light. These materials have potential applications in molecular switches, memory devices, and sensors. Understanding such phenomena requires integrating crystal field theory, exchange interactions, and lattice effects, highlighting the sophistication of transition metal solid-state chemistry.

Photoredox catalysis, enabled by transition metal complexes, has transformed synthetic methodology. Ruthenium tris-bipyridyl complexes, when irradiated with visible light, undergo metal-to-ligand charge transfer, generating excited states capable of single-electron transfers. This activation allows previously inert bonds to participate in reactions under mild conditions. Iridium complexes, with strong spin-orbit coupling, facilitate efficient intersystem crossing, enabling radical-mediated processes. Organic chemists now exploit these systems for C–H functionalization, cross-coupling, and cycloaddition reactions, showcasing the interface between coordination chemistry, photophysics, and synthetic design. The ability to harness light as a reagent exemplifies the elegance and utility of transition metal chemistry beyond thermal activation.

In medicine, transition metal complexes provide diverse therapeutic strategies. Beyond cisplatin, carboplatin and oxaliplatin, structurally modified platinum drugs, offer improved solubility and reduced toxicity while maintaining anticancer activity. Ruthenium complexes, designed to exploit selective cellular uptake and redox activity, are under investigation as next-generation chemotherapeutics. Radiometal isotopes, including technetium-99m, rhenium-186, and copper-64, enable diagnostic imaging or targeted radiotherapy, demonstrating how the chemical versatility of transition metals translates into precision medicine. Furthermore, transition metal-based enzyme mimics, or metalloenzyme models, provide insight into natural catalysis and offer potential in therapeutic and industrial applications, bridging inorganic chemistry and biomedical science.

Environmental applications of transition metals continue to expand. The use of iron-based Fenton chemistry, where Fe^{2+} catalyzes hydrogen peroxide decomposition to generate hydroxyl radicals, enables degradation of recalcitrant organic pollutants in water treatment. Manganese oxides, due to their layered structures and redox activity, participate in the removal of heavy metals and organics. Copper- and titanium-based photocatalysts harness sunlight for pollutant degradation, while transition metal-supported membranes selectively remove toxins from air and water. Such strategies exemplify sustainable chemistry, integrating fundamental transition metal behavior with environmental stewardship.

Organometallic catalysis remains a frontier for both industrial and academic research. Palladium-catalyzed cross-coupling reactions, including Suzuki-Miyaura, Heck, and Negishi reactions, form the backbone of pharmaceutical and agrochemical synthesis. These reactions exploit the oxidative addition and reductive elimination capabilities of palladium, stabilized by phosphine or N-heterocyclic carbene ligands. Nickel catalysts, often more abundant and cost-effective, are emerging as alternatives in similar transformations. Transition metal-catalyzed C–H activation, wherein inert bonds are directly functionalized, reflects the pinnacle of chemical control, minimizing waste and simplifying synthetic pathways. These developments illustrate the deep connection between electronic structure, ligand design, and practical chemical innovation.

The design of metal-organic frameworks (MOFs) and supramolecular assemblies exemplifies the extension of transition metal chemistry into complex architectures. Copper paddlewheel units and zirconium oxide clusters create robust, crystalline frameworks with tunable pore sizes and chemical functionality. MOFs serve in gas separation, catalysis, drug delivery, and chemical sensing. Beyond MOFs, polyoxometalates—clusters of transition metals linked by oxygen—demonstrate unique redox and catalytic properties, with potential in energy storage, water splitting, and oxidation reactions. The control of geometry, connectivity, and electronic environment in these assemblies embodies the essence of transition metal chemistry: the ability to orchestrate atomic interactions into functional, macroscopic systems.

Nanotechnology provides a further frontier where transition metals enable unprecedented phenomena. Gold nanoparticles, whose surface plasmon resonance produces intense colors and enhanced light-matter interactions, are used in photothermal therapy, biosensing, and imaging. Platinum, palladium, and rhodium nanoparticles catalyze hydrogenation, oxidation, and carbon-carbon bond formation with superior efficiency compared to bulk metals. Core-shell and alloy nanoparticles allow fine-tuning of electronic properties, surface reactivity, and stability. Quantum confinement effects in sub-10-nanometer clusters modify electronic energy levels, enabling applications in photocatalysis, optoelectronics, and energy conversion. These advances illustrate how fundamental understanding of d-orbital chemistry and coordination principles drives transformative technologies.

Transition metals also underpin energy technologies critical for a sustainable future. Lithium-ion batteries, sodium-ion alternatives, and solid-state systems rely on transition metal redox chemistry to reversibly store energy. Cobalt, nickel, manganese, and iron oxides serve as active cathode materials, where electronic configuration and crystal structure govern voltage, capacity, and cycling stability. In fuel cells, platinum and palladium catalyze the hydrogen oxidation and oxygen reduction reactions, while transition metal oxides support water splitting to generate hydrogen fuel. These applications exemplify the interface between chemistry, materials science, and energy engineering, highlighting the societal importance of transition metals in addressing climate change and resource challenges.

Finally, ongoing research explores single-atom catalysis, where isolated transition metal atoms anchored on supports act as highly efficient, selective catalytic centers. These systems maximize atomic utilization, reduce cost, and allow precise control over reaction pathways. Bioinspired catalysts, mimicking enzymatic metal centers such as the iron-molybdenum cofactor of nitrogenase, enable sustainable transformations under mild conditions. Computational chemistry and machine learning are increasingly integrated into transition metal research, predicting reactivity, optimizing ligand environments, and designing materials with unprecedented properties. This convergence of theory, synthesis, and application ensures that transition metals remain at the forefront of chemical science and technological innovation.

In summary, transition metals exemplify the union of fundamental chemistry and applied science. Their partially filled d orbitals impart extraordinary versatility, enabling multiple oxidation states, magnetic phenomena, vivid coloration, and catalytic prowess. Across history, from the Bronze and Iron Ages to modern electronics, energy storage, medicine, and nanotechnology, transition metals have shaped human civilization and our understanding of the natural world. Through coordination chemistry, organometallic synthesis, solid-state materials, supramolecular design, and nanotechnology, these elements continue to inspire innovation. Their enduring relevance lies not only in their intrinsic chemical properties but also in the ways chemists harness these properties to solve complex scientific, industrial, and societal challenges, ensuring that transition metals will remain central to chemistry for generations to come.

Chapter 19: Word Problems

1. A chemist is investigating the magnetic properties of a series of octahedral iron(II) complexes with ligands of varying field strength. They observe that some complexes undergo a high-spin to low-spin transition depending on temperature and ligand environment. Predict how the number of unpaired electrons, molar magnetic susceptibility, and effective magnetic moment change between the two spin states. Consider the influence of crystal field splitting energy (Δ_{oct}) relative to the pairing energy and how thermal fluctuations can induce spin-crossover behavior. Design an experiment using SQUID magnetometry, Mössbauer spectroscopy, or magnetic circular dichroism to quantitatively measure the spin transition and determine hysteresis effects under applied pressure.
2. Consider a manganese-based catalyst used in aerobic oxidation of primary alcohols. The catalyst cycles through Mn^{2+}, Mn^{3+}, and Mn^{4+} states in an octahedral ligand field. Propose a stepwise catalytic cycle, discussing how geometric flexibility, ligand electronics, and redox potentials stabilize higher oxidation states. Predict which intermediates might be rate-limiting. How could in situ spectroscopic techniques such as X-ray absorption spectroscopy (XAS), electron paramagnetic resonance (EPR), or UV–Vis stopped-flow analysis detect transient species, and what kinetic parameters could be extracted to optimize catalytic efficiency?
3. A palladium complex mediates Suzuki–Miyaura coupling between aryl bromides and phenylboronic acid. Analyze how electronic and steric properties of phosphine or NHC ligands affect oxidative addition, transmetallation, and reductive elimination steps. Predict how electron-withdrawing versus electron-donating ligands modify reaction rates, selectivity, and byproduct formation. Discuss how computational descriptors such as HOMO–LUMO gap, metal d-orbital occupancy, and natural bond orbital (NBO) charges could rationalize observed trends. Suggest experimental methods like kinetic isotope effect studies or in situ NMR to probe the mechanism quantitatively.

4. Iron-sulfur cubane clusters ([Fe_4S_4]) facilitate electron transfer in proteins. If one Fe^{2+} ion is replaced by Co^{2+}, predict changes in redox potential, electron delocalization, and cluster stability. How would this substitution impact electron transfer rates in enzymatic processes? Consider spin state, covalency, and Jahn–Teller distortions. Suggest spectroscopic approaches (EPR, Mössbauer, UV–Vis) to experimentally verify the effects on electronic structure and correlate them with biochemical efficiency.

5. A ruthenium-based photoredox catalyst promotes radical cyclizations via MLCT excitation. Analyze how ligand modification, solvent polarity, and coordination geometry influence the energy of the MLCT transition, the lifetime of the excited state, and quantum efficiency. Predict how intersystem crossing, back electron transfer, and solvent reorganization energies might affect catalysis. Suggest experimental methods (time-resolved photoluminescence, transient absorption spectroscopy) to measure excited-state lifetimes and radical generation efficiency under varying conditions.

6. A chemist designs a metal-organic framework (MOF) with copper paddlewheel nodes and bipyridyl linkers for selective CO_2 capture. Discuss how pore size, linker flexibility, and metal–ligand bond strength influence adsorption capacity and selectivity. Predict how introducing Lewis-basic sites, functional groups, or post-synthetic modifications could alter CO_2 binding, competitive adsorption of N_2 or CH_4, and thermal stability. Design adsorption isotherm experiments and in situ IR spectroscopy to quantify CO_2–MOF interactions.

7. High-T_c cuprate superconductors contain CuO_2 planes where Cu^{2+} d-orbital occupancy dictates electronic band structure and pairing mechanisms. Explain how electron correlation, hole doping, and antiferromagnetic interactions contribute to superconductivity. Predict how partial substitution with Zn^{2+} (d^{10}) disrupts Cu–O planes and reduces T_c. Suggest experimental approaches (ARPES, NMR, resistivity measurements) to quantify the effects of doping on carrier concentration, coherence length, and gap symmetry.

8. Titanium dioxide photocatalysts doped with vanadium absorb visible light due to mid-gap states. Using band theory, predict how dopant energy levels modify electron excitation, charge separation, and recombination rates. Analyze the effect

on photocatalytic degradation of organics under sunlight, considering trap states, oxygen reduction, and hydroxyl radical formation. Suggest transient photocurrent and photoluminescence experiments to quantify recombination pathways and quantum yield.

9. Chromium(VI) chromate ions oxidize primary alcohols in aqueous solution. Discuss how pH, ligand coordination, and redox potential influence the selective oxidation of primary versus secondary alcohols. Predict how complexation with ligands such as acetate or phosphate could modulate kinetics and selectivity. Suggest cyclic voltammetry and spectrophotometric monitoring to quantify reaction rates and mechanistic pathways.

10. A cobalt hydroformylation catalyst converts propene to butanal. Propose a detailed catalytic cycle, highlighting oxidative addition of CO and H_2, migratory insertion of alkene, and reductive elimination. Discuss how phosphine ligand electronics and sterics control regioselectivity (linear vs branched) and turnover frequency. Predict how varying temperature, pressure, or solvent polarity affects rate-limiting steps. Suggest kinetic modeling and isotopic labeling experiments to validate the proposed mechanism.

11. Nickel nanoparticles catalyze hydrogenation more efficiently than bulk nickel. Using surface science, explain how under-coordinated atoms at edges and corners enhance adsorption and activation of H_2. Predict the effect of particle size, shape, and support interactions on active site density and selectivity. Design experiments using TEM, XPS, and chemisorption measurements to correlate morphology with catalytic performance.

12. Ferrocene is stabilized by delocalized π-electrons. Predict how substituents on cyclopentadienyl rings influence redox potential, ligand field effects, and π-backbonding. Discuss electrochemical methods (cyclic voltammetry, square-wave voltammetry) to quantify these effects and rationalize trends in chemical stability and reactivity.

13. Iron(III) hexacyanoferrate lattices in battery cathodes involve Fe^{2+}/Fe^{3+} mixed-valence centers. Discuss electron hopping mechanisms, the role of lattice defects, particle morphology, and electrolyte composition on ionic and electronic conductivity. Predict how these factors influence rate capability, capacity retention,

and long-term cycling stability. Propose in situ conductivity and Mössbauer spectroscopy studies to investigate charge transport pathways.

14. Spinel ferrites (MFe_2O_4) exhibit magnetic ordering determined by cation distribution between tetrahedral and octahedral sites. Discuss how substitution of Co^{2+}, Ni^{2+}, or Zn^{2+} influences saturation magnetization, coercivity, and Curie temperature. Predict how synthetic parameters, annealing temperature, and particle size affect magnetic anisotropy. Suggest VSM and SQUID experiments to quantify magnetic behavior.

15. Platinum nanoparticles catalyze CO oxidation. Explain how oxygen adsorption, CO binding, and surface reconstruction affect reaction kinetics. Predict how alloying with Rh or Pd modifies d-band center, electronic structure, and catalytic performance under high-temperature, high-flow conditions. Suggest operando XAS or DRIFTS measurements to monitor reaction intermediates.

16. Molybdenum olefin metathesis catalysts mediate polymerization reactions. Discuss how metal oxidation state, ligand electronics, and steric hindrance dictate initiation, propagation, and termination rates. Predict effects of coordinating solvents or additives on active site availability. Propose NMR or IR spectroscopy experiments to monitor catalyst speciation and polymer growth in real time.

17. NiTi shape-memory alloys undergo martensitic transformations due to reversible phase changes. Explain how electron density, d-orbital bonding, and lattice distortion contribute to transformation temperature and mechanical properties. Predict how alloying with Cu or Fe affects transition temperature, hysteresis, fatigue resistance, and elasticity. Suggest DSC, XRD, and mechanical testing to characterize transformations.

18. Copper(II) sulfate pentahydrate exhibits blue color due to d–d transitions. Predict how dehydration, ligand replacement (NH_3, Cl^-), or temperature affect absorption maxima, ligand field splitting, and spectral intensity. Suggest UV–Vis spectroscopy and kinetic monitoring to quantify changes in solution or solid-state complexes.

19. Electrochemical deposition of chromium on steel is influenced by applied potential, pH, and electrolyte composition. Discuss how complexing agents like sulfate or acetate modify deposition efficiency, morphology, and oxidation state.

Predict how current density and temperature influence grain size, adhesion, and corrosion resistance. Suggest chronoamperometry, SEM, and XRD studies to optimize deposition conditions.

20. Ruthenium bipyridyl complexes in dye-sensitized solar cells inject electrons into TiO_2. Discuss how ligand substitution, metal-centered redox potential, and anchoring groups influence charge injection, recombination, and photostability. Predict how solvent polarity and electrolyte composition affect incident photon-to-current efficiency. Suggest transient absorption spectroscopy and IPCE measurements to quantify performance.

21. Vanadium pentoxide electrodes in lithium-ion batteries rely on layered structures and cation intercalation. Discuss how vanadium oxidation states, lattice flexibility, and electronic conductivity affect capacity, voltage profile, and cycling stability. Predict effects of nanostructuring, carbon composites, or surface coating on rate capability and energy density. Propose operando XRD or XAS to monitor structural changes during cycling.

22. Cobalt-containing metalloenzymes catalyze C–H activation. Using knowledge of ligand geometry, oxidation state, and electron delocalization, propose a plausible mechanism. Predict how synthetic analogues could mimic enzyme selectivity and efficiency in vitro. Suggest kinetic isotope effect studies and spectroscopic monitoring to identify intermediates.

23. Iridium photoredox complexes catalyze C–H functionalization. Predict how spin-orbit coupling, heavy-atom effects, and excited-state lifetimes affect radical generation. Suggest optimization of irradiation intensity, solvent, and temperature to maximize yield and minimize side reactions. Propose transient absorption spectroscopy or laser flash photolysis to characterize excited states.

24. Catalytic decomposition of hydrogen peroxide on MnO_2 involves surface Mn^{4+} centers. Discuss how particle size, defect density, and crystal phase influence activity. Predict the influence of pH and temperature on kinetics. Suggest in situ monitoring with oxygen evolution measurement or UV–Vis spectroscopy to quantify reaction progress.

25. Platinum-group metals in PEM fuel cells catalyze hydrogen evolution and oxygen reduction. Discuss how electronic structure, d-band center, and adsorption

energies dictate catalytic activity. Predict effects of alloying, support material, and particle morphology on overpotential and durability. Suggest rotating disk electrode and impedance spectroscopy studies to evaluate performance.

26. Nickel complexes catalyze coupling of aryl halides with organozinc reagents. Propose a detailed mechanism including oxidative addition, transmetallation, and reductive elimination. Predict effects of electron-rich versus electron-poor ligands on turnover number and selectivity. Suggest kinetic and NMR studies to probe intermediate speciation.

27. Copper-based oxygen sensors rely on phosphorescence quenching. Predict how coordination geometry, ligand field strength, and redox potential influence sensitivity. Discuss the influence of temperature, solvent polarity, and competing analytes on response time and accuracy. Suggest time-resolved photoluminescence and electrochemical measurements to quantify performance.

28. Zirconium MOFs adsorb volatile organics. Discuss how metal connectivity, ligand rigidity, and pore environment dictate selectivity. Predict how framework defects, post-synthetic functionalization, or co-ligands affect adsorption capacity and catalytic activity. Propose gas sorption isotherms and IR spectroscopy to monitor adsorption kinetics.

29. Chromium(III) forms kinetically inert octahedral complexes, while Cr^{6+} is highly oxidizing. Discuss electronic and geometric factors underlying this dichotomy. Predict how ligand substitution affects redox potential, stability, and reactivity. Suggest voltammetry and spectrophotometric monitoring to explore selective oxidation in synthetic and environmental contexts.

30. Mixed-metal oxide electrodes with Li^{+} intercalation involve Fe^{3+}/Fe^{2+} and Mn^{3+}/Mn^{4+}. Discuss how lattice strain, cation ordering, and Jahn–Teller distortions affect voltage, capacity fade, and rate capability. Predict how computational modeling could optimize electrode design. Suggest operando XRD and electrochemical measurements to track structural evolution.

31. Organopalladium complexes catalyze selective C–H activation in aromatics. Discuss how steric effects, electronic properties, and directing groups govern regioselectivity. Predict the impact of substituting N-heterocyclic carbene ligands

for phosphines on efficiency and oxidation state. Suggest kinetic studies, isotopic labeling, and in situ spectroscopy to monitor intermediates.

32. Co-doped TiO_2 photocatalysts are used for water splitting. Explain how bandgap tuning, charge mobility, and surface traps influence efficiency. Predict effects of co-catalysts and sacrificial agents on hydrogen evolution rates. Suggest transient absorption and electrochemical studies to quantify carrier lifetime and recombination.

33. Spin-crossover iron(II) complexes exhibit reversible high-spin ↔ low-spin transitions under pressure. Using ligand field theory, explain how Δ_oct and metal–ligand bond lengths control crossover temperature and hysteresis. Predict how ligand rigidity and donor strength modulate transition thermodynamics. Suggest Mössbauer and SQUID studies to characterize behavior.

34. Gold nanoparticles exhibit localized surface plasmon resonance (LSPR). Discuss how size, shape, aggregation, and dielectric environment influence resonance wavelength and intensity. Predict how to tune properties for SERS sensing or photothermal therapy. Suggest UV–Vis and dark-field microscopy to study LSPR shifts.

35. Ruthenium olefin metathesis catalysts are sensitive to coordinating solvents. Discuss how ligand competition, electron density, and sterics affect initiation and propagation. Predict decomposition pathways and methods to suppress them. Suggest in situ NMR or IR studies to monitor catalyst speciation during polymerization.

36. Mixed-valence manganese oxides in redox flow batteries exhibit electron delocalization and Jahn–Teller distortions. Discuss how lattice defects, particle size, and doping affect electron transfer kinetics, capacity, and cycling stability. Predict enhancements achievable by nanoscaling or heteroatom substitution. Suggest electrochemical impedance spectroscopy and operando XAS for analysis.

37. Palladium nanoparticles on alumina catalyze selective alkyne hydrogenation. Discuss how particle size, edge and corner atoms, and metal-support interaction influence activity and selectivity. Predict how sintering or poisoning modifies performance. Suggest in situ XPS or IR spectroscopy to identify active sites.

38. Cisplatin binds DNA via coordination to guanine N7 atoms. Discuss how ligand exchange kinetics, sterics, and electronic effects control selectivity. Predict modifications to the platinum coordination sphere that reduce toxicity while maintaining anticancer activity. Suggest DNA footprinting and NMR studies to monitor binding.

39. Nickel-based catalysts facilitate reductive amination under hydrogen. Propose a mechanism involving substrate adsorption, hydrogen activation, and imine reduction. Predict how ligands or supports alter activation energy and selectivity for primary vs. secondary amines. Suggest kinetic isotope effect experiments and in situ IR spectroscopy to track intermediates.

40. Vanadium oxides display metal-insulator transitions due to d-electron occupancy and lattice distortion. Predict how doping or strain engineering modifies transition temperature, conductivity, and electronic band structure. Suggest temperature-dependent conductivity and XRD studies to quantify effects.

41. Copper enzymes catalyze dioxygen reduction. Propose a mechanism considering redox cycling, electron delocalization, and ligand geometry. Predict how mutation of coordinating amino acids affects efficiency or selectivity. Suggest spectroscopic (EPR, stopped-flow UV–Vis) studies to identify intermediates.

42. Palladium-catalyzed Stille coupling of aryl halides and organostannanes involves oxidative addition, transmetallation, and reductive elimination. Discuss how ligand electronics, solvent polarity, and temperature influence activity. Predict factors leading to catalyst deactivation. Suggest kinetic and in situ NMR studies to probe intermediates.

43. Cobalt-phosphine complexes catalyze alkene hydroboration. Discuss how ligand bite angle, electronics, and oxidation state influence regioselectivity and turnover. Predict effects of nitrogen-based ligand substitution on product distribution and lifetime. Suggest kinetic and NMR studies for validation.

44. Iron oxide nanoparticles as MRI contrast agents depend on size, shape, surface functionalization, and magnetic anisotropy. Discuss how these factors influence relaxivity. Predict effects of aggregation, oxidation state changes, and surface chemistry on imaging performance. Suggest NMR relaxometry and TEM studies to quantify effects.

45. Permanganate oxidants degrade organic contaminants in wastewater. Discuss how pH, temperature, and competing ions affect kinetics, selectivity, and MnO_2 precipitation. Predict strategies to optimize oxidation efficiency and minimize secondary pollution. Suggest UV–Vis and ICP-OES analysis to monitor reaction and metal leaching.

46. Bimetallic Pt–Ru nanoparticles catalyze fuel cell reactions. Discuss how electronic coupling, d-band center shifts, and surface segregation influence activity and CO tolerance. Predict effects of synthesis conditions on particle homogeneity and performance. Suggest TEM, XPS, and electrochemical characterization to study structure–activity relationships.

47. Zirconium MOFs catalyze hydrolysis of organophosphates. Discuss how metal coordination, pore environment, and ligand electronics influence turnover rate and substrate specificity. Predict effects of framework defects or post-synthetic modification on activity. Suggest kinetic studies and in situ spectroscopy to quantify catalysis.

48. Iridium(III) complexes are used in OLEDs. Discuss how ligand field strength, π-conjugation, and spin-orbit coupling affect emission wavelength, quantum yield, and lifetime. Predict strategies to tune color and device stability. Suggest photophysical measurements (time-resolved emission, quantum yield) to optimize performance.

49. Vanadium oxide cathodes in sodium-ion batteries rely on layer spacing, Na^+ diffusion pathways, and mixed-valence states. Discuss effects on capacity, voltage, and rate capability. Predict improvements achievable through doping with other transition metals. Suggest operando XRD and electrochemical studies to monitor structural and electrochemical changes.

50. Single-atom iron catalysts on N-doped carbon promote oxygen reduction. Explain how coordination environment, electronic structure, and metal-support interaction dictate selectivity. Predict effects of atom dispersion and synthetic control on efficiency and durability. Suggest XAS, Mössbauer spectroscopy, and rotating ring-disk electrode studies to probe active sites.

Chapter 20

Exploring Organic Molecules

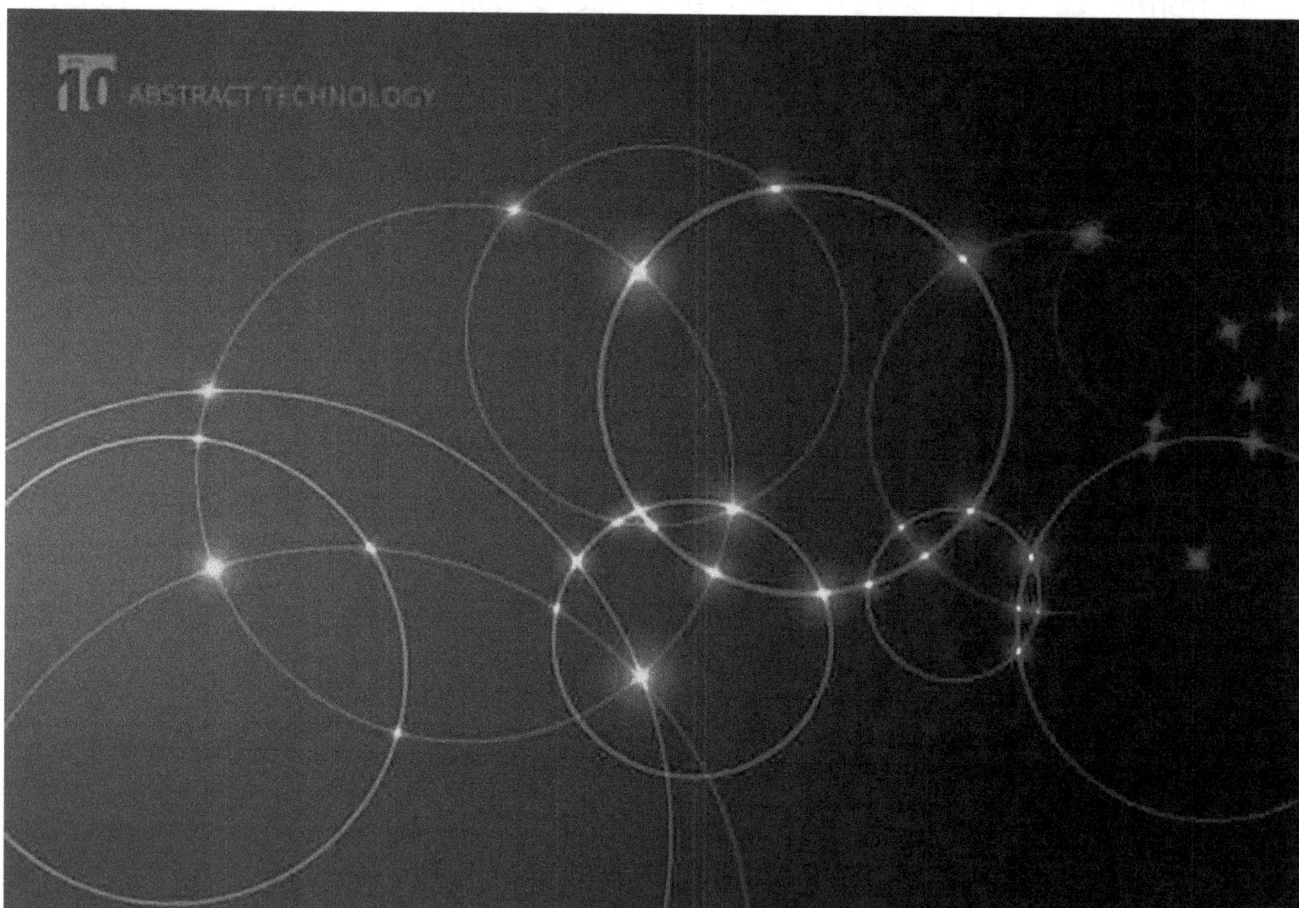

Carbon, the cornerstone of organic chemistry, is arguably the most versatile element in the periodic table. Its ability to form four covalent bonds, combined with its moderate electronegativity, enables the formation of an extraordinary variety of stable compounds. No other element matches carbon in its capacity to construct long chains, intricate branched structures, or closed rings of atoms, creating a scaffold upon which almost all life depends. The complexity and diversity of carbon compounds underpin the chemical basis of life, industrial innovation, and material science. At the most fundamental level, carbon's valence electrons allow for multiple bonding arrangements. Single bonds, formed by the overlap of sp^3 hybrid orbitals, give rise to alkanes—saturated hydrocarbons that are relatively inert and provide structural stability in fuels and raw materials. Double bonds, involving sp^2 hybrid orbitals, yield alkenes, which are more chemically reactive and form the basis of addition reactions and polymer synthesis. Triple bonds, characterized by sp hybridization, create alkynes, which are even more reactive and are critical intermediates in synthetic chemistry. The phenomenon of aromaticity, exemplified by benzene, demonstrates carbon's ability to stabilize electrons through delocalization, producing cyclic compounds with extraordinary chemical stability.

The structural versatility of carbon is matched by its functional diversity. By bonding with elements such as hydrogen, oxygen, nitrogen, sulfur, and the halogens, carbon forms functional groups that drastically alter a molecule's chemical behavior and physical properties. These functional groups, small in size yet powerful in effect, are the foundation of organic chemistry. They dictate solubility, reactivity, polarity, and biological activity, transforming simple hydrocarbons into compounds of immense industrial and biological importance. Alkanes, the simplest hydrocarbons, serve as the backbone for energy-rich materials. Methane, ethane, propane, and butane, the primary components of natural gas and liquefied petroleum gas, provide both fuel and feedstock for chemical synthesis. Their relative chemical inertness stems from the strength of the carbon-carbon and carbon-hydrogen bonds, which require significant energy input to break. Beyond energy applications, alkanes serve as lubricants, solvents, and starting materials for more complex hydrocarbons, including branched and cyclic derivatives.

Alkenes introduce unsaturation into the hydrocarbon framework. With their carbon-carbon double bonds, alkenes are highly reactive, capable of undergoing a wide array of addition reactions. Hydrogenation, halogenation, hydrohalogenation, and hydration reactions transform alkenes into functionalized derivatives, enabling industrial processes ranging from the production of plastics to the synthesis of fine chemicals. The hydrogenation of vegetable oils to produce margarine is a direct application of this chemistry, demonstrating the interface between organic reactions and everyday life. The reactivity of alkenes also makes them crucial intermediates in the synthesis of polymers, such as polyethylene and polypropylene, which form the basis of modern plastics. Alkynes, with their carbon-carbon triple bonds, are highly reactive and versatile compounds used in both industrial and laboratory synthesis. Acetylene, the simplest alkyne, serves as a fuel for welding torches and as a precursor to a variety of chemicals, including vinyl chloride for PVC production. Alkynes participate in addition reactions similar to alkenes, but their linear geometry and high electron density confer unique reactivity patterns that are exploited in the formation of conjugated and aromatic systems.

Aromatic hydrocarbons, including benzene, naphthalene, and polycyclic aromatic hydrocarbons (PAHs), occupy a special place in organic chemistry due to their resonance-stabilized electron clouds. The delocalized π-electrons confer stability and dictate characteristic chemical behavior, such as the tendency toward electrophilic substitution rather than addition reactions. Aromatic compounds are pervasive in industrial chemistry, serving as feedstocks for dyes, pharmaceuticals, resins, and explosives. Polycyclic aromatics, while industrially significant, are also environmental pollutants, with compounds like 3,4-benzopyrene recognized as potent carcinogens, highlighting the dual nature of carbon-based chemistry in both human utility and ecological impact.

The introduction of functional groups to hydrocarbon frameworks dramatically expands chemical diversity. Halides, alcohols, amines, thiols, and carbonyl derivatives transform hydrocarbons into compounds with distinct reactivity and applications. Halides, or haloalkanes, involve carbon bonded to a halogen atom. They function as intermediates in chemical synthesis, solvents, anesthetics, and refrigerants. Dichloromethane, for instance, serves as a solvent in laboratory and industrial settings, while chloroform historically found use as a surgical anesthetic. The presence of the electronegative halogen atom increases the molecule's polarity, facilitating nucleophilic substitution and elimination reactions in synthetic chemistry.

Alcohols, characterized by the hydroxyl (-OH) functional group, are highly versatile compounds. The polarity of the hydroxyl group enables hydrogen bonding, enhancing solubility in water and dictating interactions with other molecules. Ethanol, perhaps the most familiar alcohol, finds applications in beverages, medicinal formulations, and as a solvent in industrial processes. Polyhydric alcohols, such as glycerin and ethylene glycol, exhibit multiple hydroxyl groups that confer hygroscopicity, viscosity, and utility in products ranging from cosmetics to antifreeze. Alcohols also serve as key intermediates in the synthesis of esters, ethers, and other oxygen-containing functional groups, underscoring their central role in chemical transformations.

Amines, with their nitrogen-containing functional group ($-NH_2$), are both nucleophilic and basic. They play critical roles in biological systems and pharmaceuticals. Biogenic amines like adrenaline, noradrenaline, histamine, and serotonin are central to physiological regulation, affecting heart rate, immune response, and neural signaling. Synthetic amines, such as diphenhydramine and local anesthetics, exploit these properties for therapeutic purposes. The presence of nitrogen allows amines to participate in hydrogen bonding and nucleophilic substitution, making them invaluable in chemical synthesis and industrial applications. Sulfur-containing functional groups, including thiols and sulfides, contribute both chemical reactivity and sensory characteristics. Thiols, notable for their pungent odors, are employed in trace detection and odorant compounds in natural gas. Sulfides and sulfoxides are significant in pharmaceutical chemistry, acting as intermediates in drug synthesis and as biologically active compounds. For example, sulfonamides revolutionized antibacterial therapy in the early twentieth century, illustrating the pharmacological significance of sulfur in carbon-based frameworks.

The carbonyl functional group, consisting of a carbon-oxygen double bond, is perhaps the most versatile group in organic chemistry. Aldehydes, ketones, carboxylic acids, esters, and amides all derive from the carbonyl structure, yet each exhibits distinct chemical and physical properties. Aldehydes, with a carbonyl bonded to hydrogen, are highly reactive and serve as intermediates in fragrances, flavorings, and polymer chemistry. Benzaldehyde, the principal component of almond oil, exemplifies the aromatic aldehyde, prized for its olfactory and synthetic utility. Ketones, with two carbon substituents attached to the carbonyl carbon, exhibit stability and polarity that make them solvents and intermediates in organic synthesis; acetone is the prototypical example, widely used in laboratories and industry. Carboxylic acids, characterized by a carbonyl adjacent to a hydroxyl group, are acidic due to resonance stabilization of the carboxylate anion. These compounds are ubiquitous in biochemistry and industry, ranging from acetic acid in vinegar to fatty acids in lipids. Derivatives such as esters and amides expand functionality and applications: esters provide fruity aromas and flavors in foods and beverages, while amides form the backbone of proteins and pharmaceuticals, including urea and acetaminophen. Acid chlorides and anhydrides further demonstrate the chemical versatility of the carbonyl, enabling the synthesis of polymers, pharmaceuticals, and agrochemicals.

The chemical behavior of organic compounds is profoundly influenced by the interplay of functional groups. Hydrogen bonding, dipole-dipole interactions, and acid-base behavior govern solubility, boiling points, and reactivity. Alcohols, due to their hydrogen-bonding capability, are miscible with water and serve as polar solvents, while hydrocarbons lacking functional groups are hydrophobic and insoluble in polar media. Amines exhibit nucleophilicity and basicity, enabling substitution and condensation reactions that form amides, imines, and heterocyclic compounds. Carbonyl compounds participate in nucleophilic addition and condensation reactions, forming alcohols, imines, and esters, demonstrating the centrality of the carbonyl in organic synthesis.

Organic chemistry is deeply embedded in daily life. Haloalkanes act as solvents and anesthetics, providing utility in industrial and medical settings. Petroleum-derived alkanes fuel transportation and energy production, with fractional distillation yielding natural gas, gasoline, kerosene, diesel, lubricants, and asphalt. Alkenes and their derivatives underpin polymer industries, producing plastics, synthetic rubbers, and fibers. Aromatic compounds form dyes, pigments, pharmaceuticals, and flavorings, illustrating the broad scope of carbon-based chemistry. Beyond industrial utility, organic chemistry permeates biological systems. Vision relies on the photochemistry of cis-11-retinal, a conjugated aldehyde derived from vitamin A. This molecule undergoes photoisomerization to 11-trans-retinal upon light absorption, initiating a cascade of biochemical events that results in visual perception. Amines such as neurotransmitters and hormones regulate essential physiological processes, while sulfur-containing compounds influence metabolic pathways and signaling. Even flavors, fragrances, and pheromones, from limonene in citrus to bombykol in moths, demonstrate the biological and ecological significance of carbon-based molecules.

In addition to natural systems, organic chemistry shapes the design of synthetic materials and medicines. The functionalization of carbon skeletons allows chemists to tailor molecular properties for specific applications, from high-strength polymers to targeted pharmaceuticals. Aspirin, with its combination of aromatic, carboxylic acid, and ester functionalities, exemplifies the synthesis of a multifunctional therapeutic agent. Surfactants, soaps, and detergents exploit amphiphilic properties, with polar headgroups interacting with water and nonpolar tails solubilizing oils and fats. The design and optimization of such molecules illustrate the interplay between structure, function, and practical application.

Carbon's versatility extends far beyond linear, branched, or aromatic hydrocarbons. One of the most fascinating aspects of organic chemistry is the formation of heterocyclic compounds, in which carbon atoms in a ring are replaced by heteroatoms such as nitrogen, oxygen, or sulfur. These compounds are pervasive in both nature and synthetic chemistry, forming the structural backbone of nucleic acids, vitamins, hormones, and a myriad of pharmaceuticals. Heterocycles combine the rigidity of cyclic structures with the unique electronic properties of heteroatoms, resulting in molecules with diverse reactivity, polarity, and biological activity.

A prime example of a biologically critical heterocycle is pyridine, a six-membered ring containing one nitrogen atom. Pyridine is a basic, polar molecule that serves as a precursor in the synthesis of vitamins such as niacin (vitamin B3) and many pharmaceutical agents. Its electron-rich nitrogen atom allows pyridine to act as a nucleophile or a ligand in coordination chemistry, binding metal centers in catalytic processes. Similarly, pyrimidines and purines, which are fused heterocyclic systems containing both nitrogen and carbon atoms, form the building blocks of nucleic acids—DNA and RNA. Cytosine, thymine, uracil, adenine, and guanine exemplify heterocycles that encode genetic information, illustrating carbon's fundamental role in the chemistry of life.

Five-membered heterocycles such as furan, thiophene, and imidazole demonstrate the chemical richness introduced by oxygen, sulfur, and nitrogen atoms in rings. Furan, containing oxygen, is reactive toward electrophiles and serves as a precursor to pharmaceuticals and agrochemicals. Thiophene, with sulfur in the ring, exhibits aromatic character similar to benzene and is utilized in organic electronics, dyes, and conducting polymers. Imidazole, a nitrogen-containing heterocycle, is critical in biochemistry; the imidazole side chain of the amino acid histidine acts as a proton shuttle in enzyme catalysis, facilitating acid-base reactions in active sites. Histamine, a mediator of immune responses, is derived from histidine, highlighting the biological significance of nitrogen heterocycles.

Alkaloids, a large class of naturally occurring nitrogen-containing heterocycles, demonstrate the intersection of organic chemistry, biology, and pharmacology. Morphine, quinine, caffeine, nicotine, and strychnine all contain complex heterocyclic structures. Morphine's pentacyclic framework with embedded nitrogen confers analgesic properties by interacting with opioid receptors in the central nervous system. Quinine, a bicyclic nitrogen heterocycle, has historically been essential in treating malaria. Caffeine, a purine-derived alkaloid, acts as a central nervous system stimulant, highlighting how heterocycles influence physiological processes. These compounds exemplify how carbon, through the construction of heterocyclic frameworks, can mediate complex biological activities.

Vitamins and cofactors frequently incorporate heterocyclic structures. Vitamin B12, with its cobalt-containing corrin ring, is a macrocyclic heterocycle critical for DNA synthesis and red blood cell formation. Riboflavin (vitamin B2), containing an isoalloxazine ring, functions as a redox cofactor in metabolic reactions. Niacin (vitamin B3), derived from pyridine, participates in electron transfer processes as part of NAD^+/NADH and $NADP^+$/NADPH cofactors. Folic acid, with its pteridine heterocycle, facilitates one-carbon transfer reactions crucial for nucleotide biosynthesis. These examples underscore the profound influence of carbon-based heterocycles on life at the molecular level.

Carbon's role in natural products chemistry extends beyond heterocycles to include terpenes, steroids, and polyketides—structurally complex molecules derived from simple carbon building blocks. Terpenes, constructed from isoprene units (C_5H_8), form essential oils, pigments, and pheromones. Limonene, found in citrus peels, and menthol, from peppermint, illustrate the aromatic and sensory diversity achievable through carbon scaffolds. Steroids, built on a tetracyclic carbon framework, include cholesterol, testosterone, and cortisol. These molecules demonstrate how rigid carbon frameworks combined with functional groups confer highly specific biological functions. Polyketides, synthesized via alternating carbonyl and methylene units, produce antibiotics, anticancer agents, and immunosuppressants such as erythromycin, doxorubicin, and tacrolimus, highlighting the intersection of carbon chemistry with medicine.

Peptides and proteins, essential polymers in biology, are linear chains of amino acids connected through amide bonds (peptide linkages). Each amino acid contains an α-carbon bonded to a carboxyl group, an amino group, a hydrogen atom, and a distinctive side chain (R-group), which may include functional groups like hydroxyl, thiol, or aromatic rings. The side chains dictate protein structure, stability, and reactivity. Disulfide bridges between cysteine residues exemplify how carbon-linked sulfur atoms contribute to tertiary and quaternary protein structures, critical for enzymatic function, hormone signaling, and structural integrity. Proteins demonstrate carbon's ability to generate molecular complexity capable of performing precise, dynamic biological tasks.

Lipids—hydrophobic molecules with carbon-rich backbones—serve as energy stores, cellular membranes, and signaling molecules. Fatty acids, composed of long hydrocarbon chains terminated by a carboxyl group, are esterified to glycerol to form triglycerides. Phospholipids, which include phosphate-containing polar headgroups, self-assemble into bilayers, forming the structural foundation of cell membranes. Sterols, with their multi-ring carbon structures, modulate membrane fluidity and act as precursors for steroid hormones. The hydrophobic effect, driven by nonpolar carbon chains, demonstrates the physicochemical principles governing lipid assembly and function.

Carbohydrates, another critical class of carbon-based biomolecules, consist of polyhydroxylated aldehydes or ketones. Monosaccharides like glucose and fructose provide energy and serve as precursors for nucleotides, glycoproteins, and polysaccharides. Polysaccharides, including starch, cellulose, and glycogen, illustrate carbon's ability to form extensive polymeric networks with structural and storage roles. The stereochemistry of carbon centers in carbohydrates determines biological recognition, enzymatic specificity, and energy metabolism, emphasizing the importance of carbon's tetrahedral bonding in generating molecular diversity.

Nucleic acids, the information carriers of life, consist of long chains of nucleotides, each containing a sugar (ribose or deoxyribose), a phosphate group, and a nitrogenous base. The nitrogenous bases—adenine, guanine, cytosine, thymine, and uracil—are heterocycles that encode genetic information. The phosphodiester linkage between the 3′ hydroxyl of one sugar and the 5′ phosphate of the next creates a carbon-phosphate backbone that imparts stability while allowing flexibility for replication, transcription, and repair. Carbon's centrality in this structure exemplifies how the element facilitates both structural integrity and functional specificity in biomolecules.

The study of secondary metabolites, molecules not directly involved in growth or reproduction but crucial for survival and ecological interaction, further illustrates carbon's versatility. Alkaloids, terpenoids, flavonoids, and polyketides function as defense compounds, signaling molecules, and attractants. These molecules often contain complex carbon scaffolds with multiple functional groups, allowing them to interact selectively with enzymes, receptors, or other biomolecules. Morphine's analgesic effect, caffeine's stimulation of the central nervous system, and taxol's inhibition of microtubule dynamics in cancer therapy are all consequences of finely tuned carbon-based architectures.

The versatility of carbon extends far beyond natural compounds into the realm of synthetic materials and polymers, where carbon's ability to form long chains, cross-links, and diverse functional groups enables the creation of materials with tailored mechanical, thermal, and chemical properties. Polymers, macromolecules composed of repeating units (monomers), represent one of the most significant technological achievements of modern chemistry. Carbon-based polymers have revolutionized industries ranging from packaging and textiles to aerospace and medicine.

Addition polymers, also called chain-growth polymers, are formed by linking unsaturated monomers such as alkenes through addition reactions. Polyethylene, one of the simplest and most widely produced polymers, arises from the polymerization of ethylene. Its high molecular weight chains confer strength, chemical resistance, and flexibility, making it suitable for plastic bags, bottles, and piping. Polypropylene, formed from propylene monomers, is stiffer and heat-resistant, suitable for automotive components, food containers, and fibers. Polyvinyl chloride (PVC), produced from vinyl chloride, demonstrates how halogenated carbon chains impart rigidity, durability, and resistance to degradation, enabling applications in construction, flooring, and medical tubing.

Condensation polymers, formed through step-growth reactions, involve the elimination of small molecules such as water during polymerization. Polyesters, such as polyethylene terephthalate (PET), arise from the condensation of dicarboxylic acids and diols. PET's mechanical strength, clarity, and chemical resistance make it ideal for beverage bottles, textile fibers, and food packaging. Polyamides, including nylon, result from the reaction of diamines with dicarboxylic acids, producing fibers with remarkable tensile strength, elasticity, and durability. Kevlar, a para-aramid polyamide, exhibits exceptional tensile strength and heat resistance, finding applications in bulletproof vests, aerospace components, and industrial ropes. The carbon backbone of these polymers provides structural integrity, while the functional groups along the chain dictate intermolecular interactions, crystallinity, and solubility.

Elastomers, or rubber-like materials, are carbon-rich polymers capable of reversible deformation. Natural rubber, a polymer of isoprene, exhibits elasticity due to the flexible carbon-carbon backbone and the cis-configuration of double bonds, which allows chains to uncoil and recoil under stress. Vulcanization, the process of introducing sulfur cross-links, transforms rubber into a more durable, heat-resistant material by forming sulfur-carbon bonds between chains. Synthetic elastomers, such as polybutadiene, styrene-butadiene rubber (SBR), and silicone polymers, leverage carbon backbones combined with functional groups or heteroatoms to achieve tailored properties, enabling their use in tires, seals, medical devices, and adhesives.

Beyond traditional polymers, carbon allotropes—distinct structural forms of elemental carbon—demonstrate remarkable diversity in physical properties while retaining the fundamental chemical versatility of carbon. Diamond, with its three-dimensional tetrahedral network of sp^3-hybridized carbons, exhibits unmatched hardness, high thermal conductivity, and optical transparency. Its applications range from cutting and drilling tools to high-precision optics and electronics. Graphite, composed of planar sheets of sp^2-hybridized carbons arranged in hexagonal lattices, is an excellent conductor of electricity and heat while being lubricious due to weak interlayer interactions. Graphite is used in electrodes, batteries, lubricants, and as a precursor for advanced carbon materials.

Fullerenes (C_{60}, C_{70}, and higher homologues), often referred to as buckyballs, are spheroidal molecules composed entirely of carbon atoms arranged in pentagonal and hexagonal rings. Their unique geometry and electron delocalization impart exceptional stability, and they exhibit interesting electronic, optical, and chemical properties. Fullerenes have been explored in drug delivery, photovoltaics, superconductors, and nanotechnology. Carbon nanotubes, cylindrical structures of sp^2 carbon, combine high tensile strength with electrical and thermal conductivity. Their applications include composite materials, field-effect transistors, sensors, energy storage devices, and nanoscale electronics. Graphene, a single layer of carbon atoms in a two-dimensional honeycomb lattice, has revolutionized materials science due to its extraordinary strength, flexibility, conductivity, and transparency, enabling innovations in flexible electronics, membranes, and energy devices.

Carbon fibers, produced from organic precursors such as polyacrylonitrile (PAN) or pitch, are lightweight, high-strength materials with exceptional stiffness. They are widely used in aerospace, automotive, and sporting goods, where the combination of low density and mechanical strength is critical. The formation of these fibers involves pyrolysis, in which the carbon backbone undergoes reorganization to form extended sp^2 networks, demonstrating the critical role of carbon-carbon bonding in determining material properties.

Polymers and carbon-based materials in electronics and energy represent a rapidly advancing frontier. Conductive polymers, such as polyaniline and polythiophene, exploit conjugated carbon-carbon bonds to transport electrons, enabling applications in organic light-emitting diodes (OLEDs), flexible displays, and photovoltaic cells. Carbon-based anodes and cathodes in lithium-ion batteries leverage the structural and electronic properties of graphite and graphene to achieve high capacity, conductivity, and cycle stability. Supercapacitors and hydrogen storage systems also utilize carbon's unique ability to form porous, high-surface-area frameworks, demonstrating how the fundamental chemistry of carbon enables cutting-edge technology.

Carbon in composites and nanomaterials further illustrates its unparalleled versatility. Combining carbon fibers with polymer matrices produces composite materials with enhanced strength-to-weight ratios, corrosion resistance, and thermal stability, crucial for aerospace, military, and automotive applications. Carbon aerogels, ultra-light porous materials composed of interconnected carbon networks, exhibit high surface areas and low densities, enabling applications in energy storage, insulation, and environmental remediation. Carbon quantum dots and graphene derivatives expand the possibilities for bioimaging, sensing, and catalysis, highlighting the ability of carbon chemistry to bridge molecular design with macroscopic functionality.

Environmental applications of carbon-based materials underscore their practical significance. Activated carbon, with its highly porous structure, is used extensively in water purification, air filtration, and chemical adsorption. Carbon capture technologies, including functionalized graphene and carbon-based membranes, aim to mitigate greenhouse gas emissions, demonstrating the role of carbon materials in sustainable technologies. Biochar, a carbon-rich residue derived from biomass pyrolysis, enhances soil fertility and sequesters carbon, linking organic chemistry with ecological and agricultural benefits.

Carbon chemistry in medicine and drug delivery exemplifies the synergy between molecular design and therapeutic application. Carbon-based nanocarriers, including fullerenes, carbon nanotubes, and graphene derivatives, enable targeted delivery of drugs, genes, and imaging agents. Functionalization of carbon surfaces with polymers, ligands, or biomolecules tailors solubility, biocompatibility, and receptor specificity, allowing precise control over pharmacokinetics and biodistribution. These innovations demonstrate how carbon's chemical versatility supports both structural and functional roles in advanced technology and healthcare.

Chapter 20: Word Problems

1. A synthetic chemist seeks to design a novel polymer with alternating sp^2- and sp^3-hybridized carbons to achieve both rigidity and flexibility. Considering the hybridization-dependent bond angles and reactivity, propose a monomer structure capable of step-growth polymerization, predict the mechanical properties of the resulting polymer, and outline potential industrial applications. How might incorporation of polar functional groups along the polymer backbone influence solubility and thermal stability?

2. Benzene derivatives can be selectively nitrated under controlled conditions. A researcher wants to synthesize 1,3-dinitrobenzene without forming ortho-substituted byproducts. Using resonance stabilization and electronic effects of substituents, propose a stepwise reaction sequence starting from benzene, rationalize the regioselectivity, and calculate the theoretical yield given excess nitric acid.

3. Polycyclic aromatic hydrocarbons (PAHs) such as 3,4-benzopyrene are environmental pollutants with carcinogenic properties. Describe a multi-step metabolic pathway in which such a PAH is bioactivated to form DNA adducts. Discuss how the structural features of the PAH, including planarity and electron density distribution, contribute to its reactivity toward nucleophilic sites on nucleobases.

4. Ethanol can be oxidized to acetaldehyde and subsequently to acetic acid. Devise a reaction scheme using both chemical and enzymatic oxidation methods, explaining the mechanism of electron transfer in each case. Compare the kinetic and thermodynamic factors influencing each step and predict conditions under which overoxidation can be minimized.

5. A pharmaceutical chemist wants to design an analog of morphine that retains analgesic activity but reduces addictive potential. Considering the pentacyclic carbon framework and nitrogen heterocycle, propose structural modifications, justify how they would alter receptor binding, and predict any changes in pharmacokinetics or metabolism resulting from the changes in carbon functional groups.

6. Graphene exhibits extraordinary mechanical strength and electrical conductivity. A researcher plans to functionalize graphene with carboxyl and amine groups to create a hybrid material for biosensing. Discuss the chemical strategies to selectively introduce these groups without disrupting the conjugated sp^2 lattice, predict how functionalization will affect electronic properties, and suggest potential analytical applications.

7. Consider a reaction in which an alkene undergoes halogenation followed by nucleophilic substitution with a thiol to generate a thioether. Draw all possible stereoisomers, propose a mechanism that accounts for regio- and stereoselectivity, and analyze how the electronic effects of substituents on the alkene influence the reaction rate and product distribution.

8. Polyacrylonitrile (PAN) is a precursor for high-strength carbon fibers. Describe the thermal cyclization and pyrolysis processes converting PAN to sp^2-rich carbon fibers. Analyze how nitrogen atoms influence the graphitic structure formation and predict the impact of co-monomers on tensile strength, electrical conductivity, and thermal stability of the fibers.

9. Consider an alkyne undergoing a transition-metal-catalyzed coupling reaction with a haloarene to form a substituted aromatic compound. Propose a detailed catalytic cycle including oxidative addition, transmetallation, and reductive elimination. Discuss how the electronic and steric properties of the alkyne and arene influence regioselectivity, reaction rate, and yield, referencing carbon hybridization effects on bond angles.

10. Carboxylic acids can form esters via Fischer esterification. A chemist wishes to synthesize a mixed anhydride from two different carboxylic acids while minimizing self-condensation. Propose a stepwise synthetic strategy, including activation methods, discuss the influence of electronic and steric effects on reaction selectivity, and predict the relative hydrolysis rates of the resulting anhydride.

11. Design a synthetic route to transform a linear alkane into a polyfunctional molecule containing alcohol, aldehyde, and amine groups at defined positions. Rationalize the choice of protecting groups to prevent overreaction, predict regioselectivity, and evaluate how steric hindrance and electronic effects influence the reaction outcomes.

12. Alkynes can undergo cycloaddition reactions with azides to form triazoles in "click chemistry" reactions. For a given alkyne and azide substrate, predict the regioisomer formed, explain the mechanism at the molecular orbital level, and propose applications of the resulting triazole in medicinal chemistry and polymer functionalization.

13. A chemist aims to synthesize a heterocyclic compound containing both nitrogen and sulfur in a five-membered ring starting from an aldehyde and a thiolamine. Design a multi-step synthetic pathway, explain the cyclization mechanism, and predict how electron-donating or electron-withdrawing substituents on the aldehyde would affect reaction kinetics and yield.

14. The design of carbon-based nanocarriers for drug delivery requires precise functionalization of carbon nanotubes. Propose a chemical strategy for introducing hydrophilic and targeting ligands onto the nanotube surface, evaluate potential side reactions, and predict how surface chemistry will influence cellular uptake and biodistribution.

15. Polyethylene and polypropylene are addition polymers derived from alkenes. Compare the kinetics and thermodynamics of their polymerization reactions, predict the effects of temperature and pressure on chain length distribution, and discuss how branching influences crystallinity, melting point, and mechanical properties.

16. Consider the biosynthesis of cholesterol from acetyl-CoA via the mevalonate pathway. Identify the carbon-carbon bond-forming steps, predict the intermediates'

stereochemistry, and propose isotopic labeling experiments to confirm the flow of carbon atoms during the cyclization cascade.

17. Fullerenes (C_{60}) can undergo addition reactions with nucleophiles. Predict the preferred sites of nucleophilic attack, justify based on π-electron density and strain energy, and propose a functionalization strategy for generating water-soluble fullerene derivatives suitable for biomedical applications.

18. A chemist is tasked with synthesizing a polymeric material that exhibits both elastomeric and conductive properties. Propose a monomer design integrating conjugated sp^2 carbons and flexible aliphatic chains, predict the resulting polymer's mechanical and electronic properties, and suggest applications in wearable electronics or soft robotics.

19. Consider the enzymatic oxidation of a monosaccharide to a corresponding aldonic acid. Describe the carbonyl chemistry involved, predict the effect of stereochemistry on enzyme-substrate binding, and discuss how functionalization of hydroxyl groups can influence subsequent polymerization to polysaccharides.

20. A researcher aims to selectively functionalize the C5 position of a pyrimidine nucleobase without affecting other nitrogen and carbon sites. Propose a synthetic strategy, including protecting groups if necessary, explain the electronic rationale for site selectivity, and discuss potential applications in nucleic acid analogs for therapeutic use.

21. Alkaloids such as caffeine and quinine contain polycyclic nitrogen heterocycles. Devise a retrosynthetic analysis for a complex alkaloid, identifying key carbon-carbon and carbon-nitrogen bond formations, and predict challenges associated with stereocontrol and regioselectivity in the total synthesis.

22. Consider a fatty acid undergoing partial hydrogenation to yield mono- and di-unsaturated products. Analyze the stereochemistry of the resulting double bonds, predict the influence on melting point and membrane fluidity, and discuss industrial strategies to minimize trans-isomer formation during hydrogenation.

23. Sulfur-containing amino acids like cysteine form disulfide bonds in proteins. Propose a chemical mechanism for disulfide bond formation, predict how pH and redox environment influence bond stability, and analyze the effect of bond formation on tertiary and quaternary protein structure.

24. Design a stepwise synthesis for a terpenoid with multiple stereocenters starting from isoprene units. Explain the regiochemistry and stereochemistry of cyclization and rearrangement steps, and predict the impact of electronic substituents on the final product's stability and biological activity.

25. A chemist wants to prepare a conjugated polymer with alternating donor and acceptor units for use in organic photovoltaics. Propose a monomer design, explain how carbon conjugation length and substituents influence bandgap, and predict how the polymer morphology affects device efficiency.

26. Consider the oxidation of a primary alcohol to a carboxylic acid using PCC and subsequent reaction to form an amide. Propose a mechanistic pathway, evaluate competing side reactions, and discuss how solvent and temperature influence reaction rate and product selectivity.

27. Polycyclic aromatic hydrocarbons can undergo electrophilic substitution reactions. Predict the product distribution for nitration of anthracene, considering resonance structures, and explain how electron-donating and electron-withdrawing substituents influence regioselectivity.

28. Carbon nanotubes can be used as catalysts in organic reactions. Propose a mechanism for a carbon-nanotube-mediated oxidation reaction, predict the active sites on the nanotube surface, and discuss the effect of functionalization and curvature on catalytic efficiency.

29. A chemist wants to synthesize a sugar nucleotide for glycosylation reactions. Predict the reactivity of hydroxyl groups on the monosaccharide, propose protecting group strategies, and explain how carbon stereochemistry dictates the formation of glycosidic bonds.

30. The photochemistry of retinal involves cis-trans isomerization. Propose a quantum mechanical explanation for the selective isomerization of the C11=C12 bond, predict the influence of protein environment on reaction kinetics, and discuss how mutations in opsin proteins might alter visual response.

31. 31. Consider the synthesis of a carbon-based metal-organic framework (MOF) using benzene dicarboxylate ligands. Propose a self-assembly mechanism, predict how functional groups on the ligand influence porosity and gas adsorption,

and evaluate potential applications in carbon capture or catalysis.

32. Vinyl chloride undergoes polymerization to PVC. Analyze the chain-growth polymerization kinetics, predict the effect of initiator type and concentration on molecular weight distribution, and discuss the role of chlorine in determining polymer rigidity and thermal stability.

33. A chemist wants to design a fluorescent probe based on a heterocyclic aromatic framework. Propose a structure, explain how electron delocalization and substituents affect emission wavelength, and predict how functionalization for solubility and target recognition influences fluorescence efficiency.

34. Polyesters such as PET are formed by condensation reactions. Propose a kinetic model for the polymerization process, predict how monomer ratio, temperature, and catalyst type influence molecular weight and polydispersity, and discuss post-polymerization modifications to improve mechanical properties.

35. Consider the design of a carbon quantum dot for bioimaging applications. Predict how surface functional groups influence quantum confinement and emission wavelength, propose synthetic strategies to control size and shape, and evaluate biocompatibility and cellular uptake.

36. A chemist aims to synthesize a steroid derivative with selective functionalization at C17. Propose a stepwise synthetic strategy, discuss the challenges of regio- and stereocontrol on the tetracyclic carbon framework, and predict how functional groups influence receptor binding.

37. Conductive polymers such as polyaniline are proton-doped to achieve conductivity. Propose a doping mechanism, predict how the degree of polymerization and functional group substitution affects electronic properties, and discuss potential applications in energy storage and sensors.

38. A researcher wants to introduce a fluorine atom selectively into a heterocyclic compound to improve pharmaceutical properties. Predict the site of substitution, propose a synthetic strategy, and discuss the effect of fluorine on pKa, lipophilicity, and metabolic stability.

39. Nucleophilic aromatic substitution can be challenging on deactivated rings.

Design a synthetic route to introduce an amine on a highly electron-deficient aromatic ring, predict potential side reactions, and evaluate the effect of leaving group and solvent choice on reaction efficiency.

40. A chemist aims to design a biodegradable polymer based on aliphatic esters. Propose a monomer design, explain how carbon chain length and functional groups influence degradation rate and mechanical properties, and predict how copolymerization can tailor properties for biomedical applications.

41. Polycyclic nitrogen heterocycles can coordinate to transition metals in catalysis. Propose a ligand design, predict the electronic and steric effects on metal reactivity, and suggest potential catalytic transformations that exploit carbon-nitrogen frameworks.

42. Activated carbon is used in water purification. Discuss how pore size distribution, surface area, and functional groups influence adsorption of organic pollutants, and propose chemical treatments to enhance selectivity for specific contaminants.

43. A chemist wants to synthesize an enantiomerically pure amino alcohol. Propose a stereoselective synthetic pathway starting from a chiral aldehyde, predict potential racemization pathways, and evaluate how protecting groups influence stereochemical outcome.

44. Thiophene-containing polymers are used in organic electronics. Predict how the substitution pattern affects conjugation and bandgap, propose a polymerization strategy, and discuss the influence of heteroatom placement on charge transport properties.

45. Consider a reaction converting a monosaccharide into a sugar alcohol via reduction. Predict the influence of stereochemistry on the reduction outcome, propose catalysts and conditions to favor a specific diastereomer, and discuss industrial applications of sugar alcohols.

46. Carbon aerogels are synthesized via sol-gel methods. Propose a mechanism for network formation, predict how cross-linking density influences porosity and mechanical properties, and suggest functionalization strategies for energy storage or catalysis applications.

47. A chemist wants to design a carbon-based electrode for lithium-ion batteries. Predict how the microstructure and functionalization of graphite or graphene influence lithium intercalation, conductivity, and cycle stability, and propose strategies to mitigate capacity loss.

48. Alkyl halides can undergo elimination to form alkenes. Propose a reaction scheme to selectively generate a terminal versus internal alkene, explain the mechanistic rationale based on Zaitsev's and Hofmann's rules, and discuss solvent and base effects on product distribution.

49. Design a heterocyclic inhibitor targeting a kinase enzyme. Propose a carbon-based scaffold, rationalize the placement of functional groups for hydrogen bonding and hydrophobic interactions, and predict how modifications on the carbon framework alter potency and selectivity.

50. A chemist wants to synthesize a carbon-rich nanomaterial incorporating both fullerenes and carbon nanotubes to enhance mechanical and electronic properties. Propose a synthetic strategy, discuss potential challenges in interfacing the two nanostructures, and predict how carbon hybridization and functionalization influence the resulting composite's properties.

Chapter 21

A Chemical Story of Earth Science

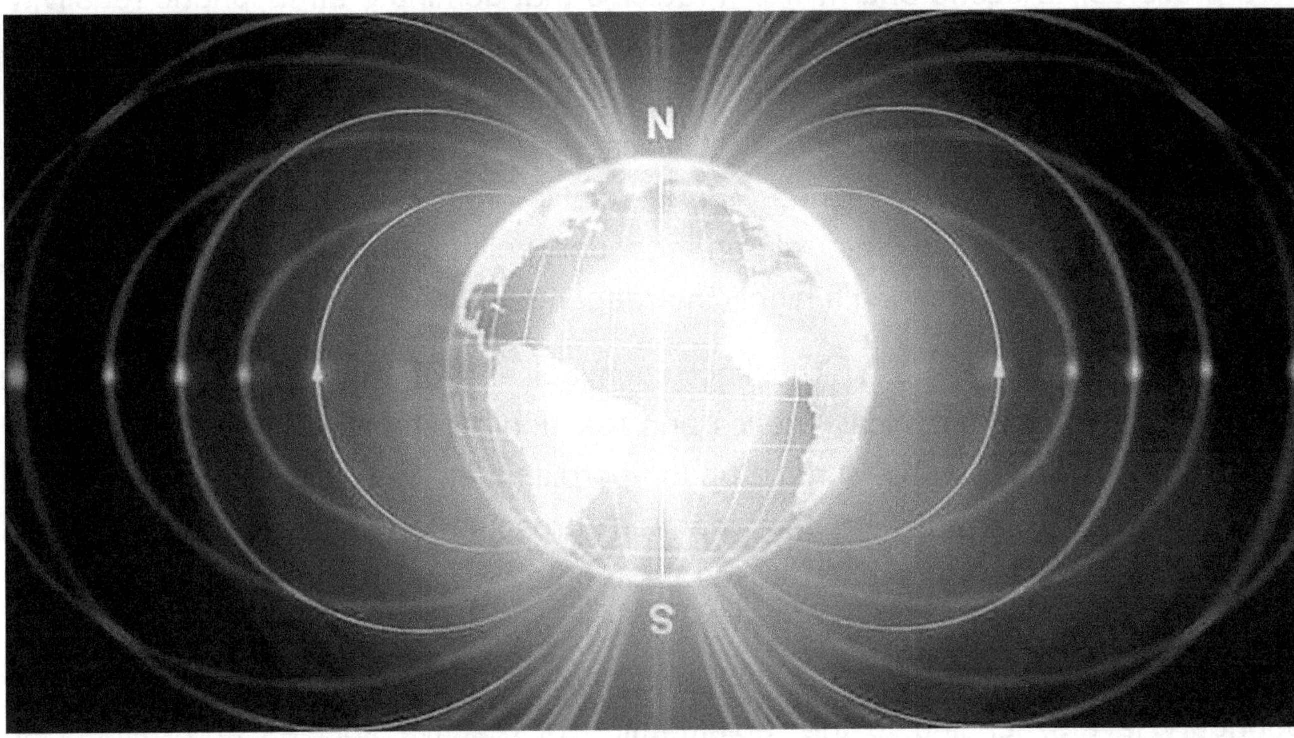

The Earth's atmosphere is not merely a passive envelope surrounding the planet; it is a chemically active, dynamically stratified medium where a multitude of processes dictate the composition and behavior of gases, aerosols, and particulates. This complex system is shaped by a delicate interplay of natural and anthropogenic factors, including volcanic eruptions, solar flux variability, ocean-atmosphere interactions, industrial emissions, combustion processes, and land-use changes. Vertical and horizontal transport mechanisms, such as convection, turbulence, boundary layer dynamics, and long-range advection, regulate the residence times and distributions of reactive species, while photochemical processes driven by ultraviolet and visible radiation dictate the transformation pathways of numerous organic and inorganic compounds.

The bulk of the atmosphere is comprised of nitrogen (~78%) and oxygen (~21%), which, despite their apparent chemical inertness under ambient conditions, are central participants in the global biogeochemical cycles. Nitrogen undergoes a series of oxidation and reduction reactions, driven both biologically through nitrification and denitrification and abiotically via lightning or high-temperature combustion. Oxygen, while critical for aerobic respiration, also participates in radical chemistry, serving as a precursor to ozone and hydroxyl radicals that dominate atmospheric reactivity. Trace gases—including argon, carbon dioxide, methane, nitrous oxide, and ozone—though present at parts-per-million or parts-per-billion concentrations, exert profound influence on climate regulation, oxidative capacity, and radiative forcing. Reactive nitrogen oxides (NO_x), sulfur oxides (SO_x), volatile organic compounds (VOCs), and halogenated species play pivotal roles in secondary pollutant formation, acid deposition, and ozone chemistry, forming a highly interdependent network of reactions.

Atmospheric layers exhibit distinct chemical and physical characteristics that profoundly influence chemical kinetics and reaction mechanisms. The troposphere, the layer extending from the surface up to approximately 8–15 kilometers depending on latitude, harbors nearly all water vapor and aerosols. Here, temperature decreases with altitude at the environmental lapse rate, promoting convective mixing and vertical transport of gases and particulate matter. Weather phenomena—including convective thunderstorms, hurricanes, and frontal systems—facilitate rapid exchange between boundary layers, leading to the redistribution of reactive species and influencing reaction rates through changes in pressure, temperature, and relative humidity. Photochemical smog, dominated by reactions among VOCs, NO_x, and sunlight, exemplifies the troposphere's capacity to produce secondary pollutants such as ozone, peroxyacyl nitrates, and aldehydes, which impact air quality and public health.

Above the troposphere lies the stratosphere, extending roughly from 15 to 50 kilometers in altitude, where temperatures increase with height due to absorption of ultraviolet radiation by ozone. Stratospheric chemistry is dominated by the formation and destruction of ozone through both the Chapman mechanism and catalytic cycles involving radicals such as OH, NO, Cl, and Br. The photodissociation of molecular oxygen (O_2) by high-energy UV photons produces highly reactive atomic oxygen, which rapidly combines with O_2 to form ozone. These ozone molecules, in turn, absorb UV radiation, preventing harmful UV-B and UV-C radiation from reaching the Earth's surface. The presence of anthropogenic halogenated compounds, such as chlorofluorocarbons and halons, introduces catalytic pathways for ozone destruction. In polar regions, stratospheric cooling and the formation of polar stratospheric clouds enable heterogeneous reactions that liberate reactive chlorine and bromine, which participate in ozone depletion events, including the Antarctic ozone hole. Stratospheric sulfate aerosols, derived from volcanic eruptions, provide additional surfaces for heterogeneous reactions, modifying ozone chemistry on regional and global scales.

The mesosphere and thermosphere, extending beyond the stratosphere, are characterized by low pressures, decreasing densities, and high-energy photochemical processes. In the mesosphere, temperatures decrease with altitude, promoting noctilucent cloud formation and meteoric particle deposition, which provide catalytic surfaces for trace reactions. The thermosphere is marked by ionization phenomena induced by extreme ultraviolet and X-ray solar radiation, creating the ionosphere, where charged particles interact with Earth's magnetic field, generating auroral emissions and influencing radio wave propagation. Within these upper layers, reactions involving atomic oxygen, molecular nitrogen, and trace species such as nitric oxide occur at rates governed by low-pressure kinetics, often proceeding via three-body recombination or radiative association mechanisms.

Ozone chemistry is intricately linked to catalytic cycles mediated by radical species. Hydroxyl radicals (OH), produced from water vapor photolysis and ozone photochemistry, act as the primary atmospheric "detergent," initiating oxidation of methane, carbon monoxide, and other hydrocarbons. Nitrogen oxides participate in both ozone production and destruction, depending on the chemical regime. In the troposphere, NO_x emissions from combustion sources drive photochemical ozone formation through the oxidation of VOCs, whereas in the stratosphere, NO_x contributes to ozone depletion through catalytic cycles. Halogen radicals, derived from anthropogenic sources, efficiently convert ozone to molecular oxygen without being consumed, illustrating the long-range environmental impact of human activities.

The greenhouse effect, an essential feature of atmospheric chemistry, is mediated by the absorption of infrared radiation by molecules possessing permanent or induced dipoles, including CO_2, CH_4, N_2O, H_2O, and halogenated compounds.

Carbon dioxide, generated through respiration, fossil fuel combustion, and volcanic degassing, is removed via photosynthesis, carbonate precipitation, and oceanic dissolution. Methane, produced biologically in anaerobic environments, reacts with hydroxyl radicals to form water and carbon dioxide, serving as a potent but short-lived greenhouse gas. Nitrous oxide, largely emitted from agricultural activities, also contributes to radiative forcing and participates in stratospheric ozone chemistry. The accumulation of these gases, coupled with feedback mechanisms such as albedo changes from ice melt and cloud formation, has led to measurable global temperature increases and altered atmospheric circulation patterns.

Acid deposition exemplifies the interplay between atmospheric chemistry and terrestrial impacts. Sulfur dioxide, emitted from fossil fuel combustion and volcanic activity, is oxidized in the presence of hydroxyl radicals to sulfuric acid, which precipitates as acid rain. Nitrogen oxides undergo similar oxidation pathways to form nitric acid. These acidic species alter soil chemistry, accelerate mineral weathering, damage vegetation, and modify aquatic ecosystems by lowering pH and solubilizing metals such as aluminum and iron. Photochemical smog, a complex mixture of ozone, aldehydes, peroxides, and nitrogen dioxide, results from sunlight-driven reactions in urban atmospheres, highlighting the interconnection between atmospheric chemistry and human health.

The hydrosphere's chemistry is inseparable from atmospheric processes. Rivers, lakes, and oceans serve as both sinks and reactors for atmospheric species. Dissolved gases—including oxygen, carbon dioxide, and nitrogen—mediate acid-base equilibria, redox reactions, and biological activity. Physical weathering releases mineral fragments into water bodies, which undergo chemical dissolution and precipitation reactions, contributing to the cycling of silica, calcium, magnesium, and other cations. Chemical weathering of silicates, carbonates, and sulfates produces bicarbonate, calcium, and sulfate ions, which regulate ocean alkalinity and carbonate equilibria. Biogeochemical cycles of carbon, nitrogen, phosphorus, and trace metals in aquatic systems are strongly influenced by thermal stratification, vertical mixing, and nutrient availability. In lakes, epilimnetic waters are rich in dissolved oxygen due to photosynthesis, while hypolimnetic waters may become anoxic, promoting anaerobic microbial processes such as denitrification and sulfate reduction.

Trace metals in aquatic systems participate in complex redox reactions and form coordination complexes with organic ligands, affecting their bioavailability and toxicity. For instance, iron undergoes alternating oxidation states between Fe^{2+} and Fe^{3+}, controlling primary productivity in iron-limited regions of the ocean. Similarly, copper, zinc, and manganese exist in multiple oxidation states, influencing enzymatic reactions and phytoplankton growth. Adsorption onto mineral surfaces, complexation with organic matter, and precipitation as sulfide or carbonate minerals modulate the cycling of these elements, highlighting the intricate balance between geochemistry and biology in aquatic environments.

Sediment formation in oceans and lakes involves the interplay of physical, chemical, and biological processes. Organic detritus from plankton, chemical precipitates such as calcium carbonate and silica, and terrigenous inputs from rivers contribute to stratified sediment layers. Redox conditions in sediments influence diagenetic reactions, including the reduction of nitrate to nitrogen gas, the transformation of sulfate to hydrogen sulfide, and the remineralization of organic carbon to carbon dioxide and methane. These reactions, in turn, feed back into the overlying water column, influencing nutrient availability and gas fluxes between the ocean and atmosphere.

Atmospheric deposition, both wet and dry, supplies essential nutrients to terrestrial and marine systems, while anthropogenic pollutants introduce heavy metals, persistent organic pollutants, and acidifying compounds. Volcanic emissions, including SO_2, CO_2, and halogens, alter atmospheric chemistry on regional and global scales, leading to episodic cooling, stratospheric aerosol formation, and transient ozone depletion events. Industrial emissions and biomass burning further perturb chemical cycles, contributing to enhanced tropospheric ozone formation, secondary organic aerosol production, and particulate matter accumulation, which affects radiative balance, cloud condensation nuclei availability, and human health.

Ultimately, the Earth's atmosphere, hydrosphere, and lithosphere constitute a highly interlinked system governed by chemical kinetics, thermodynamics, and transport phenomena. Understanding the chemical interactions among gases, liquids, and solids, both natural and anthropogenic, is fundamental to addressing global environmental challenges. From ozone photochemistry to acid deposition, from nutrient cycling to greenhouse gas accumulation, the chemistry of the Earth's systems underscores the delicate equilibrium necessary for sustaining life. Effective mitigation of climate change, protection of stratospheric ozone, preservation of water quality, and sustainable management of mineral resources all hinge on a profound comprehension of these chemical processes, their feedbacks, and their long-term consequences.

The continuous study of atmospheric and aquatic chemistry has revealed that even minor perturbations in chemical concentrations or reaction rates can cascade into significant environmental impacts. For example, slight increases in tropospheric methane enhance hydroxyl radical consumption, lengthening methane's atmospheric lifetime and amplifying radiative forcing. Similarly, perturbations in phosphorus availability can limit primary productivity in oligotrophic ocean regions, affecting the entire marine food web. Advanced modeling techniques, combining chemical kinetics, transport dynamics, and satellite observations, are now employed to predict chemical distributions, assess human impacts, and design effective remediation strategies.

Through the lens of chemistry, it becomes evident that Earth's environment is a chemically orchestrated system in perpetual flux. Interactions among radicals, trace gases, aerosols, dissolved minerals, and organic compounds shape the physical conditions, climate, and habitability of the planet. The integration of atmospheric chemistry with hydrology, geology, and ecology allows scientists to interpret past climate events, predict future trends, and formulate policies aimed at preserving planetary health. Understanding these chemical intricacies is not merely an academic pursuit—it is essential for sustaining life, protecting ecosystems, and ensuring the continuity of natural cycles in an era of unprecedented anthropogenic change.

The interaction between atmospheric chemistry and climate is particularly evident in the behavior of greenhouse gases. Carbon dioxide, methane, and nitrous oxide, while differing in atmospheric lifetime and radiative efficiency, all participate in intricate chemical and physical interactions. Carbon dioxide, for example, dissolves in water to form carbonic acid, which then equilibrates to bicarbonate and carbonate ions, influencing ocean pH, buffering capacity, and carbonate mineral precipitation. The reaction sequence, $CO_2 + H_2O \rightleftharpoons H_2CO_3 \rightleftharpoons H^+ + HCO_3^- \rightleftharpoons 2H^+ + CO_3^{2-}$, demonstrates how atmospheric CO_2 directly modifies aqueous chemistry, affecting both biological and geochemical cycles. Increased atmospheric CO_2 not only enhances radiative forcing but also drives ocean acidification, reducing carbonate ion availability for shell-forming organisms and altering the marine carbon cycle.

Methane, a more potent greenhouse gas though present at lower concentrations than CO_2, undergoes oxidation initiated by hydroxyl radicals in the troposphere: $CH_4 + OH \rightarrow CH_3 + H_2O$, followed by subsequent reactions producing formaldehyde, carbon monoxide, and eventually carbon dioxide. This oxidation pathway links methane to tropospheric ozone formation and hydroxyl radical consumption, establishing feedback loops that influence atmospheric oxidative capacity. The fate of methane in the atmosphere is therefore interwoven with both climate and air quality, illustrating the interconnectivity of chemical species across multiple domains.

Nitrous oxide, emitted primarily from microbial activity in soils and oceans, is largely inert in the troposphere but undergoes photodissociation in the stratosphere to produce nitrogen oxides, which catalyze ozone destruction: $N_2O + hv \rightarrow N_2 + O(^1D)$. Here, the dual role of N_2O as both a greenhouse gas and a contributor to stratospheric ozone depletion exemplifies the subtle but critical interdependencies among atmospheric chemical processes. The interaction of nitrogen species with halogens and ozone forms complex networks in which catalytic cycles can amplify or mitigate chemical impacts depending on the prevailing environmental conditions.

Sulfur chemistry provides another illustrative example of atmospheric-aquatic coupling. Sulfur dioxide, emitted both naturally from volcanic activity and anthropogenically through fossil fuel combustion, undergoes oxidation to sulfuric acid through multiple pathways. In the gas phase, SO_2 reacts with hydroxyl radicals to form $HOSO_2$, which rapidly converts to sulfur trioxide (SO_3) and ultimately sulfuric acid: $SO_2 + OH + M \rightarrow HOSO_2 + M$; $HOSO_2 + O_2 \rightarrow HO_2 + SO_3$; $SO_3 + H_2O \rightarrow H_2SO_4$. Aqueous-phase reactions in cloud droplets further enhance sulfate formation, leading to acid rain that deposits reactive sulfur into soils and surface waters. The resulting acidification drives mobilization of cations such as aluminum and iron, alters soil mineral equilibria, and impacts aquatic ecosystems. Sulfate aerosols also act as cloud condensation nuclei, influencing albedo, cloud microphysics, and climate feedback mechanisms.

The chemistry of nitrogen, central to both terrestrial and aquatic systems, is marked by an array of oxidation-reduction reactions. In soils and aquatic sediments, ammonium, nitrate, and nitrite participate in microbial-mediated transformations: nitrification ($NH_4^+ \rightarrow NO_2^- \rightarrow NO_3^-$) and denitrification ($NO_3^- \rightarrow NO_2^- \rightarrow NO \rightarrow N_2O \rightarrow N_2$). These reactions are tightly coupled to redox conditions, organic carbon availability, and oxygen concentration, linking atmospheric deposition of nitrogen compounds to water chemistry, eutrophication, and greenhouse gas emissions. Excess nitrogen loading into rivers and lakes from agriculture or urban runoff can stimulate phytoplankton blooms, creating hypoxic or anoxic conditions in bottom waters, further affecting the cycling of phosphorus, iron, and other trace elements.

Phosphorus, unlike nitrogen or carbon, lacks a gaseous atmospheric phase and is cycled primarily through terrestrial weathering, riverine transport, and sedimentation in aquatic systems. Chemical weathering of phosphate-bearing minerals releases orthophosphate into soils and surface waters, which is then assimilated by plants and microorganisms. Sedimentation, microbial mineralization, and resuspension processes regulate the bioavailability of phosphorus, forming a tightly coupled biogeochemical cycle with profound implications for primary productivity. In marine systems, phosphorus limitation controls the growth of phytoplankton, influencing the sequestration of carbon via the biological pump and modulating atmospheric CO_2 over geological timescales.

Organic chemistry in the atmosphere is dominated by the reactions of volatile organic compounds with radicals such as OH, NO_3, and Cl. Alkanes, alkenes, aromatics, and oxygenated VOCs undergo hydrogen abstraction, addition to double bonds, and photolytic transformations, producing peroxides, aldehydes, ketones, and organic nitrates. Secondary organic aerosol formation results from condensation of low-volatility oxidation products, influencing both visibility and climate through scattering of solar radiation and cloud nucleation. These reactions exemplify the complex interplay between radical chemistry, photochemistry, and aerosol physics in determining tropospheric composition.

Aerosols themselves provide surfaces for heterogeneous reactions that can accelerate chemical transformations otherwise limited in the gas phase. Halogen activation on sea salt particles or ice surfaces, sulfate particle-mediated oxidation of sulfur dioxide, and nitrate deposition onto dust or volcanic ash illustrate how particle chemistry can dominate certain reaction pathways. Heterogeneous chemistry is particularly important in polar stratospheric clouds, where chlorine and bromine species undergo conversion into highly reactive forms that catalyze ozone destruction during spring sunlight exposure.

Water chemistry in the hydrosphere reflects an intricate balance of dissolved gases, mineral ions, organic matter, and particulate materials. Oxygen solubility, governed by Henry's law and influenced by temperature, salinity, and pressure, dictates aerobic microbial processes, redox potential, and chemical speciation. Trace metals such as iron, manganese, copper, and zinc participate in ligand exchange, redox cycling, and enzymatic activity, regulating nutrient availability and primary productivity. Dissolved organic carbon, composed of a complex mixture of humic and fulvic acids, polysaccharides, and amino acids, interacts with metals and radicals, modulating both light absorption and chemical reactivity in aquatic systems.

Sediment chemistry is shaped by both diagenetic and lithogenic processes. In anoxic environments, microbial reduction of nitrate, sulfate, and metal oxides drives redox gradients and controls the immobilization or mobilization of elements such as phosphorus, iron, and manganese. Carbonate and silica precipitation in sediments is influenced by pH, alkalinity, and saturation states, linking ocean chemistry to atmospheric CO_2 concentrations. Trace elements, including rare earth metals, are adsorbed onto mineral surfaces, forming complexes that impact bioavailability and global biogeochemical cycling.

Atmospheric deposition, both wet and dry, delivers essential nutrients to terrestrial and marine ecosystems, while anthropogenic pollutants introduce heavy metals, halogenated compounds, and reactive nitrogen species. Riverine transport carries these chemical inputs to oceans, where they integrate into complex biogeochemical networks. Ocean circulation and mixing, including thermohaline-driven deepwater formation and upwelling events, distribute nutrients and dissolved gases, maintaining chemical gradients that sustain marine productivity. Phytoplankton, through photosynthesis and calcification, regulate carbon dioxide uptake, oxygen production, and mineral precipitation, closing the loop between atmospheric chemistry and ocean biogeochemistry.

Volcanic eruptions illustrate a natural forcing mechanism that perturbs atmospheric chemistry on both local and global scales. Emissions of sulfur dioxide, carbon dioxide, hydrogen sulfide, halogens, and particulates affect stratospheric composition, aerosol loading, and radiative forcing. Sulfate aerosols scatter sunlight, causing transient cooling, while halogen emissions contribute to episodic ozone depletion events. The interplay between volcanic inputs and anthropogenic emissions demonstrates the sensitivity of atmospheric chemistry to both natural and human-induced perturbations.

Collectively, these processes underscore the interconnectivity of chemical reactions, physical transport, and biological activity across Earth's atmosphere and hydrosphere. The continuous flux of energy and matter through photochemical, oxidative, and redox processes ensures that minor perturbations can have cascading effects, influencing climate, air quality, and ecosystem health. Modeling these systems requires integration of kinetics, thermodynamics, transport phenomena, and feedback mechanisms, providing critical insights for environmental management, policy development, and climate mitigation strategies.

Ultimately, the Earth's chemical environment exemplifies a dynamic, self-regulating system where equilibrium is constantly challenged by external forcings and internal feedbacks. From stratospheric ozone cycles to deep ocean nutrient gradients, from radical-mediated oxidation of hydrocarbons to acid deposition, the interplay of chemistry and physics defines planetary habitability. Understanding these processes in narrative, interconnected form is essential for predicting environmental change, guiding sustainable practices, and preserving the delicate balance upon which life depends.

Extreme events, whether natural or anthropogenic, highlight the sensitivity of atmospheric chemistry and its downstream impacts on aquatic and terrestrial systems. Wildfires, for instance, inject massive quantities of carbon monoxide, methane, nitrogen oxides, and particulate matter into the troposphere. The combustion of cellulose, lignin, and other plant polymers produces volatile organic compounds that rapidly react with hydroxyl radicals to form ozone and secondary organic aerosols. During prolonged fire events, the formation of pyrogenic black carbon and soot modifies the radiative balance of the atmosphere, absorbing sunlight and contributing to localized warming. These aerosols can be transported over thousands of kilometers, affecting cloud microphysics, precipitation patterns, and even snow albedo in distant polar regions, thereby coupling terrestrial fire chemistry with global climate processes.

Urban environments offer another lens through which to examine the intricacies of chemical interactions in the atmosphere. Photochemical smog, first observed in industrial cities, is a complex mixture of ozone, nitrogen dioxide, peroxyacetyl nitrates, aldehydes, and particulate matter formed under sunlight-driven reactions. NO_x emitted from vehicle exhaust interacts with volatile organic compounds to produce peroxy radicals, which then propagate radical chain reactions, leading to tropospheric ozone accumulation: $VOC + OH \rightarrow RO_2$; $RO_2 + NO \rightarrow RO + NO_2$; $NO_2 + h\nu \rightarrow NO + O$; $O + O_2 \rightarrow O_3$. The feedback between urban emissions, solar radiation, and atmospheric radicals not only affects human health but also serves as a real-time chemical laboratory illustrating radical-mediated reactions on a planetary scale.

Atmospheric deposition of pollutants into aquatic systems drives significant biogeochemical changes. Acid rain, rich in sulfuric and nitric acids, enhances mineral dissolution in soils and sediments, releasing cations such as calcium, magnesium, and aluminum. Elevated aluminum concentrations can be toxic to freshwater organisms, altering community composition and reducing biodiversity. Similarly, deposition of nitrogen compounds can trigger eutrophication in lakes and coastal waters, where excessive nutrient input promotes algal blooms that deplete dissolved oxygen through microbial respiration, forming hypoxic "dead zones." These dead zones exemplify the chemical sensitivity of aquatic systems to atmospheric inputs, demonstrating how slight alterations in deposition chemistry cascade through complex reaction networks in water and sediment.

Ocean chemistry reflects a delicate interplay between dissolved gases, nutrients, and trace metals, shaped by both physical circulation and chemical reactivity. Oxygen, produced through photosynthetic activity in the euphotic zone, decreases with depth as microbial respiration consumes organic matter. The resulting redox gradients regulate the speciation and solubility of trace metals, including iron, manganese, and copper, which cycle between reduced and oxidized forms depending on the local oxygen concentration. Iron, for example, oscillates between Fe^{2+} and Fe^{3+}, controlling the growth of phytoplankton in iron-limited regions such as the Southern Ocean. Organic ligands produced by microorganisms further stabilize metal ions in solution, affecting bioavailability and mediating trace metal chemistry on global scales. These processes illustrate the intersection of chemical speciation, biological activity, and ocean circulation in maintaining nutrient balance and supporting primary productivity.

Trace elements in sediments also undergo diagenetic transformations driven by microbial activity and chemical equilibria. Sulfate-reducing bacteria facilitate the conversion of sulfate to hydrogen sulfide under anoxic conditions, which can precipitate with iron to form iron sulfide minerals such as pyrite (FeS_2). Simultaneously, manganese and other metals undergo redox cycling, forming layered deposits that record chemical histories of ancient environments. Organic matter decomposition in sediments produces methane and carbon dioxide, which can either escape to the water column or remain sequestered in sediments, forming a chemical archive of carbon cycling and storage over geological timescales. These sedimentary reactions, linked to both water column chemistry and atmospheric feedbacks, demonstrate the continuity of chemical processes from the surface to the depths of aquatic systems.

Volcanic activity further illustrates the dynamic interplay between chemical reactions and environmental change. Major eruptions emit sulfur dioxide, carbon dioxide, halogens, and particulates into the stratosphere, where these species undergo photochemical and heterogeneous reactions. Sulfur dioxide oxidizes to sulfate aerosols, scattering sunlight and reducing surface temperatures temporarily, while halogen species catalyze ozone depletion in the polar stratosphere. The injection of fine ash particles into the troposphere can facilitate heterogeneous reactions on particle surfaces, enhancing nitrogen and sulfur cycling. Volcanic emissions thus provide a natural analog to anthropogenic pollution, demonstrating how chemical inputs influence atmospheric composition, climate forcing, and downstream biogeochemical cycles in oceans and soils.

Human activity continues to alter chemical equilibria on a global scale, particularly through the emission of long-lived greenhouse gases. The atmospheric lifetime of carbon dioxide exceeds a century, meaning that emissions today have lasting impacts on radiative forcing, ocean acidification, and carbonate equilibria. Methane's atmospheric lifetime, on the order of a decade, depends critically on hydroxyl radical abundance; perturbations in tropospheric OH concentrations can therefore amplify methane's greenhouse effect. Nitrous oxide, while less abundant, persists for over a century and contributes to both warming and stratospheric ozone depletion. These long-lived gases illustrate the temporal dimension of chemical processes, in which atmospheric kinetics, transport, and feedbacks interact over years to centuries to shape planetary chemistry.

Polar chemistry provides a striking example of seasonal and spatial variability in chemical processes. During the Antarctic winter, the formation of polar stratospheric clouds enables heterogeneous reactions that convert inert chlorine and bromine reservoirs into highly reactive radicals. In spring, sunlight photolyzes these species, triggering catalytic ozone destruction. Similarly, in the Arctic, chemical cycles are modulated by stratospheric dynamics, temperature fluctuations, and human-derived halogen emissions. These phenomena highlight the delicate balance of chemical reservoirs, radical-mediated catalysis, and environmental forcing that regulate ozone concentrations, demonstrating how minor chemical changes can produce significant climatic consequences.

Atmospheric chemistry also influences cloud formation, precipitation patterns, and radiative transfer. Aerosols act as cloud condensation nuclei, determining droplet size distribution, cloud albedo, and precipitation efficiency. Organic aerosols derived from VOC oxidation, sulfate aerosols from SO_2 oxidation, and mineral dust particles all contribute to heterogeneous chemistry in clouds. Within cloud droplets, solubilized gases undergo rapid reactions, including oxidation of sulfur dioxide to sulfate, conversion of nitrogen oxides to nitric acid, and radical-mediated organic transformations. Cloud chemistry thereby links atmospheric composition directly to hydrological cycles, illustrating the pervasive influence of chemical reactions on Earth's energy and water balance.

The intricate chemistry of oceans, sediments, and the atmosphere emphasizes the necessity of integrated modeling approaches. Chemical transport models couple reaction kinetics with atmospheric and oceanic circulation, enabling predictions of pollutant dispersion, greenhouse gas lifetimes, and feedback responses. These models incorporate photochemical, radical, heterogeneous, and redox reactions, as well as interactions with biological activity, demonstrating the inherently interdisciplinary nature of environmental chemistry. Through such modeling, scientists can anticipate the consequences of human interventions, evaluate mitigation strategies, and inform international policies aimed at preserving atmospheric and aquatic chemical integrity.

Finally, the chemical complexity of Earth's systems underscores the interconnectedness of natural and anthropogenic processes. Radical-mediated oxidation of trace gases, acidification of soils and waters, aerosol-cloud interactions, stratospheric ozone dynamics, and biogeochemical cycling of essential elements all operate on scales ranging from molecular to planetary. The delicate chemical equilibrium maintained by these processes is increasingly perturbed by industrialization, deforestation, fossil fuel combustion, and urbanization. Understanding the chemistry of Earth's atmosphere and hydrosphere, therefore, is not merely an academic exercise; it is central to sustaining life, managing resources, and mitigating environmental change. Each chemical species, reaction pathway, and transport process contributes to a larger system in which the balance of life and climate hinges on the continuous flux of matter and energy through a vast network of interacting chemical systems.

Through this integrated perspective, one appreciates that the Earth's atmosphere and hydrosphere are not static repositories but dynamic chemical environments, where feedbacks, radical-mediated reactions, and redox processes perpetually shape the planet's climate, ecosystems, and habitability. The study of these processes, in narrative and connected form, provides the framework for understanding environmental change, predicting chemical behavior under perturbation, and designing strategies for sustainable management of Earth's chemical systems. From the formation of stratospheric ozone to the cycling of trace metals in ocean sediments, the Earth's chemical environment is a testament to the intricate interplay of reaction kinetics, transport, and environmental forcing—a complex, self-regulating system whose stability is essential to life itself.

Chapter 21: Word Problems

1. In a mid-latitude urban environment during the peak of summer, the combination of high solar irradiance, vehicular emissions, and biogenic volatile organic compounds (VOCs) results in elevated tropospheric ozone concentrations. Consider the diurnal variations of NO_x and VOC emissions, the photolysis of NO_2, the daily cycle of hydroxyl (OH) radical formation, and radical propagation and termination reactions. Predict how ozone concentration varies hourly from sunrise to sunset. Discuss which chemical species (NO_x, VOCs, or radicals) limits ozone production at sunrise, at solar noon when photolysis rates are maximal, and at sunset as sunlight diminishes. Additionally, incorporate feedback effects from radical recycling, peroxy radical formation, and ozone titration by NO into your assessment. Describe how meteorological factors such as boundary layer height, temperature, and humidity might modify these diurnal patterns and discuss potential strategies for urban ozone mitigation.

2. Following a major volcanic eruption that injects approximately 10 Tg of sulfur dioxide (SO_2) into the stratosphere, consider the chemical, physical, and dynamical processes leading to the formation of sulfuric acid (H_2SO_4) aerosols. Account for gas-phase oxidation of SO_2 by hydroxyl radicals, heterogeneous reactions on the surface of pre-existing particles, coagulation, sedimentation, and stratospheric circulation patterns. Calculate the expected aerosol concentration after 30 days

and predict the changes in optical depth. Discuss how these aerosols perturb the radiative balance over the equatorial region, considering scattering of incoming solar radiation, absorption, and changes in outgoing longwave radiation. Evaluate potential impacts on surface temperatures, cloud formation, and stratospheric chemistry. Include uncertainties due to variable OH concentrations, background aerosol surface area, and eruption altitude.

3. Methane (CH_4) emissions from wetland ecosystems interact with tropospheric OH radicals in the presence of varying temperature and relative humidity. Construct a kinetic model to predict the atmospheric lifetime of methane, incorporating the feedback effects of OH depletion by carbon monoxide (CO) and VOC oxidation, the impact of temperature-dependent reaction rates, and seasonal variations in OH radical concentrations. Estimate how these processes alter methane's global warming potential over a decade. Include considerations of vertical transport, photolysis of other reactive species, and interactions with nitrogen oxides that affect radical budgets. Discuss how uncertainties in wetland emission fluxes and climate-driven changes in OH concentration propagate through the model predictions.

4. During a large-scale wildfire event, the emission inventory includes CO, CH_4, NO_x, and a wide array of VOCs. Develop a simulation of the downwind chemical evolution of the fire plume, including secondary organic aerosol (SOA) formation, tropospheric ozone production, photolysis rates, and heterogeneous reactions on particulate matter. Incorporate boundary layer mixing, diurnal changes in sunlight intensity, and variations in temperature and humidity. Predict the spatial extent and chemical composition of the plume over time, identify chemical species limiting ozone formation at various distances, and assess the potential human health impacts using exposure and toxicity considerations. Discuss feedback effects of aerosol scattering on photolysis rates and the influence of atmospheric stability on plume dispersion.

5. Nitrous oxide (N_2O) released from fertilized agricultural soils is transported to the stratosphere where photolysis produces excited oxygen atoms ($O(^1D)$) and NO radicals, catalyzing ozone depletion. Using stratospheric transport models, solar flux data, and reaction kinetics for O_3 destruction cycles, estimate the annual ozone loss resulting from a 1 Tg increase in N_2O emissions. Include the impact

of stratospheric temperature, latitude-dependent solar exposure, and background Cl and Br radicals on catalytic cycles. Evaluate the long-term implications for ozone recovery and potential feedbacks to climate through changes in ultraviolet radiation reaching the surface.

6. Urban smog formation depends on VOC speciation, NO_x ratios, intensity of sunlight, and meteorological factors such as temperature inversions and boundary layer depth. Predict the peak ground-level ozone concentrations during a heatwave, incorporating OH radical recycling, formation of peroxyacetyl nitrates (PANs) as temporary NO_x reservoirs, and nighttime NO_3 radical chemistry. Analyze how varying VOC/NO_x ratios influence the ozone production regime (VOC-limited versus NO_x-limited). Include the effects of humidity on radical lifetimes, cloud cover on photolysis rates, and horizontal transport of precursors from surrounding regions. Discuss potential mitigation strategies including traffic restrictions and emission controls.

7. A coastal upwelling event delivers nutrient-rich, iron-limited waters to the photic zone. Construct a coupled biogeochemical model to describe phytoplankton growth, iron and nitrogen limitation, dissolved oxygen consumption during microbial respiration, and methane oxidation. Predict changes in the uptake of atmospheric CO_2 by phytoplankton, the production of dimethyl sulfide (DMS) as a climate-relevant aerosol precursor, and potential feedbacks on cloud condensation nuclei formation. Consider seasonal light availability, mixing depth, and nutrient remineralization rates. Evaluate the impact on local and regional climate by linking DMS emissions to radiative forcing.

8. In polar stratospheric regions, polar stratospheric clouds (PSCs) enable heterogeneous activation of chlorine species that catalyze ozone destruction. Using detailed PSC microphysics, surface area density, temperature profiles, and concentrations of chlorine reservoir species such as HCl and $ClONO_2$, predict the onset timing, magnitude, and spatial extent of springtime ozone depletion. Include the effects of solar zenith angle, stratospheric winds, and heterogeneous reaction kinetics. Discuss uncertainties due to PSC particle composition, heterogeneous reaction efficiencies, and stratospheric mixing rates.

9. SO_2 emissions from a coal-fired power plant are transported into a humid troposphere where they dissolve in cloud droplets and undergo aqueous-phase

oxidation to sulfuric acid. Quantify the conversion rate, considering reaction kinetics, droplet pH, cloud lifetime, and potential catalysis by metal ions present in aerosols. Estimate the resulting acid deposition flux to surrounding terrestrial and aquatic ecosystems, assess potential soil acidification and lake acidification, and discuss ecological impacts on vegetation and aquatic life. Incorporate diurnal variations in cloud cover, rainfall intensity, and plume dispersion in your assessment.

10. In the mesosphere, meteoric dust provides catalytic surfaces for trace reactions involving NO and O_3. Estimate the impact of a meteoroid storm on local ozone depletion rates, taking into account three-body recombination, radiative association processes at low pressures, heterogeneous reaction efficiencies, and the vertical distribution of meteoric metals. Discuss how these reactions interact with mesospheric temperature profiles, photolysis rates, and seasonal variations in solar UV flux. Evaluate the implications for mesospheric chemistry and the long-term persistence of catalytic effects.

11. Methane oxidation in the urban troposphere produces formaldehyde, CO, and HO_2 radicals. Construct a 24-hour kinetic model incorporating reaction rates, diurnal variation of OH radicals, photolysis of formaldehyde, and the formation of secondary pollutants such as ozone and peroxyacyl nitrates (PANs). Assess the effects of varying temperature, solar flux, and VOC composition on the overall oxidative capacity of the urban atmosphere. Include feedback loops involving radical recycling, NO_x partitioning, and heterogeneous uptake on aerosols.

12. Acid rain from nitrogen and sulfur oxides alters soil geochemistry by increasing H^+ concentration, mobilizing Al^{3+} and Fe^{3+} from aluminosilicate minerals. Using rainfall pH, buffering capacity, and mineral solubility constants, model the solubilization dynamics over a month-long wet period. Predict the potential toxicological impact on freshwater biota, considering metal speciation, bioavailability, and exposure duration. Include spatial variability due to soil composition, runoff rates, and topography in your calculations.

13. Elevated atmospheric CO_2 concentrations lead to ocean acidification. Using the full carbonate equilibrium system ($CO_2 + H_2O \rightleftharpoons H_2CO_3 \rightleftharpoons H^+ + HCO_3^- \rightleftharpoons 2H^+ + CO_3^{2-}$), calculate changes in carbonate ion concentration, saturation states of

calcite and aragonite, and the shell-building capacity of marine organisms under a scenario of CO_2 doubling. Incorporate temperature and salinity dependence, buffering by bicarbonate, and feedback effects on phytoplankton productivity. Discuss implications for marine ecosystems, fisheries, and carbon sequestration.

14. During a stratospheric sulfate aerosol event, compute the scattering and absorption of solar radiation by aerosol particles, including particle size distribution, refractive index, and altitude profile. Estimate the net surface cooling, changes in stratospheric heating rates, and potential modifications to tropospheric circulation patterns. Evaluate the role of aerosol lifetime, sedimentation, and coagulation in modulating radiative forcing. Discuss uncertainties arising from particle hygroscopic growth, volcanic plume dynamics, and seasonal solar variations.

15. Urban traffic emissions release NO_x and VOCs, creating conditions for nighttime chemistry involving NO_3 radicals. Using the Leighton relationship and photostationary state assumptions, predict nocturnal NO_3 radical concentrations and assess their impact on VOC oxidation rates. Incorporate the influence of temperature inversion layers, heterogeneous uptake of NO_3 on surfaces, and the formation of reservoir species such as N_2O_5. Discuss the subsequent effects on morning ozone production and potential human exposure risks.

16. A tropical cyclone induces vigorous mixing between the boundary layer and upper tropospheric regions. Analyze the redistribution of reactive species including ozone, OH, HO_2, and NOx radicals during the storm lifecycle. Predict transient changes in local photochemistry, radical budgets, and secondary pollutant formation. Include the effects of cloud shading, precipitation scavenging, and temperature-dependent reaction rates. Evaluate the potential for long-range transport of pollutants in the cyclone outflow region.

17. In an oceanic oxygen minimum zone (OMZ), microbial denitrification reduces nitrate to N_2O. Develop a kinetic model to calculate N_2O flux to the atmosphere as a function of nitrate concentration, labile organic carbon availability, temperature, and microbial enzyme kinetics. Incorporate vertical stratification, mixing, and light-limited regions in your estimation. Discuss feedbacks to greenhouse gas forcing and the implications for regional ocean biogeochemistry.

18. Chlorofluorocarbon (CFC) emissions reach the stratosphere, where UV photolysis liberates Cl radicals that catalyze ozone destruction. Model the ClO_x catalytic cycle,

including partitioning between reactive and reservoir species, sunlight-dependent photolysis rates, and temperature-dependent reaction kinetics. Predict annual ozone loss, spatial variability across latitudes, and potential feedbacks with NO_x and BrO_x chemistry. Include uncertainties from vertical transport and stratospheric mixing processes.

19. Wildfire aerosols rich in black carbon are deposited on snow surfaces, reducing albedo and enhancing radiative heating. Quantify the resultant snowmelt acceleration by considering aerosol optical properties, snow grain size, solar zenith angle, and boundary layer mixing. Evaluate the feedback to regional hydrology and climate, including changes in surface temperature, evaporation rates, and local atmospheric stability.

20. A river receives both acidic deposition from atmospheric pollution and phosphate-rich runoff from agricultural activities. Model the coupled chemical equilibria controlling metal speciation (Fe, Al, Ca), phosphate precipitation, and pH buffering. Predict the potential for eutrophication downstream, including the timing and magnitude of algal blooms, oxygen depletion, and subsequent effects on aquatic fauna. Include hydrological transport, residence time, and seasonal variability in your assessment.

21. Volcanic halogen emissions, including chlorine and bromine species, can enhance stratospheric ozone depletion. Using measured emission rates of Cl and Br, model the chemical transformations in both gas-phase and heterogeneous reactions on sulfate aerosols. Predict the net ozone loss, considering catalytic cycles, partitioning between reactive and reservoir species, and solar flux variations at different latitudes. Include the effects of stratospheric transport, aerosol surface area evolution, and seasonal sunlight changes. Discuss uncertainties arising from halogen speciation, volcanic plume height, and particle coagulation.

22. During an intense heatwave, soil microbial activity increases, leading to elevated N_2O emissions. Estimate the impact on local radiative forcing and stratospheric ozone chemistry by coupling a soil-atmosphere emission model with tropospheric and stratospheric transport. Include OH radical depletion, stratospheric NO_x production, and lifetime considerations for N_2O. Evaluate regional versus global impacts, and discuss feedbacks to ozone recovery and greenhouse gas forcing under prolonged heatwave conditions.

23. Secondary organic aerosol (SOA) formation is strongly influenced by VOC oxidation pathways, temperature, and relative humidity. Calculate the mass yield of low-volatility products from monoterpene reactions with OH radicals under variable atmospheric conditions. Incorporate photolysis rates, radical chain propagation, and heterogeneous uptake on pre-existing aerosols. Predict the impact on aerosol optical properties, cloud condensation nuclei formation, and radiative forcing at a regional scale. Discuss uncertainties related to VOC speciation and radical budgets.

24. Phytoplankton blooms modify air-sea CO_2 fluxes and dimethyl sulfide (DMS) production. Using coupled ocean-atmosphere models, quantify the feedback effects on cloud condensation nuclei formation, regional albedo, and radiative forcing. Incorporate nutrient limitation, light availability, mixed-layer depth, and grazing by zooplankton. Evaluate the seasonal and interannual variability of these feedbacks, and discuss implications for climate models and carbon sequestration strategies.

25. An industrial plume contains SO_2, NO_x, and particulate matter. Model the downwind chemical transformations, including heterogeneous sulfate formation, photochemical ozone production, nitrate aerosol deposition, and radical propagation. Predict regional air quality impacts over time and distance, accounting for mixing layer height, meteorological conditions, and diurnal photolysis rates. Include potential effects on human health, vegetation, and regional climate through aerosol radiative forcing.

26. The formation of peroxyacetyl nitrates (PANs) in urban atmospheres depends on VOC oxidation, NO_x concentration, and ambient temperature. Compute PAN equilibrium concentrations and evaluate their role as temporary NO_x reservoirs that influence ozone formation downwind. Incorporate photolysis rates, temperature-dependent dissociation, and radical recycling. Predict how daytime and nighttime chemical regimes differ, and discuss implications for regional air pollution management.

27. Ocean stratification reduces vertical mixing of nutrients, affecting nitrogen and phosphorus cycling. Quantify the resulting impacts on primary productivity, hypolimnetic oxygen consumption, and methane oxidation rates over a seasonal

timescale. Include effects of light limitation, seasonal thermal stratification, and episodic mixing events. Evaluate implications for coastal and open-ocean biogeochemistry, including greenhouse gas fluxes and nutrient export.

28. Acidic deposition mobilizes cations from soils. Using soil mineralogy, pH-dependent solubility constants, and 50 mm rainfall scenarios, calculate the expected Al^{3+} concentration in runoff. Predict potential toxicity to freshwater organisms, accounting for bioavailable versus complexed forms of metals. Include buffering capacity, soil organic matter, and topography in your calculations, and discuss mitigation strategies for acid-sensitive watersheds.

29. Halogen activation on sea salt aerosols contributes to tropospheric ozone depletion. Model the heterogeneous reaction kinetics, including particle surface area, relative humidity, halide concentrations, and temperature effects. Estimate the resulting O_3 loss regionally and temporally, considering boundary layer dynamics, radical recycling, and sunlight-driven photolysis rates. Discuss interactions with anthropogenic NO_x emissions and the overall contribution to coastal ozone budgets.

30. NO_x emissions from a power plant interact with ambient VOCs to produce ozone. Integrate advection, turbulent mixing, and photolysis rates to predict the downwind ozone concentration profile over 100 km. Incorporate temperature inversions, solar flux variability, diurnal emission cycles, and chemical feedback from radical propagation. Assess peak ozone concentrations and potential health risks for populations living in the plume trajectory.

31. Methane oxidation by OH radicals produces formaldehyde and CO. Evaluate the effect of increased anthropogenic CO emissions on methane lifetime and hydroxyl radical availability across a continental scale. Construct a coupled kinetic box model including photochemistry, radical recycling, and boundary layer mixing. Discuss implications for tropospheric ozone, secondary organic aerosol formation, and methane's greenhouse effect under projected emission scenarios.

32. Polar spring sunlight photolyzes Cl_2 released from polar stratospheric clouds. Calculate the rate of ozone loss in Dobson units, considering sunlight intensity, temperature-dependent reaction kinetics, and partitioning of Cl radicals between reactive and reservoir forms. Include effects of stratospheric transport, PSC

surface area, and halogen abundances. Discuss uncertainties in predicting the timing, magnitude, and spatial extent of ozone depletion events.

33. Urban heat islands increase photochemical smog formation. Model the effect of elevated surface temperatures on VOC reaction rates, NO_x partitioning, and secondary aerosol formation. Include feedbacks from temperature-dependent radical lifetimes, increased evaporation of VOCs, and vertical mixing in the urban boundary layer. Assess implications for local air quality, public health, and urban climate.

34. Iron-limited regions of the ocean experience episodic dust deposition events. Quantify the impact on phytoplankton growth, CO_2 uptake, nitrate drawdown, and DMS production, incorporating ligand-mediated iron solubility and light limitation. Include dust particle solubility, seasonal stratification, and mixed-layer depth. Evaluate feedbacks on regional radiative forcing and potential impacts on atmospheric chemistry.

35. Atmospheric methane increases OH consumption, affecting tropospheric ozone production. Construct a coupled kinetic model including CH_4, OH, CO, VOCs, and NO_x to predict feedbacks on ozone, CO, and secondary organic aerosol formation under a scenario of CH_4 doubling. Incorporate diurnal variations in sunlight, temperature, and vertical mixing. Evaluate regional differences in oxidative capacity and potential climate feedbacks.

36. During a volcanic eruption, emitted SO_2 oxidizes to sulfate aerosols. Calculate the enhancement in cloud condensation nuclei (CCN) and subsequent regional albedo change. Include particle size growth through coagulation, cloud microphysics, and seasonal circulation patterns. Assess implications for radiative forcing, precipitation, and stratosphere-troposphere interactions.

37. Nitrate deposition from agricultural runoff alters coastal marine chemistry. Model coupled nitrification and denitrification processes, dissolved oxygen gradients, and potential hypoxic zone expansion over a 3-month period. Include temperature-dependent microbial kinetics, water column stratification, and seasonal variations in primary productivity. Predict ecological consequences for benthic organisms and fisheries.

38. Chlorofluorocarbons reach the stratosphere and release Cl radicals. Predict long-term changes in ozone column density over Antarctica, including recovery trends under scenarios of controlled emissions. Include seasonal sunlight variations, catalytic cycles, partitioning between reactive and reservoir species, and potential feedbacks from stratospheric temperature changes. Discuss uncertainties in modeling stratospheric transport and aerosol-mediated heterogeneous reactions.

39. Urban aerosols contain both organic and inorganic fractions. Estimate their influence on heterogeneous chemistry, cloud microphysics, and radiative forcing. Include particle hygroscopicity, surface reactivity, and light-scattering properties. Assess feedbacks on photolysis rates, radical budgets, and regional climate. Evaluate how seasonal and diurnal variations in emissions modify aerosol impacts.

40. During a heatwave, forest soils emit increased VOCs and NO_x. Calculate enhanced formation of tropospheric ozone and secondary organic aerosols, incorporating temperature-dependent reaction rates, vertical mixing, and boundary layer dynamics. Include feedback from radical propagation, PAN formation, and heterogeneous uptake. Evaluate implications for regional air quality and potential health risks.

41. Methane oxidation produces formaldehyde and eventually CO_2. Using a continental-scale box model, estimate changes in hydroxyl radical concentration and the subsequent feedback on VOC and CO oxidation rates. Include diurnal sunlight variations, temperature-dependent reaction rates, and interactions with NOx. Assess implications for regional tropospheric oxidative capacity and greenhouse gas budgets.

42. Polar stratospheric clouds facilitate Cl_2 activation. Quantify ozone destroyed per square kilometer during the onset of spring sunlight. Include PSC surface area, halogen abundance, temperature-dependent reaction kinetics, and solar zenith angle effects. Discuss uncertainties arising from cloud microphysics, halogen partitioning, and stratospheric mixing.

43. A wildfire injects pyrogenic aerosols into the upper troposphere. Model their radiative effects, cloud condensation nuclei formation, and potential modification of precipitation patterns in downwind regions. Include feedbacks on temperature, radical propagation, and secondary pollutant formation. Evaluate regional climate impacts and implications for atmospheric chemistry.

44. Acid rain modifies freshwater chemistry. Calculate the precipitation-induced mobilization of Fe^{3+}, Al^{3+}, and Ca^{2+} from soils into lakes. Predict impacts on pH, phytoplankton growth, and water quality over a one-month period. Include buffering by carbonate species, hydrological inputs, and seasonal rainfall variations. Discuss implications for aquatic ecosystems and fisheries.

45. Ocean upwelling delivers nutrients to the euphotic zone. Model coupled effects on phytoplankton productivity, CO_2 sequestration, oxygen consumption, and production of climate-relevant trace gases. Include vertical mixing, light limitation, and grazing by zooplankton. Evaluate seasonal variability and potential feedbacks on regional climate.

46. Industrial halogen emissions catalyze ozone depletion in the stratosphere. Using measured Cl and Br levels, simulate catalytic cycles, ozone loss rates, and projected recovery under gradual phase-out scenarios. Include partitioning between reactive and reservoir species, stratospheric transport, and seasonal sunlight variations. Discuss uncertainties and implications for international policy.

47. Urban photochemical smog interacts with incoming solar radiation. Quantify the effects on local temperature, ozone formation rates, and hydroxyl radical abundance. Incorporate temperature-dependent photolysis, radical recycling, VOC speciation, and boundary layer dynamics. Evaluate implications for human health, urban climate, and mitigation strategies.

48. Volcanic sulfate aerosols influence climate by scattering sunlight. Calculate surface temperature changes and modifications to cloud albedo, including particle growth, deposition rates, and seasonal circulation patterns. Incorporate aerosol microphysics, stratosphere-troposphere interactions, and uncertainties in emission height and plume composition.

49. Riverine inputs of nitrate and phosphate stimulate coastal algal blooms. Model biogeochemical consequences on dissolved oxygen, trace metal cycling, and greenhouse gas emissions over a seasonal cycle. Include nutrient limitation, hydrodynamic mixing, stratification, and microbial processes. Predict implications for eutrophication, hypoxia, and coastal fisheries.

50. A stratospheric warming event alters polar stratospheric cloud formation. Predict subsequent effects on chlorine activation, ozone depletion magnitude, and recovery trends in both polar and mid-latitude regions. Include temperature-dependent reaction kinetics, PSC microphysics, halogen partitioning, and solar flux variations. Discuss uncertainties in stratospheric circulation, chemical feedbacks, and implications for long-term ozone recovery.

Chapter 22

The World of s- and p-Block Elements

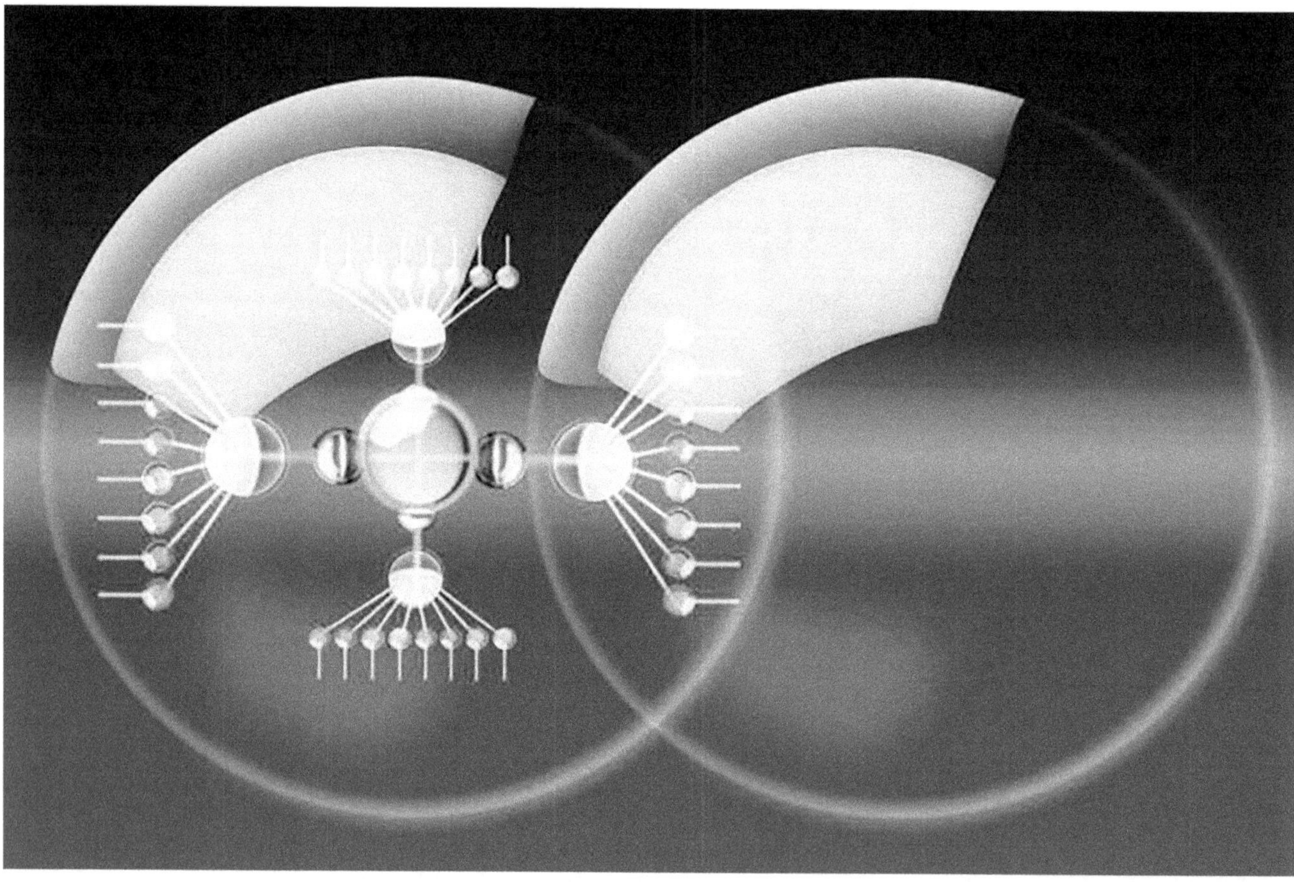

Hydrogen, the lightest and most abundant element in the universe, is an odorless and colorless gas found throughout space and within stars. On Earth, hydrogen is scarce in the atmosphere but abundant in water molecules that make up oceans and seas. Its applications are diverse, including use as rocket fuel, in the hydrogenation of fats, and in petroleum desulfurization. Helium, the second lightest element, is also an odorless, colorless, and inert gas, primarily found in its elemental form. It ranks as the second most abundant element in the universe but is rare on Earth, making up only a minor component of the atmosphere. Helium finds use in balloons, blimps, nuclear plant cooling, and leak detection.

Lithium is a soft, silver-white, and highly reactive metal that does not exist in nature in its pure form due to its rapid reaction with air and water. It is extracted from the mineral spodumene and is stored in oil or kerosene. Lithium is crucial in batteries for pacemakers and can be combined with other elements to form alloys used in space technology. Beryllium, a steel-gray, strong, lightweight, and brittle metal, is too reactive to occur freely in nature and is primarily sourced from the mineral beryl. It is used in X-ray tube windows, watch springs, and heat-conducting ceramics. Boron, existing either as a brown powder or hard black crystalline form, is rare and never found in its elemental state. It is a vital dopant for semiconductors, playing a key role in transistor manufacturing, and is also incorporated into materials such as tennis rackets and heat-resistant glass.

Carbon exists in several forms, from diamond-clear to black graphite. Though it constitutes a small fraction of Earth's crust, it is the most abundant element in nature and a fundamental component of all plants and animals. Carbon is present in coal, petroleum, perfumes, carbohydrates, proteins, and enzymes, and is used in plastics, pencils, and steel. Nitrogen, a colorless and odorless inert gas, dominates Earth's atmosphere and is essential in agricultural and industrial compounds, including ammonia, rocket fuels, and cryogenic applications. Oxygen, also a colorless gas visible as a pale blue liquid, constitutes less than half a percent of Earth's crust but is critical for life, fuel combustion, steel production, and water purification. It is continuously replenished in the atmosphere through photosynthesis.

Fluorine is a highly reactive, pale yellow gas that becomes bright yellow in liquid form. It is found in minerals such as fluorite and fluorspar and in small quantities in teeth, bones, blood, and seawater. Fluoride derived from fluorine is used in steelmaking, aluminum refining, toothpastes, and water fluoridation. Neon is a colorless, inert gas present in trace amounts in the atmosphere. It emits an orange-red glow in discharge tubes and is used in television tubes, lasers, detectors, and neon signage. Sodium is a highly reactive silvery-white metal that does not occur freely in nature and is stored under kerosene. It is found in the Earth's crust and used in alloys for nuclear reactors and batteries, as well as in streetlights. Magnesium, a shiny gray and malleable metal, is lightweight and reactive, sourced from magnesite and seawater. It has important structural applications, including in bicycle alloys.

Aluminum, a silvery-gray metal, is the most abundant metal in the Earth's crust and is derived from bauxite, found predominantly in tropical regions. It is commercially significant for window frames, foil, doorknobs, and alloys used in automobiles and airplanes. Silicon appears crystalline and reflective, often with a light blue sheen, and is the second most abundant element in Earth's crust. It is used in microchips, solar cells, and compounds such as silica, which occurs in crystalline quartz and non-crystalline flint. Phosphorus exists in various forms, including white, yellow, scarlet, red, violet, and black, and is present in phosphate rocks and fluoroapatite. It is essential in fertilizers, toothpastes, and detergents.

Sulfur is a yellow, crystalline solid found in ores and minerals, used in matches, fireworks, batteries, rubber compounds, and hair products. Chlorine, a pale yellow-green gas, is found in seawater and salt deposits and is employed for water purification, bleaching, and plastics production. Argon, a colorless, inert gas, is a minor component of the atmosphere, utilized in light bulbs, lasers, and Geiger counters. Potassium is a highly reactive, silvery-gray metal not found freely in nature, used in fertilizers, matches, gunpowder, and glass lenses. Calcium, a dull gray and soft metal, is abundant in the Earth's crust and present in teeth, bones, and sedimentary rocks. It forms compounds such as calcium carbonate, used in chalk, marble, and limestone.

Gallium, a silver-white metal, is notable for expanding when frozen and is used in quartz thermometers, tumor detection compounds, and gallium-arsenide laser diodes. Germanium, a grayish-white semiconductor, is used in transistors, wide-angle lenses, infrared night vision devices, and fiber optics. Arsenic, metallic gray, is a poor conductor found in minerals such as realgar and orpiment. It is employed in solid-state electronics and light-emitting diodes. Selenium exists in black or red forms and is used in light meters, photocopiers, and solar cells, as well as in anti-dandruff shampoos. Bromine, a red-brown element, is used in disinfectants, photographic film, and gasoline additives.

Krypton and xenon are colorless noble gases, with krypton used in neon lights and fluorescent bulbs, and xenon in ultraviolet and projection lamps. Rubidium, a gray-white metal, is stored in kerosene and used in television tubes, vacuum tubes, and heart research. Strontium, metallic and silvery-white, produces red flames in fireworks and is used in nuclear batteries, highway flares, and phosphorescent paint. Indium is a silvery, corrosion-resistant element used in alloys, solar cells, mirrors, photocells, and transistors. Tin, a silvery and malleable metal, is mainly used in alloys such as bronze, solder, and pewter, as well as in coins and organ pipes.

Antimony, a silvery-gray, poor conductor of electricity, is derived from stibnite and used in safety matches, ceramic glazes, and fire retardants. Tellurium, silvery and rare, combines with gold and is used in battery plates, percussion caps, and vulcanized rubber. Iodine, a violet nonmetallic element, is found in seaweed and brine wells, and is used as a disinfectant, in halogen lamps, and in ink pigments. Cesium, a soft silvery-gold metal, is obtained from pollucite and used in photoelectric cells and scintillation detectors. Barium, silvery-gray and abundant, is used in spark plugs, white pigments, and green fireworks.

Thallium, a silvery-white, carcinogenic, and scarce metal, is obtained as a by-product of lead and zinc refining and is used in heart muscle research. Lead, metallic gray and malleable, is obtained from galena and is used for radiation shielding, electrodes, solder, and pigments. Bismuth, a heavy, lustrous, and brittle metal, resists corrosion and is found in nature and in ores such as bismite. It is used as a catalyst in rubber production and in antacids. Polonium, a rare silvery metal, is obtained from uranium and thorium decay and finds applications in nuclear batteries and photographic film cleaning. Astatine is likely a black, metallic halogen used in alpha particle radiotherapy.

Radon, a colorless, dense noble gas, is produced from uranium and thorium decay and is used in cancer therapy and earthquake prediction. Francium, an extremely unstable metallic alkali element, is produced artificially in nuclear reactors. Radium, silvery-white and metallic, occurs in trace amounts from pitchblende and is the last of the alkaline earth elements, used in glow-in-the-dark paints and in medicine as radium chloride for cancer treatment.

The Periodic Table of Elements is a systematic chart that organizes all known elements by atomic number, reflecting their specific identity. While some elements have been known since ancient times, it was not until the late 1700s that scientists began grouping them by basic properties. In the 1800s, elements were arranged into triads and later by increasing atomic weight, leading to the Law of Octaves. Mid- to late-1800s refinements allowed larger columns and rows, though gaps remained. Eventually, elements were accurately arranged by atomic number, with new elements discovered or synthesized in laboratories filling these gaps.

The Alkali Metals of Group IA, despite their high reactivity, have become indispensable in modern technology. Lithium, for instance, has revolutionized energy storage as the core component of rechargeable lithium-ion batteries, powering everything from smartphones to electric vehicles. Sodium and potassium, while too reactive for pure-metal applications in everyday technology, play critical roles in industrial chemical processes. Sodium vapor lamps provide efficient street lighting, while potassium compounds are essential fertilizers that sustain global agriculture. Rubidium and cesium, rarer and more expensive, are used in precision instruments such as atomic clocks, where their unique atomic transitions allow timekeeping to unprecedented accuracies. Even francium, though extremely rare and radioactive, offers insight into nuclear behavior and fundamental atomic interactions.

Alkaline Earth Metals, the elements of Group IIA, are equally pivotal in both natural and engineered systems. Calcium, beyond its biological role in bone and teeth formation, finds applications in metallurgy as a reducing agent and in cement and concrete production. Magnesium, light and strong, is widely used in aerospace and automotive industries, while also functioning in chlorophyll for photosynthesis, thus sustaining life on Earth. Strontium compounds color fireworks with bright reds, while barium is used in medical imaging to highlight structures in the digestive tract. Radium, historically employed in luminescent paints, now serves as a key isotope in targeted radiation therapy, demonstrating the convergence of chemistry and medicine.

The Boron Family (Group IIIA) exemplifies versatility across technological and biological domains. Boron itself is used in the manufacture of borosilicate glass, which resists thermal shock, as well as in detergents and agriculture. Aluminum, the most familiar member, is lightweight, corrosion-resistant, and conductive, making it essential in transportation, construction, and packaging. Gallium, with its low melting point, has become vital in semiconductors and LEDs, illustrating the profound connection between periodic trends and electronic properties. Indium is employed in touchscreens and solar panels, while thallium, despite its toxicity, has specialized uses in electronics and medical imaging. These elements underscore how periodic properties—valence electrons, metallic character, and reactivity—translate directly into practical applications.

Carbon, the cornerstone of organic chemistry, extends the influence of the Periodic Table into nearly every facet of modern life. Beyond its familiar role in hydrocarbons, carbon forms the basis of polymers, advanced materials like graphene and carbon nanotubes, and even pharmaceuticals. Silicon, closely related in Group IVA, dominates the electronics industry as the primary semiconductor material, while germanium, tin, and lead have specialized applications ranging from fiber optics to soldering and radiation shielding. The interplay of chemical stability, electronic configuration, and bonding versatility allows these elements to bridge the gap between chemistry, engineering, and materials science.

Nitrogen and phosphorus, the central members of the Nitrogen Family (Group VA), are vital in biological systems and agriculture. Nitrogen, though inert in its diatomic form, becomes a life-sustaining nutrient through processes such as nitrogen fixation, which convert it into bioavailable ammonia and nitrates. Phosphorus, crucial in DNA, RNA, and ATP, also underpins fertilizers that enhance crop yields globally. Arsenic and antimony, while toxic in large quantities, serve in semiconductors and fire retardants, highlighting how controlled chemical application transforms hazards into functional technologies. Bismuth, the heaviest stable member, finds uses in cosmetics, pharmaceuticals, and low-melting alloys, combining chemical uniqueness with practical utility.

Oxygen, sulfur, selenium, and tellurium, comprising the Oxygen Family (Group VIA), illustrate the diversity of chemical roles. Oxygen is fundamental for combustion, respiration, and water formation, while sulfur underpins vulcanization of rubber, fertilizer production, and pharmaceuticals. Selenium, a trace nutrient, supports enzymatic antioxidant defenses, whereas tellurium enhances the performance of semiconductors and thermoelectric materials. Polonium, highly radioactive, serves as a neutron source in research, demonstrating how the periodic table bridges life sciences, materials science, and nuclear chemistry.

The Halogens (Group VIIA) demonstrate exceptional chemical reactivity, which chemists exploit for sanitation, medicine, and materials production. Fluorine strengthens dental enamel and enables the production of Teflon, a heat- and chemical-resistant polymer. Chlorine, historically associated with water purification, remains essential for disinfectants, PVC production, and bleaching agents. Bromine is used in flame retardants and photography, iodine supports thyroid health, and astatine, though rare and radioactive, is investigated in targeted cancer therapies. These elements exemplify how understanding periodic trends in electronegativity and reactivity enables precise manipulation of matter for human benefit.

Noble Gases (Group VIIIA), while largely chemically inert, have unique applications arising from their stability. Helium cools superconducting magnets and protects sensitive reactions from air contamination. Neon, argon, krypton, and xenon light up signage, provide inert atmospheres for welding, and enhance imaging techniques in medicine. Xenon even serves as a propellant in ion thrusters for spacecraft, demonstrating that inertness does not imply inactivity but rather controlled utility in specialized contexts. Radon, though radioactive, is studied for its role in geophysics and targeted radiation therapy.

Transition metals, spanning Groups IB through VIIIB, are the backbone of catalysis, metallurgy, and bioinorganic chemistry. Iron, cobalt, and nickel form magnets and steel alloys, copper conducts electricity, while precious metals like gold and platinum serve both decorative and industrial purposes. Their variable oxidation states and ability to form coordination complexes underpin industrial catalysis, from the Haber-Bosch process for ammonia production to hydrogenation of vegetable oils. Lanthanides and actinides, the f-block elements, enable advanced technologies such as permanent magnets, phosphors in lighting, and nuclear reactors, bridging chemistry, physics, and engineering.

In essence, the Periodic Table is a guidebook for both fundamental science and practical applications. By studying its organization, trends, and elemental characteristics, chemists can predict reactions, design materials with targeted properties, and explore the frontiers of technology and medicine. From everyday metals and gases to rare and radioactive elements, the table encapsulates the unity of chemical knowledge, illustrating the profound connection between atomic structure and the observable universe. The ongoing discovery of new elements and the development of innovative applications ensure that the Periodic Table remains a living framework, continuously guiding scientific exploration and technological advancement.

Chapter 22: Word Problems

1. Hydrogen gas is being used as a propellant in a rocket experiment designed to study high-altitude combustion efficiency. The experiment involves injecting 10 g of hydrogen into a controlled chamber and igniting it with pure oxygen, while measuring the energy release, exhaust velocity, and temperature profile. Given that the standard enthalpy of combustion of H_2 is –286 kJ/mol, calculate the total energy released and discuss how the energy density of hydrogen compares to traditional hydrocarbon fuels like RP-1. Further, consider the implications of hydrogen's low molecular weight on exhaust velocity and specific impulse, and describe safety precautions necessary due to its high flammability.

2. Helium is utilized in cooling superconducting magnets in a particle accelerator. A 2-mole sample of helium gas at 20 K is to be cooled to its boiling point of 4.2 K using a cryogenic system. Considering the specific heat capacity of helium (5.193 J/mol·K) and the latent heat of vaporization, calculate the total heat that must be removed. Additionally, evaluate why helium is preferred over liquid nitrogen despite higher costs, taking into account its low boiling point, chemical inertness, and its role in minimizing vibrations and thermal noise in superconducting systems.

3. Lithium-ion batteries are being developed for high-capacity electric vehicles. A single cell contains 0.5 moles of lithium in the form of LiC_6, which can be fully oxidized to Li^+ during discharge. Calculate the theoretical total charge that could be delivered by this amount of lithium. Compare this to a sodium-ion analog in terms of charge density, molar mass, and energy storage capacity, and discuss how the choice of metal affects battery voltage, efficiency, and thermal management in high-performance automotive applications.

4. Beryllium is used as a window material in high-energy X-ray tubes due to its low atomic number and minimal absorption. For a window 0.1 mm thick and with a density of 1.85 g/cm^3, calculate the mass per unit area. Discuss how the thickness and mechanical strength of beryllium affect X-ray transmission efficiency, thermal stress tolerance, and long-term durability. Additionally, consider the handling challenges due to beryllium's toxicity and the precautions necessary for safe manufacturing and use in medical imaging.

5. In semiconductor fabrication, boron is used to dope silicon wafers to create p-type regions. Consider a silicon wafer with an area of 10 cm^2 and thickness

0.5 mm, doped to a concentration of 10^{16} atoms/cm^3. Calculate the total number of boron atoms incorporated. Discuss how this level of doping modifies electrical conductivity, charge carrier mobility, and junction behavior, and explain why precise control over dopant concentration is critical in modern integrated circuits used in high-speed computing applications.

6. Carbon nanotubes are being tested as reinforcement fibers in advanced composite materials. A nanotube with 1 μm diameter and 1 cm length is subjected to tensile testing, with a reported tensile strength of approximately 60 GPa. Calculate the maximum load the nanotube can withstand before failure. Compare this with conventional steel wires of similar diameter, and discuss the implications for lightweight structural materials in aerospace, robotics, and high-performance sports equipment, considering both mechanical properties and potential manufacturing challenges.

7. Nitrogen fixation is a critical step in sustainable agriculture. Industrially, N_2 is converted into NH_3 via the Haber-Bosch process, while in nature, bacteria perform biological fixation. If you wish to fix nitrogen from 100 kg of air (approximately 78% N_2 by volume) using hydrogen, calculate the moles of H_2 required for complete reaction according to $N_2 + 3\,H_2 \rightarrow 2\,NH_3$. Compare energy efficiency, environmental impact, and cost between industrial catalytic and biological methods, and discuss strategies for reducing greenhouse gas emissions associated with synthetic nitrogen production.

8. Oxygen plays a critical role in steelmaking, where it is used to oxidize impurities and regulate the carbon content of molten iron. If 5 moles of O_2 react with molten iron to form Fe_2O_3, calculate the mass of oxygen consumed. Discuss how oxygen interacts with other alloying elements, how its solubility in molten metal influences microstructure, and why precise oxygen control is essential for achieving desired mechanical properties in steel used for construction or automotive applications.

9. Fluorine is highly reactive and is used industrially to produce fluorosilicates and in uranium enrichment processes. If 50 g of SiO_2 reacts with F_2 to produce SiF_4 gas, calculate the theoretical yield of SiF_4. Discuss the industrial and laboratory safety challenges associated with handling fluorine, the material compatibility issues in reactors and piping, and the role of fluorination in modifying chemical and physical properties of silicon-based materials.

10. Neon lamps are widely used for signage due to their characteristic orange-red emission at 585 nm. Calculate the energy of a single photon emitted by a neon lamp. Explain how this energy relates to electron transitions within neon atoms, discuss the quantum efficiency and intensity of light emission, and evaluate how the lamp's performance is affected by pressure, tube length, and applied voltage.

11. Sodium metal reacts violently with water to form sodium hydroxide and hydrogen gas. If 10 g of sodium is dropped into water, calculate the volume of hydrogen gas generated at STP. Compare the reactivity and energy release with potassium under identical conditions, and discuss the thermodynamic and kinetic reasons for the difference in reactivity. Consider the implications for industrial-scale alkali production and laboratory safety protocols.

12. Magnesium is valued for aerospace alloys due to its combination of low density, high strength, and corrosion resistance. A component made of magnesium weighs 2 kg and has a volume of 0.75 L. Calculate the material's density and compare it with aluminum and titanium. Discuss trade-offs in selecting magnesium for aerospace applications, considering fatigue strength, weldability, oxidation behavior, and potential use in hybrid composites for aircraft or spacecraft.

13. Aluminum extraction from bauxite using the Bayer process is energy-intensive. If 1000 kg of bauxite containing 50% Al_2O_3 is processed with 90% efficiency, calculate the mass of aluminum produced. Discuss energy requirements per kilogram of aluminum, compare them to magnesium extraction, and explain why recycled aluminum is significantly more energy-efficient, including considerations for environmental impact and industrial scaling.

14. Silicon solar cells rely on the semiconductor properties of crystalline silicon. For a 1 m^2 panel exposed to 1000 W/m^2 solar radiation, calculate the electrical power output if the cell efficiency is 18%. Discuss how impurities and crystallinity affect efficiency, the role of passivation layers, and the challenges of scaling up silicon photovoltaic technology while maintaining cost-effectiveness and performance consistency.

15. Phosphorus is a vital nutrient in fertilizers, often sourced from phosphate rock ($Ca_3(PO_4)_2$). If 50 kg of phosphate rock is processed to produce superphosphate fertilizer, calculate the mass of elemental phosphorus available for crops.

Discuss environmental consequences of phosphate overuse, including eutrophication of waterways, and strategies for sustainable phosphorus recycling in agriculture.

16. Sulfur is used in vulcanization of rubber, which improves elasticity and thermal stability. For a 100 kg batch of natural rubber, determine the mass of sulfur needed for 3% vulcanization by weight. Discuss how sulfur cross-linking affects mechanical properties, the kinetics of vulcanization, and potential environmental or health concerns associated with sulfur-containing industrial processes.

17. Chlorine is widely used to disinfect municipal water supplies. If a treatment plant treats 10,000 L of water and requires 1 ppm residual chlorine, calculate the mass of Cl_2 needed. Compare chlorine's effectiveness with other halogens such as bromine and iodine in water treatment, and discuss the formation of by-products, regulatory limits, and health implications of disinfection processes.

18. Argon provides an inert atmosphere for welding and other high-temperature processes. If 50 L of argon at STP is used per weld, calculate the mass of argon consumed in 20 welds. Discuss how argon's density, thermal conductivity, and inertness prevent oxidation of molten metals, and evaluate the economic and environmental implications of large-scale industrial argon use.

19. Potassium nitrate is used in fireworks to produce colored flames. If a pyrotechnics lab uses 100 g of KNO_3 to produce a red flame with strontium nitrate, calculate the moles of strontium required assuming complete reaction. Discuss how stoichiometry, flame temperature, and metal ion excitation determine the intensity and hue of the flame, and the safety measures necessary during mixing and ignition.

20. Calcium carbonate is used in cement production, which releases CO_2 during thermal decomposition. For a batch containing 500 kg of $CaCO_3$, calculate the mass of CO_2 released upon heating to form CaO. Discuss strategies for minimizing carbon emissions in construction materials, including alternative binders, carbon capture, and sustainable cement formulations.

21. Gallium is notable for its unusual property of expanding upon freezing. In a laboratory experiment, 100 g of gallium is cooled from its liquid state to solid at 29.76 °C. Calculate the increase in volume, given the density of liquid gallium is

6.095 g/cm³ and solid gallium 5.907 g/cm³. Discuss how this property is exploited in precise temperature measurement devices such as high-accuracy thermometers and thermal switches, and the challenges in handling and containment to prevent structural damage during solidification.

22. Germanium is an important semiconductor for infrared optics and transistors. A cubic 1 cm³ germanium crystal with a resistivity of 0.46 Ω·cm is used in a sensor application. Calculate the electrical resistance between opposite faces of the cube. Discuss how intrinsic versus doped germanium affects conductivity, the temperature dependence of resistance, and the implications for high-speed electronic and optoelectronic device performance.

23. Arsenic is employed in III-V semiconductors like GaAs for high-frequency devices. For a GaAs wafer with area 1 cm² and thickness 0.5 mm, doped with arsenic at 10^{15} atoms/cm³, calculate the total number of arsenic atoms in the wafer. Explain how such doping levels influence the electronic band structure, carrier concentration, and device behavior in lasers, LEDs, and high-speed transistors, including the effects of diffusion and activation during annealing.

24. Selenium is used in photovoltaic cells and photoelectric devices. A small solar cell of area 0.01 m² generates 0.5 W under illumination. Assuming 100% quantum efficiency, calculate the photon flux incident on the cell, given selenium's work function is 2.07 eV. Discuss why real-world efficiencies are lower due to recombination losses, surface defects, and imperfect light absorption, and explain strategies to improve selenium-based photovoltaic performance.

25. Bromine is a reactive halogen used in organic synthesis and flame retardants. In an experiment, 50 g of Br_2 is reacted with ethene in an addition reaction to form 1,2-dibromoethane. Calculate the number of moles of ethene that can react completely. Discuss the reaction kinetics, solvent effects, and safety considerations when handling bromine, as well as its industrial applications in polymers and gasoline additives.

26. Krypton is a noble gas used in fluorescent lamps. A lamp contains 0.5 L of krypton gas at STP. Calculate the number of krypton atoms in the lamp and discuss the importance of gas purity for efficient light emission. Additionally, evaluate the differences in light spectrum and intensity between krypton and neon lamps,

and the considerations for lamp lifetime and energy efficiency in industrial and commercial applications.

27. Rubidium, a highly reactive alkali metal, is stored under kerosene for safety. If 2 g of rubidium is accidentally exposed to air, calculate the energy released upon oxidation, given standard enthalpy data. Discuss the chemical reactions that occur, the formation of oxides and peroxides, and the laboratory precautions required when handling rubidium and similar highly reactive metals.

28. Strontium compounds are used to produce bright red colors in fireworks. A pyrotechnic composition contains 100 g of KNO_3 mixed with strontium nitrate for flame coloration. Calculate the moles of strontium nitrate needed for a complete reaction. Discuss the effects of flame temperature, metal ion excitation, and stoichiometry on color intensity, as well as safety protocols in handling oxidizers and fuel mixtures.

29. Indium tin oxide (ITO) is widely used in touchscreens and solar panels due to its transparency and conductivity. A coating weighs 100 g and contains 90% In_2O_3 and 10% SnO_2 by mass. Calculate the weight percentage of indium in the coating and discuss the trade-offs between conductivity, optical transparency, and mechanical robustness, as well as the role of dopants and thin-film deposition techniques.

30. Tin is used in solder alloys with lead for electronics. A 50 g solder sample contains 60% Sn and 40% Pb. Calculate the number of moles of each metal in the alloy. Discuss how the composition affects melting point, wettability, mechanical strength, and thermal cycling performance in microelectronic applications, including considerations for lead-free alternatives due to environmental regulations.

31. Antimony trioxide (Sb_2O_3) is used as a flame retardant in polymers. If 25 kg of Sb_2O_3 is incorporated into 1 ton of polymer, calculate the weight percent and discuss its influence on thermal decomposition, smoke suppression, and synergistic effects with halogenated additives. Evaluate safety, regulatory considerations, and emerging alternatives in polymer engineering.

32. Tellurium is used in thermoelectric devices to convert heat gradients into electricity. For a device containing 5 g of Te, calculate the number of moles and discuss how doping with Se or Sb modifies electrical conductivity, thermal conductivity,

and Seebeck coefficient. Evaluate applications in waste-heat recovery and low-temperature energy harvesting.

33. Iodine is used as a disinfectant in medical and industrial contexts. A 2 L solution contains 0.05 M I_2. Calculate the total mass of iodine in the solution. Discuss the chemical mechanism of antimicrobial action, differences between free iodine and iodophors, and challenges in maintaining stability, bioavailability, and safety in various applications.

34. Cesium-based atomic clocks rely on precise electronic transitions. If the cesium hyperfine transition corresponds to 9.19×10^9 Hz, calculate the energy of a single photon involved. Discuss how this property enables extreme timekeeping precision, the significance of quantum state manipulation, and the environmental and technological factors that influence clock stability.

35. Barium sulfate is widely used in medical imaging as a radiopaque contrast agent. A 500 mL suspension contains 100 g of $BaSO_4$. Calculate the molarity of barium ions and discuss the chemical and physical properties that make $BaSO_4$ suitable for X-ray imaging, including density, insolubility, and particle size considerations for safety and diagnostic effectiveness.

36. Thallium is a toxic metal with specialized applications in medical research. If a solution contains 0.01 M Tl^+, calculate the concentration in mg/L. Discuss the importance of careful handling, laboratory safety protocols, and potential alternatives for biological or electronic applications where thallium might otherwise be used.

37. Lead shields are used to protect against radiation in medical and nuclear settings. A 5 cm thick lead barrier (density 11.34 g/cm^3) is to cover 1 m^2 of wall. Calculate the total mass of lead required. Discuss concepts of half-value layer, radiation attenuation, and practical considerations for designing barriers in hospitals, laboratories, and nuclear reactors.

38. Bismuth is used in low-melting alloys for medical devices, cosmetics, and electronics. An alloy contains 50% Bi, 25% Pb, and 25% Sn by mass. For a 200 g batch, calculate the mass of each constituent. Discuss how the alloy composition affects melting point, corrosion resistance, and thermal expansion, and how these properties are leveraged in specialty applications.

39. Polonium-210 is a radioactive alpha emitter used in nuclear batteries. Calculate the decay rate in becquerels (Bq) for 1 μg of Po-210, given its half-life of 138 days. Discuss safety protocols, shielding requirements, and the mechanism by which alpha decay can be harnessed for energy generation.

40. Astatine, though extremely rare, is used experimentally in targeted alpha-particle radiotherapy. If a 0.5 MBq source is required, calculate the mass of astatine needed assuming 100% isotopic purity. Discuss the practical challenges associated with astatine's short half-life, scarcity, and safe handling in clinical settings.

41. Radon accumulation in residential basements can pose health hazards. A basement of 50 m^3 contains air with 200 Bq/m^3 of radon activity. Calculate the total activity in the room. Discuss mitigation strategies such as ventilation, sealing of foundation cracks, and air filtration, as well as the geophysical sources of radon and implications for long-term exposure.

42. Francium is produced artificially in nuclear reactors and has extremely short-lived isotopes. Estimate the number of atoms present in a 1 μg sample, and discuss why its extreme radioactivity and scarcity limit any practical applications outside fundamental nuclear research, including studies of alkali metal periodic trends and atomic structure.

43. Radium-226 decays to radon-222 and has historical use in luminescent paints. For 1 mg of Ra-226, calculate the activity in becquerels. Discuss the health hazards associated with radium exposure, current regulations, and safer alternatives in radioluminescent applications and cancer treatment.

44. Iron is used as a catalyst in the Haber-Bosch process for ammonia synthesis. If 100 moles of nitrogen are reacted with hydrogen and the reaction yield is 85%, calculate the moles of ammonia produced. Discuss how the iron catalyst facilitates nitrogen dissociation, optimizes energy efficiency, and how industrial parameters like temperature and pressure affect reaction rate and selectivity.

45. Cobalt is an essential component of $LiCoO_2$ batteries. If a battery contains 5 g of cobalt, calculate the number of moles of Co. Discuss resource sustainability, geopolitical considerations, and strategies for recycling cobalt from spent batteries, while maintaining performance in high-energy-density applications.

46. Copper wires are widely used in electrical transmission. A 2 m long wire has a cross-sectional area of 1 mm^2. Calculate its electrical resistance given the resistivity of copper ($1.68 \times 10^{-8}\ \Omega\cdot m$). Discuss how resistance impacts power loss, heating, and efficiency in power distribution, and compare copper with alternative conductive materials like aluminum or silver.

47. Nickel is employed as a hydrogenation catalyst in fine chemical synthesis. A 10 g sample of nickel is formed into nanoparticles of 5 nm diameter. Estimate the total surface area available for catalysis and discuss how nanostructuring enhances reaction rates, selectivity, and efficiency, including practical considerations for catalyst recovery and reuse.

48. Platinum is a precious metal used in catalytic converters to reduce automotive emissions. A converter contains 0.1 g of Pt. Estimate the number of active surface sites assuming 1 surface atom per 1 nm^2 and discuss how these sites facilitate oxidation of CO and hydrocarbons. Evaluate the impact of particle size, dispersion, and temperature on catalytic performance.

49. Uranium-235 is a nuclear fuel in reactors. Calculate the energy released from the fission of 1 g of U-235, assuming 200 MeV per fission. Discuss the conversion of nuclear energy to heat and then to electricity, neutron economy, and the implications for reactor design, including fuel enrichment and safety considerations.

50. Lanthanides such as neodymium are used in permanent magnets. A $Nd_2Fe_{14}B$ magnet weighs 50 g and contains 30% neodymium by mass. Calculate the mass of Nd in the magnet and discuss how the crystal structure and electron configuration contribute to high magnetic coercivity, remanence, and applications in electric motors, wind turbines, and data storage devices.

Chapter 23

The World of d- and f-Block Elements

Scandium, a silvery-white metal, remains one of the more elusive elements on Earth due to its scarcity in the crust, though it shines abundantly in the heart of stars, revealing its cosmic origin. Its lightness combined with remarkable resistance to corrosion has made it highly valuable in specialized alloys, particularly in aerospace engineering where every gram counts. The addition of scandium to aluminum produces materials that are both lightweight and strong, ideal for constructing aircraft frames and missile components. Beyond aerospace, scandium enhances the efficiency of solid-state lighting, providing brighter and more stable illumination in stadiums and large-screen televisions. In its pure metallic form, scandium can even serve as a sensitive leak detector in scientific and industrial applications, demonstrating how the unique characteristics of even trace elements can have significant technological impact.

Titanium, another silvery-gray metal, shares the lightweight nature of scandium but surpasses it in mechanical strength and thermal stability. Extracted primarily from the minerals rutile and ilmenite, titanium's resilience against corrosion and its ability to withstand extreme temperatures render it invaluable for aerospace, chemical processing, and medical applications. Rockets, jet engines, and high-performance heat exchangers frequently rely on titanium alloys to maintain integrity under stress and temperature fluctuations. In medicine, titanium is biocompatible, which allows it to form the structural basis for bone pins, joint replacements, and dental implants. Beyond industrial and medical uses, titanium compounds color paints and paper, demonstrating the versatility of this robust yet aesthetically neutral metal. Its enduring presence across multiple domains underscores the importance of metals that combine structural performance with chemical inertness.

Vanadium presents a unique profile, showing hues of blue, silver, and gray, and while it occurs in small concentrations in the Earth's crust, its extraction from minerals such as vanadinite has industrial significance. Highly resistant to corrosion, vanadium is primarily used to improve the strength and durability of steel. Even small additions of vanadium can dramatically enhance the hardness and wear resistance of steel alloys, making it indispensable for tools, construction materials, and components in jet engines. The element's ability to stabilize alloys at elevated temperatures extends its applications beyond simple reinforcement, allowing engineers to design materials capable of withstanding demanding operational environments. Vanadium compounds also find niche applications in catalysis and chemical synthesis, revealing the diverse ways in which elemental properties can be harnessed.

Chromium, recognized for its hard, brittle, and silvery metallic appearance, is primarily obtained from the mineral chromite and plays a vital role in both aesthetic and functional applications. Chromium plating imparts corrosion resistance and a mirror-like finish to automotive bumpers, appliances, and hardware, combining visual appeal with longevity. In stainless steel, chromium's contribution is paramount, preventing oxidation and extending the life of critical infrastructure and everyday tools alike. Its compounds, rich in color, find utility in pigments for paints and coatings, in audio and video recording media, and even in trace amounts in gemstones, reflecting a spectrum of human endeavors from art to technology.

Manganese, a silvery metallic element, does not occur freely in nature but is instead extracted from minerals like pyrolusite and deep-sea manganese nodules. While it corrodes in moist air, its value in metallurgy is substantial. Manganese serves as a vital alloying agent in steel production, improving hardness, tensile strength, and resistance to wear and corrosion. This makes it essential for tools, plowshares, batteries, and construction steels that face constant mechanical stress. The element's presence also plays a role in biological systems, highlighting the intersection of chemical utility with natural processes.

Iron, a lustrous metallic element and the predominant component of the Earth's core, is abundant in minerals such as hematite and magnetite. The production of iron in blast furnaces has underpinned human civilization for millennia, enabling the construction of bicycles, automobiles, bridges, machinery, and even magnets. Its versatility stems from the ability to form a wide range of alloys and compounds, from soft, ductile forms in construction to hardened steels for tools and vehicles. Iron's reactivity and abundance position it as a cornerstone of both industry and biological systems, demonstrating the profound influence a single element can have across multiple scales.

Cobalt, a hard gray metal obtained from cobaltite ore, exemplifies the synergy between corrosion resistance and magnetic properties. It is used to create alloys that resist degradation in steel, serve as sources of gamma radiation in medical and industrial applications, and form permanent magnets essential to electronics and energy technologies. Its ability to combine with other elements to form robust, functional compounds reflects the broader principle that the value of a transition metal is often determined by its interactions with others.

Nickel, another lustrous metal, is mined primarily from millerite ore and used extensively in coins, stainless steel, and rechargeable nickel-cadmium batteries.

Its resistance to corrosion and oxidation ensures longevity, while its ductility allows for diverse manufacturing processes. Nickel's capacity to form alloys with varying properties underscores its role as a fundamental element in both everyday items and high-tech applications.

Copper, celebrated for its warm red-orange metallic sheen, is obtained from chalcopyrite through roasting or electrolysis. Its excellent electrical and thermal conductivity has made it the backbone of electrical infrastructure, from household wiring to complex electronic circuits. Copper is also used in plumbing, heat exchangers, coins, and decorative arts, demonstrating the element's combination of practicality and aesthetic appeal. Its ability to alloy with other metals, such as tin to produce bronze, has historically driven technological advancement for thousands of years.

Zinc, with its silver-gray luster, is commonly sourced from sphalerite and zincblende. Its tendency to form a protective oxide layer makes it less reactive than might be expected and allows it to protect underlying metals from corrosion. Zinc finds application in batteries, gutters, pigments for rubber, and as an alloy in water and gas valves, illustrating the element's role in enhancing durability and functionality in everyday and industrial materials.

Yttrium, a rare silvery-white element, is extracted from monazite sands and has significant technological applications in modern electronics. It is used in compounds for color television screens, radar systems, superconductors, and solid-state lasers, reflecting the intersection of rarity and utility. Its ability to form stable, high-performance materials underpins advancements in imaging, communications, and energy technologies.

Zirconium, another silvery-white metal, is prized for its resistance to corrosion and high temperatures. Its applications span nuclear reactor components, space vehicle parts, catalytic converters, and even gemstones. Its chemical inertness and thermal stability make it essential in environments where failure could be catastrophic, showcasing how elemental properties determine practical applications.

Niobium, a metallic gray element, exhibits a bluish tinge upon oxidation and is obtained from columbite. It plays a critical role in high-temperature superconductivity and is added to steel alloys to enhance strength and durability, especially in pipelines and cutting tools. Molybdenum, similarly gray and metallic, is mined from molybdenite and incorporated into steel to withstand temperature and pressure extremes in aircraft engines and automobile parts, and is used in filaments and as a source of radioisotopes in medical diagnostics.

Technetium, a shiny gray metal absent in nature on Earth, appears in trace amounts in stars and nuclear reactors. Its radioactive isotopes are invaluable in medical imaging, where they trace biological pathways with minimal intrusion. Ruthenium, recovered during platinum refining, is employed as a catalyst, in eye treatments, in measuring eggshell thickness, and in corrosion-resistant alloys, demonstrating the versatile use of rare transition metals. Rhodium, a rare silvery-white metal associated with platinum ores, serves industrial catalytic processes, hardens platinum, and is used in automotive applications, reflecting the dual aesthetic and functional value of these elements.

Palladium, malleable and ductile, functions prominently in catalytic hydrogenation, dental crowns, jewelry, and anti-tumor compounds. Silver, lustrous and widely recognized, is used in mirrors, batteries, silverware, and photographic materials. Cadmium, silvery-blue, is rare, often sourced as a by-product of zinc refining, and finds application in rechargeable batteries and corrosion-resistant electroplating. Lanthanum and cerium, silvery-white rare earths, find roles in lighter flints, camera lenses, studio lighting, self-cleaning ovens, and lamp electrodes, demonstrating that even chemically similar elements can occupy widely different niches.

Praseodymium and neodymium, both metallic and silvery-white, are used in carbon arc lamp electrodes, high-strength magnets, and colored glass. Promethium, artificially produced, emits a blue glow in fluorescent light starters and nuclear batteries. Samarium contributes to high-temperature permanent magnets and specialized glasses, while europium enhances red phosphors for TVs and energy-efficient lighting. Gadolinium, terbium, and dysprosium are central to medical diagnostics, phosphors, magneto-optic alloys, and nuclear applications. Holmium, erbium, thulium, ytterbium, and lutetium offer further applications ranging from lasers to corrosion-resistant alloys, demonstrating the extensive utility of lanthanides.

Hafnium, tantalum, tungsten, and rhenium, extracted from a variety of ores, are all highly corrosion-resistant and heat-stable, serving in nuclear reactors, incandescent lamps, gas removal systems, filaments, cutting tools, and thermocouples.

Osmium, iridium, and platinum exhibit exceptional hardness, corrosion resistance, and utility in high-precision instruments, alloys, satellites, spark plugs, and catalysts. Gold and mercury, while chemically distinct, continue to influence jewelry, electronics, and measurement devices. Actinium, thorium, protactinium, uranium, neptunium, and plutonium, spanning naturally occurring and synthetic radioactive elements, underpin energy production, radiation shielding, and space exploration. Similarly, americium, curium, berkelium, californium, einsteinium, and the superheavy elements illustrate the ongoing expansion of chemical knowledge, enabling research in nuclear physics, synthesis of new elements, and practical applications such as smoke detectors and calibration tools.

The transition elements, located in the center of the Periodic Table, including the families of scandium, titanium, vanadium, chromium, manganese, iron, cobalt, nickel, the coinage metals, and the zinc family, demonstrate a wide spectrum of metallic properties, from ductility and malleability to high-temperature stability and corrosion resistance. The lanthanide and actinide series, forming additional rows below the main table, extend this diversity, each element contributing unique isotopic and chemical properties. Advanced instrumentation now allows precise measurement of atomic masses and the discovery of new elements, while the potential creation of superactinides and superlanthanides promises further expansion of the Periodic Table. These discoveries will likely unlock new compounds, materials, and technological possibilities, with each element acting as a fundamental building block for human innovation. As chemistry progresses, humanity's ability to manipulate, synthesize, and apply elements for energy, medicine, electronics, and environmental solutions will continue to expand, offering seemingly endless avenues for exploration, understanding, and practical application.

As one examines the entirety of the transition, lanthanide, and actinide series, a clear pattern emerges: each element, whether common or rare, natural or synthetic, contributes uniquely to human progress. The chemical versatility of these metals underpins modern technology, energy systems, and medical innovation. Their capacity to form stable alloys, withstand extreme conditions, and catalyze reactions allows engineers and scientists to design materials and processes that would otherwise be impossible. For example, the interplay of scandium and titanium in aerospace alloys combines lightness with durability, while vanadium and chromium strengthen steel used in construction and transportation infrastructure, showing how elemental properties can be strategically combined to optimize performance.

The rare earth metals, from lanthanum through lutetium, reveal the hidden influence of even trace elements. Though often present in minute quantities, they are essential in the production of high-strength permanent magnets, phosphors for display technologies, and lasers for industrial, medical, and military applications. Europium and terbium, for instance, are critical in producing vibrant red and green phosphors for television screens and energy-efficient lighting, exemplifying how subtle variations in electronic structure translate to large-scale technological impacts. Meanwhile, gadolinium, dysprosium, and samarium contribute to medical imaging, data storage, and high-temperature magnetic materials, bridging the gap between fundamental chemistry and real-world applications.

The actinides, particularly uranium, plutonium, and thorium, illustrate the dual nature of chemical discovery: they are both sources of tremendous energy and potential hazards. Their radioactive decay enables nuclear power generation, space exploration power sources, and specialized medical treatments, yet their handling demands meticulous control to prevent environmental contamination and ensure safety. The synthetic elements, from neptunium onward, expand the boundaries of human knowledge, pushing the limits of atomic structure and nuclear stability. These elements, though produced in minuscule quantities, allow researchers to explore the forces that hold atomic nuclei together, informing theories in nuclear physics, quantum chemistry, and even cosmology. Every new isotope synthesized adds a piece to the puzzle of atomic behavior, and potentially paves the way for future applications in medicine, materials science, and energy production.

Beyond their individual applications, these elements collectively shape entire industries and scientific disciplines. Transition metals like iron, nickel, and cobalt form the backbone of industrial manufacturing, enabling the creation of machinery, transportation systems, and infrastructure that sustain modern society. Their corrosion resistance, ductility, and catalytic properties are exploited in countless chemical reactions, including petroleum refining, fertilizer production, and polymer synthesis. The synergistic use of multiple elements in alloys demonstrates how fundamental chemical knowledge translates into practical engineering solutions, emphasizing the interplay between atomic properties and macroscopic performance.

The lanthanides and actinides further illustrate the importance of rarity and specialization. Although less abundant, their unique electronic configurations allow phenomena such as superconductivity, magneto-optical effects, and fluorescence, which have revolutionized technology. These properties underpin the miniaturization of electronic devices, the efficiency of renewable energy technologies, and the precision of medical diagnostic tools. Rare earths have also become strategically important in the global economy, as they are indispensable in smartphones, electric vehicles, wind turbines, and advanced defense systems. Their scarcity on Earth necessitates careful extraction, recycling, and innovative substitution strategies, highlighting the intersection of chemistry with geopolitics, sustainability, and resource management.

Looking forward, the continued exploration of the Periodic Table promises to expand the frontier of science and technology. The discovery of superheavy elements and theoretical superactinides could unlock new chemical behaviors, unusual bonding patterns, and unprecedented material properties. As researchers probe these extreme elements, they may uncover compounds with extraordinary stability, catalytic potential, or electronic characteristics, potentially transforming energy storage, catalysis, and materials science. Each new element not only deepens our understanding of matter but also serves as a platform for innovation, demonstrating the cumulative power of chemical knowledge built over centuries of discovery.

Ultimately, the study of these elements is not just an academic pursuit; it is a journey that connects the atomic scale with human-scale applications. From the satellites orbiting the Earth to the microprocessors in our devices, from lifesaving medical diagnostics to sustainable energy solutions, the impact of these metals is profound and pervasive. The Periodic Table, once a simple classification tool, now serves as a roadmap for technological progress and scientific inquiry. By mastering the properties of each element, understanding their interactions, and innovatively applying them in alloys, catalysts, electronics, and energy systems, humanity continues to transform raw matter into solutions for pressing global challenges.

In conclusion, the metals of the transition, lanthanide, and actinide series embody the intricate interplay between structure, reactivity, and function. They illustrate how variations in atomic size, electronic configuration, and oxidation states translate into practical utility, enabling advances in aerospace, medicine, energy, electronics, and industry. The rare and synthetic elements, although less familiar, remind us that even the most obscure corners of the Periodic Table hold untapped potential for discovery and innovation. The continuous study and application of these elements not only deepen our understanding of the natural world but also provide the foundation for future technologies that will define human progress. As chemical research advances, so too does humanity's capacity to harness the fundamental building blocks of matter, transforming the invisible architecture of atoms into tangible benefits for society, the environment, and the expansion of knowledge itself. Through this lens, the Periodic Table is more than a scientific chart—it is a testament to human curiosity, ingenuity, and the relentless pursuit of understanding the universe at its most elemental level.

Chapter 23: Word Problems

1. Scandium-aluminum alloys are prized in aerospace engineering for their exceptional combination of low density, high strength, and corrosion resistance. Imagine a scenario where an aerospace manufacturer is tasked with producing a lightweight wing spar for a new aircraft design, requiring a total mass of 200 kg of alloy that contains precisely 0.5% scandium by weight to optimize grain structure and prevent hot cracking during extrusion. Your task is to calculate the exact mass of scandium required to achieve this composition. Beyond the arithmetic, consider how the trace scandium content modifies the microstructure of the aluminum matrix, affecting its yield strength, fatigue resistance, and performance under variable thermal conditions at high altitude. How would deviations in scandium concentration impact the mechanical reliability of the spar over repeated stress cycles during flight?

2. Titanium dioxide nanoparticles are a cornerstone of dye-sensitized solar cells due to their ability to facilitate efficient electron transport and improve light absorption. Consider a photovoltaic layer containing 50 mg of TiO_2 nanoparticles deposited as a thin film 100 nm thick. Given the density of TiO_2 as 4.23 g/cm^3, determine the effective surface area coverage provided by the nanoparticles, assuming spherical particles of uniform size. In addition to the calculation, analyze how the size, surface area, and packing of nanoparticles influence electron mobility, recombination rates, and overall cell efficiency. Discuss how nanoscale engineering of TiO_2 layers can optimize light harvesting while minimizing energy losses and how thermal or mechanical stresses during fabrication might alter the film's structural and electronic properties.

3. Vanadium is a vital alloying element in high-strength steels, used to improve hardness, wear resistance, and stability at elevated temperatures. Suppose a civil engineering project requires a 5-ton structural steel beam with 0.2% vanadium by mass to withstand mechanical stress and creep deformation over decades. Calculate the exact mass of vanadium needed for the beam. Beyond this calculation, explore the metallurgical mechanisms by which vanadium carbide precipitates within the steel matrix impede dislocation motion, thereby enhancing yield strength and creep resistance. How does the distribution of vanadium affect the steel's microstructure, and what processing conditions—such as cooling rate or heat treatment—would optimize the alloy's mechanical performance in high-load environments?

4. Chromium plating is widely used to enhance both aesthetic appeal and corrosion resistance in automotive components, household appliances, and industrial hardware. Imagine a steel rod 2 m long with a diameter of 5 cm, which must be coated with a 50 μm thick layer of chromium for a protective mirror finish. Determine the mass of chromium deposited on the rod, and explain how chromium's electronic configuration and propensity to form a dense, adherent oxide layer prevent oxidative degradation of the underlying steel. Additionally, discuss the thermodynamic and kinetic factors that govern chromium deposition during electroplating, the environmental challenges associated with hexavalent chromium, and how variations in plating thickness could affect the durability of the rod under mechanical stress or exposure to corrosive environments.

5. Manganese is an essential transition metal in the electrodes of lithium-ion batteries, where its multiple oxidation states facilitate reversible redox reactions. Consider a lithium-manganese oxide cathode that contains 60% manganese by mass, used in an experimental battery system with 10 g of cathode material. Calculate the number of moles of manganese present in the sample. Then, discuss how manganese's variable oxidation states influence the electrode's electrochemical behavior, including charge-discharge capacity, voltage stability, and cycle life. Additionally, explore how structural ordering, particle size, and dopant elements affect ion diffusion pathways and electronic conductivity, highlighting the balance between maximizing energy density and maintaining long-term material stability.

6. Iron nanoparticles serve as highly efficient catalysts for industrial processes such as ammonia synthesis due to their enhanced surface area and reactivity compared

to bulk iron. Suppose a catalytic bed contains 2 kg of iron nanoparticles, each with an average radius of 50 nm. Estimate the total surface area available for catalysis, and discuss how the high surface-to-volume ratio alters adsorption-desorption kinetics for nitrogen and hydrogen molecules. Furthermore, examine the potential effects of nanoparticle aggregation, sintering at high temperature, and support material interactions on overall catalytic performance. Consider how electronic and magnetic properties of iron influence reaction pathways and how controlled nanoparticle synthesis can optimize efficiency and selectivity in large-scale industrial reactors.

7. Cobalt-chromium-nickel alloys are widely used in high-performance turbine blades due to their resistance to corrosion, oxidation, and mechanical degradation at elevated temperatures. Suppose a turbine blade has a total mass of 15 kg and requires 3 kg of cobalt incorporated into the alloy. Determine the mass fraction of cobalt in the blade, and explore how cobalt contributes to the alloy's thermal stability and magnetic properties. Discuss how the alloy's microstructure, phase composition, and grain boundary strengthening mechanisms influence fatigue resistance, creep behavior, and high-temperature corrosion resistance. Additionally, consider the manufacturing challenges in casting, directional solidification, and heat treatment needed to achieve optimal performance under extreme operating conditions.

8. Nickel-based alloys are widely employed in metal-hydride hydrogen storage tanks due to their favorable absorption and desorption kinetics. A hydrogen storage tank contains 120 kg of Ni-MH alloy capable of storing hydrogen at 1.5 wt%. Calculate the total mass of hydrogen that can be stored in the tank. Beyond the calculation, analyze the thermodynamic and kinetic considerations governing hydrogen absorption and release, including the role of metal lattice expansion, phase transitions, and pressure-temperature dependencies. Discuss how alloy composition, crystal defects, and surface catalysis influence hydrogen reversibility, storage density, and safety, highlighting the balance between material efficiency and operational reliability in renewable energy applications.

9. Copper's exceptional electrical and thermal conductivity make it a mainstay in electrical infrastructure. Consider a 100 m long copper wire in an industrial facility, carrying 50 A of current at 220 V. Using the resistivity of copper, calculate the

voltage drop and power loss along the wire. In addition to these calculations, examine how microstructural factors such as grain size, impurity content, and work hardening influence conductivity and thermal stability. Discuss the implications of long-term operation under high current, including electromigration, oxidation, and mechanical stress, and explore strategies for optimizing wire design to reduce energy losses and maintain safety in industrial power distribution systems.

10. Zinc is commonly used as a sacrificial coating to protect steel structures through galvanization. Suppose a steel beam with a total surface area of 20 m^2 is coated with a 100 μm thick zinc layer. Calculate the mass of zinc required to achieve uniform coverage. Beyond the calculation, discuss the electrochemical principles behind sacrificial protection, including the formation of zinc oxide layers and the preferential corrosion of zinc over iron. Analyze how environmental factors such as humidity, temperature, and pollutant exposure influence coating longevity, and explore processing parameters such as immersion time, bath composition, and surface preparation that ensure optimal adhesion and corrosion resistance.

11. Yttrium-doped phosphors are integral in modern LED displays and solid-state lighting due to their ability to efficiently transfer energy and produce bright, stable light. Suppose a phosphor material contains 5 mol% Y^{3+} dopants and a total mass of 1 g. Calculate the number of yttrium atoms present in the sample. Beyond the arithmetic, analyze how the introduction of Y^{3+} ions modifies the electronic band structure and facilitates energy transfer from the host lattice to the dopant ions. Discuss how particle size, crystal defects, and thermal quenching affect luminescence efficiency, and consider the implications for designing long-lasting, energy-efficient display technologies.

12. Zirconium is commonly used for cladding nuclear fuel rods because of its low neutron absorption cross-section and excellent corrosion resistance. Imagine a fuel rod 4 m long and 2 cm in diameter coated with a 1 mm thick zirconium layer. Calculate the mass of zirconium required for the cladding. In addition to the calculation, examine how zirconium's chemical inertness prevents oxidation in high-temperature, high-radiation environments. Consider the impact of hydrogen uptake, thermal expansion, and stress-induced cracking on the performance of cladding in reactor operations, and explain why zirconium alloys are favored over other transition metals in nuclear applications.

13. Niobium-titanium alloys are widely used in superconducting magnets for MRI machines and particle accelerators due to their ability to maintain zero resistance below critical temperatures. Suppose a niobium-titanium wire carries a current of 500 A below its critical temperature. Discuss how niobium content alters the electron pairing mechanism, critical current density, and magnetic field tolerance in type II superconductors. Consider the interplay between alloy composition, microstructure, and cooling conditions, and explain how defects, grain boundaries, and strain influence flux pinning and overall superconducting performance.

14. Molybdenum alloys are chosen for furnace components that must endure extreme temperatures without significant deformation. Consider a molybdenum alloy rod of length 1 m and radius 2 cm operating at 1200 °C. Calculate the expected thermal expansion of the rod given molybdenum's coefficient of thermal expansion. Analyze how molybdenum's high melting point and low expansion coefficient ensure dimensional stability under prolonged heat exposure. Additionally, discuss the role of alloying elements in enhancing creep resistance and oxidation behavior, and explore potential microstructural changes under cyclic thermal loading that may influence component longevity.

15. Technetium-99m is a radioactive isotope widely used in diagnostic imaging due to its short half-life and gamma emission properties. A hospital receives 2 GBq of Tc-99m. Calculate the number of radioactive nuclei present and their half-life in seconds. Beyond the calculation, discuss how technetium-99m's decay properties enable non-invasive imaging of organs, minimizing radiation exposure while providing high-resolution diagnostic data. Examine the chemical considerations in labeling Tc-99m to biologically active molecules and the importance of its short half-life for patient safety and timely imaging procedures.

16. Ruthenium catalysts are employed in industrial processes such as ammonia oxidation due to their high activity and selectivity. A catalytic reactor contains 500 g of ruthenium dispersed on a 50 m^2 support. Calculate the catalyst loading in g/m^2. Beyond this, explore how ruthenium's d-orbital configuration facilitates adsorption and activation of nitrogen and oxygen species. Discuss the effects of particle size, support interactions, and temperature on reaction kinetics, and consider how catalyst deactivation mechanisms such as sintering or poisoning can impact process efficiency over time.

17. Rhodium is a rare metal used extensively in automotive catalytic converters to reduce NOx emissions. Suppose a converter contains 10 g of rhodium. Calculate the molar quantity of rhodium atoms in the converter. Discuss the mechanistic steps by which rhodium facilitates the reduction of nitrogen oxides at high temperatures, considering the role of surface sites and electronic configuration. Analyze how alloying with platinum or palladium modifies catalytic activity, and explore the effects of thermal cycling and exhaust gas composition on long-term catalyst performance.

18. Palladium is a key catalyst for hydrogenation in chemical synthesis. Suppose 100 mg of palladium achieves a 95% yield in hydrogenating 2 mol of ethylene. Calculate the turnover number (TON) for this reaction. Discuss how the surface area, crystallographic facets, and nanoparticle size of palladium influence reaction kinetics and selectivity. Explore how support materials and reaction conditions such as pressure, temperature, and solvent choice impact catalyst efficiency and longevity, and consider the implications for scaling up the process industrially.

19. Silver nanoparticles are employed in antimicrobial coatings due to their ability to release Ag^+ ions, which disrupt microbial cell membranes. Consider a coating applied to a 1 m^2 surface with thickness 10 µm, containing 0.05 wt% silver. Calculate the total mass of silver in the coating. Discuss how particle size, surface chemistry, and ion release rates influence antimicrobial effectiveness. Explore the balance between maximizing biocidal activity and minimizing toxicity to human cells, and analyze how environmental factors such as humidity, pH, and temperature affect nanoparticle performance over time.

20. Cadmium is widely used in rechargeable nickel-cadmium batteries for its electrochemical reversibility. A battery contains 20 g of cadmium and operates at 1.2 V delivering 50 Ah of charge. Calculate the theoretical energy density of the battery. Beyond the numerical calculation, discuss cadmium's role in electrode reactions, including its oxidation-reduction behavior during charge and discharge. Consider the effects of cycling on electrode degradation, dendrite formation, and capacity loss, and explore strategies to enhance battery lifetime and safety while minimizing environmental hazards associated with cadmium.

21. Lanthanum alloys are widely used in metal-hydride hydrogen storage systems due to their ability to reversibly absorb hydrogen. Suppose a sample contains 2

mol of La, and each La atom can bind one H_2 molecule. Calculate the total number of hydrogen molecules the sample can theoretically store. Beyond the calculation, discuss how the crystal structure, lattice spacing, and defects influence hydrogen diffusion and storage capacity. Examine how temperature and pressure cycles affect absorption/desorption kinetics, and explore strategies to optimize alloy composition and microstructure for maximum efficiency in sustainable energy storage systems.

22. Cerium dioxide (CeO_2) is a key component in three-way catalytic converters due to its oxygen storage capacity. Imagine a catalytic converter containing 50 g of CeO_2. Calculate the molar concentration of cerium atoms. Discuss how Ce^{3+}/Ce^{4+} redox cycling facilitates oxidation and reduction reactions of exhaust gases, enhancing NO_x, CO, and hydrocarbon conversion. Consider the effects of nanoparticle size, surface area, and temperature fluctuations on catalytic performance, and explain how the incorporation of cerium into mixed oxides improves long-term durability and reaction efficiency under variable driving conditions.

23. Neodymium-iron-boron (NdFeB) magnets are essential for high-performance electric motors. A magnet weighs 1 kg with a density of 7.5 g/cm³. Calculate the volume of the magnet. Discuss how the electronic configuration of neodymium contributes to large magnetic anisotropy and high coercivity. Analyze the effects of microstructure, rare-earth content, and thermal treatment on magnetic strength, and explore challenges in maintaining magnet performance under high temperatures and mechanical stresses in electric vehicle applications.

24. Praseodymium-doped glasses are employed in fiber optics to amplify light signals in telecommunications. Consider a fiber with a cross-sectional area of 1 mm² containing 0.01 wt% Pr^{3+}. Calculate the number of praseodymium ions per meter of fiber. Beyond the arithmetic, examine how Pr^{3+} electronic transitions enable light amplification, the effects of fiber length and dopant concentration on gain, and how host glass composition influences energy transfer efficiency. Discuss potential issues such as photodarkening, thermal effects, and attenuation in long-distance optical communication systems.

25. Promethium isotopes, particularly Pm-147, are used in nuclear batteries for space and remote applications. Suppose a device contains 1 mg of Pm-147. Calculate

the decay energy released per second, considering its half-life. Analyze how the beta decay of promethium generates electrical power in miniature devices. Discuss challenges in safely containing radioactive material, maximizing energy conversion efficiency, and minimizing radiation shielding requirements in portable power sources, particularly in high-radiation or zero-gravity environments.

26. Samarium-cobalt (Sm_2Co_{17}) magnets are prized for maintaining high magnetic strength at elevated temperatures. A magnet weighs 500 g. Calculate the moles of samarium present in the magnet. Beyond the calculation, discuss how the crystal structure contributes to thermal stability and magnetic domain alignment. Examine the role of cobalt and other minor elements in enhancing coercivity, and explore the effects of manufacturing methods such as powder metallurgy and sintering on magnet performance in high-temperature industrial applications.

27. Europium-doped phosphors are widely used to produce red light in LEDs. Consider a 2 g phosphor sample containing 1 mol% Eu^{3+}. Calculate the number of europium atoms in the sample. Discuss the energy transfer processes from the host lattice to Eu^{3+} ions that result in red emission. Analyze how particle size, crystal symmetry, and dopant distribution affect luminescence efficiency and color purity. Explore applications in display technologies and energy-efficient lighting, emphasizing the role of rare-earth chemistry in modern electronics.

28. Gadolinium is used in magnetic resonance imaging (MRI) due to its strong paramagnetism. A patient receives 0.1 mol of Gd^{3+} as a contrast agent. Examine the electronic configuration of gadolinium that leads to enhanced T_1 relaxation in MRI. Discuss the chemical formulation required to safely administer gadolinium ions, including chelation and stability in biological systems. Consider the effects on image contrast, relaxivity, and patient safety, highlighting the intersection of inorganic chemistry, biophysics, and medical technology.

29. Terbium is critical in green phosphors for television and LED screens. A phosphor contains 2 mol% Tb^{3+}. Calculate the number of Tb ions in 1 g of phosphor. Beyond the calculation, analyze the mechanisms of energy transfer from the host lattice to Tb^{3+} ions that produce green emission. Discuss how crystal structure, dopant concentration, and thermal quenching affect efficiency and longevity, and explore the importance of rare-earth chemistry in high-performance display technology.

30. Dysprosium enhances coercivity in neodymium-iron-boron magnets at high temperatures. Consider a magnet with composition $Dy_{0.2}Fe_{2.8}$ weighing 200 g. Calculate the number of moles of dysprosium. Discuss how Dy incorporation strengthens magnetic anisotropy and prevents domain wall motion at elevated temperatures. Examine the trade-offs between rare-earth content, cost, and performance, particularly in electric vehicle motors and wind turbine generators operating under thermal stress.

31. Holmium-doped crystals emit at 2.1 μm, making them suitable for precise surgical lasers. Suppose a crystal contains 0.05 mol Ho^{3+} ions. Calculate the number of emitting ions. Discuss how electronic transitions in Ho^{3+} facilitate coherent infrared emission, and explore how crystal quality, dopant distribution, and thermal management influence laser efficiency. Consider clinical applications in urology and ENT surgery, where precise tissue ablation and minimal thermal damage are critical.

32. Erbium-doped fibers are used to amplify optical signals in telecommunications. A fiber contains 0.01 mol% Er^{3+} over a length of 10 km. Calculate the total number of erbium ions. Discuss how population inversion enables signal amplification, and analyze factors such as pump wavelength, fiber attenuation, and erbium concentration on gain and signal-to-noise ratio. Explore challenges in long-distance optical communication and strategies to optimize amplifier performance.

33. Thulium-doped crystals emit at 2 μm for medical and industrial laser applications. A 50 g crystal contains 0.1 mol% Tm^{3+}. Determine the number of thulium ions. Discuss the energy transfer mechanisms responsible for laser emission and analyze how doping concentration, crystal defects, and thermal effects influence output power and beam quality. Explore applications in surgery, material processing, and environmental sensing.

34. Ytterbium-doped fibers are used in high-power industrial lasers. Consider a 5 m fiber containing 0.5 mol% Yb^{3+}. Calculate the total number of ytterbium ions. Discuss how Yb^{3+} absorbs pump light and facilitates stimulated emission, and analyze how fiber design, doping uniformity, and thermal management influence laser efficiency and power scaling.

35. Lutetium-based scintillators are critical in PET scanners. A detector contains 10 g of Lu_2SiO_5. Calculate the number of lutetium atoms. Discuss how lutetium's high

atomic number and density improve gamma-ray detection efficiency. Analyze the role of crystal quality, scintillation efficiency, and electronic readout in medical imaging performance.

36. Hafnium is used in nuclear control rods for its high neutron absorption cross-section. A rod contains 1 kg of Hf. Calculate the number of hafnium nuclei and discuss how neutron capture affects reactor control. Examine the metallurgical properties required for structural integrity under high radiation and temperature conditions.

37. Tantalum's oxide layer enables high capacitance in small volumes. A capacitor contains 50 mg of Ta. Calculate the number of atoms present. Discuss how Ta_2O_5 dielectric properties allow energy storage, and analyze how grain size, purity, and anodization parameters influence breakdown voltage, capacitance stability, and long-term reliability.

38. Tungsten filaments operate at 3000 K due to high melting point and emissivity. A 0.5 g filament contains atoms with atomic mass 183.84 g/mol. Calculate the number of tungsten atoms. Discuss how filament geometry, thermal radiation, and resistivity influence light output and efficiency, and explore failure mechanisms such as evaporation and recrystallization over filament lifetime.

39. Rhenium is used to strengthen superalloys in jet engines. A component contains 2 kg of Re. Calculate the number of moles. Discuss how rhenium's high melting point and solid solution strengthening improve creep resistance and thermal stability. Analyze alloy design strategies, microstructure control, and environmental degradation under extreme mechanical and thermal loads.

40. Osmium is used for pivots and other precision components due to extreme density and hardness. Consider a 1 cm^3 cube with a density of 22.59 g/cm^3. Calculate the mass and number of atoms. Discuss how hardness, wear resistance, and corrosion stability make osmium suitable for high-precision applications in watches, navigational instruments, and aerospace components.

41. Iridium crucibles withstand extreme temperatures and corrosive environments. A 500 g crucible is used at 2500 K. Calculate the number of iridium atoms. Discuss how iridium's chemical inertness, high melting point, and thermal conductivity

make it ideal for melting reactive metals, and analyze microstructural stability under repeated heating and cooling cycles.

42. Platinum is a key catalyst in hydrogen fuel cells. A fuel cell contains 5 g of Pt. Calculate the molar quantity of platinum atoms. Discuss how Pt facilitates hydrogen oxidation and oxygen reduction, including adsorption mechanisms and electron transfer. Examine how nanoparticle size, dispersion, and support interactions influence reaction rates and durability in long-term fuel cell operation.

43. Gold nanoparticles enhance imaging due to surface plasmon resonance. A solution contains 50 mg of 10 nm diameter Au spheres. Calculate the approximate number of gold atoms per particle and total atoms. Discuss how size, shape, and surface chemistry influence optical properties, biocompatibility, and targeting in medical diagnostics.

44. Mercury's high density and thermal expansion make it ideal for precise temperature measurements. A thermometer contains 5 g of Hg. Calculate the number of mercury atoms. Discuss the thermophysical properties, phase stability, and toxicity considerations that affect thermometer performance and handling in laboratory and industrial settings.

45. U-235 is enriched to 3% in a 1000 g fuel rod. Calculate the mass and number of U-235 nuclei. Discuss how isotopic enrichment influences fission chain reaction rates, reactor design, and energy output. Explore the role of moderator materials, neutron economy, and safety considerations in maintaining controlled nuclear reactions.

46. Plutonium-239 is used in reactors and research. A reactor requires 10 kg of Pu-239. Calculate the number of nuclei. Discuss critical mass, neutron multiplication, and isotopic purity in reactor operations. Analyze handling precautions, decay heat management, and long-term storage considerations.

47. Americium-241 is used as an alpha emitter in ionization smoke detectors. A detector contains 0.9 μg of Am-241. Calculate the number of radioactive nuclei. Discuss how alpha particle ionization enables smoke detection, radiation safety, and long-term detector reliability.

48. Curium isotopes provide heat for radioisotope thermoelectric generators. A Pu-Cm alloy contains 50 g Cm-244. Calculate the number of nuclei and discuss the

energy output from radioactive decay. Analyze thermal management, shielding, and longevity in space applications where maintenance is impossible.

49. Highly radioactive transuranics like Bk and Cf are handled in microgram quantities. A sample contains 1 μg each. Calculate the number of atoms. Discuss the challenges of chemical characterization, radiological safety, and detection in research laboratories. Examine potential applications in neutron sources, nuclear chemistry, and fundamental actinide studies.

50. Superheavy elements like einsteinium are synthesized in femtogram quantities. Consider a 1 fg sample of Es-252. Estimate the number of atoms present. Discuss how mass spectrometry, nuclear reaction cross-sections, and decay pathways are used to detect these elements. Explore potential research applications in nuclear structure, quantum chemistry, and expanding the limits of the Periodic Table.

Chapter 24

Smart Healthcare Chemistry

The roots of healthcare chemistry trace back to the earliest civilizations, where humans relied on natural products to treat ailments and maintain health. Ancient healers experimented with plants, minerals, and animal products, observing their effects and refining their use over generations. The extraction of quinine from cinchona bark to combat malaria, along with the derivation of morphine from opium for pain management, represents some of the earliest documented applications of chemical principles to medicine. These discoveries laid the foundation for modern medicinal chemistry, highlighting the transformative potential of isolating bioactive compounds and understanding their mechanisms of action. As centuries passed, advances in chemical isolation techniques, structural elucidation methods, and synthetic chemistry enabled scientists to transcend reliance on natural products, facilitating the rational design of complex molecules with enhanced potency, specificity, and safety profiles.

Healthcare chemistry, in its contemporary form, integrates knowledge from diverse branches of chemistry, including organic, inorganic, analytical, and physical chemistry. Organic chemistry provides the foundation for the design and synthesis of small-molecule therapeutics, allowing chemists to manipulate molecular structures to optimize biological activity. Inorganic chemistry contributes understanding of metal-based drugs such as cisplatin, whose reactivity and coordination geometry are central to its cytotoxic properties. Analytical chemistry ensures the quality, purity, and structural verification of therapeutic compounds through precise measurements and sensitive detection techniques. Physical chemistry underpins the understanding of molecular interactions, reaction mechanisms, thermodynamics, and kinetics, all of which are critical to predicting drug behavior in biological systems. Together, these disciplines create a cohesive framework for the discovery, optimization, and application of therapeutic agents.

A central objective of healthcare chemistry is the discovery and rational design of new drugs. The process begins with the identification of a biological target—a protein, receptor, ion channel, or nucleic acid sequence—whose modulation can alter disease progression. Chemists employ a variety of strategies to design molecules capable of selectively interacting with these targets, drawing on knowledge of structure-activity relationships (SAR), stereochemistry, and electronic effects. SAR studies provide insight into how modifications of molecular structure influence binding affinity, efficacy, and safety, guiding iterative optimization. Computational chemistry and molecular modeling complement these efforts, allowing visualization of interactions at atomic resolution and enabling rational design of molecules with optimal geometry, charge distribution, and hydrophobicity to engage target sites effectively. High-throughput screening accelerates the identification of promising candidates, allowing thousands of compounds to be tested simultaneously against a given target to uncover leads for further development.

Lead optimization is a critical phase in drug development, where chemists refine chemical structures to improve pharmacological and physicochemical properties. Modifications are often subtle yet impactful, including the introduction of polar or lipophilic functional groups, isotopic labeling, and careful stereochemical adjustments. These changes can enhance solubility, stability, bioavailability, selectivity, and metabolic resilience, ultimately improving the therapeutic profile of the candidate drug. The integration of medicinal chemistry with pharmacokinetics and pharmacodynamics ensures that optimized molecules reach the intended biological targets at effective concentrations without eliciting adverse effects. Pharmacokinetics describes the absorption, distribution, metabolism, and excretion of drugs (ADME), while pharmacodynamics relates drug concentration to the resulting biological effect. Comprehensive understanding of both areas is essential for rational drug design.

Chemical modifications play a central role in influencing pharmacokinetic behavior. Polar groups can increase aqueous solubility and improve gastrointestinal absorption, while lipophilic moieties facilitate membrane permeability, aiding tissue penetration. Prodrugs, which are pharmacologically inactive precursors activated in vivo, exemplify the intersection of chemistry and pharmacokinetics, allowing selective targeting, improved bioavailability, or reduced systemic toxicity. Beyond small molecules, healthcare chemistry encompasses biologics, including monoclonal antibodies, peptides, vaccines, and nucleic acid-based therapies. These macromolecules require sophisticated chemical strategies for synthesis, conjugation, and purification, reflecting the complex interplay between chemistry and biology. Monoclonal antibodies are engineered for high specificity toward antigens, often with chemical modifications to improve stability, half-life, and reduce immunogenicity. Peptides and proteins are precisely sequenced and sometimes chemically conjugated to drugs for targeted delivery, demonstrating the versatility of chemical interventions in modern therapeutics.

Vaccines illustrate the synergy between chemical design and biological function, with formulations relying on adjuvants, stabilizers, and delivery vehicles to enhance immunogenicity. The advent of mRNA vaccines, particularly highlighted by recent global efforts against viral pandemics, underscores the importance of chemical engineering in therapeutics. Lipid nanoparticles protect fragile nucleic acids, facilitate cellular uptake, and optimize delivery, with their composition—ionizable lipids, PEGylated components, and neutral phospholipids—directly affecting efficacy and safety. Analytical chemistry is indispensable in these developments, providing precise tools for structural verification, purity assessment, and stability monitoring. High-performance liquid chromatography, mass spectrometry, nuclear magnetic resonance spectroscopy, and X-ray crystallography enable detailed chemical characterization of both small molecules and complex biologics, ensuring reproducibility and regulatory compliance.

In clinical diagnostics, analytical techniques detect disease biomarkers with high sensitivity and specificity. Mass spectrometry-based assays quantify proteins, metabolites, and nucleic acids, enabling early detection of cancers, metabolic disorders, and cardiovascular conditions. Chemical sensors and biosensors provide rapid, point-of-care diagnostics, integrating molecular recognition with electronic or optical transduction to deliver actionable data. Understanding drug metabolism is equally critical; most therapeutics are biotransformed in the liver via phase I reactions (oxidation, reduction, hydrolysis) and phase II reactions (conjugation), producing metabolites that may be active, inactive, or toxic. Chemists design molecules with metabolic stability in mind, ensuring effective delivery to target sites and minimizing adverse effects, while prodrugs can exploit enzymatic pathways for selective activation.

Toxicology, informed by chemistry, evaluates the potential harmful effects of drugs, impurities, and metabolites. In vitro assays, cell-based models, and computational predictions of structure-toxicity relationships guide the design of safer therapeutics. Chemical strategies are fundamental in the development of antimicrobial and antiviral agents, which combat infectious diseases by targeting bacterial cell walls, protein synthesis, or viral replication pathways. Modifying chemical structures can overcome resistance mechanisms, such as β-lactam ring alterations in penicillins to evade enzymatic degradation or designing nucleoside analogs to inhibit viral polymerases. Advances in genomics and proteomics enable personalized medicine, tailoring drug selection and dosing based on genetic variations that influence pharmacokinetics and pharmacodynamics, while chemical modifications enhance specificity, reduce toxicity, and improve therapeutic efficacy.

Healthcare chemistry also drives the development of companion diagnostics, assays designed to predict patient responses and optimize therapy.

Nanotechnology exemplifies the integration of chemistry with targeted delivery, where nanoparticles, liposomes, dendrimers, and polymeric micelles enable controlled release, tissue specificity, and co-delivery of multiple therapeutic agents. Surface functionalization allows attachment of targeting ligands, imaging moieties, or stimuli-responsive groups, improving precision in treatments such as cancer therapy. Nanocarriers enhance chemotherapeutic efficacy while minimizing systemic toxicity, support gene therapy, and facilitate RNA-based therapeutics, demonstrating the convergence of chemistry, biology, and materials science.

The design of diagnostic reagents and medical imaging agents further illustrates the chemical underpinnings of modern healthcare. MRI contrast agents, CT enhancers, and PET radiotracers rely on carefully tailored chemical structures to optimize imaging clarity and minimize toxicity. Gadolinium-based MRI agents exploit paramagnetic properties, while radiolabeled PET tracers must remain chemically stable yet biologically active. Chemical design also informs rapid diagnostic assays, including immunoassays and biosensors, where the choice of reagents, enzyme-substrate reactions, and detection chemistry ensures sensitivity, selectivity, and reliability.

Sustainability is an emerging priority in healthcare chemistry, with green chemistry principles guiding safer, environmentally responsible practices. Solvent selection, catalytic efficiency, energy conservation, and biodegradable delivery systems reduce environmental impact while improving scalability and safety. These sustainable approaches are increasingly integrated into pharmaceutical development, production, and waste management, reflecting a broader responsibility to human and environmental health. Regulatory oversight by agencies such as the FDA and EMA mandates comprehensive chemical characterization, stability assessments, impurity profiling, and toxicology evaluations to ensure safe and effective therapies reach patients.

Healthcare chemistry continues to evolve rapidly, propelled by advances in molecular biology, computational modeling, artificial intelligence, nanotechnology, and high-resolution analytical techniques. Emerging areas include AI-guided drug design, CRISPR-based gene editing, targeted cancer therapies, RNA therapeutics, and biosensor-enabled personalized diagnostics. By integrating chemical principles with biological insights and data-driven approaches, healthcare chemistry promises faster, safer, and more effective solutions for treating disease. Chemists play a critical role at every stage of therapeutic development, from discovery and design to delivery and diagnostics, underpinning modern medicine with rigorous scientific understanding.

Ultimately, healthcare chemistry represents a nexus of science, technology, and medicine, encompassing the synthesis, analysis, and application of therapeutic agents, diagnostic tools, and delivery systems. It provides the chemical rationale for drug action, metabolism, safety, and efficacy, ensuring that treatments are both effective and tailored to individual needs. By combining traditional chemical knowledge with innovative technologies, healthcare chemistry continues to advance our ability to prevent, diagnose, and treat disease. It fosters a deeper understanding of molecular interactions, guides the development of personalized and precision medicine, and supports sustainable pharmaceutical practices. As healthcare challenges evolve, the integration of chemistry with biology, computational science, and engineering will remain essential, shaping the future of medicine and the well-being of populations worldwide.

Healthcare chemistry's impact extends beyond the laboratory, influencing global health strategies, public policy, and personalized patient care. The integration of chemical insights into epidemiology, vaccination campaigns, and antimicrobial stewardship underscores the role of chemistry in population-level health management. Chemical analysis allows for rapid detection of infectious agents, environmental toxins, and chemical contaminants, enabling timely interventions that prevent outbreaks and reduce disease burden. For instance, water quality monitoring, air pollution assessment, and food safety testing rely heavily on analytical chemistry techniques to detect trace chemicals, microbial metabolites, or pesticide residues, directly linking chemistry to public health outcomes. These applications demonstrate that healthcare chemistry is not confined to drug design but encompasses a holistic approach to understanding and maintaining human health in complex environments.

Emerging therapies in healthcare chemistry increasingly leverage molecular engineering and synthetic biology. The development of nucleic acid-based therapeutics, such as antisense oligonucleotides, small interfering RNAs (siRNAs), and CRISPR-associated systems, depends on precise chemical modifications to improve stability, cellular uptake, and target specificity. Phosphorothioate backbones, locked nucleic acids, and chemical conjugates enhance the half-life and bioavailability of these molecules while reducing immune activation. Such chemical interventions enable previously intractable diseases to become treatable, highlighting the transformative potential of chemistry at the molecular and cellular levels. These innovations also pave the way for therapies that can be tailored to genetic profiles, furthering the goals of precision medicine and individualized treatment plans.

The integration of artificial intelligence and machine learning with healthcare chemistry is revolutionizing drug discovery and development. Computational algorithms can predict chemical properties, binding affinities, and potential toxicities of novel compounds, drastically reducing the time and cost associated with experimental screening. Machine learning models trained on extensive chemical and biological datasets can identify patterns that human researchers might overlook, accelerating lead identification and optimization. AI-guided molecular design also facilitates the creation of novel chemical scaffolds that exhibit improved efficacy, safety, and metabolic stability. By coupling these computational approaches with high-throughput synthesis and analytical technologies, healthcare chemistry is entering an era of unprecedented efficiency and innovation, bridging the gap between chemical theory and clinical application.

Another area of significant advancement is the development of multifunctional therapeutics that combine chemical, biological, and physical properties to achieve targeted and synergistic effects. Conjugates that combine drugs with imaging agents, photodynamic therapy compounds, or nanomaterials allow simultaneous diagnosis and treatment, termed theranostics. In oncology, for example, chemotherapeutic agents can be linked to fluorophores or radionuclides, permitting visualization of drug distribution while exerting cytotoxic effects. These systems rely on sophisticated chemistry to ensure stability, controlled release, and selective targeting, exemplifying the intricate interplay between chemical design and clinical efficacy. Additionally, stimuli-responsive drug delivery platforms, which release therapeutics in response to pH, temperature, enzymes, or light, are enabled by precise chemical engineering, providing temporal and spatial control over treatment.

Healthcare chemistry also contributes to regenerative medicine and tissue engineering through the design of biomaterials and scaffolds. Biocompatible polymers, hydrogels, and chemically modified matrices support cell growth, differentiation, and tissue repair. Chemical functionalization allows incorporation of growth factors, signaling molecules, and controlled degradation kinetics, creating environments that mimic natural tissue and promote healing. These chemical innovations facilitate applications ranging from wound healing and cartilage regeneration to organoids and bioartificial organs, demonstrating how chemical principles can extend the boundaries of conventional therapeutics into complex biological reconstruction.

The field of metabolomics, which relies heavily on chemical analysis, provides comprehensive insight into biochemical pathways and disease states. By profiling small molecules, lipids, amino acids, and metabolites in biological samples, chemists can identify biomarkers, understand metabolic dysregulation, and predict therapeutic responses. Mass spectrometry and nuclear magnetic resonance spectroscopy are central to these studies, enabling high-resolution, quantitative analysis of complex mixtures. Metabolomic data can inform drug design, optimize dosing strategies, and identify potential side effects, highlighting the role of chemistry in connecting molecular understanding with systemic physiology.

Sustainability and environmental considerations remain critical in contemporary healthcare chemistry. The production, use, and disposal of pharmaceuticals have ecological implications, including contamination of water systems with active drug residues, accumulation of persistent chemicals, and challenges in biodegradation. Green chemistry approaches mitigate these impacts by emphasizing atom efficiency, safer solvents, renewable feedstocks, and catalytic processes. Designing drugs with environmental stability in mind, as well as creating degradable delivery systems, ensures that therapeutic advances do not come at the expense of ecological health. These principles are increasingly mandated by regulatory bodies and are a hallmark of responsible, forward-looking chemical practice in healthcare.

Healthcare chemistry also plays a critical role in global responses to emerging diseases and health crises. Rapid development of vaccines, antivirals, and diagnostic assays during pandemics exemplifies how chemical innovation can respond to urgent societal needs. The design of broad-spectrum antiviral agents, chemical adjuvants, and modular vaccine platforms relies on an in-depth understanding of chemical reactivity, molecular recognition, and pharmacological principles. Likewise, the development of stable, transportable, and scalable therapeutics demonstrates the importance of chemical engineering in bridging laboratory discovery with real-world implementation, particularly in resource-limited settings.

Finally, the convergence of healthcare chemistry with data science, biotechnology, and materials science is redefining the landscape of modern medicine. High-dimensional chemical and biological data inform predictive modeling, precision therapy, and personalized diagnostics. Advanced imaging agents, nanocarriers, and multifunctional biomaterials rely on sophisticated chemical modifications to achieve desired biological outcomes. The iterative cycle of design, synthesis, characterization, and application ensures that chemistry remains at the forefront of innovation, translating molecular understanding into tangible health benefits. In conclusion, healthcare chemistry is an indispensable foundation of modern medicine, uniting rigorous chemical science with biological insight to develop safer, more effective, and personalized therapeutic solutions. Its continued evolution promises to address the complex challenges of disease management, global health, and sustainable pharmaceutical practice, solidifying its role as a transformative force in both science and society.

Chapter 24: Word Problems

1. Quinine, a natural alkaloid from cinchona bark, is historically crucial in treating malaria. Its activity depends on stereospecific interactions with parasite heme and enzymes. Recent studies suggest modifications at the C-8 hydroxyl can alter binding affinity and efficacy. Design a series of quinine analogs with changes at this position, considering hydrogen bonding, steric effects, and electronic distribution. Use computational docking to predict binding efficiency, then design enzyme kinetics experiments to validate these predictions, specifying substrate selection, concentration ranges, and expected readouts. Discuss potential limitations of in silico methods, complementing them with in vitro and in vivo studies, and propose experimental approaches to identify the most active enantiomer while considering pharmacokinetic implications.

2. Morphine derivatives must retain analgesic potency while minimizing addiction potential. Consider structural modifications on the aromatic ring and hydroxyl groups. Predict how these changes might influence selectivity among μ, δ, and κ opioid receptors, as well as blood-brain barrier penetration. Develop an analytical workflow using NMR, HPLC, and mass spectrometry to confirm structure, purity, and stereochemistry. Propose receptor-binding and functional assays (G-protein

activation, β-arrestin recruitment) to evaluate pharmacological effects. Discuss phase I and II metabolism, potential toxic metabolites, and strategies for lead optimization balancing potency, selectivity, and safety.

3. Cisplatin forms crosslinks with guanine in DNA, disrupting replication and transcription. For a 1000-base DNA fragment, calculate the maximum theoretical number of cisplatin molecules that can bind. Consider kinetic factors such as hydrolysis and ligand substitution in different cellular environments. Design a derivative with a modified leaving group to enhance tumor specificity and predict effects on selectivity, cytotoxicity, and side effects using thermodynamic and kinetic principles. Propose in vitro and in vivo experiments to compare DNA-binding efficiency and cytotoxicity, considering resistance mechanisms such as DNA repair or glutathione-mediated detoxification.

4. ADCs rely on selective delivery of cytotoxic drugs via antibodies. Design a series of linkers with hydrolysis rates spanning enzymatic, pH-sensitive, and redox-sensitive mechanisms. Predict how linker stability influences pharmacokinetics, therapeutic index, and off-target toxicity. Develop analytical methods (LC-MS/MS, HPLC) to quantify drug release kinetics in vitro and propose cell-based assays for cytotoxicity. Discuss challenges such as heterogeneous conjugation, variable drug-to-antibody ratios, and antibody glycosylation effects on stability and immunogenicity.

5. A liposomal mRNA vaccine requires optimal lipid composition to protect fragile nucleic acids and enhance cellular uptake. Examine how ionizable lipids, PEGylated components, and neutral phospholipids influence endosomal escape and immunogenicity. Apply thermodynamic principles to predict stability across pH and temperature variations. Propose in vitro and in vivo studies to assess uptake efficiency, antigen expression, and immune responses, and consider the role of lipid composition on biodistribution, clearance, and toxicity.

6. A small-molecule kinase inhibitor has poor aqueous solubility, limiting absorption. Design chemical modifications (e.g., polar group addition, salt formation) to enhance solubility without reducing binding affinity. Use computational models to predict binding and solubility trade-offs. Design pharmacokinetic studies to evaluate absorption, tissue distribution, and bioavailability. Discuss potential effects

on metabolism and excretion, and propose strategies to mitigate unexpected side effects.

7. After screening 500 compounds against a viral protease, develop a cheminformatics workflow using QSAR, molecular docking, and clustering algorithms to prioritize candidates. Include considerations of stereochemistry, molecular weight, logP, hydrogen bonding capacity, and predicted toxicity. Propose a stepwise experimental validation plan integrating in vitro enzyme inhibition, cytotoxicity assays, and early ADME studies. Discuss challenges in balancing computational predictions with experimental observations.

8. A poorly absorbed anti-cancer drug is proposed for prodrug development. Design chemical modifications that improve bioavailability and ensure enzymatic activation at the target site. Propose in vitro assays using liver microsomes and recombinant enzymes to quantify conversion rates. Predict the effect of modifications on plasma stability, distribution, and off-target toxicity. Include computational modeling to assess steric and electronic effects on enzyme recognition.

9. Radiolabeled PET tracers require chemical stability and high receptor specificity. Propose chemical modifications to extend circulation half-life without altering target binding. Use thermodynamic and kinetic principles to predict in vivo stability and metabolism. Design radiochemical and imaging studies to quantify biodistribution, clearance, and target engagement. Consider implications for patient safety, regulatory approval, and production scalability.

10. A peptide drug is rapidly degraded by serum proteases. Design chemical modifications (e.g., cyclization, D-amino acid substitution, PEGylation) to enhance stability and half-life. Propose LC-MS methods to quantify intact peptide in plasma over time and assess functional activity in cell-based assays. Discuss potential effects on receptor binding, immunogenicity, and toxicity.

11. An unknown metabolite of a drug appears in patient urine. Propose a workflow combining high-resolution mass spectrometry and NMR to determine its chemical structure. Predict the site of biotransformation and potential pharmacological activity. Discuss considerations of phase I and phase II metabolism, potential toxicity, and implications for dosage optimization.

12. A series of β-lactam antibiotics is being optimized to combat resistant bacteria. Describe how electronic and steric modifications to the β-lactam ring affect enzymatic hydrolysis. Propose kinetic experiments to quantify susceptibility to β-lactamases. Discuss structural strategies to overcome resistance, such as introducing steric hindrance or electron-withdrawing groups, and predict effects on pharmacokinetics and clinical efficacy.

13. Nanoparticles for targeted drug delivery must balance ligand density for optimal uptake and minimal off-target binding. Calculate the optimal density based on receptor expression and particle surface area. Propose analytical techniques (e.g., XPS, fluorescence labeling) to quantify surface coverage and predict effects on circulation time, biodistribution, and therapeutic index. Discuss trade-offs between targeting efficiency and immunogenicity.

14. Design gadolinium chelates for MRI that maximize relaxivity and minimize toxicity. Predict how electronic properties and coordination geometry influence MRI signal. Propose synthetic and analytical strategies (NMR, HPLC, relaxometry) to verify structure and function. Discuss pharmacokinetics, renal clearance, and potential side effects, including nephrogenic systemic fibrosis risk.

15. A hydrophobic anticancer drug is poorly soluble in water. Propose chemical strategies to form inclusion complexes with cyclodextrins, enhancing solubility and stability. Design dissolution studies to compare release profiles. Predict how host-guest interactions affect pharmacokinetics, biodistribution, and bioavailability, and propose analytical techniques to confirm complex formation.

16. A prodrug is activated by liver enzymes. Using Michaelis-Menten kinetics, calculate expected conversion rates given Km and Vmax values. Discuss the effect of enzyme expression variability on dosing, therapeutic efficacy, and toxicity. Design experiments to quantify activation rates and model in vivo pharmacokinetics, integrating both computational and experimental approaches.

17. Design a nucleoside analog targeting a viral polymerase. Predict how substitutions at the 2′ and 3′ positions of the ribose influence polymerase incorporation, chain termination, and resistance development. Propose assays for enzyme kinetics, viral replication inhibition, and cytotoxicity in host cells. Discuss potential metabolic modifications and pharmacokinetic implications.

18. A peptide therapeutic is PEGylated to extend circulation time. Predict effects of PEG size and attachment site on receptor binding, immunogenicity, and clearance. Design experiments to quantify PEG conjugation efficiency, plasma half-life, and functional activity. Discuss trade-offs between steric hindrance and pharmacodynamic performance.

19. Design an HPLC gradient method to separate stereoisomers of a chiral drug. Discuss how column chemistry, mobile phase composition, and temperature affect resolution. Propose quantitative strategies to measure enantiomeric excess and purity, considering regulatory requirements. Discuss implications of stereochemistry on pharmacokinetics, pharmacodynamics, and safety.

20. A drug forms an unknown metabolite adduct detected by mass spectrometry. Predict the site and mechanism of biotransformation, considering phase I and phase II reactions. Design experiments using LC-MS/MS and NMR to confirm structure. Discuss implications for activity, toxicity, and dosing strategies.

21. Design a chemotherapeutic delivery system that selectively releases the drug in the acidic tumor microenvironment. Identify chemical moieties, such as hydrazones or imines, that undergo cleavage at pH ~6.5. Propose in vitro studies using buffer systems to simulate extracellular and intracellular conditions and monitor release kinetics via HPLC or fluorescence spectroscopy. Discuss the balance between linker stability in circulation versus rapid cleavage in tumors, and predict pharmacokinetic and pharmacodynamic implications, including off-target toxicity.

22. A lipid biomarker is identified as predictive for a metabolic disease. Design chemical derivatization strategies to enhance ionization efficiency in mass spectrometry and improve detection sensitivity. Propose a workflow for sample preparation, extraction, and analysis of plasma samples. Discuss potential confounding factors such as isobaric species, matrix effects, and inter-individual metabolic variability. Evaluate how these chemical strategies contribute to diagnostic accuracy and reproducibility.

23. Design chemical modifications (e.g., 2′-O-methyl, phosphorothioate) to enhance stability of guide RNAs in CRISPR-Cas systems while maintaining Cas9 recognition and activity. Predict how modifications influence nuclease resistance, cellular uptake, and off-target effects. Propose in vitro and cell-based assays to quantify

editing efficiency, off-target events, and RNA degradation. Discuss the interplay between chemical stability, immune activation, and therapeutic efficacy.

24. AI models predict novel chemical scaffolds with high predicted binding affinity. Develop a synthetic route for the top candidate, considering stereochemistry, functional group compatibility, and scalability. Discuss potential synthetic bottlenecks and propose strategies for stereoselective reactions. Complement computational predictions with in vitro assays to validate binding, activity, and metabolic stability.

25. An mRNA vaccine formulation is sensitive to aggregation during storage. Apply physical chemistry principles to predict aggregation propensity based on ionic strength, pH, and temperature. Design experiments using dynamic light scattering and zeta potential measurements to monitor particle size and stability. Discuss how chemical modifications of lipids or mRNA sequence affect stability, cellular uptake, and immune response.

26. A radiopharmaceutical shows rapid renal clearance, limiting imaging utility. Propose chemical modifications (e.g., PEGylation, albumin-binding moieties) to extend circulation. Design preclinical imaging studies to assess biodistribution, target engagement, and clearance rates. Predict how modifications influence radiation dose to healthy tissues and contrast agent efficacy.

27. Design a small-molecule inhibitor targeting a metal-containing enzyme. Propose chelating functional groups to bind the metal cofactor, optimizing thermodynamic stability and selectivity. Predict the effect of electronic structure and ligand geometry on enzyme inhibition. Develop in vitro assays to quantify binding affinity, inhibitory kinetics, and metal displacement, and discuss potential toxicity from off-target metal chelation.

28. Design chemical strategies for covalent attachment of an enzyme to an electrode surface, preserving catalytic activity. Compare linker chemistries (e.g., amine, thiol, click chemistry) for stability, orientation, and activity. Propose electrochemical or spectrophotometric assays to quantify surface coverage and activity, and predict how immobilization affects sensitivity, response time, and reproducibility of a biosensor.

29. A nanoparticle-drug conjugate exhibits non-specific protein adsorption in plasma. Propose chemical modifications, such as PEGylation or zwitterionic coatings, to reduce protein binding. Design experiments using surface plasmon resonance or isothermal titration calorimetry to quantify adsorption. Discuss trade-offs between stealth properties, targeting efficiency, and cellular uptake.

30. A peptide-drug conjugate is internalized selectively in cancer cells. Using thermodynamic and kinetic models, predict how linker length, hydrophobicity, and charge affect cellular uptake and drug release. Design in vitro experiments with fluorescence-labeled peptides to measure internalization and endosomal escape. Discuss implications for pharmacokinetics, therapeutic index, and off-target effects.

31. An RNA-based therapeutic shows rapid degradation in serum. Propose chemical modifications (e.g., locked nucleic acids, 2′-O-methyl nucleotides) to improve stability while maintaining target specificity. Design in vitro serum stability assays, quantify half-life, and predict effects on cellular uptake and immune activation. Discuss potential challenges in scaling up modified oligonucleotide synthesis.

32. Design a multi-drug delivery platform capable of sequential release. Propose chemical strategies using stimuli-responsive linkers sensitive to pH, enzymes, or redox conditions. Develop in vitro assays to measure temporal release profiles and cytotoxicity. Discuss how linker chemistry, drug compatibility, and particle architecture influence release kinetics and therapeutic synergy.

33. A novel imaging agent aggregates due to hydrophobicity. Propose chemical modifications (e.g., addition of polar groups, salt formation, cyclodextrin inclusion) to improve solubility. Predict effects on receptor binding, biodistribution, and signal intensity. Design in vitro assays to evaluate solubility, aggregation propensity, and imaging performance.

34. Design a chemical sensor to detect a low-abundance biomarker with high specificity. Propose functional groups for molecular recognition and discuss signal transduction strategies (electrochemical, optical, or fluorescence). Design calibration experiments, evaluate limits of detection, and consider interference from other biomolecules. Discuss trade-offs between sensitivity, selectivity, and response time.

35. 35. A small-molecule drug exhibits rapid clearance. Propose chemical strategies, such as isotope labeling, steric hindrance, or metabolically stable analogs, to reduce biotransformation. Design pharmacokinetic studies to assess improved half-life and exposure. Discuss the impact of modifications on efficacy, toxicity, and off-target effects.

36. A chemotherapeutic is conjugated to a fluorophore for imaging. Predict how linker length, polarity, and conjugation site affect fluorescence efficiency, biodistribution, and therapeutic efficacy. Design analytical methods (LC-MS, fluorescence spectroscopy) to verify conjugation and stability. Discuss potential pharmacokinetic changes and clinical utility.

37. Develop a green chemistry approach for synthesizing a new drug. Propose solvent-free or catalytic pathways, energy-efficient methods, and safer reagents. Calculate atom economy, E-factor, and potential environmental impact. Discuss scalability, yield, and purity considerations for pharmaceutical production.

38. Design a vaccine adjuvant with chemical modifications to enhance immunogenicity while minimizing toxicity. Predict how variations in lipid or polymer composition influence immune activation. Propose in vitro assays using dendritic cells or macrophages, and in vivo studies to monitor cytokine response, antigen presentation, and safety.

39. An antisense oligonucleotide shows off-target binding. Propose chemical modifications to the backbone or nucleobases to improve target specificity and reduce immune activation. Design in vitro hybridization assays, RNA-seq analysis, and functional tests to quantify off-target effects and therapeutic efficacy.

40. A drug degrades under ambient light. Propose chemical modifications (e.g., substitution of labile moieties) or stabilizers (antioxidants, UV blockers) to enhance photostability. Design photostability studies with controlled light exposure, quantifying degradation kinetics using HPLC or LC-MS. Discuss implications for formulation, storage, and patient safety.

41. A liposomal anticancer agent shows differential tissue distribution. Using chemical and physical properties of lipids and drugs, model organ partitioning and predict therapeutic index. Design in vivo biodistribution studies, considering particle size,

surface charge, and circulation half-life. Discuss trade-offs between targeted delivery, clearance, and off-target toxicity.

42. Design a prodrug with ester or amide linkages. Predict enzymatic cleavage rates, in vivo stability, and drug release profiles. Design LC-MS assays to quantify conversion and pharmacokinetic studies to measure bioavailability. Discuss trade-offs between rapid activation and systemic stability.

43. AI identifies potential toxicophores in a lead compound. Propose chemical modifications to reduce toxicity while maintaining activity. Design in vitro and in vivo toxicity assays, considering reactive metabolite formation and organ-specific effects. Discuss the balance between chemical reactivity and therapeutic function.

44. Polymeric nanoparticles deliver CRISPR-Cas systems. Predict how polymer composition, molecular weight, and surface charge influence uptake, endosomal escape, and gene editing efficiency. Design in vitro assays for cellular uptake, gene editing, and cytotoxicity. Discuss strategies to balance chemical stability with biological activity.

45. A peptide shows rapid renal clearance. Suggest chemical strategies such as cyclization, fatty acid conjugation, or PEGylation to prolong half-life. Design pharmacokinetic studies in animals to evaluate improvements. Discuss trade-offs between stability, activity, and potential immunogenicity.

46. An antibody-based imaging agent accumulates in non-target tissues. Propose chemical modifications or conjugation strategies to reduce off-target binding. Predict how changes affect binding affinity, biodistribution, and imaging contrast. Design in vivo imaging studies to quantify specificity and pharmacokinetics.

47. Use isotopic labeling to trace metabolic pathways. Design a labeling strategy, predict isotopologue distribution in key metabolites, and propose LC-MS/MS methods for quantitative analysis. Discuss potential pitfalls such as isotope scrambling, matrix effects, and detection limits.

48. siRNA nanocarriers must escape endosomes efficiently. Propose chemical modifications to lipids or polymers that enhance pH-sensitive release. Design in vitro assays with fluorescent reporters to quantify endosomal escape and gene silencing efficiency. Discuss how chemical design influences efficacy and toxicity.

49. Design a nanoparticle that combines a cytotoxic drug, targeting ligand, and imaging moiety. Predict how spatial arrangement, chemical compatibility, and linker chemistry affect stability, release, and efficacy. Design analytical characterization using TEM, DLS, fluorescence, and drug release studies. Discuss pharmacokinetics, biodistribution, and therapeutic outcomes.

50. Develop a green chemistry approach for scaling up a biologic therapeutic. Optimize solvents, catalysts, and reaction conditions to maximize yield and purity while minimizing environmental impact. Design analytical and process controls to ensure bioactivity and regulatory compliance. Discuss challenges of translating lab-scale synthesis to industrial production while adhering to sustainability principles.

Chapter 25

The Elements of Dentistry

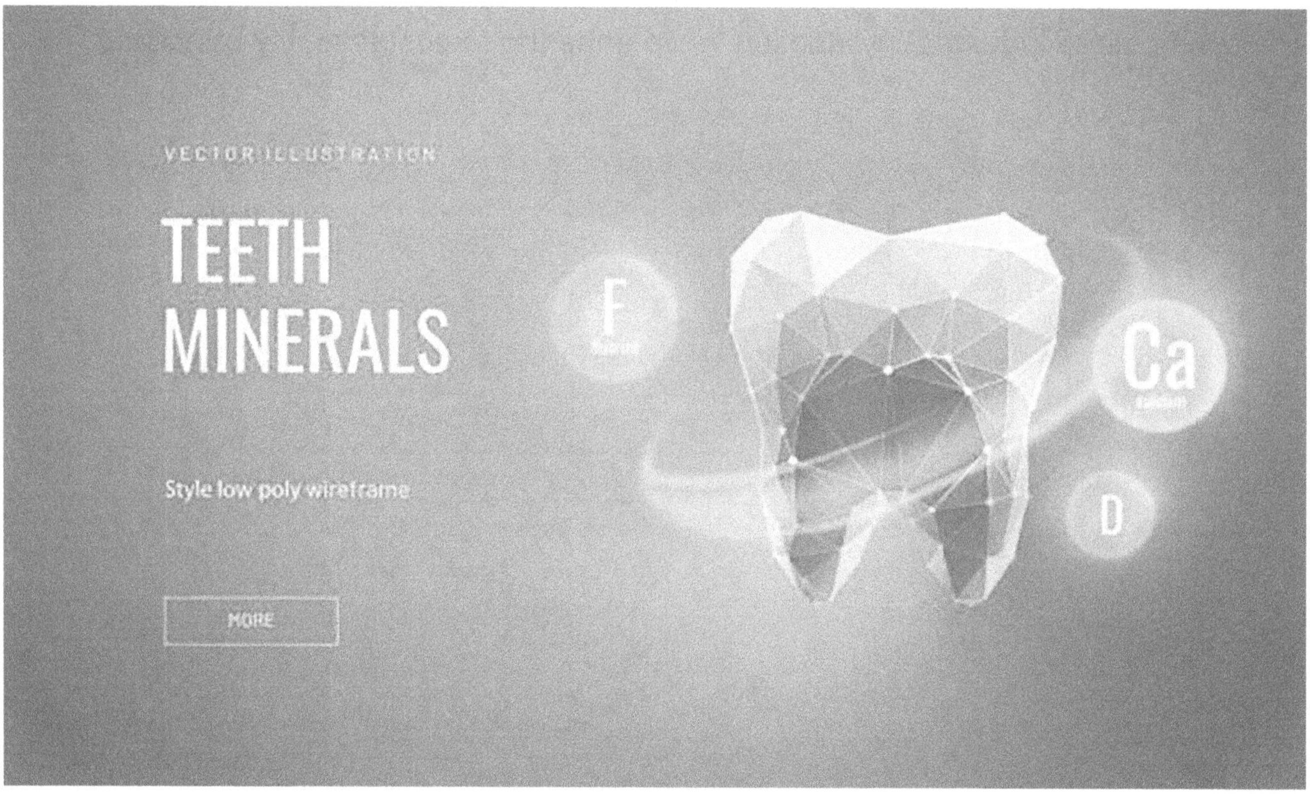

Dental chemistry, also known as oral chemistry, investigates the chemical principles underlying the health, maintenance, and treatment of teeth and oral tissues. It integrates concepts from inorganic, organic, and physical chemistry to explain processes such as tooth formation, decay, restorative materials, and preventive measures. A thorough understanding of dental chemistry is essential for dentists, researchers, and cosmetic chemists to develop effective treatments, preventive strategies, and safe materials for oral care. This field encompasses the chemistry of enamel, dentin, cementum, saliva, biofilms, dental restoratives, and therapeutic agents used in modern dentistry.

Teeth are composed of three primary tissues: enamel, dentin, and cementum. Enamel, the hardest biological material in the human body, is primarily hydroxyapatite [$Ca_{10}(PO_4)_6(OH)_2$], which provides rigidity and resistance to mechanical stress. Its highly ordered rod-like structure influences both its optical properties and mechanical strength. Beneath the enamel lies dentin, a collagen-rich matrix embedded with hydroxyapatite that combines flexibility with hardness, while cementum covers the tooth root and facilitates attachment to the periodontal ligament. The chemical properties of these tissues dictate their interactions with acids, bases, and fluoride ions. Acidic conditions in the oral cavity, often produced by bacterial metabolism, cause demineralization by dissolving hydroxyapatite, leading to dental caries, whereas remineralization occurs when calcium, phosphate, and fluoride ions redeposit into the tooth matrix, highlighting the critical chemical balance necessary to maintain tooth integrity.

Saliva, a complex aqueous solution containing electrolytes, proteins, enzymes, and organic molecules, plays a pivotal chemical role in maintaining oral health. It buffers oral pH primarily through bicarbonate (HCO_3^-), phosphate, and protein systems, preventing excessive acidity that can damage enamel. Saliva also provides calcium and phosphate ions essential for remineralization and contains antimicrobial agents, such as lysozyme, lactoferrin, and immunoglobulins, which chemically interact with bacterial surfaces to inhibit biofilm formation. Its viscosity, ionic strength, and composition are central to these protective functions, and alterations, such as reduced salivary flow or pH changes, increase susceptibility to caries, periodontal disease, and other oral pathologies.

Dental plaque is a biofilm formed by bacterial colonization on tooth surfaces, and its formation and pathogenicity are governed by biochemical interactions among bacterial extracellular polysaccharides, proteins, and salivary components. Streptococcus mutans, a key cariogenic bacterium, metabolizes dietary sugars to produce lactic acid, lowering the local pH and promoting demineralization. The chemical environment within plaque is dynamic, with gradients of oxygen, pH, and nutrients influencing bacterial metabolism and virulence. Plaque chemistry also explains the formation of calculus, or tartar, as salivary calcium and phosphate ions precipitate within the biofilm, forming hydroxyapatite or whitlockite crystals that calcify and adhere to tooth surfaces. Understanding these chemical interactions is crucial for designing effective antimicrobial agents and oral hygiene products.

Fluoride plays a central role in preventive dentistry by enhancing enamel remineralization and inhibiting demineralization. Fluoride ions (F^-) replace hydroxide ions in hydroxyapatite to form fluorapatite [$Ca_{10}(PO_4)_6F_2$], which is less soluble in acidic conditions, thereby increasing enamel resistance to acid attacks and slowing caries progression. Topical fluoride applications, such as toothpaste, mouth rinses, and varnishes, provide a direct source of F^- ions that diffuse into enamel, while systemic fluoride from water or dietary supplements incorporates into developing teeth. The effectiveness of fluoride depends on its concentration, pH, and duration of exposure, reflecting the chemical principles that guide its preventive use.

Dental restorative materials rely on distinct chemical properties to restore tooth structure. Dental amalgams, composed of mercury combined with silver, tin, and copper, set through a chemical reaction in which mercury diffuses into alloy particles, forming intermetallic compounds that are mechanically strong and corrosion-resistant. Composite resins consist of organic polymer matrices, typically bis-GMA or UDMA, reinforced with inorganic fillers such as silica or zirconia, and undergo radical-initiated polymerization, often light-activated, where the chemistry of monomer conversion, cross-linking, and filler interaction determines strength and durability. Glass ionomers combine fluoroaluminosilicate glass powders with polyacrylic acid in an acid-base reaction that releases fluoride ions while forming a chemically bonded matrix with the tooth, providing adhesion and anticariogenic properties. Dental ceramics, such as alumina and zirconia, are highly crystalline materials whose sintering process involves chemical diffusion and phase transformations to achieve chemical stability, biocompatibility, and aesthetic translucency. Dental cements, including zinc phosphate, polycarboxylate, resin-based, and glass ionomer cements, utilize chemical reactions such as acid-base reactions, polymerization, and chelation to achieve adhesion, mechanical strength, and biocompatibility. For instance, polycarboxylate cements form chelate complexes with calcium ions in the tooth, creating strong chemical bonds without requiring extensive mechanical retention, while resin adhesives employ free-radical polymerization to bond composite resins to etched enamel or dentin surfaces, exploiting chemical interactions at the interface.

Tooth whitening relies on oxidative chemical reactions that break down chromogenic molecules responsible for discoloration. Hydrogen peroxide (H_2O_2) or carbamide peroxide decomposes to release reactive oxygen species that attack conjugated double bonds in organic pigments, with the process influenced by pH, temperature, contact time, and the chemical stability of the peroxide solution. Other chemical agents incorporated into oral care products help prevent caries, control plaque, or manage gingivitis. Chlorhexidine, a cationic bisbiguanide, disrupts bacterial membranes; xylitol, a sugar alcohol, interferes with bacterial glycolysis to reduce acid production; and casein phosphopeptides (CPP-ACP) stabilize calcium and phosphate ions to promote remineralization. The effectiveness of these agents is grounded in their chemical interactions with both the oral environment and microbial metabolism.

Oral microbiology is intimately tied to chemical principles, as bacterial enzymes such as glucosyltransferases and proteases catalyze reactions that facilitate biofilm formation, tissue invasion, and acid production. Antibiotics used in dentistry, including amoxicillin and metronidazole, act by chemically inhibiting cell wall synthesis or metabolic pathways. Understanding microbial resistance mechanisms, such as β-lactamase activity and efflux pumps, is essential for designing effective treatment strategies. Advances in dental chemistry aim to develop smarter materials and therapeutics, including nanoparticles with antimicrobial or remineralizing properties, bioactive glasses that release ions to enhance tissue regeneration, enzyme-based therapeutics targeting biofilm metabolism, and regenerative dentistry approaches that leverage chemical signaling pathways to stimulate dental pulp repair and dentin formation.

Dental chemistry provides a foundational understanding of the chemical principles that govern oral health, disease prevention, and restorative dentistry. From the atomic structure of hydroxyapatite to the polymerization reactions in composite resins, chemical knowledge informs the development of preventive, therapeutic, and restorative strategies. By integrating insights from inorganic chemistry, organic chemistry, and biochemistry, dental chemistry continues to innovate, advancing materials and techniques that improve the quality of oral care and patient outcomes.

Dental biofilm formation is a dynamic chemical process influenced by both microbial metabolism and the physicochemical properties of the oral environment. Biofilms are not static; their chemical composition evolves over time. Early colonizers, such as Streptococcus sanguinis and Actinomyces species, produce extracellular polysaccharides through enzymatic reactions involving glucosyltransferases and fructosyltransferases. These enzymes catalyze the conversion of dietary sucrose into high-molecular-weight glucans and fructans, creating a sticky matrix that enhances bacterial adhesion. The chemical properties of these polysaccharides—such as charge distribution, solubility, and degree of branching—affect biofilm structure, porosity, and mechanical stability. The resultant microenvironment exhibits chemical gradients, including reduced oxygen concentration, altered pH, and accumulation of metabolic byproducts like organic acids, which further modulate bacterial physiology and interspecies interactions.

The chemical interplay between biofilm and enamel is critical in understanding dental caries. Hydroxyapatite dissolution is governed by Le Chatelier's principle: when hydrogen ions from bacterial acids shift the equilibrium, the system compensates by releasing calcium and phosphate ions from enamel. Remineralization is the reverse process, where supersaturation of calcium and phosphate ions, often aided by fluoride, drives precipitation back onto the enamel surface. Kinetic factors, such as ion diffusion rates and the presence of nucleation sites within the enamel microstructure, determine the efficiency of this chemical process. The use of chemical modeling has allowed dental researchers to predict demineralization and remineralization cycles, informing preventive strategies like fluoride varnishes and remineralizing pastes.

Saliva's chemical composition extends beyond buffering and mineral supply. It contains a diverse array of proteins, including mucins, proline-rich proteins, histatins, and statherin, each with specific chemical functionalities. Mucins, rich in glycosylated residues, contribute to saliva viscosity and facilitate the formation of the salivary pellicle, a thin proteinaceous layer on enamel surfaces. This pellicle modulates ion diffusion and acts as a selective barrier to bacterial adhesion. Statherin and proline-rich proteins bind calcium and phosphate ions, stabilizing them in supersaturated solution to prevent spontaneous precipitation, thereby maintaining a chemical environment conducive to controlled remineralization.

Fluoride chemistry is central to both preventive and therapeutic dentistry. Beyond fluorapatite formation, fluoride affects bacterial metabolism. Fluoride ions inhibit enolase, a key enzyme in the glycolytic pathway of cariogenic bacteria, reducing lactic acid production. Chemically, fluoride exists in multiple species depending on pH: free F^- ions, hydrogen fluoride (HF), or complexed fluoride, each with distinct reactivity and diffusion characteristics. Controlled delivery systems, including slow-release varnishes and fluoride-containing glass ionomer cements, exploit these chemical properties to sustain local fluoride concentrations, maintaining an environment favorable to remineralization while minimizing systemic exposure.

Restorative materials continue to evolve through detailed understanding of chemical reactions and interactions at the tooth–material interface. In composite resins, the kinetics of polymerization are influenced by radical initiator concentration, light intensity, and temperature. Incomplete polymerization leads to residual monomers, which can chemically interact with the oral environment or leach into surrounding tissues, highlighting the importance of precise chemical control. The filler particles—often silica or zirconia—are chemically silanized to improve covalent bonding with the resin matrix, enhancing mechanical properties and reducing water sorption. Additionally, the chemical stability of the resin–filler interface determines long-term color stability and resistance to hydrolytic degradation.

Glass ionomer cements exemplify the significance of acid-base chemistry in dentistry. The polyacrylic acid attacks the glass particles, leaching cations (Ca^{2+}, Al^{3+}) which cross-link with carboxylate groups to form a hardened matrix. The release of fluoride ions from the glass structure provides ongoing chemical protection against demineralization. Recent advances include resin-modified glass ionomers, which combine acid-base reactions with free-radical polymerization, creating a dual-cure chemical system with improved mechanical and aesthetic properties. Chemical tuning of the powder-to-liquid ratio, molecular weight of polyacrylic acid, and particle size distribution allows precise control over setting kinetics, ion release profiles, and final material properties.

Ceramics in dentistry rely on high-temperature chemical transformations. Alumina, zirconia, and lithium disilicate undergo sintering, a process driven by chemical diffusion of atoms across particle boundaries, densification, and phase transformations. Chemical additives, such as yttria in zirconia, stabilize the tetragonal phase, enhancing fracture toughness through a chemical–mechanical transformation mechanism. Surface chemical modifications, including glazing or acid etching, alter surface energy and wettability, improving adhesion to resin cements through chemical bonding.

Dental cements demonstrate sophisticated chemical bonding mechanisms. Polycarboxylate cements rely on chelation of calcium ions in enamel or dentin, forming strong, non-soluble complexes. Resin-based adhesives exploit both micromechanical interlocking and chemical interactions: phosphate or methacrylate functional groups form covalent bonds with hydroxyapatite or collagen, enhancing bond strength. Chemical inhibitors or stabilizers are incorporated to regulate polymerization kinetics, control shrinkage stress, and maintain long-term adhesive integrity. Understanding these reactions allows for precise manipulation of curing conditions and chemical formulations.

Tooth whitening exemplifies oxidative chemistry in oral care. Hydrogen peroxide decomposes to form hydroxyl radicals (•OH) and perhydroxyl anions (HO_2^-), highly reactive species that cleave chromophoric double bonds in organic molecules. The chemical efficiency of whitening depends on radical concentration, reaction pH, and the presence of catalytic ions, such as manganese or iron, which accelerate peroxide decomposition. Enamel chemical integrity is preserved through pH modulation and the use of remineralizing agents during or after whitening procedures. Innovations in chemical formulations aim to enhance selectivity, targeting only chromophores while minimizing collateral tissue damage.

Chemical agents in oral care extend beyond preventive applications. Chlorhexidine, with its cationic guanidinium groups, binds electrostatically to negatively charged bacterial surfaces, disrupting membrane integrity and precipitating cytoplasmic components. Its chemical interactions with salivary proteins and biofilm extracellular polymeric substances influence substantivity and antimicrobial efficacy. Polyphenols and essential oils exhibit chemical antioxidant or protein-denaturing effects that inhibit bacterial adhesion and metabolism. Xylitol, a polyhydric alcohol, competitively inhibits enzymes involved in glycolysis, reducing acid production, and is chemically stable in the oral environment, making it an effective non-cariogenic sweetener. Casein phosphopeptide–amorphous calcium phosphate (CPP-ACP) forms metastable complexes that prevent calcium phosphate precipitation while enhancing local ion supersaturation, promoting chemically driven remineralization.

Dental pharmacology and therapeutic chemistry leverage precise molecular interactions. Antibiotics used in dentistry target specific chemical pathways: β-lactams inhibit peptidoglycan cross-linking, metronidazole undergoes chemical reduction in anaerobic bacteria to generate cytotoxic radicals, and macrolides bind the bacterial ribosome to disrupt protein synthesis. Understanding the chemical basis of microbial resistance—such as hydrolysis of β-lactam rings or efflux pump-mediated ion transport—guides the design of new chemical agents and combination therapies. Enzyme inhibitors and chemical chelators are being explored to disrupt biofilm integrity or selectively modulate microbial metabolism.

Nanotechnology in dental chemistry exploits unique surface and chemical properties of nanoscale materials. Silver nanoparticles exhibit ion release chemistry that generates reactive oxygen species, disrupting bacterial membranes. Nano-hydroxyapatite mimics natural enamel chemistry, integrating seamlessly into demineralized regions and acting as a nucleation site for further mineral deposition. Bioactive glasses release calcium, phosphate, and silicate ions upon dissolution, driving localized chemical reactions that stimulate remineralization and soft tissue repair. Surface chemistry, particle size distribution, and ionic solubility are critical parameters in optimizing efficacy while ensuring biocompatibility.

Regenerative dentistry increasingly relies on chemical signaling pathways to direct cell behavior. Growth factors, chemokines, and small-molecule agonists engage receptor-mediated pathways, triggering intracellular cascades that drive dentinogenesis, pulp repair, and periodontal regeneration. Scaffold materials, often composed of biodegradable polymers or hydrogels, provide controlled chemical environments for ion release, pH modulation, and molecular signaling. Incorporating bioactive ions, such as fluoride, strontium, or zinc, further leverages chemical stimuli to enhance mineralization and cellular differentiation.

Dental material degradation is also a chemical phenomenon of practical importance. Hydrolysis, oxidation, enzymatic degradation, and acid-mediated dissolution all contribute to the long-term stability of restorative materials. Water sorption induces chemical swelling and hydrolytic breakdown of ester bonds in polymeric matrices. Oxidative radicals generated during oral inflammation can chemically modify both natural tissues and synthetic materials. Understanding these chemical pathways informs the development of materials with enhanced durability and biocompatibility.

Advanced analytical chemistry techniques underpin modern dental research. Nuclear magnetic resonance (NMR) spectroscopy, Fourier-transform infrared (FTIR) spectroscopy, Raman spectroscopy, X-ray diffraction (XRD), and scanning electron microscopy (SEM) provide detailed chemical insights into enamel composition, biofilm chemistry, and restorative material interactions. Thermodynamic and kinetic analyses allow prediction of demineralization rates, ion exchange equilibria, and polymerization dynamics. Surface chemical characterization, including X-ray photoelectron spectroscopy (XPS) and atomic force microscopy (AFM), elucidates molecular interactions at the material–tooth interface, guiding the rational design of next-generation dental materials.

The chemistry of dietary and environmental factors also impacts oral health. Sugars, acidic beverages, and chelating agents interact chemically with enamel and saliva. Acidic foods lower pH, promoting hydroxyapatite dissolution, while chelators like EDTA sequester calcium ions, hindering remineralization. Polyphenols in tea or wine chemically interact with salivary proteins, influencing pellicle formation and microbial adhesion. Understanding these chemical interactions informs dietary recommendations and the formulation of protective oral care products.

In conclusion, dental chemistry integrates inorganic, organic, physical, and analytical chemistry to understand and manipulate the complex chemical environment of the oral cavity. It encompasses mineral composition, biofilm chemistry, salivary dynamics, material science, pharmacology, nanotechnology, and regenerative approaches. By elucidating the chemical mechanisms underlying tooth formation, demineralization, remineralization, and material interactions, dental chemistry drives innovation in preventive, restorative, and therapeutic dentistry. This chemical perspective allows for the development of more effective treatments, safer materials, and personalized oral care strategies, ultimately enhancing oral health and patient outcomes.

Chapter 25: Word Problems

1. A 35-year-old patient presents with early enamel erosion, suspected to be due to reduced salivary flow caused by medication. Considering saliva as a dynamic aqueous buffer system with bicarbonate, phosphate, and protein contributions, calculate the rate of hydroxyapatite [$Ca_{10}(PO_4)_6(OH)_2$] dissolution when exposed to lactic acid produced by Streptococcus mutans in plaque. Model how the Ca^{2+} and PO_4^{3-} concentration gradients develop across the enamel surface over a 2-hour period, taking into account diffusion limitations within enamel rods. Additionally,

evaluate how topical fluoride treatments alter the kinetics by promoting fluorapatite formation, and discuss how pH buffering and local ion saturation influence demineralization dynamics.

2. During a clinical trial, fluoride varnish is applied to enamel surfaces of volunteers with moderate caries risk. Using chemical principles, calculate the theoretical increase in enamel acid resistance when hydroxide ions in hydroxyapatite are replaced by fluoride to form fluorapatite [$Ca_{10}(PO_4)_6F_2$]. Consider factors such as enamel porosity, fluoride ion mobility, and local pH. Discuss how repeated topical applications affect fluorapatite layer thickness over time and its long-term protective effect against low-pH challenges from carbohydrate metabolism.

3. Design a detailed experimental study to quantify how variations in salivary phosphate and bicarbonate concentrations affect enamel solubility under acidic conditions. Include the preparation of artificial saliva with different buffer capacities and the measurement of hydroxyapatite saturation using ion-selective electrodes. Using the Henderson-Hasselbalch equation, relate the buffer systems to the ability to maintain pH above the critical threshold for demineralization. Predict the comparative enamel dissolution for patients with normal versus low salivary buffering capacity.

4. A mature dental plaque biofilm contains multiple species, including Streptococcus mutans, Actinomyces, and Veillonella. Construct a chemical model of how pH gradients form within the biofilm due to lactic acid production and diffusion limitations. Analyze how oxygen depletion in deeper layers affects bacterial metabolism and acid production. Predict zones of enamel demineralization adjacent to the biofilm and discuss the chemical role of extracellular polysaccharides in modulating diffusion of Ca^{2+}, PO_4^{3-}, and fluoride ions into the biofilm.

5. A clinician uses a light-cured bis-GMA/UDMA composite resin for posterior restoration. Describe in detail how the concentration of radical initiator, light intensity, and penetration depth influence polymerization kinetics and monomer conversion. Predict residual monomer concentrations after 20 seconds of light curing, and discuss possible chemical interactions with saliva and biofilm. Additionally, consider the impact of oxygen inhibition at the surface on polymerization efficiency and long-term mechanical properties.

6. Design a comprehensive chemical model for fluoride ion release from a conventional glass ionomer cement in vivo. Include diffusion through the hardened matrix, ion-exchange equilibria with saliva, and the effect of pH fluctuations during meals and bacterial acid production. Predict the cumulative fluoride release over one month and discuss how sustained ion availability influences remineralization and caries prevention.

7. Explain the chelation reaction mechanism between polyacrylic acid and enamel Ca^{2+} ions in polycarboxylate cement. Calculate the theoretical bond strength of these chelates using thermodynamic parameters and evaluate the influence of salivary ions (e.g., Mg^{2+}, Na^{+}) on complex stability. Discuss how variations in powder-to-liquid ratio affect reaction kinetics, setting time, and adhesive performance in restorative procedures.

8. Hydrogen peroxide and carbamide peroxide are commonly used in whitening procedures. Design a kinetic experiment to quantify chromophore degradation in stained enamel while minimizing mineral loss. Include consideration of pH modulation, peroxide decomposition catalyzed by trace metal ions (Fe^{3+}, Mn^{2+}), and temperature effects. Predict the rate of enamel whitening and potential chemical interactions with enamel proteins or biofilm components.

9. Silver nanoparticles are incorporated into a dental varnish to prevent plaque formation. Construct a chemical kinetic model for Ag^{+} ion release in saliva, accounting for competing reactions with chloride and phosphate ions. Predict how variations in nanoparticle size, salivary ionic strength, and pH affect antimicrobial activity. Discuss implications for patients consuming a high-sugar diet that alters plaque metabolism and local acidity.

10. At an oral pH of 5.5 during a cariogenic challenge, calculate the relative proportion of fluoride in the forms of free F^{-} and hydrogen fluoride (HF). Evaluate how this speciation affects the diffusion of fluoride into enamel subsurface layers and its inhibitory effect on enolase in cariogenic bacteria. Discuss the chemical balance required to maximize remineralization while minimizing systemic fluoride absorption.

11. Nano-hydroxyapatite particles are applied to remineralize early carious lesions. Design a chemical study to determine how particle size and surface energy

affect nucleation rates on demineralized enamel. Include calculations for lattice matching with native hydroxyapatite and supersaturation thresholds for calcium and phosphate ions. Predict the optimal particle dimensions for efficient remineralization.

12. Using Gibbs free energy and solubility product data, determine whether saliva supersaturated with calcium and phosphate ions will spontaneously promote remineralization at pH 6.8. Analyze how the addition of fluoride or strontium ions modifies thermodynamic favorability and nucleation kinetics on enamel surfaces. Discuss the clinical implications for designing remineralizing pastes.

13. A bis-GMA-based resin restoration is exposed to constant saliva for one year. Estimate the rate of ester bond hydrolysis under physiological temperature (37°C) and pH fluctuations. Include contributions of salivary enzymes and water diffusion into the polymer matrix. Predict how hydrolytic degradation affects mechanical properties and ion release, and discuss strategies for improving chemical stability.

14. Streptococcus mutans converts dietary sucrose into extracellular glucans via glucosyltransferase enzymes. Calculate the expected concentration and molecular weight distribution of these glucans in a plaque biofilm given a steady sucrose flux. Predict how these polysaccharides alter the diffusion of Ca^{2+} and PO_4^{3-} ions, influence pH gradients, and modulate biofilm mechanical stability and cariogenic potential.

15. Chlorhexidine binds electrostatically to bacterial surfaces and biofilm matrix polysaccharides. Evaluate the chemical interactions governing substantivity in saliva, considering competition with proteins and divalent ions. Predict the effective concentration remaining after repeated salivary washout and its impact on antimicrobial efficacy over 12 hours.

16. During a high-sugar meal, plaque pH drops from 7.0 to 4.5. Using Le Chatelier's principle, calculate the expected shift in hydroxyapatite dissolution equilibrium. Quantify the resulting Ca^{2+} and PO_4^{3-} concentrations at the enamel surface over time, and predict the protective effect of applying fluoride immediately before or after the meal.

17. Investigate how chemical surface treatments of enamel, including acid etching, plasma modification, or silanization, modify surface energy. Calculate the interfacial

adhesion energy for phosphate-functionalized resin monomers, predict bonding strength, and evaluate the long-term effects on hydrolytic stability and resistance to enzymatic degradation by saliva or biofilm.

18. Simulate the time-dependent release of fluoride ions from a 5% NaF varnish applied to enamel. Include mass transfer through the varnish layer, salivary washout, and diffusion into subsurface enamel prisms. Predict fluoride concentrations over 24 hours and discuss implications for inhibiting bacterial metabolism and enhancing remineralization.

19. Quantify the chemical mechanism by which statherin, proline-rich proteins, and mucins stabilize calcium and phosphate ions in supersaturated saliva. Use equilibrium calculations to determine the pH threshold at which spontaneous precipitation occurs, and predict how reduced protein concentration affects enamel demineralization susceptibility during cariogenic challenges.

20. Xylitol inhibits glycolysis in Streptococcus mutans. Quantify the reduction in lactic acid production as a function of xylitol concentration using Michaelis–Menten kinetics. Predict how this chemically influences local plaque pH and hydroxyapatite dissolution, and discuss the effect of repeated exposure on bacterial metabolism over time.

21. A dentist applies a 10% carbamide peroxide gel for at-home tooth whitening. Model the chemical decomposition of carbamide peroxide into hydrogen peroxide and urea, followed by hydrogen peroxide breakdown into hydroxyl radicals (•OH) and perhydroxyl anions (HO_2^-). Quantify the radical concentration generated per minute in enamel pores, considering pH, temperature, and trace metal ion catalysis (Fe^{3+}, Mn^{2+}). Predict the selective oxidative degradation of chromophoric molecules in stained enamel while minimizing hydroxyapatite loss.

22. A silver–tin–copper dental amalgam is freshly mixed with mercury. Construct a detailed chemical model of mercury diffusion into alloy particles, formation of intermetallic compounds (Ag_2Hg_3, Cu_6Sn_5), and resultant microstructure development. Evaluate the effects of ambient temperature, humidity, and alloy composition on reaction kinetics, mechanical strength, and corrosion resistance over time.

23. Composite resins are reinforced with silanized silica fillers. Explain the chemical silanization reaction mechanism that forms covalent bonds between silica surfaces and resin monomers. Calculate the expected bond density per unit surface area and predict its influence on water sorption, hydrolytic stability, and long-term color and mechanical properties of the restoration.

24. Determine the minimum concentrations of Ca^{2+} and PO_4^{3-} ions in saliva required to maintain supersaturation with respect to hydroxyapatite at a critical pH of 5.8. Include ionic strength corrections and consider the effect of saliva proteins such as statherin and proline-rich proteins. Predict how deviations in ion concentration influence enamel demineralization rates during a high-sugar diet.

25. Bioactive glass is used to repair early enamel lesions. Model the dissolution kinetics of calcium, phosphate, and silicate ions from glass particles in saliva. Predict changes in local pH and the subsequent rate of hydroxyapatite nucleation on surrounding dentin. Discuss how particle size distribution, surface area, and saliva composition modulate ion release and remineralization efficiency.

26. Casein phosphopeptide–amorphous calcium phosphate (CPP–ACP) complexes are applied for remineralization. Design an experiment to quantify the stability of metastable calcium phosphate complexes in the presence of salivary proteins and varying pH. Predict the rate of remineralization of early enamel lesions and the effect of repeated acidic challenges.

27. Fluoroaluminosilicate glass particles react with polyacrylic acid in a glass ionomer cement. Calculate the stoichiometry of the reaction, predict fluoride ion release over the first 24 hours, and determine matrix hardness. Discuss the effect of particle size, acid concentration, and powder-to-liquid ratio on the chemical reaction kinetics and long-term protective effects.

28. Investigate the chemical origin of natural enamel fluorescence, focusing on protein and chromophore contributions. Predict how early demineralization alters the electronic environment of chromophores and the resulting changes in fluorescence intensity. Consider the effect of dietary stains, hydrogen peroxide, and enamel porosity on these chemical interactions.

29. Using standard Gibbs free energy data, calculate the equilibrium constant for chelation of enamel Ca^{2+} ions by polyacrylic acid chains. Include competition from

Mg^{2+} and Na^{+} in saliva, and predict how variations in temperature, pH, and ionic strength influence chelation efficiency and adhesion strength of the cement.

30. Composite restorations are exposed to reactive oxygen species generated during oral inflammation. Model the oxidative chemical degradation of polymer matrices, including free radical attack on ester and urethane bonds. Predict the rate of mechanical property loss, color changes, and monomer leaching over time, and propose chemical strategies to improve polymer durability.

31. Simulate cycles of demineralization and remineralization during meals using chemical kinetics. Quantify net mineral gain or loss over 24 hours in enamel treated with fluoride toothpaste. Include diffusion limitations, salivary ion concentrations, supersaturation, and competitive adsorption of ions on enamel surfaces. Discuss the influence of pH fluctuations, fluoride availability, and biofilm presence.

32. Yttria-stabilized tetragonal zirconia is used in all-ceramic crowns. Explain the chemical mechanism by which yttria stabilizes the tetragonal phase and prevents spontaneous tetragonal-to-monoclinic transformation. Calculate the energy barrier for phase transformation and predict the impact on fracture toughness and long-term mechanical reliability.

33. Design a polymeric delivery system for antimicrobial agents in periodontal pockets with pH ~6.5. Describe the chemical mechanism of pH-triggered drug release, including protonation/deprotonation of polymer functional groups. Model drug diffusion, local concentration profiles, and the expected impact on biofilm reduction. Discuss potential side reactions with salivary components.

34. Fluoride ions inhibit enolase in cariogenic bacteria. Develop a kinetic model incorporating Michaelis–Menten parameters, fluoride concentration, and local pH. Predict the reduction in lactic acid production over time and its effect on plaque acidogenicity and enamel demineralization. Include chemical considerations for fluoride speciation (F^- vs. HF).

35. Dental calculus forms from the precipitation of calcium phosphate salts in plaque. Model the chemical equilibrium between salivary Ca^{2+} and PO_4^{3-} ions under varying pH, presence of bacterial nucleation sites, and plaque polysaccharide matrix. Predict the rate of mineral deposition and discuss how oral hygiene or chelating agents affect this process.

36. Compare the chemical setting kinetics of zinc phosphate cement and polycarboxylate cement. Include acid-base reactions, ion release, exothermicity, and matrix formation. Predict mechanical strength development over time and evaluate the influence of salivary moisture, temperature, and powder-to-liquid ratio on setting chemistry.

37. Enamel contains carbonate and magnesium substitutions in the hydroxyapatite lattice. Quantify how these substitutions alter solubility, mechanical strength, and susceptibility to acidic dissolution. Predict the chemical consequences for patients with high-carbonate diets or systemic mineral imbalances.

38. Salivary mucins and glycoproteins form a pellicle on enamel surfaces. Analyze the chemical interactions governing adsorption, ion binding, and modulation of enamel surface charge. Predict how these interactions influence bacterial adhesion, diffusion of Ca^{2+}/PO_4^{3-} ions, and enamel protection during cariogenic challenges.

39. Salivary lysozyme and lactoferrin contribute to antimicrobial defense. Quantify reaction rates of bacterial cell wall hydrolysis under physiologic ionic strength. Predict the combined chemical effect of enzyme concentration, pH, and salivary ionic composition on plaque biofilm reduction.

40. Given measured concentrations of Ca^{2+} and PO_4^{3-} in patient saliva, calculate the ion activity product and compare it with hydroxyapatite solubility product. Determine whether demineralization or remineralization is thermodynamically favored, and discuss how fluoride supplementation alters this equilibrium.

41. Hydroquinone acts as a polymerization inhibitor in resin composites. Explain the chemical mechanism of radical quenching, predict its effect on polymerization kinetics, degree of monomer conversion, and final mechanical properties. Include the effect of inhibitor concentration and temperature on the reaction.

42. Design an in vitro simulation of repeated acid attacks due to high-sucrose diets. Include chemical kinetics of hydroxyapatite dissolution, diffusion of ions, salivary buffering, and the effect of fluoride or calcium phosphate supplementation. Predict cumulative enamel mineral loss over a week.

43. During peroxide-based whitening, hydroxyl radicals may modify salivary mucins. Quantify the chemical alteration of functional groups, predict changes in viscosity

and biofilm resistance, and discuss strategies to minimize protein oxidation while maximizing whitening efficacy.

44. Construct a chemical kinetic model of sucrose metabolism by S. mutans, incorporating lactic acid and glucan production rates. Predict local pH changes and hydroxyapatite dissolution rates. Include enzyme kinetics, substrate availability, and diffusion limitations within plaque biofilm microenvironments.

45. Calculate the photochemical energy absorbed by a 2 mm layer of composite resin during curing at 470 nm wavelength. Predict radical formation rates, degree of polymerization, and the depth of cure as a function of light intensity and exposure time. Include chemical quenching effects of oxygen at the surface.

46. Model the diffusion coefficient of fluoride ions into enamel prisms during topical application. Predict time-dependent fluoride concentration profiles in subsurface layers under pH 6.0 and discuss implications for cariogenic inhibition and remineralization.

47. EDTA chelates calcium in saliva. Assess how this affects hydroxyapatite solubility and remineralization potential during acidic challenges. Predict enamel mineral loss over a 24-hour period with intermittent EDTA exposure and discuss chemical interactions with fluoride.

48. During in-office peroxide whitening, predict the steady-state concentration of hydroxyl radicals in enamel pores. Calculate reaction rates for chromophore oxidation versus hydroxyapatite damage. Discuss how pH modulation, radical scavengers, and peroxide concentration influence selective whitening chemistry.

49. Model the substitution of sodium ions by protons in a bioactive glass implant immersed in saliva. Predict the resulting local pH increase, calcium phosphate precipitation, and hydroxyapatite nucleation rate. Include the chemical effect of competing ions such as Mg^{2+} and PO_4^{3-}.

50. Integrate chemical kinetics of demineralization, remineralization, bacterial metabolism, fluoride availability, and salivary buffering into a computational model. Predict net enamel mineral loss over six months in a patient with hyposalivation. Include chemical interactions between fluoride, calcium, phosphate, and biofilm polysaccharides, and propose intervention strategies based on model outcomes.

Chapter 26

Power of Forensics Chemistry

Forensic chemistry is the meticulous application of chemical principles to the investigation of crime and civil disputes. It serves as the bridge between the scientific and legal worlds, providing concrete evidence that can corroborate or refute testimonies, link suspects to criminal acts, and establish the identity or origin of materials. Every substance, whether visible or microscopic, carries a chemical signature.

Forensic chemists analyze these signatures using advanced techniques, transforming trace evidence into compelling, legally admissible data. The discipline requires a profound understanding of chemistry, keen observational skills, and rigorous methodological standards.

Historically, forensic chemistry has evolved alongside the broader field of analytical chemistry. In the 19th century, chemists such as Mathieu Orfila pioneered toxicology, analyzing poisons in biological samples to solve criminal cases. Orfila's work in detecting arsenic in exhumed bodies marked the beginning of modern forensic toxicology. Over time, the field has expanded to include the chemical analysis of drugs, explosives, fibers, paints, inks, and even environmental pollutants. Today, forensic chemistry combines traditional wet-chemical methods with cutting-edge instrumentation such as mass spectrometry, nuclear magnetic resonance, and electron microscopy to uncover details invisible to the naked eye.

A major component of forensic chemistry is biological evidence analysis. Biological samples—blood, hair, saliva, sweat, and tissues—contain rich chemical information that can link individuals to crime scenes. Blood, for example, can reveal alcohol, drugs, or poisons through chromatographic and spectrometric techniques. Gas chromatography-mass spectrometry (GC-MS) is widely used to separate and identify volatile organic compounds in biological matrices, while liquid chromatography-mass spectrometry (LC-MS) enables the analysis of non-volatile, polar substances. Proteins and nucleic acids extracted from these samples also provide critical information; forensic DNA analysis relies on precise extraction and amplification of genetic material, allowing the identification of individuals with extraordinary accuracy.

Toxicology is a specialized branch within forensic chemistry that focuses on the detection of toxins, drugs, and other harmful chemicals in biological systems. Forensic toxicologists assess whether chemicals have contributed to death, impairment, or criminal activity. Heavy metals such as arsenic, lead, and mercury are detected at trace levels using techniques like inductively coupled plasma mass spectrometry (ICP-MS). Organic toxins, including plant alkaloids, opioids, and synthetic drugs, are often analyzed using GC-MS or LC-MS. Understanding the chemical metabolism of these compounds is essential; some substances degrade rapidly in biological systems, requiring chemists to interpret their presence in light of metabolic processes.

The analysis of controlled substances is another central area. Illicit drugs, including cocaine, heroin, methamphetamine, and designer cannabinoids, often exist as complex mixtures. Analytical techniques such as infrared spectroscopy (IR), nuclear magnetic resonance (NMR), and X-ray diffraction (XRD) are used to determine chemical composition, purity, and structural features. Isotopic profiling can sometimes identify the geographical origin of a sample, providing critical investigative leads. Chemists also analyze precursor chemicals and adulterants, which can reveal manufacturing methods or distribution networks.

Explosives and post-blast residues require highly specialized forensic analysis. Explosive materials—TNT, RDX, PETN, and improvised devices—contain energetic compounds that can leave microscopic residues on surfaces or in debris. Detection techniques include ion mobility spectrometry (IMS), LC-MS, and capillary electrophoresis. By analyzing chemical residues, forensic chemists can reconstruct the type of explosive used, the method of detonation, and sometimes the source of the explosive. Chemical reactions involved in detonation, including oxidation-reduction processes and rapid decomposition, are studied to understand residue formation and to prevent secondary hazards during analysis.

Fire and arson investigations exemplify the intersection of chemistry and forensic interpretation. Fires often destroy evidence, complicating investigations. Chemists analyze debris to detect accelerants like gasoline, kerosene, or alcohol-based liquids, often using headspace gas chromatography to identify volatile compounds. Combustion chemistry is crucial for understanding how fires spread and for distinguishing accidental from deliberate ignition. Thermal decomposition reactions, pyrolysis of materials, and interaction of fire with chemical accelerants are carefully studied to reconstruct events. For example, a trace of naphthalene on charred flooring might indicate gasoline use, while its absence could suggest spontaneous combustion or electrical fault.

Trace evidence analysis focuses on materials transferred in minuscule amounts, including fibers, paints, glass, soil, and cosmetic residues. Each carries a unique chemical signature. Fibers can be analyzed using Fourier-transform infrared spectroscopy (FTIR) to determine polymer type and dye composition. Glass fragments can be characterized by their elemental composition through scanning electron microscopy (SEM) with energy-dispersive X-ray spectroscopy (EDS). Soil analysis often involves elemental profiling to link samples from a suspect to a specific location. The underlying principle is Locard's Exchange Principle, which states that every contact leaves a trace. Chemical analysis transforms these tiny transfers into critical evidence.

The chemistry of documents and inks is also significant. Forensic document examination often requires chemical analysis to detect alterations, forgeries, or discrepancies in inks and papers. Ink compositions vary due to dyes, pigments, and solvents, which can be differentiated using thin-layer chromatography (TLC) or Raman spectroscopy. Paper, depending on the manufacturing process, contains different fillers and bleaching agents, which can be analyzed chemically to establish consistency across documents. Even adhesives in envelopes or stamps can be characterized, linking documents to sources or revealing tampering.

Forensic chemistry also applies to environmental and industrial investigations. Environmental forensic chemistry deals with the identification of pollutants, hazardous chemicals, and contamination sources. Heavy metals, persistent organic pollutants, and industrial chemicals in soil, water, or air can be traced back to factories, waste disposal sites, or illegal dumping. Analytical techniques must consider chemical stability, degradation products, and interactions with environmental matrices. Similarly, industrial chemical analyses might be necessary in workplace accidents or product liability cases, where the chemical composition of materials is central to determining cause and responsibility.

Analytical instrumentation is the backbone of forensic chemistry. Chromatography, spectroscopy, and microscopy are routinely employed to detect, identify, and quantify substances. GC-MS combines separation and identification for volatile organic compounds. LC-MS/MS allows for high-resolution detection of complex, non-volatile molecules, including metabolites and synthetic drugs. Nuclear magnetic resonance (NMR) provides structural details of organic compounds, while atomic absorption spectroscopy (AAS) and ICP-MS detect metals at trace levels. Advanced microscopy, including SEM and transmission electron microscopy (TEM), reveals morphology and elemental composition at the nanoscale. Chemometric and statistical tools are increasingly applied to interpret complex datasets, detect patterns, and support decision-making.

Emerging trends in forensic chemistry include portable detection devices, which allow rapid on-site analysis. Handheld Raman spectrometers, portable gas chromatographs, and ion mobility spectrometers enable immediate assessment of chemical substances at crime scenes, reducing sample degradation and accelerating investigations. Computational chemistry and machine learning tools assist in predicting chemical behavior, interpreting mass spectra, and matching unknown substances with reference databases. These technologies expand the capabilities of forensic chemists, making the field faster, more accurate, and increasingly proactive.

Standards, validation, and quality control are critical to forensic credibility. Every method must be validated for accuracy, precision, sensitivity, and specificity. Reference materials and calibration standards ensure analytical reliability. Rigorous documentation, adherence to chain-of-custody procedures, and standardized protocols are essential, especially when results will be scrutinized in legal proceedings. Chemical stability, environmental degradation, and potential contamination must always be considered in data interpretation. The forensic chemist must balance analytical rigor with practical constraints while ensuring that evidence is scientifically defensible.

Case studies illustrate the impact of forensic chemistry. In the famous Lindbergh kidnapping case (1932), chemical analysis of wood and tools helped link a suspect to the crime scene. In modern times, GC-MS analysis of fentanyl analogues has been critical in tracing overdose cases to illicit drug networks. Similarly, arson investigations often rely on chemical detection of accelerants to secure convictions, while DNA extracted from trace biological evidence has exonerated wrongly accused individuals, underscoring the life-altering importance of forensic chemistry.

Finally, forensic chemistry is not just a laboratory pursuit—it is a critical component of the criminal justice system. Expert chemists testify in court, explaining complex analyses in understandable terms. They provide clarity and context, translating molecular-level observations into actionable evidence. The combination of scientific rigor, investigative acumen, and legal responsibility defines the field. As technology advances, forensic chemistry will continue to expand, providing ever more precise and insightful tools for justice.

Forensic chemistry, while rooted in core chemical principles, is distinguished by its rigorous approach to evidence. Every sample analyzed—whether a fiber, a droplet of blood, or a residue from an explosive device—must be treated not only as a chemical entity but also as a piece of a broader investigative puzzle.

This dual perspective demands precision at every stage: collection, preservation, analysis, and interpretation. The principles of analytical chemistry, including chromatography, spectroscopy, and mass spectrometry, are employed not in isolation but in concert to maximize reliability and minimize uncertainty. Even seemingly straightforward measurements, such as determining the concentration of a drug in a biological sample, require careful consideration of matrix effects, potential interferences, and the limits of detection. In this way, forensic chemistry exemplifies the marriage of theoretical knowledge with practical problem-solving, revealing information that is often inaccessible through other means.

A central challenge in forensic chemistry is the detection and interpretation of trace evidence. Substances left at a crime scene—microscopic particles of paint, soil, or gunshot residue—carry unique chemical signatures that can connect objects, people, or locations. Trace analysis relies heavily on the sensitivity and selectivity of modern instrumentation, yet it also depends on a deep understanding of chemical behavior. For instance, the solubility, volatility, and reactivity of a substance can determine whether it is recoverable from a complex matrix such as fabric or soil. Forensic chemists must therefore consider both the intrinsic properties of analytes and the environmental conditions to which they have been exposed. This intricate balancing of chemical knowledge and methodological rigor ensures that conclusions drawn from trace evidence are not only accurate but legally defensible.

The field of toxicology, a critical branch of forensic chemistry, illustrates the profound societal impact of chemical science. By analyzing the presence and effects of drugs, poisons, and metabolic byproducts in biological samples, forensic chemists can reconstruct events leading to injury or death. These analyses often involve highly complex matrices—blood, urine, tissue—where analytes exist at trace levels and may be subject to post-mortem changes. Advanced chromatographic and spectrometric techniques, often coupled with bioanalytical methods, enable precise identification and quantification of these substances. Beyond the laboratory, the interpretation of toxicological results requires knowledge of pharmacokinetics, metabolism, and chemical interactions, translating raw data into actionable insights for investigators, coroners, and courts. Through this process, chemistry becomes a tool not merely for measurement but for understanding human behavior and its consequences.

Forensic chemists also confront the challenge of unknown or novel substances, particularly in the context of synthetic drugs, industrial chemicals, or improvised explosives. The rapid pace of chemical innovation means that new compounds frequently appear on the market or at crime scenes before standardized testing protocols exist. Here, the chemist's ability to deduce molecular structure from limited data becomes essential. Techniques such as nuclear magnetic resonance spectroscopy, high-resolution mass spectrometry, and infrared spectroscopy allow for the elucidation of unknown substances, while computational chemistry provides predictive models for their reactivity and potential hazards. This capacity to identify and interpret novel chemicals underscores the adaptive nature of forensic chemistry, bridging the gap between emergent threats and established analytical methodology.

Environmental forensics is another dimension where chemistry reveals truths hidden in complex systems. Contaminants such as heavy metals, persistent organic pollutants, and industrial effluents can provide evidence linking individuals or corporations to environmental harm. Chemical fingerprinting—using isotopic ratios, elemental profiles, or molecular markers—enables the tracing of pollutants to specific sources, while kinetic and thermodynamic modeling helps reconstruct timelines of contamination. Here, forensic chemistry intersects with ecology, regulatory science, and public health, demonstrating the discipline's reach beyond the confines of traditional criminal investigations. By quantifying exposure and identifying causative agents, chemists provide critical evidence that informs both remediation efforts and legal accountability.

The evolution of forensic chemistry has been inextricably linked to advances in instrumentation and methodology. Techniques that were once laborious or impossible are now routine, yet the underlying chemical principles remain constant. Mass spectrometry continues to rely on ionization, fragmentation, and mass-to-charge ratios; chromatographic separation still depends on differential affinities and partitioning; spectroscopy remains grounded in the interaction of light with matter. What has changed is the precision, sensitivity, and speed with which these methods can be applied, allowing forensic chemists to tackle problems of increasing complexity. This evolution reflects a broader truth in chemistry: that advances in instrumentation expand the horizons of inquiry, but fundamental chemical knowledge is the lens through which observations are interpreted.

In conclusion, forensic chemistry embodies the convergence of science, technology, and societal need. Its practitioners operate at the interface of molecules and meaning, translating subtle chemical signatures into narratives that can clarify events, assign responsibility, and uphold justice. The discipline demonstrates the universality and versatility of chemical principles, showing that whether analyzing a drop of blood, a fiber, or a complex mixture of environmental contaminants, chemistry provides a framework for understanding and explaining the world. In doing so, forensic chemistry not only advances scientific knowledge but also affirms the ethical imperative of using science to serve truth, transparency, and the public good.

Ultimately, the future of forensic chemistry will continue to be shaped by innovation, interdisciplinarity, and rigorous standards of evidence. Emerging fields such as metabolomics, nanotechnology, and machine learning promise to enhance the sensitivity and interpretive power of chemical analyses, while global collaboration ensures that methodologies remain standardized and reproducible. Yet even as new tools emerge, the core mission remains unchanged: to reveal the invisible, interpret the unknown, and provide objective evidence that informs decisions critical to society. Through this lens, forensic chemistry stands as a testament to the enduring power of chemical science—not only to illuminate the molecular world but to illuminate truth itself.

Chapter 26: Word Problems

1. Investigators recover a blood sample from a hit-and-run scene that has been exposed to sunlight for several hours. When the forensic laboratory attempts to quantify ethanol and a panel of common drugs using GC-MS, the chromatograms reveal several unexpected degradation peaks. As the forensic chemist, explain how photodegradation, matrix effects, and the volatility of analytes could affect quantitation and determine which analytical approach or internal standard strategy would best ensure reliability in court.

2. A body exhumed after three years is suspected to contain traces of arsenic. The local humidity and soil composition are known to mobilize heavy metals. You must determine whether arsenic detected by ICP-MS originates from poisoning or post-burial soil migration. Discuss how isotopic ratios, tissue distribution patterns, and comparative soil analyses could be interpreted to address the evidentiary question.

3. A suitcase containing a white powder is seized at an international airport. Preliminary IR spectroscopy suggests a mixture of cocaine, a cutting agent, and an unknown impurity that does not match reference spectra. Propose how NMR, HR-MS, and chemometric analysis could be used together to elucidate the impurity's identity and assess whether it indicates a specific manufacturing route or geographic origin.

4. An arson incident destroys a warehouse storing paints, solvents, and lubricants. Headspace GC of charred debris reveals a mixture of volatile compounds, some of which are common paint solvents. Explain how a forensic chemist should differentiate an accelerant from background industrial chemicals and how pyrolysis products and burn patterns influence the chemical interpretation.

5. A soil sample found in a suspect's car is compared with soil collected from an illegal dumping site contaminated with heavy metals and industrial solvents. SEM-EDS elemental profiles show close similarity but not exact matches. Explain how statistical comparison, heterogeneity within soil horizons, and environmental weathering should be factored into determining whether the soils likely share a common origin.

6. Investigators seize gelatin capsules containing a novel synthetic opioid. LC-MS reveals the parent ion, but daughter-ion fragmentation patterns do not match existing databases. Explain how fragmentation pathways, computational chemistry predictions, and derivatization strategies could help identify the compound's structure well enough to classify it legally.

7. A partially burned document is recovered from a suspect's fireplace. Raman analysis of the remaining writing shows spectral contributions from two ink types layered on one another. Explain how ink solubility behavior, laser-induced fluorescence, and PCA analysis of spectral data could help determine whether the document was altered before burning.

8. A victim's clothing contains microscopic blue fibers. FTIR analysis indicates a polyester base with a dye profile matching several commercial brands. Explain how trace-level dye component ratios, manufacturing additives, and fiber cross-section morphology might narrow the fiber source sufficiently to be probative.

9. Post-blast debris from an improvised explosive device shows nitrate-based residues, but LC-MS analysis detects multiple organic stabilizers not typically associated with commercial explosives. Discuss how the chemist should interpret these stabilizers in terms of homemade synthesis routes, contamination, or secondary reactions during detonation.

10. A forensic toxicologist detects a weak LC-MS signal corresponding to a benzodiazepine metabolite in blood drawn 48 hours after alleged drug-facilitated assault. The defendant argues the compound is an environmental contaminant. Explain how metabolic half-life, matrix suppression, and authentic standards could be used to support or refute this claim.

11. A clandestine drug lab is raided, and mixtures of solvents are found in open containers. GC-MS reveals multiple isotopologues of a precursor chemical, suggesting synthesis involving different reagent batches. Explain how isotopic fingerprinting could help trace supply chains and whether environmental evaporation might distort isotopic ratios.

12. A wine bottle is suspected to be poisoned with a plant-derived alkaloid. The beverage matrix contains sugars, tannins, alcohol, and complex organic acids. Discuss how LC-MS/MS method development—including ionization mode choice and calibration matrix selection—affects the detection and quantification of the alkaloid.

13. A piece of glass retrieved from a suspect's jacket is compared with fragments from a shattered window at a burglary scene. SEM-EDS analysis reveals nearly identical elemental profiles except for minor variations in trace strontium. Evaluate how melt-batch variability, manufacturing tolerances, and particle transfer probabilities influence evidential interpretation.

14. Investigators recover a USB drive coated with a sticky residue believed to originate from an adhesive used in counterfeit tax stamps. FTIR reveals a polymeric base, but the additive profile is unclear. Explain how GC-MS of extractables, pyrolysis analysis, and reference libraries could be combined to determine whether the adhesive belongs to a particular counterfeit production batch.

15. Blood stains on carpet are treated with luminol during investigation. Later, LC-MS toxicology testing on the preserved sample reveals unexpected oxidized

drug metabolites. Explain how presumptive test reagents might chemically alter analytes and how chain-of-custody documentation should address this when reporting findings.

16. A river near a factory shows elevated levels of a persistent organic pollutant. Samples collected downstream reveal degradation products not found upstream. Explain how environmental forensics uses reaction kinetics, photolysis pathways, and EPA-approved chemical fingerprints to determine whether the factory is the likely pollution source.
17. A decomposed body yields negligible intact drug but substantial amounts of a known metabolite. Tissue pH had shifted post-mortem due to putrefaction. Discuss how the chemist should infer possible antemortem drug concentration using knowledge of post-mortem redistribution and metabolite stability.
18. Investigators analyze soot from an apartment fire suspected to involve electrical malfunction. GC-MS detects phthalates from melted plastic insulation alongside small amounts of aromatic hydrocarbons typical of accelerants. Explain how combustion temperatures, pyrolysis chemistry, and substrate effects help differentiate accidental from intentional ignition.
19. A clandestine lab synthesizes a designer cannabinoid precursor. IR and NMR spectra show overlapping signals from multiple byproducts. Discuss how multivariate curve resolution, 2D-NMR, and chromatographic purification analysis could separate and identify the relevant chemical species.
20. A rare heavy metal is detected in a poisoning case. ICP-MS isotope ratios match a known industrial waste site. Evaluate how transport through water distribution systems, environmental dilution, and metal-complex formation could influence whether this isotopic match is meaningful or coincidental.
21. An evidence technician swabs the interior door handle of a vehicle suspected to be involved in a kidnapping. LC-MS detects trace amounts of a sedative whose degradation products depend strongly on pH. Because the swab solution used in collection was slightly acidic, interpret how the chemist should evaluate whether detected metabolites represent authentic biological transfer or degradation artifacts introduced during evidence collection.

22. Pieces of duct tape are found at a crime scene and on a suspect's clothing. FTIR analysis shows nearly identical adhesive spectra, but GC-MS of volatile components reveals slight variations in plasticizer content. Explain how manufacturing batch variability, storage history, and environmental exposure could create such differences and how they influence evidential significance.

23. Post-blast debris from a suspected homemade explosive reveals inorganic oxidizer crystals contaminated with soil particles. Ion chromatography shows ratios of nitrate, chloride, and sulfate that differ from the same oxidizer's commercial formulations. Discuss how environmental contamination, device construction, and pre-blast storage conditions might modify these ratios and how to interpret them.

24. A corpse discovered in a shallow grave contains traces of organophosphate pesticides. As organophosphates hydrolyze quickly in moist environments, evaluate how the chemist can use metabolite profiles, soil composition, and mass-spectral ion ratios to determine whether exposure was ante- or post-mortem.

25. A luxury handbag is suspected to be counterfeit. Microscopic paint chips from the logo are analyzed using SEM-EDS and show a pigment mixture inconsistent with genuine items. Discuss how elemental fingerprinting, binder chemistry, and trace impurity analysis can identify manufacturing origin and support the authenticity assessment.

26. A hit-and-run victim's clothing contains tiny glass fragments. Refractive-index measurements match automotive window glass, but LA-ICP-MS detects minor rare-earth elements not typical for that model year. Explain how recycling practices, aftermarket replacements, or contamination could explain this anomaly and how the chemist should report it.

27. A sealed container labeled as "olive oil" is suspected of concealing a hydrophobic drug. NMR reveals unusual peaks for long-chain esters, and GC-MS shows several non-food-grade impurities. Explain how forensic chemists use lipid-matrix effects, solubility predictions, and impurity profiles to conclude whether the container was used for drug transport.

28. A suspect claims that trace cocaine found on their clothing results from environmental contamination on public transportation. By comparing isotopic

ratios, impurity profiles, and particle morphology, explain how the forensic chemist can differentiate incidental contamination from deliberate handling.

29. A fugitive's abandoned car contains a fuel canister with traces of an unusual gasoline additive. Headspace GC-MS identifies alkylates and oxygenates in ratios uncommon to local fuel suppliers. Evaluate how fuel-supply logistics, seasonal formulations, and evaporation rates affect whether this chemical evidence can link the suspect's movements to a specific region.

30. A jewelry store robbery leaves behind a tiny plastic fragment near a broken display case. FTIR suggests polycarbonate, but Raman spectroscopy shows peaks consistent with a flame retardant used only in certain industrial products. Explain how mixed-material formulations and additive migration can complicate polymer identification.

31. In a suspected poisoning case, only a small volume of vitreous humor remains available for testing. LC-MS/MS detects a drug at trace levels near the limit of quantification. Discuss how matrix stability, dilution strategies, and calibration curve selection affect whether the reported result can withstand legal scrutiny.

32. An arson suspect claims the fire was caused by spontaneous ignition of oily rags. GC-MS of charred debris reveals signatures of petroleum distillates consistent with certain accelerants but also with some types of wood-finishing oils. Explain how combustion chemistry, distillation profiles, and substrate interactions help differentiate these sources.

33. A cold case reopens when hairs found on old evidence are analyzed with modern proteomics. Mass spectrometry identifies hair-shaft proteins unique to certain ethnic or genetic groups, though DNA is too degraded for standard profiling. Discuss how protein stability, environmental exposure, and bioinformatic classification affect the evidentiary value.

34. A forged painting is suspected because pigment analysis shows titanium dioxide nanoparticles inconsistent with the alleged age. Evaluate how XRD, SEM, and binder-chemistry analysis can confirm modern materials and distinguish restoration artifacts from deliberate forgery.

35. A water sample taken near a chemical plant shows elevated mercury with a distinctive isotopic pattern. The company claims natural geological sources are

responsible. Explain how isotope fractionation, sediment-water partitioning, and background regional profiles can determine whether the pollution originates from the plant.

36. A mail bomb investigation reveals traces of polymers and oxidizers on the envelope adhesive. Because the envelope passed through multiple sorting machines, investigators question whether residue is transfer from machinery. Explain how wear-debris chemistry, adhesive composition, and transfer likelihood studies inform the analysis.

37. A body retrieved from a lake shows minimal intact nicotine but abundant hydrolysis products. Temperature logs show the lake was unusually warm for several weeks. Discuss how environmental hydrolysis, microbial activity, and drug partitioning in water affect interpretation of toxicology results.

38. A suspect claims a positive test for a controlled substance resulted from consuming herbal tea containing a plant with chemically similar alkaloids. Using mass-spectral fragmentation, retention-time stability, and metabolite specificities, evaluate how a forensic chemist distinguishes true drug exposure from legal plant consumption.

39. Particles recovered from a shooting scene show ambiguous GSR signatures; SEM-EDS detects lead, barium, and antimony, but in ratios inconsistent with standard ammunition. Discuss how primer-formulation variation, environmental contamination, and firearm modifications affect interpretation of atypical GSR profiles.

40. An antique manuscript is alleged to be centuries old, but ink-dating analysis reveals multiple layers of ink with different solvent compositions. Explain how evaporative loss, solvent-residue ratios, and paper-fiber chemistry help determine whether parts of the manuscript were created much later.

41. A clandestine lab dump site shows soil stained with a mixture of organic solvents. GC-MS reveals several ketone degradation products not typical of the parent chemicals. Evaluate how soil moisture, microbial activity, and photolysis could generate these products and how they might indicate timeline of dumping.

42. A mass-casualty poisoning incident involves a toxin that rapidly binds to serum proteins. Only small amounts of free toxin remain measurable. Discuss how protein-

binding kinetics, deconjugation methods, and target-protein mass spectrometry are used to reconstruct exposure levels.

43. A suspect's boots contain dried mud that visually resembles soil from a protected wildlife reserve. ICP-MS reveals matching rare-earth patterns, but XRD shows subtle mineral differences. Discuss how geochemical heterogeneity, microstratification, and depositional history determine whether the match is significant.

44. A package of pills purchased online is advertised as a dietary supplement but contains several unlisted synthetic stimulants. NMR shows overlapping aromatic signals, while LC-HR-MS detects multiple isomers. Explain how chromatographic resolution strategies and chemometric clustering can classify the pill's true contents.

45. A vehicle fire investigation reveals melted wiring insulation containing brominated flame retardants. Trace accelerant residues appear in the same samples. Explain how to discern whether flame-retardant decomposition could mimic accelerant signatures and how thermal degradation profiles help clarify the cause.

46. A kidnapping suspect is linked to a ransom note printed on a color laser printer. Analysis of toner particles using SEM-EDS reveals chemical composition consistent with several printer models. Evaluate how polymer binders, pigment ratios, and fuser-temperature artifacts help narrow the printer source.

47. A smuggling operation uses ceramic figurines to conceal powdered drugs. XRD of ceramic fragments reveals inconsistent firing temperatures. Discuss how mineral phase transitions, glaze-chemistry anomalies, and porosity measurements can indicate tampering or concealment cavities within the figurines.

48. A decomposed sample of biological tissue contains a mixture of endogenous metabolites and possible exogenous toxins. GC-MS analysis shows overlapping peaks and coelution. Explain how derivatization chemistry, selective ion monitoring, and matrix-matched standards improve interpretation.

49. A wildlife-poisoning investigation finds bait laced with a pesticide that degrades rapidly in sunlight. Photolysis experiments show multiple possible degradation pathways. Discuss how chromatographic peak-ratio changes, intermediate-product stability, and sunlight-exposure modeling can identify when the bait was prepared.

50. A suspect's phone case contains microscopic paint smears. FTIR identifies a urethane-based coating, while Raman suggests metallic flakes typical of certain automotive paints. Evaluate how paint-layer stratigraphy, binder chemistry, and spectral library matching can link the paint to a specific vehicle.

Chapter 27

FIREFIGHTING CHEMISTRY SAVES LIVES

Fire has fascinated humans for millennia, serving as a tool of survival, a catalyst for technological progress, and a symbol of both destruction and transformation.

At its essence, fire is a chemical phenomenon: a rapid oxidation process in which a fuel reacts with an oxidizing agent, releasing energy in the form of heat, light, and reaction products. Unlike a simple thermal event, fire represents a complex interplay of molecular interactions, chemical kinetics, and thermodynamic principles. It is a self-sustaining chemical system in which microscopic reactions propagate through macroscopic media, demonstrating emergent behaviors that are both predictable and chaotic. Understanding fire in its full chemical complexity requires examining the detailed mechanisms of bond-breaking and bond-forming processes, radical intermediates, energy distribution, and the influence of environmental factors such as pressure, humidity, and oxygen availability.

Combustion, the chemical heart of fire, is fundamentally an exothermic redox reaction. Hydrocarbon fuels such as methane, ethane, propane, and complex petroleum derivatives undergo oxidation when exposed to an adequate source of energy and oxygen. The simplest stoichiometric reactions, such as $CH_4 + 2\ O_2 \rightarrow CO_2 + 2\ H_2O$, represent idealized scenarios; real combustion is far more intricate. Incomplete combustion occurs when oxygen supply is limited or mixing is imperfect, leading to the formation of carbon monoxide (CO), unburned hydrocarbons, soot, and other intermediates. These byproducts are not merely environmental concerns—they provide insight into the reaction mechanisms at play, revealing the stepwise propagation of radicals and energy through molecular networks. Fire is thus a highly sensitive system in which slight variations in fuel composition, oxygen concentration, and temperature can dramatically alter outcomes.

At the molecular level, the energy released during combustion arises from the difference in bond energies between reactants and products. Breaking the C–H and O=O bonds in hydrocarbons and oxygen requires input energy, but the formation of C=O and O–H bonds in carbon dioxide and water releases greater energy, resulting in a net exothermic process. This energy manifests as thermal radiation, visible light, and kinetic motion of gaseous molecules. Chemical kinetics plays a critical role, governed by chain-branching reactions involving reactive radicals such as hydroxyl (•OH), hydrogen (•H), and oxygen (•O). These radicals are highly reactive intermediates that propagate the reaction by abstracting hydrogen atoms, forming new radicals, and sustaining the combustion cycle. Radical recombination and termination events modulate flame stability, intensity, and the likelihood of quenching.

Ignition, the onset of combustion, is dictated by the fuel's chemical and physical properties. The ignition temperature is the threshold at which sufficient kinetic energy allows molecules to overcome the activation energy barrier, initiating rapid radical formation. Factors such as molecular structure, phase state, particle size, and ambient conditions influence this threshold. Finely divided solids, like sawdust or powdered metals, ignite at lower temperatures than bulk materials due to their larger surface area, which facilitates oxygen contact and heat transfer. Liquids with low flash points, such as ethanol or gasoline, produce flammable vapors that ignite readily. Gaseous fuels, being already dispersed at the molecular level, often exhibit highly sensitive ignition behaviors. Understanding these ignition dynamics is critical for fire prevention, safety engineering, and the design of controlled combustion systems.

Thermodynamics governs the progression and intensity of fire. The release of chemical energy elevates the temperature of adjacent materials, accelerating reaction rates—a process often described as thermal runaway. The interplay of heat transfer mechanisms, including conduction through solids, convection of hot gases, and radiation across gaps, determines fire propagation in both natural and built environments. In forest ecosystems, convective currents carry embers and preheat distant fuels, contributing to the rapid spread of wildfires. In buildings, radiation through walls and ceilings can ignite flammable materials even without direct flame contact. Accurate thermodynamic modeling of these processes enables the design of fire-resistant structures, sprinkler systems, and firebreaks, highlighting the importance of chemistry in practical safety solutions.

The products of combustion have profound environmental, chemical, and health implications. Complete combustion produces carbon dioxide and water, but incomplete reactions yield carbon monoxide, nitrogen oxides, volatile organic compounds, polycyclic aromatic hydrocarbons, and particulate matter. Soot formation, a consequence of carbon-rich intermediates aggregating during incomplete oxidation, not only pollutes air but also affects flame color, temperature distribution, and radiative heat transfer. Analyzing combustion products provides chemists with diagnostic tools for assessing reaction efficiency, pollutant formation, and the environmental footprint of fires. The study of these emissions has driven advances in cleaner combustion technologies, including catalytic converters, optimized burner designs, and alternative fuels that minimize harmful byproducts.

Fire chemistry extends beyond hydrocarbon combustion. Metals such as magnesium, aluminum, and titanium can burn with extreme exothermicity, producing intense light and heat. These reactions are harnessed in pyrotechnics, metal cutting, and incendiary applications, illustrating the diversity of chemical processes that constitute fire. Furthermore, unconventional oxidizers such as fluorine, chlorine, and permanganates can initiate combustion, demonstrating the versatility of redox chemistry. The energetic interactions of metals with oxidizing agents are dictated by thermodynamic potentials, reaction kinetics, and surface reactivity, highlighting the role of chemical principles in predicting and controlling these energetic events.

Flame coloration and luminosity reveal additional layers of molecular complexity. Flame color arises from the electronic excitation of atoms and molecules within the hot gas. Sodium ions emit bright yellow light, copper compounds produce green-blue hues, and strontium yields red emissions. These phenomena result from electron transitions between energy levels, with photon emission at characteristic wavelengths. Molecular vibrations, collisions, and blackbody radiation contribute to the visible and infrared spectrum, allowing chemists to interpret flame diagnostics and create controlled visual effects in fireworks and chemical demonstrations. The chemistry of light emission thus bridges molecular processes and observable phenomena.

Fire suppression relies on chemical and physical principles to interrupt combustion. Removing fuel, oxygen, or heat destabilizes the chain reaction. Chemical retardants often act by neutralizing radicals, thereby terminating the propagation steps in the flame. Water functions both as a cooling agent and as a medium that displaces oxygen from the combustion zone. Other suppression methods include foam, inert gases, and halogenated compounds, each designed to interfere with specific molecular processes. Advanced fire suppression strategies are informed by a deep understanding of combustion chemistry, radical kinetics, and thermodynamic equilibria, particularly in high-risk industrial and forested environments.

Combustion chemistry is central to energy production, from power plants to internal combustion engines. Optimizing fuel efficiency, minimizing pollutant formation, and controlling reaction kinetics are critical objectives. Alternative fuels, including biofuels and hydrogen, introduce new molecular pathways, requiring detailed studies of ignition, flame propagation, and pollutant profiles. Catalysts, additives, and combustion modeling allow for precise control over energy output while mitigating environmental harm. Understanding the chemical underpinnings of combustion facilitates the transition to more sustainable energy systems and informs climate science, linking molecular reactions to global societal outcomes.

Fire chemistry also intersects with forensic science. Arson investigations leverage knowledge of combustion residues, accelerant chemistry, and reaction kinetics to reconstruct fire events. Specific fuels and accelerants leave identifiable molecular signatures, detectable through chromatographic and spectroscopic techniques. Understanding the timing, sequence, and environmental conditions of combustion enables investigators to distinguish accidental fires from deliberate acts. This intersection of chemistry and law illustrates how fundamental molecular understanding can serve practical societal needs, providing evidence and guiding policy.

Fire behavior in complex environments highlights emergent chemical phenomena. In wildfires, heterogeneous fuels, variable moisture content, and turbulent airflows create dynamic chemical systems where radical chemistry, heat transfer, and environmental factors interact nonlinearly. Laboratory studies using small-scale combustion models help elucidate mechanisms, but real-world fires often display stochastic behaviors. Computational modeling of combustion chemistry and fluid dynamics provides predictive insights, aiding in disaster management, firefighting strategy, and policy development. Here, molecular understanding translates directly into societal safety and environmental protection.

At the interface of material science, chemistry of fire informs the design of fire-resistant polymers, coatings, and structural materials. Flame-retardant additives, intumescent layers, and heat-resistant composites exploit chemical reactions to absorb energy, release inert gases, or form protective char layers. The selection and optimization of these materials require understanding chemical stability, radical scavenging mechanisms, and thermal decomposition pathways. Advances in materials chemistry, informed by fire science, enhance safety across industries, from construction to aerospace.

Furthermore, fire serves as a model system for studying nonequilibrium chemical dynamics. Combustion exhibits oscillatory reactions, chain branching, and feedback loops, making it a rich platform for exploring fundamental principles of chemical kinetics and thermodynamics. Studies of laminar and turbulent flames, premixed versus diffusion flames, and catalytic combustion provide insights into molecular transport, energy partitioning, and reaction network behavior. Fire chemistry thus contributes not only to applied science but also to the broader understanding of complex chemical systems.

Finally, the cultural, ecological, and technological significance of fire underscores the importance of chemistry in shaping human interactions with this elemental force. Fire drives nutrient cycling in ecosystems, enables industrial processes, and has been central to human evolution. Its study integrates physical chemistry, environmental science, materials chemistry, and forensic investigation, demonstrating the interconnectedness of chemical knowledge with societal needs. By probing the molecular foundations of fire, chemists continue to enhance our ability to harness, control, and coexist with one of nature's most dynamic and transformative phenomena.

The study of fire, when viewed through the lens of modern chemistry, highlights not only the reactive transformations of matter but also the principles of energy conservation, molecular dynamics, and chemical equilibria. Fire serves as a vivid illustration of how chemical systems respond to thermodynamic potentials and kinetic constraints. The rapid oxidation of fuels demonstrates how energy landscapes govern reaction pathways: molecules traverse potential energy surfaces, form transient intermediates, and release energy when more stable products are achieved. These principles, while vividly observable in flames, are universal across chemical phenomena. By analyzing fire, chemists gain insight into fundamental mechanisms that extend to catalysis, materials degradation, and biochemical energy transduction.

Radical chemistry, a central feature of combustion, also exemplifies broader chemical concepts such as chain reactions, reaction branching, and oscillatory dynamics. The propagation of radicals in a flame mirrors processes in atmospheric chemistry, polymerization, and enzymatic catalysis, revealing the universality of reaction network principles. The balance between chain propagation and termination events dictates stability, efficiency, and reaction selectivity. Incomplete combustion, often viewed as a problem, becomes a laboratory for studying intermediate species, electron transfer processes, and molecular rearrangements, deepening our understanding of reactive chemistry. These insights demonstrate that even destructive chemical processes like fire are fundamentally instructive, providing frameworks for predicting, controlling, and harnessing reactivity in other chemical contexts.

Fire also exemplifies the coupling between chemical kinetics and transport phenomena. Heat, mass, and momentum transfer interact with reaction networks to produce emergent behaviors. Convection spreads radicals and energy, conduction preheats adjacent materials, and radiation provides long-range thermal feedback. The study of these coupled processes informs not only fire science but also chemical engineering, where reactors must balance heat release, mixing, and product formation. Understanding these principles allows chemists to design safer, more efficient combustion systems and to translate lessons from fire into industrial, environmental, and energy applications.

Environmental chemistry benefits directly from the study of fire. The emissions of particulate matter, carbon monoxide, nitrogen oxides, and volatile organic compounds illustrate the intersection of combustion chemistry with atmospheric processes. Radical chemistry in flames is mirrored in the atmosphere, where free radicals mediate oxidation, smog formation, and ozone depletion. By characterizing combustion products and understanding their formation pathways, chemists can develop cleaner fuels, optimize reaction conditions, and design mitigation strategies that reduce environmental impact. In this way, the chemistry of fire extends beyond immediate heat and light to influence air quality, climate systems, and public health.

The practical applications of fire chemistry span multiple disciplines. In materials science, knowledge of combustion and thermal degradation informs the design of fire-resistant polymers, coatings, and structural composites. Radical scavengers, char-forming agents, and heat-absorbing additives are carefully engineered to interrupt chemical propagation pathways, illustrating the application of fundamental chemistry in protective technologies. In energy science, controlled combustion of fossil fuels and alternative energy carriers depends on precise chemical understanding to maximize energy yield while minimizing pollutants. Catalysts, burners, and chemical additives optimize reaction kinetics, demonstrating how molecular-level insight translates into macroscopic efficiency.

Emerging research further underscores fire as a model system for studying complex chemical networks. Modern computational methods allow simulation of radical propagation, flame dynamics, and multi-phase interactions, providing predictive power over systems previously too complex to analyze. Quantum chemical calculations illuminate electronic structures of reactive intermediates, while spectroscopic techniques detect transient species in real-time. These advances deepen understanding of chemical reactivity, revealing subtleties in energy transfer, electron dynamics, and reaction branching that inform both theoretical and applied chemistry. The study of fire thus operates at the intersection of classical thermodynamics, molecular kinetics, and quantum mechanics, exemplifying the interconnected nature of modern chemical science.

Fire also illuminates chemical safety and risk management. Understanding ignition thresholds, reaction rates, and propagation mechanisms enables chemists and engineers to predict hazardous conditions, design safer chemical processes, and implement effective suppression strategies. The interplay between molecular chemistry and environmental factors highlights the importance of system-level thinking: a seemingly minor change in oxygen concentration, fuel geometry, or ambient temperature can dramatically alter reaction outcomes. This sensitivity underscores the broader principle that chemical systems must be understood not only at the molecular scale but also in their environmental context.

Finally, the study of fire reinforces chemistry's dual role as both a tool for innovation and a lens for understanding natural phenomena. Fire embodies the conversion of chemical energy into multiple forms—thermal, kinetic, and radiant—demonstrating the universality of energy transformation principles. It provides insight into reaction mechanisms, radical chemistry, thermodynamic constraints, and environmental interactions. By studying fire, chemists refine predictive models, develop practical applications, and advance our understanding of the physical world. Fire is not merely a destructive force but a dynamic chemical laboratory, revealing fundamental truths about matter, energy, and reactivity.

In conclusion, fire chemistry exemplifies the breadth and depth of chemical science. From molecular bond dynamics to system-level emergent behavior, from energy conversion to environmental impact, the study of fire integrates physical chemistry, materials science, environmental chemistry, and applied technology. It challenges chemists to understand rapid, complex, and highly energetic systems, while providing a framework for innovation in safety, energy, and materials design. By observing and analyzing the flame, chemists confront the delicate balance between order and chaos inherent in chemical reactions, ultimately demonstrating how the principles of chemistry govern both natural phenomena and human ingenuity. Fire is not only a window into the reactive world of molecules—it is a testament to the transformative power of chemistry itself.

Chapter 27: Word Problems

1. Investigators recover a blood sample from a hit-and-run scene that has been exposed to sunlight for several hours. When the forensic laboratory attempts to quantify ethanol and a panel of common drugs using GC-MS, the chromatograms reveal several unexpected degradation peaks. As the forensic chemist, explain how photodegradation, matrix effects, and the volatility of analytes could affect quantitation and determine which analytical approach or internal standard strategy would best ensure reliability in court.

2. A body exhumed after three years is suspected to contain traces of arsenic. The local humidity and soil composition are known to mobilize heavy metals. You must determine whether arsenic detected by ICP-MS originates from poisoning or post-burial soil migration. Discuss how isotopic ratios, tissue distribution patterns, and comparative soil analyses could be interpreted to address the evidentiary question.

3. A suitcase containing a white powder is seized at an international airport. Preliminary IR spectroscopy suggests a mixture of cocaine, a cutting agent, and an unknown impurity that does not match reference spectra. Propose how NMR, HR-MS, and chemometric analysis could be used together to elucidate the impurity's identity and assess whether it indicates a specific manufacturing route or geographic origin.

4. An arson incident destroys a warehouse storing paints, solvents, and lubricants. Headspace GC of charred debris reveals a mixture of volatile compounds, some of which are common paint solvents. Explain how a forensic chemist should differentiate an accelerant from background industrial chemicals and how pyrolysis products and burn patterns influence the chemical interpretation.

5. A soil sample found in a suspect's car is compared with soil collected from an illegal dumping site contaminated with heavy metals and industrial solvents. SEM-EDS elemental profiles show close similarity but not exact matches. Explain how statistical comparison, heterogeneity within soil horizons, and environmental weathering should be factored into determining whether the soils likely share a common origin.

6. Investigators seize gelatin capsules containing a novel synthetic opioid. LC-MS reveals the parent ion, but daughter-ion fragmentation patterns do not match

existing databases. Explain how fragmentation pathways, computational chemistry predictions, and derivatization strategies could help identify the compound's structure well enough to classify it legally.

7. A partially burned document is recovered from a suspect's fireplace. Raman analysis of the remaining writing shows spectral contributions from two ink types layered on one another. Explain how ink solubility behavior, laser-induced fluorescence, and PCA analysis of spectral data could help determine whether the document was altered before burning.

8. A victim's clothing contains microscopic blue fibers. FTIR analysis indicates a polyester base with a dye profile matching several commercial brands. Explain how trace-level dye component ratios, manufacturing additives, and fiber cross-section morphology might narrow the fiber source sufficiently to be probative.

9. Post-blast debris from an improvised explosive device shows nitrate-based residues, but LC-MS analysis detects multiple organic stabilizers not typically associated with commercial explosives. Discuss how the chemist should interpret these stabilizers in terms of homemade synthesis routes, contamination, or secondary reactions during detonation.

10. A forensic toxicologist detects a weak LC-MS signal corresponding to a benzodiazepine metabolite in blood drawn 48 hours after alleged drug-facilitated assault. The defendant argues the compound is an environmental contaminant. Explain how metabolic half-life, matrix suppression, and authentic standards could be used to support or refute this claim.

11. A clandestine drug lab is raided, and mixtures of solvents are found in open containers. GC-MS reveals multiple isotopologues of a precursor chemical, suggesting synthesis involving different reagent batches. Explain how isotopic fingerprinting could help trace supply chains and whether environmental evaporation might distort isotopic ratios.

12. A wine bottle is suspected to be poisoned with a plant-derived alkaloid. The beverage matrix contains sugars, tannins, alcohol, and complex organic acids. Discuss how LC-MS/MS method development—including ionization mode choice and calibration matrix selection—affects the detection and quantification of the alkaloid.

13. A piece of glass retrieved from a suspect's jacket is compared with fragments from a shattered window at a burglary scene. SEM-EDS analysis reveals nearly identical elemental profiles except for minor variations in trace strontium. Evaluate how melt-batch variability, manufacturing tolerances, and particle transfer probabilities influence evidential interpretation.

14. Investigators recover a USB drive coated with a sticky residue believed to originate from an adhesive used in counterfeit tax stamps. FTIR reveals a polymeric base, but the additive profile is unclear. Explain how GC-MS of extractables, pyrolysis analysis, and reference libraries could be combined to determine whether the adhesive belongs to a particular counterfeit production batch.

15. Blood stains on carpet are treated with luminol during investigation. Later, LC-MS toxicology testing on the preserved sample reveals unexpected oxidized drug metabolites. Explain how presumptive test reagents might chemically alter analytes and how chain-of-custody documentation should address this when reporting findings.

16. A river near a factory shows elevated levels of a persistent organic pollutant. Samples collected downstream reveal degradation products not found upstream. Explain how environmental forensics uses reaction kinetics, photolysis pathways, and EPA-approved chemical fingerprints to determine whether the factory is the likely pollution source.

17. A decomposed body yields negligible intact drug but substantial amounts of a known metabolite. Tissue pH had shifted post-mortem due to putrefaction. Discuss how the chemist should infer possible antemortem drug concentration using knowledge of post-mortem redistribution and metabolite stability.

18. Investigators analyze soot from an apartment fire suspected to involve electrical malfunction. GC-MS detects phthalates from melted plastic insulation alongside small amounts of aromatic hydrocarbons typical of accelerants. Explain how combustion temperatures, pyrolysis chemistry, and substrate effects help differentiate accidental from intentional ignition.

19. A clandestine lab synthesizes a designer cannabinoid precursor. IR and NMR spectra show overlapping signals from multiple byproducts. Discuss how

multivariate curve resolution, 2D-NMR, and chromatographic purification analysis could separate and identify the relevant chemical species.

20. A rare heavy metal is detected in a poisoning case. ICP-MS isotope ratios match a known industrial waste site. Evaluate how transport through water distribution systems, environmental dilution, and metal-complex formation could influence whether this isotopic match is meaningful or coincidental.

21. An evidence technician swabs the interior door handle of a vehicle suspected to be involved in a kidnapping. LC-MS detects trace amounts of a sedative whose degradation products depend strongly on pH. Because the swab solution used in collection was slightly acidic, interpret how the chemist should evaluate whether detected metabolites represent authentic biological transfer or degradation artifacts introduced during evidence collection.

22. Pieces of duct tape are found at a crime scene and on a suspect's clothing. FTIR analysis shows nearly identical adhesive spectra, but GC-MS of volatile components reveals slight variations in plasticizer content. Explain how manufacturing batch variability, storage history, and environmental exposure could create such differences and how they influence evidential significance.

23. Post-blast debris from a suspected homemade explosive reveals inorganic oxidizer crystals contaminated with soil particles. Ion chromatography shows ratios of nitrate, chloride, and sulfate that differ from the same oxidizer's commercial formulations. Discuss how environmental contamination, device construction, and pre-blast storage conditions might modify these ratios and how to interpret them.

24. A corpse discovered in a shallow grave contains traces of organophosphate pesticides. As organophosphates hydrolyze quickly in moist environments, evaluate how the chemist can use metabolite profiles, soil composition, and mass-spectral ion ratios to determine whether exposure was ante- or post-mortem.

25. A luxury handbag is suspected to be counterfeit. Microscopic paint chips from the logo are analyzed using SEM-EDS and show a pigment mixture inconsistent with genuine items. Discuss how elemental fingerprinting, binder chemistry, and trace impurity analysis can identify manufacturing origin and support the authenticity assessment.

26. A hit-and-run victim's clothing contains tiny glass fragments. Refractive-index measurements match automotive window glass, but LA-ICP-MS detects minor rare-earth elements not typical for that model year. Explain how recycling practices, aftermarket replacements, or contamination could explain this anomaly and how the chemist should report it.

27. A sealed container labeled as "olive oil" is suspected of concealing a hydrophobic drug. NMR reveals unusual peaks for long-chain esters, and GC-MS shows several non-food-grade impurities. Explain how forensic chemists use lipid-matrix effects, solubility predictions, and impurity profiles to conclude whether the container was used for drug transport.

28. A suspect claims that trace cocaine found on their clothing results from environmental contamination on public transportation. By comparing isotopic ratios, impurity profiles, and particle morphology, explain how the forensic chemist can differentiate incidental contamination from deliberate handling.

29. A fugitive's abandoned car contains a fuel canister with traces of an unusual gasoline additive. Headspace GC-MS identifies alkylates and oxygenates in ratios uncommon to local fuel suppliers. Evaluate how fuel-supply logistics, seasonal formulations, and evaporation rates affect whether this chemical evidence can link the suspect's movements to a specific region.

30. A jewelry store robbery leaves behind a tiny plastic fragment near a broken display case. FTIR suggests polycarbonate, but Raman spectroscopy shows peaks consistent with a flame retardant used only in certain industrial products. Explain how mixed-material formulations and additive migration can complicate polymer identification.

31. In a suspected poisoning case, only a small volume of vitreous humor remains available for testing. LC-MS/MS detects a drug at trace levels near the limit of quantification. Discuss how matrix stability, dilution strategies, and calibration curve selection affect whether the reported result can withstand legal scrutiny.

32. An arson suspect claims the fire was caused by spontaneous ignition of oily rags. GC-MS of charred debris reveals signatures of petroleum distillates consistent with certain accelerants but also with some types of wood-finishing oils.

Explain how combustion chemistry, distillation profiles, and substrate interactions help differentiate these sources.

33. A cold case reopens when hairs found on old evidence are analyzed with modern proteomics. Mass spectrometry identifies hair-shaft proteins unique to certain ethnic or genetic groups, though DNA is too degraded for standard profiling. Discuss how protein stability, environmental exposure, and bioinformatic classification affect the evidentiary value.

34. A forged painting is suspected because pigment analysis shows titanium dioxide nanoparticles inconsistent with the alleged age. Evaluate how XRD, SEM, and binder-chemistry analysis can confirm modern materials and distinguish restoration artifacts from deliberate forgery.

35. A water sample taken near a chemical plant shows elevated mercury with a distinctive isotopic pattern. The company claims natural geological sources are responsible. Explain how isotope fractionation, sediment-water partitioning, and background regional profiles can determine whether the pollution originates from the plant.

36. A mail bomb investigation reveals traces of polymers and oxidizers on the envelope adhesive. Because the envelope passed through multiple sorting machines, investigators question whether residue is transfer from machinery. Explain how wear-debris chemistry, adhesive composition, and transfer likelihood studies inform the analysis.

37. A body retrieved from a lake shows minimal intact nicotine but abundant hydrolysis products. Temperature logs show the lake was unusually warm for several weeks. Discuss how environmental hydrolysis, microbial activity, and drug partitioning in water affect interpretation of toxicology results.

38. A suspect claims a positive test for a controlled substance resulted from consuming herbal tea containing a plant with chemically similar alkaloids. Using mass-spectral fragmentation, retention-time stability, and metabolite specificities, evaluate how a forensic chemist distinguishes true drug exposure from legal plant consumption.

39. Particles recovered from a shooting scene show ambiguous GSR signatures; SEM-EDS detects lead, barium, and antimony, but in ratios inconsistent with

standard ammunition. Discuss how primer-formulation variation, environmental contamination, and firearm modifications affect interpretation of atypical GSR profiles.

40. An antique manuscript is alleged to be centuries old, but ink-dating analysis reveals multiple layers of ink with different solvent compositions. Explain how evaporative loss, solvent-residue ratios, and paper-fiber chemistry help determine whether parts of the manuscript were created much later.

41. A clandestine lab dump site shows soil stained with a mixture of organic solvents. GC-MS reveals several ketone degradation products not typical of the parent chemicals. Evaluate how soil moisture, microbial activity, and photolysis could generate these products and how they might indicate timeline of dumping.

42. A mass-casualty poisoning incident involves a toxin that rapidly binds to serum proteins. Only small amounts of free toxin remain measurable. Discuss how protein-binding kinetics, deconjugation methods, and target-protein mass spectrometry are used to reconstruct exposure levels.

43. A suspect's boots contain dried mud that visually resembles soil from a protected wildlife reserve. ICP-MS reveals matching rare-earth patterns, but XRD shows subtle mineral differences. Discuss how geochemical heterogeneity, microstratification, and depositional history determine whether the match is significant.

44. A package of pills purchased online is advertised as a dietary supplement but contains several unlisted synthetic stimulants. NMR shows overlapping aromatic signals, while LC-HR-MS detects multiple isomers. Explain how chromatographic resolution strategies and chemometric clustering can classify the pill's true contents.

45. A vehicle fire investigation reveals melted wiring insulation containing brominated flame retardants. Trace accelerant residues appear in the same samples. Explain how to discern whether flame-retardant decomposition could mimic accelerant signatures and how thermal degradation profiles help clarify the cause.

46. A kidnapping suspect is linked to a ransom note printed on a color laser printer. Analysis of toner particles using SEM-EDS reveals chemical composition consistent with several printer models. Evaluate how polymer binders, pigment ratios, and fuser-temperature artifacts help narrow the printer source.

47. A smuggling operation uses ceramic figurines to conceal powdered drugs. XRD of ceramic fragments reveals inconsistent firing temperatures. Discuss how mineral phase transitions, glaze-chemistry anomalies, and porosity measurements can indicate tampering or concealment cavities within the figurines.

48. A decomposed sample of biological tissue contains a mixture of endogenous metabolites and possible exogenous toxins. GC-MS analysis shows overlapping peaks and coelution. Explain how derivatization chemistry, selective ion monitoring, and matrix-matched standards improve interpretation.

49. A wildlife-poisoning investigation finds bait laced with a pesticide that degrades rapidly in sunlight. Photolysis experiments show multiple possible degradation pathways. Discuss how chromatographic peak-ratio changes, intermediate-product stability, and sunlight-exposure modeling can identify when the bait was prepared.

50. A suspect's phone case contains microscopic paint smears. FTIR identifies a urethane-based coating, while Raman suggests metallic flakes typical of certain automotive paints. Evaluate how paint-layer stratigraphy, binder chemistry, and spectral library matching can link the paint to a specific vehicle.

Chapter 28

CHEMISTRY BEHIND WATER INNOVATION

Water's molecular behavior becomes even more revealing when examined through the lens of dynamic fluctuations and transient structural motifs that emerge within its hydrogen-bonded network. Although often depicted as a static lattice of tetrahedrally arranged molecules, liquid water is in fact a rapidly shifting ensemble of bond formation and rupture events occurring on femto- to picosecond timescales.

These dynamics influence everything from solvation thermodynamics to the rates of acid-base reactions and the transport of ions. In recent years, researchers have exploited ultrafast spectroscopy and neutron scattering to probe these fluctuations, uncovering subtle heterogeneities in water's structure. Such insights are propelling innovations in water treatment where controlling local hydration environments can dramatically improve sorption capacity, membrane selectivity, and catalytic efficiency. Materials engineered to mimic or stabilize specific hydrogen-bond environments—such as biomimetic channels or structured polymeric networks—can selectively facilitate or hinder the passage of salts, organic solutes, or gases. This approach reflects a broader trend in water chemistry: rather than merely responding to contaminants, modern technologies increasingly manipulate the molecular landscape in which water's interactions unfold.

The anomalous properties of water near its freezing point continue to inspire chemical strategies for thermal management and resource preservation. Water's maximum density at approximately 4°C arises from the competition between hydrogen-bonded open networks and more compact disordered structures. Understanding this balance has enabled the design of cryoprotectants, phase-change materials, and antifreeze coatings that modify ice nucleation pathways. In water purification, this knowledge informs freeze desalination technologies that exploit selective exclusion of ions during ice crystal growth. Advances in nucleation chemistry now allow researchers to manipulate the morphology, purity, and stability of ice phases through additives that preferentially bind to specific crystal faces. These innovations hold promise for energy-efficient desalination in cold regions and for preserving sensitive biological samples. In nature, organisms have long used such chemistry, producing peptides and polysaccharides that inhibit ice growth or alter water's dynamics. By replicating these strategies, engineers hope to create low-energy purification methods that operate under conditions where conventional thermal or membrane processes are inefficient.

Water's solvent capabilities extend far beyond simple dissolution; they govern reaction equilibria, molecular recognition, and the assembly of macromolecules. Hydrophobic interactions, long attributed to entropic effects, arise because water organizes itself around nonpolar surfaces, minimizing disruptions to its hydrogen-bonded network. This principle is relevant in water treatment where surfactants, micelles, and emulsifiers are used to isolate hydrophobic pollutants, enabling their removal through flotation, adsorption, or oxidation. Innovations in supramolecular chemistry are expanding the toolbox for manipulating solvation structures. Cyclodextrins, calixarenes, and cucurbiturils can encapsulate organics with remarkable selectivity, offering biodegradable alternatives to activated carbon. These host–guest systems operate through a balance of hydrophobic forces, hydrogen bonding, and van der Waals interactions, underscoring how subtle shifts in water structure influence macroscopic outcomes. Meanwhile, advances in computational chemistry have made it possible to quantify hydration free energies with high accuracy, allowing chemists to design molecules that dissolve, aggregate, or separate in predictable ways. Such predictive control is transforming fields ranging from drug formulation to industrial solvent recovery, reaffirming water's dual role as both a medium and active participant in chemical transformations.

Acid-base equilibria in water reveal the profound connection between molecular structure and macroscopic function. The autoionization constant of water, though small, sets the stage for a wide spectrum of proton-transfer reactions central to environmental and industrial water chemistry. Proton mobility in water occurs primarily through the Grotthuss mechanism, a rapid relay process where hydrogen bonds reorganize to shuttle protons without requiring net molecular displacement. This mechanism underpins the high conductivity of acidic solutions and influences electrochemical water purification methods such as capacitive deionization and proton-exchange membrane processes. Advances in computational modeling have allowed researchers to visualize proton hopping pathways and identify how solutes, interfaces, or confinement within nanopores perturb these processes. For example, proton conduction in reverse osmosis membranes depends on controlling hydration levels within ionic clusters that determine membrane charge and permeability. Insights into such proton dynamics are enabling the development of membranes that self-adjust local pH, modulate surface charge, or prevent fouling by repelling acidic or basic species. The interplay of proton transfer with surface chemistry remains a fertile area for innovation, especially in designing materials that mimic the precise proton management used by biological systems such as aquaporins and proton pumps.

Redox chemistry in aqueous environments is not only vital for pollutant degradation but also for regulating the speciation, mobility, and toxicity of inorganic contaminants. Heavy metals such as chromium, arsenic, and mercury shift between oxidation states that differ dramatically in solubility and environmental impact. Reductive precipitation, oxidative dissolution, and complexation reactions determine whether a contaminant is immobilized or transported through ecosystems. Modern water treatment increasingly targets these redox pathways using selective electron donors or acceptors, including engineered microbes, functionalized nanoparticles, and photocatalytic surfaces. Semiconductor photocatalysts, when illuminated, produce electron–hole pairs that initiate oxidative or reductive transformations of pollutants. Modifying the band structure through doping or heterojunction formation allows chemists to tune reactivity, improve quantum efficiency, and reduce the energy required for activation. Coupling such photocatalysts with sunlight-driven processes offers a sustainable route to degrade persistent pollutants like pharmaceuticals, dyes, and endocrine disruptors. Electrochemical oxidation systems similarly exploit controlled potentials to generate radicals that mineralize organics or break down microplastics into more manageable fragments. These methods represent a shift toward reagent-free water treatment, where the driving force is electrical or photonic rather than chemical, reducing waste and operational hazards.

Desalination technologies continue to evolve through deeper understanding of ion hydration, membrane transport, and interfacial chemistry. Reverse osmosis remains the dominant method, but its performance is limited by fouling, energy consumption, and the trade-off between permeability and selectivity. New membrane designs incorporate hydrophilic coatings, zwitterionic polymers, or nanoporous fillers such as graphene oxide to create pathways that facilitate water flow while blocking ions. These materials leverage the fact that water diffuses more quickly along hydrophilic surfaces with disrupted hydrogen bonding, whereas ions are hindered by their tightly bound hydration shells. By designing nanochannels with specific chemical functionalities, engineers can manipulate dehydration barriers and steric constraints to achieve unprecedented selectivity. Forward osmosis, capacitive deionization, and membrane distillation also benefit from advances in interfacial chemistry. In forward osmosis, the osmotic pressure difference is controlled by draw solutes engineered to be easily separated and regenerated, often using temperature-responsive or magnetic materials. Capacitive deionization relies on the reversible adsorption of ions onto charged surfaces, requiring optimized electrode materials with high surface area, fast ion diffusion, and chemical stability. Integrated hybrid systems combine multiple desalination methods to improve energy efficiency and allow recovery of valuable minerals from brine, demonstrating how chemical insight drives both performance and sustainability.

Hard water management exemplifies how even simple inorganic chemistry plays a central role in maintaining water infrastructure. Calcium and magnesium ions, though not harmful to health, form insoluble carbonates and sulfates that foul pipes, boilers, and industrial systems. Understanding the equilibria governing carbonate solubility, influenced by temperature, pH, and dissolved carbon dioxide, enables precise control of scaling. Ion exchange resins replace hardness ions with sodium or hydrogen ions, but advances in resin chemistry now favor materials with chelating ligands that selectively capture Ca^{2+} and Mg^{2+} even in the presence of competing ions. Such ligands may exploit differences in ionic radius, coordination geometry, or hydration enthalpy. Antiscalants, often based on phosphonates or polycarboxylates, inhibit crystal growth by binding to mineral surfaces and disrupting lattice formation. Modern research focuses on biodegradable alternatives that reduce environmental impact while maintaining effectiveness. Meanwhile, magnetic and electrochemical strategies continue to be explored, aiming to manipulate nucleation kinetics without chemical additives. Though controversial in their efficacy, these approaches highlight the continual search for solutions grounded in fundamental aqueous chemistry.

Filtration technologies benefit immensely from materials designed with molecular precision. Adsorbent surfaces functionalized with amino, carboxyl, thiol, or sulfonate groups interact selectively with metal ions, organics, and emerging contaminants such as PFAS. Metal–organic frameworks (MOFs) offer tunable pore sizes, high surface areas, and customizable chemical functionality, making them promising candidates for capturing specific pollutants through coordination chemistry or size-exclusion mechanisms. Graphene oxide adsorbents leverage π–π stacking interactions and oxygen-containing groups to remove dyes, pharmaceuticals, and heavy metals. Bio-inspired polymers, such as those mimicking mussel adhesive proteins, adhere strongly to surfaces and resist fouling, enhancing membrane longevity. Innovations in controlling pore architecture at the nanoscale are leading to filters that not only trap contaminants but also selectively catalyze their degradation, merging adsorption and catalysis into a single platform. Such multifunctional materials represent the next generation of water purification systems capable of continuous operation with minimal maintenance.

Catalysis in aqueous environments involves complex interplay among solvent structure, reactant solvation, mass transfer, and surface chemistry. Heterogeneous catalysts used for water treatment must remain stable under diverse pH, ionic strength, and oxidative conditions. Nanoparticles such as zero-valent iron, manganese oxides, and ceria exhibit high reactivity due to large surface-area-to-volume ratios and abundant active sites. Immobilizing enzymes on polymeric or inorganic supports allows biologically inspired catalysis to operate under harsh environmental conditions, extending enzyme lifetime and enabling selective degradation of pollutants that resist conventional treatments. Photocatalysis continues to advance through the engineering of materials with tailored bandgaps that maximize absorption of visible light, enhancing solar-driven water purification. Coupling photocatalysts to co-catalysts that facilitate charge separation or surface reactions further improves efficiency. Understanding electron transfer at the catalyst–water interface is essential to optimizing these systems, and ongoing research strives to reveal how solvation dynamics, adsorbate orientation, and surface hydration layers influence reactivity.

Membrane technology is undergoing a revolution driven by molecular-level control of surface properties and internal pore structure. Hydrophilic modifications reduce fouling by creating hydration layers that repel organic matter and limit microbial adhesion. Zwitterionic coatings, composed of balanced positive and negative charges, attract a tightly bound water layer that provides superior antifouling performance. Nanoparticle integration—using silica, silver, titanium dioxide, or carbon derivatives—enhances mechanical strength, antimicrobial activity, and permeability. Advances in block copolymer self-assembly allow membranes to be fabricated with uniform pore sizes and anisotropic channels that mimic biological structures. Stimuli-responsive membranes, which alter porosity or charge distribution in response to pH, temperature, or electric fields, promise real-time adaptability in water treatment systems. Such membranes could dynamically adjust to fluctuations in feed water composition, reducing fouling, improving selectivity, and extending operational lifetimes.

Nanomaterials continue to redefine the boundaries of water chemistry. Silver nanoparticles provide microbial inhibition through controlled release of Ag^+ ions, while titanium dioxide nanoparticles catalyze oxidation reactions under UV or visible light. Carbon nanotubes and graphene derivatives offer exceptional surface areas, electrical conductivity, and mechanical strength, enabling applications in filtration, sensing, and electrocatalysis. Understanding quantum size effects, surface charge, and aggregation behavior in water is crucial to harnessing these materials effectively and safely. Nanomaterials introduce unique challenges due to their interactions with natural organic matter, proteins, and cell membranes. Researchers are developing coatings and functional groups that modulate nanoparticle stability, toxicity, and reactivity, aiming to create materials that maximize performance while minimizing environmental impact. Nano-enabled water treatment systems illustrate how manipulating matter at the smallest scales can address large-scale problems.

Resource recovery from wastewater integrates water chemistry with circular economy principles. Chemical precipitation, ion exchange, electrodialysis, and membrane processes can recover phosphorus, nitrogen, rare earth elements, and precious metals. Electrochemical methods allow selective extraction through controlled potentials that target specific redox states. For example, ammonium can be converted to gaseous ammonia at electrodes, enabling capture and reuse as fertilizer. Phosphorus can be precipitated as struvite using controlled pH and magnesium additions. Microbial fuel cells convert organic contaminants into electrical energy while simultaneously treating wastewater. Advances in reactor design and electrode materials are improving efficiency and enabling recovery of valuable resources that would otherwise be lost. Such systems require careful management of reaction conditions to avoid unwanted byproducts, illustrating the importance of precision chemistry in sustainable water management.

Analytical chemistry continues to expand the limits of detection and resolution, providing tools to monitor contaminants at trace levels and guide adaptive treatment strategies. Spectroscopic methods, including Raman, infrared, and nuclear magnetic resonance, reveal structural information about solutes, water clusters, and reaction intermediates. Chromatographic techniques separate complex mixtures, allowing identification of emerging contaminants and their transformation products. Electrochemical sensors provide real-time monitoring of pH, redox potential, ion concentration, and organic pollutants, enabling dynamic control of treatment processes. Innovations in microfluidics and lab-on-a-chip devices are reducing the cost and complexity of water analysis, making advanced diagnostics accessible in remote or resource-limited settings. Biosensors employ enzymes, antibodies, or nucleic acid probes to detect pathogens and toxins with high specificity.

The integration of these analytical tools with digital systems and machine learning supports predictive maintenance, automated treatment, and early detection of contamination events.

Hydrogen production through electrolysis links water chemistry directly to the global shift toward renewable energy. Understanding electrode kinetics, proton transport, and catalyst behavior is central to improving electrolyzer efficiency and durability. Proton exchange membrane (PEM) electrolyzers rely on acidic environments and noble metal catalysts, while alkaline electrolyzers operate with cheaper materials but require careful management of hydroxide transport. Anion exchange membranes represent a hybrid approach, combining alkaline operation with membrane-based separation, but require materials with high chemical stability. Overpotential reduction is a key challenge, and researchers are developing catalysts that lower the energy barrier for water splitting, including transition metal phosphides, nitrides, and oxides. The oxygen evolution reaction, in particular, suffers from sluggish kinetics, necessitating catalysts with optimized surface states, electronic structures, and reaction pathways. Coupling water electrolysis with renewable energy sources such as solar or wind power demands systems that can operate under variable conditions, adding further complexity. Innovations in this area underscore how advancements in water chemistry contribute directly to sustainable energy infrastructure.

Water's influence on biological systems reveals the deep connections between chemistry and ecology. The solubility, speciation, and bioavailability of contaminants determine their toxicity and environmental fate. Metal ions can complex with organic matter, forming species that are either more or less bioavailable depending on ligand identity and environmental conditions. Organic pollutants may undergo hydrolysis, photolysis, or biodegradation, each governed by aqueous chemistry. Bioremediation strategies harness microbes that metabolize contaminants, often requiring precise control of pH, redox conditions, and nutrient availability. Enzymatic pathways can be optimized through chemical additives that enhance microbial activity or stabilize reactive intermediates. Chemical modeling of aquatic systems integrates thermodynamics, kinetics, and transport phenomena to predict how pollutants behave at ecosystem scales. These models support environmental risk assessments, regulatory frameworks, and habitat restoration efforts, illustrating the importance of chemistry in safeguarding natural water resources.

Future innovations in water chemistry are increasingly driven by computational tools that simulate molecular behavior with unprecedented accuracy.

Quantum chemistry methods allow prediction of reaction pathways, solvation energies, and spectral properties. Molecular dynamics simulations reveal how water molecules reorganize around solutes, surfaces, and interfaces. Machine learning accelerates the discovery of new materials, identifying promising membrane polymers, catalysts, and adsorbents from vast chemical spaces. These computational approaches enable rational design rather than trial-and-error experimentation, reducing costs and accelerating the development of water technologies. Integrating computational predictions with experimental validation moves the field toward materials and processes that achieve optimal performance with minimal environmental footprint. Sustainability remains the overarching goal, emphasizing renewable energy inputs, minimal chemical waste, and recyclability of components. The vision for future water systems includes smart infrastructure capable of autonomous operation, resource recovery, and adaptive treatment tailored to local conditions.

As water chemistry continues to advance, its foundational role becomes ever clearer. From the molecular intricacies of hydrogen bonding and solvation to the macroscopic engineering of membranes and reactors, every innovation depends on understanding how water interacts with matter. The convergence of chemistry, materials science, biology, engineering, and computation promises transformative progress in addressing global water challenges. Through deeper molecular insight and thoughtful integration of emerging technologies, it becomes possible to build water systems that are not only effective and efficient but also sustainable and resilient, ensuring that this essential molecule remains accessible to future generations.

Chapter 28: Word Problems

1. Researchers performing ultrafast pump–probe IR spectroscopy on liquid water introduce a solute that transiently disrupts tetrahedral ordering. The spectrum reveals a 40% increase in the population of weakly hydrogen-bonded OH groups during the first 300 fs after the pump pulse. If this fluctuation correlates with a 15% increase in local solvation entropy for an anionic pollutant, formulate how these dynamic structural changes would alter its hydration free energy and predict whether adsorption to a hydrophobic membrane surface becomes more or less favorable. Justify quantitatively or qualitatively based on thermodynamic relationships.

2. A polymer membrane mimics the confined geometry of biological channels by stabilizing linear rather than tetrahedral water chains. Diffusion measurements indicate that proton conductivity through the membrane increases threefold, while Na^+ and Ca^{2+} transport decrease by 70% and 85%, respectively. Given the timescales of hydrogen-bond rearrangements (ps range), analyze how the membrane's altered hydration environment enables this selectivity and predict the effect of further narrowing pore diameters by 0.2 nm.

3. In a water treatment reactor, a sorbent designed to stabilize low-coordination water clusters around its surface shows a 20% increase in adsorption capacity for a hydrophobic organic contaminant. If neutron scattering indicates suppressed hydrogen-bond switching rates near the sorbent, formulate a mechanistic explanation linking dynamic fluctuations to sorption thermodynamics, and calculate whether the distribution coefficient would increase or decrease if temperature were raised by 15°C.

4. A freeze-desalination system cools seawater from 5°C to –3°C in 20 minutes while using a peptide additive that selectively binds to the prismatic face of ice crystals. If the resulting ice excludes 98.4% of Na^+ and 99.1% of Mg^{2+}, but only 92% of boron due to its smaller hydration shell, develop an equation describing how binding affinity of the additive to a specific ice face influences solute rejection efficiency.

5. A cryoprotectant mixture is engineered to shift water's density maximum from 4.0°C to 6.3°C. The additive stabilizes compact hydrogen-bonded motifs over open tetrahedral ones. Using this altered density behavior, determine how phase-change enthalpy and nucleation rates would shift in a biological-sample freezer operating at –10°C and predict the resulting effects on intracellular ice formation.

6. A contaminated stream contains a hydrophobic pesticide with a hydration free energy of +33 kJ/mol. After adding a surfactant, dynamic light scattering shows micelles with a critical micelle concentration (CMC) of 0.8 mM encapsulate 94% of the pesticide. Given the entropy penalty of water structuring around the pollutant, calculate whether increasing temperature by 10°C will improve or hinder micellar capture and justify by analyzing hydrophobic hydration thermodynamics.

7. Cyclodextrins with cavity diameters of 0.7–0.8 nm are deployed in a pilot treatment

system to remove a pharmaceutical contaminant whose molecular diameter is 0.65 nm. If computational modeling shows that encapsulation stabilizes by −18 kJ/mol relative to hydration, predict how changes in pH (affecting hydrogen bonding) or ionic strength (affecting cavity water displacement) would modulate uptake efficiency.

8. A desalination membrane containing nanoscale ionic clusters exhibits significant proton mobility governed by the Grotthuss mechanism. When the membrane dehydrates by 5%, proton conductivity drops by 60%. Develop a model relating hydrogen-bond connectivity to proton hopping rates and predict how adding sulfonic acid groups at fixed intervals would alter local hydration, proton mobility, and membrane fouling resistance.

9. A membrane engineered to self-adjust surface charge includes functional groups that locally catalyze water autoionization. During operation, feedwater pH fluctuates from 6.0 to 8.5. Estimate how the local proton concentration near the membrane surface changes relative to bulk, and evaluate the resulting shifts in rejection of weak acids whose degree of deprotonation depends sensitively on pH.

10. A groundwater remediation reactor uses zero-valent iron nanoparticles to reduce Cr(VI) to Cr(III), forming insoluble hydroxides. If the electron transfer rate doubles when dissolved organic matter (DOM) concentrations increase due to ligand-bridging effects, determine how DOM's redox behavior and complexation chemistry jointly influence chromium mobility and the risk of remobilization during pH changes.

11. A TiO_2 photocatalyst modified with N-doping shifts its band gap from 3.2 to 2.6 eV, improving visible-light absorption. When treating dye-laden wastewater at 25°C, the catalyst shows a 45% higher degradation rate. Analyze how electron–hole recombination rates change due to doping and predict the effect of constructing a heterojunction with a second semiconductor of conduction band 0.3 eV lower.

12. In an electrochemical reactor, controlled potentials generate hydroxyl radicals ($\cdot OH$) that break down polyethylene microplastics into carboxylic acids. If radical generation efficiency is limited by mass transfer, calculate how decreasing solution ionic strength affects electric field distribution, radical production, and polymer fragmentation rates.

13. A GO-based membrane features nanochannels 0.9 nm wide and functional groups with variable oxidation states. Ion rejection increases when environmental pH rises from 6 to 9. Propose a detailed mechanism relating hydration shell stability, surface charge density, and channel narrowness to explain why larger multivalent ions are disproportionately rejected.

14. A thermoresponsive polymer used as a draw solute collapses when heated above 32°C, reducing osmotic pressure by 40%. Given an FO system operating cyclically between 25°C and 40°C, calculate water flux changes during each cycle and discuss regeneration efficiency and solute leakage as functions of polymer–water hydrogen-bond dynamics.

15. A CDI system uses carbon aerogels with pore sizes centered at 1.2 nm. Surface charging increases ion uptake until voltage reaches 1.4 V, where co-ion repulsion reduces capacity. Using Poisson–Boltzmann theory, predict how the ion-adsorption isotherm changes when pore size is narrowed to 0.8 nm and solution salinity doubles.

16. In a geothermal plant, $CaCO_3$ scaling occurs when water at 60°C containing 4 mM Ca^{2+} and 15 mM bicarbonate undergoes degassing. Model the shift in carbonate speciation as CO_2 escapes and determine whether a chelating resin with preference for octahedral coordination will outperform a conventional Na-form ion-exchange resin.

17. A biodegradable polycarboxylate antiscalant shows strong affinity for the {104} face of calcite. If surface adsorption decreases growth rate by 75%, evaluate how introducing Mg^{2+} (which incorporates poorly into calcite) affects lattice distortion, inhibition efficiency, and long-term stability.

18. A Zn-based MOF with 1.1-nm pores captures PFOS via both hydrophobic interactions and electrostatic attraction to $-SO_3^-$ groups. Given that competing organic matter occupies 30% of pore volume, derive how breakthrough time changes when humidity rises (increasing competitive hydration) and when the MOF is functionalized with fluorinated groups.

19. MnO_2 nanoparticles catalyze oxidation of phenolic contaminants through surface redox cycling. If nanoparticle aggregation increases particle size from 10 to 40

nm, calculate the resulting loss in catalytic surface area and estimate the effect on reaction turnover frequency.

20. A biomimetic membrane incorporates peptide channels that pass water at 10^8 molecules per second but block ions. If ionic leakage increases when pore-lining residues become partially protonated at pH 5, quantify how protonation changes channel electrostatics and predict the overall membrane selectivity at varying pH.

21. A water treatment facility introduces silver nanoparticles (AgNPs) for microbial control. Upon entering a river system rich in humic substances, the AgNPs rapidly acquire an organic corona that reduces Ag^+ release by 55%. Using colloidal stability principles and ligand-exchange equilibria, predict how corona formation alters particle aggregation, antimicrobial efficacy, and long-term fate. Evaluate how changes in ionic strength during seasonal runoff might influence environmental mobility.

22. During solar-driven purification, TiO_2 nanoparticles experience fluctuating pH (4–9) and ionic strength. Over time, zeta potential measurements reveal a shift toward the isoelectric point, increasing aggregation. Considering surface hydroxylation, protonation–deprotonation equilibria, and electric double-layer compression, model how these variables collectively govern catalyst lifetime and quantum efficiency.

23. An electrodialysis system captures NH_4^+ by converting it to NH_3(g) at a cathode. If electrode kinetics accelerate at high pH but membrane transport slows due to reduced hydration of ammonium, analyze the competing effects to determine overall recovery efficiency. Predict how adjusting the applied potential influences Faradaic efficiency and byproduct formation.

24. A reactor recovering phosphorus forms struvite ($MgNH_4PO_4 \cdot 6H_2O$) when Mg^{2+} and phosphate reach supersaturation at pH 8.5. If competing Ca^{2+} levels rise, forming Ca-phosphates preferentially, derive how ion speciation, nucleation kinetics, and solubility product differences alter precipitation selectivity. Predict operational changes needed to maintain high struvite purity.

25. A microbial fuel cell treating organic-rich wastewater shows declining voltage output as biofilm thickness increases. Given that proton transport through the

biofilm becomes rate-limiting, use the Grotthuss mechanism and pH gradients to assess how biofilm hydration, redox mediator diffusion, and surface charge affect electron recovery efficiency.

26. A microfluidic chip integrates Raman spectroscopy and electrochemical sensing to detect an endocrine-disrupting compound at 10^{-12} M. If flow rate increases shear stress that rearranges hydration layers around the analyte, determine how this influences Raman cross-section and electrode adsorption kinetics. Propose design changes to maintain sensitivity.

27. High-field NMR detects slow-exchanging water clusters in a desalination concentrate stream. If paramagnetic ions broaden spectra by shortening relaxation times, calculate the extent to which solvation-shell rearrangements can still be resolved. Discuss how changes in magnetic susceptibility from trace Fe^{3+} or Mn^{2+} alter structural interpretations.

28. During treatment of oilfield wastewater, Raman mapping reveals microdomains rich in hydrophobic contaminants surrounded by structured water. With temperature increasing from 15°C to 45°C, these domains shrink. Explain how hydrophobic hydration and hydrogen-bond disruption contribute to domain collapse and predict how this affects oxidation processes.

29. A freeze-concentration desalination unit forms dendritic ice at −6°C. Introducing a polysaccharide additive produces slab-like crystals instead. Evaluate how selective binding to ice faces, hydrogen-bond templating, and thermal gradients control crystal morphology and estimate the effect on solute entrapment and energy efficiency.

30. A machine-learning model trained on DFT-calculated hydration energies predicts solubilities of emerging pollutants with ±2 kJ/mol error. If a new pollutant has extensive π-conjugation and polarizable substituents, analyze whether the model may underpredict or overpredict solvation energy. Propose additional descriptors needed for accurate prediction.

31. A membrane distillation module shows reduced flux when hydrophobic pores partially flood due to surfactants. Analyze how changes in liquid–vapor interfacial curvature, capillary pressure, and temperature polarization influence transport.

Predict whether increasing membrane surface roughness improves or worsens wetting stability.

32. A hybrid system combining forward osmosis, nanofiltration, and electrodialysis aims to maximize water recovery while extracting lithium. If FO concentrates salts that later cause NF scaling, evaluate how inter-stage chemistry (pH, redox state, saturation index) governs scaling propensity. Assess how ED's ion-selective extraction alters downstream equilibria.

33. A seawater RO plant experiences fouling when humic acids adsorb onto the polyamide layer. Given that humics adopt different conformations at pH 6 versus pH 9, describe how charge density, hydration, and conformation influence fouling thickness and permeability decline. Quantify how adding divalent cations changes the fouling layer's cohesion.

34. A PEM electrolyzer operating at 80°C degrades due to sulfonic acid group desulfonation, lowering proton conductivity by 30%. Considering hydration-shell collapse and polymer-chain rearrangement, predict how water content affects conductivity recovery and estimate the change in overpotential required to maintain hydrogen production rate.

35. An alkaline electrolyzer using NiFe-oxide catalysts experiences performance losses at high current densities due to limited OH^- transport. Model how ionic strength, viscosity, and electric-field distribution at the electrode interface influence mass-transfer limitations. Suggest catalyst-layer design modifications to mitigate concentration polarization.

36. A visible-light photocatalyst treating brackish water suffers reduced activity in high-SO_4^{2-} environments. Evaluate whether sulfate acts primarily as a radical scavenger, surface adsorbate, or charge-transfer inhibitor, and determine how ionic strength influences conduction-band electron availability.

37. Carbon nanotube electrodes detect trace Pb^{2+} via differential pulse voltammetry. If natural organic matter binds Pb^{2+} with high affinity, reducing free-ion concentration, predict how sensor response shifts. Analyze how modifying nanotube surface chemistry could enhance detection of strongly complexed metal ions.

38. A chelating resin featuring iminodiacetate groups selectively removes Cu^{2+} in the presence of Ca^{2+} and Mg^{2+}. Given that Cu^{2+} prefers square planar or Jahn–Teller

distorted octahedral environments, describe how resin geometry and hydration-shell structure impart selectivity. Predict how resin performance changes at low pH when proton competition increases.

39. A graphene-supported Co_3O_4 catalyst oxidizes pharmaceuticals in water using applied potentials of 1.5–2.0 V. If functional groups on graphene modulate electron transfer rates, derive how changing oxygen content alters catalytic turnover. Predict how catalyst stability changes in the presence of persistent radicals.

40. A laccase-modified polymer mimicking mussel proteins enhances fouling resistance. When exposed to oily wastewater, the hydrated adhesive layer repels hydrophobic droplets. Evaluate how hydration-layer structuring, interfacial tension, and polymer cross-link density control antifouling performance under variable shear conditions.

41. A batch reactor uses ultra-small (2–3 nm) ceria nanoparticles for oxidative degradation of chlorinated solvents. If enlarging particles to 8 nm reduces oxygen-vacancy concentration by 40%, calculate how this affects redox cycling capacity and pollutant degradation kinetics. Discuss trade-offs between reactivity and stability.

42. A lake impacted by mining waste contains As(V) and As(III), with NOM controlling redox and complexation equilibria. As sunlight drives photoreduction of Fe(III)–organic complexes, determine how this shifts arsenic speciation and mobility. Predict how seasonal mixing events alter vertical concentration gradients.

43. A pesticide degrades via base-catalyzed hydrolysis with rate dependent on hydroxide activity rather than concentration. Using activity coefficients that change with ionic strength, compare degradation rates in freshwater versus saline water and determine which environmental conditions accelerate breakdown.

44. In a contaminated aquifer, microbial consortia degrade hydrocarbons under mildly reducing conditions. If adding nitrate shifts redox potentials upward, enabling denitrification, assess how this changes degradation pathways, microbial competition, and community structure. Predict effects on hydrocarbon half-lives.

45. A pharmaceutical undergoes direct photolysis with quantum yield dependent on water clustering around excited states. If DOM both shields light and forms

reactive triplet states, determine whether net photolysis increases or decreases in DOM-rich water. Evaluate wavelength-dependent effects.

46. MD simulations of Li^+, Na^+, and K^+ show divergent hydration-shell residence times. Given that a graphene-oxide membrane selectively rejects Li^+ despite its small size, explain how dehydration penalties, channel polarity, and free-energy barriers dictate selectivity. Predict how functional group density modifies rejection order.

47. A soil column contains both hydrophobic organics and heavy metals. As water percolates, organics partition into soil organic matter while metals adsorb to clays. If redox gradients develop with depth, analyze how these gradients alter metal speciation and organic biodegradation rates. Predict breakthrough curves under flooded conditions.

48. A neural network trained on thousands of polymer structures predicts permeability–selectivity relationships. When applied to zwitterionic membranes, predictions deviate due to hydration-layer effects not represented in training data. Identify missing physical descriptors and propose how to improve model generalizability.

49. A pH-responsive membrane switches from neutral to positively charged when pH drops below 6.0, altering rejection of anionic dyes by 70%. Model how protonation alters Debye length, surface potential, and hydration-layer thickness. Predict membrane behavior during rapid pH oscillations.

50. A "smart" pilot facility uses real-time sensors, computational chemistry predictions, and adaptive controls. When sudden contamination increases ionic strength and introduces a hydrophobic pollutant, the system must select between oxidation, adsorption, or membrane processes. Formulate a decision-making framework using solvation thermodynamics, redox potential windows, and membrane selectivity to justify the optimal treatment pathway under dynamically changing conditions.

Chapter 29

Chemical Art of Masterpieces

The chemistry of art conservation deepens further when one examines the dynamic relationship between pigments, binders, varnishes, environmental stressors, and the intrinsic chemical reactivity of materials over decades or centuries. Paintings are not static objects; their surfaces remain chemically active long after completion. Oxygen, moisture, pollutants, and light continuously interact with organic and inorganic components, causing subtle or dramatic transformations. Even the most stable pigments undergo slow but measurable changes when exposed to fluctuating relative humidity or reactive atmospheric species such as ozone, nitrogen oxides, and sulfur-containing gases. As organic binders oxidize, cross-link, or hydrolyze, their optical properties shift, altering saturation, gloss, and translucency. These microscopic changes accumulate gradually, contributing to the patina of age or, in more severe cases, to visible deterioration. The capacity of conservation science to diagnose these transformations relies on modern analytical tools—infrared spectroscopy, Raman spectroscopy, X-ray fluorescence, gas chromatography–mass spectrometry, and optical microscopy—each revealing signatures of chemical change that guide restoration decisions. While artists of the past selected materials for their beauty, workability, and cultural symbolism, conservators must now interpret the aging pathways of these same materials by understanding how chemistry governs their long-term stability.

Light-induced degradation remains one of the most pervasive challenges in preserving paintings. Photons, especially in the ultraviolet and blue region, carry enough energy to break chemical bonds in organic pigments and binders. When chromophores absorb light, they may undergo photoreduction, photooxidation, or irreversible cleavage, resulting in fading or discoloration. Natural organic colorants—such as carmine from cochineal insects, indigo from plants, or madder lake derived from anthraquinone structures—are particularly prone to photochemical degradation. Their molecular conjugation systems, responsible for their vivid hues, are delicate and easily disrupted by bond scission or oxidation. In contrast, inorganic pigments like ultramarine, iron oxides, and cadmium sulfide exhibit greater lightfastness, though they, too, can undergo alterations. Ultramarine, for example, can fade under acidic conditions as sulfur species in its aluminosilicate cage oxidize or leach out. Cadmium yellow may oxidize to cadmium sulfate, lightening the paint and altering texture. Conservation scientists mitigate these risks by controlling illumination levels, using UV-filtered lighting, and performing periodic colorimetry to monitor subtle drifts in chromatic values over time.

Temperature and humidity contribute their own complex influences. Organic binders—whether proteins, oils, resins, or polymers—expand and contract in response to humidity changes, producing mechanical stresses that can cause craquelure or delamination. Oil-based films, formed through oxidative polymerization of drying oils such as linseed or walnut oil, continue to cross-link over the centuries, gradually becoming more brittle. In cold, dry conditions, these films may shrink and crack, while in warm, humid environments, hydrolysis can weaken the polymer network and lead to blooming, efflorescence, or softening. Water-sensitive pigments, such as those based on copper acetates or certain iron compounds, may dissolve or undergo hydration–dehydration cycles that alter color or structural integrity. Meanwhile, canvas supports react differently from wood or metal supports, producing differential tensions that translate into warping or distortions. Museums counteract these effects with climate-controlled storage and display environments calibrated to minimize mechanical and chemical strain. The interplay of material science and environmental engineering becomes crucial, ensuring that paintings experience as little variation as possible in their microclimates.

Beyond environmental chemistry, the internal chemistry of paints reveals intricate stories of composition and transformation. Many pigments exist in multiple polymorphs or oxidation states, each with distinct colors and stabilities. For instance, the iron oxide spectrum ranges from yellow goethite to red hematite and black magnetite, with transformations between these species occurring through dehydration or redox processes. Copper-based pigments such as verdigris, malachite, and azurite are susceptible to acid–base reactions and CO_2-induced conversions. Verdigris, a basic copper acetate, may convert into brownish-black copper oxides or carbonates, altering the intended green. Vermilion, composed of mercuric sulfide, can darken by forming metallic mercury or convert into chlorine-containing mercury salts under exposure to light and humidity, especially in the presence of chloride ions. Even titanium white, a modern pigment prized for stability, exists as both rutile and anatase; the latter can catalyze photochemical reactions that accelerate degradation of adjacent organic binders. These transformations underscore the importance of analytical characterization when planning treatments, as visual inspection alone cannot reveal the specific molecular changes responsible for deterioration.

Binding media—ranging from egg yolk in tempera paintings to drying oils in Renaissance and Baroque works to synthetic acrylics in contemporary art—play equally important roles in determining a painting's chemical fate. Egg tempera dries through physical evaporation followed by protein denaturation and cross-linking, resulting in a hard, matte surface that resists yellowing but remains sensitive to water and high humidity. Oil paints, conversely, cure by autoxidation, producing complex polymer networks shaped by fatty acid composition. Linseed oil, rich in polyunsaturated linolenic and linoleic acids, forms strong films but tends to yellow over time as chromophoric oxidation products accumulate. Walnut and poppy oils yellow less but form softer coatings. Acrylic paints, introduced in the mid-20th century, consist of polymer dispersions that coalesce into films as water evaporates. Their plasticizers and surfactants, however, can migrate to surfaces, creating sticky or glossy residues that attract dust. Acrylics are also prone to softening under moderate heat and may become brittle with age as polymer chains undergo scission or loss of plasticizers. The chemistry of these binders determines not only optical qualities but also vulnerability to solvents, making conservation cleaning a highly delicate operation requiring tailored methods for each medium.

Varnishes add another layer of complexity. Natural resins such as dammar, mastic, and copal have long been used to impart gloss, saturation, and surface protection, but they oxidize and yellow significantly over time. Their triterpenoid or polymeric structures react readily with oxygen and UV light, forming chromophoric degradation products that obscure the artist's original palette. Synthetic varnishes, including ketone resins, acrylic resins, and modern aliphatic urethanes, offer greater stability and reversibility, yet no material is entirely immune to oxidation or environmental stress. Conservation philosophy emphasizes reversibility to ensure that future conservators can safely remove or modify treatments. This has led to intensive research into solvent mixtures, gel-based cleaning systems, and microemulsions that can remove aged varnish layers with minimal disruption to underlying paints. The chemical principles governing solubility, diffusion, polymer swelling, and selective solvency underpin every decision in such processes.

Furthermore, the interface between pigment particles and binders is a domain of molecular interaction often overlooked in traditional art history but essential in conservation science. Pigments dispersed within a medium may catalyze or inhibit oxidation, influence mechanical strength, or alter the binder's rheology. Lead white, for example, can accelerate oil curing by catalyzing peroxide formation, resulting in unusually robust paint films. Conversely, zinc white can produce brittle, crack-prone layers due to the formation of zinc soaps—metal carboxylates formed when fatty acids from the binder react with zinc ions. These soaps may migrate, crystallize, or create protrusions on the paint surface known as "art acne." Metal soap formation is not restricted to zinc; it can occur with lead, calcium, copper, and other metal-containing pigments, reshaping the internal structure of the paint over time. Understanding these reactions allows conservators to anticipate potential failure modes in historic works and develop strategies to stabilize or mitigate them.

Glazes and transparent paint layers represent another chemically fascinating aspect of artistic technique. Many masters, including Titian and Rembrandt, relied on thin layers of transparent pigments to build depth, luminosity, and subtle color transitions. These glazes often contain organic dyes precipitated onto inorganic substrates, forming lake pigments that are inherently less stable than their mineral counterparts. The interaction between dye molecules and their substrates involves ionic bonding, coordination chemistry, and van der Waals forces, all of which influence color strength and stability. Over time, environmental acids or humidity may disrupt these interactions, detaching dye molecules and causing fading. Additionally, glazes are highly sensitive to cleaning solvents and require meticulous care to avoid swelling, leaching, or solubilizing their organic components. Conservation of glazed passages thus demands an intimate understanding of dye chemistry, binder solubility parameters, and microenvironmental factors that affect aging.

The chemistry of supports—canvas, wood, paper, metal, or composite materials—also contributes significantly to the long-term behavior of artworks. Wooden panels, common in early European painting, respond strongly to humidity, expanding and contracting anisotropically according to grain direction. This movement can shear paint layers, causing cracks or flaking. Canvas supports, often made from linen or cotton, age through oxidation and hydrolysis of cellulose fibers, becoming brittle or slack depending on conditions. Ground layers applied to supports—typically containing chalk, gypsum, or lead white bound in animal glue or oil—serve as intermediaries that distribute mechanical stresses, but they are themselves susceptible to shrinkage, embrittlement, and chemical alteration. Metal supports such as copper panels offer stability but may corrode, producing metal salts that migrate into paint layers. Each support material introduces its own chemical narrative, influencing choices in display, storage, and conservation treatment.

Glass and ceramic materials, though often considered more durable, exhibit their own aging processes rooted in chemical kinetics and materials science. Glass, despite its reputation for permanence, can undergo leaching of alkali ions when exposed to moisture, forming hydrated silica layers that manifest as iridescence or crizzling. Ceramic glazes may craze due to thermal expansion mismatches between glaze and body, allowing moisture and pollutants to infiltrate microcracks. Transition metal ions incorporated into these materials produce striking colors through d–d transitions or charge-transfer processes, but their oxidation states may shift over time, subtly modifying hue. Understanding these processes enriches both conservation approaches and artistic practice, emphasizing that even the most seemingly inert materials possess dynamic chemical lives.

Within museums and conservation laboratories, analytical chemistry plays a crucial role in uncovering these hidden histories. Techniques such as Fourier-transform infrared spectroscopy reveal functional groups in binders and varnishes, while Raman spectroscopy identifies pigments based on molecular vibrations. X-ray diffraction provides crystalline phase information, distinguishing calcite from aragonite or identifying corrosion products on metals. Scanning electron microscopy paired with energy-dispersive X-ray spectroscopy allows visualization of pigment particles at the micron scale, revealing their morphology and elemental composition. Chromatographic methods detect aging products in oils or natural resins, mapping oxidative pathways and polymeric rearrangements. These tools enable conservators to reconstruct the material choices of artists, diagnose deterioration mechanisms, and design tailored interventions based on chemical evidence rather than assumption.

As the chemistry of painting materials becomes better understood, new questions arise concerning the ethics and philosophy of restoration. Should a painting be returned to its presumed original appearance, or should the patina of age—also a product of chemistry—be preserved as part of the artwork's history? Some pigments originally bright and vivid have darkened irreversibly due to chemical transformation; attempting to "correct" these changes through retouching risks misrepresenting the artist's original intent while suppressing the artwork's material narrative. Similarly, removing discolored varnish layers restores saturation and contrast but may erase the softening effects of time that have shaped the viewer's experience for generations. Chemistry cannot answer these philosophical questions but provides the factual foundation needed for informed decision-making, enabling conservators, curators, and historians to weigh aesthetic, historical, and material considerations holistically.

The creation of new conservation materials is another frontier where chemistry and art intersect. Scientists develop consolidants, adhesives, varnishes, and cleaning systems that combine stability, reversibility, and minimal visual impact. Innovations such as solvent-free gels, nanostructured dispersions, and enzyme-based cleaning agents demonstrate how modern chemistry can address problems that traditional solvents and resins cannot solve safely. Nanomaterials, including calcium hydroxide nanoparticles used to consolidate frescoes or deacidify paper, offer unprecedented control over reactivity and penetration. However, the long-term behavior of cutting-edge materials must be studied carefully to avoid unintended consequences in decades to come. Conservation science thus demands not only chemical expertise but also humility, acknowledging the uncertainties inherent in predicting material behavior over extended timescales.

The emotional and symbolic dimensions of color—so central to artistic expression—expand further when viewed through a chemical lens. The spectral signature of a pigment is not merely a physical property but a mediator of psychological and cultural meaning. Ultramarine, derived historically from the semiprecious stone lapis lazuli, carried religious significance and economic value, shaping artistic choices in medieval painting. Its blue arises from sulfur species trapped within a silicate lattice, an elegant interplay of crystal chemistry and color symbolism. The warm reds of cinnabar or cadmium pigments derive from mercury and cadmium's unique electronic structures, yet their beauty coexists with toxicity, reminding us that materials possess both aesthetic and chemical identities. Green pigments such as malachite or viridian evoke nature and vitality but may undergo transformations that alter hue, demonstrating the fragility of visual meaning when mediated by reactive materials. Understanding these connections enriches appreciation of artworks by linking their visual impact to the fundamental chemistry underlying their creation.

Even contemporary art, often employing synthetic polymers, industrial pigments, and unconventional materials, faces chemical challenges that echo older traditions while introducing new complexities. Modern pigments may incorporate organic molecules engineered for high chromatic purity, yet these molecules can undergo photobleaching or oxidative stress. Acrylic paints contain surfactants that migrate unpredictably, and flexible supports introduce mechanical stresses unfamiliar to traditional conservation frameworks. Mixed-media artworks combine metals, plastics, natural fibers, and electronic components, each with unique aging behavior. The chemical heterogeneity of such works tests the limits of conservation science, requiring interdisciplinary approaches to anticipate interactions and devise preventive measures.

In this ever-evolving dialogue between art and chemistry, one theme becomes clear: the preservation of cultural heritage depends on understanding matter at every scale—from electronic transitions defining color to macroscopic behaviors of materials under environmental stress. Artists manipulate materials intuitively or deliberately, harnessing chemical principles to achieve expressive goals. Conservators, in turn, decode these material systems, guided by analytical techniques and chemical reasoning. Museums provide stable conditions that mediate the relentless influence of light, temperature, humidity, and pollutants. Chemists contribute knowledge about reaction mechanisms, materials properties, and degradation pathways. Together, these perspectives establish a vibrant, collaborative field where science illuminates art and art inspires scientific inquiry.

Thus, the conservation and restoration of artworks is not merely a technical endeavor but an intellectual and cultural one, grounded in the rigorous application of chemical understanding. Each pigment particle, polymer chain, or crystalline lattice carries within it a story of creation, transformation, and survival. By studying these stories, conservators preserve not only the physical objects but also the human narratives encoded within them—the creativity of the artist, the historical journey of the object, and the shared cultural values that give art its profound meaning. Chemistry, far from being a distant discipline, becomes an essential lens through which the visual world is interpreted, protected, and celebrated. Through this union of art and science, future generations may continue to experience the beauty and significance of artworks that have endured because their material essence was respected, studied, and preserved.

Chapter 29: Word Problems

1. A 17th-century oil painting kept in fluctuating RH (35–75%) exhibits progressive craquelure over 30 years. Analytical data show increased formation of zinc soaps in passages containing zinc white mixed with linseed oil. Propose a quantitative model describing how humidity-driven fatty-acid migration combined with pigment-catalyzed oxidative polymerization accelerates crack formation. Identify which

rate laws and diffusion equations best describe each contributing process and determine which environmental parameter would yield the greatest reduction in crack-propagation rate.

2. A museum discovers that an oil-tempera mixed-technique work displays unexpected darkening in red glaze layers composed of madder lake on alum substrates. FTIR shows loss of carboxylate binding modes, and Raman indicates partial anthraquinone cleavage. Develop a mechanistic pathway explaining the degradation, factoring in local pH shifts caused by adjacent proteinaceous binder aging. Predict whether adding a microclimate buffer would slow or accelerate the observed fading, giving chemical justification.

3. An installation using cadmium yellow, ultramarine, and acrylic medium shows localized chalking under blue-rich LED lighting. Colorimetry reveals a consistent ΔE^* of ~2.1 annually. Considering the photochemistry of cadmium sulfide and polymer chain scission in acrylic films, estimate the relative contributions of photoxidation versus photocatalysis initiated by anatase TiO_2 particles elsewhere in the artwork. Recommend lighting spectra and intensities to minimize further damage.

4. An 18th-century panel painting undergoes warping when displayed near an HVAC vent. Wood moisture content cycles between 6% and 13%, while the oil paint contains lead-rich ground layers. Using anisotropic expansion coefficients for hardwoods, develop a stress-distribution model predicting where delamination will occur first. Explain how lead-catalyzed polymer cross-linking in the paint alters the mechanical response compared to a zinc-based ground.

5. A conservator detects blooming on a varnished Renaissance painting. GC–MS identifies oxidized triterpenoids from aged dammar; SEM–EDS shows microcrystalline zinc soaps protruding from the paint surface. Create a step-wise hypothesis describing how varnish oxidation products migrate into upper paint layers and interact with metal carboxylate phases. Predict solubility parameters needed for a gel-based cleaning system that removes bloom without destabilizing zinc soaps.

6. An ultramarine blue section in a medieval manuscript illuminates exhibits fading after exposure to acidic atmospheric pollutants. XRF identifies sulfur depletion,

and XRD indicates partial breakdown of the aluminosilicate lattice. Construct a kinetic model incorporating both proton diffusion and redox degradation of sulfur species. Provide estimates for pollutant concentrations required to halve ultramarine chromatic intensity in a century.

7. A contemporary acrylic artwork shows surfactant migration that leaves glossy patches. FTIR reveals increased free-water absorption bands near the surface. Propose a thermodynamic explanation linking plasticizer volatilization to surfactant migration, and estimate the activation energy of the migration process using Arrhenius plots derived from artificially aged samples.

8. An oil painting with linseed-based binder shows progressive yellowing despite UV-filtered lighting. GC–MS detects dicarboxylic acids and conjugated ketones. Using known autoxidation pathways of linolenic acid, propose a quantitative mechanism describing chromophore accumulation. Calculate the expected spectral shift in absorbance if the concentration of conjugated trienes doubles.

9. A late-19th-century watercolor using indigo experiences localized photobleaching in high-illumination cases. Using the blue-light absorption coefficients for indigo, derive the quantum yield of photodegradation needed to produce the observed fading rate. Suggest an illumination limit that would extend the pigment life by at least 200 years.

10. A canvas painting exhibits slackening and tightening cycles that mismatch the expansion behavior of an oil ground. Using the viscoelastic properties of oxidized cellulose, model stress transfer at the ground–canvas interface. Predict the long-term outcome if RH fluctuations remain ±10% and explain the chemical mechanism of cellulose depolymerization contributing to embrittlement.

11. A mummy portrait painted on wood with copper-based pigments shows green areas turning brown-black. XRD reveals formation of copper oxide and basic copper carbonate. Describe the competing equilibria between verdigris and atmospheric CO_2/H_2O. Determine whether increasing pH slightly (by microenvironment control) would stabilize or accelerate the transformation.

12. A study of aged mastic varnish shows increased absorption at 420 nm. Using known oxidation rates for mastic's triterpenoids, estimate the average oxidation

state of carbon in the polymer network and relate it to the observed yellowing. Propose a reversible varnish alternative with superior photostability and justify why its polymer structure resists chromophore formation.

13. A panel shows protrusions ("art acne") in regions containing zinc white. FTIR reveals long-chain zinc carboxylates; cross-sections show migration channels. Using Fick's second law, model the migration of free fatty acids from the oil matrix. Identify environmental thresholds where soap formation becomes diffusion-limited versus reaction-limited.
14. An acrylic painting softens under moderate heat during exhibition. DSC analysis shows a decrease in Tg by 8°C due to plasticizer presence. Predict how long-term plasticizer loss will shift Tg over decades and how this would change risk of microcracking under low temperatures.
15. A lacquered wooden artifact shows iridescence due to glasslike layers undergoing alkali leaching. ICP–MS data show sodium depletion from the top 150 nm. Explain diffusion-controlled leaching kinetics and calculate time required for complete loss of alkalis from the outer 1 μm under 60% RH.
16. A copper panel painting develops green corrosion products after being stored at 75% RH. Analyze the electrochemical series to determine whether formation of malachite or brochantite is more favorable. Discuss how organic binder residues might shift redox equilibria.
17. A fresco consolidated with $Ca(OH)_2$ nanoparticles shows uneven strengthening. Using solubility and carbonation kinetics, predict depth of penetration based on particle size and pore architecture. Recommend an optimal relative humidity for controlled carbonation.
18. A Rembrandt glaze of organic lake pigment on oil shows solvent sensitivity during cleaning tests. Propose a solvency-parameter-based selection of microemulsion components that discriminate between resin varnish and lake-rich glaze. Justify choices using Hansen solubility parameters.
19. An artwork combining aluminum foil, acrylic medium, and copper wires undergoes galvanic corrosion when RH > 60%. Construct a galvanic series–based model predicting corrosion currents and estimate coating thickness needed to inhibit electrochemical interaction for a decade.

20. A painting's lead-white passages darken, forming lead sulfide, due to hydrogen sulfide exposure. Calculate expected PbS formation given known H_2S concentrations and propose a reversible optical correction using photocatalytic reconversion chemistry.

21. A multi-layered oil painting has a hidden earlier composition revealed by X-radiography; the lower layer contains vermilion that has partially converted to metallic mercury under chloride-rich exposure. Design an analytical campaign combining micro-Raman, µXRF, and TEM-EDS to map mercury migration. Propose a thermodynamic rationale for chloride-facilitated conversion and suggest a stabilization protocol that minimizes further mercury coalescence.

22. A 20th-century mixed-media work includes cellulose acetate film adhered over painted paper. The acetate emits acetic acid that catalyzes hydrolysis in adjacent gelatin-based media. Develop a coupled-reactor model describing acetate diffusion, acid hydrolysis rate constants for gelatin, and expected mechanical weakening over 50 years at 25°C. Recommend mitigation strategies based on barrier coatings.

23. An archaeological mural's pigment palette includes both azurite and orpiment. Over decades, orpiment (As2S3) undergoes photodecomposition producing arsenic oxides which catalyze local darkening. Using solid-state reaction kinetics, model the rate of As2S3 photodegradation under museum lighting and assess health/safety implications for handling and treatment of the mural.

24. A study of a modern painting shows formation of surface efflorescence composed of sulfur-containing salts. Ion chromatography reveals ammonium and sulfate as major ions. Trace the likely sources of these ions in urban environments and propose an atmospheric chemistry model to estimate deposition rates on museum exteriors compared with interior galleries.

25. An 18th-century egg tempera panel exhibits localized water sensitivity in glazes containing aluminum lakes. Microscale solvent uptake experiments reveal anomalously high swelling. Construct a diffusion-swelling model for proteinaceous binders with ionic lake particles, accounting for osmotic effects, and predict solvent exposure times that will remain safe for the glaze during conservation cleaning.

26. A contemporary installation uses light-activated photochromic dyes embedded in polymer films. Over accelerated photostability tests, dyes degrade by N-dealkylation producing colored byproducts that migrate to the surface. Propose an analytical strategy to identify byproducts (LC–MS/MS workflow), determine migration coefficients, and design polymer matrices that suppress dye fragmentation pathways.

27. A copper-based green pigment in a historic painting forms basic copper chlorides after decades in a chloride-rich microenvironment. Use Pourbaix diagrams to explain stable phases as a function of pH and potential. Assess whether poulticing with deionized water risks converting stable basic carbonates into more mobile chloride complexes.

28. A gypsum ground on canvas shows sulfate transport into the wooden stretcher producing iron–sulfate corrosion stains. Model coupled ion transport and redox reactions that generate iron sulfates at the wood interface and propose both preventative humidity control strategies and chemical barriers to halt progression.

29. An artist's use of metallic leaf (gold leaf) over a bole ground exhibits localized lifting due to interface delamination. Analysis shows migration of organic acids from the bole (animal glue with additives) reacting with metallic impurities to form microscopic intermetallic corrosion. Develop a mechanical-chemical model that couples adhesive failure to corrosion product growth and design a reversible consolidation strategy.

30. A late-medieval illuminated folio with organic lake pigments displays microscale fading after centuries. Analyze the possible roles of trace metal catalysis, pH excursions during humidity cycles, and complexation changes in the lake substrate. Propose an experimental plan using micro-sampling, colorimetry, and electrochemical probes to test hypotheses.

31. An oil painting exhibits localized brittleness where lead white was heavily applied over an aged varnish layer. Nanoindentation maps reveal higher modulus values correlating with steep gradients in soap concentration. Construct a chemo-mechanical model for how lead soaps alter cross-link density and how that affects energy dissipation during mechanical deformation.

32. A museum is evaluating LED lighting retrofit options. Using known photochemical action spectra of common organic colorants (madder, indigo, carmine), design a weighted lighting index that balances visual acuity with pigment-specific photodamage rates to recommend spectral power distributions and allowable lux-hours per year.

33. A restoration uses an enzyme-based gel to remove proteinaceous adhesives from canvas verso. Enzyme kinetics reveal Michaelis–Menten behavior; however, immobilization on a gel affects substrate accessibility. Model the effective reaction rate and diffusion-limited uptake, and propose optimal gel thickness and enzyme loading for safe removal with minimal residual activity.

34. A modern artwork contains polymeric foam elements that off-gas volatile organic compounds accelerating adjacent pigment fade. Using gas-phase diffusion and reaction kinetics, model the concentration profile of emitted VOCs in a gallery space and evaluate active-sorbent strategies (charcoal, molecular sieves) to lower effective exposure to photochemically reactive VOCs.

35. A polychrome wooden sculpture shows flaking where ground layers contain basic lead carbonate and exposed iron-fasteners have corroded. Using galvanic coupling and lead–iron interactions, predict migration of lead into corrosion layers and its effect on porosity. Recommend electrochemical and chemical treatments compatible with polychrome preservation.

36. A conservation team evaluates two synthetic varnishes: an acrylic copolymer and an aliphatic polyurethane. Using accelerated oxidation tests and spectrophotometric yellowing indices, compare expected lifetimes and reversibility. Provide a decision matrix incorporating chemical stability, solvent compatibility for removal, and potential interactions with underlying binders.

37. A set of mixed-media works includes photographic gelatin silver prints adhered to polymer supports using pressure-sensitive adhesives. Over time, silver image density decreases adjacent to adhesive lines. Formulate a hypothesis about silver complexation or halide migration induced by adhesive components, and design an experiment using TOF-SIMS and μFTIR to map species distributions.

38. A conservator observes cracking in a contemporary painting where titanium white (anatase-rich) was used adjacent to organic red dyes. The anatase may

catalyze photodegradation. Using semiconductor photocatalysis principles, estimate electron–hole pair generation rates under standard gallery illumination and correlate to observed dye fade rates. Propose pigment segregation strategies for future practice.

39. A large canvas shows differential fading across stitched repair lines; the conservation stitch used a polyester thread that altered local microclimate. Model heat and moisture transport across the stitch region compared to surrounding canvas and show how this results in altered ageing kinetics of nearby pigments.

40. An archaeological glass bead assemblage shows crizzling and iridescence due to alkali leaching and silica rehydration. Derive an activation-energy-based model for leaching under burial porewater chemistries differing in pH and ionic strength. Predict conditions that maximize surface iridescence, and suggest minimal-intervention stabilization.

41. A painting restored in the 1970s with synthetic adhesives shows exudation of adhesive in warm galleries. Thermal analysis suggests a Tg near ambient. Model time-dependent viscous flow of adhesive under cyclic temperature conditions and recommend display temperature controls to prevent further exudation without degrading pigments.

42. A polychrome terracotta reliquary displays sulfate-induced efflorescence after exposure to sulfate-rich groundwater during a disastrous flood. Model salt crystallization pressure in pore spaces and predict the stress threshold that would cause micro-cracking in the ceramic matrix. Recommend desalination strategies that consider dissolution kinetics and the risk of swelling.

43. A conservator must choose a consolidant for flaking paint where the ground includes gypsum and protein binder. Compare small-molecule alkoxysilanes, acrylic consolidants, and calcium hydroxide nanoparticle consolidants for compatibility, penetration depth, reversibility, and long-term chemical interactions with the gypsum phase.

44. A modern mural sprayed with aerosol paints shows layered interactions where solvent-borne resins dissolve lower layers during reapplication. Model solvent diffusion and polymer swelling during overlaying processes to determine minimal flash-off times required to avoid interlayer mixing for three common solvent types.

45. A set of watercolors on wove paper shows staining from iron-gall ink migration. Using electromigration and capillary models, predict rates of iron and sulfate transport through paper fibers during high humidity events. Design a localized treatment that immobilizes iron while preserving optical appearance.

46. A textile artwork uses mordanted natural dyes (aluminum, iron mordants) and shows variation in lightfastness across mordant types. Propose how coordination chemistry differences (ligand field strength, geometry) influence photostability, and design an accelerated light-aging study to quantify relative lifetimes for each mordant-dye complex.

47. A composite artwork includes embedded electronic components that heat during operation, causing local thermal aging of adjacent polymer paints. Model thermal diffusion, polymer oxidation kinetics as a function of temperature, and predict service cycles that keep chemical aging within acceptable limits for 25 years.

48. A copper-containing green pigment shows transgranular cracking associated with recrystallization of metal soaps. Using crystal-growth models and thermodynamic solubility, explain how soap crystallization exerts mechanical forces and propose a stabilization protocol that reduces further crystal growth without solvating paint layers.

49. A conservator investigating a fresco finds microbial colonization contributing to pigment alteration. Cultures reveal heterotrophic bacteria producing organic acids and pigment-reducing metabolites. Propose an integrated biocide and environmental control plan that eradicates microbes while minimizing chemical damage to original pigments and binders.

50. A major museum plans to loan a collection of modern mixed-media works internationally. Prepare a chemical risk-assessment framework that evaluates relative vulnerability of works to transport stresses (temperature/humidity swings, light exposure, vibration) and proposes packaging and climate control specifications to reduce probabilistic failure risk to below 1% per annum.

Chapter 30

Cosmetics at the Molecular Level

Cosmetic chemistry is a fascinating field dedicated to understanding the fundamental principles behind the formulation of beauty and personal care products. The purpose of studying this discipline is not merely academic; it equips individuals with the knowledge to make informed decisions about the products they use daily. Many women, for instance, reach for lipsticks, mascaras, or foundations without fully understanding the chemical compositions of these items. Makeup, in essence, is a carefully designed mixture of chemicals, and knowing how these substances interact with the skin, hair, or nails allows for safer, more effective use. By understanding cosmetic chemistry, one can select products that enhance natural beauty while minimizing potential side effects, such as rashes, acne, or eye irritation.

The primary function of cosmetic ingredients is to enhance or refine the appearance of skin, lips, lashes, and other facial and bodily features. However, applying makeup is not always straightforward, as improper use can lead to adverse reactions. A fundamental step in responsible makeup application is reading ingredient labels, which the United States Food and Drug Administration requires to be listed in descending order of concentration. Ingredients that constitute less than one percent of a product, such as fragrances or colorants, appear after the primary components. Consumers must remain vigilant to avoid irritation, infections, or other health complications, and they should recognize that expensive products are not inherently superior to more affordable alternatives.

Primers are among the first cosmetic products applied, serving as a preparatory layer that enhances the adherence and longevity of subsequent makeup. Foundation primers, for example, can function as moisturizers, control oiliness, and even provide antioxidant protection through ingredients like vitamins A, C, and E, grape seed extract, or green tea extract. Primers vary in composition, including water-based, silicone-based, and mineral-based formulations, each with unique properties. Eye primers smooth the eyelid surface, intensify eyeshadow colors, and prevent creasing, while lip primers prepare the lips for lipstick, preventing smudging and prolonging wear. Mascara primers enhance lash volume and length before the application of mascara, ensuring a longer-lasting effect. Botanical extracts such as lavender, rose, and aloe are sometimes added to primers to soothe and refresh the skin.

Lipsticks combine waxes, oils, and pigments to add color, texture, and protection to the lips. Waxes such as beeswax, carnauba, and candelilla give structure, while oils like castor oil, olive oil, and lanolin provide shine and smooth application. Pigments, both organic and inorganic, determine the wide array of lipstick shades, from pinks to reds, and specialized finishes such as matte, crème, sheer, or glossy are achieved by varying the proportion of oils, waxes, and fillers. Chemistry techniques like thin-layer chromatography can even separate pigments in lipstick, revealing the intricate composition of these seemingly simple products.

Concealers and foundations work together to create an even skin tone. Concealers, richer in pigments than foundations, mask dark circles, blemishes, and other imperfections, while foundations provide coverage ranging from light to full, depending on the desired effect. Foundations themselves come in a variety of formulations—oil-based, water-based, powder-based, and silicone-based—each suited for different skin types and conditions. Face powders, blushes, and bronzers complement these products by setting makeup, adding color, and enhancing facial structure. Blushes typically contain red talcum powder and colorants, while bronzers deepen the complexion and create a natural sun-kissed effect.

Mascara enhances the eyes by darkening, thickening, and lengthening lashes, with formulations ranging from water-resistant to regular varieties. Ingredients such as carbon black, iron oxides, waxes, oils, and preservatives determine the performance, texture, and durability of mascara, while fibers like nylon or rayon can add length or curl. Proper hygiene is essential, as old or contaminated mascara may harbor bacteria and cause infections or allergic reactions.

Color is one of the most crucial aspects of makeup, achieved through pigments and dyes derived from natural, synthetic, or animal sources. Common colorants include titanium dioxide, iron oxides, mica, coal tar dyes, and carmine, and they can be combined with bases, bulking agents, and waxes to produce long-lasting, visually appealing products. Bases, whether oily, waxy, or emulsified mixtures, provide the medium for colorants to adhere to the skin or lips. Bulking agents like talc and French chalk ensure even application, while waxes and thickeners impart structural integrity and texture.

Active ingredients, such as alpha hydroxy acids, salicylic acid, and benzoyl peroxide, interact directly with the skin to improve its appearance, fight acne, or regulate oiliness. Emollients, moisturizers, and essential fatty acids protect, hydrate, and nourish the skin, maintaining softness and elasticity. Sunscreens filter harmful UV rays, while preservatives prevent microbial contamination and chemical degradation. Emulsions, surfactants, and additives ensure stability, consistency, and safety in cosmetic formulations.

Effective skin care is intertwined with cosmetic use. Cleansing, exfoliating, toning, moisturizing, mask treatments, and sun protection are critical practices tailored to specific skin types—normal, dry, oily, combination, or sensitive. Proper application, regular hygiene, careful selection of ingredients, and attention to potential allergens are all vital to achieving optimal results. Reading labels, testing new products on small areas, and avoiding expired or shared cosmetics are key safety precautions.

Ultimately, cosmetic chemistry is an essential discipline for anyone seeking to understand, use, or create makeup responsibly. Knowledge of primers, lipsticks, concealers, foundations, powders, blushes, bronzers, mascara, colors, bases, thickeners, active ingredients, emulsions, moisturizers, sunscreens, preservatives, surfactants, and emollients empowers users to make informed choices, enhancing beauty while prioritizing health. Coupled with diligent skin care, this understanding allows individuals to maintain youthful, healthy, and radiant appearances, making cosmetic chemistry both a science and an art in the pursuit of personal care excellence.

In this broader setting, cosmetics chemistry becomes another arena where chemical matter expresses its informational capacities through formulation behavior, interfacial structuring, and dynamic molecular responses to the biological environment. Modern cosmetic systems are no longer passive mixtures but adaptive assemblies whose performance depends on how ingredients collectively encode hydration, protection, color, texture, and sensory experience. At the molecular level, the stability of an emulsion, the persistence of a pigment, or the controlled release of an active ingredient emerges from a choreography of intermolecular attractions, phase equilibria, and microstructural transitions. Cosmetics thus exemplify a subtle form of chemical computation: ingredients “decide” their spatial organization and functional contributions by following thermodynamic cues and kinetic pathways, transforming a simple mixture into a responsive material calibrated to human skin’s complexities.

As research deepens, cosmetic formulations increasingly rely on an understanding of self-assembly as a central design principle. Surfactant molecules arrange themselves into micelles, vesicles, liquid-crystalline phases, and lamellar structures depending on concentration, ionic strength, and temperature. These supramolecular arrangements can encapsulate hydrophobic or hydrophilic components, modulate viscosity, and influence tactile properties such as slip, spreadability, and cushion. The interplay between hydrophilic-lipophilic balance, critical micelle concentration, packing geometry, and interfacial curvature becomes a guiding framework, allowing chemists to craft emulsions that remain stable over months or respond to shear forces during application. Even subtle molecular adjustments—branching, degree of saturation, headgroup charge—alter these assemblies, demonstrating how cosmetic performance is encoded in molecular architecture rather than imposed externally.

Pigments and colorants in cosmetics reveal another layer of chemical sophistication, where optical properties arise from electronic transitions, crystal packing, particle morphology, and surface treatments. Organic chromophores rely on conjugated π-systems whose absorption spectra depend on substituents, resonance stabilization, and electron density distribution. Inorganic pigments derive color from d–d transitions, charge-transfer processes, or band-gap absorption in crystalline lattices. Meanwhile, physically structured pigments exploit interference, diffraction, and scattering to generate iridescence or soft-focus effects. The choice and modification of pigments—through coatings, dispersion aids, or surface functionalization—determine not only shade and vibrancy but also longevity, radiance, and interactions with sebum and moisture. These systems illustrate how materials "compute" visual outcomes through structural organization, linking aesthetics directly to molecular and particulate chemistry.

Fragrance chemistry brings yet another dimension of complexity, rooted in volatility, molecular recognition, and emergent olfactory perceptions. Small, semi-volatile organic molecules diffuse through air and bind transiently to olfactory receptors, triggering neural signals that the brain assembles into scent profiles. The performance of a fragrance within a cosmetic matrix depends on its partitioning behavior, vapor pressure, solubility, and interactions with oils, waxes, polymers, and surfactants. Fixatives, encapsulation technologies, and controlled-release matrices modulate evaporation dynamics, allowing fragrance accords to unfold over time in top, middle, and base notes. Microencapsulation systems—whether polymeric shells, cyclodextrin complexes, or lipid-based carriers—demonstrate how temporal sensory experiences emerge from chemical microenvironments and molecular mobility.

Skin biology adds yet another computational layer: the skin barrier's structure determines how cosmetic ingredients partition, diffuse, and transform upon contact. The stratum corneum, with its lipid bilayer stacks and corneocyte matrix, acts as a selective membrane whose permeability depends on molecular size, polarity, hydrogen bonding ability, and conformational flexibility. Cosmetic actives must navigate this landscape, where enzymatic processes, hydration gradients, and lipid phases determine their fate.

Whether hydrating agents draw moisture through osmotic mechanisms, antioxidants quench reactive species through electron-transfer pathways, or UV filters dissipate absorbed radiation via internal conversion, each function emerges from molecular behavior in a living, dynamic milieu. Cosmetics thereby interface with biologically embedded chemical networks, shaping outcomes through gentle manipulations of equilibrium and kinetics that respect skin's functional integrity.

Polymers play a central role in modern cosmetics as structural and sensory modulators. Synthetic and biobased polymers create films, gels, viscosifying networks, and sensory-modifying matrices. Their behavior depends on chain conformation, charge distribution, cross-linking density, and interactions with water and oils. Vinyl polymers, cellulose derivatives, silicones, and protein-based polymers each create distinct rheological signatures—ranging from thixotropic flow to elastic gelation—allowing formulators to tune how a product spreads, grips, melts on contact, or leaves a residue. These polymers participate in subtle forms of chemical information transfer: hydration levels alter their chain mobility and network architecture, which in turn influences feel and active release. Even the tactile qualities consumers describe subjectively—silky, creamy, lightweight, velvety—arise from the interplay of polymer physics and molecular interactions.

The science of skin feel itself increasingly relies on quantitative structure–sensory relationships, which attempt to correlate tactile perceptions with measurable attributes such as modulus, yield stress, slip coefficient, and interfacial lubrication behavior. Silicones illustrate this beautifully: their flexible Si–O backbones, low surface tension, and controlled volatility create characteristic sensory signatures. Cyclic silicones impart rapid spread and a fleeting, dry touch as they evaporate, while linear silicones contribute glide, softness, or film formation depending on chain length. Substituted silicones add further dimensions: phenyl groups influence refractive index and radiance; polyether groups tune hydrophilicity; crosslinked networks create elastomeric textures. These phenomena highlight how minute chemical modifications translate into macroscopic user experience.

Cosmetics also rely on photochemistry, especially in the domain of sun protection. UV filters function by absorbing radiation and dissipating the energy harmlessly as heat via non-radiative decay pathways. Their photostability—the capacity to cycle through excitation and relaxation without degrading—is essential for sustained protection.

Molecular structures with conjugated aromatic systems and electron-donating or withdrawing substituents modulate absorption maxima and stability. Formulations must also consider intermolecular energy transfer, potential for photodegradation, and interactions between organic and inorganic filters. Inorganic UV filters such as titanium dioxide and zinc oxide introduce additional complexity through band-gap transitions, particle size effects, surface coatings, and scattering behavior. The balance between transparency, protection, and photochemical robustness emerges from a lattice of molecular and particulate considerations.

In anti-aging and skin-health cosmetics, active ingredients demonstrate how chemical reactivity intersects with biological signaling. Antioxidants such as ascorbates, tocopherols, flavonoids, and polyphenols neutralize reactive oxygen species through electron or hydrogen atom transfer processes governed by redox potentials and resonance stabilization. Humectants like glycerol, hyaluronic acid fragments, and certain sugar derivatives modulate hydration by binding water through hydrogen bonding networks. Peptides act via structural motifs that influence receptor interactions or modulate enzymatic cascades, with activity tied to their conformational stability, sequence charge distribution, and susceptibility to degradation. Even encapsulation methods—liposomes, polymeric nanoparticles, cross-linked hydrogel beads—use chemical confinement to regulate diffusion, protect sensitive molecules, and direct their release profiles.

Cosmetic preservation strategies illustrate another subtle interplay of chemistry, microbiology, and material behavior. Preservatives act by interfering with microbial membrane integrity, enzyme function, or metabolic pathways, yet their performance depends on pH, partition coefficients, binding affinities, and compatibility with other ingredients. Multifunctional additives such as certain glycols, organic acids, and chelators participate indirectly by altering water activity, disrupting biofilm formation, or sequestering metal ions that catalyze unwanted reactions. Preservation efficacy emerges from the collective behavior of formulation components, where the microenvironment modulates the balance between active antimicrobial species and those sequestered or neutralized by interactions.

Emerging “smart” cosmetics embody the future direction of the field, where chemical systems respond dynamically to stimuli such as temperature, pH, ionic strength, or mechanical stress. Thermoresponsive polymers adjust viscosity with skin temperature; pH-responsive encapsulation matrices release actives in slightly acidic skin environments; mechanochromic pigments shift optical properties under friction. These transformations do not require complex machinery—only the intrinsic capacity of molecules to reorganize according to changes in energetic landscapes. In this sense, smart cosmetics behave like molecular computers, interpreting physical cues as inputs and producing sensory or functional outcomes as outputs.

Sustainable cosmetics add yet another dimension, where renewable feedstocks, biodegradable polymers, fermentation-derived surfactants, and green solvent systems reshape the chemical toolkit. Life cycle analysis informs ingredient selection, considering upstream synthesis impacts, biodegradation pathways, and interactions with environmental matrices. Green chemistry principles drive innovation in mild, efficient reactions and inherently safe molecular designs. Here too, chemical systems process information—this time ecological, regulatory, and biological—by embedding environmental considerations into molecular structure itself.

Across all these domains, cosmetics chemistry offers a lens through which to view chemical matter as an active participant in functional design. Emulsions, gels, fragrances, pigments, and active delivery systems all behave as dynamic networks whose properties emerge not from isolated components but from collective organization. The field increasingly recognizes that performance cannot be reduced to a list of ingredients; it arises from how those ingredients negotiate interactions, self-assemble, respond to stimuli, and adapt to biological surfaces. Cosmetics thus become a microcosm of the broader trend in chemical research: a shift from static formulations to dynamic, information-rich materials capable of processing environmental cues and generating tailored responses.

As this shift deepens, the boundaries between formulation chemistry, soft matter physics, skin biology, and materials science continue to blur. The future of cosmetics lies not in single miracle molecules but in orchestrated systems whose emergent properties stem from multi-scale interactions—from atomic-level bonding to macroscopic sensory perception. In embracing complexity, the field aligns with the wider trajectory of chemical science, where matter is understood as both structured and structuring, both receptive and expressive. Cosmetics chemistry, in all its subtlety, demonstrates that even everyday materials participate in the grand interplay of chemical information, energy landscapes, and dynamic organization that shapes the material world.

Chapter 30: Word Problems

1. A cosmetic chemist is developing a water-in-silicone foundation that must remain stable for at least 12 months at 40 °C. The formulation contains 30 % cyclopentasiloxane, 15 % dimethicone, 10 % silicone crosspolymer gel, 40 % aqueous phase, and 5 % pigment dispersion treated with a hydrophobic coating. During accelerated testing, the chemist observes phase separation when the foundation is subjected to repeated shear cycles mimicking consumer use. Considering the hydrophilic–lipophilic balance (HLB), interfacial curvature, and the critical micelle concentration of the surfactant system employed, determine how the chemist should adjust the emulsifier blend to prevent coalescence and whether the silicone phase requires altered polymer cross-linking density to maintain structural integrity under stress.

2. A lipstick manufacturer wants to create a matte formula using a combination of candelilla wax, beeswax, hydrogenated polyisobutene, and silica microspheres. However, the current formula produces drag on application, likely due to the crystalline structure of the wax network and insufficient oil phase penetration. Analyze how adjusting wax ratios, incorporating esters with specific branching patterns, or altering pigment surface treatment might reduce yield stress while maintaining the matte optical profile. Calculate the expected changes in viscosity if wax crystallization is reduced by 20 %.

3. A research team is studying how the microstructure of an oil-in-water emulsion primer changes when subjected to skin temperature (32–34 °C) and mechanical spreading. The primer contains a lamellar liquid-crystal surfactant system, humectants, and silicone elastomers. Using concepts of phase transitions, interfacial rheology, and lamellar spacing, evaluate how the lamellar arrangement reorganizes during application and predict whether active encapsulated molecules (vitamin C derivatives) will be prematurely released or remain trapped within the lamellae until evaporation of the volatile components shifts osmotic balance.

4. A new mascara formula incorporates nylon fibers, carnauba wax, iron oxides, film-forming acrylic polymers, and a water-resistant emulsifier. After several weeks on the market, users report smudging and reduced curl retention in humid climates. Investigate how humidity affects polymer glass transition temperature

(Tg), capillary condensation on fibers, and wax phase deformation. Propose modifications to polymer chain length, cross-linking, or wax melting points that would maintain lash lift under 90 % relative humidity.

5. A skin-tint foundation uses pigment-encapsulated silica microspheres that burst upon shear, releasing color gradually for a "tone-adjusting" effect. During scale-up, the manufacturer notices that shear exposure during mixing prematurely breaks the capsules, altering the final shade. Considering capsule shell polymer chemistry, shear-thinning behavior, and mechanical energy transfer, devise strategies to adjust processing parameters (RPM, impeller type, batch temperature) or modify shell crosslink density so encapsulation remains intact until consumer application.

6. A sunscreen includes avobenzone, octocrylene, Tinosorb S, and zinc oxide nanoparticles. Under simulated solar exposure, the formulation shows declining UVA protection due to avobenzone photodegradation. Examine how electron-donating substituents on avobenzone, triplet-state quenching by octocrylene, and surface-coated ZnO particle charge influence photostability. Determine whether adjusting pH, encapsulating avobenzone in liposomes, or modifying ZnO coatings will best prevent radical formation and maintain long-term SPF stability.

7. A cosmetic brand formulates a botanical serum containing niacinamide, ferulic acid, hyaluronic acid fragments, and green tea polyphenols. Over time, the serum darkens, suggesting oxidative degradation. Using redox potentials, pKa values, and hydrogen-bonding networks, analyze how niacinamide reacts with acidic components, how polyphenols undergo oxidation, and how water activity influences stability. Design a buffer system or encapsulation method that minimizes browning while preserving bioactivity.

8. A blush formulation uses mica coated with iron oxides and silicone-treated talc to create a soft-focus effect via light scattering. However, consumer feedback indicates inconsistent opacity across different skin types. Evaluate how sebum composition, skin moisture levels, and refractive index mismatches modify pigment dispersion and scattering efficiency. Determine how altering particle size distribution, coating chemistry, or emollient volatility can create a more uniform appearance.

9. A brand seeks to replace synthetic surfactants in a cleanser with fermentation-derived biosurfactants to improve sustainability. Yet the cleanser now produces

insufficient foam and fails to remove oily makeup. Assess how biosurfactant packing parameter, headgroup charge, and critical aggregation concentration influence micelle size and solubilization capacity. Propose how co-surfactants, electrolytes, or polymers could rebuild cleansing performance without sacrificing biodegradability.

10. A “smart” moisturizer uses thermoresponsive polymers that increase viscosity at skin temperature. Consumers report that the moisturizer feels sticky in warm climates. Examine how polymer lower critical solution temperature (LCST), hydrophilic–hydrophobic balance, and hydrogen bonding with humectants affect this sticky sensation. Recommend modifications to polymer architecture or hydration modulation that reduce tackiness without losing the temperature-triggered thickening effect.

11. A high-coverage concealer contains dispersed titanium dioxide and iron oxides in a silicone-based medium. Under electron microscopy, pigments form flocculated clusters that reduce opacity and create streaks during application. Investigate how surface charge, steric stabilization, and dispersant molecular weight affect pigment dispersion. Predict how altering particle coatings or adding small-molecule dispersants will modify the colloidal stability.

12. A fragrance-infused hair serum uses cyclodextrin inclusion complexes to prolong scent release. At high humidity, the fragrance evaporates too quickly. Using host–guest chemistry, vapor pressure shifts, and water-competition interactions, determine how humidity alters fragrance binding affinity inside cyclodextrin cavities and evaluate whether hydrophobic substituents or polymeric encapsulation would slow evaporation more effectively.

13. A sheet mask uses a hydrogel matrix containing peptides, glycerin, and ceramide nanoparticles. During wear, users notice uneven serum distribution on the skin. Apply principles of diffusion, surface tension, and hydrogel network porosity to explain how hydrogel–skin interactions affect release kinetics. Propose adjustments to crosslinking or the osmotic gradient to ensure uniform active delivery.

14. A matte foundation thickened with a cellulose derivative begins to pill when layered over silicone primers. Using polymer–silicone incompatibility, interfacial energy, and chain mobility concepts, evaluate why pilling occurs. Suggest modifications to

the polymer substitution degree, primer silicone chain length, or formulation order of application that would eliminate the issue.

15. A brow gel uses a combination of PVP, acrylate copolymers, and waxes to create long-lasting hold. In cold climates, the gel becomes brittle and flakes off. Considering polymer glass transitions, wax crystallinity, and moisture-plasticization effects, determine what adjustments to polymer ratios or wax types could maintain flexibility across temperatures.
16. A tinted sunscreen uses encapsulated organic UV filters within lipid nanoparticles to reduce skin irritation. However, the encapsulation dramatically increases viscosity, making the product difficult to spread. Analyze how nanoparticle volume fraction, electrostatic repulsion, and particle–polymer interactions contribute to rheological thickening. Propose how to reduce viscosity while keeping encapsulation efficiency high.
17. A setting spray includes alcohol, film-forming polymers, and humectants. Over time, nozzle clogging occurs due to polymer aggregation. Using polymer solubility parameters, volatility-driven concentration shifts, and pH-dependent polymer ionization, determine how to prevent clogging without reducing film integrity.
18. An anti-aging cream contains retinol encapsulated in multilamellar liposomes. Under stress, such as vibration during shipping, the liposomes leak prematurely. Considering lamellar bending modulus, membrane fluidity, and stabilizing cholesterol ratios, evaluate how to redesign the liposomal architecture for mechanical robustness.
19. A stick highlighter uses volatile silicones and ester oils to create a smooth glide. At high temperatures, the stick softens excessively. Assess how van der Waals interactions in the wax/oil network, silicone evaporation, and oil solubility contribute to collapse. Calculate how altering wax melting point distribution can preserve structure up to 45 °C.
20. A transparent lip gloss uses hydrogenated polyisobutene and isohexadecane, but clouding appears when stored below 10 °C. Apply phase behavior principles, including solubility curves, branching effects, and crystallization kinetics, to determine why haze forms and how to maintain clarity at low temperatures.

21. A color-correcting concealer relies on a microemulsion containing a high concentration of iron-oxide pigments and ethoxylated emulsifiers to achieve a smooth skin finish; however, clinical panels report that the color shifts under different lighting conditions (incandescent vs. LED). Using Mie scattering theory, pigment refractive index, particle aspect ratio, and the role of thin-film interference from layered application, analyze the origin of metamerism in this product and propose formulation changes—such as particle-size tuning, use of diffractive vs. absorbing pigments, or refractive-index-matching oils—that would reduce perceptible color shift across common light sources.

22. A biodegradable polymeric microcapsule is proposed to deliver benzoyl peroxide in a leave-on acne product; the polymer degrades enzymatically in sebum-rich microenvironments. Considering polymer backbone chemistry, ester hydrolysis kinetics, sebum lipase activity, and diffusion-limited release, model the expected release profile in sebaceous vs. non-sebaceous regions and determine the monomer composition and molecular weight range that would yield a burst release within 2–6 hours in high-lipase conditions while remaining intact for at least 48 hours on dry skin.

23. A liquid eyeliner uses carbon black dispersed in a waterborne acrylic matrix. After storage at elevated temperature, the product shows increased viscosity and syneresis of low-molecular-weight components. Apply principles of polymer phase separation, glass transition shift due to plasticizer loss, and pigment–polymer interactions to explain these observations; then recommend stabilizing additives, humectant strategies, or packaging changes that minimize syneresis without compromising fast-drying performance.

24. A company is developing mechanochromic nail polish that changes hue when stressed (bent) using spiropyran-containing polymers embedded in a nitrocellulose lacquer. Evaluate how polymer Tg, dye aggregation, and matrix polarity affect the reversibility and fatigue life of the color change through repeated flexing, and design an experiment to quantify color recovery and the number of cycles to 50 % intensity loss under realistic mechanical strains.

25. A novel cleanser uses proteolytic enzymes to remove keratin plugs in pores; however, irritation is observed in sensitive skin cohorts. Using enzyme kinetics,

substrate selectivity, pH-dependent activity, and skin barrier disruption models, determine how to tune enzyme concentration, immobilization (e.g., encapsulation in pH-triggered carriers), or use protective co-actives to maximize comedone clearance while keeping transepidermal water loss and inflammatory markers within acceptable safety thresholds.

26. A luxury foundation incorporates nacre-like platelets to produce a pearlescent glow. During formulation, the manufacturer experiences sedimentation of platelets despite a thixotropic rheology modifier. Analyze hindered settling in concentrated suspensions, particle aspect ratio effects on rheology, and the balance between yield stress and spreadability required to prevent sedimentation yet maintain a luxurious glide; propose a strategy combining network-forming polymers and surface functionalization to stabilize the platelet dispersion long-term.

27. A matte-to-dewy hybrid primer uses a pH-responsive polymer that collapses at skin pH to expose hydrophobic sites, thereby reducing sebum uptake. The effect is inconsistent among users with differing skin pH (4.1–6.0). Using Henderson–Hasselbalch relationships, ionization-dependent polymer conformation, and interfacial energy considerations, predict how polymer pKa and side-chain hydrophobicity should be tuned to produce consistent collapse across typical skin pH ranges while avoiding irritation from surfactant-like behavior.

28. A refillable compact uses solid emulsions (emulsion gels) to deliver creamy blush that must withstand hundreds of open/close cycles without microbial contamination. Considering water activity, preservative partitioning between solid and liquid domains, and the diffusion rates of antimicrobial agents within a semi-solid matrix, assess the minimum preservative concentration and the most effective preservative systems that maintain preservative efficacy against mixed flora while meeting consumer-preference constraints for "preservative-light" labels.

29. A product contains photochromic dyes to give sun-exposed lip products a tint that deepens outdoors. Over months, the photochromic response weakens. Examine the role of dye photobleaching, oxygen permeability of the lip matrix, and photostabilizer partitioning in the formulation. Recommend molecular design changes (e.g., stabilized spiropyran analogs), antioxidant co-formulants, or microencapsulation approaches to prolong reversible photochromism without introducing systemic toxicity concerns.

30. A spray-on bronzer relies on volatile solvents to leave finely dispersed silica and iron-oxide pigments on skin. Some users report inhalation discomfort and pulmonary concerns during application. Using aerosol droplet-size distribution, respiratory deposition models, and particle agglomeration behavior in solvent-laden plumes, quantify the fraction of respirable particles (<10 μm) produced by a typical pump sprayer and propose particle engineering (granulometry, binder choice) and packaging redesigns (non-aerosol pumps, mousse formats) to mitigate inhalation risk while preserving application uniformity.

31. A multifunctional BB cream includes encapsulated retinoid esters and water-dispersible titanium dioxide. Unexpectedly, the retinoid acts as a pro-oxidant when mixed with the inorganic filter under UV exposure. Using redox chemistry, interfacial electron transfer, and surface-catalyzed photochemistry, diagnose potential pathways for reactive oxygen species generation and determine whether antioxidant co-encapsulation, opaque surface coatings on TiO_2, or spatial separation via incompatible microphases would most effectively prevent oxidative degradation and irritation.

32. A formulator wants to design a hair-conditioning polymer that provides detangling yet is easily washed out. They consider polyquaterniums with varying charge densities and side-chain lengths. Using polymer–hair surface interaction models, Debye screening in hard-water conditions, and friction coefficient measurements, evaluate which degree of quaternization and molecular weight range will maximize transient adsorption (for conditioning) while enabling desorption with surfactant shampooing, and how chelators or counterions in the rinse affect performance.

33. A plant-based cosmetic line replaces mineral oils with renewable esters derived from fatty acids. In cold climates, products develop waxy bloom and reduced spreadability. Assess crystallization behavior of mixed esters, polymorphism, and cloud point phenomena; propose strategies such as branching in ester structure, use of liquid esters with low melting points, or microemulsion structuring to retain spreadability and prevent visible bloom down to −10 °C.

34. A high-luminance eyeshadow uses pearlescent pigments and a low-viscosity solvent system. Users report fallout (loose particles detaching) after application. Consider adhesive forces between pigment and binder, tackiness, and plasticizer

migration. Propose a formulation redesign that increases particle adhesion (e.g., polymeric tackifiers, rheology modifiers that create a soft film) while maintaining quick setting and avoiding a heavy feel on the eyelid—also quantify how much tackifier (wt%) would be needed to increase the work of adhesion by 30 %.

35. A sunscreen intended for children must maximize spreadability and maintain inorganic filter transparency on light skin tones. However, nanoparticles of ZnO produce a slight whitening. Using particle-size-dependent scattering (Rayleigh/Mie regimes), surface coatings to reduce refractive-index contrast, and agglomeration suppression strategies, calculate the median particle diameter required to render ZnO effectively transparent at visible wavelengths while still providing broad-spectrum UV attenuation, and discuss acceptable coating chemistries that preserve photoprotection.

36. A polyol-based humectant blend (glycerin, propanediol, and a sugar alcohol) in a moisturizer shows unexpectedly high stickiness when combined with a silicone oil. Using concepts of hygroscopicity, water-plasticization of polymer networks, and interfacial rheology between hydrophilic humectants and silicone phases, explain the mechanistic origin of stickiness and propose formulation pathways—such as inclusion of volatile esters, chain-length tailored polyethers, or partitioning agents—to reduce tack while preserving hydration efficacy.

37. A mascara preservative system uses an organic acid blend that is active at low pH; however, when the formula includes cationic conditioning polymers, preservative efficacy diminishes. Analyze how polymer binding sequesters protonated acid species, altering free concentration and antimicrobial activity, and propose a multifaceted preservation strategy combining chelators, multifunctional preservatives, and formulation pH adjustments that maintain efficacy without destabilizing the polymer or inducing ocular irritation.

38. An anti-pollution serum includes nanoparticle scavengers (functionalized ceria) to neutralize reactive oxygen species on the skin surface. Investigate the balance between catalytic antioxidant activity, particle cellular uptake, and potential pro-oxidant effects at high local concentrations. Using surface functionalization to modulate cellular interactions, design a safe ceria functionalization strategy (ligand density, hydrophilicity) that maximizes extracellular ROS scavenging while minimizing transdermal penetration and intracellular accumulation.

39. A color-stable lip stain aims to resist transfer through strong film formation but still needs to feel non-drying. The team considers elastic crosslinked polymer networks that fracture under shear to deposit color yet heal to retain film integrity. Drawing on fracture mechanics, thixotropic recovery, and network relaxation times, propose polymer backbones and crosslink densities that allow shear-thinning deposition during application but rapid elastic recovery to minimize transfer—also outline tests to measure transfer-resistance quantitatively.

40. A biodegradable glitter uses coated cellulose particles instead of microplastics. During product aging, the coating cracks, releasing hydrophilic cellulose that absorbs moisture and changes optical performance. Assess coating selection (e.g., acrylic vs. silicone coatings), mechanical mismatch due to thermal expansion and modulus differences, and the importance of coating thickness to prevent microcracking. Recommend a coating system and processing conditions that preserve optical reflectivity and moisture resistance across a wide temperature range.

41. A sunscreen developer notices that encapsulated octyl methoxycinnamate (OMC) migrates from micellar reservoirs into silicone-based primers when both are layered, reducing SPF. Using partition coefficients, micelle–oil phase equilibria, and kinetics of surfactant exchange, model the transfer of OMC between aqueous micelles and silicone matrices and recommend encapsulation strategies (lipid vesicles, polymeric nanoparticles with specific surface chemistries) or primer composition constraints that retain OMC in the sunscreen layer.

42. A hair-color formulation uses oxidative dyes that rely on small-molecule precursors and an alkaline developer. The client wants gentler formulations with lower ammonia equivalents, instead employing monoethanolamine (MEA). Evaluate how MEA modifies hair cuticle swelling, dye penetration, and developer kinetics compared to ammonia, and determine whether using MEA at reduced concentrations with co-solvents or penetration enhancers can achieve similar permanent coloration efficiency while reducing perceived odor and irritation.

43. A novel deodorant relies on pH-responsive polymer beads that sequester malodorous volatile fatty acids at neutral pH and release them for enzymatic breakdown in a second, acidic microenvironment. Analyze the thermodynamics

of binding and release, the influence of sweat composition on bead swelling, and how competing ions or enzymes in the axillary microflora will affect sequestration. Propose bead chemistries and multi-stage activation strategies that retain odorants during perspiration peaks yet enable biodegradation between uses.

44. A sunscreen's photostability assay shows that mixing two organic filters produces a synergistic photodegradation pathway leading to reactive intermediates. Use frontier molecular orbital theory, excited-state energy transfer, and intersystem crossing concepts to analyze how energy transfer between chromophores could sensitize degradation and propose pairwise filter choices or antioxidant additives that quench triplet states and minimize deleterious cross-coupling reactions.

45. A manufacturer wants to formulate a zero-waste solid shampoo bar incorporating protein hydrolysates, clay, and cationic surfactants. The bar dissolves unevenly and forms chalky deposits on hard water. Using solubility products, ion-exchange with Ca^{2+}/Mg^{2+}, and surfactant precipitation mechanisms, diagnose the cause of chalkiness and propose chelating agents, counterion strategies, or substitute surfactant chemistries that maintain performance in a broad range of water hardness levels while maintaining biodegradability.

46. A biodegradable film-former for nail polish is designed from a bio-based polymer with pendant hydroxyls. During curing, the film shows microcracking due to shrinkage stress. Apply crosslinking density calculations, volumetric shrinkage estimations, and modulus development curves to propose a co-monomer selection or plasticizer system that reduces shrinkage-induced stress and enhances fracture toughness without compromising gloss or adhesion.

47. A long-wear liquid lipstick uses a coalescing polymer that forms a continuous film on the lip surface. In hot, humid climates the film becomes tacky and attracts dust/pollen. Analyze how environmental humidity affects the glass transition, water plasticization, and surface energy that govern tackiness. Recommend polymer chemistries (e.g., hydrophobic backbones with low hygroscopicity), inclusion of matte silica fillers, or volatile co-solvents to mitigate tack while preserving long-wear film integrity.

48. A skincare device–compatible serum is formulated to be iontophoresis-friendly, containing charged peptides and low-ionic-strength buffers. During device use,

peptide aggregation and electrode fouling occur. Using electrochemistry, ionic flux, and isoelectric point considerations, determine buffer systems, peptide modifications (e.g., PEGylation), or electrode materials/coatings that minimize aggregation and fouling while maximizing transdermal peptide delivery under low current densities.

49. A pigment dispersion company is investigating how plasma treatment of mica substrates alters adhesion and color intensity in pearlescent cosmetics. Analyze how plasma-induced surface oxidation changes surface energy, silanol density, and subsequent coupling-agent chemistry (e.g., silanes), and predict the optimal plasma conditions and silane selection that yield maximal pigment wetting by oil-phase vehicles and improved durability on skin without compromising biocompatibility.

50. A sustainable cosmetics startup aims to replace conventional preservatives with a multifunctional glycol system and a low-pH formulation to reduce synthetic preservative load. Microbiological challenge tests, however, show slow growth of osmophilic yeasts. Integrating osmotic stress theory, microbial adaptation mechanisms, and preservative partitioning into complex formulations, propose a preservation strategy that combines low-pH with small-molecule chelators, selective natural antimicrobials, and water-activity lowering excipients to robustly control yeasts while aligning with green-chemistry goals and regulatory constraints.

Chapter 31

MOLECULAR COMPUTING & CHEMICAL LOGIC GATES

Molecular computing occupies a remarkable frontier at the intersection of chemistry, information theory, materials science, and nanotechnology. It aims to process, store, and transform information using molecules rather than silicon-based electronics. While electronic computers rely on voltages, currents, and semiconductor junctions to represent logical states, molecular computing uses chemical concentrations, conformational states, reaction networks, and molecular interactions to encode the equivalents of “0,” “1,” and complex logical operations. This perspective builds upon a profound conceptual shift: that information is not an abstract construct independent of matter, but rather is embodied in the material transformations and energetic landscapes of chemical species. Chemical reactions inherently transform molecular states in a directional and quantifiable manner, making chemistry a natural substrate for computation.

The foundational principle of molecular computing is that molecules can be designed to undergo transformations that mirror the behavior of logical operations. A logic gate, in its most general form, is a device that converts defined inputs into a predictable output according to Boolean logic. In chemistry, such inputs may correspond to concentration thresholds, binding events, pH changes, redox potentials, or the presence of orthogonal molecular species. The outputs—changes in fluorescence, colorimetric shifts, product formation, or catalytic activation—can be measured or transduced into further chemical events. When a chemical system is engineered such that these outputs can serve as inputs to subsequent reactions, cascading logic becomes possible, enabling the assembly of molecular circuits capable of meaningful information processing.

The earliest conceptual developments in chemical logic explored simple binary operations based on measurable analytical signals. For example, photochromic molecules can switch between two structures when irradiated with light of different wavelengths. These structures correspond to distinct absorption spectra, effectively encoding digital states. Such systems demonstrated that by controlling the light input, one could reliably toggle molecular states. Yet this early work represented only the most rudimentary form of molecular logic. True computation requires not just switching between states, but also the conditional dependence of outputs on combinations of inputs—AND, OR, NOT, NAND, NOR, XOR, and other logic functions. The transformation from mere switching behavior to conditional logic demanded precise control over reaction kinetics and molecular interactions.

One crucial avenue of development emerged through supramolecular chemistry. Host–guest systems—crown ethers, cyclodextrins, calixarenes, cucurbiturils, and rotaxanes—demonstrated that complexation equilibria could serve as logic primitives.

In a supramolecular context, a receptor might bind a particular ion or small molecule, causing a structural or electronic change that alters spectroscopic signatures. If binding is conditional on the simultaneous presence of two guests, the system performs an AND operation. Conversely, systems where either of two guests trigger the receptor response perform an OR operation. Because supramolecular interactions are highly tunable through functional-group modification, they provide a rich platform for encoding diverse logic functions with fine control over sensitivity, selectivity, and output modalities.

Another major branch of molecular computing lies in DNA-based logic systems, where information is encoded not simply in chemical identity, but in sequence-specific interactions. DNA computing exploits base-pair recognition, strand displacement, and enzymatic catalysis to implement logic gates that rival the complexity of electronic circuits, albeit operating on slower timescales. However, the significance of DNA systems to chemical logic extends beyond biology-inspired computation. DNA exemplifies a general principle: that molecular information storage can be intrinsically coupled to molecular information processing. The pairing of structure and reactivity, inherent to nucleic acids, helps illuminate how purely synthetic molecules can also serve as programmable information-processing units.

Even beyond supramolecular and nucleic-acid frameworks, reaction networks themselves can perform computation. Chemical reaction networks (CRNs) can be designed to exhibit dynamic behaviors such as oscillation, bistability, excitability, and threshold responses. These behaviors can be translated into logical operations by associating specific dynamical states with digital values. For instance, the Belousov–Zhabotinsky (BZ) reaction oscillates in concentration and color, and has been harnessed to implement spatial logic gates in reaction–diffusion media. Patterns that propagate, collide, and annihilate can serve as logic signals in a manner akin to gliders in cellular automata. The chemistry of CRNs demonstrates that computation is not limited to discrete molecular transformations; it can also arise from emergent reaction dynamics and self-organizing behavior.

Molecular logic gates can be broadly categorized based on their mechanisms. The simplest category includes **binding-based gates**, where inputs modulate receptor–ligand interactions. In these systems, the output often involves optical signals such as fluorescence enhancement or quenching, making them useful for molecular diagnostics and sensing. For example, a receptor may fluoresce only when both protons (acting as acid inputs) and a metal ion are present, thus functioning as a molecular AND gate. These gates are typically reversible, fast, and amenable to real-time monitoring, though they may lack the temporal integration needed for more complex computation.

A second category encompasses catalysis-based gates, where inputs alter the rate or pathway of a chemical reaction. Catalysts may be activated, inhibited, or modulated by multiple inputs, enabling highly non-linear responses. Catalytic logic gates offer amplified outputs, since catalytic turnover can convert small input signals into large product quantities. This property makes them ideal for sensing applications where signal amplification is desirable. However, catalytic systems often suffer from irreversible consumption of substrates, limiting reusability unless designed for reversible catalysis.

A third category involves structural-switching gates, exemplified by molecular machines, foldamers, and conformationally switchable polymers. These systems rely on the fact that molecular conformation dictates reactivity and function. An input may cause a molecule to fold into a catalytically active state, act as a mechanical switch, or expose a functional group. By engineering multiple independent switching mechanisms, molecules can implement multi-input logic behaviors. The challenge in these systems lies in ensuring orthogonality among inputs, such that each one selectively triggers its designated conformational change without interfering with other parts of the molecule.

Logic gates may also operate via redox-based mechanisms, in which inputs shift the oxidation state of a metal center or organic moiety. Redox chemistry offers precise control over electronic properties and reactivity, enabling gates with electrochemical readouts. These gates are particularly attractive for integration with electronic devices, potentially enabling hybrid molecular–electronic systems. Such integration may bridge the speed and signal-processing advantages of electronics with the nanoscale programmability of chemical systems.

In the context of information theory, the concept of noise is central to the reliability of logical operations. Chemical systems inherently exhibit thermal fluctuations, concentration variability, and stochastic reaction events. Thus, the design of molecular logic gates must account for noise tolerance, thresholding, and signal restoration. In traditional digital electronics, transistors provide gain and signal restoration; molecular systems, by contrast, often lack natural amplifying components unless catalytic or redox processes provide the necessary gain. Some molecular systems implement chemical thresholding by regulating outputs according to sigmoidal or nonlinear response curves. Cooperative binding, autocatalysis, and allosteric regulation can all contribute to threshold behavior that filters weak signals and enhances computational fidelity.

The design of chemical circuits requires a deliberate orchestration of reaction kinetics. Unlike electronic circuits where signals propagate at speeds near the speed of light, chemical signals propagate through diffusion and molecular movement, leading to timescales ranging from seconds to hours. As a result, the architecture of molecular circuits can differ significantly from conventional electronics. Serial logic operations may require long waiting times, whereas parallel operations are more natural in chemical systems. For instance, if multiple logic gates are embedded within a solution or on a surface, they may operate concurrently as long as diffusion supplies the necessary inputs to each. Spatial organization, such as compartmentalization inside vesicles, microdroplets, or solid-state matrices, can dramatically alter computational behavior by modulating transport, reducing cross-talk, or establishing spatial gradients.

Reaction–diffusion computing offers a compelling demonstration of spatially distributed computation. In such systems, chemicals diffuse through a medium while reacting locally, generating waves, fronts, and patterns that can represent logical states. When two reaction fronts meet, they may annihilate or interact in structured ways that resemble logical operations. In excitable media, a single trigger can propagate through the entire system, enabling global signaling. This architecture differs fundamentally from conventional logic circuits; it is inherently parallel, analog, and continuous. Yet it can still implement discrete operations by defining thresholds for wave amplitude, timing, or spatial patterns. Such spatial computing may hold promise for massively parallel problem-solving, self-regulating materials, and autonomous chemical systems.

Modern molecular computing increasingly seeks to emulate biological information processing. Biological systems use reaction networks, feedback loops, signal amplification, and compartmentalization to perform sophisticated computations essential for life—cell differentiation, immune response, circadian rhythm regulation, and metabolic control. These biological networks demonstrate that chemical logic need not be limited to simple Boolean operations. Instead, it can incorporate fuzzy logic, probabilistic inference, analog computation, and learning-like behavior. Synthetic chemists now aim to create artificial systems that mimic these biological strategies, enabling chemical materials that autonomously adapt, respond, and self-regulate.

One critical element borrowed from biology is feedback control. Positive feedback produces bistability—systems that toggle between two stable states, forming the basis of memory units or flip-flop circuits in molecular form. Negative feedback stabilizes outputs and reduces noise, improving computational fidelity. Oscillatory networks that combine both types of feedback can serve as clocks, timing modules, or rhythmic signal generators, enabling temporal logic operations. These features broaden the computational capabilities of chemical systems beyond simple feed-forward logic gates.

Energy flow is another essential concept. Every computation requires energy dissipation to overcome thermodynamic constraints. In electronic systems, energy dissipation occurs via electrical resistance. In chemical systems, dissipation manifests in irreversible reactions, fuel consumption, or conformational transitions. Designing chemical logic gates that operate near thermodynamic equilibrium is difficult because reversible systems lack the driving force to produce sharp thresholds or robust signal amplification. Thus, many molecular logic systems incorporate chemical fuels—such as ATP, redox agents, or synthetic high-energy species—to sustain non-equilibrium behavior. These energy-driven systems can perform more sophisticated computations, albeit at the cost of fuel turnover.

Advancements in materials chemistry have further expanded the scope of molecular logic. Responsive polymers, hydrogels, metal–organic frameworks, and covalent organic frameworks now host chemical logic gates embedded in macroscopic materials. These materials can change mechanical properties, permeability, optical characteristics, or catalytic activity in response to chemical inputs, behaving as programmable materials. For instance, a hydrogel might swell only when two specific metal ions are present, performing an AND operation at the macroscopic scale. Such materials blur the boundary between computation and function, making chemical logic not merely a means of manipulating information but also a tool for designing adaptive materials.

Perhaps one of the most transformative developments is the emergence of interfaced chemical–electronic systems, where molecular logic gates communicate with traditional electronics. Chemical inputs may modulate electronic signals via chemically responsive transistors, conductive polymers, or molecular junctions. Conversely, electronic inputs may trigger chemical reactions through electrochemical control. This bidirectional interface opens pathways to molecular sensors, smart diagnostics, environmental monitoring systems, and hybrid computational devices that integrate chemical reactivity with electronic speed.

The ultimate vision of molecular computing extends to bottom-up construction of computing architectures where molecules serve not only as logic elements but also as wiring, memory, and processing units. Such architectures could theoretically reach densities impossible for silicon-based systems, as molecules occupy nanometer-scale volumes. While practical implementation remains distant, the conceptual foundations are maturing rapidly, informed by lessons from nanotechnology, supramolecular chemistry, and synthetic biology.

Chemical computing also offers unique advantages in environments where traditional electronics fail—high-radiation environments, extreme temperatures, biological systems, or highly miniaturized spaces. Chemical logic gates can operate in aqueous environments, inside living cells, or under conditions where electronic components degrade. This resilience suggests potential applications in biosensing, drug delivery, autonomous therapeutic systems, and smart materials that react to chemical signatures.

Yet challenges remain substantial. Chemical computation is slower, more stochastic, and often less reliable than electronic computation. Cross-reactivity and unintended interactions pose design difficulties. Achieving modularity—where one chemical logic gate can be reliably connected to another without interference—is a persistent challenge. Standardization of signal magnitudes, thresholds, and output modalities is still in its infancy. Nevertheless, the field is progressing toward more modular systems by using orthogonal chemistries, compartmentalization strategies, and reaction engineering analogous to modular design in synthetic biology.

In reflecting on the future of molecular computing, it is clear that the goal is not simply to replicate electronic computers at a molecular scale. Rather, the aim is to develop a new paradigm of computation that exploits the strengths of chemical systems: parallelism, intrinsic sensing, self-assembly, and stimulus-responsive behavior. Chemical logic gates represent the building blocks of this paradigm, but the true potential lies in integrating these gates into autonomous, adaptive networks capable of performing complex tasks. Whether in materials that heal themselves, chemical robots that navigate microenvironments, or smart therapeutics that compute molecular cues before acting, molecular computing promises to transform how we conceptualize information processing in the chemical world.

In this broader chemical landscape, the maturation of molecular computing reflects a deeper truth about matter: information is not an abstract construct imposed on chemical systems, but an emergent property of their interactions, states, and transformations. As research moves toward increasingly integrated chemical circuits, the boundary between structure and function becomes fluid—molecules both encode and process information through their conformational dynamics, reaction pathways, and collective organization. This convergence reveals computation as an intrinsic capacity of chemical matter rather than a foreign layer added to it. The challenge ahead lies in mastering this capacity, not by forcing chemical systems into rigid computational architectures, but by embracing their nonlinearities, stochasticity, and thermodynamic constraints. Such an approach broadens the meaning of "programming," shifting it from the manipulation of digital symbols to the design of energetic landscapes and network topologies that guide reactivity and self-organization. As these principles solidify, molecular computing evolves from a futuristic concept into a foundational framework for designing responsive materials, dynamic catalysts, and adaptive nanosystems—each demonstrating how information processing, at its core, is a chemical phenomenon woven into the fabric of matter itself.

Chapter 31: Word Problems

1. A supramolecular AND gate uses a host molecule H that fluoresces only when simultaneously bound to proton P (K_a = 1.2 × 10^5 M^{-1}) and metal ion M^{2+} (K_a = 6.0 × 10^4 M^{-1}). In a buffered solution at pH 6.5, the protonated fraction of H is measured to be 0.35. If the initial concentrations of H, P, and M^{2+} are each 10 μM, determine the equilibrium fraction of H molecules simultaneously bound to both

inputs and evaluate whether the fluorescence output will exceed the experimental detection threshold of 15% complex formation.

2. A structural-switching foldamer used as an XOR gate transitions to state S_1 when exposed to ligand A (K_d = 8 μM) and to state S_2 when exposed to ligand B (K_d = 20 μM). When both ligands are present, steric interference prevents either state from forming, leaving the foldamer in the non-fluorescent state S_0. If A is added at 12 μM and B at 30 μM, calculate the distribution of states and determine whether the XOR condition (exactly one ligand bound) is satisfied sufficiently to produce >60% fluorescent output.

3. A reaction–diffusion OR gate is constructed from a BZ-type medium whose excitation fronts propagate at 0.7 mm/s. Two input points separated by 5 mm are triggered simultaneously. The output zone is 2 mm downstream from the midpoint. Assuming circular propagation fronts, compute the time window within which the fronts must reach the output zone to generate a superthreshold signal. Evaluate whether the gate maintains reliability if input triggering varies by ±0.5 s.

4. A DNA strand-displacement logic gate uses a 40-nt duplex with a 7-nt toehold. Reaction kinetics follow $k = 2.3 \times 10^5\ M^{-1}\ s^{-1}$. For an AND gate requiring sequential displacement by strands A and B, each added to 100 nM, estimate the time required to reach 90% displacement completion at 25°C. Assess whether the gate can operate within a 10-minute computation step constraint for a larger molecular circuit.

5. A catalytic NOR gate uses a metal complex whose catalytic activity is suppressed by either ligand X or ligand Y. In the absence of both ligands, turnover frequency is 80 s^{-1}; binding of either ligand reduces it to 2 s^{-1}. If an output signal requires accumulation of 0.5 μmol of product in 30 seconds within a 1 mL reactor, determine the maximum allowable concentrations of X and Y such that the NOR condition (both absent) triggers product above threshold.

6. A hydrogel-based AND gate swells when both ion inputs I_1 and I_2 bind to cross-linking sites. Swelling follows $\Delta V = k[I_1][I_2]$, where $k = 4.0 \times 10^2\ L^2\ mol^{-2}$. If detection requires a volume change of at least 0.3 mL and the hydrogel initially occupies 1.0 mL, calculate the minimum input concentrations needed when the system is dosed with equal concentrations of the two ions.

7. In a redox-based NAND gate, oxidation of a metal center ($M^{2+} \rightarrow M^{3+}$) suppresses an electrochemical current. Inputs E_1 and E_2 both increase the redox potential by +40 mV. The oxidation threshold is +60 mV above baseline. Determine whether the system oxidizes (undesirable) when both inputs are present, one input is present, or neither is present, given that thermal fluctuations contribute ±10 mV noise.

8. A bistable chemical flip-flop stores a bit as high or low concentration of species A. Positive feedback is governed by rate law $d[A]/dt = k_1[A] - k_2[A]^2 + I$, where I is the input stimulus. For $k_1 = 0.8\ s^{-1}$ and $k_2 = 0.04\ M^{-1}\ s^{-1}$, determine the minimum I required to push the system from the low-A state (0.5 mM) to the high-A state (>5 mM) within 20 seconds.

9. A reversible supramolecular NOT gate binds an inhibitor Q that quenches fluorescence. The system is designed so that the presence of input X displaces Q, restoring fluorescence. If Q has $K_a = 5 \times 10^6\ M^{-1}$ and X competes with K_d = 40 μM, determine the concentration of X required to restore >90% fluorescence when Q is initially at 5 μM.

10. A catalytic amplifier integrated into a molecular computing circuit must offset signal attenuation across three prior logic gates. Each upstream gate reduces the effective concentration of the signal molecule by 35%. If the amplifier gate must regenerate the signal to at least 80% of its original concentration, calculate the minimum catalytic turnover number (TON) required when 2 μM substrate is available.

11. A DNA-based OR gate relies on two toeholds, each 5 nt, that independently initiate strand displacement. Given $k = 8 \times 10^4\ M^{-1}\ s^{-1}$ per pathway and competing leak reactions at $2 \times 10^3\ M^{-1}\ s^{-1}$, determine whether adding 50 nM of each input strand can ensure >95% activation within 200 s without excessive false-positive activation.

12. A host–guest XOR gate requires that the host binds G_1 or G_2 but not both due to steric blocking. If binding free energies are −28 and −25 kJ/mol for G_1 and G_2, respectively, calculate the population distribution at 310 K when each guest is added at 20 μM, and determine if XOR output exceeds 70% fidelity.

13. A BZ reaction-based logic system uses wave collisions to encode AND operations. If wave amplitude decreases exponentially with distance ($A = A_0e^{(-x/3\ mm)}$), compute the minimal spacing of input zones such that both waves retain ≥30% amplitude at the collision point required to generate an output wave.

14. A foldamer logic gate exhibits cooperative conformational switching described by Hill coefficient n = 2.3. If ligand binding follows $\theta = [L]^n/(Kd^n+[L]^n)$ with K_d = 15 µM, calculate the ligand concentration at which the gate transitions from <10% to >90% activation and discuss threshold sharpness.

15. In a catalytic AND gate, substrate conversion requires two cofactors C_1 and C_2 that bind sequentially. Their binding rate constants are 1.1×10^5 and $6.5 \times 10^4\ M^{-1}\ s^{-1}$. If each cofactor is present at 5 µM, compute the expected lag time before full catalytic activation begins.

16. A proton-sensitive logic gate operates in a microfluidic channel where diffusion coefficient $D = 7 \times 10^{-6}\ cm^2/s$. If a proton pulse representing input "1" must reach a downstream gate 800 µm away within 3 seconds, determine whether diffusion alone suffices or whether convective flow is required.

17. A supramolecular OR gate uses two orthogonal metal ions, each inducing a color change. Due to cross-reactivity, adding both ions results in partial quenching of each color pathway. If individual inputs produce 100 a.u. absorbance but both together yield only 150 a.u., determine whether the gate still meets OR logic thresholds (≥80 a.u.).

18. A redox-controlled polymer undergoes a conductivity shift of two orders of magnitude at potential +0.2 V. Two chemical inputs raise potential by +0.1 V each, but background drift lowers it by 0.05 V. Determine whether the polymer transitions to the high-conductivity state and how reliably it acts as an AND gate.

19. A surface-anchored reaction-diffusion network implements NAND by dissipating reactive fronts on contact with inhibitor regions. If the inhibitor reaction consumes fronts at rate $k = 3.0\ s^{-1}$ and fronts arrive at intervals of 1.5 s, determine whether inhibitory capacity will saturate and lead to logic failure.

20. A DNA hairpin-based NAND gate requires two separate strand displacements to open fully. Each intermediate has ΔG° = +6 kJ/mol. If the toehold domain 1

releases −12 kJ/mol and domain 2 releases −10 kJ/mol, compute the overall thermodynamic driving force for complete opening.

21. A memory element based on bistable pH shifts toggles between states at pH 5.0 and 7.0. Buffer capacity β = 0.02 mol/L·pH. Determine how many μmol of acid/base must be added to flip between states in a 2-mL microcompartment.

22. A supramolecular AND gate requires simultaneous binding of S_1 and S_2, but S_1 binding shifts the K_d of S_2 from 10 μM to 200 μM via allostery. Determine whether a 20 μM dosing of each substrate can realistically trigger AND output.

23. A foldamer switch has two orthogonal conformational triggers, heat and ligand L. Heating to 50°C gives 30% activation; adding L at 15 μM gives 70% activation. However, simultaneous application gives only 60% due to pathway interference. Determine whether this system can serve as an XOR gate.

24. A cyclic catalytic reaction used as a timing module oscillates with period 100 s. Environmental noise introduces fluctuations of ±12 s. If a computation requires three sequential oscillations to coordinate gating events, determine cumulative timing uncertainty.

25. A hybrid chemical–electronic gate uses a molecular transistor that responds to redox changes with modulation factor 0.8 μA per 10 mV. Compute expected signal swing when ligand binding raises redox potential by 45 mV.

26. A reaction network displays bistability when parameter α exceeds 0.5 in the equation $dX/dt = \alpha X - X^3$. Determine whether a perturbation increasing X from 0.1 to 0.3 is sufficient to push the system to the high-X state when α = 0.62.

27. A supramolecular receptor has two binding sites that must both be occupied to trigger a fluorescence turn-on event. If sequential binding shows negative cooperativity with $K_{d_2} = 8K_{d_1}$, compute site occupancies when total ligand concentration is twice Kd_1.

28. A catalytic XOR gate exhibits reaction pathways A and B that compete for substrate S. If kA[S] and kB[S] govern outputs and XOR requires dominance of exactly one pathway, determine the necessary kA:kB ratio to ensure <10% crosstalk when both activators are present.

29. A redox-active logic gate uses NADH/NAD^+ cycling with catalytic amplification. If each cycle yields 50 NADH molecules and input strength determines number of cycles (0, 3, or 7), compute total signal output and whether amplification can preserve digital fidelity.

30. A DNA-origami logic circuit places three AND gates in a row, spaced 30 nm apart. If local diffusion of strands is limited to $D = 1 \times 10^{-8}$ cm^2/s, compute average diffusion time between gates and evaluate whether signal propagation fits within a 5-minute computation.

31. A reaction–diffusion system implements threshold detection via autocatalysis with rate law $v = k[A]^2$. If detection threshold is 0.2 mM and initial A is 0.05 mM, compute time needed to reach threshold when $k = 4 \times 10^3$ M^{-1} s^{-1}.

32. A light-controlled NOT gate uses a photochromic switch with quantum yield 0.35. If 1 mW of blue light produces 1.2×10^{15} photons/s and each molecule requires 20 absorbed photons to switch, estimate switching time for a 3 μM solution in 100 μL.

33. A molecular robot uses a chemical sensor to compute whether the sum of two analyte concentrations exceeds 100 nM. If sensor response is nonlinear and follows output $O = ([A]+[B])^2 / (K^2+([A]+[B])^2)$ with K = 80 nM, determine minimal (A,B) pairs that trigger output >0.8.

34. A metal–organic framework hosting molecular logic gates restricts diffusion by factor 0.1 compared to bulk. If a catalytic AND gate requires substrate mobility of at least 2×10^{-6} cm^2/s, determine whether MOF confinement permits operation.

35. A hydrogel logic material changes opacity according to cross-link cleavage induced by inputs X and Y. Cleavage rate = k[X][Y], with $k = 5 \times 10^3$ L mol^{-1} s^{-1}. Determine opacity change after 10 s when [X] = 100 μM and [Y] = 50 μM.

36. A redox-based memory element stores a bit as oxidation state Fe^{2+}/Fe^{3+}. If spontaneous re-reduction by impurities occurs at rate 0.004 s^{-1}, compute retention time before 50% loss.

37. A DNA catalytic network computes AND using two allosteric activators. Activation energy barrier decreases from 65 kJ/mol to 52 kJ/mol when both inputs bind. Calculate rate enhancement at 298 K.

38. A supramolecular OR gate must detect either guest G_1 or G_2. G_1 binds in 1 ms, G_2 in 20 ms. If computation step must complete in <10 ms, determine which inputs are valid for real-time operation.

39. A polymeric conformational switch follows Arrhenius kinetics with E_a = 80 kJ/mol. If heating from 25°C to 60°C increases activation rate by factor f, calculate f and evaluate viability for thermal logic.

40. A reaction–diffusion multiplier performs analog multiplication by interacting wave intensities. If intensities decay as $I = I_0 e^{(-t/\tau)}$ with τ = 12 s, determine output intensity for input waves triggered 4 seconds apart.

41. A catalytic AND gate embedded in a matrix requires substrate diffusion radius ≥100 μm before reaction. If diffusion coefficient is 3×10^{-6} cm^2/s, compute time to reach required radius.

42. A signaling hydrogel stores information spatially as ion gradients. If gradient loss follows $dC/dt = -0.03C$, compute half-life of stored information.

43. A metal-centered XOR gate requires redox states M^{2+} (bright) and M^{3+} (dim). Inputs X and Y oxidize at 30 mV and 45 mV respectively, but simultaneous addition drives over-oxidation to M^{4+}. Determine whether concentrations must be tuned to preserve XOR behavior.

44. A bistable autocatalytic system computes NOT by switching between high and low A concentrations. If autocatalytic amplification factor β = 4.2 and decay rate γ = 1.1 s^{-1}, determine conditions needed to sustain high-A state in absence of input.

45. A DNA-enzyme hybrid gate shows Michaelis–Menten kinetics with K_M = 90 μM and V_{max} = 5 μM/s. Determine substrate levels that differentiate between logical "0" (low activity) and "1" (high activity) when threshold is 2 μM/s.

46. A nanoparticle-based AND gate uses plasmonic coupling that increases absorbance by factor 3 when aggregation exceeds 40% clustering. Given binding probability $P = [L] / (K_d+[L])$ with K_d = 12 μM, find ligand concentration required.

47. A hydrogel reaction network creates oscillatory color changes with period T = 70 s. If two inputs shift T by +5 s and −3 s, respectively, determine net effect when both inputs are present and whether system can function as a temporal XOR gate.

48. A catalytic cascade simulates AND–OR logic by sequential substrate channels. If channel efficiencies are 0.6 and 0.7, compute overall probability that output exceeds detection threshold of 0.3.

49. A reaction network mimicking fuzzy logic produces output O between 0 and 1. If $O = w_1A + w_2B$ with weights 0.4 and 0.7, determine allowed input regions for $O > 0.8$ and interpret as a molecular decision boundary.

50. A molecular circuit operating under non-equilibrium conditions consumes chemical fuel F at rate proportional to number of active gates. If each gate requires 2.5×10^{-10} mol/s and a 100-gate circuit runs for 30 minutes, compute total fuel used and discuss feasibility for autonomous operation.

Chapter 32

CHEMISTRY IN EXTREME ENVIRONMENTS

Chemistry in extreme environments reveals the profound plasticity of matter when the governing conditions of temperature, pressure, chemical composition, and energy flux deviate far from the narrow range humans typically experience. Most chemical knowledge comes from experiments conducted at or near ambient conditions, yet these represent only a tiny sliver of the possible physical states of matter in the universe.

The majority of planetary and interstellar environments operate under extremes that reshape chemical principles so thoroughly that behaviors assumed to be universal become contingent. Extreme environments—deep ocean trenches, volcanic fields, permafrost layers, hydrothermal vents, nuclear reactors, planetary surfaces bathed in radiation, the upper atmosphere, and extraterrestrial oceans—function as natural laboratories that stretch and challenge our understanding of chemical mechanisms. They expose how matter rearranges, stabilizes, and transforms when pushed into unfamiliar thermodynamic regimes, revealing not only the robustness of chemical laws but also their immense versatility. Chemistry under extreme conditions is not a fringe curiosity; it is a fundamental domain of study that informs planetary formation, materials design, the origin of life, and the habitability of worlds beyond Earth. Through extreme chemistry, scientists observe and model how atomic and molecular structures evolve when subjected to forces and energies that exceed the thresholds tolerated by biological organisms and most materials. By studying these boundaries, we reveal the spectrum of possible chemical architectures that matter can adopt across the cosmos.

Temperature is among the most significant parameters defining extreme chemistry. Elevated temperatures, such as those found in volcanic magma conduits, geothermal fields, and the deep lithosphere, shift the balance between enthalpy and entropy. Thermal energy disrupts noncovalent interactions—including hydrogen bonding, ionic attractions, and van der Waals forces—allowing only the strongest covalent frameworks to remain intact. At sufficiently high temperatures, solids melt, liquids vaporize, and gases ionize into plasmas where electrons detach from atoms. In supercritical water, accessible at pressures above 218 atm and temperatures beyond 374 °C, the fluid loses the polarity that normally organizes aqueous chemistry. Supercritical water dissolves nonpolar compounds readily, behaves as both an oxidant and reductant under appropriate conditions, and participates in rapid hydrolysis. Organic molecules degrade into simple species like CO_2, CO, H_2, and CH_4, often catalyzed by mineral surfaces. Meanwhile, silicate minerals undergo dissolution and reprecipitation, forming high-temperature clays, zeolites, and metal-bearing phases that do not form at lower temperatures. These transformations drive the enrichment of certain metals, the cycling of sulfur, and the redistribution of carbon within Earth's crust. The temperature dependence of phase stability explains the formation of polymorphs such as quartz, coesite, and stishovite, each stable only under distinct thermodynamic windows. Thermal extremes thus reshape not only molecular configurations but the solid-state architecture of entire mineral systems.

Low temperatures, by contrast, slow kinetic processes but stabilize otherwise transient intermediates. Cryogenic environments—whether Arctic permafrost, polar ice sheets, the surfaces of icy moons, or interstellar molecular clouds—host a chemistry that unfolds over years, centuries, or even millions of years. In these settings, water exists as crystalline or amorphous ice, often formed under nonequilibrium conditions. Amorphous solid water traps gases, ions, and organic molecules within its irregular structure, creating micro-compartments where reactions occur via localized tunneling events or photolysis. Ultraviolet photons penetrating interstellar ices initiate reactions that produce amino acids, nucleobases, alcohols, aldehydes, and polycyclic organics. On the surfaces of moons like Europa, cosmic radiation interacts with ice to generate oxidants such as hydrogen peroxide and molecular oxygen, which may later mix into underlying oceans. Meanwhile, the stability of clathrate hydrates—the crystalline cages capable of trapping gases like methane, nitrogen, or carbon dioxide—depends sensitively on temperature. In deep, cold environments, these clathrates form massive reservoirs that influence planetary climate, geochemistry, and the stability of potential biospheres. Cryogenic conditions thus provide a distinct chemistry dominated by slow diffusion, quantum effects, radiolysis, and entrapment-driven reactivity.

Pressure introduces another axis of environmental extremity. Earth's interior is dominated by pressures that exceed hundreds of thousands of atmospheres, compressing materials into dense phases with bonding characteristics that defy ambient expectations. Even at the depths of the Mariana Trench, pressures exceed 1,100 atm, influencing solubility equilibria, reaction kinetics, and biomolecular conformations. High pressure favors reactions that decrease volume, shifting equilibria and enabling compounds to form that would otherwise be unstable. For example, under high pressures, noble gases form stable oxides and clathrates, hydrogen enters metallic states, and simple salts adopt polymeric or metallic structures. Water assumes high-pressure ice phases such as Ice VII, which may exist deep within terrestrial and extrasolar planets. Deep-sea organisms counteract the volume changes associated with pressure by accumulating piezolytes—small molecules that stabilize protein structure by reducing compressibility. Protein folding landscapes shift dramatically under pressure, with hydrophobic interactions becoming weakened as water molecules infiltrate protein cores. Pressure-induced denaturation thus represents a unique challenge that biological systems must overcome. In inorganic systems, high-pressure experiments reveal new classes of materials, including superhard phases, ultra-dense electrides, and unexpected stoichiometries such as $NaCl_3$ or $NaCl_7$. These discoveries reshape our understanding of bonding and mineralogy under planetary interior conditions.

The extremes of acidity and alkalinity illustrate the degree to which pH modulates chemical behavior. Highly acidic environments occur wherever sulfur, chlorine, or nitrogen oxides interact with water, forming potent acids that dissolve mineral matrices, mobilize metals, and drive redox reactions. Volcanic fumaroles and acid-sulfate geothermal areas, often at pH values below 1, are environments where iron, copper, and aluminum are solubilized into complex ionic networks. In these environments, silicate minerals slowly dissolve, forming clay minerals or silica gels that precipitate in patterns dictated by pH, temperature, and fluid flow. Acidophilic microorganisms use robust membranes enriched with cyclized lipids to prevent proton influx, and their proteins feature surface residues that resist protonation. Conversely, strongly alkaline environments such as soda lakes or serpentinization systems promote rapid nucleophilic reactions and create conditions under which organic molecules undergo hydrolysis or condensation depending on the functional groups present. Alkaliphilic organisms rely on sodium gradients instead of proton gradients for bioenergetics and maintain enzyme stability through precise control of surface charge distributions. Highly alkaline fluids produced by serpentinization—sometimes exceeding pH 12—generate hydrogen gas and reduced carbon species that may have fueled ancient metabolic pathways on early Earth. The juxtaposition of acid and base extremes in different environmental niches highlights the chemical diversity of planetary processes.

Salinity exerts profound chemical effects by altering water activity and ion-mediated interactions. In hypersaline environments such as salt flats, deep brines, or evaporitic basins, ionic strength can reach levels that drastically impact equilibrium constants, solubility, reaction rates, and molecular stability. High salinity promotes the formation of brine-inclusion minerals that trap gases, microbes, or organic molecules for geological timescales. The physical structure of water changes at high ionic strengths, with ions organizing hydration shells that reduce the solvent's dielectric constant. Precipitation sequences during evaporation reveal the order in which minerals crystallize from complex solutions, producing layered deposits with characteristic mineral assemblages. Halophilic organisms alter the chemical composition of their intracellular fluid, accumulating potassium ions or synthesizing osmoprotective organics to counteract osmotic stress. Their proteins use acidic residues to remain soluble in salty conditions, while their enzymes exhibit low isoelectric points that preserve activity at high ionic strengths. The interaction between salinity and organic chemistry extends to prebiotic contexts, where concentration effects in evaporating pools may have played a role in polymer formation and the emergence of catalytic macromolecules.

Radiation presents one of the most transformative environmental extremes by inducing ionization, excitation, and radical formation in molecules. In the upper atmosphere, solar radiation drives photochemical cycles that produce ozone, nitrate, and reactive oxygen species. These radicals influence climate, atmospheric stability, and the breakdown of pollutants. On planetary surfaces without strong magnetic fields, cosmic rays and solar wind particles interact with regolith, forming oxidants like perchlorates, which dominate the surface chemistry of Mars. Radiolysis of water in nuclear reactors generates a host of reactive species—hydroxyl radicals, solvated electrons, and hydrogen atoms—that attack metal cladding, degrade organic materials, and complicate fuel-cycle chemistry. Organisms such as Deinococcus radiodurans survive intense radiation exposure by compacting their DNA, using highly efficient repair enzymes, and accumulating manganese-based antioxidants that scavenge reactive oxygen species. Radiation induces structural changes in minerals, such as color centers in halite or damage tracks in zircon, which geologists use to date geological processes. Radiation chemistry thus spans a continuum from atmospheric photochemistry to nuclear materials design.

Desiccation introduces a chemical regime characterized by the absence of bulk water, forcing reactions to proceed via alternative mechanisms. Dry environments such as deserts, salt crusts, volcanic ash layers, and interplanetary dust grains lack liquid water but possess abundant catalytic mineral surfaces. Solid-state chemistry dominates these settings, with reactions occurring through surface adsorption, limited diffusion, or photochemical excitation. Minerals such as clays bind organic molecules and create confined spaces where temperature fluctuations or irradiation drive condensation reactions that form oligomers or polymeric species. The absence of water can stabilize otherwise labile molecules, preserve biomolecular structures for millennia, or enable radicals to persist long enough to undergo secondary reactions. In biological contexts, desiccation-tolerant organisms rely on glass-forming sugars or vitrified cytoplasmic matrices that immobilize structures and protect them from damage. Prebiotic chemistry research suggests that wet–dry cycles may have been crucial for the formation of early nucleic acids or peptides by alternating between concentration and mobility phases. Desiccation chemistry thus has implications ranging from planetary science to the origin of biochemical macromolecules.

The deep subsurface combines multiple extremes—high pressure, variable salinity, limited nutrient availability, restricted water activity, and occasional high temperatures. These environments harbor slow, long-lived geochemical reactions between fluids and host rocks. Serpentinization, a reaction between water and ultramafic rock, produces hydrogen, methane, and small organic molecules, creating redox gradients that support microbial ecosystems. Radiolysis of water by radioactive minerals generates hydrogen peroxide, dissolved hydrogen, and oxygen species that can drive subsurface metabolism even in complete darkness. Mineral surfaces act as catalysts for electron transfer, hydrocarbon generation, or carbon fixation. Deep subsurface microbes use metabolic pathways that rely on iron oxidation, sulfate reduction, methanogenesis, or anaerobic methane oxidation. Their survival over millions of years suggests that chemical ecosystems can persist independently of sunlight, driven entirely by geochemical energy. The deep biosphere illustrates how chemical conditions shape the limits of life and provides analogs for potential extraterrestrial ecosystems.

The behavior of biomolecules under extreme conditions reflects the interplay between chemical stability and evolutionary adaptation. Proteins in thermophiles exhibit enhanced hydrophobic packing, rigidified secondary structures, and salt bridge networks that maintain stability at high temperatures. Psychrophilic proteins show increased flexibility, reduced core density, and active-site architectures tuned to low thermal energy. Acidophiles and alkaliphiles maintain membrane integrity through lipid structures that resist proton or hydroxide permeation. Some extremophiles incorporate unusual amino acids or metal cofactors that stabilize redox enzymes under extreme conditions. DNA stability is influenced by salt concentration, temperature, and radiation, with extremophiles employing protective binding proteins or alternative repair mechanisms. These adaptations have practical applications in biotechnology, including the use of thermostable enzymes for industrial reactions, extremophile-derived pigments for radiation shielding, and membrane components for designing robust liposomes or nanocarriers.

Materials science in extreme environments provides a bridge between natural phenomena and engineered solutions. High-temperature alloys, ceramic composites, refractory materials, and high-entropy alloys exhibit stability under thermal shock, corrosive media, and radiation. Ceramic matrix composites withstand conditions in jet engines or hypersonic vehicles, while alloys designed for deep-sea pipelines resist both pressure and corrosive brines. In nuclear reactors, materials face neutron bombardment, radiolytic radicals, thermal gradients, and mechanical stresses, necessitating designs that prevent embrittlement, swelling, or phase degradation. Synthetic materials often draw inspiration from natural extremophile systems, such as thermophilic enzyme stability, biomineral resilience, or radiation-resistant pigments. By understanding how matter behaves in extreme natural contexts, engineers design materials capable of performing in challenging industrial or extraterrestrial environments.

Planetary chemistry extends the concept of extreme environments beyond Earth. Titan's methane and ethane lakes host an exotic solvent environment with radically different solubility properties than water. Reaction kinetics proceed slowly but generate complex organic aerosols through atmospheric photochemistry. Venus's surface, with temperatures near 465 °C and pressures over 90 atm, is dominated by supercritical CO_2 chemistry that dissolves or alters mineral surfaces and drives carbonate-silicate cycles. Mars exhibits perchlorate-rich soils, oxidizing conditions, and freezing temperatures, leading to unusual aqueous chemistries that may episodically support brine flows. Icy moons such as Europa and Enceladus host high-pressure ocean worlds with hydrothermal activity at the seafloor, producing chemical gradients that could support life. Interstellar chemistry occurs in cold, low-pressure environments where complex organic molecules form within icy dust grains subjected to UV or cosmic ray irradiation. Each environment demonstrates how chemical systems evolve under conditions far removed from terrestrial norms.

Across all extremes, a consistent thread emerges: chemical principles remain universal, but their specific manifestations depend intimately on environmental context. Thermodynamics dictates which phases and reactions are possible under given conditions, while kinetics determines the rates and feasibility of transformations. Solvation structures, molecular conformations, redox potentials, and stability windows all shift depending on environmental parameters. Extreme environments make these dependencies explicit, offering clear examples of how matter reorganizes itself under stress. Studying chemistry under extremes advances our understanding of planetary evolution, the limits of life, the formation of minerals, the design of materials, and the origins of biological complexity. It reveals that the diversity of chemical behavior across the universe is vast, and that the familiar chemistry of Earth represents only one narrow chapter in a much broader cosmic narrative. By embracing the study of these environments, scientists uncover mechanisms that operate across planetary systems, support potential ecosystems, and inspire technological innovations that allow human exploration of places once deemed unreachable.

Chemistry at extremes ultimately demonstrates that the universe operates through a unified set of principles that adapt fluidly to context, revealing new structural and reactive landscapes when pressured, cooled, irradiated, or energized beyond everyday bounds. As conditions diverge from the terrestrial norm, the familiar hierarchy of intermolecular forces can invert, bonding patterns reorganize, and unexpected molecular symmetries emerge. Reaction coordinates bend as quantum tunneling becomes more prominent at low temperatures, while high-energy environments reshape potential energy surfaces, enabling otherwise inaccessible transition states. Even elemental behavior shifts: noble gases form stable compounds under compression; hydrogen metallizes; carbon adopts novel frameworks; and metals assume oxidation states unknown at ambient conditions.

These phenomena compel scientists to revisit assumptions about chemical "rules," recognizing them as context-dependent tendencies rather than absolutes. At the same time, extreme chemistry reveals deep continuity: the same mathematical formalisms—partition functions, rate equations, wavefunctions, and thermodynamic potentials—remain valid even when matter behaves in seemingly alien ways. By pushing systems into unconventional regimes, researchers illuminate the full scope of chemical possibility, refine theoretical models, and develop predictive tools that encompass both familiar and exotic domains. This expanding picture strengthens the connection between laboratory chemistry and the vast diversity of cosmic environments, from the mantles of icy moons to the atmospheres of ultrahot exoplanets. As humanity ventures farther into space and deeper into the frontiers of materials science and planetary exploration, the insights gained from extreme chemistry will continue to guide innovation, expand our understanding of matter, and reveal the fundamental continuity that links the smallest molecular interactions to the grand architecture of the universe.

Chapter 32: Word Problems

1. A silicate melt ascending through a volcanic conduit reaches 1200 °C and experiences pressures decreasing from 2 kbar to 0.5 kbar. A geochemist notices that an organic contaminant introduced by drilling equipment is rapidly decomposing into CO_2, CO, CH_4, and H_2. If the rate of decomposition doubles for every 60 °C increase and the initial decomposition rate at 900 °C was 3.0×10^{-3} s^{-1}, estimate the decomposition rate at conduit temperatures and determine whether hydrothermal minerals forming along the conduit walls could catalyze oxidation of CH_4 before it exits to the surface.

2. Inside a geothermal power plant, water accidentally superheats to 400 °C at 300 atm, creating supercritical water in contact with stainless steel piping. The fluid begins dissolving nonpolar lubricants and oxidizing metal ions released from pipe corrosion. Given that SCW simultaneously allows hydrolysis and oxidation, design a reaction pathway showing how a C_{12} alkane lubricant could degrade into CO_2, H_2, and organic acids, and calculate the enthalpy change if metal oxides catalyze the breakdown steps.

3. A research team subjects pure SiO_2 to a controlled heating schedule where temperature increases from 500 °C to 1500 °C at ambient pressure. Using the known stability fields of quartz, tridymite, cristobalite, and melt, determine the sequence of polymorph transitions expected and compute the Gibbs free-energy differences driving each transition. Assess whether any metastable retention of quartz would persist upon quenching.

4. An astronomer simulating a molecular cloud cools a water–methanol–ammonia mixture to 20 K under high vacuum. UV radiation induces formation of small organics including amino acids at a measured rate of 1 molecule cm^{-2} s^{-1}. If ices persist for 10^6 years with a surface area of 10^{22} cm^2 per cloud region, estimate the total yield of glycine and evaluate whether entrapment in amorphous solid water affects diffusion-limited reaction progress.

5. A permafrost core exhibits patches where methane clathrate is present at −15 °C and 40 atm. Over decades of warming to −5 °C, clathrates begin to dissociate. Using the Clapeyron slope for methane clathrate stability (dP/dT ≈ 4 atm/°C), determine the pressure drop required for dissociation and estimate methane release per cubic meter of permafrost.

6. A submersible descends into a trench at 11,000 m depth (≈1100 atm). Water samples show formation of high-pressure Ice VI inclusions in supercooled droplets. Determine the free-energy change needed for Ice VI nucleation at this depth and whether the presence of dissolved salts would shift the phase boundary enough to suppress ice formation.

7. Experimentalists compress hydrogen to several Mbar using a diamond anvil. Electrical conductivity increases dramatically once hydrogen enters a metallic state. Given an initial molar volume of 14 cm^3 mol^{-1} at ambient conditions shrinking to 2 cm^3 mol^{-1} at metallization, compute the work done on the sample and discuss how quantum delocalization of protons could alter predicted bonding behavior.

8. Xenon is observed forming XeO_3-like species at ≈200 kbar in mantle-analog experiments. If the volume decreases by 15% upon oxidation, predict the equilibrium shift based on Le Châtelier's principle and estimate the associated pressure–volume contribution to the Gibbs free energy of reaction.

9. A deep-sea microbe's enzyme denatures at 800 atm due to water intrusion into its hydrophobic core. Assuming volume change of unfolding ΔV = 25 mL/mol, calculate the pressure-dependent contribution to ΔG and identify piezolyte concentrations required to restore folding equilibrium at trench pressures.

10. A sodium–ammonia compound forms an electride under 50 GPa, trapping electrons in lattice cavities. If compressibility reduces the lattice volume by 40%, determine the electron density in cavities and evaluate whether such phases could exist in the mantles of icy exoplanets.

11. A fumarole located in an active volcanic field emits acidic gases, forming solutions with pH ≈ 0.5 that percolate through surrounding basaltic rock. The basalt contains Ca-rich plagioclase and minor pyroxenes. Considering that proton-promoted dissolution has a rate constant k = 1.5 × 10−6 mol m−2 s − 1per unit H+, estimate the dissolution rates of plagioclase over daily and seasonal timescales, assuming typical mineral surface areas. Discuss the extent to which silica released from the dissolution could supersaturate the downstream fluids and whether secondary silica gels are likely to precipitate under varying temperature and ionic-strength conditions. Evaluate how gas composition, flow rates, and surface passivation might alter these predictions.

12. In a serpentinizing ultramafic system, fluids generated at pH ≈ 12.5 and temperatures near 200 °C react with CO_2-rich water. If the reduction of CO_2 to CH_4 (the Sabatier reaction) is thermodynamically favorable under these conditions, calculate the Gibbs free energy change, ΔG , accounting for realistic concentrations of H_2, CH_4, and CO_2. Predict which mineral assemblages, such as magnetite, brucite, or awaruite, could catalyze CH_4 formation and how reaction kinetics depend on fluid-rock ratios. Consider how variable pressure, trace metals, and silica activity influence carbon speciation and the stability of carbonate versus methane phases.

13. A halophilic enzyme functions optimally in 5 M KCl brines, where the water activity is significantly lowered and dielectric constant is reduced. Using Debye-Hückel or extended ion-interaction models, determine the activity coefficients for charged residues in the enzyme's active site. Predict how changes in ionic strength and dielectric constant shift the stabilization energies of acidic and basic residues that maintain protein solubility, and discuss implications for enzyme folding, aggregation, and catalytic turnover in extreme saline environments.

14. A hypersaline basin evaporates 2 meters of brine containing Na^+, Mg^{2+}, Ca^{2+}, SO_4^{2-}, and Cl^- over a summer season. Using published solubility products and assuming equilibrium conditions at 25 °C, compute the expected sequence of mineral precipitation (gypsum → halite → Mg-salts, etc.) and estimate the thickness of each mineral layer. Evaluate how brine convection, temperature gradients, and minor trace ions (e.g., K^+, Li^+) might modify this idealized sequence, and discuss implications for natural evaporite deposition and resource recovery.

15. A nuclear reactor's cooling pool is irradiated at an absorbed dose rate 106 Gy yr−1. Using radiolytic yields (G-values) for water, calculate the steady-state concentrations of hydroxyl radicals (OH•), solvated electrons (e⁻_aq), and hydrogen atoms (H•). Assess whether these reactive species are sufficient to accelerate corrosion of zirconium alloy cladding. Discuss feedback mechanisms such as radical recombination, scavenging by dissolved gases, and temperature effects on radical lifetimes.

16. Perchlorate-rich Martian regolith is exposed to ultraviolet radiation that generates reactive oxygen species (ROS). Estimate decomposition rates for amino acids deposited on the surface over diurnal and seasonal cycles. Evaluate whether

transient brines could briefly stabilize organic molecules before oxidizing them, and discuss how mineral surfaces (clays, salts) could catalyze or inhibit these photochemical reactions.

17. A spacecraft observes a significant decrease in stratospheric ozone concentrations following solar flare events. Using the Chapman mechanism, model ozone photochemistry under enhanced UV flux, incorporating reactions with NO, OH, and HO_2 radicals. Calculate the amount of NO and HO required to trigger the observed catalytic ozone destruction cycles and predict how these cycles might vary with altitude, temperature, and background ozone concentration.

18. Zircon crystals retain alpha-recoil damage tracks over geologic timescales. If a sample experiences 10^9 alpha decays per gram over 2 billion years, calculate the expected defect density. Consider partial annealing at 300 °C and model how such thermal histories reduce lattice defects. Discuss the implications for U-Pb geochronology and the reliability of zircon as a chronometer in high-temperature environments.

19. Volcanic ash surfaces coated with clay experience repetitive dry-wet cycles in hydrothermal or early Earth environments. Assess whether condensation of amino acids into dipeptides is thermodynamically favored during the dry phase and calculate cumulative yields assuming a 2% conversion per cycle over 10,000 cycles. Discuss how ionic strength, mineral catalysis, and fluctuating pH could enhance or inhibit peptide formation.

20. On a cold interstellar dust grain at 280 K with no bulk water, radicals generated by UV photolysis persist long enough to undergo secondary reactions. Estimate radical lifetimes in the absence of solvation and determine whether these conditions facilitate polymerization of formaldehyde into polyoxymethylene. Consider how surface morphology, radical recombination, and low-temperature diffusion influence polymer yields.

21. Deep granite formations produce 5×10^4 molecules of H_2 per gram per year via radiolysis. Estimate the microbial biomass that could be supported by this hydrogen flux and calculate whether sulfate-reducing pathways remain energetically favorable under low electron acceptor availability. Include effects of temperature, fluid flux, and porosity on energy-limited microbial growth rates.

22. A fault zone exposes ultramafic rocks to circulating water, generating CH_4 and small organics. Model carbon-isotope fractionation during abiotic CH_4 formation using equilibrium and kinetic isotope effects, and compare expected $\delta^{13}C$ signatures to those typical of microbial methanogenesis. Discuss uncertainties introduced by variable temperature, water-rock ratio, and fluid residence time.

23. A thermophilic enzyme containing a Mo–Fe cofactor operates at 110 °C. Determine how redox potential varies with temperature and evaluate whether cofactor geometry might distort under thermal stress, affecting electron-transfer kinetics. Predict how these changes influence overall catalytic efficiency and substrate specificity in high-temperature environments.

24. A cold-adapted enzyme shows a dramatic increase in turnover rate between −5 °C and +5 °C. Using Arrhenius analysis, calculate activation energy and assess how increased protein flexibility impacts enthalpy–entropy compensation. Discuss implications for enzyme stability, substrate binding, and evolutionary adaptation to extreme cold.

25. Given hydroxide concentrations corresponding to pH 12, estimate proton influx rates into normal versus alkaliphilic membranes. Determine the sodium-gradient requirements needed to sustain ATP synthesis under alkaline stress. Analyze how membrane composition, buffering capacity, and electrochemical gradients modulate bioenergetic efficiency.

26. Deinococcus radiodurans stores manganese antioxidants at ~100× normal cellular concentrations. Model suppression of OH•-mediated DNA damage and calculate Mn^{2+}/DNA binding ratios required to minimize strand breaks. Discuss how compartmentalization and redox cycling affect overall cellular protection against ionizing radiation.

27. A nickel-based superalloy exposed to a 1500 °C atmospheric plasma forms protective oxides. Predict the equilibrium oxide-layer composition using Ellingham diagrams and assess whether evaporation of CrO_3 or NiO destabilizes the protective layer. Evaluate the impact on high-temperature creep, oxidation resistance, and service lifetime.

28. A ceramic-matrix composite must withstand repeated thermal shocks up to 1800 °C. Calculate thermal-shock resistance using material toughness, thermal expansion

coefficients, and heat capacity. Assess suitability for repeated atmospheric entry or industrial refractory applications, considering microcrack formation and oxidation effects.

29. Deepwater pipelines at 4000 m depth experience both high pressure and saline corrosion. Estimate corrosion rates assuming oxygen solubility increases under hydrostatic pressure. Determine whether alloying with molybdenum reduces pitting and discuss how flow rate, temperature, and biofouling modify degradation.

30. Reactor structural components accumulate helium through nuclear transmutation. Compute swelling percentages using helium bubble formation energies and neutron fluence over 20 years. Evaluate mechanical property degradation and potential mitigation strategies through material selection or annealing cycles.

31. A probe dissolves organic solids in Titan's methane–ethane lakes at 94 K. Using solubility parameters, determine which classes of organics (aromatics, nitriles, hydrocarbons) will dissolve. Calculate rates of slow photochemical polymerization in the lakes and discuss the formation of macromolecular organics under cryogenic conditions.

32. At 465 °C and 92 atm, supercritical CO_2 interacts with basaltic rock. Model olivine dissolution and carbonate precipitation, estimating reaction rates given CO_2 diffusivity and surface-area constraints. Evaluate feedbacks from evolving pH, secondary mineral formation, and CO_2 saturation on long-term carbon sequestration efficiency.

33. Perchlorate-rich brines on Mars depress the freezing point to −70 °C. Compute water activity under variable temperature cycles and assess whether transient melts could persist for hours under typical Martian diurnal humidity and temperature conditions. Discuss implications for habitability and potential brine-driven geochemical processes.

34. Europa's ocean experiences pressures of 1–2 GPa at the seafloor. Determine the stability fields of Ice VI and VII and evaluate how hydrothermal venting could produce chemical gradients sufficient to support hypothetical metabolism. Consider phase transitions, salt partitioning, and buoyancy-driven circulation.

35. Interstellar dust grains at 10 K catalyze hydrogenation of CO to form methanol. Using quantum tunneling probabilities and grain surface site densities, calculate

reaction rates and expected methanol abundance per grain. Discuss the influence of dust composition, grain size, and cosmic ray flux on prebiotic molecule formation.

36. An ultrahot Jupiter receives intense UV flux, driving H_2O photodissociation and formation of metal oxides in the upper atmosphere. Model chemical equilibrium and compute metal oxide abundances at temperatures exceeding 2000 K. Evaluate the impact on spectral signatures and atmospheric opacity.

37. A plasma torch operating at 8000 K ionizes N_2 and O_2. Compute ionization fractions using the Saha equation and assess whether NO formation is thermodynamically or kinetically favored under these conditions. Consider implications for plasma diagnostics and high-temperature combustion processes.

38. A reaction between H-atoms and CO occurs at 15 K via quantum tunneling. Given barrier width, barrier height, and particle mass, calculate tunneling probability and overall rate constant. Discuss how tunneling dominates over thermal activation at such low temperatures and the effect on interstellar chemistry.

39. $NaCl_7$ forms at 40 GPa. Propose possible electron configurations, calculate lattice enthalpy, and explain how extreme pressure stabilizes this unusual stoichiometry. Consider implications for high-pressure chemistry and exotic ionic compounds.

40. Amorphous ice traps CH_4, N_2, CO, and organics. Determine the energy released when amorphous ice warms to crystalline ice and calculate how this influences volatile escape in cometary activity. Discuss implications for early solar system chemistry and comet surface evolution.

41. A subseafloor basalt microbe tolerates 500 atm, pH 10, and 80 °C. Model combined effects of pressure, alkalinity, and temperature on protein folding free energy and membrane permeability. Predict adaptive strategies for maintaining bioenergetics under such polyextremophilic conditions.

42. Over 10^6 years, radiolysis produces oxidants that alter Fe-bearing minerals. Calculate redox front migration rates and expected mineralogical changes in the host rock. Include diffusion-limited transport, fluid flow, and secondary mineral precipitation.

43. Supercritical CO_2 is used for rare-earth extraction and is doped with chelating ligands. Model solubility enhancements and calculate ligand binding energies

under supercritical conditions. Evaluate effects of temperature, pressure, and co-solvent presence on extraction efficiency.

44. A high-entropy alloy is tested at 1600 °C and 3 GPa. Predict stability of multi-component phases and calculate contributions of configurational entropy to free energy. Discuss the likelihood of phase separation or intermetallic formation under extreme conditions.

45. A rocky exoplanet with mantle temperatures of 3000 K undergoes partial ionization of silicates. Estimate free charge carrier concentration and discuss consequences for electrical conductivity and possible magnetic field generation. Consider how pressure, melt fraction, and composition affect ionization equilibrium.

46. A meteor impact generates transient pressures of 50 GPa for milliseconds. Compute formation probabilities for high-pressure phases such as stishovite. Evaluate whether impact heating allows survival of pre-existing organic molecules and potential synthesis of new compounds.

47. A lunar regolith sample experiences extreme UV flux. Calculate the photolysis rate of adsorbed water molecules and determine whether resulting OH radicals can recombine to form H_2O or escape into space. Discuss how regolith composition and porosity influence radical lifetimes.

48. Evaporating brines concentrate phosphate, promoting nucleotide condensation. Estimate polymer length distributions under fluctuating ionic strengths and temperature cycles. Analyze how these conditions could favor prebiotic polymer assembly in early Earth or extraterrestrial environments.

49. Cosmic rays penetrate Europa's ice crust generating H_2O_2. Compute oxidant flux into the ocean and evaluate the potential redox energy available to hypothetical microbial metabolisms. Discuss constraints imposed by diffusion rates, ice thickness, and radiation shielding.

50. A cross-environment simulation models chemical reactions across temperatures from 10 K to 3000 K and pressures from 10^{-6} atm to 100 GPa. Using identical partition-function and Gibbs-potential formalisms, calculate shifts in reaction equilibria for three contrasting regimes: interstellar ices, Earth-surface aqueous conditions, and deep-mantle silicates. Discuss how universal thermodynamic laws remain valid despite drastically different chemical outcomes.

Chapter 33

Quantum Biology and Chemical Mechanisms

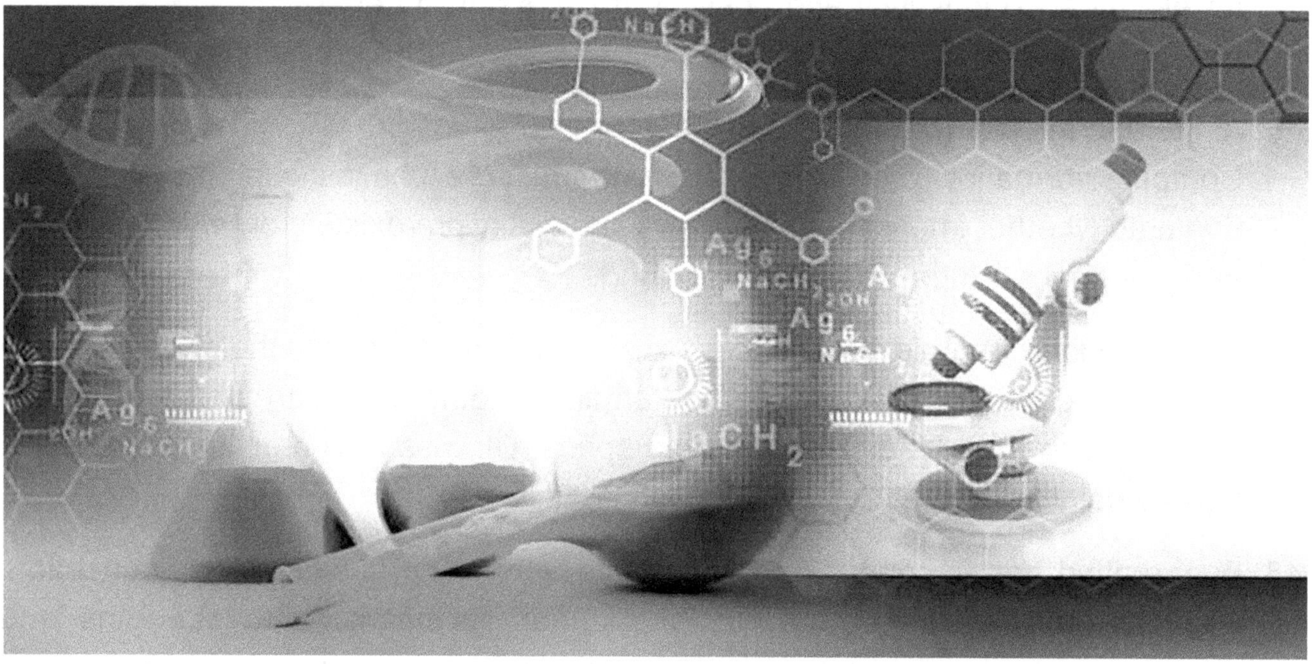

Quantum biology is an emerging interdisciplinary field that bridges the gap between the fundamental principles of quantum mechanics and the complex molecular machinery of living systems. For much of the 20th century, the consensus among biologists and chemists was that quantum phenomena—superposition, entanglement, and tunneling—would be irrelevant to life. The reasoning was straightforward: biological systems are "warm, wet, and noisy," conditions that promote rapid decoherence and effectively destroy delicate quantum states. Under such conditions, classical thermodynamic and kinetic models were assumed to suffice in explaining the chemical and biochemical behavior of macromolecules. Yet, increasing experimental and theoretical evidence has revealed that quantum effects not only occur in living systems but are often exploited to optimize efficiency, sensitivity, and reaction specificity. Photosynthesis exhibits near-perfect energy transfer efficiency, enzymes catalyze reactions at rates far exceeding predictions from classical transition state theory, and certain birds appear to detect geomagnetic fields through quantum spin-dependent processes. These phenomena suggest that evolution has shaped biological molecules to leverage quantum effects rather than simply tolerate them. Quantum biology, therefore, represents a paradigm shift in understanding how life orchestrates chemical processes, demonstrating that quantum mechanics is not restricted to subatomic systems but extends meaningfully into the realm of biochemistry.

Central to the field is the recognition that biological systems are not merely passive recipients of quantum mechanical rules but actively manipulate quantum states through structural, vibrational, and electronic mechanisms. Proteins, nucleic acids, and pigments provide highly organized microenvironments in which quantum coherence can persist, tunneling barriers can be optimized, and energy landscapes can be sculpted to guide reaction pathways. Unlike artificial quantum systems, which often require extreme isolation and cryogenic conditions, living organisms achieve functional quantum effects at physiological temperatures, using molecular architecture and environmental interactions to manage decoherence. This approach has profound implications for our understanding of chemical reactivity, energy transfer, and sensory perception in biological contexts.

Quantum mechanics provides the foundation for understanding the behavior of particles at molecular and submolecular scales, including electrons, protons, and photons. The wavefunction, Ψ, encodes the probability amplitude for the position and momentum of particles, and the evolution of the wavefunction is governed by the Schrödinger equation. Energy levels are quantized, meaning that molecules and molecular complexes occupy discrete states rather than continuous energy distributions. Superposition allows particles to exist in multiple states simultaneously, while entanglement creates correlations between spatially separated particles, such that the state of one particle instantaneously influences the state of another. These principles, long considered relevant primarily to physics and chemistry under controlled laboratory conditions, are now understood to influence biological function.

Tunneling, the quantum mechanical ability of a particle to traverse a potential energy barrier lower than its classical energy, is especially significant in hydrogen and electron transfer reactions. Protons and electrons can move between donor and acceptor sites without acquiring the energy needed to surmount the barrier classically. Decoherence, the loss of quantum coherence due to environmental interactions, limits the timescales over which superposition and entanglement are maintained. Biological systems appear to balance the competing demands of coherence and decoherence, using structured environments and vibrational couplings to maintain quantum behavior for functionally relevant timescales. Spin chemistry, a branch of quantum chemistry that focuses on the role of electron spin states in chemical reactions, provides a framework for understanding radical pair mechanisms and magnetoreception.

Photosynthetic organisms capture solar energy with remarkable efficiency, converting photons into chemical energy with minimal losses. Classical models such as Förster resonance energy transfer (FRET) describe energy migration as incoherent hopping between pigment molecules, mediated by dipole–dipole interactions. However, experimental observations using two-dimensional electronic spectroscopy have revealed oscillatory signals indicative of quantum coherence, where excitonic energy is delocalized over multiple pigments simultaneously. This coherence persists for hundreds of femtoseconds, a surprisingly long duration given the thermal noise present in biological systems.

The Fenna–Matthews–Olson (FMO) complex, a well-studied light-harvesting protein in green sulfur bacteria, exemplifies how protein structure can guide quantum dynamics. Pigments are embedded within a protein scaffold whose vibrational modes interact with electronic excitations. Rather than simply causing decoherence, these vibrations create a form of environment-assisted quantum transport (ENAQT), enabling excitons to explore multiple pathways and reach the reaction center with near-unity efficiency. Quantum coherence thus works synergistically with classical thermal relaxation, optimizing energy funneling. The implications extend beyond biology: bio-inspired materials designed to mimic coherent energy transfer may enable more efficient solar energy devices.

Enzymes catalyze chemical reactions by lowering activation barriers, but in reactions involving light atoms such as hydrogen, quantum tunneling can contribute significantly. Proton-coupled electron transfer (PCET) is a common mechanism where electron and proton transfers are coupled either sequentially or concertedly. Enzymes such as soybean lipoxygenase, methylamine dehydrogenase, and aromatic amine dehydrogenase exhibit hydrogen transfer rates that far exceed classical transition state theory predictions. Kinetic isotope effects, in which hydrogen is replaced by deuterium, often reveal larger-than-expected differences in rate constants, providing strong evidence for tunneling contributions.

Protein structure plays a critical role in facilitating tunneling. Enzymes optimize donor–acceptor distances, align hydrogen bonds, and modulate vibrational frequencies to create "tunneling-ready" conformations. Dynamic fluctuations of protein domains ensure that these favorable arrangements occur transiently but sufficiently often to enhance reaction rates. Computational approaches, including path-integral molecular dynamics and hybrid quantum/classical simulations, allow the mapping of tunneling pathways and provide quantitative insights into enzyme-catalyzed reactions. These studies reveal that enzymes do not simply lower energy barriers; they actively shape the quantum landscape of the reaction coordinate.

DNA stability relies on precise hydrogen bonding between nucleobases. Quantum mechanical proton tunneling can induce tautomeric shifts, temporarily altering base-pairing patterns. Such events, although rare, may contribute to spontaneous mutations during replication. Proton tunneling occurs on timescales much shorter than thermal fluctuations, and local environmental factors, such as hydration shells and vibrational couplings, influence its probability. Computational models suggest that tautomeric forms stabilized transiently during DNA replication may escape proofreading mechanisms, generating mismatches that become permanent mutations. These findings illustrate how quantum effects, though subtle, have profound implications for genetic fidelity. RNA folding may also involve quantum-assisted base pairing, where electron delocalization and proton tunneling influence the formation of secondary and tertiary structures. While less studied than DNA, these processes highlight the potential role of quantum dynamics in shaping the conformational landscape of nucleic acids, with implications for ribozymes, RNA-based regulation, and prebiotic chemistry.

Certain migratory birds navigate using Earth's magnetic field, a phenomenon explained by the radical pair mechanism in photoreceptive proteins called cryptochromes. Light excitation generates radical pairs with entangled electron spins, existing in singlet or triplet states. The interconversion between these spin states is influenced by weak magnetic fields, altering the chemical fate of the radicals and producing signals that can be interpreted by the nervous system.

The persistence of coherence in radical pairs, despite thermal noise, is facilitated by the protein microenvironment, which regulates hyperfine interactions and limits decoherence. Spin chemistry models, including spin Hamiltonians and density matrix approaches, allow quantitative predictions of magnetic sensitivity and align well with behavioral observations in birds. This remarkable sensitivity demonstrates that quantum entanglement and spin dynamics can be exploited by biological systems for macroscopic function.

The olfactory system may utilize quantum effects through vibrationally assisted electron tunneling. In this hypothesis, odorant molecules modulate electron transfer within receptors based on their vibrational spectra, rather than shape alone. Although experimental evidence remains mixed, studies with deuterated odorants suggest that some aspects of smell may involve inelastic electron transfer. This proposed mechanism represents a novel form of chemical recognition that combines molecular vibrations with quantum electron dynamics. Additional sensory processes, including vision and taste, may also involve subtle quantum effects, although these remain largely speculative.

Microtubules, components of the cytoskeleton, have been proposed to support coherent dipole oscillations and quantum processing in neurons. While these ideas are controversial, they emphasize the potential for macromolecular structures to maintain quantum states. Modeling such coherence requires consideration of dielectric environments, vibrational couplings, and decoherence rates. Even if functional quantum processing in microtubules is ultimately disproven, these investigations highlight the broader principle that biological macromolecules can create microenvironments that modulate quantum dynamics.

Biological systems maintain quantum effects by structuring their environments to preserve coherence and enhance tunneling. Proteins provide rigid yet dynamic scaffolds, positioning chromophores, hydrogen donors, and radical pairs optimally. Vibronic couplings allow environmental vibrations to assist rather than disrupt quantum states. Conformational gating enables selective tunneling, and electrostatic interactions stabilize charged intermediates. Across photosynthesis, enzymatic catalysis, and magnetoreception, organisms demonstrate that quantum phenomena are not passively tolerated but actively managed through structural and dynamic strategies.

Evidence suggests that natural selection may have favored molecular architectures that optimize quantum behavior. Photosynthetic pigments display spectral tuning aligned with coherence-supported energy transport. Enzymes maintain donor–acceptor geometries that maximize tunneling probability. Cryptochromes maintain radical pair coherence on timescales ideal for magnetic sensing. These observations imply that quantum mechanics is not merely an unavoidable background rule but a functional tool shaped by evolutionary pressures. Quantum biology, therefore, represents an intersection of chemical physics and evolutionary biology, where quantum optimization confers selective advantages.

Quantum biological mechanisms inspire a new generation of technologies. Artificial photosynthetic devices emulate coherent energy transfer to improve solar energy capture. Radical pair mechanisms inspire organic quantum sensors capable of detecting weak magnetic fields. Enzymatic tunneling principles guide the design of catalytic materials with enhanced proton or electron transfer. Advances in quantum computational chemistry allow simulation of complex biological processes, paving the way for biomimetic quantum materials and chemical devices. Understanding how life exploits quantum mechanics provides both a scientific foundation and an engineering blueprint for future innovations.

Quantum biology reveals that quantum mechanics profoundly shapes chemical and biological processes, from energy transfer to catalysis, from genetic stability to sensory perception. Biological systems maintain, regulate, and exploit quantum states through structural design, dynamic fluctuations, and vibrational coupling. By integrating quantum mechanics with chemical mechanisms, researchers gain new insights into the principles of life and the design of bio-inspired technologies. The field challenges classical assumptions and demonstrates that quantum phenomena, far from being confined to physics labs, are intimately woven into the fabric of living systems.

Building on these insights, chemistry gains a renewed appreciation for the ways molecular systems orchestrate delicate quantum interactions under ambient conditions. At physiological temperatures, quantum coherence and tunneling may appear improbable, yet biomolecules demonstrate architectures that subtly enhance these phenomena. Proteins, pigments, and nucleic acids seem to create local environments that soften energetic barriers, align electronic states, and coordinate motions in ways that collectively elevate the probability of quantum events. This capacity suggests that chemical reactivity in living systems cannot always be reduced to classical collision theory or simplified potential surfaces; instead, it emerges from a carefully tuned interplay of structure, dynamics, and the probabilistic nature of quantum states.

In catalysis, for example, enzymes achieve extraordinary rate enhancements by stabilizing transition states and shaping reaction pathways. Quantum biology adds a deeper perspective: that vibrationally assisted tunneling, proton transfer through delocalized networks, and dynamic coupling between electronic and nuclear motions may be crucial contributors to enzymatic efficiency. These quantum contributions do not replace classical models but refine them, providing an additional layer of explanatory power. They also highlight that evolution may have selected for molecular motifs that support coherent vibrational modes or charge-transfer pathways, improving catalytic precision in ways that surpass conventional chemical strategies. Understanding these design principles opens new opportunities for chemists to engineer catalysts that exploit similar quantum-enabled behaviors.

The study of charge and energy transfer in photosynthetic and photochemical systems further reinforces the significance of quantum effects. Light-harvesting complexes display remarkable efficiency, guiding excitons across pigment networks with minimal energy loss. Coherence allows these excitons to sample multiple pathways simultaneously, finding optimal routes that would be far less accessible through classical hopping alone. This perspective reframes chemical energy conversion as not merely a sequence of discrete steps but as a phenomenon shaped by a subtle balance between coherence and decoherence. Artificial photosynthetic systems and photovoltaic technologies may benefit from these insights, inspiring designs that preserve coherence long enough to enhance efficiency without requiring the pristine environments of physics laboratories.

Genetic stability may also hinge upon quantum-level interactions. Proton tunneling within DNA base pairs can transiently reshape tautomeric forms, influencing replication fidelity. While rare, such events have far-reaching consequences, potentially contributing to mutation rates and evolutionary dynamics. Chemistry traditionally treats hydrogen bonding and base pairing as largely classical interactions, yet quantum biology reveals that even the foundations of genetic encoding may involve fleeting quantum fluctuations. By integrating these perspectives, chemists gain a richer understanding of molecular recognition, hydrogen-bond stability, and the subtle forces governing nucleic acid behavior.

Olfaction, magnetoreception, and other sensory processes extend the relevance of quantum phenomena beyond molecular chemistry into organism-level behavior. In olfaction, vibrational theory proposes that odorant recognition involves not only shape complementarity but also quantum-assisted electron transfer facilitated by characteristic vibrational frequencies. In magnetoreception, spin-dependent reactions within cryptochrome proteins may allow birds to sense Earth's magnetic field, involving radical-pair mechanisms that are exquisitely sensitive to quantum spin dynamics. These examples demonstrate that chemistry, even when embedded within complex biological systems, remains deeply influenced by quantum principles that operate at the smallest scales.

The convergence of quantum mechanics with chemistry encourages a reevaluation of long-held assumptions regarding molecular behavior. It suggests that biological systems have evolved to take advantage of quantum opportunities rather than merely contend with quantum unpredictability. Molecular flexibility, dynamic disorder, and fluctuating environments—once considered challenges—may actually be essential components that sustain coherence, enable tunneling, and modulate reaction rates. Such features transform our understanding of the chemical landscape inside living cells, portraying it as a dynamic arena where classical and quantum principles continuously intersect.

As experimental and computational methods advance, chemists can probe quantum effects with unprecedented resolution. Ultrafast spectroscopy, high-level quantum simulations, and precision isotope studies reveal how energy landscapes shift over femtosecond timescales, how electrons and protons move through complex scaffolds, and how quantum correlations persist despite thermal noise. These tools allow researchers to validate theoretical predictions, uncover hidden mechanisms, and build models that bridge the gap between atomic-level events and macroscopic function. The resulting knowledge enriches not only quantum biology but the broader field of chemistry, offering insights applicable to materials science, catalysis, energy storage, and molecular engineering.

Ultimately, quantum biology encourages chemists to adopt a more holistic view of molecular systems, one that recognizes the limitations of strictly classical interpretations while embracing the nuanced contributions of quantum behavior. It reveals that the boundary between the quantum and classical worlds is not an abrupt divide but a continuum shaped by environmental interactions, structural constraints, and dynamic processes. By understanding how biology navigates this continuum, chemists gain powerful new strategies for designing molecules, reactions, and materials that harness quantum principles with precision and purpose.

In this way, quantum biology serves as a bridge between fundamental physics and practical chemistry, illuminating the hidden quantum threads that underpin life's chemical tapestry. It pushes the boundaries of what is considered possible under everyday conditions, demonstrating that quantum behavior is not a rare phenomenon but a foundational aspect of molecular function. As research continues, the synergy between quantum mechanics and chemistry will deepen our understanding of biological complexity and inspire innovative technologies rooted in the same principles that nature has refined over billions of years.

Chapter 33: Word Problems

1. In photosynthetic organisms such as green sulfur bacteria, energy transfer in the Fenna–Matthews–Olson (FMO) complex occurs with near-unity efficiency despite the presence of a thermally noisy protein environment. Using the framework of environment-assisted quantum transport (ENAQT), explain how vibrational modes of the protein scaffold facilitate exciton migration between bacteriochlorophylls. Describe how coherence and decoherence interplay to optimize energy flow. Propose a computational approach, combining quantum master equations and molecular dynamics simulations, to quantify the relative contributions of coherent versus incoherent pathways. Include considerations of spectral density, temperature dependence, and how static disorder in site energies influences energy funneling.

2. Hydrogen tunneling in enzymes, exemplified by soybean lipoxygenase, leads to reaction rates exceeding classical transition state theory predictions. Given an experimentally measured kinetic isotope effect (KIE) of 7 when hydrogen is replaced by deuterium, estimate the fraction of the reaction rate attributable to quantum tunneling. Discuss how protein dynamics, including fluctuations in

donor–acceptor distances and vibrational gating motions, modulate tunneling probabilities. Suggest a combined experimental and computational approach to resolve the time scales of conformational motions contributing to tunneling.

3. Radical pairs formed in cryptochrome proteins mediate magnetoreception in migratory birds. Formulate a spin Hamiltonian for a radical pair including Zeeman and hyperfine interactions, and predict how variations in hyperfine couplings influence singlet–triplet interconversion rates under Earth-strength magnetic fields. Analyze how nuclear spin orientation, electron spin relaxation, and protein microenvironment fluctuations modulate magnetic sensitivity. Propose computational simulations using density matrix formalism to explore radical pair dynamics and design behavioral experiments to test predictions.

4. Vibrationally assisted electron tunneling has been proposed as a mechanism in olfactory receptors, potentially explaining odor discrimination. Design an experiment in which odorants are isotopically substituted (e.g., hydrogen → deuterium) to test whether receptor activation depends on vibrational frequencies. Predict how electron transfer rates and resulting receptor signaling would be modified, and discuss how these changes could influence odor perception and discrimination thresholds. Consider control experiments to separate classical binding effects from quantum tunneling contributions.

5. Quantum coherence in light-harvesting complexes can persist for hundreds of femtoseconds, even at physiological temperatures. Calculate the minimum dephasing time required for coherence to meaningfully influence energy transfer between pigments. Discuss the impact of environmental noise, spectral overlap of donor–acceptor sites, and exciton delocalization. Evaluate implications for designing artificial photosynthetic devices with optimized coherence lifetimes and efficiency, including considerations of pigment spacing, orientation, and coupling to vibrational modes.

6. DNA base pairs can undergo proton tunneling between canonical and tautomeric forms, potentially inducing mutations. Using a one-dimensional potential energy model, estimate the probability of a proton tunneling event given a barrier height of 0.5 eV and a donor–acceptor distance of ~1.8 Å. Discuss how the resulting tautomeric states could lead to mispairing during DNA replication and the possible

contribution of tunneling to spontaneous mutation rates. Suggest computational simulations (e.g., path-integral molecular dynamics) to quantify tunneling effects under physiological conditions.

7. RNA folding involves both electron delocalization and proton tunneling, which can influence secondary and tertiary structure formation. Using principles of quantum chemistry, describe how proton tunneling could modulate hydrogen bond formation and helix stability. Propose a computational framework integrating density functional theory and molecular dynamics to simulate folding pathways, including tunneling contributions, and predict how isotope substitution or environmental changes could alter folding kinetics.

8. Consider a hypothetical enzyme whose donor–acceptor hydrogen distance fluctuates between 2.5–3.0 Å due to thermal motion. Using quantum tunneling theory, estimate how these fluctuations affect the reaction rate, including the exponential sensitivity of tunneling probability to distance. Suggest structural or dynamic features in the enzyme that could stabilize tunneling-ready conformations, such as hydrogen bond networks or conformational gating motions, and describe experimental strategies to validate these features.

9. In photosynthetic complexes, excitons can delocalize over multiple pigments, enabling faster energy transfer. Construct a tight-binding Hamiltonian to model the excitonic energy landscape and quantify the effect of coherence on exciton transfer pathways. Compare results with incoherent hopping models, and discuss how pigment site energy disorder, exciton-phonon coupling, and environmental noise influence the relative efficiencies of coherent versus incoherent transport.

10. A radical pair in a cryptochrome protein has a lifetime of ~10 μs. Estimate the minimum coherence time required to detect Earth-strength magnetic fields, taking into account singlet–triplet interconversion rates and spin relaxation. Discuss how protein microenvironments, such as polarity, rigidity, and vibrational couplings, contribute to preserving entanglement, and propose simulations to predict behavioral consequences for bird navigation.

11. Proton-coupled electron transfer (PCET) in methylamine dehydrogenase can proceed through sequential or concerted pathways. Construct a detailed reaction coordinate diagram including both classical over-the-barrier pathways and

tunneling contributions. Predict isotope effects for hydrogen versus deuterium, considering both tunneling enhancement and kinetic barriers. Discuss how environmental fluctuations, such as protein vibrations and solvent reorganization, influence the relative contributions of the sequential and concerted pathways.

12. In bio-inspired photovoltaic systems, excitonic energy must travel efficiently across a network of pigment molecules. Design a computational simulation to compare classical Förster resonance energy transfer (FRET) with quantum coherent energy transfer under ambient conditions, incorporating site energy disorder, thermal noise, and decoherence rates. Predict conditions under which quantum coherence provides a measurable efficiency advantage and suggest design principles for synthetic light-harvesting networks.

13. The Schrödinger equation governs electron and proton dynamics in enzymatic reactions. Solve a one-dimensional tunneling problem for a proton across a rectangular potential barrier of height 0.4 eV and width 0.8 Å. Estimate the tunneling probability and rate, and discuss how changes in barrier height, width, and environmental fluctuations (e.g., protein vibrations) modulate tunneling efficiency. Compare with classical transition state theory predictions.

14. Consider a protein scaffold that stabilizes radical pairs for magnetoreception. Model how variations in vibrational coupling between the protein and radicals affect decoherence rates. Predict how increased or decreased decoherence influences the sensitivity of the radical pair mechanism to Earth's magnetic field, and suggest potential behavioral consequences in migratory birds. Design computational simulations combining quantum spin dynamics and molecular vibrations to validate your predictions.

15. Vibrationally assisted tunneling in enzymes relies on coupling nuclear motion with electronic transitions. Propose a path-integral molecular dynamics simulation to quantify how thermal fluctuations enhance tunneling rates. Consider isotopic substitution, donor–acceptor distance distributions, and protein conformational gating. Discuss how your results could be experimentally validated via kinetic isotope effects or temperature-dependent rate measurements.

16. Energy transfer in the FMO complex exhibits oscillatory signals in two-dimensional electronic spectroscopy experiments. Interpret these oscillations in terms of

quantum coherence, exciton delocalization, and environmental decoherence. Propose how such measurements could be used to extract coherence times, site energies, and couplings between pigments. Predict how temperature changes or protein mutations would modify the oscillatory signatures.

17. Design a DNA mutation that increases the flexibility of hydrogen bonds in a base pair. Predict how this alteration would affect the frequency and probability of proton tunneling events. Discuss potential consequences for replication fidelity, spontaneous mutagenesis, and evolutionary implications. Suggest experimental and computational strategies to measure altered tunneling rates and structural dynamics.

18. Proton tunneling can influence enzymatic catalysis in aromatic amine dehydrogenase. Given a barrier height of 0.3 eV and a donor–acceptor distance of 1.6 Å, calculate the tunneling rate for a hydrogen atom. Discuss how protein dynamics, such as conformational gating and vibrational modes, modify the tunneling contribution to the overall reaction rate. Suggest kinetic isotope experiments to validate theoretical predictions.

19. Spin chemistry predicts that weak magnetic fields can influence radical pair reactions. Simulate the effect of a 50 μT magnetic field on singlet–triplet interconversion using a density matrix formalism. Predict how changes in interconversion rates alter reaction yields and discuss implications for biological magnetoreception. Consider the effects of hyperfine coupling, spin relaxation, and protein environment on field sensitivity.

20. Consider an olfactory receptor that allows electron transfer influenced by vibrational modes of the odorant molecule. Model the tunneling probability for two odorants with different vibrational frequencies. Predict how this difference affects receptor activation and odor discrimination efficiency. Propose an experimental design using isotopically labeled odorants to test your predictions, and discuss potential confounding factors such as classical binding affinity differences.

21. Quantum coherence in microtubules has been proposed to influence neuronal signaling, possibly contributing to information processing. Calculate the decoherence time for dipole oscillations in a microtubule segment at physiological temperature (310 K), considering thermal fluctuations and environmental interactions. Discuss

the feasibility of sustained quantum processing in neurons and the implications for neurobiological models of consciousness. Suggest experimental approaches, such as ultrafast spectroscopy or quantum dot probes, to measure coherence in cytoskeletal structures.

22. Proton tunneling in DNA can generate transient tautomeric states that may lead to mispairing. Using a symmetric double-well potential to model proton transfer between canonical and tautomeric states, estimate the energy splitting and the probability of a tunneling event during replication. Discuss how local environment factors—such as hydration shells, ionic strength, and DNA sequence context—affect tunneling likelihood and potential mutagenesis.

23. Enzyme-catalyzed hydrogen transfer often exhibits temperature-dependent tunneling. Derive an expression for the temperature dependence of tunneling rates using quantum transition state theory, and compare it to classical Arrhenius behavior. Analyze how protein dynamics, donor–acceptor distance fluctuations, and vibrational coupling modify the apparent activation energy. Suggest experimental measurements across a range of temperatures to quantify tunneling contributions and validate the model.

24. Photosynthetic pigments are spectrally tuned to maximize coherent energy transfer. Using exciton theory, calculate how mismatches in site energies between donor and acceptor pigments affect transfer efficiency. Include the effects of site energy disorder, thermal fluctuations, and pigment–pigment coupling. Propose design principles for artificial light-harvesting systems that optimize energy funneling under realistic environmental noise.

25. In radical pair magnetoreception, conformational fluctuations in proteins modulate hyperfine interactions and spin dynamics. Model how these fluctuations influence the sensitivity of the radical pair mechanism to weak magnetic fields. Discuss the role of protein rigidity, local polarity, and vibrational modes in preserving coherence. Suggest computational experiments combining molecular dynamics and spin density calculations to predict field-dependent reaction yields.

26. Proton tunneling can influence RNA folding kinetics by facilitating hydrogen bond formation and rearrangements. Develop a kinetic model that incorporates tunneling probabilities into folding pathways. Predict how changes in hydration,

ionic environment, or isotope substitution would affect folding rates and the population of secondary structures. Suggest experimental techniques, such as temperature-jump spectroscopy or NMR, to validate the model.

27. A light-harvesting complex exhibits environment-assisted quantum transport (ENAQT) with exciton lifetimes of ~200 fs. Calculate the probability that an exciton reaches the reaction center via multiple coherent pathways. Include effects of dephasing, site energy disorder, and pigment–protein coupling. Discuss implications for artificial systems and the potential optimization of pigment networks to maximize transfer efficiency.

28. Enzymatic proton-coupled electron transfer (PCET) reactions involve simultaneous motion of electrons and protons. Using quantum transition state theory, estimate the rate enhancement due to tunneling for a hydrogen atom compared to classical over-the-barrier predictions. Consider donor–acceptor distances, vibrational gating, and isotope effects. Suggest experimental or computational methods to isolate the tunneling contribution in complex enzymatic systems.

29. Vibrationally assisted tunneling may occur in olfactory receptors, affecting electron transfer rates. Propose an experimental test using isotopically labeled odorants (H → D) to measure tunneling contributions. Predict how receptor activation and odor discrimination would change. Discuss confounding factors, including classical binding affinities, conformational flexibility, and receptor heterogeneity, and propose strategies to control for these effects.

30. Quantum effects can stabilize tautomeric forms of DNA, potentially influencing mutation rates. Using density functional theory (DFT), predict the energy barrier for proton transfer between canonical and tautomeric states. Quantify the likelihood of tautomer formation under physiological temperature and discuss the potential impact on replication fidelity. Suggest experiments or simulations to test the influence of sequence context and environmental factors on proton transfer.

31. In the FMO complex, excitonic coherence can span multiple pigments, but thermal fluctuations reduce coherence times. Model the impact of site energy disorder and vibrational noise on coherence decay. Quantify how reduced coherence affects energy transfer efficiency and predict which pigments serve as critical nodes for optimized transport. Suggest experimental strategies, such as two-dimensional electronic spectroscopy, to measure coherence lifetimes in vivo.

32. 32. Radical pair spin entanglement is influenced by hyperfine couplings to nearby nuclei. Design a computational experiment using density matrix simulations to predict singlet–triplet ratios under varying nuclear spin environments. Analyze how changes in isotopic composition, protein fluctuations, or magnetic field strength alter entanglement and the resulting chemical yields. Discuss implications for animal magnetoreception.
33. Enzymatic tunneling is highly sensitive to donor–acceptor alignment. Propose a molecular dynamics simulation protocol to identify conformational states that maximize tunneling probability. Include analysis of hydrogen bond networks, local vibrational modes, and conformational gating. Predict how mutations or solvent conditions could modulate the rate of hydrogen transfer.
34. Proton tunneling in RNA ribozymes can be influenced by local hydration shells and ion binding. Estimate how water-mediated hydrogen bonding and ionic screening modify the effective barrier height and reaction rates. Suggest experimental or computational approaches to quantify the impact of hydration on tunneling-mediated catalysis.
35. Energy transfer in photosynthetic systems can be enhanced by quantum coherence. Calculate the difference in transfer efficiency between coherent exciton migration and incoherent hopping for a pigment network of 10 coupled sites. Include effects of site energy disorder, decoherence rates, and temperature. Discuss strategies to experimentally distinguish coherent versus incoherent contributions.
36. Quantum spin dynamics in radical pairs can modulate chemical yields. Using a density matrix approach, simulate the effect of weak magnetic fields (e.g., 50 μT) on singlet–triplet interconversion. Include contributions from hyperfine interactions, spin relaxation, and environmental fluctuations. Predict field-dependent yield changes relevant to magnetoreception.
37. Vibrational modes can assist tunneling in enzymatic reactions by coupling nuclear motion to electronic transitions. Propose an experiment using isotopic substitution ($H \rightarrow D$) to measure the vibrational contribution to reaction rates. Predict how tunneling enhancement would change with temperature and donor–acceptor distance. Discuss experimental observables that can quantify vibrationally assisted tunneling.

38. Proton tunneling in DNA may contribute to mutagenesis. Using a one-dimensional potential energy model, compute the tunneling frequency for a proton in a hydrogen bond between base pairs. Assess how local fluctuations, hydration, and base stacking influence tunneling probability and the likelihood of mispairing during replication.

39. Coherent energy transfer in photosynthetic complexes can be modeled using a tight-binding Hamiltonian. Predict how introducing disorder in site energies (e.g., due to thermal fluctuations or mutations) affects exciton migration pathways and overall energy transfer efficiency. Propose strategies to enhance robustness of coherent transport in synthetic systems.

40. Radical pair mechanisms in magnetoreception rely on entangled electron spins. Construct a model to predict the effect of decoherence on directional sensitivity in migratory birds. Include protein microenvironment factors, such as polarity and vibrational modes, and predict thresholds for geomagnetic field detection. Suggest behavioral experiments to validate model predictions.

41. Enzymes maintain specific donor–acceptor geometries to optimize proton tunneling. Design a protein engineering strategy aimed at enhancing proton transfer rates through geometric preorganization. Predict how modifications to hydrogen bond networks, active site rigidity, and conformational flexibility would influence tunneling probability. Suggest computational and experimental tests of your design.

42. Quantum effects in olfaction may involve inelastic electron transfer through odorant molecules. Propose a spectroscopic method (e.g., ultrafast pump–probe or inelastic electron tunneling spectroscopy) to detect tunneling-mediated electron flow in receptors. Discuss expected spectral signatures and how isotopic substitution would alter the observed signals.

43. Consider microtubule-associated dipole oscillations as potential quantum information carriers. Using a harmonic oscillator model, calculate decoherence rates at physiological temperatures, incorporating environmental noise and thermal fluctuations. Discuss implications for the feasibility of quantum processing in cellular systems and suggest experimental approaches to measure coherence.

44. Proton tunneling in DNA base pairs can cause rare mutations. Model the probability distribution of tautomeric states during replication using a double-well potential framework. Predict mutation frequency as a function of barrier height, donor–acceptor distance, and environmental factors such as hydration or ionic strength. Compare with observed mutation rates in vivo.

45. Environment-assisted quantum transport (ENAQT) enhances exciton transport in light-harvesting complexes. Develop a kinetic model incorporating both coherent and incoherent hopping pathways. Predict overall energy transfer rates as a function of dephasing, temperature, and pigment network connectivity. Suggest experimental tests to validate the contributions of coherence to transport efficiency.

46. Quantum coherence in photosynthesis persists despite thermal noise. Using a Lindblad master equation, simulate decoherence dynamics for a three-pigment subsystem. Predict how varying environmental coupling strengths, site energy disorder, and temperature alter coherence times and energy transfer efficiency. Discuss implications for designing bio-inspired energy systems.

47. Proton-coupled electron transfer (PCET) in enzymes is sensitive to isotope substitution. Predict the effect of replacing hydrogen with tritium on reaction rates and tunneling probabilities. Include contributions from mass-dependent barrier penetration, zero-point energy shifts, and protein dynamics. Suggest experimental validation via kinetic isotope effect measurements.

48. Radical pair reactions depend on the local protein microenvironment. Model how altering the polarity or dielectric constant of the binding pocket affects singlet–triplet interconversion rates and overall reaction yields. Discuss how this could be exploited to tune magnetoreception sensitivity in biological or synthetic systems.

49. Vibrationally assisted tunneling may occur in olfactory receptors. Estimate the probability of electron transfer for a molecule with a vibrational frequency of 1500 cm^{-1}. Discuss how isotopic substitution ($H \rightarrow D$) would modify tunneling probability and receptor activation, and propose a combined experimental and computational approach to validate predictions.

50. Consider the evolution of quantum-optimized molecular architectures in enzymes and photosynthetic proteins. Propose a comparative study across species to

identify structural features that enhance tunneling efficiency and energy transfer. Include analysis of hydrogen bond networks, donor–acceptor distances, pigment arrangements, and protein dynamics. Discuss how evolutionary pressures may have selected for quantum-facilitating features, and suggest experiments to quantify their functional significance.

Chapter 34

Self-Healing Materials at the Molecular Level

The field of self-healing materials has undergone an extraordinary transformation over roughly the past two decades, driven by advances in dynamic covalent chemistry, supramolecular chemistry, and composite materials engineering that collectively aim to emulate the adaptive, repairable characteristics of biological tissues. Early milestones in the field focused heavily on reversible covalent reactions, with the Diels–Alder (DA) cycloaddition between furan (as diene) and maleimide (as dienophile) serving as one of the most influential and widely explored molecular platforms. In such polymers, multifunctional furan and maleimide monomers are crosslinked to form a three-dimensional network that can reversibly dissociate at elevated temperatures through the retro-Diels–Alder reaction, yielding free reactive groups that can subsequently recombine upon cooling. This reversible chemistry introduced a fundamentally new capability to synthetic polymer networks: the ability to selectively break and reform crosslinks in response to thermal stimuli, enabling reprocessing, remodeling, and repair of microstructural damage in materials previously considered permanent thermosets.

Over time, researchers began refining this DA-based paradigm, realizing that the full utility of reversible networks depends critically on the balance between bond strength, bond dynamics, molecular mobility, and the structural architecture of the network. One of the major achievements came from manipulating the stoichiometric ratio of maleimide-to-furan groups to deliberately incorporate free pendant groups that enhance mobility and increase reaction probability across damaged surfaces. In several studies, off-stoichiometric ratios enabled rapid self-healing under unusually mild conditions—even at or near room temperature—thereby overcoming one of the original limitations of DA systems, which often required temperatures of 80–140 °C to activate the retro-DA bond cleavage. Such breakthroughs demonstrated that reversible networks need not be restricted to high-temperature applications but could instead function under everyday ambient conditions, broadening their relevance to consumer products, wearable electronics, soft robotics, biomedical devices, and flexible packaging.

Yet these improvements highlighted a persistent challenge: the trade-off between dynamic responsiveness and mechanical robustness. Lowering the crosslink density or increasing segmental mobility promotes efficient healing by allowing reactive groups to diffuse and re-encounter one another at a damaged interface. However, these same structural modifications reduce stiffness, thermal resistance, creep resistance, and overall mechanical strength—properties that are essential in load-bearing or high-performance applications. Studies on polyketone-based thermosets modified with furan groups revealed that adjusting the density of grafted furan moieties and selecting bis-maleimide crosslinkers with appropriate rigidity allowed fine-tuning of glass transition temperatures, elastic moduli, and healing behavior. Some formulations achieved full healing within minutes under moderate heat (e.g., 120 °C), even demonstrating shape-memory capabilities in which a programmed temporary shape could be recovered concurrently with the healing event. These examples confirmed that controlled network design enables a rich interplay between dynamic healing chemistry and functional mechanical performance.

As the field matured, researchers began to explore hybrid strategies that combine dynamic covalent networks with reinforcing fillers in order to overcome the inherent trade-offs between mechanical strength and healing efficiency. One prominent approach involved integrating nanocellulose fibers—ultra-high-aspect-ratio, stiff, renewable nanomaterials—into reversible polyurethane-based networks crosslinked through furan–maleimide DA bonds. The resulting composites demonstrated significantly enhanced elastic modulus and tensile strength due to the reinforcing nanofibers, while still maintaining the reprocessability and self-healing capabilities conferred by the DA covalent chemistry. Such systems address a central challenge in applying self-healing materials to real engineering contexts: ensuring that the material is sufficiently robust for daily use or structural performance while still retaining the ability to autonomously repair microcracks, surface damage, or internal defects. This composite strategy, which leverages both molecular reversibility and nanoscale reinforcement, exemplifies the trend toward multifunctional hybrid materials that align with sustainability goals by incorporating bio-derived components.

Parallel to dynamic covalent systems, supramolecular self-healing materials emerged as a powerful complementary avenue. Supramolecular interactions—including hydrogen bonding, metal–ligand coordination, π–π stacking, halogen bonding, host–guest complexation, and dynamic electrostatic pairings—offer a wide range of reversible interactions that can break and reform without the need for high temperatures or harsh stimuli. These interactions typically proceed rapidly and efficiently under ambient conditions, making them ideal for applications requiring spontaneous or low-energy repair. Among the most important motifs are strong, directional hydrogen-bonding pairs such as ureidopyrimidinone (UPy), which can form quadruple hydrogen bonds with high association constants. Polymers decorated with such motifs can heal within minutes simply by bringing fractured surfaces into contact, allowing the interfacial hydrogen-bond network to reconstitute. In so-called "smart rubbers," these supramolecular interactions can even mimic biological tissue behavior, enabling soft materials to fuse back together at room temperature with minimal external stimulus. While these materials excel in fast and mild-condition healing, they often trade off high-load performance, exhibiting lower creep resistance, reduced modulus, or vulnerability to humidity, which competes for hydrogen-bonding sites.

The desire to harness both the strength of covalent networks and the responsiveness of supramolecular systems has led to the development of dual-dynamic or hybrid healing networks. These materials incorporate two or more dynamic mechanisms, such as combining Diels–Alder chemistry with disulfide metathesis, imine exchange, boronic ester dynamics, or metal–ligand interactions. A dual-dynamic polymer might, for instance, heal via DA bond reformation under thermal stimulation but also repair through light-activated disulfide exchange under UV irradiation. This multi-modality expands the operational versatility of self-healing systems: a single material can heal in multiple environments, under different stimuli, or at different time scales. However, hybrid systems introduce additional complexity. Competing dynamic mechanisms may interfere with each other, reactive groups may degrade under environmental exposure, and network fatigue over repeated healing cycles can become more pronounced. Designing such systems requires careful consideration of orthogonality among dynamic bonds, stimulus selectivity, activation energies, and the physical mobility needed for each healing mode.

Composite materials represent another frontier, particularly in the context of engineering applications such as aerospace, automotive, civil infrastructure, and protective coatings. Embedding self-healing interleaves within carbon-fiber composites illustrates how dynamic chemistry can be incorporated at structural scales. Electrospun interleaves containing DA-based healing agents have been shown to mitigate crack propagation in laminated composites, reducing delamination and improving fatigue resistance under cyclic loading. Modeling work on such systems treats cracks as disruptions in the structural manifold and healing as a restorative process that re-establishes continuity across plies. This blend of materials design and computational modeling provides insight into how healing kinetics, interleaf distribution, and interfacial adhesion influence structural durability. Yet translating self-healing chemistry to composite manufacturing environments remains challenging. High-temperature curing conditions can prematurely trigger dynamic reactions, and the addition of healing agents must not compromise fiber–matrix bonding, which is essential for load transfer and mechanical performance.

At the molecular level, the central challenge in designing self-healing systems is striking a balance between stability and reversibility. Bonds must be robust enough to prevent catastrophic failure under mechanical load, yet sufficiently labile to enable healing under accessible stimuli. Diels–Alder chemistry, disulfide bonds, imine bonds, boronic esters, transesterification reactions, and metal coordination bonds all have tunable dynamics governed by activation energies, pH dependence, redox states, or light sensitivity. The design process must account for how polymer chain mobility, crosslink density, backbone flexibility, and free volume affect the likelihood of reactive group encounters during healing. Achieving high healing efficiency often requires introducing softer, more flexible segments to increase segmental mobility, but these modifications inevitably influence modulus, creep, and thermal performance. Thus, designing highly dynamic networks is an exercise in balancing competing property requirements where the optimal chemistry depends on the target application environment.

Activation conditions represent another critical design consideration. While early dynamic covalent networks required substantial heat to trigger healing, newer architectures demonstrate healing under ambient or slightly elevated temperatures, opening possibilities for everyday-use materials. However, ambient-temperature healing often proceeds more slowly due to limited chain mobility, especially if the polymer is below its glass transition temperature or contains crystalline domains that impede diffusion. Achieving robust healing at low temperatures typically requires high mobility, low crosslink density, plasticizers, or solvent-assisted swelling, each of which brings potential compromises in mechanical or environmental stability. Thus, tuning the thermal and mechanical properties to match the operating range of the dynamic chemistry remains a delicate balancing act.

An increasingly influential part of the field involves computational modeling. Molecular dynamics simulations, Monte Carlo methods, finite-element modeling, and non-Euclidean elastic theories provide insight into how bond exchange rates, chain diffusion, network topology, and damage morphology influence healing efficiency. Some computational frameworks treat polymer networks as evolving manifolds where damage introduces geometric distortions and healing restores local compatibility. These models can predict how healing improves fracture toughness, slows crack propagation, or affects cyclic fatigue behavior. With advances in multiscale modeling and machine-learning-assisted material prediction, computational approaches are becoming indispensable tools for designing next-generation self-healing systems that meet multiple performance targets simultaneously.

Looking toward future directions, one of the most exciting frontiers is the development of autonomic self-healing materials—systems that sense damage and initiate repair without external intervention. This vision involves integrating dynamic covalent networks with stimuli-responsive motifs and molecular sensors capable of detecting stress, chemical changes, or mechanical disruption. Possible sensing mechanisms include fluorescence turn-on near crack tips, mechanochromic units that undergo chemical transformation under stress, redox-sensitive linkages that trigger bond exchange upon oxidation, or embedded liquid microcapsules that rupture upon damage and release healing agents. Coupling sensing and healing functions brings materials closer to biological analogues like skin or muscle, which continuously monitor, respond, and repair themselves. Autonomic systems could revolutionize structural materials, soft electronics, artificial organs, and responsive coatings.

Sustainability and recyclability have also become central motivations. Traditional thermosets are notoriously difficult to recycle because of their permanent crosslinked networks. Dynamic covalent chemistries—particularly transesterification-based vitrimers, DA networks, and associative exchange mechanisms—offer pathways to reprocess and reshape materials multiple times. When combined with bio-derived backbones such as cellulose, lignin derivatives, natural rubber, chitosan, or polypeptides, these materials can significantly reduce environmental impact while maintaining advanced functionality. Several recent studies demonstrate promising combinations of renewability, mechanical robustness, and dynamic repair chemistry within the same material, suggesting that the next generation of self-healing polymers may simultaneously address performance, longevity, and sustainability.

Despite promising progress, major challenges remain. Long-term durability under repeated healing cycles is not yet fully understood, and many systems show diminishing recovery efficiency after multiple cycles due to irreversible side reactions, oxidation of functional groups, microstructural densification, or diffusion limitations. Environmental factors—such as humidity, UV radiation, oxidative conditions, or mechanical fatigue—can degrade dynamic bonds or interfere with healing processes. Furthermore, scaling laboratory-scale dynamic networks to industrial manufacturing processes requires overcoming issues such as maintaining uniform distribution of dynamic functionalities, preventing premature reaction during cure, and ensuring robust performance across temperature and humidity ranges. Standardized testing protocols are also needed—

current definitions of healing efficiency vary widely, making it difficult to compare materials across studies. Metrics for tensile strength recovery, fracture toughness recovery, fatigue life extension, and long-term cyclic stability must be harmonized to accelerate industrial adoption.

Several case studies illustrate recurring principles in the design of self-healing materials. In one example, elastomeric DA polymers synthesized with off-stoichiometric maleimide-to-furan ratios demonstrated room-temperature healing capabilities. When severed and pressed back together at approximately 25 °C, the fractured surfaces reconnected sufficiently within minutes to support modest loads, and full healing over several hours restored mechanical properties close to their original values. This performance was achieved by incorporating excess furan groups that increased mobility and availability of reactive sites at the healing interface. However, this same design led to a slight compromise in dynamic mechanical properties due to reduced crosslink density—an example of the universal trade-off between healing efficiency and structural performance.

A second case involved polyketone-based thermosets functionalized with furan groups and crosslinked with aromatic bis-maleimides. By adjusting the furan grafting ratio, researchers achieved tunable thermomechanical behavior, rapid high-temperature healing (as fast as five minutes at 120 °C), and notable shape-memory functionality. These materials demonstrated that dynamic covalent networks can simultaneously deliver high strength, thermal stability, and rapid healability when designed with appropriate backbone rigidity and crosslinker structure.

A third example illustrated how reinforcing fillers can enhance the performance of dynamic networks. Polyurethane elastomers containing DA bonds were reinforced with nanocellulose fibers, producing composites with improved toughness and mechanical integrity while retaining thermal reversibility and healing capacity. Structural analysis confirmed successful DA crosslink formation and uniform filler dispersion, underscoring the value of integrating nanostructural reinforcement with repairable chemistries.

Together, these examples reveal several general design principles. Network architecture governs healing kinetics and trade-offs between stiffness and mobility. The choice of dynamic bond chemistry dictates the required stimulus and influences the practicality of healing under realistic conditions. Backbone selection and composite design allow mechanical properties to be tuned independently of healing chemistry. Finally, application-specific requirements determine the optimal balance between healing speed, mechanical strength, environmental resilience, and manufacturability.

As self-healing research continues to expand, the field is converging toward multifunctional, multimodal systems that unify dynamic covalent chemistry, supramolecular interactions, embedded microcapsules, and sensing capabilities within a single material platform. Advances in additive manufacturing, high-throughput polymer synthesis, machine learning, and data-driven materials optimization are expected to further accelerate the discovery and deployment of next-generation self-healing materials. The long-term vision is clear: synthetic materials that can autonomously monitor their structural health, repair themselves repeatedly, maintain high performance throughout extended lifetimes, and ultimately reduce maintenance costs, environmental impact, and material waste. These innovations bring us closer to a future in which synthetic materials demonstrate lifespans and adaptive behaviors previously reserved for nature.

Chapter 34: Word Problems

1. A polymer network is crosslinked via furan–maleimide Diels–Alder (DA) chemistry. If the retro-DA reaction occurs at 120 °C and regenerates free reactive groups, calculate how the healing efficiency would change if the polymer is instead heated at 90 °C for the same duration. Consider the activation energy for the retro-DA reaction, the Arrhenius dependence of reaction rates on temperature, and the influence of polymer segmental mobility on reaction probability. Discuss how incomplete bond dissociation could limit healing efficiency and suggest strategies to optimize thermal healing without degrading the polymer backbone.

2. A polyketone-based thermoset contains 20% furan-modified monomers and is crosslinked with a rigid bis-maleimide. Predict how increasing the furan grafting density to 35% would influence the glass transition temperature (Tg), elastic modulus, and healing kinetics. Consider the effect of increased crosslink density on segmental mobility, steric hindrance during DA bond reformation, and possible changes in network homogeneity. Discuss the trade-offs between improved healing rates and potential embrittlement due to higher crosslink density, including implications for load-bearing applications.

3. Consider a DA-based polymer designed to heal at room temperature. If the maleimide-to-furan stoichiometry is off-balance, explain how the presence of excess furan groups influences segmental mobility, interfacial reaction probability, and the formation of dangling chains. Analyze how this stoichiometric imbalance affects mechanical recovery, diffusion-limited kinetics at the healing interface, and the thermodynamic driving force for bond reformation. Propose experimental methods to quantify healing efficiency under nonstoichiometric conditions.

4. A soft robotic actuator is constructed using a supramolecular polymer decorated with quadruple hydrogen-bonding ureidopyrimidinone (UPy) units. If the actuator develops microcracks under cyclic loading, discuss the molecular mechanisms that enable autonomous healing at ambient temperature, including bond dissociation-reassociation, chain diffusion, and local relaxation. Estimate the timescale of recovery based on typical hydrogen-bond lifetimes, polymer viscosity, and network topology. Predict how varying UPy density and polymer molecular weight would alter healing kinetics and actuator performance.

5. A DA-based network is reinforced with nanocellulose fibers. Describe how the incorporation of these high-aspect-ratio fillers affects tensile strength, elastic modulus, fracture toughness, and healing efficiency. Consider fiber–matrix interfacial adhesion, restriction of polymer segmental mobility, and potential steric hindrance to DA bond reformation. Propose an optimal fiber loading that balances mechanical robustness with repairability, and discuss how fiber orientation and dispersion influence both properties.

6. You are designing a hybrid self-healing polymer that combines DA chemistry and disulfide exchange. Identify potential interferences between these mechanisms during simultaneous activation, including competitive bond reformation, radical-mediated side reactions, and thermal or photochemical incompatibilities. Suggest strategies to maintain orthogonality of bond dynamics, such as sequential activation, selective catalysts, or environmental control, and analyze how these strategies would impact healing efficiency, mechanical performance, and fatigue resistance.

7. A laminated carbon-fiber composite contains an electrospun DA interleaf. Under cyclic loading, a delamination occurs. Explain how the interleaf chemistry restores

structural integrity through retro-DA and forward DA reactions, and calculate the potential increase in fatigue life if healing is 80% efficient. Incorporate stress distribution models, interfacial fracture mechanics, and potential diffusion limitations of reactive groups. Discuss how interleaf thickness, crosslink density, and environmental conditions (temperature, humidity) influence healing performance.

8. A polymer network relies on hydrogen bonding for healing. Exposure to high humidity reduces healing efficiency. Explain the underlying molecular interactions, including competition between water molecules and donor/acceptor sites, disruption of network dynamics, and alteration of chain mobility. Propose strategies to mitigate moisture interference without compromising mechanical properties, such as backbone hydrophobization, incorporation of moisture-resistant supramolecular motifs, or protective coatings. Analyze the trade-offs between environmental stability and healing speed.

9. A dual-dynamic polymer heals via DA bond reformation at 120 °C and disulfide exchange under UV light. Devise an experimental protocol to quantify the individual contributions of each mechanism to total healing efficiency after simultaneous damage. Consider selective activation conditions, kinetic monitoring via spectroscopic methods, and mechanical recovery measurements. Discuss potential coupling effects, competition between mechanisms, and how to decouple the contributions for precise quantification.

10. A polymer with boronic ester dynamic bonds is exposed to an oxidizing environment. Predict how redox-induced bond exchange affects healing kinetics and mechanical stability, considering activation energy, bond lifetime, and network relaxation. Analyze possible side reactions, such as boronic acid hydrolysis or backbone oxidation, and suggest molecular design strategies to enhance oxidative stability while retaining self-healing functionality.

11. You are tasked with designing a self-healing coating for aerospace applications. Outline the choice of dynamic chemistry (DA, disulfide, boronic ester, or supramolecular), reinforcing fillers, and crosslink density to balance high-load performance, fatigue resistance, and healing speed under operational temperature ranges. Discuss environmental challenges, such as UV exposure, temperature cycling, and moisture, and propose design criteria to ensure consistent healing performance throughout the coating lifecycle.

12. In a room-temperature DA polymer, the healing interface is pressed together after severing. Model the diffusion-limited reaction kinetics that dictate the timescale for full mechanical recovery. Include considerations of free volume, chain mobility, reactive group concentration, and the effect of applied pressure on interface contact. Quantitatively estimate the time required for full recovery and discuss how altering polymer molecular weight or plasticizer content would influence the process.

13. A polyketone-based thermoset exhibits shape-memory behavior concurrent with healing. Explain the molecular mechanism by which temporary shapes are recovered during thermal activation, including the interplay between crosslink density, segmental mobility, and DA bond reformation. Analyze how the kinetics of retro-DA and forward DA reactions influence the simultaneous recovery of shape and mechanical integrity, and predict optimal conditions for maximal recovery without causing network degradation.

14. Consider a supramolecular polymer network relying on metal–ligand coordination for healing. Propose a method to quantitatively measure the association constants of the motifs, such as isothermal titration calorimetry or spectroscopic titration, and relate them to observed self-healing times. Discuss how ligand geometry, metal ion identity, and network topology affect both binding strength and dynamic bond exchange rates, and predict implications for macroscopic healing efficiency.

15. You aim to scale a DA-based polymer to industrial production. Identify challenges related to premature retro-DA reactions during high-temperature curing, including loss of crosslink density, network heterogeneity, and degradation of reactive groups. Propose solutions, such as controlled heating ramps, protective solvents, or encapsulated reactive groups, and analyze how these strategies affect final mechanical properties, healing efficiency, and production throughput.

16. A composite polymer contains 5 wt% nanocellulose and demonstrates both healing and high modulus. If fiber alignment is random rather than unidirectional, predict the effect on mechanical anisotropy, stress distribution, and healing uniformity. Discuss how fiber orientation influences local segmental mobility, interfacial reaction kinetics, and fracture propagation, and propose strategies to optimize fiber alignment while maintaining process scalability.

17. A dual-dynamic network combines DA and imine bonds. Discuss how environmental factors (pH, temperature, humidity) could selectively activate or deactivate one mechanism, and predict the implications for healing under realistic operating conditions. Consider the effect of competing hydrolysis or oxidation reactions on network stability, and propose design modifications to maximize healing efficiency while preserving mechanical performance across a broad range of conditions.

18. A dynamic covalent polymer is below its glass transition temperature (Tg). Discuss how low chain mobility limits healing efficiency, including the effect on diffusion of reactive groups, bond reformation kinetics, and interfacial contact. Suggest molecular modifications, such as flexible spacers, plasticizers, or supramolecular additives, to improve repair at sub-Tg conditions, and evaluate their influence on modulus, creep resistance, and thermal stability.

19. Consider a polymer network where segmental mobility is increased to enhance healing. Quantitatively analyze how reducing crosslink density influences creep resistance, elastic modulus, and fracture toughness. Propose an optimal compromise for load-bearing applications, balancing rapid healing with sufficient mechanical integrity. Include discussion of time-temperature superposition principles and the effect of network defects on macroscopic properties.

20. A polymer with quadruple hydrogen-bonding motifs is submerged in water. Predict the competition between water molecules and hydrogen-bonding sites, including partial disruption of noncovalent interactions and reduction of effective crosslink density. Propose strategies, such as backbone hydrophobization, sacrificial hydrogen-bonding motifs, or encapsulation, to retain healing capability under aqueous conditions. Discuss the impact on mechanical performance and long-term durability.

21. A DA polymer network is exposed to repeated healing cycles. Discuss the factors that lead to degradation of functional groups over multiple cycles, including thermal degradation, hydrolysis, and irreversible bond scission. Calculate the expected decrease in tensile recovery over ten cycles using a simple kinetic model, and propose strategies to extend the functional lifetime, such as using more thermally stable maleimide analogs or incorporating protective fillers.

22. You are designing a polymeric soft electronic device. Evaluate whether dynamic covalent or supramolecular healing chemistries are more suitable for ambient-

temperature repair. Justify your selection based on reaction kinetics, reversibility, mechanical compliance, and device operational constraints, and discuss potential trade-offs in terms of healing speed, fatigue resistance, and long-term reliability under repeated deformation.

23. A self-healing polymer incorporates light-sensitive disulfide linkages. Devise an experiment to measure the activation energy for photoinduced healing, considering light intensity, wavelength, polymer thickness, and irradiation time. Discuss how photophysical properties, absorption coefficients, and potential side reactions influence the measured healing rates, and propose methods to decouple thermal and photochemical contributions.

24. In a hybrid polymer network, competing dynamic mechanisms reduce overall healing efficiency. Propose a computational modeling approach, such as kinetic Monte Carlo or molecular dynamics simulations, to predict optimal ratios of each dynamic bond type that maximize recovery. Include considerations of network topology, reaction kinetics, bond lifetime distributions, and environmental conditions. Validate predictions through hypothetical experimental scenarios.

25. A carbon-fiber laminate contains a DA-based interleaf. Use finite-element modeling to simulate crack propagation and healing. Determine stress distribution changes before and after interleaf activation, and predict the influence of interleaf thickness, crosslink density, and fiber–matrix adhesion on recovery of load-bearing capacity. Discuss implications for fatigue life and safety margins in aerospace applications.

26. A polymer contains DA and boronic ester bonds. Exposure to acidic conditions selectively accelerates boronic ester exchange. Predict the impact on composite material integrity over prolonged acid exposure, including potential softening, reduction in elastic modulus, and changes in healing kinetics. Propose mitigation strategies, such as protective coatings or pH-buffered environments, to maintain both mechanical performance and self-healing capability.

27. A self-healing material is intended for wearable electronics and must heal at ~25 °C. Analyze how free-volume enhancement, polymer flexibility, and segmental mobility contribute to reaction probability at the healing interface. Consider diffusion-limited kinetics, interfacial adhesion, and possible mechanical constraints imposed by device architecture. Propose molecular design principles to optimize healing

without compromising electrical performance or mechanical compliance.

28. Consider a polymer network reinforced with nanocellulose fibers. Evaluate the effect of fiber aspect ratio on modulus, fracture toughness, crack bridging, and self-healing kinetics. Discuss how high-aspect-ratio fibers may enhance load transfer but impede reactive group mobility, and propose an optimal fiber geometry that balances mechanical robustness with dynamic bond accessibility.

29. A supramolecular polymer relies on host–guest complexation for self-healing. Design an experiment to measure association/dissociation rates of host–guest pairs, using spectroscopic, calorimetric, or rheological techniques, and correlate these rates with observed macroscopic healing times. Analyze how network topology, polymer molecular weight, and guest mobility influence the kinetics and efficiency of repair.

30. You are designing a dual-dynamic polymer with DA and disulfide bonds. Calculate the optimal stimulus conditions, such as temperature and irradiation intensity, where both mechanisms contribute synergistically without interfering. Consider kinetic parameters, bond lifetime, and network relaxation, and propose a strategy to maximize mechanical recovery while minimizing potential side reactions or premature degradation.

31. A polyketone thermoset exhibits rapid high-temperature healing. Discuss how backbone rigidity influences elastic modulus, glass transition temperature, creep resistance, and the rate of DA bond reformation. Analyze the interplay between polymer stiffness, segmental mobility, and network relaxation, and predict the optimal balance of rigidity and flexibility for fast yet mechanically robust healing.

32. A polymer is engineered for autonomic self-healing using fluorescence turn-on sensors that respond to bond scission. Propose a quantitative method to relate sensor activation to the onset of healing and degree of damage, including calibration procedures, imaging protocols, and mechanical testing correlations. Discuss potential limitations arising from sensor fatigue, photobleaching, or incomplete reporting of microdamage.

33. A composite material incorporates self-healing interleaves in aerospace laminates. Evaluate how interleaf thickness, spacing, and crosslink density affect crack

propagation, stress redistribution, and healing efficiency. Consider both mechanical and environmental factors, such as temperature fluctuations, humidity, and repeated loading, and propose optimal design parameters to maximize structural reliability.

34. A DA polymer exhibits shape-memory healing. Using thermomechanical principles, predict the energy required to induce shape recovery while simultaneously reforming DA crosslinks. Analyze how crosslink density, segmental mobility, and thermal ramp rate influence both recovery efficiency and mechanical integrity, and propose design strategies to minimize energy consumption while maximizing repair performance.

35. A dual-dynamic network includes imine and disulfide bonds. Discuss how environmental oxidation selectively affects disulfides, and predict the impact on long-term healing cycles, bond reformation rates, and mechanical recovery. Suggest molecular strategies to protect sensitive bonds or exploit selective activation for stimuli-responsive healing.

36. A supramolecular polymer displays rapid room-temperature healing via halogen bonding. Quantify the effect of halogen bond strength, donor–acceptor geometry, and network topology on healing kinetics and mechanical recovery. Predict how environmental factors, such as humidity or temperature, influence bond stability and propose strategies to maintain consistent healing performance.

37. You are tasked with designing a vitrimer with transesterification exchange. Calculate how crosslink density, activation energy, and catalyst loading influence reprocessing temperature, healing rate, and long-term mechanical stability. Discuss the trade-offs between thermal accessibility of exchange reactions and network robustness, and propose design principles for scalable industrial applications.

38. A polymer network is engineered with off-stoichiometric DA chemistry. Analyze the effect of excess reactive groups on mechanical properties, chain mobility, network topology, and healing efficiency. Include consideration of dangling chains, steric hindrance, and potential entanglements, and propose strategies to tune stoichiometry for optimal macroscopic performance.

39. A hybrid polymer contains light-activated disulfide bonds and thermal DA bonds.

Predict the sequence of activation under simultaneous UV and heat stimuli, the possible interactions between mechanisms, and their effect on multicycle healing. Propose experimental protocols to decouple mechanisms and evaluate synergistic effects on recovery of mechanical properties.

40. A supramolecular polymer is exposed to mechanical fatigue. Using molecular dynamics and diffusion concepts, explain how chain mobility, bond reformation, and network relaxation influence fatigue life extension. Quantify the expected improvement in cyclic durability as a function of bond lifetime and polymer viscosity, and discuss practical limitations for load-bearing applications.

41. A composite polymer is reinforced with bio-derived fibers. Evaluate the trade-off between environmental sustainability and mechanical performance, considering healing kinetics, modulus, fracture toughness, and network compatibility. Discuss strategies to optimize fiber–matrix interactions, network design, and processing conditions for high-performance, eco-friendly self-healing materials.

42. A DA-based polymer is designed to heal under moderate heat (80–140 °C). Construct a kinetic model incorporating retro-DA and forward DA reactions to estimate the minimum healing time required for full mechanical recovery at 100 °C. Include Arrhenius parameters, polymer segmental mobility, and diffusion-limited reaction effects, and validate your approach with hypothetical experimental scenarios.

43. A polymer network contains both rigid and flexible segments. Discuss how segmental heterogeneity influences healing interface accessibility, bond reformation probability, and overall fracture toughness. Consider the effect of local chain mobility, stress distribution, and reactive group density on healing efficiency, and propose strategies to optimize network design for combined mechanical performance and repairability.

44. A supramolecular polymer exhibits healing via π–π stacking. Quantify the relationship between stacking interaction energy, chain diffusion, and rate of macroscopic repair. Discuss the influence of polymer backbone rigidity, aromatic content, and environmental conditions (solvent, temperature) on the effectiveness of healing, and propose strategies to enhance bond accessibility without sacrificing mechanical robustness.

45. You are tasked with scaling self-healing polymers for industrial manufacturing. Identify potential process-induced defects, such as incomplete crosslinking, heterogeneity, or premature bond dissociation, and propose design criteria to maintain uniform dynamic functionality. Discuss processing conditions, such as temperature profiles, solvent removal, and curing kinetics, and analyze their effect on final healing performance.

46. A hybrid network incorporates metal–ligand coordination and DA chemistry. Discuss how selective ligand exchange could enable stimulus-responsive healing, including competitive binding, network relaxation, and kinetic control. Predict potential interferences between mechanisms and propose molecular design strategies to maintain orthogonality while maximizing recovery efficiency.

47. A polymer is engineered to self-heal in ambient conditions. Model the temperature dependence of healing rate using Arrhenius principles, incorporating the effect of polymer Tg, segmental mobility, and free-volume distribution. Predict healing timescales at various temperatures and propose modifications, such as plasticizers or flexible linkers, to accelerate repair while retaining mechanical integrity.

48. A composite material with electrospun DA interleaves experiences microcrack formation. Using continuum mechanics and finite-element concepts, predict stress redistribution after healing and evaluate the effect on structural integrity. Consider the influence of interleaf placement, thickness, crosslink density, and polymer modulus on macroscopic mechanical recovery.

49. A supramolecular polymer exhibits humidity-dependent healing. Design a water-resistant variant by modifying hydrogen-bond density, backbone hydrophobicity, and network topology. Predict its mechanical performance, healing efficiency, and environmental stability, and analyze trade-offs between water resistance and dynamic bond accessibility for effective repair.

50. A polymer incorporates sensing motifs that trigger autonomic self-healing. Develop a computational framework to predict the relationship between damage severity, sensor activation threshold, and healing kinetics. Include consideration of signal transduction, network relaxation, diffusion of reactive species, and environmental factors, and propose experimental validation strategies to quantify predictive accuracy.

Chapter 35

CHEMOMETRICS: DATA-DRIVEN CHEMISTRY

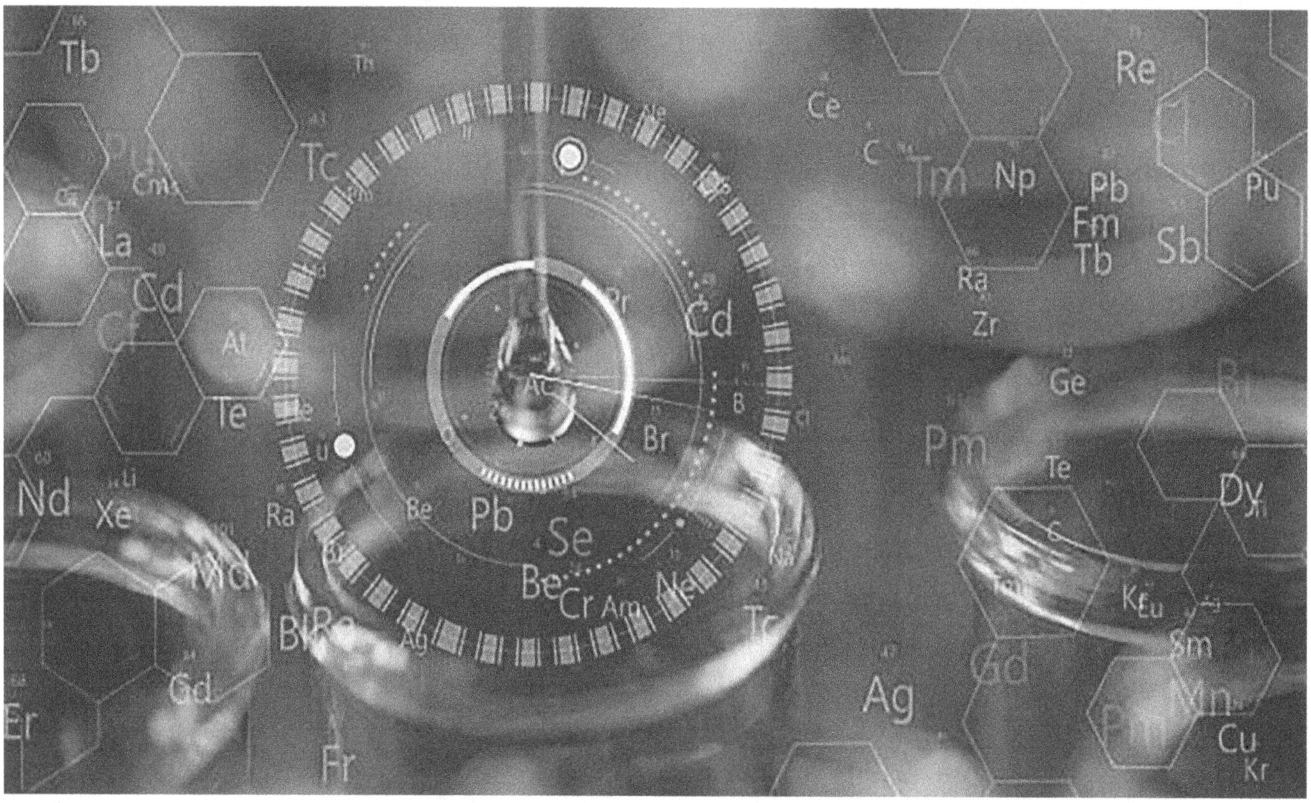

Chemometrics has found profound applications in pharmaceutical research and development, where the complexity of chemical and biological systems necessitates advanced data-driven approaches. A notable case study involves the optimization of a multi-step synthetic route for an active pharmaceutical ingredient (API) with several chiral centers. Traditional trial-and-error experimentation in such systems can be prohibitively time-consuming and costly, often requiring hundreds of reaction variations to achieve optimal yield and stereoselectivity. By applying chemometric techniques such as design of experiments (DoE) combined with multivariate analysis, researchers can systematically evaluate the influence of multiple factors simultaneously. In one example, reaction temperature, solvent composition, catalyst loading, and reaction time were varied according to a factorial experimental design. The resulting dataset, comprising hundreds of reaction outcomes, was analyzed using partial least squares regression to model the relationship between process parameters and product yield. The chemometric model revealed subtle interactions, such as a synergistic effect between solvent polarity and catalyst concentration, which would have been difficult to detect using univariate analysis. By predicting the optimal conditions computationally and validating experimentally, the team reduced experimental effort by over 60% while achieving consistently high yields and enantiomeric excess. This case highlights how chemometrics transforms pharmaceutical synthesis from an empirical endeavor into a rational, predictive process.

In environmental chemistry, chemometrics plays a crucial role in monitoring and understanding complex pollutant systems. For example, the assessment of water quality in a river impacted by industrial effluents often involves multivariate measurements, including heavy metal concentrations, organic contaminants, pH, dissolved oxygen, and microbial load. A case study of a river contaminated with textile industry effluents employed principal component analysis to reduce the dimensionality of the dataset and identify patterns of pollution. PCA revealed that variations in heavy metal concentrations were strongly correlated with the proximity to industrial discharge points, whereas organic contaminant levels were influenced by seasonal variations and upstream agricultural activities. Hierarchical clustering further classified monitoring sites into groups with similar pollution profiles, enabling targeted remediation strategies. Beyond descriptive analysis, partial least squares regression models were constructed to predict the concentration of key pollutants based on spectroscopic measurements, facilitating rapid in-field monitoring using portable UV-Vis and near-infrared (NIR) instruments. This approach not only accelerated environmental assessment but also enabled proactive regulatory intervention, demonstrating the tangible societal impact of chemometrics.

Materials science provides another rich domain for chemometric application, particularly in the design and characterization of functional materials. A striking case involves the development of perovskite-based solar cells, where the performance of the device depends on a delicate balance of compositional, structural, and processing parameters. Researchers faced a combinatorial explosion of possible formulations, as variations in cation composition, halide ratios, solvent additives, and annealing conditions could each influence the resulting photovoltaic efficiency. To navigate this complexity, chemometric approaches such as multivariate curve resolution and response surface modeling were employed. Spectroscopic and X-ray diffraction data for hundreds of samples were analyzed to identify latent variables corresponding to crystallinity, defect density, and phase purity. PLS regression was then used to correlate these structural features with device efficiency. The resulting model not only predicted the most promising compositions but also provided mechanistic insights, revealing that certain halide combinations promoted defect passivation, leading to improved charge carrier lifetimes. By integrating chemometrics with high-throughput experimentation, the researchers were able to rapidly optimize the material composition, ultimately achieving record-breaking solar cell efficiencies. This example illustrates how data-driven analysis can accelerate the discovery and optimization of complex functional materials.

In chemical process optimization, chemometrics enables real-time monitoring and control, enhancing efficiency, safety, and product quality. One instructive case study is the implementation of multivariate statistical process control (MSPC) in the continuous production of fine chemicals. In this process, multiple sensors monitored temperature, pressure, flow rate, reactant concentration, and spectroscopic signatures. Traditional univariate control charts failed to capture the interplay among these variables, leading to occasional off-spec batches. By applying PCA-based MSPC, engineers created a multivariate model representing the normal operational space of the process. Deviations from this space were detected in real time, triggering automatic adjustments to maintain optimal conditions. For instance, the model identified a previously unnoticed correlation between feedstock moisture content and reactor temperature, allowing preemptive adjustment of heating profiles. This not only reduced product variability but also minimized waste and energy consumption. The success of this approach demonstrates the transformative potential of chemometrics in industrial chemistry, where data-driven control leads to more robust, efficient, and sustainable processes.

In the field of metabolomics, chemometrics enables the analysis of extremely complex biological matrices containing thousands of metabolites with varying abundances and chemical properties. A case study in the investigation of plant secondary metabolites involved NMR and mass spectrometric profiling of hundreds of plant extracts under different growth conditions. Multivariate techniques, including PCA and orthogonal PLS, were applied to identify patterns in metabolite composition associated with environmental stressors, such as drought or pathogen exposure. Chemometric models revealed that specific metabolites, previously considered minor components, played a central role in stress response pathways. Furthermore, clustering analyses allowed the classification of plant cultivars based on their metabolic profiles, facilitating breeding programs aimed at improving stress tolerance. The integration of chemometrics with high-resolution analytical techniques in metabolomics exemplifies the power of data-driven chemistry to uncover biological insights that would be otherwise inaccessible.

Chemometric analysis is also pivotal in the characterization of polymers and complex macromolecular systems. In a study of polymer blends designed for biomedical applications, near-infrared spectroscopy was employed to monitor phase composition and molecular interactions in real time. A large dataset encompassing spectra from blends with varying monomer ratios and processing conditions was analyzed using multivariate curve resolution, which separated overlapping spectral contributions into individual component profiles. This allowed the identification of subtle compositional heterogeneity and provided quantitative information on phase segregation. Subsequent correlation of chemometric results with mechanical testing data revealed that specific microstructural features governed tensile strength and elasticity, guiding the rational design of polymer blends with tailored properties. This case illustrates the ability of chemometrics to link molecular-level information to macroscopic performance, bridging the gap between characterization and materials engineering.

In pharmaceutical quality control, chemometrics enhances both speed and reliability. One example involves the use of near-infrared spectroscopy coupled with PLS regression to predict active ingredient concentrations in tablets without the need for destructive testing. The calibration model, trained on a diverse set of tablet compositions and manufacturing conditions, allowed rapid prediction of dosage uniformity across production batches. Outliers were detected efficiently, reducing the risk of substandard or non-compliant products reaching the market.

Beyond simple quantification, chemometric models also provided insights into the sources of variability, such as mixing efficiency or granule size distribution, enabling process improvements and regulatory compliance. The case highlights the dual role of chemometrics in ensuring both analytical precision and operational efficiency in pharmaceutical manufacturing.

Another significant application is in catalytic reaction development, where chemometrics aids in optimizing complex catalytic systems with multiple co-factors and competing reaction pathways. For instance, in the development of asymmetric hydrogenation catalysts, researchers faced a vast combinatorial space of ligands, metal precursors, solvents, and reaction conditions. By systematically varying these parameters using a designed experimental matrix and applying PLS regression to correlate conditions with enantiomeric excess and yield, chemometric analysis identified subtle non-linear interactions that enhanced selectivity. Subsequent iterative experimentation guided by the model allowed researchers to achieve superior catalytic performance with fewer experiments than traditional methods. This demonstrates how chemometrics can provide both predictive power and mechanistic insight, streamlining catalyst development and reducing experimental burden.

The versatility of chemometrics extends to emerging fields such as environmental metabolomics and exposomics, where human exposure to complex chemical mixtures is studied. For example, in assessing the impact of urban pollution on human health, large datasets comprising hundreds of biomarkers, pollutants, and lifestyle variables were analyzed using PCA and hierarchical clustering. The models revealed associations between exposure profiles and specific metabolic alterations, highlighting previously unknown pathways of pollutant-induced stress. Such insights inform public health interventions and guide regulatory policies. Chemometric integration of high-dimensional environmental and biological data underscores the ability of data-driven chemistry to address pressing societal challenges.

Across all these domains, chemometrics relies heavily on careful data curation, preprocessing, and validation. Case studies consistently emphasize that the predictive power of chemometric models depends on the quality and representativeness of the input data. Techniques such as cross-validation, bootstrapping, and external validation sets ensure model robustness, while preprocessing methods like normalization, baseline correction, and noise reduction improve signal fidelity. The case studies collectively illustrate that chemometrics is not merely a set of statistical tools but a comprehensive framework for translating raw chemical data into actionable knowledge, spanning discovery, process optimization, quality control, and environmental assessment.

Chemometrics continues to expand its influence as chemical systems become more complex, multidimensional, and tightly integrated with real-time sensing technologies. Modern analytical chemistry no longer deals with isolated measurements but with elaborate spatiotemporal datasets emerging from hyphenated spectroscopies, multi-omics workflows, multisensor arrays, autonomous platforms, and high-throughput experimentation. As a result, chemometric methodologies have evolved beyond traditional multivariate statistics into a unified chemical data science ecosystem that incorporates machine learning, digital twins, uncertainty quantification, and explainable modeling. This evolution is reshaping how chemists interrogate matter, interpret signals, and design new systems with molecular-level precision. In high-dimensional spectroscopy, chemometrics provides the tools needed to extract meaningful information from massive datasets, including multi-wavelength absorption cubes, hyperspectral images, multidimensional NMR matrices, and ultrafast pump–probe measurements. Multiway and tensor-based approaches such as PARAFAC, Tucker decomposition, and multilinear PLS preserve structural relationships in spectral–temporal–spatial data, enabling the resolution of overlapping species, identification of latent chemical components, and quantification of analytes in complex matrices. Sparse and regularized spectroscopic models, incorporating penalties like LASSO, ridge regression, and elastic nets, help manage the curse of dimensionality while enhancing interpretability. These methods improve vibrational feature extraction, maintain calibration robustness across instruments, and facilitate compressed sensing strategies that reconstruct full spectra from undersampled measurements—a capability especially transformative in NMR, mass spectrometry imaging, and infrared microscopy.

The integration of chemometrics with robotics and algorithmic decision-making has accelerated the rise of autonomous and self-optimizing laboratories. Closed-loop experimentation platforms use chemometric analyses combined with Bayesian optimization, Gaussian process regression, and reinforcement learning to iteratively design and refine experiments. Such systems have dramatically enhanced reaction discovery, catalyst screening, and polymer optimization, often converging on ideal conditions far more rapidly than human-guided workflows. Real-time process analytical technology (PAT) further extends chemometrics into continuous manufacturing and flow chemistry, where in-line NIR, Raman, UV–Vis, and mass spectrometric probes generate continuous data streams. Multivariate calibration and adaptive PLS models translate these signals into quantitative reaction metrics, while multivariate statistical process control (MSPC) detects anomalies, identifies deviations from design space, and maintains product quality. Chemometrics, therefore, forms a core component of Quality-by-Design (QbD) strategies in pharmaceutical development and scalable chemical manufacturing.

Beyond instrumentation, chemometrics has become a foundational tool for mechanistic elucidation and kinetic modeling. Global-fitting strategies supported by singular value decomposition, PCA, and multilinear factorization reveal hidden intermediates, resolve parallel versus sequential pathways, and extract microkinetic parameters otherwise inaccessible through direct measurement. These techniques are increasingly complemented by machine learning models that classify kinetic profiles, infer rate laws, or suggest plausible mechanistic hypotheses based on patterns learned from literature and experimental databases. Reaction network inference is a rapidly advancing domain in which chemometric algorithms reconstruct complex chemical reaction networks directly from time-series datasets. Applications include deciphering catalytic cycles, mapping polymerization pathways, analyzing combustion chemistry, and quantifying metabolic flux in biological systems. When integrated with digital twins, these methods allow chemists to simulate reaction behavior under diverse conditions, predict bottlenecks, and optimize processes before performing physical experiments.

In materials science, chemometrics accelerates discovery by linking high-throughput characterization data to properties and design principles. Massive datasets generated from XRD, XPS, solid-state NMR, AFM, and electron microscopy often contain latent structural information buried beneath noise and experimental variability. Techniques like non-negative matrix factorization, independent component analysis, clustering, and unsupervised manifold learning reveal phase fractions, compositional gradients, defect densities, nanodomain distributions, and other structural features that correlate with material performance. Predictive models including kernel ridge regression, PLS, support vector machines, and neural networks then map these descriptors to functional properties such as conductivity, catalytic activity, mechanical toughness, or optical response. The resulting materials informatics frameworks enable accelerated discovery of battery electrolytes, perovskite absorbers, nanostructured catalysts, and biomimetic polymers by learning from both experimental results and computed descriptors derived from quantum chemistry.

Environmental and geochemical applications of chemometrics also continue to grow as global monitoring networks produce increasingly large and complex datasets. Multisensor fusion integrates data from remote sensing instruments, atmospheric monitoring devices, water quality probes, and soil analysis platforms. Multivariate calibration corrects for environmental interferences, sensor drift, and matrix effects, enabling robust quantification of pollutants across diverse conditions. Cluster analysis, factor analysis, PCA, and isotopic mixing models play a central role in source attribution, helping identify anthropogenic versus natural pollutant origins, reconstruct atmospheric transport pathways, and determine the provenance and degradation states of microplastics. Spatial chemometric methods incorporate geostatistical tools like kriging and variogram analysis, making it possible to map chemical distributions across landscapes and correlate them with ecological, geological, or anthropogenic processes.

In multi-omics and systems biology, chemometrics enables the integration and interpretation of high-dimensional data from LC–MS, GC–MS, NMR, and FTIR metabolomics platforms. Dimensionality reduction techniques isolate relevant features from thousands of variables, while peak deconvolution, baseline correction, and retention time alignment ensure analytical consistency. Supervised models such as OPLS-DA, discriminant analysis, and hierarchical PLS identify biomarkers, characterize metabolic states, and distinguish between disease phenotypes, dietary effects, or toxicological exposures. Network-based chemometric approaches examine correlations and covariance structures to reveal enzyme bottlenecks, feedback loops, channeling behaviors, and regulatory motifs within metabolic networks, bridging chemistry, biochemistry, and computational biology into a unified systems-level framework.

Looking forward, chemometrics is evolving toward models that are explainable, predictive, autonomous, and deeply integrated with chemical theory. Explainable AI tools such as SHAP values, local surrogate models, and counterfactual reasoning help chemists understand why a model makes a particular prediction by linking outcomes to specific spectral regions, structural descriptors, or mechanistic features. Hybrid chemometric–quantum models incorporate descriptors derived from quantum chemistry—electron density maps, orbital-based fingerprints, solvation energies, and conformational landscapes—to improve predictions of reactivity, stability, and spectral behavior. These approaches combine the strengths of physics-based and data-driven modeling, particularly in reaction prediction and molecular design. Ultimately, the convergence of chemometrics with automation, robotics, real-time analytics, and machine learning is ushering in an era of fully autonomous chemical discovery. These systems will not only conduct experiments but will also interpret data, propose hypotheses, optimize conditions, and design molecules or materials in an iterative, self-directed manner. As such, chemometrics stands poised to become the computational backbone of chemical research, ensuring that data-driven insight, predictive modeling, and automated decision-making remain central to the future of both fundamental chemistry and industrial innovation.

Chapter 35: Word Problems

1. A pharmaceutical research team is optimizing a multi-step asymmetric synthesis of an active pharmaceutical ingredient (API) containing three stereocenters. The team systematically varies temperature, solvent polarity, catalyst type and loading, and reaction time in order to maximize both overall yield and enantiomeric excess. Using partial least squares (PLS) regression on the collected dataset, they observe a subtle but significant interaction between solvent polarity and catalyst concentration that influences enantioselectivity in an unexpected way. Discuss how the chemometric model can be used to guide further experimental design, such as suggesting new combinations of reaction conditions, prioritizing experiments that reduce uncertainty in poorly explored regions, and potentially predicting conditions that achieve high yield and high stereoselectivity without resorting to exhaustive trial-and-error screening.

2. In a design-of-experiments (DoE) study for pharmaceutical synthesis, hundreds of reactions are carried out across multiple factors, including temperature, pressure, reagent ratios, and catalyst types. The resulting dataset contains significant noise arising from instrument variability, pipetting errors, and minor

environmental fluctuations. Prior to applying multivariate analyses like PLS or PCA, what preprocessing steps—such as centering, scaling, normalization, outlier detection, and noise filtering—should be considered, and how do these steps improve model reliability, interpretability, and the ability to uncover true underlying chemical relationships?

3. During a river water quality assessment, principal component analysis (PCA) identifies patterns of heavy metal concentrations that correlate strongly with nearby industrial discharge points. Explain how hierarchical clustering could be applied to group sampling locations or pollutants into meaningful clusters for designing targeted remediation strategies. Consider how dendrograms could reveal pollutant source similarities, inform site prioritization for clean-up, and guide regulatory decisions while minimizing unnecessary interventions.

4. A chemometrician is developing a PLS regression model to predict concentrations of pollutants in water based on UV-Vis spectra. Some of the spectral datasets exhibit baseline drift, caused by instrumental instability or light scattering. Discuss preprocessing approaches such as baseline correction, Savitzky–Golay smoothing, multiplicative scatter correction, and derivative spectra, explaining how each method addresses baseline drift and improves predictive accuracy and model robustness.

5. Researchers working on perovskite solar cells collect hundreds of X-ray diffraction (XRD) patterns and complementary spectroscopic measurements across multiple synthesis conditions. Their goal is to identify latent variables that reflect crystallinity, defect density, and phase purity. Explain how multivariate curve resolution (MCR) can be applied to deconvolute overlapping features, quantify contributions from different phases, and relate structural variations to device efficiency, potentially enabling rapid optimization of processing conditions.

6. In the optimization of a continuous chemical process, PCA-based multivariate statistical process control (MSPC) identifies a correlation between feedstock moisture and reactor temperature fluctuations. Propose a strategy for real-time process adjustments that accounts for these correlations, such as closed-loop feedback, predictive correction models, or adaptive control charts, to maintain product consistency and reduce the risk of off-specification batches.

7. Plant metabolomics data reveal that minor metabolites, often below standard detection thresholds, play critical roles in stress response pathways. Discuss how orthogonal PLS (O-PLS) and hierarchical clustering can be used to extract these subtle signals, distinguish meaningful biological variations from noise, and classify plant cultivars according to stress tolerance. Include considerations for data normalization, scaling, and cross-validation in ensuring reproducible metabolite identification for breeding programs.

8. A polymer blend for biomedical applications is analyzed using near-infrared (NIR) spectroscopy, where overlapping spectra of different monomer units complicate interpretation. Explain how multivariate curve resolution can separate individual spectral contributions, quantify component concentrations, and link these spectral features to mechanical properties such as tensile strength and elasticity. Discuss challenges in ill-posed solutions and strategies for incorporating constraints based on known chemical structure.

9. During pharmaceutical tablet production, NIR spectroscopy is used to predict active pharmaceutical ingredient (API) content across production batches. Describe how cross-validation and external validation can assess model performance, prevent overfitting, and ensure robustness across batches with varying excipient compositions and minor process variations. Include discussion of statistical metrics such as RMSEP, R^2, and bias for evaluating predictive accuracy.

10. In asymmetric hydrogenation catalyst optimization, ligand and solvent combinations exhibit complex, non-linear interactions affecting enantiomeric excess. How can chemometric approaches like response surface modeling, PLS, and interaction term analysis help identify key factors, predict optimal conditions, and reduce the number of experimental runs, while maintaining interpretability of mechanistic effects?

11. Urban pollution metabolomics generates high-dimensional datasets comprising hundreds of biomarkers. PCA and hierarchical clustering reveal metabolic alterations associated with specific exposure profiles. Discuss how these chemometric insights could inform public health policy, for instance, by identifying subpopulations at higher risk, guiding urban planning, and prioritizing targeted interventions to reduce exposure to specific pollutants.

12. In a multi-sensor environmental monitoring study, data from pH sensors, turbidity meters, and metal detectors exhibit widely varying signal-to-noise ratios and sampling frequencies. Explain how normalization, baseline correction, and noise-reduction techniques can improve model reliability in subsequent chemometric analyses, ensuring that environmental trends are captured accurately and spurious variations are minimized.

13. A chemometric analysis integrates multiway NMR and hyperspectral imaging data collected from heterogeneous biological samples. Compare tensor decomposition methods, such as PARAFAC and Tucker decomposition, in preserving spatial, spectral, and temporal correlations. Discuss how these approaches enable identification of subtle chemical patterns that would be missed by conventional two-dimensional PCA or PLS.

14. Sparse regression models with LASSO penalties are applied to high-dimensional NIR datasets to predict chemical properties of complex mixtures. Explain how enforcing sparsity reduces overfitting, facilitates interpretability by selecting relevant spectral regions, and allows chemists to link specific absorption bands to molecular features or impurities, particularly when the number of predictors far exceeds the number of observations.

15. Closed-loop autonomous laboratories combine chemometric analysis with Bayesian optimization to iteratively design experiments. Describe how this approach accelerates reaction discovery compared to traditional sequential experimentation by focusing on regions of the design space with high uncertainty or predicted optimal outcomes, while continuously updating models with new data to refine predictions.

16. In continuous flow synthesis, in-line NIR and Raman probes generate large streams of multivariate spectral data. Discuss how adaptive PLS models and MSPC techniques can detect deviations from expected behavior in real time, trigger alarms or automated process adjustments, and maintain product quality within the defined design space.

17. For kinetic modeling of complex reactions, singular value decomposition (SVD) is applied to time-resolved spectroscopic data. Explain how SVD can reveal hidden intermediates, distinguish sequential versus parallel pathways, and provide

insight into reaction mechanisms, particularly when direct observation of species is challenging due to spectral overlap or low concentration.

18. Chemometric analysis of polymerization pathways utilizes multilinear factorization techniques. Discuss how these methods can extract microkinetic parameters, such as rate constants and chain-length distributions, from multidimensional datasets, thereby improving mechanistic understanding and enabling prediction of polymer properties under varying reaction conditions.

19. In materials informatics, kernel ridge regression is used to link XRD, AFM, and chemical composition descriptors to functional properties such as electronic conductivity and mechanical strength. How does this chemometric approach facilitate rational materials selection, guide synthesis, and prioritize experimental validation of promising candidates?

20. Environmental chemometrics involves integrating soil, water, and atmospheric data to assess pollutant sources. Describe how cluster analysis combined with isotopic mixing models can distinguish anthropogenic from natural contributions, identify spatial patterns of contamination, and inform remediation or regulatory strategies.

21. Multi-omics metabolomics datasets often contain thousands of variables with strong correlations and noise. Explain how OPLS-DA can isolate biomarkers indicative of disease states, classify phenotypes, and control for confounding factors while maintaining model interpretability in the context of high-dimensional data.

22. A chemometrician applies SHAP (SHapley Additive exPlanations) values to explain a predictive model for reaction yield. Discuss how linking model outcomes to spectral regions or chemical descriptors enhances mechanistic insight, builds trust in model predictions, and informs targeted experimentation.

23. Hybrid chemometric-quantum models predict molecular stability using both electron density descriptors and experimental spectral data. Explain how combining physics-based quantum chemical features with data-driven chemometric models improves predictive accuracy and allows rationalization of trends in chemical reactivity.

24. Autonomous chemical laboratories iteratively propose and test novel reactions. Discuss how reinforcement learning, integrated with chemometric modeling, can guide optimization in complex catalytic systems by balancing exploration of new chemical space with exploitation of known high-performing conditions.

25. A digital twin simulates a multistep reaction network under varying conditions. Explain how chemometric techniques can analyze the simulated datasets to identify reaction bottlenecks, predict optimal operating windows, and reduce experimental effort while maintaining confidence in predicted yields and selectivities.

26. Non-negative matrix factorization (NMF) is applied to XPS and NMR datasets collected from nanostructured catalysts. Describe how NMF can reveal latent features such as defect density, surface composition gradients, and active site distributions, providing insights for targeted catalyst design.

27. Multi-sensor fusion combines remote sensing data, in-field probes, and laboratory analyses for environmental monitoring. How does multivariate calibration correct for sensor interferences, harmonize measurements, and enable reliable quantification of pollutants across different spatial and temporal scales?

28. Chemometric analysis is applied to hyperspectral images of heterogeneous samples containing overlapping chemical species. Explain how multilinear PLS and tensor-based approaches can deconvolute spectral mixtures, identify spatially resolved chemical distributions, and provide quantitative maps of component concentrations.

29. A materials scientist develops predictive models to optimize battery electrolytes. Discuss the role of supervised machine learning, feature selection, and chemometric modeling in connecting compositional descriptors to ionic conductivity, electrochemical stability, and thermal properties, facilitating accelerated electrolyte design.

30. In assessing river pollution, geostatistical kriging is applied to chemometric datasets integrating chemical concentrations and environmental parameters. Explain how spatial correlations are incorporated to produce high-resolution pollutant distribution maps, guide monitoring strategies, and target remediation interventions.

31. Multi-omics integration in systems biology relies on chemometric network analysis. Describe how covariance structures, correlation networks, and partial correlation models can reveal enzyme bottlenecks, feedback loops, and regulatory motifs in complex metabolic pathways.

32. For FTIR metabolomics datasets, preprocessing steps such as peak deconvolution, retention alignment, and baseline correction are crucial. Discuss how these steps affect downstream supervised chemometric modeling, including PLS and OPLS, by reducing noise, correcting drift, and improving quantitative accuracy.

33. Chemometrics is used to infer reaction networks from time-series datasets obtained from catalytic or polymerization reactions. Explain how multivariate modeling distinguishes between catalytic cycles, polymerization pathways, and combustion mechanisms by revealing correlated spectral or concentration patterns over time.

34. A chemometric study examines the effect of mixed solvent compositions on a multi-step organic synthesis. Explain how response surface modeling can predict optimal solvent ratios that maximize reaction yield, selectivity, and solubility while minimizing byproducts, and discuss strategies for experimental validation.

35. In perovskite solar cell research, PLS regression correlates structural defects detected by XRD and microscopy with device efficiency. Describe how the model can reveal which halide or additive combinations improve defect passivation and inform rational material engineering strategies.

36. High-throughput screening produces hundreds of polymer blend samples characterized by NIR and mechanical testing. Discuss how multivariate curve resolution links spectral features to macroscopic properties such as tensile strength, flexibility, and thermal stability, and how this guides formulation optimization.

37. Chemometric analysis in pharmaceutical manufacturing identifies outlier tablet batches. Explain how hierarchical clustering complements PLS by grouping similar batches, identifying anomalous patterns, and guiding quality control interventions before products reach consumers.

38. An environmental metabolomics study investigates the impact of urban pollution on human metabolism. Describe how hierarchical clustering can detect subpopulations with distinct exposure profiles, uncover metabolic pathways affected by pollutants, and inform targeted public health interventions.

39. In reaction optimization, Gaussian process regression guides closed-loop experimentation. Explain how chemometric reasoning balances exploration (testing uncertain regions) and exploitation (refining promising conditions) to efficiently identify optimal reaction parameters while reducing experimental load.

40. NMR metabolomics datasets are often undersampled due to instrument time constraints. Discuss how compressed sensing combined with sparse chemometric models can reconstruct full spectra, maintain spectral resolution, and extract meaningful metabolite information for downstream analysis.

41. A multivariate chemometric model monitors catalyst performance in real time during a continuous reaction. Explain how PCA-based MSPC can detect deviations, identify root causes of drift or deactivation, and suggest preemptive adjustments to maintain reaction efficiency.

42. In polymeric biomaterials, spectroscopic data often show overlapping signals from multiple monomers and additives. Discuss how multivariate curve resolution quantifies individual components, guides formulation tuning, and links compositional variation to mechanical and thermal properties.

43. High-dimensional environmental datasets often include temperature, pH, heavy metals, organic contaminants, and nutrient levels. Explain how PCA reduces dimensionality while preserving patterns relevant to pollution assessment, regulatory compliance, and ecosystem health monitoring.

44. A chemometric study integrates spectroscopic data and computational quantum chemistry descriptors. Describe how hybrid models enhance prediction of reaction kinetics and thermodynamics compared to conventional regression by combining mechanistic insight with data-driven correlations.

45. Multiway tensor analysis is applied to ultrafast pump-probe spectroscopy of transient chemical intermediates. Explain how preserving temporal, spectral, and spatial correlations improves identification of short-lived species, elucidates reaction pathways, and enables mechanistic interpretation.

46. Chemometric models guide autonomous labs in predicting outcomes of novel chemical reactions. Discuss how explainable AI techniques provide transparency in predictions, increase user confidence, and allow chemists to rationalize model-suggested experimental designs.

47. In perovskite research, spectral data from hundreds of samples are analyzed to understand how additive concentrations, crystallinity, and processing conditions affect device efficiency. Discuss how multivariate regression reveals interactions and informs materials engineering strategies.

48. Chemometric analysis links LC-MS metabolomics data with environmental exposure metrics. Explain how supervised learning approaches can detect subtle associations between pollutants and metabolic perturbations, accounting for confounding variables and inter-individual variability.

49. In continuous pharmaceutical manufacturing, multivariate calibration of NIR spectra is required to monitor product quality. Discuss the role of adaptive PLS in compensating for instrument drift, process variability, and changes in raw material batches, ensuring consistent product specifications.

50. A chemometric pipeline integrates robotics, real-time analytics, and reinforcement learning to autonomously optimize complex chemical and material systems. Explain how these elements work synergistically, from data acquisition to model updating to experimental suggestion, to achieve efficient, autonomous discovery and optimization of target properties.

Chapter 36

Astrochemistry: Complex Molecules in Space

The chemical evolution of interstellar clouds can be visualized as an intricate network of reactions, where simple molecules act as precursors to more complex species. For instance, the transformation of CO into methanol and related organics follows a sequential pathway on icy dust grains: CO first reacts with hydrogen atoms to form HCO, which further hydrogenates to formaldehyde (H_2CO), and eventually produces methanol (CH_3OH). This process is not linear; side reactions generate radicals such as CH_3O and CH_2OH, which can recombine to yield methyl formate ($HCOOCH_3$) and dimethyl ether (CH_3OCH_3), both of which have been detected in hot molecular cores. In parallel, gas-phase ion-molecule reactions complement grain-surface chemistry. For example, CH_5^+ can transfer a proton to neutral molecules like NH_3 or H_2O, forming NH_4^+ or H_3O^+, which subsequently recombine to generate neutral species. These intertwined networks, operating under low temperatures and pressures, demonstrate how molecular complexity can emerge without requiring conventional laboratory conditions.

Prebiotic chemistry in interstellar space is increasingly supported by the detection of amino acids and sugar precursors. Glycine, the simplest amino acid, has been tentatively identified in the Sgr B2(N) region, while glycolaldehyde and ethylene glycol ($HOCH_2CH_2OH$) have been confirmed in hot molecular cores. Laboratory simulations indicate that such molecules can form through the radical-radical recombination of simpler species on icy grains, facilitated by UV photolysis and cosmic-ray irradiation. For example, the combination of NH_2 and CH_2OH radicals can produce amino-acetonitrile, a precursor to glycine. Similarly, formaldehyde polymerization on ice surfaces can yield sugars or sugar-like compounds. These processes highlight that the fundamental chemical precursors of life are not restricted to planets but are widespread across the cosmos, potentially delivered to planetary surfaces via comets, asteroids, and dust grains.

Comparative studies of molecular clouds reveal the sensitivity of chemistry to environmental conditions. Cold, quiescent clouds like TMC-1 favor the formation of long carbon chains and cyanopolyynes, while warmer regions such as Orion KL, with temperatures exceeding 100 K, promote the sublimation of ices and the prevalence of more complex organics. Dense clouds shielded from UV radiation preserve fragile molecules, whereas regions exposed to high-energy photons experience active photochemistry that generates radicals and ions. Observations of L483 and L1527, two low-mass star-forming cores, demonstrate contrasting chemical profiles: L483 is rich in carbon-chain species, whereas L1527 exhibits higher abundances of oxygen-containing organics like methanol and formamide. These variations underscore that molecular synthesis is intimately tied to physical parameters, including temperature, density, radiation flux, and the dynamical evolution of the cloud.

The interplay between gas-phase and grain-surface chemistry can be further illustrated by sulfur- and nitrogen-bearing molecules. H_2S, formed via hydrogenation of sulfur atoms on ices, can be photodissociated to yield HS and S radicals, which participate in the formation of SO_2, OCS, and more complex sulfur organics. Similarly, the formation of nitrogenous molecules such as NH_3, HCN, and CH_3CN involves both surface hydrogenation and gas-phase ion-neutral reactions. The coexistence of multiple pathways enhances chemical diversity and allows the formation of species that are otherwise kinetically inaccessible under interstellar conditions. Observational studies using ALMA have detected gradients of sulfur- and nitrogen-bearing molecules in protoplanetary disks, revealing that these networks are active not only in clouds but also in the environments where planets form.

Isotopic labeling studies provide additional insight into chemical mechanisms. For instance, laboratory experiments using ^{13}C-labeled and ^{15}N-labeled precursors demonstrate that isotopic substitution can influence reaction branching ratios on ice surfaces. Such studies are critical for interpreting astronomical observations, where isotopic ratios in molecules like HCN, H_2CO, and NH vary across clouds and disks. The measured $^{12}C/^{13}C$ and $^{14}N/^{15}N$ ratios in Sgr B2(N) and Orion KL reveal both inheritance from earlier cloud stages and in situ fractionation, providing a chemical "fingerprint" of molecular evolution. Similarly, deuterium enrichment in molecules like D_2CO and CD_3OH indicates that cold, dense conditions favor isotopic substitution, offering clues about the molecular history of solar system material, including comets and meteorites.

The formation of aromatic and heterocyclic compounds illustrates the increasing complexity achievable in interstellar chemistry. Polycyclic aromatic hydrocarbons (PAHs) serve as intermediates in the synthesis of nitrogen-containing heterocycles such as quinoline and pyrimidine, which are of direct prebiotic relevance. Laboratory simulations show that UV-irradiated ices containing PAHs and simple organics can generate nitrogen-substituted rings, potentially leading to the formation of nucleobase analogs. Similarly, fullerenes and carbon chains can fragment under irradiation to provide reactive species that further enrich the chemical inventory. The presence of such molecules in protoplanetary disks suggests that planets may inherit preassembled organic structures capable of contributing to early biochemistry.

Recent observations of exoplanetary atmospheres provide an additional perspective on astrochemical evolution. Spectroscopy of planets like HD 189733b and WASP-12b reveals the presence of water, carbon monoxide, methane, and hydrogen cyanide, indicating ongoing thermochemical and photochemical processes. Variations in molecular abundances between planets of similar mass and stellar irradiation suggest that disk chemistry, migration history, and stellar activity influence atmospheric composition. Moreover, the detection of hazes composed of complex organic aerosols hints at the in situ synthesis of prebiotic species in planetary atmospheres, extending the concept of chemical complexity beyond interstellar clouds to fully formed planetary environments.

Laboratory studies complement observational data by elucidating reaction mechanisms under interstellar analog conditions. Ultra-high vacuum chambers and cryogenic surfaces simulate the low pressures and temperatures of molecular clouds, while photon and particle irradiation mimic the effects of UV and cosmic rays. Experiments have demonstrated that radicals generated in ices can recombine to form sugars, amino acid precursors, and nucleobases, confirming theoretical predictions. For instance, the formation of formamide (NH2CHO) from NH3 and H2CO on icy surfaces has been observed under simulated interstellar irradiation, providing a plausible pathway to biologically relevant molecules. These findings underscore the synergy between laboratory chemistry, theoretical modeling, and observational astronomy in unraveling the origins of molecular complexity.

Theoretical modeling remains indispensable for understanding the evolution of complex molecular networks. Kinetic simulations integrate thousands of reactions, including gas-phase ion-molecule processes, surface-mediated hydrogenation, radical recombination, and photodissociation, to predict molecular abundances as a function of cloud age, density, and temperature. Models of hot molecular cores, for example, reproduce the observed abundances of methyl formate, ethyl cyanide, and dimethyl ether by incorporating both surface and gas-phase pathways, demonstrating the interplay between different reaction mechanisms. Similarly, simulations of protoplanetary disks reveal chemical stratification, with ices preserved in the midplane and photochemically processed species dominating surface layers, offering predictive frameworks for interpreting observational surveys.

The detection of molecules in comets and meteorites provides compelling evidence for the transfer of complex interstellar organics to planetary surfaces. Analysis of the Murchison meteorite revealed amino acids, purines, pyrimidines, and sugar derivatives, many of which are identical or closely related to molecules observed in molecular clouds. Cometary studies, including the Rosetta mission to comet 67P/Churyumov-Gerasimenko, detected glycine, methylamine, and phosphorus-bearing compounds, indicating that prebiotic chemistry can survive incorporation into small solar system bodies. These findings highlight a continuity of molecular evolution from interstellar clouds to planetary environments, reinforcing the notion that the building blocks of life are cosmically widespread.

Astrochemistry also provides insights into the chemical potential of extreme environments, such as regions subjected to shocks, outflows, and intense radiation. Shock-induced heating liberates ices from grain surfaces and triggers endothermic reactions that form molecules like SiO, H_2CO, and complex organics. Observations of molecular outflows in Orion BN/KL show enhanced abundances of COMs along shock fronts, illustrating the dynamic interplay between physical processes and chemical evolution. In addition, X-ray irradiation from young stellar objects ionizes surrounding gas, promoting the formation of radicals and ions that feed into complex reaction networks, demonstrating that chemical complexity is often driven by energetic interactions between stars and their natal clouds.

The astrochemical evolution of the universe is ultimately a story of progressive complexity, linking simple atoms to molecules capable of seeding life. From the primordial hydrogen and helium created in the Big Bang, through nucleosynthesis in stars that produces heavier elements, chemistry in cold molecular clouds and protoplanetary disks transforms these elements into a rich tapestry of organics. The detection of sugars, amino acid precursors, nucleobase analogs, and large carbonaceous molecules indicates that prebiotic chemistry is a natural consequence of interstellar evolution. By studying the mechanisms, pathways, and environments that produce these molecules, astrochemistry illuminates the chemical continuity between stars, planets, and life itself.

The synthesis of complex molecules in space has profound implications for astrobiology. If prebiotic molecules are common throughout the galaxy, then the raw materials for life are likely available on planets orbiting other stars. The transport of these molecules via comets, asteroids, and interplanetary dust may deliver them to planetary surfaces, providing the molecular scaffolding for the emergence of biology. Comparative studies of molecular clouds, protoplanetary disks, and planetary atmospheres suggest that chemical pathways are influenced by local conditions, but the fundamental processes are universal, emphasizing the broad relevance of astrochemistry to understanding the origin and distribution of life in the cosmos.

Astrochemistry, at its core, seeks to understand the chemical processes that govern the cosmos, from the dense cores of molecular clouds to the tenuous environments of diffuse interstellar space. One of the central challenges in the field is elucidating the formation and survival of molecules under extreme conditions—ultra-low temperatures (10–100 K), high-energy radiation fields, and low densities (often $<10^3$ particles/cm^3). Under such conditions, classical chemistry does not apply straightforwardly; instead, quantum tunneling, radiative association, and surface-mediated reactions on dust grains dominate molecular synthesis. For example, hydrogen molecules—the most abundant molecular species in the universe—form predominantly on the surfaces of submicron silicate or carbonaceous grains through a combination of physisorption, diffusion, and recombination, processes that cannot proceed efficiently in the gas phase at such low densities. This underscores the interplay between solid-state chemistry and gas-phase reactions, a hallmark of astrochemical networks.

The complexity of interstellar chemistry becomes particularly evident in the formation of organic molecules. Observations have revealed the presence of molecules ranging from simple diatomics like CO, CN, and OH to more complex species such as methanol (CH_3OH), formamide (NH_2CHO), and even amino acid precursors. Gas-phase reactions often initiate radical-radical processes, ion-molecule reactions, and photodissociation-induced syntheses, while grain-surface chemistry allows for sequential hydrogenation and complexation reactions. Quantum chemical calculations are increasingly critical to modeling reaction rates and pathways, particularly for reactions with high activation barriers or involving transient intermediates. The synergy of experimental data, such as cryogenic ice analog studies, with computational modeling allows for accurate simulations of these intricate reaction networks, which in turn inform astronomical observations.

Isotopic fractionation provides an additional window into cosmic chemical evolution. Variations in the ratios of isotopes such as $^{12}C/^{13}C$, $^{16}O/^{18}O$, and D/H across different molecular species reveal not only the chemical history of interstellar environments but also the physical conditions prevalent during molecular formation. Deuterium enrichment, for instance, occurs preferentially at low temperatures via exothermic ion-molecule exchange reactions, producing observable signatures in molecules like HDO, DCO^+, and DCN. Such isotopic fingerprints are invaluable for tracing the evolutionary pathways of molecular clouds and for drawing parallels between interstellar chemistry and the isotopic composition of meteorites, comets, and planetary atmospheres.

Prebiotic chemistry in space is one of the most tantalizing aspects of astrochemistry, linking molecular evolution to the potential origins of life. Laboratory simulations have demonstrated that simple ices composed of H_2O, CO, NH_3, and CH_3OH, when exposed to ultraviolet irradiation or energetic particles, can produce amino acids, nucleobases, and other biologically relevant molecules. These findings suggest that the molecular precursors to life may be widespread in the cosmos, delivered to nascent planets through comets, asteroids, or interplanetary dust. Observational campaigns have increasingly focused on detecting these prebiotic molecules in star-forming regions, protoplanetary disks, and even the cold envelopes of evolved stars, providing empirical support for the hypothesis that the building blocks of life are cosmically ubiquitous.

The chemistry of protoplanetary disks represents another frontier of astrochemical research. Within these disks, gradients of temperature, density, and irradiation create distinct chemical niches. The inner, warmer regions favor gas-phase reactions and high-temperature condensates, while the colder outer disk regions allow for extensive ice chemistry and complex organic molecule formation. Observations with ALMA and JWST have revealed ring-like structures enriched in molecules such as CO, H_2CO, and CH_3CN, hinting at chemical zoning that may influence planetesimal and planet formation. In particular, the snowlines—boundaries where volatiles condense into ices—play a critical role in determining the chemical composition of emerging planetary bodies, influencing everything from water delivery to the organic inventory of nascent planets.

Astrochemical modeling increasingly relies on multi-scale simulations that integrate hydrodynamics, radiative transfer, and chemical kinetics. These models allow researchers to predict molecular abundances, isotopic ratios, and reaction efficiencies across environments ranging from diffuse clouds to dense cores. Coupling these models with observational data provides stringent tests of chemical networks and identifies gaps in our understanding, such as the formation pathways of complex organic molecules or the survival of fragile species in high-radiation environments. Machine learning and artificial intelligence are also beginning to play a role in pattern recognition within large spectral datasets, accelerating the identification of unknown molecules and revealing previously hidden chemical correlations.

Finally, astrochemistry extends its implications to planetary atmospheres and exoplanetary science. The detection of molecules such as water vapor, methane, carbon dioxide, and ammonia in exoplanet atmospheres informs models of planetary formation, migration, and potential habitability. High-resolution spectroscopy allows for the identification of minor species and isotopologues, revealing processes such as photochemistry, atmospheric escape, and surface-atmosphere interactions. These studies create a bridge between molecular astrophysics and astrobiology, connecting the chemical evolution of the interstellar medium with the prospects for life beyond Earth.

In conclusion, astrochemistry is a profoundly interdisciplinary field, linking quantum chemistry, laboratory astrophysics, observational astronomy, and computational modeling to understand the molecular universe. As instrumentation and analytical techniques continue to advance, from ALMA's millimeter-wave sensitivity to JWST's infrared spectroscopic precision, the cosmic inventory of molecules will expand dramatically. Each discovery—from the simplest diatomics to complex prebiotic organics—provides critical insight into how chemical complexity emerges from primordial matter, laying the foundation for understanding the origins of planets, life, and the broader narrative of cosmic evolution. The chemistry of the cosmos is not merely a catalog of molecules but a dynamic, evolving network that informs our understanding of where we come from and the universal processes that shape life itself.

Chapter 36: Word Problems

1. Consider the sequential hydrogenation of CO on icy dust grains, which leads to the formation of HCO, H_2CO, and ultimately CH_3OH. Design a comprehensive kinetic model that incorporates not only the main hydrogenation pathway but also competing radical recombination processes that generate CH_3O and CH_2OH radicals. Using this model, calculate the relative abundances of CH_3OH and methyl formate ($HCOOCH_3$) after a period of 10^5 years, assuming a constant temperature of 10 K and a hydrogen atom flux of 10^4 atoms cm^{-2} s^{-1}. Discuss how variations in radical diffusion rates and surface coverage might influence these abundances.

2. A molecular cloud contains CO ice with a thickness of 100 monolayers. Assuming

a hydrogenation rate coefficient of 10^{-3} s^{-1} per adsorbed species, estimate the timescale required for 50% conversion of CO to formaldehyde (H_2CO). Further, evaluate how cosmic-ray-induced UV photons might alter this timescale by either promoting radical formation or photodissociation, and discuss the broader implications for prebiotic molecule formation in dense clouds.

3. Using the radical-radical recombination mechanism, propose a detailed reaction network for producing dimethyl ether (CH_3OCH_3) from methanol-derived radicals. Quantify the expected branching ratios if CH_3O recombines with CH_3 versus CH_2OH with CH_3O, assuming equal radical diffusion rates. Include discussion of the kinetic bottlenecks and potential effects of surface temperature and ice morphology on these branching ratios.

4. Evaluate the effect of temperature on the mobility of hydrogen atoms adsorbed on icy grains. Calculate the diffusion length of H atoms at 10 K and 50 K over a 10^4 s interval, assuming an activation energy for hopping of 100 K. Discuss the implications of this mobility for the efficiency of methanol formation, considering both Langmuir-Hinshelwood and Eley-Rideal reaction mechanisms, as well as the possibility of tunneling at very low temperatures.

5. Consider a dense molecular cloud where both gas-phase CH_5^+ reactions and ice-surface chemistry contribute to NH_4^+ formation. Derive expressions for the steady-state abundance of NH_4^+ that account for both pathways, including recombination with electrons. Discuss how variations in electron density, temperature, and cosmic-ray ionization rates might shift the relative contributions of gas-phase versus surface-mediated processes.

6. Laboratory simulations indicate that NH_2 + CH_2OH radicals can combine to form amino-acetonitrile. If NH_2 radicals have a density of 10^{-6} cm^{-3} and CH_2OH radicals have a density of 10^{-7} cm^{-3} in a molecular cloud, calculate the reaction rate and estimate the timescale required to reach 1% conversion. Extend the discussion to include possible competing reactions and the effects of radical trapping on ice surfaces.

7. Glycolaldehyde has been detected in hot molecular cores, suggesting formation on icy grains. Construct a plausible surface reaction mechanism starting from formaldehyde monomers and calculate the activation energy barrier, assuming

Langmuir-Hinshelwood kinetics. Evaluate how thermal processing, UV irradiation, and radical mobility could influence the overall formation efficiency and observed abundances in astrophysical environments.

8. Compare the formation efficiencies of glycine via ice-surface radical recombination versus gas-phase ion-molecule reactions. Quantify the relative abundances under cold (10 K) and warm (100 K) conditions, assuming rate constants of 10^{-12} cm^3 s^{-1} for surface reactions and 10^{-9} cm^3 s^{-1} for gas-phase processes. Discuss the implications for prebiotic chemistry in different astrophysical environments.

9. Analyze the competition between amino-acetonitrile formation and radical recombination pathways leading to methyl formate. Using a simplified kinetic network, determine the conditions under which one pathway dominates over the other. Justify your assumptions regarding radical concentrations, surface coverage, and temperature, and evaluate how observational constraints might inform the relative importance of these pathways in interstellar clouds.

10. Model the UV photolysis of H_2O-CO-NH_3 ice mixtures leading to the formation of amino acids. Estimate the fraction of initial CO that is converted into biologically relevant products after exposure to a UV dose of 10^{18} photons cm^{-2}, assuming a quantum yield of 0.01. Discuss how ice thickness, photon penetration depth, and radical recombination efficiencies affect the predicted yields.

11. Cold, quiescent clouds favor cyanopolyyne formation. Calculate the expected column density of HC_7N in TMC-1, given a gas density of 10^4 cm^{-3} and a temperature of 10 K over 10^5 years. Include simple gas-phase ion-neutral reaction pathways in your estimation, and consider the effects of freeze-out and depletion on long carbon-chain species.

12. Compare chemical abundances between the carbon-chain-rich source L483 and the oxygen-rich source L1527. Propose a model that links temperature, UV shielding, and ice sublimation to the observed molecular distributions, and discuss how differences in cloud age and evolutionary stage might further influence the chemistry.

13. In Orion KL, where temperatures exceed 100 K, ices sublimate into the gas phase. Estimate the enhancement in methanol abundance if sublimation increases

surface coverage by 50%. Discuss how gas-phase reactions might further modify abundances and how observational spectroscopy can be used to validate such models.

14. Quantitatively analyze the effect of UV radiation on fragile molecular species in exposed molecular clouds. Calculate the photodissociation rate for formaldehyde (H_2CO) given a photon flux of 10^8 cm^{-2} s^{-1} and a cross-section of 10^{-17} cm^2. Evaluate how self-shielding, dust extinction, and ice-mantle protection may alter this rate in realistic interstellar conditions.

15. Derive an expression for the ratio of carbon-chain to oxygen-containing organics in a molecular cloud as a function of temperature and density. Compare the model predictions with observations of L1527 and L483 and discuss how such ratios provide insights into cloud evolution and chemical pathways.

16. H_2S forms via surface hydrogenation of sulfur atoms on icy grains. Compute the equilibrium abundance of HS and S radicals under UV irradiation, assuming a dissociation rate of 10^{-5} s^{-1}. Discuss the balance between photodissociation, hydrogenation, and radical recombination, and its implications for sulfur chemistry in dense clouds.

17. OCS forms via the reaction HS + CO. Given an HS density of 10^{-7} cm^{-3} and CO of 10^{-4} cm^{-3}, estimate the OCS formation rate and the timescale to reach 50% of the observed abundance in a dense cloud. Consider competing pathways and potential destruction mechanisms in your calculation.

18. NH_3 and CH_3CN are produced via both surface hydrogenation and gas-phase reactions. Construct a coupled kinetic model to estimate their steady-state abundances under cold (~20 K) and warm (~100 K) cloud conditions. Discuss how variations in surface coverage, temperature, and UV flux might shift the relative importance of each pathway.

19. ALMA observations detect radial and vertical gradients of sulfur-bearing species (e.g., CS, H_2S, SO) in protoplanetary disks surrounding T Tauri and Herbig Ae/Be stars. Develop a reaction–diffusion model that quantitatively reproduces these gradients by incorporating radial and vertical temperature profiles, disk density structures, molecular diffusion coefficients, adsorption/desorption rates on dust

grains, and reactive fluxes. Consider both gas-phase chemistry and grain-surface reactions, including possible sulfur sequestration in refractory compounds. Evaluate how varying disk turbulence (α-viscosity parameter) and UV irradiation alter sulfur chemistry, and propose specific observational tests using line intensities, isotopologue ratios, and excitation temperatures to validate your model.

20. Discuss in detail how competing surface-mediated reactions and gas-phase pathways contribute to the chemical diversity observed in molecular clouds. Use explicit examples such as the formation of NH_3 via successive hydrogenation of N atoms on icy grains and the gas-phase synthesis of HCN through CN + H_2 reactions. Include kinetic bottlenecks, branching ratios, and the role of radical mobility and tunneling at low temperatures (~10–50 K). Analyze the influence of environmental parameters such as gas density, dust-to-gas ratio, cosmic-ray ionization rate, and temperature gradients, and explore how competition between these pathways leads to time-dependent abundances that can be traced with molecular line surveys.

21. Laboratory experiments using ^{13}C-labeled CO have demonstrated modifications in reaction branching ratios during H-atom addition on icy surfaces. Calculate the expected $^{12}C/^{13}C$ isotopic ratio in H_2CO after 10^4 s at 15 K, given an experimentally determined fractionation factor of 1.2. Incorporate surface coverage, radical encounter probability, and possible photolytic effects on fractionation. Discuss the implications of these isotopic fractionation patterns for interpreting interstellar formaldehyde and methanol observations, and explore how such measurements constrain chemical evolution models in dense cores.

22. Deuterium enrichment in methanol (CH_3OH) occurs via sequential hydrogenation/ deuteration of CO on grain surfaces. Estimate the D/H ratio in methanol formed at 10 K, assuming a gas-phase D/H ratio of 0.01, taking into account reaction barriers, tunneling probabilities, and diffusion-limited radical encounters. Discuss how the resulting fractionation patterns in CH_2DOH, CHD_2OH, and CD_3OH can serve as tracers of cold molecular cloud conditions, including insights into grain surface coverage, ice morphology, and the duration of cold phases before star formation.

23. Model the inheritance of $^{14}N/^{15}N$ isotopic ratios from molecular clouds to protoplanetary disks. Examine the sensitivity of NH_3, N_2, and HCN isotopic ratios

to in situ fractionation versus inheritance from earlier stages. Include isotope-selective photodissociation, gas-grain exchange, and potential fractionation during disk accretion. Propose observational diagnostics, such as comparing CN/HCN isotopologue ratios at different disk radii, to constrain chemical evolution from clouds to forming planetary systems.

24. Interpret deuterium fractionation in D_2CO observed in protostellar envelope L1527. Calculate the expected D_2CO/H_2CO ratio in a cold (~10 K), dense (~10^5 cm^{-3}) environment, considering sequential hydrogenation of CO, radical diffusion on ice surfaces, and competition with H/D abstraction reactions. Compare your calculation with the observed ratio (~0.1), and evaluate how variations in radical mobility, ice layering, and reaction branching influence the final isotopic ratio.

25. Evaluate the effects of isotopic substitution (e.g., D vs H) on radical mobility on icy grain surfaces. Discuss the consequences for the formation of prebiotic species such as amino-acetonitrile and formamide, including modifications to reaction kinetics, tunneling rates, and branching ratios. Consider how these isotope-dependent effects influence the spatial distribution and temporal evolution of complex organics in cold dense cores and protostellar envelopes.

26. Polycyclic aromatic hydrocarbons (PAHs) are hypothesized precursors to nitrogen-containing heterocycles such as quinoline. Propose a UV-induced reaction pathway that involves photodissociation, radical formation, and addition reactions. Calculate the expected yield for a PAH population density of 10^{-6} cm^{-3} exposed to a photon flux of 10^8 cm^{-2} s^{-1}, incorporating reaction cross sections, quantum yields, and radical lifetimes. Discuss the roles of excited-state chemistry, radical intermediates, and potential quenching pathways in controlling product formation.

27. Laboratory simulations of ices containing CH_4, NH_3, and PAHs exposed to UV irradiation produce nitrogen-substituted rings. Estimate the probability of forming a pyrimidine analog over a period of 10^5 years in a cold cloud environment, accounting for radical generation rates, diffusion coefficients on ice, recombination efficiencies, and competing radical reactions. Include discussion on how ice porosity and photon penetration depth influence overall yields.

28. Fullerenes such as C_{60} undergo fragmentation under energetic irradiation (UV photons, cosmic rays). Model the production rate of reactive carbon radicals

from C_{60} subjected to a cosmic-ray flux of 1 particle $cm^{-2}\ s^{-1}$. Discuss how these radicals contribute to the formation of complex organic molecules, considering subsequent radical-radical recombination, addition to unsaturated molecules, and incorporation into larger carbonaceous networks.

29. Quantify the potential contribution of aromatic molecules to prebiotic nucleobase formation in protoplanetary disks. Include the effect of radial temperature gradients, ice sublimation zones, radical diffusion barriers, and recombination rates. Evaluate how such spatial chemical differentiation might produce localized reservoirs of nitrogen-containing organics that could be incorporated into planetesimals and comets.

30. Compare the stability of heterocyclic molecules (e.g., pyrimidines, purines) versus linear carbon-chain molecules (e.g., HC_5N) under UV and cosmic-ray irradiation in hot molecular cores. Discuss how molecular structure, conjugation, and functional groups influence photostability and radical formation. Analyze potential pathways leading to prebiotic chemistry and the likelihood of survival of these molecules during early star formation.

31. Spectroscopic observations of the exoplanet HD 189733b reveal CH_4 and HCN in its atmosphere. Using chemical equilibrium models at 1000 K and 1 bar, estimate the CH_4/HCN ratio as a function of C/O ratio, elemental abundances, and vertical temperature gradients. Discuss how deviations from equilibrium due to photochemistry or vertical mixing could alter observable ratios.

32. Consider the effect of intense stellar irradiation on the atmosphere of WASP-12b. Calculate the photodissociation rate of H_2O for a UV flux of 10^{13} photons $cm^{-2}\ s^{-1}$, taking into account absorption cross sections, shielding effects, and photochemical reaction branching. Discuss implications for atmospheric composition, observability of key molecules, and constraints on potential biosignatures or prebiotic precursors.

33. Discuss how disk chemistry and planetary migration history influence exoplanet atmospheric composition. Construct a simplified reaction network linking disk abundances of CO, H_2O, NH_3, CH_4, and organics to observed atmospheric species. Analyze how inward or outward migration and associated accretion of icy planetesimals modify the chemical inventory, and propose observational signatures that could trace migration pathways.

34. Predict the formation of complex organic aerosols in exoplanetary atmospheres (e.g., hot Jupiters, Titan analogs). Estimate particle size distributions given condensation rates of 10^{-12} cm^3 s^{-1}, vertical mixing velocities, and coagulation timescales. Discuss implications for atmospheric opacity, spectral features, radiative transfer, and climate feedback processes, emphasizing the interplay between chemistry, microphysics, and dynamics.

35. Evaluate chemical pathways by which interstellar organics (e.g., amino acids, nucleobase precursors) survive incorporation into planetary atmospheres. Include both photochemical (UV, X-ray) and thermal processes during accretion. Discuss how such pathways influence delivery efficiency, chemical modifications, and the potential for prebiotic chemistry on early Earth-like planets.

36. Design an ultra-high vacuum laboratory experiment to produce formamide (NH_2CHO) via NH_3 + H_2CO reactions on ice surfaces. Calculate the expected product yield for a 10-monolayer ice under UV photon flux of 10^{17} cm^{-2} s^{-1}, incorporating radical diffusion coefficients, recombination efficiency, photon absorption depth, and surface coverage. Discuss how experimental results can be extrapolated to interstellar and protoplanetary disk conditions.

37. Model cryogenic radical recombination leading to sugar formation from formaldehyde (H_2CO) on icy grains. Use a Monte Carlo approach to simulate radical encounters and calculate the fraction of H_2CO converted to glycolaldehyde over 10^5 s. Analyze the influence of ice morphology, radical diffusion barriers, ice porosity, and energy dissipation pathways on reaction efficiency.

38. Evaluate the effect of photon energy on radical formation in ice analogs. Calculate the threshold photon energy required to dissociate CH_3OH into CH_2OH and H radicals. Discuss how the distribution of photon energies in interstellar space affects the overall radical pool, reaction branching, and formation of more complex molecules.

39. Compare laboratory yields of amino acid precursors produced under UV versus cosmic-ray irradiation of ice analogs. Quantify energy efficiency in terms of molecules formed per eV absorbed. Evaluate how these different energy sources contribute to prebiotic chemistry in various astrophysical environments, including cold molecular clouds, hot cores, and protoplanetary disks.

40. Using laboratory reaction rate data for ice chemistry, extrapolate rates to interstellar conditions (10 K, 10^3 cm^{-3}). Discuss assumptions involved in scaling, potential sources of error, and limitations arising from differences in surface coverage, cosmic-ray flux, and tunneling effects. Propose methods to validate such extrapolations using observations or isotopic fractionation measurements.

41. Construct a kinetic simulation incorporating over 1000 reactions relevant to a hot molecular core. Predict time-dependent abundances of methyl formate ($HCOOCH_3$), ethyl cyanide (C_2H_5CN), and dimethyl ether (CH_3OCH_3) over 10^6 years, including gas-phase and grain-surface chemistry. Assess how uncertainties in rate constants, branching ratios, and desorption mechanisms affect predicted abundances and compare with observational constraints from ALMA and single-dish surveys.

42. Model chemical stratification in a protoplanetary disk. Predict radial and vertical abundances of H_2O, CH_3OH, and CO in the midplane versus disk surface, accounting for temperature gradients, UV irradiation, cosmic-ray ionization, adsorption/desorption, and vertical mixing. Discuss implications for planetesimal formation, ice composition, and delivery of volatiles to forming planets.

43. Use Monte Carlo or rate-equation simulations to study radical diffusion on grain surfaces. Calculate encounter probabilities for NH_2 and CH_2OH radicals as a function of surface coverage, diffusion barriers, and ice temperature. Analyze how these factors influence the formation of complex organics such as amino alcohols, amino acids, and prebiotic precursors.

44. Compare gas-phase versus surface-mediated pathways for complex organic molecule formation in cold molecular clouds. Determine under which density and temperature regimes each mechanism dominates, and evaluate the implications for observed abundances and isotopic fractionation patterns. Include discussion on bottlenecks, radical lifetimes, and the effect of ice layering.

45. Evaluate the impact of photodissociation on molecular abundances in a protoplanetary disk. Incorporate self-shielding effects for CO and H_2O, radial and vertical stellar radiation fields, and disk geometry. Discuss how photochemistry influences chemical stratification, the survival of complex organics, and observable emission lines in infrared and millimeter wavelengths.

46. Analyze amino acid abundances measured in the Murchison meteorite. Estimate the fraction of interstellar organics preserved during incorporation, accounting for thermal processing, aqueous alteration, and photodegradation. Discuss implications for the delivery of prebiotic molecules to early Earth and the isotopic signatures expected for comparison with laboratory ice analogs.

47. Using Rosetta mission data, calculate the delivery rate of glycine to early Earth via comet 67P, assuming a cometary mass flux of 10^8 kg/year. Compare contributions from comets versus asteroids in seeding prebiotic molecules. Include discussion on survival during atmospheric entry, degradation processes, and cumulative delivery over Earth's early history.

48. Model shock-induced chemistry in molecular outflows and protostellar jets. Predict enhancements of SiO, H_2CO, and complex organic molecules along shock fronts, incorporating shock velocity, gas density, temperature profiles, and sputtering from dust grains. Discuss observational diagnostics for tracing shock chemistry and implications for complex molecule formation in star-forming regions.

49. Evaluate the contribution of X-ray irradiation from young stellar objects to radical and ion production in surrounding molecular clouds. Quantify its effect on HCN and other nitrogen-bearing species, considering ionization rates, secondary UV photons, and resulting chemical networks. Discuss implications for nitrogen isotopic fractionation and the formation of prebiotic molecules in protoplanetary environments.

50. Synthesize astrochemical evidence for the cosmic ubiquity of prebiotic molecules. Propose a probabilistic model for delivery to habitable planets, incorporating formation rates, survival probabilities during transport (via comets, asteroids, and interplanetary dust), and environmental factors such as UV flux and planetary atmospheres. Evaluate how this framework could constrain the likelihood of prebiotic chemistry occurring elsewhere in the galaxy.

Chapter 37

Chemistry of the Microbiome

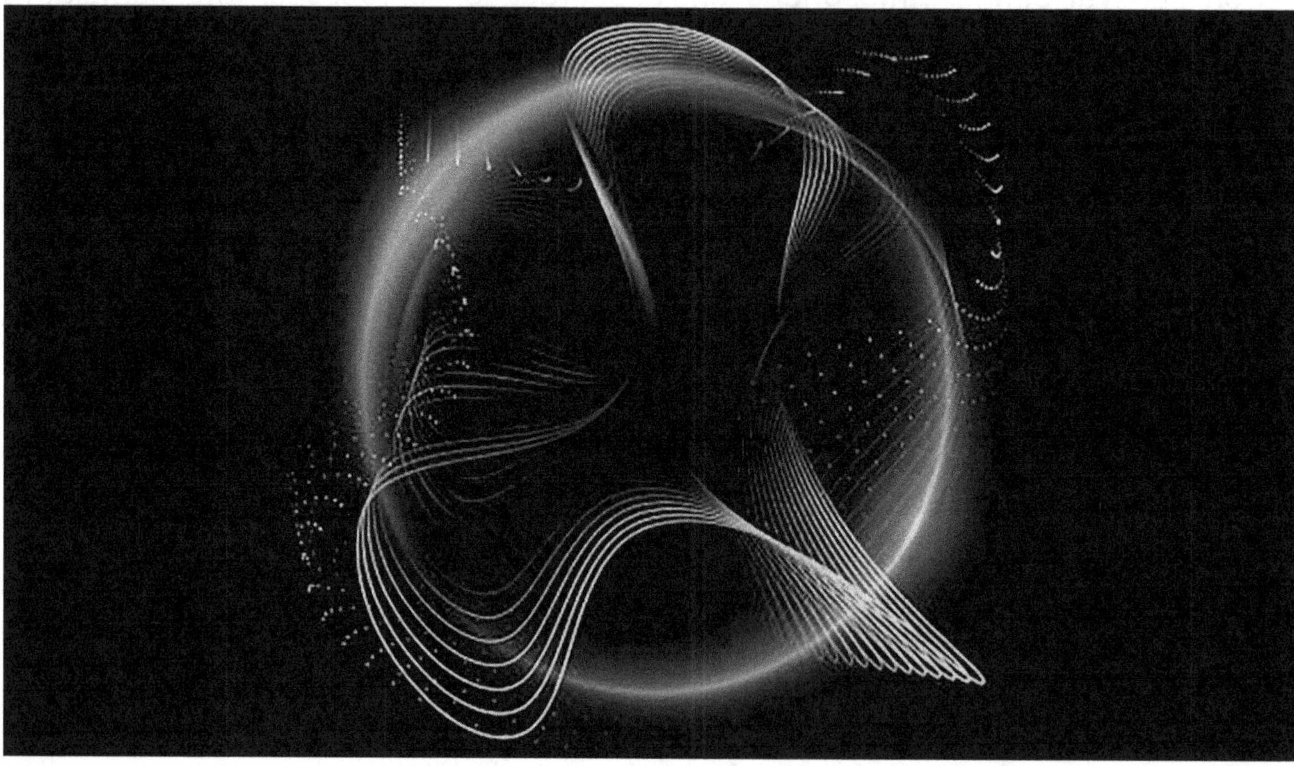

The human microbiome, comprising trillions of microorganisms including bacteria, viruses, fungi, and archaea, inhabits virtually every surface of the body, from the gut to the skin, mouth, and respiratory tract. Its chemical landscape is extraordinarily complex, shaped by intricate interactions between microbial metabolism, host biochemistry, and environmental factors. Central to understanding the microbiome is the recognition that microbial activity is not merely a biological phenomenon but a chemical ecosystem. Microorganisms secrete a diverse array of metabolites, enzymes, and signaling molecules that modulate host physiology, influence immune responses, and even impact neurological function. These chemical mediators range from short-chain fatty acids and bile acids to secondary metabolites such as phenolic compounds and microbial-derived neurotransmitters, all of which collectively define the chemical signature of the microbiome.

At the molecular level, microbial metabolism is a fascinating interplay of canonical and specialized biochemical pathways. Gut bacteria, for instance, ferment indigestible polysaccharides to produce short-chain fatty acids, including acetate, propionate, and butyrate. These molecules serve multiple chemical roles: they are energy sources for colonocytes, modulate gene expression through histone acetylation, and act as signaling molecules affecting host lipid and glucose metabolism. The chemistry of SCFAs is underscored by their carboxylate functionality, which allows them to participate in acid-base reactions, hydrogen bonding, and enzymatic transformations within both microbial and host cells. Beyond SCFAs, gut microbes modify bile acids via deconjugation, oxidation, and epimerization reactions, producing secondary bile acids with distinct chemical properties that influence the solubilization of lipids and the activation of nuclear receptors involved in metabolism and immunity.

Microbial enzymes are chemical powerhouses that catalyze reactions inaccessible to host enzymes. Glycoside hydrolases, polysaccharide lyases, and proteases expand the chemical repertoire of the microbiome, breaking down complex carbohydrates, proteins, and xenobiotics into smaller, absorbable units. These transformations often involve precise acid-base catalysis, nucleophilic attacks, and redox reactions, highlighting the sophisticated chemical logic embedded in microbial metabolism. For example, the cleavage of β-glycosidic bonds in dietary fibers involves protonation and stabilization of transition states, generating monosaccharides that feed both microbial and host metabolic pathways. Similarly, microbial reductive dehalogenation reactions detoxify halogenated compounds in the gut, illustrating the chemical versatility and adaptability of microbial communities.

The chemistry of inter-microbial communication, known as quorum sensing, reveals another layer of chemical complexity. Microbes release and detect small signaling molecules such as acyl-homoserine lactones in Gram-negative bacteria, oligopeptides in Gram-positive bacteria, and autoinducer-2 molecules shared across diverse species. These molecules engage in reversible chemical interactions with receptor proteins, triggering transcriptional responses that coordinate collective behaviors, including biofilm formation, virulence factor production, and secondary metabolite synthesis. Quorum sensing molecules often possess electrophilic or nucleophilic centers that allow covalent and non-covalent interactions with target proteins, underscoring the precise chemistry underlying microbial sociality.

A particularly intriguing chemical dimension of the microbiome lies in its interaction with xenobiotics and pharmaceuticals. Microbial enzymes such as β-glucuronidases, nitroreductases, and azoreductases chemically transform drugs and dietary compounds, modulating their bioavailability, efficacy, and toxicity. For instance, the microbial reduction of the cardiac drug digoxin produces inactive metabolites, while the metabolism of the chemotherapeutic agent irinotecan by bacterial β-glucuronidases generates toxic compounds responsible for intestinal side effects. These interactions highlight a complex chemical dialogue between host, microbiome, and exogenous compounds, where enzyme catalysis, substrate specificity, and microenvironmental pH govern the fate of chemical entities within the gut.

The skin microbiome provides a distinct chemical environment shaped by lipid-rich secretions, sweat-derived metabolites, and host-derived antimicrobial peptides. Cutaneous microbes metabolize sebum components, such as triglycerides and wax esters, generating free fatty acids that maintain acidic pH and inhibit pathogenic colonization. The chemical interplay between microbial enzymes and skin lipids involves hydrolysis, oxidation, and esterification reactions, producing bioactive lipids that influence inflammation and barrier function. Moreover, microbial metabolites such as indoles, phenols, and sulfur-containing compounds contribute to the characteristic odor of the skin, demonstrating the chemical specificity of host-microbe interactions at the molecular level.

In the oral cavity, microbial chemistry orchestrates the formation and maintenance of dental biofilms. Polysaccharide synthesis by Streptococcus mutans and other oral microbes drives biofilm adhesion and structural integrity, while acid production from carbohydrate fermentation promotes enamel demineralization. The chemical gradient within biofilms, including pH and redox potential, shapes microbial composition and inter-species interactions. Extracellular polymeric substances, composed of polysaccharides, proteins, and extracellular DNA, provide a chemically dynamic matrix mediating diffusion of nutrients, signaling molecules, and antimicrobial agents, exemplifying the microenvironmental chemistry of microbial communities.

Advances in analytical chemistry have revolutionized our understanding of the microbiome's chemical repertoire. Mass spectrometry, nuclear magnetic resonance, and high-performance liquid chromatography enable the identification and quantification of thousands of microbial metabolites, while isotope labeling experiments reveal metabolic fluxes and chemical transformations. These chemical insights are complemented by computational metabolomics, which models metabolic networks and predicts chemical interactions within microbial consortia. Such approaches illuminate how chemical structure, reactivity, and thermodynamics govern the behavior of microbial communities in health and disease.

The chemical ecology of the microbiome extends to its role in host immunity. Microbial-derived metabolites modulate immune cell differentiation, cytokine production, and barrier integrity. Butyrate, for example, induces regulatory T cell differentiation through epigenetic modification, while secondary bile acids interact with farnesoid X receptors to suppress pro-inflammatory pathways. These chemical mediators act through precise molecular interactions involving hydrogen bonding, hydrophobic effects, and electrostatic complementarity, illustrating the importance of chemistry in host-microbe signaling.

Diet is a primary determinant of microbiome chemistry. The chemical composition of dietary fibers, polyphenols, proteins, and lipids dictates microbial metabolism, metabolite production, and community composition. Polyphenols undergo microbial biotransformation through hydrolysis, reduction, methylation, and dehydroxylation reactions, producing bioactive compounds with antioxidant, anti-inflammatory, and neuroprotective properties. The chemical diversity of dietary components thus shapes microbial chemical output, linking nutrition, microbial metabolism, and host physiology in a complex network of chemical interactions.

Environmental factors such as oxygen availability, pH, and redox potential further influence the chemical behavior of the microbiome. Facultative anaerobes and obligate anaerobes occupy distinct niches, reflecting the chemical constraints imposed by oxygen gradients. Microbial production of hydrogen sulfide, nitric oxide, and reactive oxygen species modulates redox chemistry, impacting both microbial survival and host cellular responses. These reactive chemical species participate in signaling, defense, and tissue remodeling, highlighting the interplay of chemical reactivity and biological function within microbial ecosystems.

Emerging research on the gut-brain axis underscores the role of microbial chemistry in neurological function. Microbes synthesize neurotransmitters such as gamma-aminobutyric acid, serotonin, and dopamine precursors, which influence neuronal signaling, mood, and cognition. These molecules engage in receptor-mediated chemical interactions, modulate ion channel activity, and participate in synaptic neurotransmission, illustrating the profound chemical interconnection between microbial metabolism and host neurophysiology. Furthermore, microbial modulation of tryptophan metabolism impacts the production of kynurenine and indole derivatives, linking microbial chemical activity to neuroinflammation and neurodegenerative disease pathways.

Chemical diversity within the microbiome also encompasses secondary metabolites with antimicrobial, antiviral, and anticancer properties. Polyketides, non-ribosomal peptides, and ribosomally synthesized peptides exemplify chemically complex molecules that confer competitive advantage within microbial communities while exerting profound effects on the host. Biosynthesis of these molecules involves sophisticated enzymatic chemistry, including chain elongation, cyclization, oxidation-reduction, and glycosylation reactions, revealing the depth of chemical innovation encoded in microbial genomes.

The chemical plasticity of the microbiome allows rapid adaptation to perturbations such as antibiotic exposure, dietary shifts, and pathogen invasion. Microbial communities reorganize through chemical signaling, metabolic rewiring, and horizontal gene transfer, altering their chemical output to restore ecological balance. Antibiotic-induced shifts illustrate how chemical pressures select for resistant strains, which possess enzymes capable of hydrolyzing, modifying, or effluxing antibiotic molecules. The chemistry of resistance, including β-lactam hydrolysis and aminoglycoside modification, exemplifies the dynamic interplay between microbial metabolism and chemical survival strategies.

Emerging metabolomic and metagenomic approaches integrate chemical and genetic data, enabling systems-level understanding of microbiome function. Network analysis reveals correlations between specific metabolites and microbial taxa, while isotope tracing elucidates chemical flux through microbial and host pathways. Such integrative chemistry enables the identification of biomarkers for disease, prediction of therapeutic responses, and rational design of interventions that modulate microbial chemical activity. These strategies underscore the potential of microbiome chemistry to inform precision medicine, nutritional strategies, and microbial therapeutics.

Finally, synthetic biology offers opportunities to harness and redesign microbial chemistry. Engineered microbes can be programmed to produce specific metabolites, degrade toxins, or modulate host signaling. Chemical considerations, including substrate specificity, reaction kinetics, and metabolite stability, are central to the design of such systems. By manipulating microbial chemistry, it becomes possible to create living therapeutics, personalized nutrition strategies, and bioactive molecules with unprecedented precision, highlighting the transformative potential of understanding and leveraging the chemistry of the microbiome.

The chemical landscape of the microbiome is defined by the vast diversity of metabolites produced and transformed by microbial communities. These metabolites include small molecules such as short-chain fatty acids (SCFAs), branched-chain amino acids, bile acid derivatives, polyamines, and an array of secondary metabolites, each playing distinct roles in host physiology and microbial ecology. SCFAs, including acetate, propionate, and butyrate, are particularly noteworthy for their multifaceted influence: they serve as energy substrates for colonocytes, act as signaling molecules through G-protein-coupled receptors (GPCRs), and modulate gene expression via epigenetic mechanisms such as histone deacetylase inhibition. The production of these metabolites is not uniform but depends intricately on microbial composition, substrate availability, and environmental context, reflecting a dynamic chemical equilibrium within the gut ecosystem.

Beyond SCFAs, microbial metabolism exerts profound effects on the host through the chemical modification of dietary components and host-derived molecules. For instance, microbial bile salt hydrolases catalyze the deconjugation of primary bile acids, converting them into secondary bile acids that influence lipid digestion, cholesterol metabolism, and signaling pathways such as the farnesoid X receptor (FXR) axis. Similarly, microbial conversion of dietary tryptophan produces indole derivatives that serve as ligands for the aryl hydrocarbon receptor (AhR), impacting mucosal immunity, intestinal barrier integrity, and systemic inflammatory responses. The complexity of these interactions illustrates the microbiome not as a passive collection of organisms but as a chemically active network that integrates diet, host metabolism, and environmental signals.

Enzymatic transformations within the microbiome extend to xenobiotic metabolism, wherein microbial enzymes modify drugs, dietary polyphenols, and environmental chemicals. Reductive, hydrolytic, and conjugative reactions catalyzed by microbial enzymes can activate, inactivate, or produce novel bioactive compounds from these substrates, often influencing pharmacokinetics and therapeutic efficacy. For example, microbial β-glucuronidases can deconjugate drug metabolites, impacting enterohepatic circulation and systemic exposure. Polyphenol-derived metabolites, generated by microbial enzymes, may exert antioxidant, anti-inflammatory, or cardiometabolic benefits, highlighting the microbiome as a critical mediator of nutraceutical activity. The chemical consequences of these reactions underscore the microbiome's capacity to expand the host's metabolic repertoire beyond its intrinsic enzymatic potential.

Inter-microbial chemical communication adds another dimension to microbiome chemistry. Quorum sensing molecules, such as N-acyl homoserine lactones and autoinducing peptides, regulate gene expression across microbial populations, coordinating behaviors such as biofilm formation, virulence factor production, and resource competition. These signaling molecules often intersect with host pathways, influencing immune responses, epithelial integrity, and even neuronal signaling. Similarly, the production of bacteriocins, secondary metabolites with antimicrobial activity, mediates competitive interactions that shape microbial community structure. This chemical crosstalk reflects a dual role: microbes interact among themselves through a chemically orchestrated network while simultaneously modulating host physiology, creating a layered and interconnected chemical ecosystem.

The interplay between microbial metabolism and host neurochemistry is a particularly active area of research, commonly framed within the concept of the gut–brain axis. Microbial production and transformation of neurotransmitters, including gamma-aminobutyric acid (GABA), serotonin, dopamine, and catecholamines, exemplify this chemical connection. For instance, specific gut microbes convert dietary tryptophan into serotonin precursors, influencing both local gastrointestinal function and systemic neurochemical signaling. Similarly, microbial metabolites may cross the intestinal barrier or influence vagal nerve signaling, providing a chemical conduit between the microbiome and central nervous system. The microbiome thus functions as a bioactive chemical organ, capable of modulating neurological processes through complex molecular interactions.

Host–microbiome co-metabolism exemplifies the integration of microbial and host chemistry. Metabolic handoffs, wherein microbial metabolites serve as substrates for host enzymatic pathways, are widespread. Butyrate, for example, is produced exclusively by microbes but fuels colonocyte energy metabolism and influences systemic lipid homeostasis. Secondary bile acids generated by microbial action modulate hepatic gene expression, while microbial transformation of dietary polyphenols can enhance systemic antioxidant capacity. Such cooperative metabolic networks highlight the chemical interdependence between host and microbiome, where perturbations in microbial chemistry can propagate systemic physiological consequences.

Emerging technologies are expanding our capacity to decode microbiome chemistry with unprecedented resolution. Mass spectrometry-based metabolomics, nuclear magnetic resonance spectroscopy, and isotope tracing allow the identification and quantification of thousands of metabolites simultaneously. Coupled with metagenomic, metatranscriptomic, and metaproteomic data, these approaches enable the mapping of chemical flux through microbial communities, revealing enzyme–substrate relationships, co-metabolic pathways, and host–microbiome chemical interactions. Computational modeling, network analysis, and machine learning further allow predictions of metabolic outcomes, guiding targeted interventions and the rational design of probiotics, prebiotics, and postbiotics.

The therapeutic potential of manipulating microbiome chemistry is vast. Strategies include dietary modulation to favor beneficial metabolite production, engineered microbes capable of producing therapeutic compounds, and small-molecule inhibitors targeting specific microbial enzymes. Prebiotic fibers can selectively enhance SCFA-producing bacteria, whereas engineered probiotics may secrete anti-inflammatory metabolites or neurotransmitter precursors. Furthermore, understanding microbial drug metabolism facilitates personalized medicine, optimizing dosing and minimizing adverse effects. Microbiome-based chemical interventions thus represent a convergence of molecular biology, chemistry, and clinical therapeutics.

Beyond human health, the chemical principles governing the microbiome have broader ecological and industrial implications. Microbial consortia in soil, water, and engineered environments similarly rely on metabolite-mediated interactions, influencing nutrient cycling, pollutant degradation, and bioenergy production. Harnessing these microbial chemical networks enables the design of biotechnological applications, from biofertilizers and bioremediation strategies to microbial cell factories for high-value chemicals. The principles derived from human microbiome chemistry thus have universal applicability, demonstrating the centrality of chemical interactions in microbial ecology and function.

In summary, the chemistry of the microbiome is a multidimensional tapestry encompassing metabolite production, enzymatic transformation, signaling, and host–microbe co-metabolism. It integrates environmental cues, dietary inputs, microbial genetics, and host physiology into a dynamic, chemically mediated system. Continued elucidation of this molecular complexity will enhance our understanding of health and disease, enable rational design of microbiome-targeted therapeutics, and provide blueprints for harnessing microbial chemistry in ecological and industrial contexts. In essence, the microbiome constitutes a living chemical network, its molecular products and transformations forming the foundation of a symbiotic relationship as chemically rich and diverse as life itself.

Chapter 37: Word Problems

1. A researcher is studying how different types of dietary fiber affect short-chain fatty acid (SCFA) production in the human gut. They provide colonocyte cultures with fibers rich in cellulose, inulin, and resistant starch, then measure acetate, propionate, and butyrate concentrations over 48 hours. Considering the chemical structure of these fibers, microbial enzyme specificity, and the roles of SCFAs as energy sources, signaling molecules, and epigenetic modulators, explain how fiber composition might influence SCFA profiles and host metabolic effects.

2. In a patient with disrupted lipid metabolism, gut microbiome analysis reveals a marked reduction in bacterial populations responsible for producing secondary bile acids. Describe in detail the chemical transformations that convert primary bile acids synthesized in the liver (e.g., cholic acid, chenodeoxycholic acid) into secondary bile acids (e.g., deoxycholic acid, lithocholic acid), including enzymatic deconjugation of taurine or glycine residues, 7α-dehydroxylation, and epimerization reactions at the C3 and C7 positions. Discuss how a reduction in secondary bile acid concentrations could alter lipid micelle formation, impair solubilization of

dietary lipids and fat-soluble vitamins, and influence enterohepatic circulation. Further, analyze how changes in secondary bile acid profiles can modulate nuclear receptor signaling, particularly the farnesoid X receptor (FXR) and Takeda G-protein receptor 5 (TGR5), and predict systemic metabolic consequences, including impacts on glucose homeostasis, hepatic lipid metabolism, and inflammatory pathways.

3. β-Glycosidase enzymes produced by gut microbes cleave β-glycosidic linkages in complex dietary fibers such as cellulose, resistant starches, and oligosaccharides. Detail the chemical mechanism of this hydrolysis, including protonation of the glycosidic oxygen, nucleophilic attack by a water molecule or active-site residue, formation of the oxocarbenium ion-like transition state, and stabilization through hydrogen bonding and electrostatic interactions within the enzyme active site. Explain how active-site architecture and substrate-binding residues dictate specificity for β-linkages versus α-linkages. Predict the consequences of inhibiting microbial β-glycosidases on the availability of monosaccharides to both the microbiome and host epithelial cells, considering downstream effects on fermentation pathways, short-chain fatty acid (SCFA) production, and energy harvesting.

4. Skin-resident microbes metabolize triglycerides present in sebum into free fatty acids via lipase-catalyzed hydrolysis of ester bonds. Describe the chemical sequence of triglyceride hydrolysis, detailing nucleophilic attack at the carbonyl carbon, formation of tetrahedral intermediates, and release of mono- and diglycerides as well as free fatty acids. Discuss how the resulting acidification of the skin microenvironment alters the pH-dependent activity of both commensal and pathogenic microbes, modifies competitive microbial interactions, and influences enzymatic kinetics. Additionally, analyze how free fatty acids interact chemically with microbial cell membranes, destabilizing lipid bilayers, forming micellar complexes, and exerting antimicrobial effects. Include considerations of chain length, degree of unsaturation, and hydrophobicity in determining potency.

5. Streptococcus mutans produces extracellular polysaccharides (EPS) that organize into dense biofilms with steep microgradients of pH, redox potential, and nutrient availability. Examine how these chemical gradients influence local microbial composition, facilitating niches for acid-tolerant species while suppressing others.

Discuss the role of EPS in mediating diffusion of metabolites, quorum-sensing molecules, and signaling compounds, thereby shaping interspecies communication. Explain the chemical basis for acid-induced enamel demineralization, emphasizing hydroxyapatite dissolution kinetics under fluctuating local pH, and explore how biofilm EPS modulates microenvironmental buffering, ionic strength, and enamel protection or susceptibility.

6. Gram-negative gut bacteria employ N-acyl homoserine lactones (AHLs) as quorum-sensing molecules that bind reversibly to receptor proteins to coordinate population-level behaviors such as biofilm formation, virulence factor expression, and motility. Describe the chemical features of AHLs that enable receptor binding, including the electrophilic lactone carbonyl, the hydrogen bond donor/acceptor pattern of the amide, and the hydrophobic tail for receptor pocket interactions. Predict how hydrolysis or lactonolysis of the AHL ring would disrupt molecular recognition, attenuate signal transduction, and reduce biofilm formation, considering both kinetic and thermodynamic aspects of ligand-receptor interactions.

7. Gut microbes metabolize dietary tryptophan into indole derivatives, including indole-3-acetic acid, indole-3-propionic acid, and indole-3-aldehyde, which interact with the aryl hydrocarbon receptor (AhR) to modulate host immune responses. Outline the chemical steps of tryptophan conversion, such as deamination, oxidative decarboxylation, and hydroxylation, and explain how variations in gut pH, redox state, or cofactor availability could alter enzymatic activity, reaction rates, and metabolite distribution. Discuss the molecular interactions that stabilize ligand-receptor complexes, including hydrogen bonding, π–π stacking, and hydrophobic interactions, and predict how changes in metabolite stereochemistry or electronic properties could affect AhR activation.

8. Certain gut microbes express β-glucuronidases that hydrolyze glucuronide-conjugated drug metabolites, altering drug pharmacokinetics and systemic exposure. If enzyme activity increases by 50%, quantitatively analyze the potential changes in substrate turnover rates, enterohepatic recycling, and plasma drug concentrations. Explain the chemical principles underlying substrate recognition, such as complementarity to glucuronic acid moieties, hydrogen bonding, and nucleophilic attack at the anomeric carbon. Discuss how local gut environment factors, including pH, ionic strength, and cofactor availability, influence enzymatic efficiency and drug metabolism outcomes.

9. Butyrate, a short-chain fatty acid produced by microbial fermentation of dietary fibers, serves both as an energy source for colonocytes via β-oxidation and as an epigenetic modulator through inhibition of histone deacetylases (HDACs). Compare and contrast these chemical roles by detailing the oxidation reactions that generate acetyl-CoA and the molecular interactions that allow butyrate to chelate the Zn^{2+} ion in HDAC active sites. Discuss how fluctuations in butyrate concentrations may influence systemic lipid and glucose metabolism, colonocyte energy balance, and chromatin remodeling, integrating enzymatic kinetics, receptor-mediated signaling, and metabolite transport considerations.

10. Some gut microbes perform reductive dehalogenation of halogenated xenobiotics, transforming them into less toxic or more excretable compounds. Describe the chemical mechanism, emphasizing electron transfer from redox cofactors, nucleophilic attack on halogenated carbons, and stepwise removal of halogens. Predict which types of carbon-halogen bonds (C–Cl, C–Br, C–I) are most susceptible based on bond dissociation energies and electron-withdrawing effects. Analyze how the resulting products influence host detoxification, microbial community interactions, and potential accumulation of intermediate metabolites with differing reactivities or toxicities.

11. Gut microbes convert dietary tyrosine into dopamine precursors, influencing host neurotransmission. Describe in detail the chemical transformations from tyrosine to L-DOPA via tyrosine hydroxylation, followed by decarboxylation to dopamine. Include the roles of microbial enzymes such as tyrosine hydroxylase-like oxygenases and aromatic amino acid decarboxylases, the cofactors involved (e.g., tetrahydrobiopterin, PLP), and the electronic changes during hydroxylation and decarboxylation steps. Predict how microbial enzyme activity, pH, and redox potential in the gut lumen modulate local dopamine concentrations and the extent of systemic absorption, affecting enteric and central nervous system signaling.

12. Dietary polyphenols are metabolized by gut bacteria through hydrolysis, reduction, methylation, dehydroxylation, and ring cleavage, generating bioactive derivatives with altered chemical properties. Describe the stepwise chemical transformations, detailing nucleophilic aromatic substitution, redox reactions, and enzymatic methyl transfer. Discuss how structural modifications influence solubility, antioxidant potential, metal-chelating ability, and interactions with host enzymes or receptors,

and predict how these changes might impact systemic inflammatory responses, cardiovascular health, or neural signaling.

13. Short-chain fatty acids (SCFAs) such as acetate, propionate, and butyrate act as ligands for G-protein-coupled receptors (GPCRs) in the gut. Explain in detail how the carboxylate functional group forms hydrogen bonds and ionic interactions with receptor binding sites, emphasizing the orientation of the functional groups, protonation state at physiological pH, and hydrophobic interactions with receptor pockets. Predict how variations in SCFA chain length or substitution affect receptor affinity, downstream signal transduction, and systemic metabolic and immune responses.

14. After broad-spectrum antibiotic treatment, a patient shows markedly reduced propionate production. Analyze in detail how the depletion of specific bacterial populations alters polysaccharide fermentation, considering the substrate specificity of microbial enzymes and the chemical pathways leading to SCFA formation. Describe the mechanistic basis for decreased propionate yield, including potential bottlenecks in methylmalonyl-CoA pathways and succinate conversion, and predict downstream effects on colonocyte energy supply, gut barrier function, and systemic metabolic homeostasis.

15. Explain the chemical mechanism, including the formation of the Schiff base intermediate between glutamate and the cofactor, proton transfer events, and decarboxylation kinetics. Predict how variations in gut pH, cofactor availability, or microbial redox state could influence enzymatic efficiency and the levels of GABA, affecting local inhibitory signaling in enteric neurons and systemic modulation of host nervous system function.

16. Describe the chemical basis of bile acid-FXR interactions, including hydrogen bonding, hydrophobic contacts with receptor pockets, and conformational changes upon ligand binding. Predict how alterations in hydroxylation pattern, epimerization, or conjugation of bile acids could influence receptor binding affinity, signaling strength, and downstream transcriptional regulation of metabolic genes.

17. Explain the chemical mechanism of thiol oxidation, including electron transfer steps, formation of disulfides or sulfur-centered radicals, and redox cycling. Discuss the potential effects on microbial survival, oxidative stress responses, host redox signaling, and modulation of cellular thiol pools.

18. Using principles of acid-base chemistry and solubility equilibria, calculate how local pH changes influence hydroxyapatite dissolution, ionic saturation, and enamel solubility constants. Predict how gradients of acid production within the biofilm affect colonization by aciduric microbes versus more sensitive species, and consider feedback effects on microbial community composition and dental caries progression.

19. Discuss in detail how these chemical features enable specific interactions with receptor proteins, considering hydrogen bonding, dipole interactions, and steric complementarity. Predict how the introduction of small-molecule analogs or covalent inhibitors could block quorum sensing, perturb biofilm formation, and attenuate microbial virulence.

20. Describe the chemical mechanism in detail, including nucleophilic attack on the amide carbonyl, formation of the tetrahedral intermediate, acyl-enzyme complex formation, and hydrolysis to regenerate the free enzyme. Analyze the role of active-site residues such as Ser70 or Lys73 in catalysis, and discuss how environmental factors like pH, temperature, and metal ion availability influence enzymatic efficiency and clinical resistance outcomes.

21. Outline the chemical pathway from cysteine or methionine to volatile thiols (e.g., 3-methyl-mercaptan, hydrogen sulfide), detailing enzymatic cleavage, decarboxylation, and reduction steps. Predict how modulation of enzymatic activity, cofactor availability, or substrate concentration could reduce odor production while preserving normal microbial physiology.

22. Explain the chemical basis of electrostatic interactions between the positively charged amine groups and the negatively charged phosphate backbone of DNA or RNA. Discuss how these interactions stabilize nucleic acid structures, influence transcriptional regulation, and modulate chromatin compaction, while considering pH-dependent protonation states and ionic shielding effects.

23. Describe the detailed chemical reaction mechanism, including nucleophilic attack at the anomeric carbon of the glucuronide, formation and stabilization of oxocarbenium-like transition states, stereochemical considerations, and enzyme-substrate specificity. Discuss the influence of pH, cofactor availability, and active site residues on reaction kinetics. Evaluate strategies to mitigate toxicity, such as

selective β-glucuronidase inhibitors or chemical modifications of the glucuronide linkage, and predict potential effects on systemic drug metabolism, enterohepatic circulation, and therapeutic efficacy.

24. Predict the chemical reactivity of H_2S with metalloproteins, including heme-containing enzymes and iron-sulfur clusters, as well as with redox-sensitive cysteine thiols. Describe possible reaction pathways such as ligand exchange, metal reduction, or persulfide formation, incorporating thermodynamic and kinetic considerations. Discuss the consequences of these interactions for microbial survival under oxidative stress, modulation of host enzymatic activity, regulation of vascular tone, and inflammatory signaling, including potential feedback mechanisms between H_2S production and redox homeostasis.

25. Using mechanistic and stoichiometric reasoning, predict SCFA output from specific fiber types, considering enzyme specificity, pathway branch points, and competition for reducing equivalents. Discuss how variations in SCFA profiles influence colonocyte energy metabolism, luminal pH modulation, microbial cross-feeding networks, and local redox balance. Include thermodynamic constraints on pathway fluxes and potential regulatory effects of microbial signaling molecules on fermentation rates.

26. Describe the chemical mechanism, including electron transfer from reduced cofactors (e.g., flavins, ferredoxins) to the substrate, nucleophilic attack on halogenated carbons, and stepwise removal of halogens. Predict substrate selectivity based on C–Cl, C–Br, and C–I bond strengths, polarizability, and electronic effects. Analyze potential impacts on host detoxification pathways, microbial ecological dynamics, and the formation of reactive or partially dehalogenated intermediates that may influence gut or systemic chemistry.

27. Describe the detailed chemical mechanism, emphasizing pyridoxal phosphate (PLP) cofactor participation, Schiff base formation, proton transfers, and decarboxylation kinetics. Predict how variations in gut pH, PLP availability, and microbial redox potential affect GABA synthesis. Discuss implications for local gut-neural signaling, host enteric nervous system modulation, and potential systemic neurological effects mediated by GABA absorption or signaling.

28. Discuss the chemical principles underlying substrate specificity, including acyl chain length, stereochemistry, and active site accessibility. Describe the stepwise

mechanism of ester bond cleavage, formation of acyl-enzyme intermediates, and hydrolysis to release FFAs. Analyze how the profile of released FFAs affects antimicrobial activity, pH modulation, microbial competition, and colonization dynamics. Predict systemic consequences if microbial lipase activity is dysregulated.

29. Outline the chemical steps involved, including malonyl-CoA condensation, decarboxylation, ketoreduction, dehydration, cyclization, and post-synthetic modifications. Predict how mutations in polyketide synthases alter chemical structure, functional group placement, stereochemistry, and bioactivity. Discuss implications for host-microbe interactions, pharmacological activity, and metabolic stability.

30. Explain how electron transfer, hydroxylation, and epimerization modify bile acid chemical properties. Predict effects on nuclear receptor signaling (FXR, TGR5), bile acid solubility, enterohepatic recirculation, lipid emulsification, and systemic metabolic regulation. Include discussion of how gut redox potential, oxygen gradients, and microbial enzyme specificity influence reaction rates and metabolite distributions.

31. Describe the chemical mechanism in detail, emphasizing PLP-dependent Schiff base formation, proton abstraction, and decarboxylation kinetics. Predict downstream effects on host serotonergic signaling, gastrointestinal motility, and potential systemic neuroactive consequences. Include considerations of microbial population density, cofactor availability, and local pH on metabolite flux.

32. Discuss how EPS chemical composition—including polysaccharides, proteins, and extracellular DNA—affects diffusion coefficients, electrostatic interactions, hydration, and molecular binding. Predict impacts on microbial competition, resilience to environmental stressors, biofilm thickness, and spatial organization. Include thermodynamic and kinetic considerations of nutrient transport and metabolite sequestration.

33. Model the receptor-ligand interactions chemically, detailing hydrogen bonding, ionic interactions, van der Waals forces, and active site complementarity. Predict how SCFA structural variations, receptor mutations, or local microenvironmental factors modulate signaling outputs affecting systemic lipid and glucose metabolism.

Include discussion of downstream metabolic feedback loops and cofactor-dependent enzymatic regulation.

34. Describe the chemical steps, including electron transfer from microbial reductases, sequential reduction of nitrate to nitrite and NO, and formation of reactive intermediates. Predict effects on gut microbial ecology, redox homeostasis, and host signaling pathways, including vasodilation, immune modulation, and inflammation. Include kinetic and thermodynamic constraints on intermediate accumulation.

35. Discuss chemical considerations in their design, including substrate specificity, enzyme kinetics, cofactor stability, metabolite solubility, chemical reactivity in gut conditions, and avoidance of reactive intermediates. Predict how these factors influence metabolite yield, bioavailability, host safety, and off-target chemical interactions. Include consideration of competitive microbial metabolism and chemical degradation in vivo.

36. Explain the molecular interactions, including hydrogen bonding, π–π stacking, hydrophobic contacts, and electronic complementarity that stabilize ligand-receptor complexes. Predict how chemical modifications or analogs influence receptor activation, downstream transcriptional programs, and immune modulation. Include thermodynamic considerations of binding affinity and potential metabolic stability in the gut.

37. Describe the chemical transformations, detailing enzyme-catalyzed electronic rearrangements, cofactor requirements (e.g., SAM for methylation, tetrahydrobiopterin for hydroxylation), and reaction kinetics. Predict impacts on host neurological signaling, systemic catecholamine levels, and local redox state in the gut. Discuss potential feedback mechanisms from host metabolism or diet.

38. Explain the chemical mechanisms of resistance, including enzymatic hydrolysis, modification of antibiotic functional groups, target protection, and efflux. Discuss how horizontal gene transfer propagates these chemical resistance traits and alters microbial population dynamics, metabolic interactions, and chemical fluxes in the gut environment.

39. Propose chemical strategies—including derivatization, isotopic labeling, selective ionization, and fragmentation—to distinguish SCFA isomers and quantify their

concentrations accurately. Discuss how chemical properties such as polarity, acidity, volatility, and ionizability influence detection, quantitation, and interpretation of metabolic fluxes.

40. Discuss the chemical transformations involved, including elimination reactions, keto-enol tautomerism, and rearrangements, and analyze how changes in hydroxylation patterns affect antioxidant potential, receptor interactions, metal chelation, and host health outcomes. Include kinetic considerations and enzyme-substrate specificity.

41. Select a microbial metabolite and describe in detail the chemical interactions stabilizing its binding to a host receptor, including orientation, stereochemistry, and electronic effects. Predict how structural modifications alter binding affinity and efficacy, and analyze potential downstream physiological or signaling consequences.

42. Explain chemical mechanisms by which reactive oxygen and nitrogen species interact with proteins, nucleic acids, and other redox-sensitive targets, altering enzymatic activity, signaling pathways, and microbial survival. Predict effects on immune modulation, host-pathogen interactions, and chemical homeostasis in the gut microenvironment.

43. Describe chemical modifications such as methylation, reduction, and conjugation, and predict their effects on metabolite structure, receptor binding, solubility, bioavailability, and antioxidant capacity. Discuss downstream chemical signaling consequences for vascular function, endothelial integrity, and systemic oxidative stress.

44. Detail the chemical steps, including peptide bond rearrangements, heterocycle formation, oxidative modifications, and stereochemical control. Predict how these changes affect peptide stability, bioactivity, receptor interactions, and chemical reactivity in the host environment.

45. Discuss chemical factors that govern metabolite formation, including substrate reactivity, enzyme specificity, pH, redox potential, and cofactor availability. Predict chemical strategies to mitigate harmful effects, such as enzyme inhibition, substrate modification, or alteration of gut redox conditions.

46. Using thermodynamic principles, compare energy yields from butyrate versus glucose, considering ΔG values for β-oxidation, glycolysis, and mitochondrial electron transport. Discuss implications for colonocyte metabolism, energy homeostasis, redox balance, and response to microbial metabolite fluctuations.

47. Model chemical fluxes and metabolite concentrations to predict pathway rerouting to maintain essential functions, considering substrate competition, cofactor availability, redox balance, and interspecies metabolic interactions. Include predictions for transient accumulation of intermediates and chemical stress responses.

48. Design a chemical experiment using isotopically labeled compounds to monitor conversion steps, intermediate formation, and systemic absorption. Include discussion of detectable chemical transformations via mass spectrometry or NMR, kinetic isotope effects, and reaction mechanisms.

49. Discuss the influence of stereochemistry, functional group orientation, hydrogen bonding, hydrophobic interactions, and electronic properties on binding affinity and receptor activation. Predict downstream chemical signaling outcomes, immune modulation, and potential metabolic feedback loops.

50. Create a chemical model linking microbial enzyme activity, metabolite concentrations, diffusion, receptor interactions, and host neuronal responses. Include local gut effects, systemic circulation, potential feedback regulation, and cofactor-dependent reaction kinetics. Predict chemical constraints and bottlenecks that may influence neuroactive metabolite availability and signaling fidelity.

Chapter 38

Sustainable Molecular Design

Recent industrial initiatives exemplify the translation of sustainable molecular design principles into large-scale chemical production. The synthesis of ibuprofen serves as a paradigmatic case. Traditional routes relied on multi-step syntheses that generated significant waste and employed stoichiometric amounts of reagents. Modern approaches, notably the BHC (Boots-Hoechst-Celanese) route, utilize a three-step, one-pot process with highly selective catalytic steps that minimize waste and improve atom economy. This redesign illustrates how careful molecular-level planning, informed by catalysis and reaction engineering, can achieve substantial environmental and economic benefits. Furthermore, the use of green solvents such as ethanol or water in these transformations exemplifies solvent-conscious design, demonstrating the alignment of synthetic efficiency with environmental responsibility.

In the realm of polymer chemistry, sustainable molecular design has catalyzed the development of recyclable plastics. Poly(lactic acid) (PLA), derived from renewable feedstocks such as corn starch or sugarcane, exemplifies a material engineered for both performance and biodegradability. Its stereocontrolled polymerization allows for tunable mechanical properties, while its ester linkages ensure hydrolytic degradability under composting conditions. Industrial adoption of PLA in packaging and biomedical applications demonstrates the feasibility of translating molecular design principles into commercially viable materials. Advanced research focuses on enhancing thermal and mechanical stability without compromising biodegradability, employing copolymerization strategies and incorporation of renewable additives. These innovations reflect the broader trend of integrating life-cycle considerations at the molecular design stage.

Catalysis-driven sustainable transformations have also gained prominence in fine chemical production. The hydrogenation of biomass-derived platform molecules, such as furfural or levulinic acid, into value-added chemicals illustrates the synergy between molecular design and renewable feedstocks. Heterogeneous catalysts, including supported metals like Ni, Cu, or Pd, allow selective reduction under mild conditions, producing bio-based solvents, fuels, and intermediates with minimal byproduct formation. Mechanistic studies reveal that ligand modification and support engineering can dramatically influence reaction selectivity and catalyst lifetime. Such insights demonstrate the importance of molecular-level design in optimizing catalytic systems, enabling processes that are both environmentally benign and industrially scalable.

Photoredox catalysis offers another frontier in sustainable molecular design, providing energy-efficient pathways for bond formation. Recent studies have demonstrated the use of visible-light-activated photocatalysts to drive C–C and C–N bond formation under ambient conditions. Organic dyes, transition-metal complexes, and semiconductor nanoparticles serve as photocatalysts that absorb visible light, triggering selective transformations without the need for harsh reagents or elevated temperatures. For example, photoredox-mediated cross-coupling reactions have enabled the functionalization of pharmaceuticals and agrochemicals with reduced energy input and waste generation. The molecular design of both photocatalysts and substrates is critical, requiring tuning of redox potentials, excited-state lifetimes, and reaction kinetics to achieve high efficiency and selectivity.

Enzyme-catalyzed reactions continue to demonstrate transformative potential in sustainable molecular design. Lipases, oxidoreductases, and transaminases facilitate highly selective transformations in aqueous or solvent-free conditions, often outperforming conventional synthetic routes in both efficiency and stereocontrol. Industrial examples include the synthesis of statin precursors via transaminase-mediated amination and the production of biodiesel using lipase-catalyzed transesterification of triglycerides. Protein engineering and directed evolution have expanded the substrate scope and enhanced the stability of these enzymes, enabling their integration into multi-step chemoenzymatic sequences. These approaches exemplify the integration of molecular design, catalysis, and systems thinking, achieving high performance with minimal environmental impact.

Sustainable molecular design in pharmaceuticals also addresses the environmental persistence of drugs. Active pharmaceutical ingredients (APIs) such as fluoroquinolones and sulfonamides, though therapeutically valuable, are often resistant to biodegradation and accumulate in aquatic ecosystems. Researchers are employing molecular modifications, such as cleavable ester or amide linkages, to generate prodrugs that retain pharmacological activity but degrade into harmless metabolites post-excretion. Computational toxicology and predictive degradation models guide these designs, ensuring that modifications do not compromise efficacy. These examples illustrate the necessity of integrating chemical functionality, environmental chemistry, and computational insights at the molecular design stage.

Supramolecular and self-assembling systems offer additional opportunities for sustainability. Metal–organic frameworks (MOFs) designed for gas storage and catalysis demonstrate that porosity, stability, and catalytic activity can be tuned at the molecular level while minimizing energy-intensive processing. For instance, MOFs with biodegradable organic linkers or water-stable metal nodes provide high performance without introducing persistent pollutants. Similarly, dynamic covalent assemblies and host–guest systems are being engineered to capture and release pollutants, serve as recyclable catalysts, or facilitate controlled drug delivery. These systems highlight the intersection of molecular precision, functionality, and environmental responsibility.

Electrochemical transformations represent a growing frontier in sustainable molecular design, enabling selective oxidation and reduction reactions without stoichiometric chemical reagents. For example, electrochemical CO_2 reduction to formic acid or methanol has been achieved using molecularly engineered catalysts that stabilize reactive intermediates and minimize overpotential. Similarly, anodic oxidation processes can replace hazardous chemical oxidants in fine chemical synthesis, generating benign byproducts such as hydrogen gas. Molecular design of both the catalyst and the substrate interface is critical, with ligand engineering, electron transfer pathways, and surface chemistry dictating efficiency and selectivity. By integrating renewable electricity sources, these approaches further reduce the carbon footprint of chemical manufacturing.

Life-cycle assessment (LCA) and sustainability metrics increasingly guide molecular design. Tools such as E-factor, process mass intensity (PMI), and green chemistry metrics quantify waste generation, energy consumption, and overall environmental impact. Incorporating these metrics during the design phase enables chemists to identify the most impactful modifications, prioritize atom economy, and evaluate alternative reaction pathways. For example, the replacement of toxic halogenated solvents with biodegradable alternatives may increase reaction time slightly but significantly improve environmental outcomes. Quantitative sustainability assessment fosters iterative design, creating molecules and processes that meet both functional and ecological criteria.

Circularity in chemical manufacturing is exemplified by the valorization of waste streams. Industrial processes generating glycerol, lignin, or CO_2 as byproducts can be reimagined as feedstocks for new chemical entities. Glycerol, abundant in biodiesel production, has been transformed into solvents, polymers, and pharmaceuticals through catalytically guided reactions, demonstrating how molecular design can close material loops. Similarly, lignin, a complex polyphenolic polymer, is being depolymerized and chemically modified to produce high-value aromatic compounds, integrating sustainable design with renewable feedstocks. These examples illustrate the practical implementation of circular principles in molecular design, emphasizing resource efficiency and environmental responsibility.

Emerging research explores adaptive and stimuli-responsive molecules to enhance sustainability. Materials incorporating dynamic covalent bonds or supramolecular interactions can self-heal, reorganize, or degrade under controlled conditions, extending functional lifespan and reducing waste. For instance, polymers with reversible Diels–Alder linkages exhibit reversible crosslinking, enabling material recycling without harsh chemical treatments. Similarly, photoresponsive molecules allow on-demand degradation or functional switching, aligning performance with environmental safety. These approaches demonstrate how sophisticated molecular engineering can simultaneously address functionality, efficiency, and sustainability challenges.

The integration of machine learning and artificial intelligence (AI) into molecular design further accelerates sustainable innovation. By analyzing vast datasets of reaction outcomes, AI algorithms can predict optimal reaction conditions, catalyst structures, and molecular scaffolds with high efficiency and low environmental impact. Generative models allow the design of novel molecules with predefined properties, such as biodegradability, low toxicity, or high catalytic efficiency. Coupling AI predictions with automated synthesis platforms enables rapid prototyping of sustainable molecules, reducing experimental burden, chemical consumption, and associated waste. This convergence of computational power and chemical insight represents a transformative tool for sustainable molecular design.

Educational initiatives and industry-academia partnerships are critical to embedding sustainable molecular design principles into practice. Universities are increasingly incorporating green chemistry and sustainability metrics into core curricula, training chemists to consider environmental, economic, and social factors from the outset of molecular design. Industrial consortia, such as those in the pharmaceutical and materials sectors, foster collaboration in developing scalable, green processes that meet regulatory and sustainability benchmarks. Case studies from these collaborations, including solvent reduction in large-scale pharmaceutical syntheses and the design of recyclable polymer coatings, provide concrete examples of molecular design principles applied at scale.

Future directions in sustainable molecular design are increasingly guided by the principle of atom economy, emphasizing reactions that maximize the incorporation of all atoms from starting materials into the final product while minimizing waste. This approach encourages chemists to rethink traditional synthetic routes, prioritizing catalytic over stoichiometric processes and designing transformations that reduce the need for hazardous solvents, heavy metals, and energy-intensive conditions. Green chemistry metrics, such as the E-factor, process mass intensity (PMI), and reaction mass efficiency, are becoming essential tools for quantifying the sustainability of molecular syntheses, enabling researchers to systematically compare, optimize, and innovate toward more sustainable chemical processes.

The development of renewable feedstocks is another critical frontier in sustainable molecular design. Biomass, carbon dioxide, and other abundant, non-fossil resources are being converted into high-value chemicals through both biological and chemical transformations. Chemoenzymatic cascades exemplify this integration, combining the selectivity of enzymes with the robustness of chemical catalysis to achieve complex molecular architectures under mild, energy-efficient conditions. Similarly, electrocatalytic strategies leverage electricity—ideally sourced from renewable energy—to drive redox transformations with minimal chemical waste, opening pathways for carbon-neutral or carbon-negative production of key intermediates. These innovations position chemistry at the forefront of a global shift toward carbon-conscious molecular manufacturing.

Circularity in molecular design extends beyond feedstock considerations to encompass the full life cycle of a chemical product. Molecular structures are now being engineered for recyclability, degradability, and upcycling, ensuring that end-of-life materials can be reintegrated into the production cycle rather than disposed of. Polymers with programmable depolymerization pathways, self-healing materials, and dynamic covalent networks exemplify this paradigm, enabling chemical products to maintain functional performance while reducing environmental impact. Advanced analytical techniques, including real-time mass spectrometry, spectroscopy, and computational modeling, support the rational design of these circular molecules, allowing chemists to predict and optimize degradation pathways, stability profiles, and recyclability metrics before scale-up.

The integration of stimuli-responsive functionalities represents a transformative direction in sustainable molecular design. Molecules and materials that respond to light, pH, redox potential, or mechanical stress can enable precision-controlled reactivity, targeted delivery, and adaptive performance. In pharmaceuticals, this approach minimizes off-target effects and reduces the dosage required, while in materials science, it supports adaptive coatings, sensors, and self-regulating systems. By incorporating responsiveness into molecular architecture, chemists can create smart molecules that dynamically interact with their environment, enhancing efficiency and reducing resource consumption across diverse applications.

Computational chemistry, machine learning, and artificial intelligence are becoming indispensable in guiding sustainable molecular design. Predictive models can evaluate the environmental impact, energy requirements, and toxicity of hypothetical molecules before laboratory synthesis, enabling the rational selection of candidates that balance performance with sustainability. High-throughput virtual screening, molecular dynamics simulations, and multi-scale modeling integrate seamlessly with experimental workflows, accelerating discovery while minimizing resource consumption. In addition, AI-driven retrosynthetic analysis allows chemists to identify synthetic routes that are not only efficient but also environmentally responsible, bridging the gap between theoretical design and practical implementation.

The rise of bioinspired and biomimetic strategies is further expanding the toolkit for sustainable chemistry. By emulating nature's molecular logic—such as enzyme catalysis, metabolic pathways, and hierarchical self-assembly—chemists can achieve transformations that are highly selective, energy-efficient, and compatible with aqueous or mild conditions. Synthetic analogs of natural cofactors, enzyme-mimicking catalysts, and artificial metabolic networks exemplify this trend, allowing for scalable production of complex molecules while minimizing chemical hazards. Such bioinspired strategies often dovetail with circularity and renewable feedstock initiatives, creating a coherent framework for sustainable molecular innovation.

Global regulatory frameworks and market dynamics are increasingly aligning with these scientific advances. Governments, international organizations, and industry stakeholders are setting ambitious targets for green chemistry adoption, waste reduction, and carbon neutrality, incentivizing the development of sustainable molecules and processes. Compliance with these standards often requires a holistic approach, encompassing process safety, environmental impact, life-cycle assessment, and socio-economic considerations. Companies and research institutions that integrate sustainable molecular design early in development gain both competitive advantage and long-term resilience, highlighting the convergence of chemistry with strategic planning and policy.

Education and workforce development play an essential role in sustaining this momentum. Curricula in chemistry are evolving to emphasize not only classical principles but also sustainability metrics, green synthesis, life-cycle thinking, and interdisciplinary collaboration. Training the next generation of chemists to think in terms of environmental responsibility, systems-level design, and technological innovation ensures that sustainable molecular design becomes a core aspect of chemical practice rather than a peripheral concern.

Looking ahead, the future of sustainable molecular design lies in integrated, multi-scale approaches. Molecular innovations will increasingly be evaluated in the context of system-level sustainability, considering interactions between chemical production, energy use, ecological impact, and human health. Collaboration across chemistry, materials science, chemical engineering, biology, and data science will be crucial for realizing molecules and materials that are not only high-performing but also environmentally harmonious and socially responsible. By embracing this holistic framework, chemists can position their discipline as a cornerstone of a sustainable technological society, driving innovation that harmonizes human needs with planetary stewardship.

Ultimately, sustainable molecular design represents a paradigm shift: a movement away from linear, resource-intensive chemistry toward dynamic, circular, and responsive molecular systems. The convergence of advanced catalysis, renewable feedstocks, computational design, bioinspired strategies, and life-cycle thinking establishes a new chemical ethos, where efficiency, performance, and environmental responsibility are inseparable. As the chemical community continues to adopt, refine, and expand these principles, the discipline will not only meet the pressing challenges of sustainability but also unlock new avenues for creativity, innovation, and transformative impact across technology, healthcare, energy, and materials. In this vision, chemists emerge not only as molecular architects but also as stewards of a sustainable future, demonstrating that the science of molecules can serve both human progress and the planet.

Chapter 38: Word Problems

1. In the synthesis of ibuprofen via the BHC (Boots–Höchst–Ciba) route, which involves three highly selective catalytic steps including Friedel–Crafts acylation, 1,2-aryl migration, and hydrolysis, chemists seek to further reduce environmental impact by replacing one of the catalysts with a more active, recyclable alternative. Discuss the molecular-level criteria for catalyst selection, including electronic effects, ligand environment, and active site accessibility. Evaluate how changes in catalyst activity and selectivity could alter reaction kinetics, energy requirements, atom economy, byproduct formation, and scale-up feasibility, and propose a systematic methodology to assess these parameters experimentally and computationally.

2. Poly(lactic acid) (PLA) exhibits tunable mechanical and thermal properties depending on the stereochemistry of the lactide monomers used. Design a hypothetical scenario where adjusting the L- to D-lactide ratio optimizes PLA for biomedical implant applications, such as bone scaffolds or resorbable sutures. Incorporate considerations of hydrolytic degradation rates, crystallinity, mechanical modulus, and thermal stability. Discuss potential trade-offs between environmental life-cycle impacts (including production energy, compostability, and post-use degradation) and performance requirements, and propose experimental and computational methods to predict optimal stereocontrol.

3. Biomass-derived furfural can be selectively hydrogenated into value-added chemicals such as furfuryl alcohol or tetrahydrofurfuryl alcohol using transition metal catalysts (Ni, Cu, Pd). Compare how ligand modification, support choice, and metal dispersion influence catalytic selectivity, turnover frequency, and byproduct formation. Predict how these molecular-level changes could affect yield, energy efficiency, catalyst recyclability, and environmental footprint when scaling the process to an industrial level.

4. Photoredox catalysis enables visible-light-driven C–C bond formation through single-electron transfer processes. For a substrate containing multiple redox-active functional groups, propose a detailed strategy to select both an appropriate photocatalyst and light source to maximize chemoselectivity. Consider molecular orbital alignment, excited-state lifetimes, solvent polarity, and competing side

reactions. Evaluate how this approach can minimize energy consumption, hazardous waste generation, and photodegradation of the product.

5. Enzyme-catalyzed transesterification of triglycerides into biodiesel is often performed under solvent-free conditions using lipases. Discuss how protein engineering (such as site-directed mutagenesis, directed evolution, or immobilization strategies) could expand substrate specificity, enhance thermal and solvent stability, and improve catalytic efficiency. Analyze the resulting effects on process sustainability, cost-effectiveness, waste reduction, and potential regulatory compliance for large-scale biodiesel production.

6. Fluoroquinolones, as persistent antibiotics, pose environmental risks due to biodegradation resistance. Propose a molecular modification strategy to design a prodrug that retains antimicrobial activity but degrades into environmentally benign metabolites. Incorporate computational toxicology, predictive degradation modeling, and structure–activity relationship analysis to guide your design. Evaluate trade-offs between pharmacokinetics, therapeutic efficacy, and environmental fate.

7. Metal–organic frameworks (MOFs) are being developed for CO_2 capture using biodegradable organic linkers and water-stable metal nodes. Analyze how molecular-level tuning of porosity, linker chemistry (hydrophobicity, functional groups), and node coordination affects adsorption capacity, selectivity, thermal and chemical stability, and ease of synthesis. Discuss life-cycle considerations, including environmental impact of production, potential recyclability, and end-of-life disposal.

8. Electrochemical reduction of CO_2 to formic acid can be enhanced using molecularly engineered catalysts. Describe how ligand design, electron-transfer pathways, and coordination environment can lower overpotential, improve selectivity, increase turnover frequency, and reduce energy consumption. Include consideration of reactor design, mass transport, electrolyte choice, and scalability to pilot-scale operations.

9. When evaluating two synthetic routes to a fine chemical—one using conventional stoichiometric oxidants and another employing electrochemical oxidation—construct a quantitative framework based on green chemistry metrics such as

E-factor, process mass intensity (PMI), energy input, and safety profiles. Analyze molecular-level reactivity, waste formation, and reaction conditions to recommend the optimal route for environmental and economic performance.

10. Glycerol, a major byproduct of biodiesel production, can be converted into high-value chemicals such as acrolein, glycerol carbonate, or dihydroxyacetone. Propose a catalytic pathway, considering molecular-level transformations, reaction kinetics, catalyst selection (heterogeneous vs homogeneous), and thermodynamic constraints. Evaluate how these factors align with circular economy principles, process sustainability, and energy efficiency.

11. Polymers containing reversible Diels–Alder linkages are being developed for recyclable coatings. Predict how crosslink density, bond reversibility, and reaction kinetics influence mechanical performance, healing efficiency, thermal and chemical stability, and overall environmental footprint. Include considerations for processing, recyclability, and end-of-life degradation.

12. Machine learning algorithms are increasingly used to predict optimal catalysts for photoredox reactions. Design a comprehensive workflow integrating experimental kinetic data, computational modeling, and sustainability metrics (e.g., energy input, waste generation, solvent usage) to iteratively improve molecular design. Discuss challenges in data quality, model transferability, and incorporation of multi-objective optimization.

13. Industrial PLA production may incorporate renewable additives to improve thermal stability or mechanical properties. Evaluate the trade-offs between enhanced performance, biodegradability, process energy requirements, extrusion temperature, and environmental impact. Propose molecular strategies to optimize additive incorporation while maintaining sustainable production practices.

14. A transaminase-mediated synthesis of statin precursors is being optimized for stereoselectivity. Discuss how enzyme immobilization, substrate concentration, cofactor regeneration, and reaction medium composition can influence activity, selectivity, and process scalability. Evaluate the interplay between molecular-level enzyme modifications and overall green chemistry metrics.

15. In solvent-conscious reaction design, replacing dichloromethane with water introduces challenges in solubility, phase behavior, and reaction kinetics.

Propose molecular strategies, such as surfactant use, co-solvents, or catalyst modification, to overcome these limitations while preserving reaction yield, selectivity, and environmental sustainability.

16. Dynamic covalent assemblies are employed for selective heavy metal capture from wastewater. Discuss how reversible bond design, host–guest complementarity, network topology, and environmental stability influence capture efficiency, recyclability, and process robustness under variable conditions.

17. A visible-light-activated photocatalyst is being engineered for C–N bond formation in agrochemical synthesis. Analyze how molecular tuning of redox potentials, excited-state lifetimes, and absorption spectra affects reaction efficiency, selectivity, energy input, and process scalability.

18. Scaling PLA for high-temperature packaging applications requires enhanced thermal stability. Propose a copolymerization strategy using renewable monomers (e.g., lactide and bio-derived cyclic esters), considering stereochemistry, crystallinity, mechanical properties, and compostability. Assess trade-offs in production energy, post-use biodegradability, and industrial processing.

19. Electrochemical anodic oxidation offers an alternative to hazardous chemical oxidants in pharmaceutical synthesis. Design a molecular-level experiment to evaluate catalyst stability, substrate compatibility, overpotential requirements, and energy efficiency. Incorporate reaction monitoring, mechanistic understanding, and green chemistry assessment.

20. Lignin depolymerization generates aromatic monomers for polymer synthesis. Formulate a catalytic strategy that maximizes monomer yield, minimizes side reactions, and ensures economic and environmental viability. Include consideration of catalyst design, solvent selection, and process scalability.

21. Photoresponsive polymers allow on-demand degradation upon exposure to light. Develop a detailed model linking light wavelength, intensity, and exposure time to polymer degradation kinetics, considering photochemical reaction pathways, bond dissociation energies, and polymer morphology. Analyze how these factors influence polymer lifetime, recycling potential, and environmental impact when deployed in real-world applications.

22. AI-generated molecular scaffolds are evaluated for both biodegradability and catalytic efficiency. Construct a multi-criteria assessment framework integrating predicted molecular properties, synthetic feasibility, energy input, catalyst turnover, and environmental impact. Discuss strategies for iterative optimization, including feedback loops between computational predictions and experimental validation.

23. Solvent replacement is proposed in a large-scale pharmaceutical synthesis to improve safety and reduce environmental burden. Discuss how changes in polarity, hydrogen-bonding capacity, and dielectric constant influence reaction mechanisms, solubility of reactants and intermediates, reaction kinetics, yield, and side-product formation. Propose molecular strategies to mitigate potential challenges while maximizing sustainability metrics such as PMI and E-factor.

24. A chemoenzymatic cascade converts biomass-derived intermediates into a complex target molecule. Analyze how enzyme compatibility, intermediate stability, sequential reaction order, and molecular-level interactions influence overall atom economy, yield, and environmental footprint. Propose strategies for enzyme immobilization, cofactor recycling, and reaction medium optimization to enhance both efficiency and sustainability.

25. Circularity principles are applied to CO_2 valorization through catalytic conversion into polymerizable monomers. Design a molecular-level catalytic cycle, considering energy input, catalyst turnover, molecular selectivity, reaction kinetics, and thermodynamic constraints. Evaluate how catalyst design, reactor operation, and process integration influence overall carbon efficiency and environmental performance.

26. Self-healing polymers rely on dynamic covalent bonds such as imines, boronate esters, or Diels–Alder adducts. Predict how bond dissociation energies, bond exchange rates, network connectivity, and polymer chain mobility influence repair efficiency, mechanical durability, and life-cycle sustainability. Discuss trade-offs between material robustness and recyclability.

27. A biocatalyst is engineered to functionalize a pharmaceutical molecule under mild, ambient conditions. Evaluate how pH, temperature, substrate hydrophobicity, and enzyme active-site microenvironment influence conversion efficiency, selectivity, and waste minimization. Discuss implications for process scalability, cost, and regulatory compliance.

28. Life-cycle assessment identifies energy-intensive steps in polymer synthesis. Propose molecular modifications or alternative monomers to reduce overall carbon footprint without compromising functional performance, thermal and mechanical properties, or end-of-life recyclability. Discuss how predictive modeling can guide sustainable monomer design.

29. Stimuli-responsive molecules are designed to degrade upon mechanical stress. Formulate a reaction model linking stress intensity, bond cleavage rate, and subsequent molecular degradation pathways. Evaluate how structural features such as bond strength, steric environment, and polymer architecture affect environmental impact reduction and material lifetime.

30. Photoredox-mediated cross-coupling is scaled to a 100 L reactor for industrial production. Predict how photon flux, catalyst concentration, substrate solubility, and reactor geometry influence reaction yield, energy efficiency, and scalability. Discuss strategies to optimize light penetration, reduce side reactions, and maintain process reproducibility.

31. A renewable feedstock is converted into a high-value intermediate using electrocatalysis. Analyze how electrode material, applied potential, electrolyte composition, and mass transport influence selectivity, overpotential, energy efficiency, and catalyst stability. Discuss how molecular-level modifications to the electrocatalyst could enhance turnover frequency and sustainability.

32. Computational models predict degradation pathways of environmental contaminants. Construct a workflow linking molecular structure, predicted metabolites, and ecosystem toxicity to guide the design of safer molecules. Discuss integration of quantum chemical calculations, machine learning, and experimental validation to optimize environmental safety.

33. MOFs are designed for dual applications of gas storage and catalysis. Evaluate how metal node selection, linker hydrophobicity, pore size, and framework topology affect adsorption efficiency, catalytic activity, and material recyclability. Discuss trade-offs between structural stability, scalability, and environmental footprint.

34. A circular economy approach proposes upcycling polymer waste into reusable monomers. Propose a depolymerization and repolymerization pathway, considering

catalyst selection, reaction temperature, energy input, and byproduct management. Assess environmental trade-offs, industrial feasibility, and life-cycle sustainability.

35. AI-assisted retrosynthetic analysis identifies a greener synthetic route for a fine chemical. Compare predicted atom economy, E-factor, solvent usage, energy demand, and safety profile between conventional and AI-suggested routes. Discuss how molecular design and reaction mechanism understanding can improve sustainability outcomes.

36. Bioinspired catalysts are developed to mimic enzymatic selectivity for multi-step oxidation reactions. Design an experimental plan to evaluate turnover frequency, regioselectivity, enantioselectivity, and environmental performance. Include consideration of ligand environment, redox potentials, and reaction conditions to optimize both efficiency and sustainability.

37. Light-activated polymers allow spatially controlled crosslinking for applications in coatings and adhesives. Discuss how polymer architecture, photophysical properties, and light absorption profile influence energy efficiency, functional performance, and recyclability. Analyze molecular design strategies to optimize both reaction kinetics and environmental impact.

38. Directed evolution is used to improve enzyme activity for stereoselective synthesis. Predict how mutation sites, substrate docking, active-site flexibility, and cofactor interactions influence conversion rates, selectivity, and overall environmental metrics. Discuss integration with computational modeling to accelerate enzyme optimization.

39. A prodrug is engineered to degrade after excretion to reduce environmental impact. Formulate a predictive model linking molecular structure, enzymatic cleavage rate, metabolite stability, and environmental degradation. Evaluate how this approach balances pharmacological efficacy with ecological safety.

40. Supramolecular systems capture CO_2 and release it under controlled conditions. Evaluate how binding constants, pore accessibility, framework robustness, and molecular interactions influence system efficiency, selectivity, and long-term sustainability. Discuss potential industrial applications and regeneration strategies.

41. High-throughput screening identifies green solvents for a catalytic reaction.

Design a scoring system integrating solvent polarity, toxicity, biodegradability, cost, reaction performance, and energy efficiency. Discuss how molecular descriptors and predictive models can guide solvent selection and process optimization.

42. Polymer depolymerization pathways are being optimized for chemical recycling. Predict how monomer purity, reaction temperature, catalyst turnover, and network topology influence circularity, energy consumption, and scalability. Discuss strategies to minimize waste and maximize polymer recovery.

43. Electrochemical CO_2 reduction is coupled to a renewable energy source. Construct a model to quantify carbon savings, energy efficiency, catalyst lifetime, and process scalability over prolonged operation. Include considerations of electrode design, electrolyte composition, and molecular-level catalyst optimization.

44. Dynamic covalent networks are proposed for self-healing coatings. Assess how bond reversibility, network topology, polymer chain mobility, and environmental exposure (temperature, moisture, light) affect performance, durability, and sustainability metrics. Discuss trade-offs between material robustness and recyclability.

45. Photocatalytic reactions are adapted for pharmaceutical functionalization. Determine how substituent electronics, solvent choice, light intensity, and photocatalyst structure influence reaction efficiency, selectivity, energy consumption, and waste generation. Evaluate scalability and environmental impact.

46. AI predicts optimal biocatalysts for converting lignin-derived compounds. Develop a decision tree integrating predicted activity, stability, selectivity, environmental footprint, and synthetic accessibility to prioritize candidates. Discuss iterative cycles of computational prediction and experimental validation.

47. Circular PLA-based packaging is being scaled to industrial production. Discuss how molecular modifications (e.g., stereochemistry, copolymer incorporation), processing conditions (extrusion, molding), and composting standards influence product performance, thermal and mechanical stability, and overall life-cycle sustainability.

48. A bioinspired metabolic network is engineered for carbon-negative chemical production. Predict how pathway flux, enzyme expression levels, substrate availability, and cofactor regeneration influence atom economy, process efficiency,

and environmental impact. Discuss optimization strategies at both molecular and systems levels.

49. Smart molecules respond to multiple stimuli simultaneously, such as light, pH, and redox potential. Propose a molecular design framework where these stimuli collectively control chemical activity, and evaluate the sustainability advantages of such multifunctional systems compared to conventional single-stimulus materials.

50. Considering the full life cycle of a chemical product, design a molecular scaffold that balances functional performance, recyclability, biodegradability, and regulatory compliance. Describe an iterative computational-experimental approach to optimize properties, including molecular modeling, high-throughput synthesis, predictive toxicity assessment, and process-scale sustainability evaluation.

Chapter 39

Chemical Metamaterials

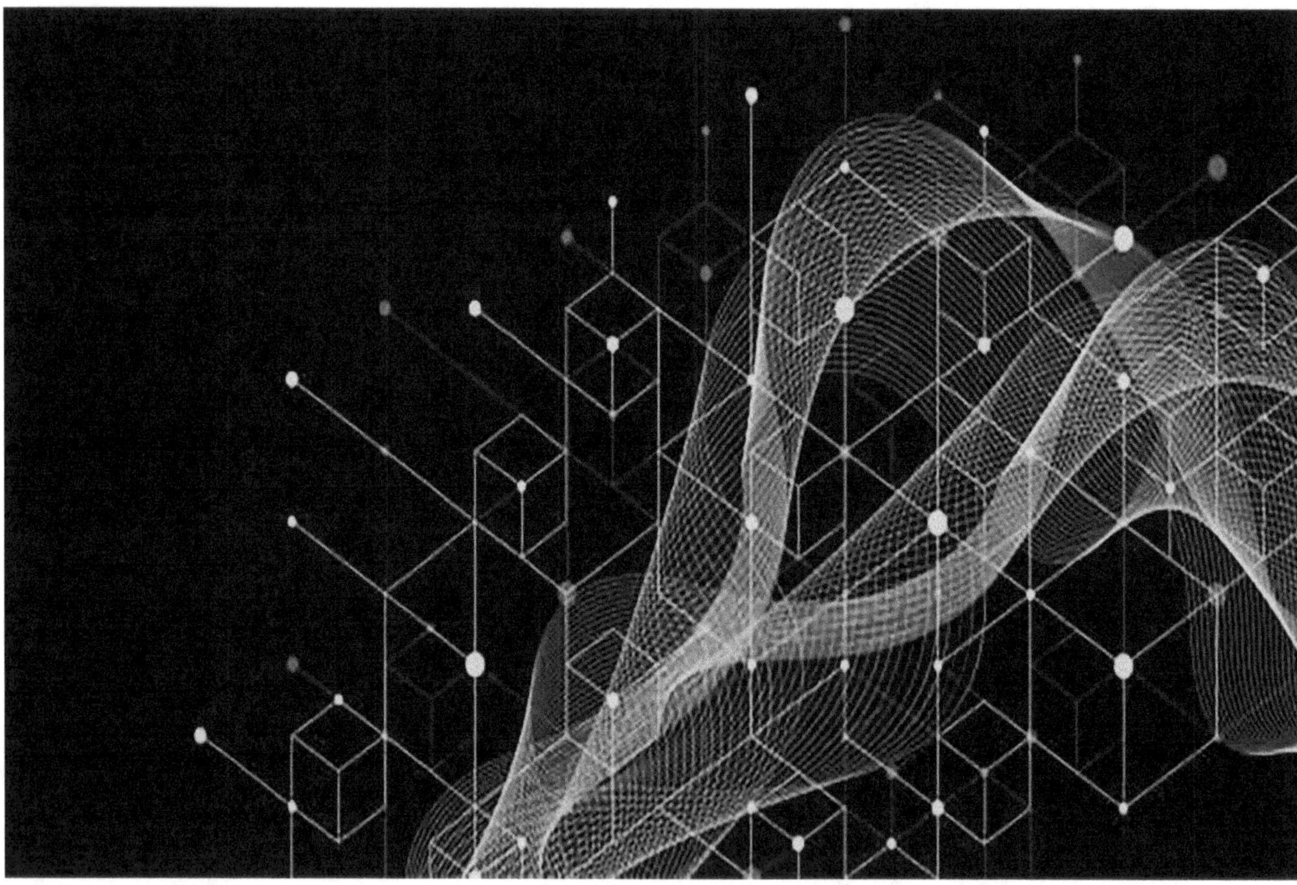

Chemical metamaterials represent a revolutionary class of engineered substances whose remarkable properties emerge not merely from their chemical composition but also from their meticulously designed architectures at multiple length scales. Unlike conventional materials, whose bulk behaviors are dominated by intrinsic atomic and molecular characteristics, chemical metamaterials exploit the synergy between molecular chemistry and hierarchical structuring to achieve properties that are absent in nature. The foundation of chemical metamaterial design lies in the strategic arrangement of molecular building blocks into periodic, quasi-periodic, or aperiodic lattices, often with features at nanometer, micrometer, and sometimes even macroscopic scales. These hierarchies enable unprecedented control over the propagation of electromagnetic waves, acoustic vibrations, heat flow, and chemical reactivity. The concept of metamaterials first arose in the context of electromagnetic wave manipulation, notably with photonic crystals and negative-refractive-index materials, but its extension into chemistry leverages the vast toolbox of molecular interactions, supramolecular self-assembly, and responsive functionalities, creating materials whose behavior is programmable, tunable, and adaptive.

The early developments in chemical metamaterials drew inspiration from natural photonic structures, such as butterfly wings and opal-like minerals, which exhibit iridescence due to nanoscale periodicity rather than pigment. Chemists recognized that by designing artificial analogs with precise molecular composition and controlled nanoscale patterning, it was possible to tailor light-matter interactions, including reflection, refraction, diffraction, and absorption, with a level of precision unattainable in bulk materials. This realization spurred the development of bottom-up synthetic strategies, in which molecular units, such as organic ligands, metal-organic frameworks (MOFs), covalent organic frameworks (COFs), dendrimers, and nanoparticles, self-assemble into well-defined periodic or hierarchical arrangements. Complementary top-down techniques, such as lithography, nanoimprint patterning, and focused ion beam etching, enable integration of these molecularly designed structures into larger-scale architectures. The combination of bottom-up and top-down approaches allows the fabrication of chemical metamaterials with both precise chemical identity and macroscopic control, bridging the gap between molecular design and functional application.

At the molecular scale, the choice of building block is critical. Metal-organic frameworks, for instance, provide a versatile platform where transition metal nodes are connected via rigid or flexible organic linkers, forming lattices with tunable porosity, electronic properties, and mechanical strength. By functionalizing linkers with photoresponsive, redox-active, or catalytic groups, researchers can imbue these lattices with dynamic functionalities, enabling reversible modulation of optical, electronic, or chemical responses. Covalent organic frameworks, in contrast, rely entirely on robust covalent linkages to form stable, extended networks with predictable pore sizes and electronic structures. Dendritic macromolecules offer a third strategy, in which branched architectures provide high surface area, tunable functional group density, and spatially defined reactive sites. Nanoparticles, including noble metals, semiconductors, and magnetic materials, act as both structural motifs and functional centers, whose plasmonic, photonic, or magnetic properties can be harnessed and tuned through size, shape, composition, and interparticle spacing.

The optical properties of chemical metamaterials are among the most extensively explored. Plasmonic metamaterials, for example, exploit the collective oscillations of electrons at the surfaces of metal nanoparticles to achieve localized surface plasmon resonances, resulting in strong light confinement and enhancement of electromagnetic fields at the nanoscale. By arranging these nanoparticles into periodic lattices embedded in dielectric or polymer matrices, chemists can create materials with negative refractive indices, hyperbolic dispersion, and subwavelength focusing capabilities. The introduction of stimuli-responsive molecules, such as azobenzenes or spiropyrans, allows dynamic tuning of optical properties in response to light, temperature, or chemical triggers, enabling smart windows, reconfigurable optical filters, and adaptive lenses. Similarly, photonic MOFs, which combine periodic dielectric structures with molecular functionality, can manipulate the propagation of specific wavelengths of light while facilitating photochemical transformations within their pores. These systems exemplify how molecular design and hierarchical structure converge to produce emergent optical phenomena.

Beyond optics, chemical metamaterials exhibit extraordinary electronic and magnetic behaviors. Spin-crossover complexes, radical-containing polymers, and conductive MOFs can switch between distinct electronic or magnetic states under external stimuli such as temperature, light, or pressure. For example, iron(II)-based MOFs can reversibly toggle between high-spin and low-spin configurations, dramatically altering their magnetic susceptibility, electrical conductivity, and dielectric response. Radical polymers, in which unpaired electrons are delocalized across conjugated backbones, can exhibit tunable conductivity and magnetism, especially when organized into periodic lattices. Chemically doped graphene-based metamaterials and layered transition metal dichalcogenides extend this concept to two-dimensional systems, where doping, defect engineering, and interlayer stacking produce tailored electronic band structures. The combination of molecular-level control, supramolecular assembly, and nanoscale periodicity enables electronic and magnetic metamaterials that operate far beyond the limitations of traditional bulk materials.

Mechanical properties of chemical metamaterials also reveal novel phenomena. By programming flexibility, anisotropy, and auxetic behavior at the molecular and nanoscale levels, chemists can design materials that respond to mechanical stress in counterintuitive ways. Auxetic networks, which expand laterally under tension, are realized by re-entrant lattice motifs in polymers or coordination networks. Dynamic covalent bonds, hydrogen bonding, metal-ligand interactions, and π-π stacking serve as reversible sacrificial linkages that allow materials to absorb mechanical energy, self-heal, and undergo programmable shape transformations. For instance, polyrotaxane-based metamaterials exploit sliding crosslinks to combine high stiffness with elasticity, while shape-memory polymers functionalized with reversible chemical groups can alter their macroscopic geometry in response to temperature or light. The integration of molecular chemistry and topological design thus provides precise control over mechanical responses, opening pathways for soft robotics, wearable devices, and biomedical implants.

Chemical metamaterials can be chemically active, serving as both structural components and reaction media. Catalytically active metamaterials integrate nanoscale catalysts into periodic lattices to enhance reaction efficiency, selectivity, and specificity. Plasmonic metamaterials, which concentrate electromagnetic fields into nanoscale hot spots, can drive photochemical reactions with high local energy density, facilitating transformations that would otherwise require extreme conditions. MOFs and COFs act as nanoscale reactors, spatially confining reactants to enhance multistep catalytic cascades. Supramolecular metamaterials utilize host-guest chemistry, hydrogen bonding, and electrostatic interactions to organize catalytic sites with nanometer precision, enabling cooperative effects and emergent reactivity. This intersection of chemical reactivity and metamaterial design exemplifies a paradigm in which the material itself actively participates in transformations rather than merely supporting them.

Energy-related applications are a natural extension of chemical metamaterials. Photonic and plasmonic structures can concentrate light and heat to enhance solar energy conversion, while catalytic metamaterials enable efficient hydrogen production, CO2 reduction, and fuel cell operation. Electrochemical metamaterials, including redox-active MOFs and conductive polymers, exhibit enhanced ion transport, surface area, and electronic conductivity, improving battery and supercapacitor performance. Hierarchical structuring at the nanoscale ensures precise control over pathways for charge carriers, ions, and photons, resulting in devices with superior efficiency, selectivity, and longevity. Stimuli-responsive metamaterials can further adapt to environmental conditions, modulating energy capture or storage dynamically, demonstrating a new class of smart energy materials.

Characterization of chemical metamaterials requires a combination of techniques to probe structure, composition, and functionality across multiple scales. Electron microscopy, including high-resolution TEM and cryo-EM, reveals nanoscale periodicity and defect distribution. Atomic force microscopy and scanning tunneling microscopy provide topographical and electronic mapping at molecular resolution. Spectroscopic methods, including NMR, IR, Raman, UV-Vis, and X-ray absorption spectroscopy, elucidate chemical composition, bonding interactions, and electronic states. Time-resolved spectroscopy and in situ monitoring enable observation of dynamic processes, such as stimuli-induced structural changes, catalytic turnover, and optical switching. These advanced characterization strategies are critical for understanding the structure-property relationships that underpin the functionality of chemical metamaterials and for guiding iterative design improvements.

Case studies illustrate the breadth and potential of chemical metamaterials. In optical metamaterials, silver and gold nanoparticle arrays embedded in polymer matrices have been engineered to produce negative refractive indices at visible wavelengths, demonstrating superlensing and subwavelength imaging. Plasmonic MOFs functionalized with photochromic ligands achieve reversible modulation of optical absorption, serving as light-responsive filters. In electronic systems, graphene-MOF hybrids exhibit tunable conductivity and photodetection capabilities, while radical-based polymers show switchable magnetism when organized into periodic lattices. Mechanically, polyrotaxane-based networks demonstrate programmable stiffness and reversible deformation, while dendritic hydrogels with sacrificial bonds achieve self-healing after mechanical damage. Catalytically, plasmonic metamaterials drive selective hydrogenation reactions under ambient conditions, and MOF-based systems catalyze multi-step organic transformations with spatial control over active sites. These examples underscore the interplay of chemistry, structure, and functionality in producing emergent properties unattainable in conventional materials.

Challenges remain in the field of chemical metamaterials. Synthesis and fabrication at scale require precise control over molecular identity, periodicity, and defect distribution, which is often difficult to achieve consistently. Environmental stability, including resistance to moisture, temperature fluctuations, and chemical degradation, is essential for practical applications but can be compromised by the complexity of the architectures. Integration of dynamic and responsive functionalities can further complicate stability and reproducibility. Addressing these challenges involves innovative synthetic strategies, including self-healing chemistries, dynamic covalent networks, hierarchical fabrication combining lithography with molecular self-assembly, and post-synthetic functionalization to enhance stability. Advances in machine learning and predictive modeling are increasingly being applied to accelerate design and optimize functional outcomes, reducing trial-and-error approaches and enabling rational development of next-generation metamaterials.

The future of chemical metamaterials lies in the convergence of chemistry, computation, and automation. Artificial intelligence and machine learning can predict molecular motifs, lattice structures, and emergent properties, guiding targeted synthesis. Robotic platforms can execute complex multistep syntheses and hierarchical assembly with high precision. Integration with dynamic feedback systems allows metamaterials to sense environmental conditions and self-adjust their optical, electronic, mechanical, or chemical behavior in real time. Biologically inspired strategies, including peptide-based assembly, DNA origami, and enzyme-catalyzed synthesis, provide pathways for self-healing, adaptive, and energy-autonomous materials. The ultimate goal is a new class of programmable matter, where molecular composition and hierarchical structure are co-designed to achieve complex functionality, blurring the line between synthetic and natural systems.

Chemical metamaterials thus epitomize the fusion of chemistry, physics, and materials engineering, demonstrating how molecular precision and structural innovation can produce phenomena beyond the capabilities of conventional materials. Through careful design of building blocks, control of hierarchical architecture, and incorporation of dynamic chemical functionalities, these materials exhibit tunable optical, electronic, mechanical, and catalytic behaviors. The combination of experimental, computational, and theoretical approaches continues to expand the possibilities of chemical metamaterials, offering transformative potential in energy, photonics, biomedicine, quantum technologies, and beyond. As understanding deepens and fabrication techniques advance, chemical metamaterials are poised to redefine our conception of what materials can do, providing platforms with properties limited only by imagination and molecular creativity.

Building upon the foundational concepts of chemical metamaterials, one encounters a rich landscape of molecular and supramolecular design strategies that enable the tailoring of macroscopic properties from the nanoscale. Central to this endeavor is the concept of programmable chemistry, wherein specific functional groups, ligands, or reactive motifs are judiciously arranged to encode desired interactions. For instance, coordination-driven self-assembly allows for the precise formation of discrete cages, frameworks, and extended lattices, whose geometric and electronic features can be systematically modulated. Similarly, covalent organic frameworks (COFs) and metal–organic frameworks (MOFs) provide modular platforms in which pore size, topology, and chemical environment are finely controllable, directly influencing guest–host interactions, catalytic performance, and transport phenomena. These frameworks serve not merely as static scaffolds but as dynamic chemical metamaterials, capable of adapting to external stimuli such as pH, light, electric fields, or mechanical stress.

A defining characteristic of chemical metamaterials is the integration of emergent functionality arising from collective molecular interactions. Unlike conventional materials, where properties are largely determined by bulk composition, metamaterials leverage structural organization across multiple scales to generate unconventional behavior. Optical metamaterials, for example, exploit subwavelength structuring to achieve negative refractive indices, cloaking effects, or hyperbolic dispersion, but when these architectures are functionalized chemically—via photoresponsive azobenzenes, redox-active centers, or plasmonic nanoparticles—the material gains active tunability, enabling real-time control over light–matter interactions. Similarly, electronic metamaterials benefit from molecular doping, charge-transfer complexes, or π-conjugated frameworks that produce anisotropic conductivity, topological states, or non-linear responses unattainable in homogeneous solids.

The field of mechanically adaptive metamaterials is rapidly expanding, with chemistry providing avenues to encode responsiveness at the molecular level.

Polymers incorporating reversible covalent bonds, supramolecular crosslinks, or mechanophores can be engineered into architected lattices or cellular solids whose stiffness, toughness, or shape memory evolves under external forces. Chemical gradients and spatial patterning further allow for the design of metamaterials with directional or gradient responses, inspiring applications in soft robotics, impact mitigation, and biomedical implants. Importantly, computational chemistry, molecular dynamics, and finite-element modeling play a critical role in predicting emergent behavior, linking molecular-scale interactions to macroscale mechanical properties, and guiding experimental realization.

Catalysis represents another domain where chemical metamaterials are redefining functional paradigms. By structuring active sites with precise spatial and electronic environments, researchers can create metamaterials with enzyme-like selectivity and cooperative effects, amplifying reaction rates or steering reaction pathways toward desired products. Porous networks with hierarchically tuned chemical heterogeneity not only enhance substrate accessibility but also introduce synergistic effects reminiscent of allosteric regulation in biological systems. The dynamic incorporation of stimuli-responsive ligands further allows catalytic metamaterials to operate under controllable on/off states, effectively merging material science with reaction engineering in a programmable manner.

Energy applications exemplify the transformative potential of chemical metamaterials. In photovoltaics and light-harvesting systems, arranging chromophores, quantum dots, or perovskite domains with precise nanoscale periodicity enables enhanced absorption, exciton migration, and charge separation. Chemical functionalization can further stabilize these architectures or introduce self-healing pathways, improving operational lifetimes. Similarly, in energy storage, ion-conductive metamaterials exploit chemically defined channels and interfacial architectures to achieve high selectivity, rapid ion transport, and suppression of dendritic growth, addressing longstanding challenges in battery and supercapacitor technology.

Beyond static properties, the temporal dimension of chemical metamaterials offers a frontier of adaptive, reconfigurable behavior. Molecular switches, dynamic covalent bonds, and out-of-equilibrium chemical reactions permit the construction of materials that evolve in response to environmental cues or programmed chemical flux. Such systems blur the line between matter and function, allowing for chemical intelligence, self-organization, and emergent computation embedded within the material itself. Examples include self-oscillating gels, responsive photonic crystals, and reaction–diffusion-based patterns, which illustrate how molecular chemistry can encode spatiotemporal behavior across scales.

Integration with quantum technologies represents a particularly exciting horizon. Chemical metamaterials can host qubits or spin centers with precise spatial arrangements, enabling engineered interactions that preserve coherence and allow for tailored entanglement networks. Molecular design strategies facilitate tunable energy landscapes, spin–orbit coupling, and phonon confinement, providing avenues to control decoherence pathways and optimize quantum gate operations. In parallel, chemically structured photonic metamaterials can manipulate light at the nanoscale to interface with quantum emitters, creating hybrid platforms for quantum communication, sensing, and computation.

The fabrication of chemical metamaterials increasingly relies on synergistic approaches combining bottom-up and top-down methodologies. Self-assembly, template-guided synthesis, and molecular printing enable precise control of chemical composition and nanoscale architecture, while lithography, additive manufacturing, and microfluidics provide macroscopic patterning and scalability. The convergence of these approaches allows for hierarchical materials in which each length scale—atomic, molecular, mesoscale, and macroscopic—is engineered for a specific functional contribution, resulting in materials with unprecedented multifunctionality.

Finally, the future of chemical metamaterials will be shaped by interdisciplinary integration, where advances in synthetic chemistry, computational modeling, physics, materials engineering, and bioinspiration converge. Ethical and environmental considerations, including sustainability, recyclability, and biocompatibility, will become integral to the design paradigm, ensuring that these advanced materials contribute positively to societal needs. As the limits of chemical creativity expand, metamaterials stand to redefine not only technological capabilities but also our conceptual understanding of materiality itself, transforming the interplay between molecules, structures, and emergent phenomena into a platform for next-generation innovation.

Chapter 39: Word Problems

1. A research team fabricates a plasmonic metamaterial composed of gold nanoparticles (NPs) arranged in a 3D MOF-derived lattice with 15 nm interparticle spacing. When illuminated with 520 nm light, the material exhibits a strong localized surface plasmon resonance (LSPR) peak redshift of 18 nm. If the same lattice is thermally expanded so the spacing increases to 18 nm, predict how the plasmonic coupling strength and resonance wavelength will change, and justify your reasoning using electrodynamic coupling principles.

2. An azobenzene-functionalized COF is designed to switch between trans and cis states when exposed to UV and visible light, modulating pore diameter from 1.5 nm (trans) to 0.9 nm (cis). If a catalytic reaction relies on substrate diffusion through these pores, estimate how the reaction rate would change under continuous UV irradiation and explain the mechanistic basis for this change.

3. A photonic MOF is constructed with alternating high- and low-refractive-index layers, forming a 1D Bragg stack. When a guest molecule infiltrates the pores and increases the refractive index of the low-index layer from 1.3 to 1.45, analyze how the photonic bandgap shifts and how this could be exploited for chemical sensing.

4. A conductive MOF containing Fe(II) centers undergoes spin-crossover from low-spin to high-spin at 330 K, decreasing electrical conductivity by two orders of magnitude. Develop a thermodynamic model explaining how external pressure could shift the transition temperature and predict the resulting effect on conductivity at ambient conditions.

5. A self-healing polymeric metamaterial contains reversible Diels–Alder bonds that break at 120 °C and reform upon cooling. During cyclic mechanical loading, the material's storage modulus drops by 40%. Explain how the density and kinetics of reversible bonds influence both the mechanical fatigue behavior and the healing efficiency.

6. A hierarchical catalytic metamaterial includes plasmonic Ag NPs embedded within a COF containing Brønsted acid sites. Under green-light illumination, plasmonic heating increases the local temperature by ~35 °C. Predict how this localized temperature rise affects catalytic turnover frequency, and explain how plasmonic fields might also alter reaction selectivity independent of heating.

7. A graphene–MOF hybrid metamaterial exhibits anisotropic electron mobility, with mobility parallel to MOF channels being 4× higher than perpendicular pathways. Propose an experimental method to quantify this anisotropy and explain how molecular orientation within the MOF contributes to directional charge transport.

8. A nanoparticle-based mechanical metamaterial displays auxetic behavior due to re-entrant lattice geometry. If the Poisson ratio is measured as –0.25 during tensile testing, derive how volumetric strain evolves under elongation and explain how nanoscale architectural motifs enable macroscopic auxeticity.

9. A plasmonic catalyst consisting of Au nanorods embedded in a porous polymer drives a photocatalytic hydrogenation reaction. When the nanorod aspect ratio increases from 3.2 to 4.8, the longitudinal plasmon resonance shifts by 90 nm. Discuss how this shift impacts reaction rate under broadband illumination and determine whether optimizing absorption cross-section or hot-electron generation is more important.

10. A doped COF with redox-active linkers undergoes charge transfer upon exposure to NO_2 gas, changing its electrical resistivity. If the COF exhibits a linear response up to 50 ppm but saturates beyond that concentration, derive a model describing adsorption-induced electronic changes and explain the saturation mechanism.

11. A photonic crystal assembled from colloidal silica spheres displays iridescence due to its periodicity. When infiltrated with an organic monomer and polymerized in situ, the lattice spacing expands by 4%. Calculate the expected shift in reflected peak wavelength and comment on structural vs. chemical contributions to optical behavior.

12. A shape-memory chemical metamaterial containing spiropyran motifs changes stiffness when exposed to UV light. If the Young's modulus decreases 5× upon switching, analyze how this property could be exploited to design reconfigurable soft robots and what constraints molecular switching imposes on cycling stability.

13. A catalytic MOF with two distinct pore environments (0.8 nm and 1.4 nm) catalyzes a multistep cascade reaction in which intermediate B must diffuse from small pores to large pores. Develop a diffusion-reaction model describing how pore connectivity affects overall reaction yield.

14. A 2D transition metal dichalcogenide (TMD) metamaterial is engineered with periodic sulfur vacancies. These vacancies create mid-gap states enabling strong photoluminescence quenching. Predict how controlled chemical passivation of these vacancies would influence electronic band structure and optoelectronic performance.

15. A self-oscillating hydrogel metamaterial operates using a Belousov–Zhabotinsky reaction embedded in a polymer network. Describe how chemical oscillations couple to mechanical deformation and propose a method for tuning oscillation frequency via molecular design.

16. A reversible covalent COF used for gas separation undergoes framework breathing when exposed to CO_2, altering pore volume by 20%. Analyze how enthalpic vs. entropic contributions determine the selectivity of CO_2 over N_2, and model the energetics driving structural expansion.

17. A metamaterial battery electrode uses a hierarchical carbon–MOF composite to maximize Li-ion flux. If the MOF pore channels provide 3× higher ionic conductivity than the carbon matrix, determine how the relative volume fractions of each component influence overall electrode impedance.

18. A polyrotaxane-based metamaterial combines sliding ring crosslinks with covalent backbones. When stretched at high strain rates, frictional interactions between rings dominate. Explain how molecular topology governs energy dissipation and propose conditions in which entropic elasticity becomes significant.

19. A molecularly doped polymer metamaterial exhibits nonlinear dielectric behavior at high electric fields. Construct a model describing field-induced alignment or ion migration effects, and predict how altering dopant dipole moment would modify the dielectric constant.

20. A supramolecular photonic gel incorporates host–guest complexes that reversibly dissociate under pH changes. If dissociation reduces refractive index contrast between domains, analyze how this transition impacts photonic bandgap width and suggest sensing applications.\

21. A chiral MOF-based metamaterial rotates plane-polarized light due to asymmetrically arranged ligands. When ligand substitution alters the helical pitch

of the lattice, determine how optical rotation magnitude changes and whether birefringence or molecular chirality contributes more strongly.

22. A plasmonic–catalytic metamaterial uses Cu_2O nanocubes aligned in a photonic crystal. Under 600 nm illumination, hot carriers drive CO_2 reduction. Explain how photonic bandgap alignment enhances quantum efficiency and predict how disorder affects performance.
23. A flexible electronic metamaterial built from radical-containing polymers displays conductivity switching under pressure. Model the relationship between chain packing density and conductive pathway formation, and propose how hydrogen bonding strength could modulate switching threshold.
24. A nanoporous COF containing π-stacked chromophores promotes exciton migration. If exciton diffusion length increases upon increasing crystallinity, evaluate how synthetic defect control influences energy transport and optical gain.
25. A chemical cloaking metamaterial designed for IR absorption uses a gradient of metal–organic clusters. Derive how spatial refractive index gradients bend IR radiation and propose how chemical composition gradients can be fabricated at scale.
26. A catalytic metamaterial for ammonia synthesis uses Fe sites spaced periodically inside a COF, lowering activation barriers cooperatively. Explain how spatial arrangement induces cooperative catalysis and compare predicted rates to isolated Fe nanoparticles.
27. A battery separator metamaterial uses aligned sulfonated polymer channels to suppress dendrite growth. Develop a mechanistic model showing how channel anisotropy influences electric field distribution and dendrite stability.
28. A magnetically responsive polymer–nanoparticle metamaterial changes stiffness under alternating magnetic fields. Analyze how particle volume fraction and ligand shell thickness influence magneto-mechanical coupling.
29. A quantum metamaterial hosts molecular spin centers arranged with 2 nm spacing. Derive how dipole–dipole interactions affect coherence times and suggest chemical strategies to suppress decoherence.

30. A thermal metamaterial uses hierarchically organized MOF particles with extremely low thermal conductivity. Explain how pore size distribution and interparticle contacts combine to produce near-amorphous heat flow behavior.
31. A sunlight-harvesting metamaterial integrates perovskite nanocrystals into a photonic scaffold. Predict how photonic recycling improves external quantum efficiency and identify degradation pathways specific to hybrid structures.
32. A reaction–diffusion metamaterial creates spatial patterns through catalytic COF nodes. Describe how chemical feedback loops generate Turing patterns and predict how pore size affects wavelength of emergent features.
33. A humidity-responsive metamaterial includes hygroscopic ionic groups that swell anisotropically. Model the relationship between humidity, osmotic pressure, and directional swelling to design humidity-controlled actuators.
34. A neural-mimetic chemical metamaterial uses ion-conducting channels and redox-active moieties to emulate synaptic plasticity. Propose how temporal chemical inputs adjust conductivity and create memory-like behavior.
35. A 3D-printed architected COF composite incorporates rotating molecular hinges. Explain how hinge torsional flexibility impacts macroscopic deformation modes under compression.
36. A catalytic plasmonic metamaterial accelerates Suzuki coupling via hot-electron injection into Pd sites. Evaluate how interfacial electronic alignment determines rate enhancement and propose how linker chemistry tunes charge transfer.
37. A metamaterial acoustic absorber embeds soft polymer domains within a rigid COF matrix. Analyze how chemical identity and domain geometry determine resonant absorption frequencies.
38. A metamaterial membrane for ion separation uses crown ether functional groups periodically spaced. Predict how ion size and hydration energy influence selectivity, and relate spacing periodicity to hopping energy barriers.
39. A transparent metamaterial coating uses index-matched MOF–polymer composites. Calculate how nanoparticle volume fraction must be adjusted to minimize scattering at visible wavelengths.

40. A mechanochemical metamaterial includes force-activated mechanophores that fluoresce under stress. Explain how stress concentration at molecular junctions determines activation thresholds.

41. A multiferroic chemical metamaterial integrates magneto-ionic polymers and ferroelectric nanoparticles. Derive how ionic migration under electric fields modulates magnetization.

42. A nanopatterned COF photonic lattice exhibits slow-light modes that enhance photocatalysis. Predict how reduced group velocity modifies reaction rates at embedded catalytic sites.

43. A self-sorting supramolecular metamaterial forms two lattices from similar building blocks. Model how kinetic vs. thermodynamic control determines domain purity and structural order.

44. A chemical gradient metamaterial stores chemical energy in spatially patterned redox couples. Evaluate how controlled diffusion and reaction kinetics enable programmable energy release.

45. A peptide-based metamaterial uses β-sheet stacking to create rigid nanofibers. Assess how hydrogen bonding directionality influences modulus and fracture toughness.

46. A light-driven COF rotary motor array generates macroscopic torque upon irradiation. Determine how molecular torque per motor scales to bulk angular momentum of the entire material.

47. A porous catalytic metamaterial integrates acid–base bifunctional sites at fixed distances. Predict how distance influences cooperative catalysis and determine optimal spacing.

48. A dielectric metamaterial with tunable permittivity incorporates polarizable metal clusters. Explain how cluster aggregation state affects polarization response under AC fields.

49. A chemical metamaterial for gas sensing uses photochromic linkers to modulate optical absorption upon analyte binding. Develop a spectroscopic calibration model relating analyte concentration to absorbance change.

50. A quantum-optical metamaterial embeds molecular emitters in a photonic crystal cavity. Analyze how Purcell enhancement modifies emission lifetime and propose how chemical functionalization tunes cavity–emitter coupling strength.

Chapter 40

Neurochemistry and Molecular Cognition

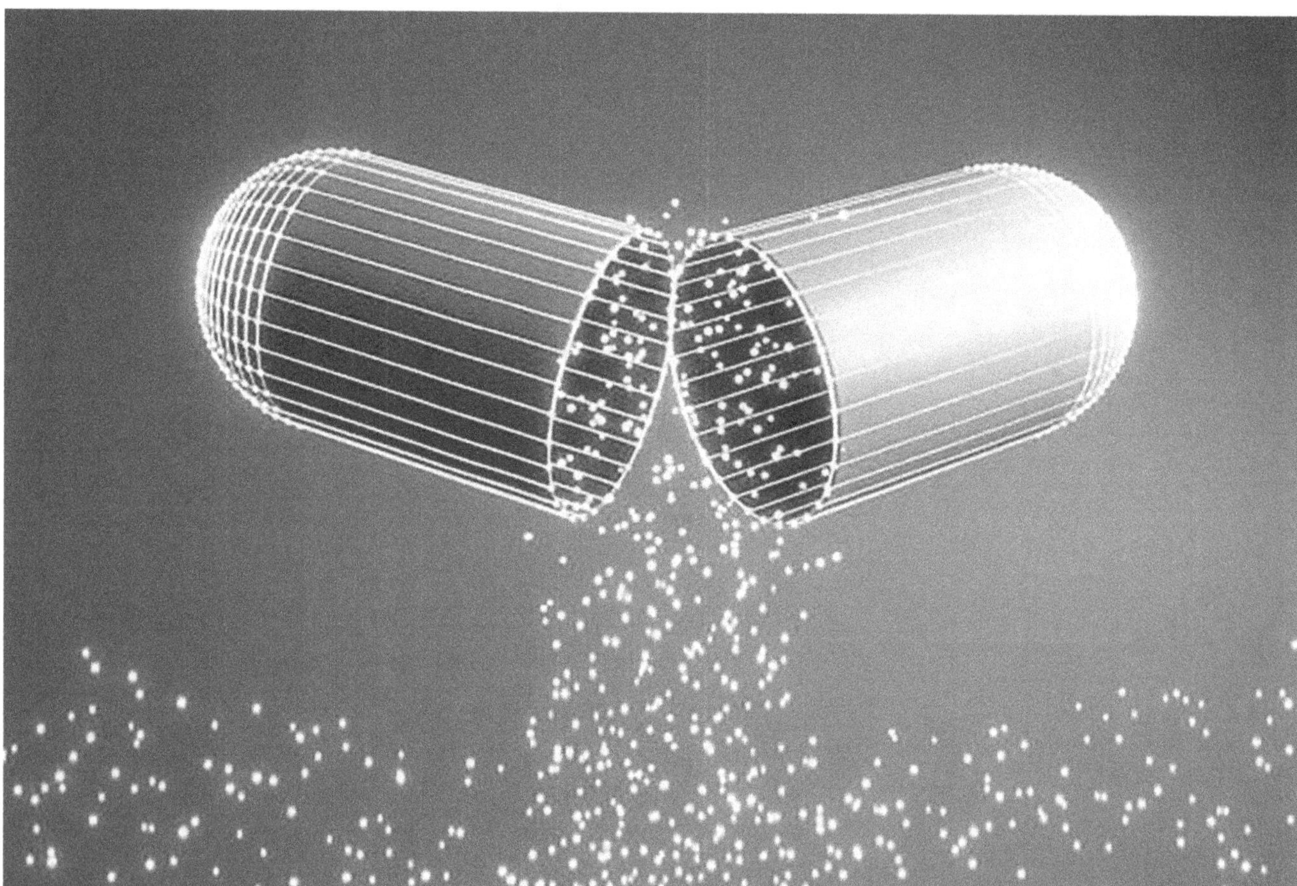

Neurochemistry, at its core, represents the intricate intersection of chemistry and the nervous system, encompassing the molecular foundations of cognition, memory, perception, and behavior. The human brain, comprising roughly 86 billion neurons, functions as a highly complex chemical network where information is encoded, transmitted, and modulated through precise molecular interactions.

Neurons communicate through a combination of electrical impulses and chemical signaling, with neurotransmitters serving as primary mediators. The synthesis, release, reception, and degradation of neurotransmitters are governed by a variety of enzymatic processes and receptor-mediated pathways, reflecting the sophistication of neurochemical control over cognitive processes.

Central to molecular cognition is the role of neurotransmitters such as glutamate, gamma-aminobutyric acid (GABA), dopamine, serotonin, acetylcholine, and norepinephrine. Glutamate, the predominant excitatory neurotransmitter in the mammalian central nervous system, exerts its effects through ionotropic receptors such as AMPA, NMDA, and kainate receptors, and metabotropic glutamate receptors, each linked to distinct intracellular signaling cascades. NMDA receptors, in particular, are critical for synaptic plasticity, long-term potentiation (LTP), and memory consolidation, as their voltage-dependent magnesium block allows for coincidence detection that integrates presynaptic activity with postsynaptic depolarization. The calcium influx through NMDA channels triggers downstream pathways involving calmodulin-dependent protein kinase II (CaMKII), protein kinase C (PKC), and mitogen-activated protein kinase (MAPK), ultimately influencing gene transcription through CREB (cAMP response element-binding protein), which mediates synaptic growth and the stabilization of memory traces.

GABA, in contrast, provides inhibitory control through GABA_A and GABA_B receptors, which modulate chloride and potassium ion fluxes, respectively, hyperpolarizing neurons and shaping network excitability. The balance between excitatory and inhibitory neurotransmission is critical for cognitive fidelity, preventing pathological hyperexcitability while allowing for plastic adaptation during learning. Dopaminergic signaling, originating primarily from the ventral tegmental area and substantia nigra, regulates reward-based learning, motivation, and executive function. Dopamine receptors are classified into D1-like and D2-like families, each coupled to distinct G-protein-mediated pathways. D1-like receptors activate adenylate cyclase, increasing intracellular cyclic AMP and stimulating protein kinase A (PKA), whereas D2-like receptors inhibit adenylate cyclase, modulating the phosphorylation state of ion channels and transcription factors. The interplay between these signaling pathways underlies reinforcement learning, decision-making, and working memory processes.

Serotonergic systems, distributed across the raphe nuclei, influence mood, cognition, and perception, acting via multiple receptor subtypes (5-HT1 through

5-HT7). Serotonin modulates excitatory and inhibitory circuits, neurogenesis, and synaptic plasticity, often through intracellular cascades involving phospholipase C, PKC, and intracellular calcium signaling. Similarly, cholinergic neurons, particularly those originating in the basal forebrain and projecting to cortical regions, modulate attention, memory encoding, and sensory processing through nicotinic and muscarinic receptors, engaging both ionotropic and G-protein coupled mechanisms. Acetylcholine's regulation of hippocampal theta rhythms illustrates its role in temporal coding and information integration in cognition.

The neurochemical basis of cognition is not limited to classical neurotransmitters. Neuromodulators, neuropeptides, and gaseous transmitters such as nitric oxide contribute to fine-tuning synaptic efficacy and network dynamics. Nitric oxide, synthesized from L-arginine by nitric oxide synthase (NOS), diffuses freely across membranes, activating soluble guanylate cyclase and increasing cyclic GMP levels, which in turn modulate ion channels, neurotransmitter release, and synaptic plasticity. Endocannabinoids, such as anandamide and 2-arachidonoylglycerol, provide retrograde signaling by binding to presynaptic CB1 receptors, suppressing neurotransmitter release, and shaping long-term synaptic modifications critical for learning and memory.

At the cellular level, the molecular machinery of cognition involves the dynamic regulation of ion channels, transporters, and intracellular signaling networks. Voltage-gated sodium and potassium channels dictate action potential initiation and propagation, while calcium channels orchestrate activity-dependent neurotransmitter release at synaptic terminals. Calcium, serving as a versatile second messenger, activates numerous pathways including calpain-mediated cytoskeletal remodeling, calcium/calmodulin-dependent kinases, and phosphatases like calcineurin, which collectively influence dendritic spine morphology, receptor trafficking, and synaptic efficacy. The precise spatial and temporal regulation of calcium transients is fundamental to synaptic integration and Hebbian plasticity, wherein coincident activity strengthens or weakens specific synaptic connections, encoding memory at the molecular level.

Beyond single synapses, cognition emerges from the coordinated activity of neuronal ensembles and microcircuits, whose chemical interactions are mediated by astrocytes and microglia. Astrocytes, through their tripartite synapse interactions, regulate neurotransmitter uptake, release gliotransmitters, and modulate extracellular ion concentrations, thereby influencing synaptic plasticity and network homeostasis. Microglia, through the release of cytokines, growth factors, and synaptic pruning mechanisms, contribute to structural remodeling, refinement of neural circuits, and neuroimmune modulation of cognition. The chemical dialogue between neurons and glia integrates metabolic support, trophic signaling, and synaptic modulation, illustrating that cognition is inherently a systems-level chemical phenomenon.

At the genomic and epigenetic level, molecular cognition is further influenced by transcriptional and translational regulation, which allows neurons to adapt to sustained activity and environmental stimuli. Immediate early genes such as c-fos, Arc, and Egr1 are rapidly induced by synaptic activity, triggering protein synthesis-dependent LTP and synaptic consolidation. Epigenetic modifications, including DNA methylation, histone acetylation, and non-coding RNA regulation, shape neuronal gene expression patterns, modulating synaptic strength and long-term memory formation. These molecular mechanisms reveal that cognitive processes are encoded not only in synaptic networks but also in dynamic chemical modifications that bridge experience with genomic expression.

Molecular studies of cognition also emphasize the importance of protein-protein interactions in synaptic signaling. Scaffolding proteins such as PSD-95, Homer, and Shank organize receptor complexes, anchor signaling molecules, and facilitate cross-talk between glutamatergic and dopaminergic pathways. The dynamic assembly and disassembly of these protein networks are modulated by phosphorylation, palmitoylation, and ubiquitination, which provide rapid yet reversible control over synaptic strength. Protein kinases, phosphatases, and proteases interact in a finely tuned chemical symphony that allows neurons to encode, retrieve, and update information with high fidelity.

Recent advances in neurochemical imaging and optogenetics have revealed the temporal and spatial complexity of molecular cognition, allowing the direct observation of neurotransmitter dynamics, receptor activation, and intracellular signaling in vivo. Techniques such as two-photon microscopy, genetically encoded calcium indicators, and optogenetic control of specific signaling pathways provide unprecedented resolution of the chemical events underlying learning, memory, and decision-making. These technologies have demonstrated that cognition emerges from the integration of molecular signals across scales, from synapses to networks, and highlight the importance of both transient and sustained chemical changes in neural computation.

Cognitive dysfunction, whether in neurodegenerative diseases, psychiatric disorders, or acute injury, often arises from perturbations in neurochemical networks. Alzheimer's disease, for example, is characterized by deficits in cholinergic transmission, dysregulated glutamatergic signaling, and pathological accumulation of amyloid-β and tau proteins, which disrupt synaptic chemistry and plasticity. Parkinson's disease features degeneration of dopaminergic neurons in the substantia nigra, leading to impaired basal ganglia circuitry and altered synaptic signaling. Schizophrenia and major depressive disorder have been linked to imbalances in dopaminergic, serotonergic, and glutamatergic systems, reflecting how subtle chemical dysregulation can profoundly affect cognition, perception, and behavior. Understanding these pathologies at the molecular level informs the rational design of pharmacological interventions aimed at restoring chemical homeostasis and cognitive function.

Pharmacological modulation of cognition relies on the precise targeting of receptors, transporters, and enzymes involved in neurotransmitter metabolism. Agents such as selective serotonin reuptake inhibitors (SSRIs), acetylcholinesterase inhibitors, NMDA receptor antagonists, and dopamine agonists exemplify the chemical strategies used to enhance, restore, or suppress specific neurochemical pathways. Moreover, emerging therapies, including gene therapy, RNA-based modulation, and optogenetic pharmacology, aim to correct molecular dysfunctions at their source, offering targeted interventions for cognitive enhancement and neuroprotection.

Molecular cognition also intersects with neuroenergetics and metabolic chemistry. The high energy demands of synaptic transmission require tightly regulated ATP production, glucose metabolism, and mitochondrial function. Neurons rely on oxidative phosphorylation and glycolytic flux to maintain ionic gradients, recycle neurotransmitters, and support protein synthesis. Metabolic intermediates such as acetyl-CoA not only fuel energy production but also serve as substrates for acetylcholine synthesis, histone acetylation, and lipid signaling, directly linking cellular metabolism with chemical modulation of cognition.

Finally, the chemical basis of cognition extends to emergent phenomena such as consciousness, decision-making, and creativity. Although these processes remain challenging to define in purely chemical terms, studies of network oscillations, neuromodulatory interactions, and synaptic integration suggest that cognition arises from the coordinated activity of molecularly diverse signaling pathways, dynamically interacting across spatial and temporal scales. Molecular models of cognition increasingly emphasize the probabilistic and context-dependent nature of neural computation, in which chemical signals encode not only deterministic responses but also stochastic variability, allowing for adaptive behavior in complex environments.

Neurochemistry and molecular cognition represent an intricate tapestry of chemical interactions that underlie the human mind. From neurotransmitter synthesis and receptor activation to intracellular signaling, glial modulation, gene expression, and metabolic integration, cognition emerges as a chemically orchestrated process. Advancements in molecular neuroscience, imaging, and computational modeling continue to deepen our understanding of how chemical networks encode, transmit, and adapt information, providing a foundation for both fundamental insights into human cognition and translational strategies for addressing cognitive disorders. The convergence of chemistry, biology, and computation at the molecular level underscores the central role of chemical principles in shaping thought, memory, and behavior, revealing the brain as both a chemical engine and a molecular information processor.

Neurochemistry encompasses a remarkable diversity of molecular entities, ranging from classical neurotransmitters like glutamate, GABA, dopamine, serotonin, and acetylcholine, to neuropeptides, endocannabinoids, and neuromodulators that fine-tune synaptic and extrasynaptic signaling. Each of these chemical messengers is synthesized through precise enzymatic cascades, often tightly regulated by cofactors, allosteric modulators, and subcellular compartmentalization. For instance, the biosynthesis of dopamine requires tyrosine hydroxylase-mediated hydroxylation to L-DOPA, followed by aromatic L-amino acid decarboxylase action, processes that are sensitive to cellular redox state and availability of tetrahydrobiopterin. The flux through these pathways not only determines neurotransmitter levels but also modulates neuronal excitability, plasticity, and the probability of synaptic vesicle release.

Receptor-ligand interactions constitute the chemical interface through which extracellular signals are translated into intracellular responses. Ionotropic receptors, such as NMDA and AMPA glutamate receptors, exhibit ligand-gated conformational changes that open ion channels, while metabotropic G-protein-coupled receptors (GPCRs) initiate second-messenger cascades, modulating cyclic AMP, phosphoinositides, and calcium signaling. The spatiotemporal dynamics of these signaling molecules are influenced by scaffolding proteins, lipid microdomains, and subcellular compartmentalization, demonstrating that cognition relies not merely on the presence of molecules but on their precise chemical choreography. Understanding the kinetics, thermodynamics, and structural determinants of these receptor-ligand interactions has opened avenues for rational drug design targeting cognitive disorders, psychiatric disease, and neurodegeneration.

Intracellular signaling networks translate receptor activation into gene expression, synaptic remodeling, and metabolic adaptation. Kinase cascades, phosphatases, and ubiquitin-mediated proteolysis provide reversible chemical switches that control the functional state of neurons. Calcium ions serve as universal second messengers, linking membrane excitability to transcriptional programs via calmodulin-dependent kinases and calcineurin pathways. The redox environment, influenced by mitochondrial activity and reactive oxygen species, further modulates signaling fidelity, enzyme activity, and synaptic plasticity. Chemically, the neuron is a highly sensitive microreactor, where minor shifts in ion gradients, phosphorylation states, or metabolic flux can lead to substantial changes in functional output.

Glial cells, once considered passive support elements, are now recognized as active chemical participants in cognitive processes. Astrocytes regulate neurotransmitter clearance and recycling, modulate extracellular potassium and pH, and release gliotransmitters such as ATP, D-serine, and glutamate. Microglia contribute to synaptic pruning and immune surveillance through the release of cytokines, chemokines, and reactive nitrogen and oxygen species. Oligodendrocytes, by modulating myelin composition and lipid metabolism, influence the conduction velocity of action potentials, thereby affecting the temporal precision of synaptic integration. The chemical interplay between neurons and glia demonstrates that cognition arises from a multicellular chemical symphony rather than isolated neuronal signaling.

At the level of systems chemistry, metabolic integration plays a central role in cognition. Glucose metabolism, oxidative phosphorylation, and lactate shuttling between astrocytes and neurons provide energy substrates necessary for synaptic transmission and plasticity. Neurotransmitter synthesis consumes specific metabolites, linking metabolic state to signaling capacity. Amino acid availability, fatty acid oxidation, and one-carbon metabolism can influence epigenetic landscapes, affecting histone and DNA modifications that encode long-term memory. In this sense, cognitive function is a chemical phenotype emerging from the integration of metabolic, signaling, and structural networks.

Molecular cognition also depends on the precise temporal coding of chemical signals. Short-term plasticity relies on rapid vesicular release and receptor desensitization, while long-term potentiation and depression involve sustained chemical modifications of receptors, ion channels, and cytoskeletal proteins. Protein phosphorylation, SUMOylation, nitrosylation, and glycosylation serve as chemical memory marks, maintaining synaptic strength and network connectivity over hours, days, or even a lifetime. Computational modeling of these modifications at the molecular level reveals how signal integration, noise filtering, and information storage can be understood in terms of chemical reaction networks and probabilistic kinetics.

Emerging technologies, such as super-resolution imaging, optogenetics, and chemogenetics, allow direct visualization and manipulation of chemical events in living neural circuits. Fluorescent and luminescent chemical probes track ions, metabolites, and second messengers with subcellular resolution. Photoactivatable ligands and designer receptors enable spatiotemporal control of neurotransmission, permitting dissection of chemical causality in cognitive behaviors. These tools reveal that cognition is not only a product of which molecules are present but also of their precise timing, localization, and chemical context.

Chemical dysregulation underlies a spectrum of cognitive disorders. Parkinson's disease arises from dopaminergic neuron degeneration, Alzheimer's disease from amyloid-beta aggregation and tau hyperphosphorylation, and schizophrenia from imbalances in glutamatergic and dopaminergic signaling. Therapeutic strategies increasingly target these molecular pathways, including small-molecule enzyme inhibitors, receptor modulators, and metabolite replacement therapies. Even subtle shifts in chemical homeostasis—altered pH, redox potential, or neurotransmitter ratios—can precipitate profound cognitive deficits, highlighting the sensitivity of the brain's chemical networks.

Epigenetic and neurochemical integration further expands the landscape of molecular cognition. DNA methylation, histone modification, and noncoding RNAs respond to neuronal activity and environmental stimuli, linking chemical signaling to stable alterations in gene expression. Neurotransmitter-dependent epigenetic changes can reinforce synaptic connectivity, thereby embedding experiences within molecular memory traces. Metabolites such as S-adenosylmethionine, acetyl-CoA, and NAD^+ function as both metabolic intermediates and cofactors for chromatin-modifying enzymes, directly connecting cellular chemistry to the storage of cognitive information.

Finally, artificial intelligence and computational chemistry are increasingly applied to model the chemical underpinnings of cognition. Molecular dynamics simulations, reaction network modeling, and machine learning analyses of chemical data sets enable prediction of emergent cognitive behaviors from molecular interactions. These approaches reveal design principles of chemical information processing, offering insights into memory storage, learning algorithms, and potentially even artificial cognition. By decoding the brain as a chemical information processor, neurochemistry provides a framework for understanding not only how we think but how chemical logic itself can be harnessed to create intelligent systems.

In conclusion, neurochemistry and molecular cognition exemplify a multidimensional integration of chemistry, biology, and computation. Cognition arises from the orchestration of neurotransmitter synthesis, receptor activation, intracellular signaling, glial modulation, metabolic flux, and epigenetic regulation. Each molecular interaction contributes to the emergent properties of thought, memory, and behavior, revealing the brain as a chemical engine of unparalleled complexity. Continued exploration at the interface of chemical biology, systems neuroscience, and computational modeling promises to unravel the molecular logic of the mind, guiding both therapeutic innovation and fundamental understanding of human cognition.

Chapter 40: Word Problems

1. A researcher stimulates hippocampal CA1 neurons repeatedly to induce LTP and measures CaMKII autophosphorylation rates. She finds that blocking NMDA receptors reduces CaMKII activation by 85%, but intracellular calcium remains elevated due to L-type calcium channels. Explain why CaMKII phosphorylation still fails to reach threshold and calculate the proportional contribution of NMDA-mediated calcium influx to CaMKII activation if basal activation is 5%, L-type channels alone yield 10%, and full LTP requires 70%.

2. During an experiment on dopaminergic VTA neurons, a pharmacological agent selectively inhibits D2-like receptor coupling to Gi proteins. Firing rates increase by 25%, and cAMP levels double. If baseline cAMP is 40 μM, determine the new concentration and describe how the disruption alters reinforcement learning models.

3. A glial depletion model reduces astrocytic glutamate uptake by 40%, causing spillover onto extrasynaptic NMDA receptors. If synaptic glutamate concentration decays with a time constant $\tau = 2$ ms normally, but τ increases to 8 ms after depletion, calculate the percentage increase in extracellular dwell time and predict its effect on LTP/LTD balance.

4. A mutant PSD-95 variant reduces AMPA receptor anchoring efficiency by 35%. If the neuron initially expresses 900 synaptic AMPA receptors but compensates by increasing receptor insertion by 20%, determine the final number of synaptic receptors and discuss how this affects synaptic scaling.

5. In cortical interneurons, GABA_A receptor chloride reversal potential shifts from –70 mV to –60 mV due to altered KCC2 expression. Compute the change in inhibitory postsynaptic potential amplitude if the driving force was originally 20 mV and correlate this to increased seizure susceptibility.

6. A serotonin-producing neuron experiences 50% reduction in tryptophan hydroxylase activity. If serotonin synthesis rate correlates linearly and baseline release is 120 nM per stimulus, estimate the new release level and explain how 5-HT1A autoreceptor feedback might further modulate firing rate.

7. Endocannabinoid retrograde signaling is impaired because DAG lipase activity decreases by 60%. If 2-AG normally suppresses presynaptic release probability from 0.45 to 0.25, compute the new release probability and evaluate consequences for short-term depression.

8. A neuron's NMDA receptor Mg^{2+} block is experimentally lowered so that channel open probability increases by 40% at resting potential. Predict how this would influence coincidence detection, and calculate relative Ca^{2+} entry if baseline is 1.0 arbitrary units.

9. A microglial knockout model lacks complement receptor C3R, reducing synaptic pruning efficiency by 75%. Given a developing cortical region with 12,000 synapses

that normally stabilizes to 8,000, estimate the number after knockout and interpret network-level effects.

10. An acetylcholinesterase inhibitor increases acetylcholine lifetime in the synaptic cleft from 1 ms to 4 ms. If muscarinic receptor activation probability increases with the square root of lifetime, calculate the fold change and discuss implications for memory encoding.
11. An antioxidant drug reduces neuronal ROS levels by 30%, increasing mitochondrial ATP output by 18%. If baseline ATP is 2.5 mM, compute the new level and evaluate how altered redox signaling influences kinases involved in synaptic plasticity.
12. Blocking CREB phosphorylation reduces transcription of immediate early genes by 50%. If 700 Arc mRNA molecules per neuron were typical after stimulation, estimate new levels and describe the expected long-term effect on spine stabilization.
13. A GPCR mutation alters Gq-coupling efficiency, reducing IP3 production by 40% when stimulated. If intracellular calcium release depends linearly on IP3 concentration and normally increases by 200 nM, compute the new increase and consider effects on PKC activation.
14. A neuronal culture is exposed to a glutamate transporter inhibitor that raises extracellular glutamate by 15 µM. If NMDA receptor activation probability increases by 2% per µM of glutamate, compute the increase and evaluate excitotoxic risk.
15. An AMPA receptor undergoes a Q/R site mutation that increases Ca^{2+} permeability by 300%. If baseline Ca^{2+} influx per event is 0.2 pA, determine the new value and discuss how this may shift the neuron toward Ca^{2+}-dependent plasticity or toxicity.
16. A dopaminergic neuron exposed to chronic D1 agonist treatment shows a 45% increase in PKA activity. If baseline PKA activity is 3.2 units (arbitrary), calculate the new activity level. Explain how sustained PKA elevation may alter DARPP-32 phosphorylation and influence working memory circuits in the prefrontal cortex.
17. A mutation in CaMKII reduces its affinity for Ca^{2+}/calmodulin by 55%. If the dissociation constant (Kd) was originally 5 nM, compute the new Kd and describe how this change affects the threshold for LTP induction during theta-burst stimulation.

18. A mouse line lacking NMDA receptors in interneurons exhibits increased pyramidal neuron firing synchrony. If inhibitory postsynaptic current (IPSC) timing jitter normally ranges ±3 ms but increases to ±9 ms, calculate the percentage increase in timing variability and reason how this disrupts gamma oscillations.

19. Blocking GABA_B receptors results in decreased activation of inwardly rectifying potassium channels, raising neuronal input resistance by 25%. If baseline resistance is 180 MΩ, determine the new value and predict how EPSP amplitude changes given Ohm's law.

20. A hippocampal slice treated with an NO synthase inhibitor shows a 60% reduction in cGMP production. If cGMP-dependent protein kinase activity drops proportionally and baseline activity is 90 units, compute the new value. Discuss the impact on retrograde plasticity mechanisms.

21. A neuron receives two converging synaptic inputs whose glutamate release probabilities are 0.30 and 0.50. After induction of presynaptic LTP, probabilities increase by 40% for both synapses. Calculate new probabilities and analyze how this reshapes synaptic competition during Hebbian plasticity.

22. An astrocytic lactate-shuttle blockade reduces neuronal ATP availability by 15%. If the neuron previously devoted 65% of its ATP budget to Na^+/K^+ ATPase function, determine the percentage of total ATP now available for other processes. Discuss consequences for action potential reliability.

23. An mGluR5 antagonist lowers intracellular DAG levels by 70%. If DAG-dependent PKC signaling originally elevated AMPA receptor insertion by 25%, estimate the new insertion percentage. Interpret effects on metabotropic-induced LTD.

24. A neuron expresses a hyperactive form of calcineurin, producing a 2.4-fold increase in dephosphorylation rate of synaptic proteins. If the baseline dephosphorylation rate was 0.8 events/s, calculate the new rate. Evaluate the likelihood of LTD dominance in this system.

25. During oxidative stress, glutathione levels fall by 50%, leading to 35% higher ROS accumulation. If ROS normally oxidize 4% of redox-sensitive cysteine residues in ion channels, determine the new oxidation percentage and relate this to altered firing thresholds.

26. A neuron with reduced Shank3 expression shows a 30% decline in mGluR–NMDA receptor cross-talk. If this cross-talk normally enhances Ca^{2+} influx by 0.5 μM, compute the decreased enhancement and discuss ASD-related implications for synaptic integration.

27. A cholinergic deficit reduces cortical nicotinic receptor activation by 40%. If attentional performance correlates linearly with activation and baseline performance is indexed at 100, estimate the new performance index and describe effects on prefrontal theta modulation.

28. Blocking voltage-gated calcium channels reduces neurotransmitter release probability from 0.6 to 0.22. Calculate the percentage reduction and discuss how this impacts spike-timing-dependent plasticity (STDP) curves.

29. Acetyl-CoA availability is lowered by 25% in a metabolic manipulation. If acetyl-CoA contributes equally to energy metabolism and histone acetylation, determine the proportional reduction in histone acetylation substrates and relate this to memory consolidation deficits.

30. An NMDA receptor antagonist increases EPSP rise time from 1.5 ms to 4.2 ms. Compute the fold change and explain how slowed EPSPs affect temporal summation during high-frequency bursts.

31. Serotonin transporter (SERT) overexpression reduces extracellular 5-HT by 65%. If postsynaptic receptor activation drops from 0.80 to 0.30 probability, compute the correlation ratio between extracellular 5-HT and receptor activation. Discuss mood-related consequences.

32. A dendritic spine shrinks from 0.12 μm^3 to 0.07 μm^3 following LTD induction. Calculate the percentage volume decrease and analyze the expected change in AMPA receptor surface area.

33. An optogenetic stimulation protocol increases intracellular calcium peaks from 300 nM to 550 nM. Compute the percentage increase and provide reasoning on how CaMKII versus calcineurin balance shifts.

34. An interneuron's GABA synthesis rate declines by 20% due to reduced glutamic acid decarboxylase activity. If synaptic inhibitory charge transfer per event

decreases proportionally from 45 pC, determine the new value and the likely effect on cortical excitability.

35. Excitatory synapses in the prefrontal cortex exhibit a 150% increase in spontaneous glutamate release under stress. If frequency rises from 2 Hz to 5 Hz, verify the percentage increase and explore implications for stress-related working memory impairment.

36. Tau hyperphosphorylation reduces microtubule stability by 48%. If axonal transport velocity declines from 2.5 μm/s, calculate the new velocity and discuss the downstream effects on synaptic vesicle replenishment.

37. An inhibitory neuromodulator reduces cAMP levels by 70%. If PKA activity decreases accordingly and baseline activity is 4.8 units, determine the new value and describe consequences for CREB-dependent transcription.

38. A cortical region loses 20% of its oligodendrocytes, decreasing conduction velocity from 45 m/s to 32 m/s. Compute the percentage reduction and interpret timing effects on multisensory integration.

39. A mutation in a voltage-gated sodium channel slows activation kinetics such that action potential threshold shifts from –55 mV to –48 mV. Determine the change in threshold and explain effects on population coding.

40. D-serine release from astrocytes decreases by 60%, reducing NMDA co-agonist site occupancy. If occupancy falls from 0.75 to 0.30, calculate the proportional decrease and evaluate effects on spike-timing plasticity.

41. During neuroinflammation, cytokine release increases glutamate release probability by 30%. If baseline probability is 0.25, compute the new value and predict excitotoxic risk.

42. Enhanced mitochondrial biogenesis increases ATP availability by 22%. If the neuron originally produced 3.6 mM ATP, compute the new concentration and analyze consequences for synaptic vesicle cycling.

43. Blocking astrocytic K^+ buffering increases extracellular K^+ from 3 mM to 7.5 mM. Calculate the fold increase and explain how this affects membrane potential and propensity for spreading depolarization.

44. Loss of BDNF reduces TrkB receptor signaling by 55%. If downstream MAPK activation drops from 100 units to a reduced value, compute it and discuss long-term consequences for dendritic arborization.

45. A neuron increases its G-protein turnover rate by 2.5-fold under chronic drug exposure. If baseline turnover is 0.4 cycles/s, determine the new rate and interpret adaptation within GPCR desensitization pathways.

46. A synapse's vesicle recycling time increases from 80 ms to 140 ms due to clathrin pathway disruption. Calculate the percentage increase and predict limitations on sustained high-frequency firing.

47. An AMPA receptor potentiator increases single-channel conductance from 10 pS to 16 pS. Compute the percentage increase and evaluate the expected EPSC amplitude enhancement.

48. Under prolonged learning tasks, Arc-mediated endocytosis removes 18% of synaptic AMPA receptors. If a synapse originally contained 620 receptors, compute the remaining number and explore implications for metaplasticity.

49. A neuron's intrinsic firing frequency increases from 12 Hz to 18 Hz after modulation of HCN channels. Calculate the percentage increase and describe how enhanced Ih current affects temporal encoding.

50. A pharmacological agent increases histone acetylation by 45%, raising expression of memory-associated genes from 280 mRNA copies to a higher value. Compute the new copy number and explain how this shifts the balance from transient to stable memory traces.

Chapter 41

Molecular Gastronomy

Molecular gastronomy, as a scientific discipline, represents the convergence of chemistry, physics, and culinary art, providing a framework to understand the molecular transformations that occur during food preparation, cooking, and consumption. It is not merely concerned with aesthetics or flavor combinations; rather, it investigates the underlying chemical, physical, and biochemical mechanisms that dictate texture, flavor release, aroma development, and visual appeal. The field emerged from a desire to apply rigorous scientific methods to understand traditional cooking phenomena while enabling the design of novel culinary experiences. At its core, molecular gastronomy explores how macromolecules, including proteins, carbohydrates, and lipids, interact with solvents, ions, and other food components under specific conditions of temperature, pH, mechanical stress, and chemical environment. By studying these interactions, chefs and scientists alike can manipulate food properties at the molecular level, producing textures, flavors, and presentations that would be unattainable with conventional techniques.

Proteins serve as a central component in the molecular understanding of food. They are responsible for structure, gelation, foam stabilization, and enzymatic activity in culinary systems. Proteins undergo denaturation when exposed to heat, pH shifts, mechanical forces, or chemical additives, resulting in the disruption of secondary, tertiary, and quaternary structures. This exposes hydrophobic residues and sulfhydryl groups, leading to aggregation, cross-linking, and the formation of stable gels. In egg-based systems, the precise denaturation of ovalbumin and ovotransferrin during thermal processing allows the transformation of liquid egg whites into firm meringues or custards, depending on temperature and agitation. Similarly, casein micelles in milk, when subjected to controlled heat and enzymatic action, aggregate to form cheese curds, demonstrating the importance of protein chemistry in structuring foods. Enzymes such as transglutaminase facilitate covalent cross-linking between lysine and glutamine residues in proteins, creating novel textures and shapes without altering the intrinsic flavor profile, exemplifying how chemical modification at the molecular level can be harnessed for culinary innovation.

Polysaccharides, including starches, cellulose derivatives, and hydrocolloids, provide additional means for manipulating texture and viscosity. Starch gelatinization occurs when granules absorb water and swell upon heating, breaking intermolecular hydrogen bonds and leaching amylose into the surrounding medium.

This process converts a semi-crystalline granular structure into a viscous gel, whose consistency depends on the amylose-to-amylopectin ratio, temperature, and water content. Hydrocolloids such as agar, carrageenan, xanthan gum, and pectin extend the culinary toolkit, enabling gelation, encapsulation, and stabilization of emulsions and foams. Spherification, a signature technique in molecular gastronomy, relies on the ionic cross-linking of alginate with divalent cations such as calcium to form a thin gel membrane around a liquid core, producing spheres that burst in the mouth. This process exemplifies the application of polymer chemistry principles in controlling the structure and mouthfeel of food at the microscale. Moreover, complex coacervation between oppositely charged polysaccharides or proteins can be exploited to encapsulate flavor molecules or bioactive compounds, allowing controlled release during consumption.

Lipids and lipid-based systems are equally critical to molecular gastronomy due to their role in flavor solubilization, texture, and mouthfeel. Emulsions, which are dispersions of immiscible liquids stabilized by surfactants or amphiphilic molecules, demonstrate how chemistry can be used to engineer food systems. Traditional mayonnaise relies on lecithin from egg yolk to stabilize oil droplets within an aqueous phase, forming a viscoelastic network. Advanced techniques, such as high-pressure homogenization, microfluidization, or ultrasonic emulsification, reduce droplet size to the nanometer scale, producing nanoemulsions with enhanced stability and modified organoleptic properties. The stabilization of foams, the gas-liquid analog of emulsions, employs proteins, polysaccharides, or phospholipids to reduce surface tension at the air-water interface, resulting in lightweight, aerated textures. Techniques such as whipping, foaming with siphons, or using surfactant-stabilized foams illustrate the application of interfacial chemistry to culinary innovation.

The Maillard reaction, a non-enzymatic browning process, exemplifies the chemical complexity underlying flavor development in cooking. It occurs when reducing sugars react with amino acids, forming Schiff bases that rearrange into Amadori compounds and subsequently undergo a cascade of dehydration, fragmentation, and condensation reactions to produce melanoidins and hundreds of volatile flavor molecules. The reaction is highly sensitive to temperature, pH, water activity, and the nature of the reactants, enabling precise modulation of flavor, color, and aroma. Caramelization, the thermal decomposition of sugars, generates complementary flavors through the formation of furans, aldehydes, and ketones. Understanding the thermodynamics and kinetics of these reactions allows molecular gastronomists to control sensory outcomes, producing complex flavors without the development of off-notes such as bitterness or acridness.

Phase transitions, such as melting, freezing, and glass formation, offer additional routes for texture modification. Sous-vide cooking exemplifies the application of precise thermal control to influence protein denaturation, enzymatic activity, and moisture retention. By maintaining food at sub-denaturing temperatures for extended periods, collagen can hydrolyze slowly into gelatin, resulting in tender meats with retained juiciness. Cryogenic techniques, employing liquid nitrogen or carbon dioxide, exploit rapid freezing to arrest molecular motion, producing novel textures such as brittle foams, frozen powders, or granular ice structures. Freeze-drying and lyophilization remove water under vacuum at low temperatures, preserving volatile flavors and producing crisp, lightweight textures suitable for deconstructed culinary presentations.

Flavor perception in molecular gastronomy is fundamentally chemical, arising from the interaction of volatile and non-volatile molecules with olfactory and gustatory receptors. Aroma compounds, often small hydrophobic molecules, volatilize during cooking or upon consumption and bind to olfactory receptors, generating complex perceptions of flavor. Taste is mediated by the interaction of soluble molecules with receptor proteins on the tongue, where sweet, salty, bitter, umami, and sour sensations are detected. Techniques such as encapsulation in cyclodextrins, liposomes, or hydrocolloids allow controlled release of flavor molecules, producing delayed or staged flavor delivery. The manipulation of pH, ionic strength, polarity, or redox environment can further modulate flavor perception, enabling the design of sensory experiences that defy conventional expectations. For example, altering pH can shift the perception of acidity or sweetness, while ionic gradients can enhance or suppress specific flavor notes through receptor-mediated mechanisms.

Analytical chemistry plays a crucial role in molecular gastronomy by providing methods to quantify, characterize, and monitor molecular transformations in food systems. Spectroscopic techniques, including nuclear magnetic resonance (NMR), mass spectrometry, infrared spectroscopy, and ultraviolet-visible spectroscopy, allow the identification and quantification of flavor compounds, Maillard intermediates, and lipid oxidation products. Chromatography, including gas chromatography and high-performance liquid chromatography, separates complex mixtures of volatile and non-volatile molecules, enabling precise monitoring of flavor development or degradation. Rheology provides insights into texture, gel strength, viscoelastic properties, and flow behavior, informing the rational design of foams, gels, and emulsions. Microscopy, including scanning electron microscopy and confocal laser scanning microscopy, reveals structural organization at micro- and nanoscales, guiding the engineering of food constructs with controlled porosity, layering, and encapsulation.

Enzymatic processes are harnessed to produce specific chemical modifications and flavor developments. Proteases, lipases, and carbohydrases catalyze selective hydrolysis, generating peptides, free fatty acids, or sugars that influence taste, aroma, and texture. For instance, proteolytic tenderization of meat using papain or bromelain modifies the protein matrix to increase tenderness without compromising flavor integrity. Enzyme-assisted flavor release can convert inactive precursors into active aroma compounds, enabling controlled development of flavor during cooking or at the moment of consumption. Transglutaminase-mediated protein cross-linking enables the formation of edible networks with customized textures, supporting innovative presentations in modern cuisine.

Redox reactions and pH-driven color changes provide additional avenues for chemical creativity. Anthocyanin pigments in fruits and vegetables exhibit pH-dependent chromatic shifts, enabling chefs to create visually dynamic dishes that respond to acidity or alkalinity. Oxidation-reduction reactions, sometimes catalyzed by enzymes or transition metals, can alter pigment structure or flavor compound profiles, producing dramatic changes in both appearance and taste. Such chemical manipulation underscores the ability of molecular gastronomy to transform food into an interactive, dynamic system rather than a static substrate.

Self-assembly and supramolecular chemistry have been adapted to culinary applications to create structures that mimic biological forms or exhibit emergent properties. Amphiphilic molecules, including surfactants or designer lipids, can organize into micelles, vesicles, or bilayers that encapsulate flavors or bioactive molecules. Polysaccharide-protein complexes can self-organize into layered gels, lattices, or filaments, producing textures and forms that are not achievable through conventional cooking. These approaches highlight the integration of nanochemistry and materials science into culinary design, enabling the construction of food at scales ranging from molecular to macroscopic.

Safety and nutritional considerations intersect with molecular gastronomy, as chemical reactions can produce unintended byproducts or alter bioavailability. Lipid oxidation, for instance, can generate off-flavors and potentially harmful aldehydes if not controlled. High-temperature Maillard reactions, if pushed beyond optimal conditions, may produce acrylamide, a neurotoxic and potentially carcinogenic compound. The judicious application of chemical principles allows chefs to minimize these risks while maximizing flavor, texture, and aesthetic impact. Molecular gastronomy also informs functional food design, where chemical modification or encapsulation can enhance nutrient stability, bioavailability, or targeted delivery of vitamins, minerals, and bioactive compounds.

Historical and contemporary developments in molecular gastronomy illustrate the iterative relationship between chemistry and culinary art. Early experiments with foams, gels, and emulsions provided empirical evidence of chemical phenomena that were later rationalized using modern chemistry. Contemporary chefs, in collaboration with food scientists, have extended these principles to create multi-sensory experiences, deconstructed dishes, and edible displays that challenge traditional conceptions of cuisine. Techniques such as spherification, foaming under controlled conditions, sous-vide cooking, and cryogenic freezing are now widely used, demonstrating the reproducibility and scalability of molecular gastronomy when grounded in chemical understanding.

Future directions in molecular gastronomy emphasize the integration of synthetic biology, computational modeling, and advanced analytical techniques. Designer enzymes and genetically encoded flavor precursors allow precise temporal control of flavor and texture. Computational chemistry models predict reaction kinetics, flavor release, and textural outcomes under varying conditions, guiding rational design of dishes before empirical testing. The combination of chemical insight, sensory science, and technological innovation promises to expand the frontiers of culinary creativity while deepening our understanding of the molecular basis of taste, aroma, and texture.

Molecular gastronomy is a multidimensional discipline that merges chemistry, physics, biochemistry, and culinary art to manipulate food at the molecular level. From protein denaturation and polysaccharide gelation to lipid emulsification, Maillard chemistry, enzymatic modification, and self-assembly, molecular gastronomy transforms the preparation, presentation, and perception of food. Analytical techniques provide rigorous methods to study and optimize these processes, while chemical principles inform both safety and creativity. Through its systematic application of molecular science, molecular gastronomy has elevated cooking from a traditional practice to a laboratory of sensory exploration, innovation, and scientific discovery. By understanding and harnessing the chemistry of food, practitioners create novel textures, flavors, and visual effects, offering both intellectual and sensory satisfaction. This discipline exemplifies the profound interrelationship between science and art, demonstrating that culinary innovation is fundamentally a chemical endeavor, guided by molecular insight and a spirit of experimentation.

At the core of molecular gastronomy is the precise control of chemical transformations in food. Proteins, for instance, are not merely sources of nutrition but versatile molecular scaffolds whose structure and function can be manipulated through heat, pH, ionic strength, or enzymatic activity. Denaturation, the unfolding of protein tertiary and quaternary structures, can be tuned to generate diverse textures—from the creaminess of custards to the elasticity of gels. Similarly, the Maillard reaction, a non-enzymatic browning process between reducing sugars and amino acids, is a chemical symphony that produces hundreds of volatile flavor compounds. By controlling temperature, time, water activity, and reactant ratios, chefs and scientists can favor the formation of specific aroma molecules, balancing savory, caramelized, and roasted notes.

Polysaccharides, another cornerstone of molecular gastronomy, enable the construction of gels, foams, and encapsulated systems through molecular self-assembly and network formation. Agar, carrageenan, xanthan gum, and pectin each possess distinct molecular architectures that dictate gel strength, elasticity, and water retention. The interplay of ionic crosslinking, hydrogen bonding, and chain entanglement allows practitioners to design textures that challenge conventional expectations, such as spheres that burst with liquid upon biting or foams that collapse into creamy sauces. Understanding these chemical principles permits not only creative expression but also consistency and reproducibility in culinary preparation.

Lipids, often relegated to the role of flavor carriers, are equally dynamic in molecular gastronomy. Their amphiphilic nature facilitates the formation of emulsions, microemulsions, and lipid-based colloids that modify mouthfeel and stabilize multi-phase systems. By leveraging the principles of interfacial chemistry and surfactant behavior, chefs can construct stable vinaigrettes, airy mousses, or encapsulated flavor droplets that release aromatics in a controlled manner. The precise manipulation of lipid phase transitions—melting, crystallization, and polymorphism—also informs the creation of chocolate tempering, butter creams, and specialty frozen desserts, where texture and gloss depend critically on molecular alignment.

Enzymatic catalysis adds yet another layer of chemical sophistication. Enzymes such as transglutaminase, bromelain, and lipase enable selective bond formation or cleavage under mild conditions, circumventing harsh heating or chemical modification. Transglutaminase, for example, crosslinks glutamine and lysine residues in proteins, producing novel textures such as cohesive meat gels or restructured fish fillets. Bromelain, a protease derived from pineapple, tenderizes proteins by hydrolyzing peptide bonds, whereas lipases selectively liberate or restructure fatty acids to enhance flavor. By applying enzyme kinetics, substrate specificity, and reaction engineering, practitioners transform raw ingredients into complex, multi-textural experiences.

Analytical chemistry is indispensable in molecular gastronomy for both fundamental understanding and practical optimization. Techniques such as gas chromatography–mass spectrometry (GC-MS), high-performance liquid chromatography (HPLC), nuclear magnetic resonance (NMR), and spectroscopy allow quantification of volatile compounds, identification of reaction intermediates, and monitoring of structural transformations. Rheology and texture analysis provide quantitative measures of gel strength, elasticity, and viscosity, linking molecular interactions to sensory perception. These data-driven insights guide chefs in designing flavor profiles and textures with scientific precision, bridging the gap between experimentation and reproducibility.

Beyond mechanics and flavor, molecular gastronomy probes the sensory chemistry of taste and aroma. Umami perception, driven by free glutamate and nucleotides, can be enhanced or suppressed through enzymatic treatment or Maillard reaction control. Aroma compounds, often present at trace concentrations, influence flavor perception disproportionately, a phenomenon quantified through odor thresholds and gas chromatography–olfactometry. By understanding chemical synergy, volatility, and partitioning between phases, chefs can engineer dishes that maximize olfactory and gustatory impact, creating experiences that engage multiple senses simultaneously.

Furthermore, molecular gastronomy intersects with nutrition and health chemistry. Encapsulation techniques protect sensitive bioactive compounds—vitamins, antioxidants, or probiotics—during cooking and digestion, enhancing bioavailability. Acid-base chemistry and enzyme-mediated modifications can reduce antinutritional factors, improve digestibility, or modulate glycemic response. By integrating chemical principles with food science, molecular gastronomy demonstrates that culinary innovation can align with health and functional objectives without compromising sensory appeal.

Finally, molecular gastronomy serves as a platform for experimentation in chemical creativity. By applying principles such as phase behavior, molecular recognition, and reaction kinetics, practitioners invent novel forms—spherified liquids, edible films, smoke-infused gels, and deconstructed dishes—that challenge conventional culinary boundaries. This fusion of science and art exemplifies chemistry's power to transform experience, revealing the invisible molecular underpinnings of taste, aroma, and texture. As the discipline evolves, the integration of computational modeling, predictive chemistry, and high-throughput experimentation promises to expand the frontiers of what is possible at the intersection of molecules and gastronomy.

In essence, molecular gastronomy illuminates the chemistry of the everyday, translating molecular interactions into sensory delight. By harnessing chemical principles systematically, it transforms cooking into an investigative and expressive science, demonstrating that the kitchen can be as much a laboratory as it is a creative space. Through the continued interplay of molecular insight, culinary curiosity, and technological innovation, molecular gastronomy will undoubtedly continue to redefine how we understand, experience, and celebrate food.

Chapter 41: Word Problems

1. A custard mixture containing 8% egg protein is heated at 78 °C for 20 minutes to achieve a specific gel firmness. Ovalbumin in the mixture denatures according to first-order kinetics with a known rate constant of 0.12 min^{-1} at this temperature. The chef wants to know what fraction of protein remains folded after cooking and whether this level of unfolding is sufficient to create a stable gel network. Because denaturation exposes hydrophobic residues that promote aggregation, the fraction of unfolded protein directly correlates with firmness. Calculate the fraction of native protein remaining after 20 minutes and predict how this degree of denaturation affects the custard's final texture.

2. During fresh cheese production, milk is acidified from pH 6.6 to 5.0, reducing casein micelle charge and promoting aggregation. Assume the surface charge density decreases approximately linearly with pH. The cheesemaker wants to estimate the relative reduction in electrostatic repulsion among micelles and predict the resulting curd firmness. Because lower net charge allows micelles to collide and stick, texture becomes more cohesive. Calculate the proportional change in charge and explain how this influences the mechanical strength of curds.

3. A 3 cm-thick meat portion is cooked sous-vide at 60 °C for 6 hours, allowing collagen to slowly hydrolyze. The activation energy for collagen breakdown is 75 kJ/mol. The chef considers raising the bath temperature to 65 °C to accelerate tenderization without overcooking proteins. Using the Arrhenius equation, estimate how much faster hydrolysis proceeds at 65 °C, and predict how the altered rate affects tenderness. Discuss whether this temperature change risks excessive moisture loss.

4. A starch slurry with 60% water gelatinizes at 72 °C. When 20% sucrose is added, the observed gelatinization temperature increases to 78 °C. Sucrose competes for available water, reducing water activity and requiring more heat for granule swelling. Using known water-binding behavior of sucrose, estimate the reduction in "free" water responsible for this shift. Explain how this change affects the viscosity and stability of the final sauce.

5. In basic spherification, calcium ions diffuse into a droplet of 2% sodium alginate, forming a membrane that thickens with time. Given a Ca^{2+} diffusion coefficient of 0.8×10^{-9} m^2/s, the chef wants to estimate the membrane thickness after 60 seconds. The thickness determines how the sphere bursts in the mouth: too thick and the sphere becomes rubbery, too thin and it ruptures prematurely. Using diffusion length approximations, calculate the gel layer thickness. Based on the result, explain how manipulating ion concentration and exposure time changes texture.

6. A citrus foam stabilized with egg-white proteins exhibits a surface tension drop from 72 mN/m to 50 mN/m, enabling bubble formation. However, acidity and shear stress can destabilize the foam. The chef wants to know whether the protein concentration is sufficient to maintain stability during plating. Given that

minimally stable foams require surface tension below 55 mN/m, evaluate the foam's robustness. Predict how small pH adjustments might improve or weaken stability.

7. A nanoemulsion with 120 nm droplets appears slightly cloudy due to light scattering. The chef contemplates lowering the droplet diameter to 60 nm using higher homogenization pressure. Using Rayleigh scattering principles, estimate how turbidity would change if droplet size is halved. Explain how this affects visual clarity and mouthfeel. Recommend whether the extra processing is worthwhile for the intended presentation.

8. A nutty aroma compound forms optimally at 140 °C during the Maillard reaction. When heating is increased to 160 °C, the chef notices bitter notes. Using knowledge of Maillard reaction branching pathways, estimate how increasing temperature may shift intermediates toward degradation products. Predict whether adjusting water activity or reactant ratios could help preserve the desired aroma. Evaluate whether lower temperatures might also reduce overall browning intensity.

9. A syrup is reduced from 20% water to 5% water, accelerating caramelization. Because caramelization requires low water activity while Maillard browning requires more water, the chef wants to understand which pathway dominates. Using qualitative reaction-rate reasoning, estimate how reducing water changes the onset temperature. Predict how this shift influences the final flavor profile and color. Suggest ways to balance caramel and Maillard notes in the same dish.

10. A mayonnaise begins breaking near 65 °C, resulting in oil separation. Lecithin's phase transition temperature is around 70 °C, and heating increases droplet collision rates. The chef wants to determine why destabilization occurs slightly below lecithin's transition. Explain how increased kinetic energy and decreased viscosity contribute to coalescence. Predict how switching to oils of different polarity would alter thermal stability.

11. A sauce thickened with 0.5% xanthan gum shows viscosity decreasing from 1200 cP at low shear to 200 cP at high shear. The chef wants to connect this rheological behavior to mouthfeel. Explain how polymer chain alignment causes shear thinning and how this helps sauces feel rich at rest but smooth during eating. Discuss how altering polymer concentration changes perceived thickness. Estimate the sensory impact of doubling xanthan concentration.

12. Steaming milk at 65 °C reduces average casein micelle size by 15%, improving microfoam stability. The barista wants to understand how this size change affects the foam's density and smoothness. Explain the relationship between micelle size, film formation, and bubble stability. Predict whether increasing steaming time would continue to improve microfoam or lead to collapse. Evaluate whether non-dairy alternatives behave similarly.

13. A marinade containing 50 U/mL bromelain increases meat tenderness by 30% after two hours. Assuming Michaelis–Menten kinetics, doubling the enzyme concentration does not necessarily double tenderization rate. Estimate the expected increase in reaction rate at high substrate concentration. Explain how over-tenderization might occur if the enzyme is too concentrated. Predict the ideal marinade duration for balanced texture.

14. Two new peaks appear in GC-MS chromatograms when mushrooms are seared at high temperature. These peaks increase with heat intensity. Given that mushrooms contain sulfur-rich amino acids, identify likely classes of volatiles responsible for the new peaks. Predict how changes in cooking temperature influence peak intensity. Explain how controlling moisture during searing alters compound formation.

15. A frying oil repeatedly heated to 180 °C begins forming peroxides that degrade flavor. Oxygen diffusion drives the oxidation rate. The chef considers switching to a more saturated fat to reduce breakdown. Estimate how saturation affects oxidation kinetics and predict the impact on long-term flavor quality. Discuss whether antioxidant additives could further improve stability.

16. Chocolate tempered to maintain Form V crystals melts around 34 °C. If the chocolatier accidentally heats it to 36 °C, some Form V transitions to the more brittle Form VI. Using relative thermodynamic stability, estimate the fraction likely to convert. Explain how this affects gloss, snap, and melt-in-mouth quality. Recommend recovery steps to re-establish Form V dominance.

17. A berry sauce turns from bright red to purple when baking soda is added. Anthocyanin pigments undergo structural changes across pH ranges. Estimate the approximate pH shift needed for this visible transition. Explain the molecular basis for color change and its effect on perceived flavor. Recommend alternatives for color modification without altering taste drastically.

18. A foam loses 20% volume in 10 minutes due to liquid drainage between bubbles. The chef wants to understand how film thinning impacts stability. Using the Lucas–Washburn model qualitatively, estimate the thinning rate and identify dominant variables. Predict how adding polysaccharides slows drainage. Suggest formulation changes for longer-lasting plating foams.

19. A 1.5% agar gel has measured gel strength of 800 g/cm². The chef wants to predict gel strength at 1.0% agar for a softer texture. Assuming approximately linear scaling, estimate the new gel strength. Explain why agar gels exhibit thermal hysteresis between melting and setting. Describe how this affects serving temperature flexibility.

20. Liposomes encapsulate a flavor compound with logP = 2.5 at 65% efficiency. The chef wants to understand how bilayer rigidity influences release during eating. Explain how lipid composition changes permeability at 37 °C. Predict how modifying bilayer saturation affects flavor timing. Estimate whether increasing rigidity would noticeably lower the perceived aroma intensity.

21. A fruit purée loses moisture from 90% to 5% during freeze-drying. Sublimation rate depends strongly on chamber pressure. The chef wants to choose a pressure low enough to preserve volatiles but high enough for reasonable drying time. Estimate how reducing pressure affects sublimation temperature and volatile retention. Recommend an approximate pressure range for optimal quality.

22. Adding 15% sucrose lowers the freezing point of an ice-cream base. The chef wants to estimate the approximate temperature shift using colligative principles. Explain how this affects ice-crystal size and smoothness. Predict textural differences if sugar concentration is increased to 20%. Discuss how stabilizers interact with freezing-point depression.

23. A homogenizer at 200 MPa reduces droplet size from 800 nm to 200 nm. The chef considers increasing pressure to 300 MPa. Using empirical scaling relations, estimate the new droplet size. Explain how smaller droplets affect opacity, viscosity, and flavor release. Evaluate whether the additional pressure is justified for sensory gains.

24. An edible film 50 μm thick releases vanillin aroma fully over 2 hours. Using Fick's law qualitatively, estimate the diffusion coefficient required for such release.

Explain how adjusting polymer matrix density alters release timing. Predict how temperature during serving influences aroma burst. Recommend film compositions for multi-phase flavor release.

25. Two oppositely charged polysaccharides form coacervate capsules at pH 4.5. Raising the pH to 6.0 stops capsule formation. Using charge-density relationships, explain why coacervation collapses at higher pH. Estimate how much charge reduction occurs. Predict whether salt addition could restore capsule formation.

26. Apples treated with lemon juice brown 40% more slowly during storage. Lower pH and added antioxidants inhibit polyphenol oxidase (PPO). Estimate the degree of PPO inhibition needed to reduce browning by this amount. Explain how enzymatic kinetics govern browning speed. Propose complementary techniques for further reduction.

27. A glycosidase enzyme liberates floral aromatics from non-volatile glycosides during dessert consumption. At 37 °C, release occurs steadily over 10 minutes. Using basic enzyme-kinetic reasoning, estimate reaction rate needed for this timing. Explain how temperature shifts modulate release speed. Predict how encapsulating the enzyme might extend aroma delivery.

28. Mixing dough for 15 minutes forms a strong gluten network through hydration and disulfide-bond formation. Doubling the shear rate may increase elasticity but also cause over-development. Estimate how increased shear affects gluten alignment. Explain the role of mechanical energy in gluten crosslinking. Predict consequences for crumb structure in the baked product.

29. Immersion of a 2 cm mousse sphere in liquid nitrogen produces rapid freezing and small ice crystals. The chef wants to estimate cooling rate based on thermal conductivity and sphere radius. Explain why rapid cooling leads to smoother textures. Predict issues if the sphere is much larger. Suggest thawing strategies to maintain texture.

30. High-methoxyl pectins with DE 70% require high sugar conditions for gelation, whereas DE 30% forms low-methoxyl gels with calcium. Compare expected gel strength for identical concentrations of each. Explain how DE influences crosslink type and gel firmness. Predict sensory differences between the two gels. Recommend applications suited to each type.

31. A lemon aroma encapsulated in β-cyclodextrin releases when heated above 60 °C due to weakened host–guest interactions. The chef wants to estimate the energy required for this release. Explain how humidity affects equilibrium between bound and free aroma. Predict whether encapsulation improves aroma persistence during cooking. Recommend dish types best suited to such thermal release.

32. NMR analysis reveals that fat crystals in a stored chocolate shift from the β′ form to the more stable β form. This transition alters texture. Estimate how crystal size changes during this transformation. Explain how this affects snap and mouthfeel. Suggest storage conditions to limit polymorphic transitions.

33. A foam stabilized by lecithin and saponin exhibits unexpected stability. Mixed micelles form when both surfactants are above their CMCs. Estimate how micelle composition affects surface elasticity. Predict whether adding more of either surfactant improves or worsens stability. Explain why synergy occurs in some ratios but not others.

34. A layered dessert consisting of mousse above a crust exhibits sogginess after 6 hours. Water migrates from the high-moisture mousse into the low-moisture crust. Using diffusion concepts, estimate the water-migration timescale. Explain how fat layers or hydrophobic barriers slow moisture transfer. Recommend layering adjustments to extend shelf life.

35. Acrylamide formation increases significantly above 160 °C. The chef lowers frying temperature from 170 °C to 150 °C to reduce acrylamide. Estimate how reaction rate decreases based on typical thermal sensitivity of Maillard-related pathways. Explain how the lower temperature impacts texture and surface browning. Suggest trade-offs between safety and crispness.

36. Ultrasonic emulsification at 20 kHz creates cavitation bubbles that fragment oil droplets. The chef wants to estimate the smallest achievable droplet size given a known energy density. Explain how cavitation energy correlates with droplet diameter. Predict effects on emulsion stability and opacity. Suggest when ultrasonic methods are preferable to pressure homogenization.

37. Alginate crosslinks more strongly with Ca^{2+} than with Mg^{2+}. If Mg^{2+} concentration doubles while Ca^{2+} is halved, gel firmness decreases. Estimate the proportional

reduction in crosslinking efficiency. Explain how ion radius and charge density influence gel rigidity. Predict whether adding more alginate compensates for weaker crosslinking.

38. Egg-white foams contain proteins with denaturation temperatures between 62 °C and 85 °C. Heating to 75 °C partially denatures some proteins while fully denaturing others. Estimate the fraction of the total protein population affected. Explain how partial denaturation contributes to foam stability. Predict why overheating collapses the foam.

39. A hydrophobic aroma compound dissolves readily into lipid micelles. The chef wants to understand how micelle size affects volatility and perceived aroma intensity. Explain how partitioning between aqueous and lipid phases governs aromatic release. Estimate whether reducing micelle diameter enhances aroma perception. Predict consequences for emulsified sauces.

40. Fresh herbs lose about half their aroma when heated at 90 °C for 5 minutes. Volatile degradation competes with evaporation. Estimate the volatility-driven loss rate. Explain how gentler extraction methods such as cold infusion reduce loss. Recommend hybrid techniques for maximum aromatic retention.

41. A stew simmered at 90 °C for 4 hours softens significantly as collagen converts to gelatin. Using a time-temperature integral, estimate collagen breakdown extent. Explain how slow heating affects moisture retention. Predict how extending simmering time alters both texture and flavor concentration. Discuss diminishing returns for very long cooks.

42. A carbonated foam is stable at low temperature due to high CO_2 solubility. When warmed by 15 °C, bubbles expand and destabilize. Use Henry's law qualitatively to estimate the decrease in solubility. Explain how gas release affects foam texture. Recommend serving-temperature strategies for stability.

43. Protein solubility in broths depends on ionic strength. Increasing salt may initially improve clarity (salting-in) but eventually cause precipitation (salting-out). Estimate the ionic-strength threshold for the transition. Explain how this affects mouthfeel and appearance. Suggest seasoning strategies that avoid turbidity.

44. Pressure cooking raises water's boiling point, increasing reaction rates. Doubling the pressure significantly accelerates both Maillard pathways and caramelization.

Estimate how much reaction rate increases based on the boiling-point shift. Explain how this intensifies color and flavor. Predict when pressure cooking may produce undesirable darkening.

45. A 250-bloom gelatin forms a firm gel at 2%. Lowering concentration to 1% weakens the gel noticeably. Estimate the new gel strength using typical concentration-strength relationships. Explain how triple-helix formation contributes to elasticity. Predict how chilling time affects final firmness.

46. A rosemary-infused oil stored in light for 30 days shows increased peroxide values. Phenolics in rosemary can be either antioxidants or pro-oxidants depending on conditions. Estimate the oxidative-rate increase due to light exposure. Explain how metal ions influence peroxide formation. Suggest ideal storage methods.

47. Liquid chocolate behaves as a yield-stress, shear-thinning fluid. Using Herschel–Bulkley parameters, the chef wants to predict coating thickness during enrobing. Explain how applied shear determines final flow. Estimate coating variation with small changes in shear stress. Discuss how fat content alters flow behavior.

48. A sorbet freezes too quickly, forming large crystals. Stabilizers can control nucleation and slow crystal growth. Estimate how adding 0.2% stabilizer changes freezing curve slope. Explain how slower freezing improves smoothness. Predict whether alcohol addition could enhance texture.

49. A dessert uses two gel layers: a soft pectin gel and a firm agar gel. When eaten, each layer dissolves at different rates. Estimate dissolution times based on gel-network density. Explain how this controls sequential flavor release. Recommend modifications for a more dramatic multi-phase sensory effect.

50. A chef adjusts amino-acid ratios to favor formation of specific Maillard volatiles. Using provided reaction-rate constants, modeling predicts shifts in volatile composition. Estimate the expected increase in the targeted aroma compound. Explain how computational tools reduce trial-and-error. Recommend experimental conditions to validate predictions.

Chapter 42

Chemical Aspects of Climate Engineering

Climate engineering, also referred to as geoengineering, encompasses a set of deliberate interventions in the Earth's climate system aimed at mitigating the adverse impacts of anthropogenic climate change. These interventions are grounded fundamentally in chemistry, from the molecular reactions that regulate atmospheric composition to the design of engineered materials that modify radiative balance, carbon cycling, and chemical equilibria in the oceans and atmosphere. The chemical underpinnings of climate engineering can be broadly divided into two categories: solar radiation management (SRM), which aims to reflect or scatter incoming solar radiation to reduce global temperatures, and carbon dioxide removal (CDR), which seeks to extract CO_2 from the atmosphere or enhance its long-term sequestration. Both categories rely heavily on understanding reaction kinetics, photochemistry, aerosol chemistry, and heterogeneous chemical processes, as well as advanced materials chemistry for designing efficient and environmentally compatible systems.

Solar radiation management strategies exploit the chemical properties of particles and gases to increase the Earth's albedo. Stratospheric aerosol injection (SAI) is one of the most studied SRM approaches. It involves the introduction of fine particles, typically derived from sulfur dioxide (SO_2), into the stratosphere, where they undergo oxidation to form sulfuric acid (H_2SO_4). This acid nucleates to form sulfate aerosol particles capable of scattering incoming solar radiation. The conversion of SO_2 to H_2SO_4 is influenced by stratospheric photochemistry, involving reactions with hydroxyl radicals (OH) generated through the photolysis of water vapor. The nucleation, growth, and coagulation kinetics of sulfate aerosols are strongly dependent on temperature profiles, water vapor content, and the presence of trace gases such as ammonia, nitrogen oxides, and halogens. Additionally, these aerosols act as reactive surfaces for heterogeneous chemical reactions, including those that catalyze ozone depletion, highlighting the need for a sophisticated chemical understanding to predict both climate benefits and unintended consequences.

Beyond sulfates, engineered aerosols composed of alumina (Al_2O_3), titanium dioxide (TiO_2), or diamond nanoparticles have been proposed. These particles are selected for their high reflectivity and chemical stability under stratospheric ultraviolet (UV) radiation. The surface chemistry of these materials, including potential reactions with ozone, nitrogen oxides, and water vapor, affects their atmospheric lifetime, scattering efficiency, and potential side reactions. Nanostructuring and surface functionalization can enhance optical properties or reduce aggregation, but introduce additional chemical considerations such as photocatalytic activity, surface oxidation, or dissolution under acidic conditions. Understanding the interplay between surface chemistry, particle morphology, and atmospheric reactions is therefore central to designing aerosols for SRM that are both effective and environmentally safe.

Marine cloud brightening (MCB) represents an alternative SRM approach, leveraging the chemistry of sea salt aerosols. Seawater contains dissolved salts, primarily sodium chloride, along with magnesium, calcium, and trace metals. When seawater is atomized and injected into the marine boundary layer, water evaporates to leave fine salt particles that act as cloud condensation nuclei (CCN), increasing cloud reflectivity. The chemical composition of these particles determines their hygroscopic growth, surface tension, and CCN activity. Organic compounds, including dimethyl sulfide (DMS) and other organosulfur species emitted by marine phytoplankton, can coat particle surfaces or participate in atmospheric oxidation reactions, altering cloud microphysics and radiative properties. Additionally, heterogeneous reactions on sea salt aerosols can release reactive halogens, influencing tropospheric ozone chemistry. A comprehensive chemical understanding of these multiphase reactions is critical for modeling MCB effectiveness and avoiding unintended atmospheric effects.

Carbon dioxide removal strategies hinge on chemical processes capable of sequestering CO_2 either directly from air or indirectly through enhancement of natural carbon sinks. Direct air capture (DAC) technologies commonly employ aqueous amine solutions, such as monoethanolamine (MEA), to chemically absorb CO_2. The reaction forms carbamates and bicarbonates, with reaction kinetics depending on temperature, solvent concentration, and mass transfer rates. However, amine solutions are prone to oxidative degradation, forming byproducts such as nitrosamines and ammonia, which pose environmental and operational challenges. Solid sorbents, including functionalized metal-organic frameworks (MOFs), amine-grafted silica, and ion-exchange resins, offer alternatives with tunable CO_2 binding affinities. The chemical design of these sorbents focuses on active site optimization, pore chemistry, moisture stability, and regeneration efficiency. Thermal, pressure-swing, or electrochemical regeneration of sorbents involves reversibility of chemical equilibria, often requiring precise control over pH, temperature, and partial pressure of CO_2. Molecular-level understanding of sorbent chemistry, including adsorption energetics and surface reaction mechanisms, is fundamental for scaling DAC systems economically and sustainably.

Ocean-based carbon dioxide removal strategies exploit the inherent chemical equilibria of carbonate and silicate systems. Ocean alkalinity enhancement involves adding basic minerals, such as olivine (Mg_2SiO_4) or lime (CaO), to seawater to facilitate the formation of bicarbonate and carbonate ions. CO_2 dissolves in seawater to form carbonic acid, which reacts with added bases to sequester carbon in stable forms. The kinetics of silicate mineral weathering are influenced by particle surface area, ocean temperature, pH, and the presence of trace metal ions. The chemical transformations in these systems are complex, as secondary reactions can alter the speciation of metals, influence nutrient availability, and affect marine biogeochemistry. Laboratory and field experiments, coupled with geochemical modeling, are essential for quantifying rates of carbon sequestration and evaluating potential ecological impacts. Additionally, controlled electrochemical enhancement of seawater alkalinity has been proposed, utilizing redox reactions to increase bicarbonate concentrations while producing hydrogen as a co-product, integrating chemical engineering principles with electrochemistry for efficient CDR.

Biochar production represents another chemical pathway for CO_2 sequestration. Pyrolysis of biomass under limited oxygen produces stable, carbon-rich materials with extensive aromatic structures resistant to degradation. The chemical structure of biochar, including aromaticity, functional groups, and mineral content, determines its longevity in soils, its ability to adsorb nutrients or pollutants, and its influence on soil chemistry. Chemical modifications of biochar, such as oxidation or nitrogen doping, can enhance CO_2 adsorption, nutrient retention, and cation exchange capacity, illustrating the interplay between chemical engineering and environmental chemistry in CDR strategies.

A critical component across all climate engineering methods is the need to understand and model chemical reactions in multicomponent, dynamic environments. For SRM, stratospheric chemistry models incorporate photolysis rates, catalytic cycles, heterogeneous reactions, and transport of aerosols and gases. Accurate modeling requires fundamental chemical data, including reaction rate constants, activation energies, solubility, and aerosol surface reactivity. For CDR, chemical models must account for gas-liquid-solid equilibria, transport phenomena, reaction thermodynamics, and kinetics in both the atmosphere and ocean. Uncertainties in chemical parameters can propagate into predictions of climate impacts, necessitating laboratory experiments, field campaigns, and advanced spectroscopic monitoring to validate models. Multi-scale modeling approaches, integrating quantum chemical calculations for reaction mechanisms with mesoscale and global atmospheric models, are increasingly employed to optimize climate engineering strategies.

Unintended chemical consequences are a major concern in geoengineering. Sulfate aerosols, while reflecting sunlight, can catalyze stratospheric ozone destruction via heterogeneous halogen reactions, enhancing the concentration of reactive chlorine and bromine species. Ocean alkalinity enhancement may mobilize trace metals such as nickel or chromium, potentially altering marine microbial ecology and biogeochemical cycles. Engineered nanoparticles for SRM may undergo photochemical transformation, releasing reactive radicals or interacting with natural organic matter. Comprehensive chemical risk assessments, including thermodynamic, kinetic, and toxicological analyses, are necessary to evaluate these secondary effects and guide the selection of safe, sustainable intervention strategies.

Material chemistry plays an increasingly prominent role in climate engineering. Advanced photochemical nanoparticles can catalyze the reduction of CO_2 in atmospheric or oceanic environments, converting it to fuels or stable carbonates. These processes require precise control of surface chemistry, electronic structure, and reaction pathways to maximize conversion efficiency and selectivity while minimizing unwanted side reactions. Hybrid systems, combining SRM and CDR approaches, present additional chemical complexity, with simultaneous multiphase reactions, photochemistry, and aerosol-cloud interactions. Designing such systems requires interdisciplinary expertise, bridging materials chemistry, atmospheric chemistry, and chemical engineering principles.

Analytical chemistry is indispensable for monitoring climate engineering interventions. Techniques such as mass spectrometry, spectroscopy, chromatography, and remote sensing are used to detect and quantify atmospheric aerosols, trace gases, and oceanic chemical changes. High-resolution chemical analysis allows real-time tracking of engineered particles, verification of CO_2 removal rates, and identification of potentially hazardous byproducts. Chemical monitoring is also critical for regulatory compliance, ensuring that interventions remain within safe operational bounds and mitigating environmental risk.

Photochemistry underpins many climate engineering processes. In the stratosphere, ultraviolet radiation drives the photolysis of ozone, water, and nitrogen oxides, generating radicals that catalyze aerosol transformations and ozone depletion. In oceans, photochemical reactions can influence dissolved organic matter, trace metals, and carbonate equilibria, altering the efficiency of carbon sequestration.

Understanding these reactions requires integrating quantum chemical calculations with experimental photochemistry to predict reaction pathways and radical lifetimes under varying environmental conditions.

The environmental fate of engineered chemicals is a key consideration. Sulfate aerosols eventually sediment and can contribute to acid deposition, influencing terrestrial and aquatic chemistry. Alkaline minerals added to oceans gradually dissolve, modifying pH and potentially affecting nutrient cycles. Engineered nanoparticles may aggregate, dissolve, or undergo surface transformations, altering their chemical reactivity and mobility. Predicting these outcomes relies on chemical kinetics, thermodynamics, and environmental transport models, highlighting the centrality of chemistry in assessing both efficacy and ecological impact.

Emerging research increasingly explores synergistic combinations of SRM and CDR. For example, stratospheric aerosols may be engineered to facilitate heterogeneous CO_2 reduction reactions, simultaneously reflecting solar radiation and promoting chemical sequestration. Similarly, marine cloud brightening may be coupled with ocean alkalinity enhancement, integrating aerosol chemistry with seawater chemistry to maximize radiative and carbon capture effects. These hybrid strategies require detailed chemical understanding of interfacial reactions, multiphase processes, and feedback mechanisms, emphasizing the interdisciplinary nature of chemical research in climate engineering.

Thermodynamics and kinetics govern all chemical interventions in climate engineering. For SRM, particle scattering efficiency and lifetime depend on nucleation thermodynamics, coagulation kinetics, and heterogeneous chemical reactions. For CDR, CO_2 absorption, mineral dissolution, and catalytic conversion are dictated by reaction energetics and rate laws. Accurate determination of activation barriers, equilibrium constants, and reaction orders is essential for predicting system performance and optimizing intervention design. Computational chemistry, including density functional theory and molecular dynamics simulations, provides insights into reaction pathways, binding energies, and solvation effects, complementing experimental studies.

Chemical innovation is also driving the development of safer, more efficient materials. Functionalized nanoparticles with tailored optical properties, sorbents with high CO_2 selectivity, and biochar with enhanced surface chemistry represent the forefront of climate engineering research. Rational design requires a deep understanding of surface chemistry, electronic structure, and reaction mechanisms, allowing for the tuning of chemical reactivity, stability, and environmental compatibility. Collaboration between chemists, material scientists, and atmospheric modelers is critical to ensure that chemical design translates into real-world climate impact.

Ultimately, the success and safety of climate engineering depend on rigorous chemical understanding. The interactions of engineered particles, gases, and solvents with the atmosphere and oceans are governed by reaction kinetics, thermodynamics, photochemistry, and surface chemistry. Laboratory experiments, field studies, and computational models are essential for elucidating these interactions, predicting environmental outcomes, and informing policy. Chemical principles not only guide the design of interventions but also frame regulatory, ethical, and sustainability considerations.

The future of climate engineering will increasingly rely on interdisciplinary chemical research. Advances in computational chemistry, high-throughput material synthesis, and in-situ analytical techniques will enhance our ability to design, monitor, and control interventions. Understanding the detailed chemical mechanisms of both intended and unintended reactions will remain central to developing effective, safe, and socially responsible strategies for mitigating climate change. By integrating chemical insights with atmospheric, oceanic, and environmental sciences, researchers can advance climate engineering from conceptual proposals to scientifically robust, actionable solutions that address the pressing challenges of global warming.

The chemical dimension of climate engineering extends across multiple scales, from molecular-level reactions to ecosystem-wide processes. At the molecular level, understanding the thermodynamic and kinetic properties of greenhouse gases, aerosols, and reactive intermediates is critical. For instance, the design of novel carbon capture materials hinges on precise knowledge of adsorption energetics, chemical selectivity, and stability under fluctuating environmental conditions. Functionalized metal-organic frameworks (MOFs), covalent organic frameworks (COFs), and porous polymers exemplify materials whose performance depends on subtle chemical interactions between gas molecules and the engineered pore environment. Computational modeling and machine learning approaches are increasingly deployed to predict these interactions before materials are synthesized, accelerating the identification of promising candidates while minimizing experimental trial and error.

Beyond carbon capture, chemical strategies for solar radiation management involve complex atmospheric chemistry. Sulfate aerosols, reflective nanoparticles, and halocarbon compounds each interact with solar radiation and atmospheric constituents in ways that can produce both cooling effects and unintended chemical byproducts. The oxidation pathways, photolysis reactions, and catalytic cycles that govern the atmospheric lifetimes of these species require detailed elucidation. Moreover, the interactions of engineered chemicals with naturally occurring radicals, ozone, and volatile organic compounds can initiate cascading chemical reactions with potentially far-reaching ecological consequences. Developing predictive, chemistry-based models for these processes is essential not only for evaluating efficacy but also for assessing environmental safety and long-term impact.

Ocean-based climate engineering also presents unique chemical challenges. Approaches such as ocean alkalinization or nutrient fertilization rely on precise manipulation of carbonate chemistry and biogeochemical cycles. Introducing alkaline compounds to seawater alters pH, carbonate equilibria, and the solubility of carbon dioxide, affecting both carbon sequestration efficiency and marine ecosystems. Similarly, adding iron or other micronutrients to stimulate phytoplankton growth initiates a cascade of chemical and biological interactions, including the production of dimethyl sulfide and other sulfur-containing compounds that influence cloud formation and atmospheric chemistry. Understanding these reactions at a mechanistic, molecular level is essential to predicting both the short-term outcomes and long-term feedbacks of oceanic interventions.

Advances in analytical chemistry are integral to these efforts. High-resolution mass spectrometry, laser-based spectroscopy, and synchrotron techniques enable the detection of trace intermediates and reactive species that are pivotal to climate engineering reactions. Real-time monitoring of chemical transformations in the field allows researchers to track deviations from predicted pathways and rapidly adjust interventions. Additionally, isotopic labeling and chemical tracers can help disentangle complex reaction networks, providing insight into the fate of introduced substances and their interactions with natural chemical cycles. By combining these experimental tools with computational simulations, chemists can construct comprehensive reaction models that inform safe and effective deployment strategies.

Interdisciplinary collaboration remains a cornerstone of progress in climate chemistry. Integrating chemical research with atmospheric science, oceanography, ecology, and policy ensures that interventions are grounded in mechanistic understanding while remaining responsive to societal and environmental concerns. Ethical considerations, regulatory frameworks, and environmental justice must guide experimental design, field trials, and potential implementation, requiring chemists to communicate complex reaction chemistry in a way that informs broader decision-making. The development of standardized protocols for chemical risk assessment, long-term monitoring, and mitigation of unintended consequences will be essential to ensuring that climate engineering strategies do not create more problems than they solve.

In conclusion, the future of climate engineering is inseparable from advances in chemical science. The ability to predict, manipulate, and monitor chemical processes with precision will define the efficacy and safety of interventions designed to mitigate global warming. By leveraging molecular-level insights, innovative materials, sophisticated analytical techniques, and interdisciplinary collaboration, chemists are poised to transform climate engineering from a largely theoretical concept into a suite of practical, controllable, and responsible solutions. As global climate challenges intensify, the role of chemistry in guiding evidence-based strategies will become ever more central, bridging the gap between human ingenuity and the complex chemical networks of the Earth system.

Chapter 42: Word Problems

1. Stratospheric aerosol injection (SAI) uses sulfur dioxide (SO_2) to form sulfate aerosols that scatter sunlight. SO_2 reacts with hydroxyl radicals (OH) to produce sulfuric acid (H_2SO_4), which nucleates into aerosols. If the OH concentration is 1×10^6 molecules/cm^3 and the rate constant is 1.2×10^{-12} cm^3 molecule^{-1} s^{-1}, calculate the initial sulfate formation rate per cm^3. Discuss how doubling OH affects nucleation and the climate cooling potential, and note potential side reactions such as ozone depletion on sulfate surfaces.

2. Alumina (Al_2O_3) nanoparticles are considered for reflective stratospheric aerosols due to their high refractive index and UV stability. These particles can react

with ozone on their surfaces, which could affect particle lifetime and ozone concentrations. For a particle with 1 m^2 surface area and a reaction rate of 10^{-12} cm^3/s at an O_3 concentration of 10^{12} molecules/cm^3, estimate the total ozone molecules reacted over a week. Consider how particle size, surface defects, and hydroxyl groups influence reactivity, and discuss trade-offs between scattering efficiency and potential ozone depletion.

3. Marine cloud brightening sprays seawater to produce salt aerosols as cloud condensation nuclei. For 1 L of 0.6 M NaCl seawater with 80% evaporation, estimate the number of CCN formed. Discuss how particle size, solute composition, and organic coatings influence droplet activation, cloud albedo, and radiative forcing, including effects of particle coagulation and redistribution of organics.

4. Direct air capture (DAC) using aqueous monoethanolamine (MEA) solutions relies on the chemical absorption of atmospheric CO_2, forming carbamates via nucleophilic attack on CO_2 by the amine group. Given a 2 M MEA solution, an atmospheric CO_2 concentration of 400 ppm, and a carbamate equilibrium constant of 5 M^{-1}, calculate the carbamate concentration at equilibrium. Expand your analysis to discuss how variations in ambient temperature, partial pressure of CO_2, and oxidative degradation of MEA influence the overall CO_2 capture efficiency, solvent longevity, and energy requirements for thermal regeneration. Include considerations of side reactions, potential formation of degradation byproducts such as nitrosamines, and how these chemical processes could limit repeated solvent cycling in long-term DAC deployment.

5. Ocean alkalinity enhancement involves the addition of olivine (Mg_2SiO_4) particles to seawater to accelerate CO_2 sequestration via mineral weathering. With a measured dissolution rate of 10^{-9} mol/cm^2/s and a total particle surface area of 50 m^2, estimate the time required to sequester 1 ton of CO_2 in 10 m^3 of seawater. Discuss the influence of temperature, particle size distribution, surface passivation, trace metal content, and seawater buffering capacity on the kinetics of dissolution. Consider how these chemical and physical factors affect local carbonate chemistry, saturation indices, and potential ecological impacts, such as shifts in pH-sensitive marine biota and trace metal bioavailability.

6. Pyrolysis of biomass at 450 °C produces a mixture of biochar, bio-oil, and syngas. For 500 kg of feedstock yielding 30% mass as biochar, calculate the carbon content

retained in the biochar fraction. Discuss how the degree of aromaticity, functional group composition, and porosity of biochar influence its chemical stability in soil, nutrient retention capabilities, and potential for additional CO_2 adsorption. Evaluate chemical strategies such as surface functionalization, oxidation, or activation to enhance long-term carbon sequestration, considering potential trade-offs between reactivity and structural stability in diverse soil environments.

7. Sulfate aerosols catalyze ozone-depleting reactions involving halogen species in the stratosphere. In an aerosol layer containing 10 ppb ClO and 5 ppb BrO, estimate the total ozone loss over 24 hours, assuming relevant catalytic cycles. Discuss how aerosol surface area, particle acidity, composition, and phase state affect catalytic efficiency. Include a consideration of competing atmospheric processes, such as NO_x-mediated reactions, and evaluate the trade-offs between enhanced scattering for cooling and chemical destruction of stratospheric ozone, including seasonal and latitudinal variability in chemical reactivity.

8. Titanium dioxide (TiO_2) nanoparticles absorb ultraviolet (UV) light and generate reactive oxygen species (ROS) via photocatalysis. Given a 3.2 eV bandgap and 350 nm irradiation, calculate the fraction of UV photons absorbed by a single particle and estimate the number of hydroxyl radicals generated per particle per second. Discuss how nanoparticle size, shape, surface coatings, and aggregation influence both scattering efficiency and ROS-mediated atmospheric reactions. Consider potential side reactions, such as oxidation of volatile organic compounds, interactions with existing aerosols, and implications for stratospheric chemical equilibria.

9. Amine-functionalized silica captures CO_2 through surface carbamate formation. Given a surface area of 500 m^2/g and 2 mmol/g of accessible amine sites, calculate the maximum adsorption capacity per gram of sorbent. Discuss how relative humidity, competitive adsorption of water molecules, chemical degradation of amines, and cyclic thermal regeneration affect CO_2 capture efficiency and sorbent lifetime. Include considerations of diffusion limitations, reaction kinetics on porous surfaces, and strategies to enhance durability and reduce energy consumption in practical DAC applications.

10. Ocean iron fertilization stimulates phytoplankton growth, leading to the production of dimethyl sulfide (DMS), which contributes to cloud condensation nuclei

formation. For 1 kg C/m^2 of phytoplankton biomass distributed over 1 ha, with 0.1% of carbon converted to DMS, estimate the total DMS released. Discuss the role of microbial oxidation, photochemical transformations, and atmospheric transport on DMS lifetime. Evaluate how sulfate aerosol formation from DMS affects cloud droplet number concentration, albedo, and feedbacks to radiative forcing, including potential chemical and ecological side effects of large-scale fertilization.

11. Stratospheric sulfate aerosol particles coagulate over time, altering size distributions and optical properties. For 0.1 μm particles at a concentration of $10^8/cm^3$ with a coagulation rate constant of 10^{-15} m^3/s, estimate the timescale for the average particle radius to double. Discuss how coagulation impacts particle lifetime, scattering efficiency, and overall solar radiation management effectiveness. Include considerations of particle shape, fractal aggregation, heterogeneous chemical reactions, and seasonal variations in stratospheric dynamics.

12. Addition of calcium oxide (CaO) to seawater increases alkalinity and shifts the carbonate-bicarbonate equilibrium. Calculate the resulting pH change for a 1 g/L CaO addition and discuss how seawater buffering limits alkalinity enhancement, thereby affecting CO_2 sequestration efficiency. Evaluate potential consequences for local marine chemistry, including carbonate saturation, calcium carbonate precipitation, and impacts on marine ecosystems sensitive to pH variations.

13. Biochar adsorbs ammonium via cation exchange mechanisms. For biochar with a cation exchange capacity (CEC) of 150 meq/100 g, calculate the maximum NH_4^+ adsorption per kilogram of biochar. Discuss how biochar surface chemistry, porosity, functional groups, and particle size influence nutrient retention, soil chemical dynamics, and interactions with other cations such as K^+, Ca^{2+}, and Mg^{2+}. Consider implications for agricultural productivity and nutrient leaching mitigation.

14. Hybrid solar radiation management (SRM) and carbon dioxide reduction nanoparticles can simultaneously scatter sunlight and catalyze CO_2 conversion. For a UV flux of 200 W/m^2, particle cross-section of 10^{-12} m^2, and quantum yield of 0.01, estimate the CO_2 conversion rate per particle. Discuss the challenges of scaling such systems, including side reactions that could produce undesirable byproducts, photochemical stability, particle aggregation, and the environmental safety of widespread deployment.

15. The gas-phase oxidation of SO_2 to H_2SO_4 in the stratosphere is temperature-dependent. If the reaction rate doubles when temperature increases from 210 K to 220 K, calculate the activation energy using the Arrhenius equation. Discuss implications for seasonal and latitudinal variability in sulfate aerosol formation, atmospheric lifetime, and the radiative forcing contribution of aerosols. Include chemical interactions with halogens and water vapor that may modify reaction pathways.

16. Trace ammonia reacts with sulfate aerosols to form ammonium bisulfate (NH_4HSO_4) in the atmosphere. Using known Henry's law constants and reaction rates, estimate the fraction of sulfate converted in a layer containing 5 ppb NH_3. Discuss how this chemical transformation alters aerosol particle size distribution, composition, hygroscopicity, and radiative properties, and how feedbacks might influence cloud microphysics and local atmospheric chemistry.

17. In a pressure-swing adsorption (PSA) DAC system designed to capture 1 ton of CO_2 per cycle, calculate the energy required for sorbent regeneration if the operating pressure swings from 1 atm to 0.1 atm, assuming ideal gas behavior. Discuss factors that limit energy efficiency, including thermal losses, sorbent degradation, competitive adsorption of water, and cyclic chemical stability, as well as implications for scaling PSA technology to megaton CO_2 capture levels.

18. Diamond nanoparticles embedded in acidic stratospheric aerosols (pH ~3) may undergo surface oxidation over time. Assess whether this oxidation significantly decreases reflectivity over six months, considering reaction kinetics, particle size, surface area-to-volume ratio, and chemical stability of sp^3 carbon surfaces. Discuss implications for long-term solar radiation management performance and potential byproducts that could influence stratospheric chemistry.

19. Köhler theory predicts cloud droplet activation based on aerosol size and solute properties. If particle size or solute composition changes, estimate the effect on cloud droplet number concentration. Discuss how chemical composition, hygroscopicity, and surface-active species influence droplet formation, cloud albedo, and overall radiative forcing. Consider feedbacks with atmospheric chemistry, including interactions with trace gases and secondary aerosol formation.

20. Silicate weathering using olivine (Mg_2SiO_4) is strongly temperature-dependent.

Compare reaction rates at 15 °C and 25 °C using Arrhenius parameters. Discuss consequences for oceanic CO_2 sequestration efficiency, chemical limitations imposed by seawater saturation states, and interactions with trace metals and phosphate. Include consideration of local ecosystem impacts due to alkalinity changes and nutrient release.

21. TiO_2 nanoparticles in the stratosphere degrade under UV light, with a decomposition rate constant of 10^{-13} s^{-1}. Estimate their atmospheric lifetime and discuss how particle size, surface coatings, and aggregation affect both chemical stability and scattering efficiency. Include analysis of heterogeneous reactions with ozone, water vapor, and other atmospheric oxidants.

22. Using computational chemistry methods, calculate the CO_2 binding energy to an amine-functionalized metal-organic framework (MOF). Predict how competing adsorption of water vapor and other gases such as O_2 or N_2 influences adsorption energetics and sorbent performance. Discuss implications for DAC system design, including material selection, regeneration conditions, and chemical stability under cyclic operation.

23. Calculate the sulfate aerosol optical depth required to reduce incoming solar radiation by 1%. Discuss chemical trade-offs, including potential ozone depletion, acid deposition, aerosol particle lifetime, and heterogeneous chemistry. Evaluate implications for radiative forcing and climate feedbacks in different stratospheric conditions.

24. Biochar oxidation in soil follows first-order kinetics at 298 K. Estimate carbon loss over one year and discuss how functional groups, moisture, temperature, microbial activity, and redox conditions influence chemical stability. Include considerations of environmental factors that affect long-term carbon sequestration potential in soils with varying mineralogy and microbial communities.

25. If marine cloud brightening (MCB) increases cloud droplet number concentration by 50%, calculate the expected change in cloud albedo using Twomey's formula. Discuss how particle chemistry, size distribution, solute composition, and atmospheric dynamics influence cloud microphysics, radiative effects, and climate feedbacks.

26. A MOF module with 100 m^2 surface area captures CO_2 at a 0.04% atmospheric concentration, following a Langmuir isotherm with K = 100 atm^{-1}. Determine CO_2 uptake and discuss how temperature, water vapor, and chemical limitations affect sorbent efficiency, regeneration cycles, and operational stability.

27. Evaluate olivine solubility in seawater containing 0.5 mM Fe^{2+} at pH 8.2 and 20 °C. Predict how trace metals influence silicate dissolution rates, carbonate formation chemistry, and overall CO_2 sequestration efficiency. Discuss potential secondary precipitation of Fe minerals and chemical feedbacks on seawater buffering capacity.

28. Sulfate aerosols mediate heterogeneous reactions producing reactive chlorine. Estimate total Cl atoms generated per m^2 of aerosol surface over 24 hours if Cl_2 and HOCl concentrations are 1 ppb and 5 ppb, respectively. Discuss implications for stratospheric ozone chemistry, catalytic ozone loss, and interactions with other halogen species under varying temperature and humidity conditions.

29. Sea salt aerosol halides affect tropospheric ozone through reactions such as $Cl + O_3 \rightarrow ClO + O_2$. Estimate ozone loss in a 1 km^3 air parcel containing 10^9 particles/cm^3 over 12 hours, and discuss chemical feedbacks, including interactions with clouds, photochemistry, and competing reactions with nitrogen oxides.

30. In a dual-gas DAC system absorbing CO_2 and SO_2 with aqueous amines, model competitive reactions using known equilibrium constants, solubility limits, and reaction kinetics. Predict capture efficiency for each gas and potential formation of byproducts, including sulfamate and carbamate species, that could affect system performance and environmental safety. Discuss solvent degradation pathways and energy implications for regeneration.

31. Electrochemical ocean alkalinity enhancement produces bicarbonate ions and hydrogen gas. Calculate the energy efficiency per mole of CO_2 sequestered if the electrolysis cell operates at 2 V with 80% current efficiency. Discuss competing side reactions, seawater buffering limitations, electrode stability, and scaling considerations for large-scale carbon removal.

32. Nanoparticle aggregation reduces scattering efficiency in SRM applications. For fractal aggregates with dimension 2.5, starting at a 0.05 μm radius, estimate

changes in albedo as aggregation proceeds over several months. Discuss how coagulation, fractal morphology, and chemical interactions with atmospheric gases influence particle lifetime and radiative forcing.

33. Photolysis of water vapor generates radicals in the stratosphere. Calculate the steady-state hydroxyl radical concentration under 250 nm UV light given known photolysis rates and radical lifetimes. Discuss implications for aerosol oxidation, SO_2 conversion to sulfate, and secondary chemical reactions affecting stratospheric composition.

34. Amine-grafted silica sorbents require thermal regeneration to release captured CO_2. Determine the optimal desorption temperature that maximizes CO_2 release while minimizing degradation of amine functional groups. Discuss chemical trade-offs, energy input, sorbent longevity, and operational considerations for DAC systems.

35. Cloud droplets oxidize SO_2 to sulfate via diffusion-limited kinetics. Calculate the half-life for SO_2 conversion in a 10 μm droplet at 15 °C. Discuss how droplet chemistry, oxidation rates, aerosol nucleation, and growth influence radiative efficiency and cloud microphysics, including potential feedbacks on local and global climate.

36. Biochar with 70% carbon retention is used for CO_2 sequestration. Calculate the mass required to sequester 10 tons of CO_2 annually. Discuss chemical modifications, surface functionalization, and environmental conditions that could improve adsorption capacity, carbon stability, and nutrient retention for long-term sequestration in different soil types.

37. Photochemical transformations of nanoparticles in the atmosphere can alter lifetime, scattering efficiency, and side reactions. Model oxidation rates, radical generation, and chemical interactions, discussing their impact on SRM effectiveness, atmospheric chemistry, and potential unintended environmental consequences.

38. Adding 1 Mt of CaO to 1 km³ of seawater increases bicarbonate concentrations. Calculate the theoretical maximum HCO_3^- concentration and discuss limits imposed by carbonate equilibria, solubility constraints, and pH buffering. Consider

implications for ocean CO_2 uptake, local chemistry, and potential effects on marine organisms.

39. Nitrosamine formation occurs in MEA-based DAC systems under oxygen exposure. Estimate nitrosamine production if O_2 rises from 0 to 10 ppm, considering reaction kinetics and environmental toxicity. Discuss chemical strategies to minimize hazardous byproducts, including antioxidant addition, solvent formulation, and operational controls.

40. A hybrid SRM/CDR system has a CO_2 reduction rate proportional to $[CO_2]^1[UV]^{0.5}$. Predict how the rate changes if atmospheric CO_2 concentration doubles and UV intensity increases by 50%. Discuss chemical and environmental constraints, such as quantum yield limitations, photochemical side reactions, and scaling challenges for large-scale implementation.

41. Sulfate aerosol deposition is influenced by particle size and sedimentation velocity. Calculate the deposition rate for 0.1 μm particles at 220 K, and discuss how sedimentation dynamics affect aerosol lifetime, radiative forcing, and long-term climate impacts. Include considerations of turbulent mixing, coagulation, and stratosphere-troposphere exchange.

42. Temperature affects CO_2 adsorption on metal-organic frameworks. Using the Van't Hoff relation with $\Delta H = -40$ kJ/mol, calculate the change in CO_2 uptake when temperature increases from 25 °C to 35 °C. Discuss chemical and operational implications for DAC performance, sorbent regeneration, and thermal stability.

43. TiO_2 nanoparticles generate hydroxyl radicals under 300 nm irradiation. Calculate the radical generation rate per m^2 of aerosol layer under continuous UV exposure. Discuss how these radicals influence both SRM effectiveness and atmospheric chemical reactions, including oxidation of SO_2 and volatile organic compounds.

44. Biochar surface oxidation increases cation exchange capacity (CEC). Using first-order kinetics, estimate CEC enhancement after one year at 298 K. Discuss chemical factors controlling functional group formation, environmental influences, and implications for soil nutrient retention and long-term carbon sequestration.

45. Addition of 1 M $Mg(OH)_2$ to seawater accelerates bicarbonate formation. Calculate the HCO_3^- production rate given CO_2 solubility and reaction kinetics, and discuss

how chemical equilibria, precipitation limits, and seawater buffering constrain CO_2 sequestration efficiency. Evaluate potential ecological impacts of large-scale implementation.

46. Photochemical CO_2 reduction in stratospheric aerosols requires a specific quantum yield. Determine the quantum yield needed to reduce CO_2 by 0.1 ppm over one year under realistic UV flux. Discuss chemical, engineering, and environmental feasibility for large-scale implementation, including stability of photoreactive particles and side reaction pathways.

47. Mixed sulfate and TiO_2 particles nucleate and coagulate in the stratosphere. Model the evolution of particle size distribution over three months and discuss how chemical interactions between particle types affect SRM efficiency, particle lifetime, and stratospheric chemistry, including heterogeneous oxidation reactions and photochemical stability.

48. Organic coatings on sea salt aerosols alter hygroscopic growth. Estimate changes in cloud condensation nuclei (CCN) activation for a 5% organic coating and discuss chemical mechanisms affecting droplet formation, cloud microphysics, radiative properties, and feedbacks to regional and global climate.

49. Water vapor competes with CO_2 for adsorption sites in MOFs used for DAC. Predict the reduction in CO_2 uptake at 50% relative humidity using known adsorption constants. Discuss implications for sorbent design, thermal and pressure regeneration strategies, chemical stability, and operational efficiency in realistic atmospheric conditions.

50. Design a reaction sequence that combines stratospheric cooling with CO_2 sequestration. Integrate kinetic and thermodynamic data to evaluate chemical feasibility, environmental risks, and operational constraints for large-scale deployment. Consider particle stability, side reactions, feedbacks on climate and atmospheric chemistry, and potential interactions with natural aerosol populations and cloud systems.

Chapter 43

Chemically Activated Energy Storage

Chemically activated energy storage represents a frontier in modern chemistry, where energy is captured, stored, and released through precise chemical processes rather than through purely physical means. The essence of these systems lies in the storage of energy in chemical bonds, metastable intermediates, or electronically excited states that can be triggered to release energy on demand. Unlike conventional electrochemical batteries or mechanical storage devices, chemically activated systems allow the harnessing of intrinsic molecular energy, providing potentially higher energy densities and versatile release modalities. At the molecular level, the challenge is to balance the energy stored in chemical bonds with the stability required to prevent uncontrolled reactions. Compounds must be sufficiently energetic to provide meaningful storage but chemically stable enough for handling, transport, and integration into devices. The selection and design of such molecules demand a deep understanding of thermodynamics, kinetics, electronic structure, and reaction mechanisms, as all these factors govern both energy density and controllability.

Energetic small molecules form the foundation of many chemically activated storage systems. Nitrogen-rich compounds, including azides, tetrazoles, and polynitrogen species, possess substantial enthalpies of formation due to the inherent strain in N–N bonds. Decomposition of these compounds can release large amounts of energy per unit mass, rivaling conventional fuels. For example, the controlled decomposition of hydrazoic acid or nitrogen-rich heterocycles has been investigated for micropropulsion and energy-dense chemical batteries. The design of such molecules involves balancing bond strain and steric hindrance to prevent premature decomposition while maintaining rapid energy release kinetics when triggered. Computational methods, including density functional theory (DFT) and ab initio molecular dynamics, allow chemists to predict stability, reaction pathways, and energetic output, guiding the synthesis of safer yet high-energy molecules. Similar strategies apply to oxygen- and hydrogen-rich molecules such as peroxides and hydrazines, where the presence of high-energy bonds facilitates storage but introduces significant handling risks. Encapsulation, solid supports, and covalent stabilization have been explored to mitigate these challenges, highlighting the interplay of molecular chemistry and materials engineering.

Redox chemistry is central to many chemically activated energy storage systems. Redox-active organic molecules such as quinones, viologens, phenazines, and phenothiazines undergo reversible electron-transfer processes, storing energy in the form of oxidized or reduced states. The energy density of such systems depends on the redox potential, electron stoichiometry, and reversibility of the electron-transfer processes. Substituents and conjugation patterns can tune redox potentials and stabilize charged intermediates, allowing fine control over energy storage and release. Redox flow batteries exemplify the practical application of this concept, where soluble redox-active species circulate through electrochemical cells. Molecular design in these systems targets both high solubility and stability to maximize volumetric energy density while minimizing side reactions that can degrade performance. Polyoxometalates, for instance, offer multi-electron redox capacities, allowing the storage of several electrons per molecule with high stability, bridging molecular chemistry and scalable device engineering.

Catalysis is intimately connected to chemically activated energy storage, governing the rates, selectivity, and reversibility of energy release. Transition metal complexes, particularly those of ruthenium, cobalt, nickel, and iron, have been widely explored for catalyzing multi-electron redox processes and reversible chemical transformations. Metal-mediated hydrogen storage illustrates this principle: metal hydrides absorb hydrogen under mild conditions and release it upon heating or catalysis. The coordination environment, ligand design, and electronic structure of the metal dictate both the thermodynamics and kinetics of hydrogen absorption and desorption. Advances in ligand design, including pincer-type ligands and frustrated Lewis pair systems, have enhanced reversibility and lowered activation barriers, enabling practical energy storage at ambient conditions. The integration of homogeneous catalysts with solid-state supports, such as MOFs or porous polymers, further improves stability and facilitates process scalability, highlighting the synergistic relationship between molecular chemistry and materials design.

Thermochemical energy storage leverages reversible chemical reactions to store energy as enthalpy, providing a pathway for integrating chemical storage with heat management systems. Systems such as metal oxide hydration/dehydration cycles, ammonium salt decomposition, and carbonate-bicarbonate conversions demonstrate the potential for high-density storage with tunable release profiles. Thermodynamic analysis based on Gibbs free energy, enthalpy, and entropy changes guides the selection of reactions that are both efficient and reversible. Kinetics play a pivotal role in determining the speed and controllability of energy release, requiring precise engineering of catalysts, reaction media, and thermal management. For example, magnesium oxide hydration to magnesium hydroxide offers a high enthalpy change with relatively benign reactants, whereas lithium hydroxide hydration provides a faster kinetic pathway suitable for rapid energy release in smaller-scale applications. The integration of such reactions with solar thermal or waste heat sources allows sustainable energy storage, transforming intermittent energy inputs into chemically stable forms for later utilization.

Polymer-based chemically activated systems expand the design space for molecular energy storage. Polymers incorporating high-energy functional groups such as azides, peroxides, strained rings, and nitroalkanes provide both energy density and mechanical stability. The polymer backbone allows spatial control over energy distribution, mitigating hazards associated with concentrated high-energy molecules. Cross-linked networks, dendrimers, and supramolecular assemblies enhance stability while allowing tunable release characteristics. Stimuli-responsive polymers, which undergo structural or chemical changes in response to heat, light, or chemical triggers, offer programmable energy storage capabilities. For instance, azobenzene-functionalized polymers can store light energy in cis-trans isomerization processes, releasing it thermally or photochemically on demand. These systems illustrate the convergence of synthetic chemistry, polymer science, and materials engineering in the design of safe, high-density energy storage platforms.

Photochemical energy storage harnesses solar energy to create high-energy metastable isomers. Molecules such as azobenzenes, norbornadienes, and fulvalenes undergo reversible photochemical transformations, storing energy in strained or electronically excited states. The efficiency of these systems depends on absorption spectra, quantum yield, thermal relaxation rates, and photochemical fatigue resistance. Molecular engineering, including rigidification of backbones, introduction of electron-donating or -withdrawing substituents, and optimization of conjugation length, enhances energy density and lifetime. Hybrid approaches combining photochemistry with catalysis or redox mediation expand the scope of chemically activated energy storage, enabling the capture and release of solar energy with high temporal and spatial control.

Bioinspired chemically activated storage provides insights from nature's molecular solutions to energy storage. High-energy phosphate bonds in ATP, redox cofactors such as NADH and FADH2, and proton gradients across membranes illustrate highly efficient natural energy storage mechanisms. Synthetic analogues, including enzyme-mimetic catalysts, molecular redox carriers, and biomimetic energy transfer networks, translate these concepts into artificial systems. Proton-coupled electron transfer, molecular recognition, and multistep reaction cascades in biological systems inform the design of energy storage molecules with high specificity, reversibility, and minimal energy loss. Such biomimetic approaches bridge fundamental chemistry and applied energy storage, demonstrating the value of interdisciplinary inspiration.

Advanced chemical storage systems increasingly rely on computational and data-driven design. Quantum chemical calculations predict electronic structure, bond strain, and reaction pathways, enabling rational design of high-energy molecules. Machine learning and high-throughput experimentation accelerate discovery by screening large chemical spaces for candidates with optimal energy density, stability, and reaction kinetics. Coupled with automated synthesis and in situ characterization, these approaches shorten development cycles and reduce experimental trial-and-error, making the discovery of new chemically activated storage materials more systematic and predictive.

Real-world implementation of chemically activated energy storage requires attention to safety, scalability, and sustainability. High-energy compounds must be stabilized against decomposition, compatible with storage and containment materials, and environmentally benign. Encapsulation, polymer matrices, and covalent functionalization mitigate hazards, while life-cycle assessment informs material selection and process design. Integration with renewable energy sources, such as solar, wind, and geothermal input, leverages chemically activated storage as a bridge between intermittent energy supply and continuous demand. Emerging applications range from grid-scale storage and microactuation to space propulsion, where energy density and controllability are critical.

Hybrid systems combining chemical, electrochemical, and mechanical energy storage are emerging as particularly promising. For example, redox-active polymers integrated with electrochemical cells provide rapid charge-discharge capabilities while storing high-energy molecules safely. Similarly, thermochemical systems coupled with heat engines or phase-change materials optimize energy management by combining chemical storage with thermal regulation. These hybrid approaches highlight the versatility of chemically activated storage, enabling tailored solutions for diverse applications ranging from portable electronics to renewable energy infrastructure.

Frontiers in the field continue to explore extreme energy-density compounds, including polynitrogen species, strained cyclic molecules, high-oxidation-state metal complexes, and multi-electron redox centers. The exploration of these compounds is guided by computational chemistry, catalysis, and materials science, balancing energy density, reversibility, and safety. Emerging strategies also include adaptive and self-healing materials that respond to environmental stimuli, dynamically controlling energy storage and release in response to temperature, light, or chemical triggers. Such smart systems point toward a future where chemically activated energy storage is not only efficient and dense but also programmable and resilient.

Ultimately, chemically activated energy storage exemplifies the transformative power of chemistry in energy technology. By storing energy in molecular bonds and intermediates, these systems provide solutions beyond the limitations of conventional batteries, capacitors, or mechanical storage. Advances in molecular design, catalysis, polymer engineering, photochemistry, bioinspiration, and computational design converge to create systems with higher energy densities, faster response times, and enhanced safety profiles. As global energy demands grow and renewable energy sources proliferate, chemically activated storage provides a versatile and sustainable pathway, bridging fundamental chemistry with practical applications and redefining the landscape of energy storage science.

Building on the concept of chemically activated energy storage, the field increasingly emphasizes the molecular-level control of energy capture and release. Traditional energy storage systems, such as lithium-ion batteries or pumped hydroelectric storage, rely on bulk electrochemical or mechanical processes that impose intrinsic limits on energy density, scalability, and response time. In contrast, chemically activated systems operate at the level of discrete molecular bonds and reactive intermediates, allowing for a precision of energy management that is unprecedented. For instance, the development of high-energy organometallic compounds or metastable chemical species enables rapid energy release on demand, effectively functioning as molecular springs that can store and deploy energy with minimal losses. These approaches not only push the boundaries of energy density but also enable novel forms of energy delivery that are tunable across temporal scales—from milliseconds to hours—depending on the application.

A key strategy in this domain is the harnessing of redox-active molecules, which can reversibly accept or donate electrons while storing significant amounts of energy. Organic redox molecules, such as quinones, phenazines, and viologens, are being explored for their ability to function in both aqueous and non-aqueous environments, offering scalable and environmentally benign alternatives to heavy-metal-based systems. Moreover, polymeric redox materials extend these concepts by integrating multiple redox centers into flexible, processable frameworks, enabling high volumetric energy densities without the safety concerns associated with metallic electrodes. Advances in synthetic chemistry allow precise tailoring of redox potentials, solubility, and kinetics, thereby optimizing performance for applications ranging from grid-scale energy storage to portable electronics. Computational modeling plays an increasingly vital role in this design process, allowing chemists to predict the thermodynamic and kinetic behaviors of candidate molecules before experimental synthesis, saving both time and resources while accelerating discovery.

Catalysis also represents a central pillar in chemically activated energy storage, particularly in systems where chemical transformations govern energy uptake or release. For example, hydrogen storage technologies rely on catalytic dehydrogenation and hydrogenation reactions that store energy in the form of chemical bonds. Transition-metal complexes, metal-organic frameworks (MOFs), and nanostructured catalysts have all demonstrated remarkable efficiency in mediating these reactions, lowering activation barriers, and enhancing cycle lifetimes. Similarly, photoactivated catalytic systems, which combine photochemistry with catalysis, provide mechanisms for directly converting solar energy into chemically stored energy, creating an integrated platform for sustainable energy storage. These photochemical approaches are particularly promising in the development of solar fuels, where sunlight drives the generation of high-energy molecules such as hydrogen, formic acid, or methanol from water or carbon dioxide. The challenge lies in designing catalysts and molecular systems that not only exhibit high conversion efficiencies but also maintain stability over extended operational cycles, a requirement that demands a deep understanding of both chemical thermodynamics and reaction kinetics.

Bioinspiration offers another rich avenue for innovation in chemically activated energy storage. Nature demonstrates remarkable efficiency in storing and mobilizing energy through molecular intermediates, from adenosine triphosphate (ATP) in cellular metabolism to the light-driven proton gradients that power photosynthesis. By mimicking these biological strategies, chemists can design synthetic analogues that exploit hierarchical structures, self-assembly, and feedback regulation to improve energy storage performance. For instance, supramolecular assemblies can encapsulate reactive species to prevent premature energy release, while enzymatic mimetics or peptide-based catalysts can enable highly selective reaction pathways that reduce energy losses. Furthermore, integrating bioinspired approaches with synthetic polymers or inorganic frameworks offers the potential for hybrid systems that combine the high efficiency of biological processes with the robustness and scalability of engineered materials.

Photochemistry itself continues to expand the frontier of energy storage by enabling the direct capture of light energy and its transformation into chemically stored potential. Molecules such as azobenzenes, diarylethenes, and other photochromic compounds undergo reversible structural changes upon light absorption, effectively acting as molecular switches that store photon energy as chemical strain. When triggered, these molecules release the stored energy through a controlled thermal or photochemical reaction, providing a platform for light-driven actuation, energy release, or on-demand power generation. Coupling these systems with advanced materials, such as conductive polymers or carbon-based scaffolds, enhances charge transfer, energy retention, and material stability, creating hybrid architectures that blur the boundaries between photochemistry, materials science, and chemical engineering.

Finally, the integration of computational chemistry, machine learning, and high-throughput experimental methods is transforming the discovery and optimization of chemically activated energy storage systems. Quantum mechanical calculations and molecular dynamics simulations allow detailed mapping of reaction pathways, energy landscapes, and electronic properties, guiding the rational design of molecules with targeted storage capacities and kinetic profiles. Machine learning algorithms can further analyze vast chemical spaces, identifying correlations and predicting promising candidates that might otherwise remain undiscovered. This convergence of theory, computation, and experiment not only accelerates the development of new storage materials but also provides fundamental insights into the principles governing chemical energy transformation, bridging molecular understanding with practical engineering solutions.

In conclusion, chemically activated energy storage exemplifies a paradigm shift in energy technology, leveraging the principles of chemistry to achieve precise, efficient, and sustainable energy management. By combining innovations in molecular design, catalysis, polymer engineering, photochemistry, bioinspiration, and computational modeling, these systems transcend the limitations of conventional approaches, offering new opportunities for grid stabilization, portable power, and renewable energy integration. As global energy demands intensify and environmental constraints become more pressing, the ongoing exploration of chemically activated energy storage underscores the vital role of chemistry as both a fundamental science and a driver of technological innovation, redefining the very landscape of energy storage in the twenty-first century.

Chapter 43: Word Problems

1. A chemist is tasked with designing a nitrogen-rich energetic molecule intended for micropropulsion applications, where the compound must release approximately 500 kJ/mol of energy upon decomposition while maintaining kinetic stability at ambient conditions (298 K). Considering both bond strain (e.g., N–N, N–O bonds) and steric hindrance effects, propose a molecular design strategy that balances high energy content with safety. Explain how density functional theory (DFT) calculations could be employed to predict both the decomposition energy and kinetic barriers, including transition state searches and vibrational analysis. Discuss how electronic structure parameters such as HOMO–LUMO gap, natural bond orbital charges, and bond orders can inform predictions of stability, and propose experimental validation methods such as calorimetry or gas evolution measurements.

2. A hydrogen storage molecule is being developed for portable power devices and must reversibly absorb molecular hydrogen under mild temperature and pressure conditions. Assuming the molecule has a molar mass of 120 g/mol and can bind two moles of H_2 per mole of carrier, calculate the theoretical hydrogen storage density in weight percent. Discuss how thermodynamic parameters (ΔH, ΔS) dictate storage reversibility, and how catalyst selection influences both kinetics of absorption/desorption and operational temperature. Consider structural

modifications, ligand effects, or metal center selection if a transition metal-based system is used, and explain how computational modeling could predict optimal hydrogen binding energy for practical operation.

3. A redox-active organic molecule undergoes a reversible two-electron reduction with a standard potential of 0.8 V versus SHE. Compute the Gibbs free energy change per mole during reduction and evaluate the energy storage capacity. Discuss how substitution on the aromatic or heterocyclic backbone could modulate redox potential while preserving reversibility, considering electron-withdrawing and electron-donating groups. Propose how solvent, ionic strength, and proton-coupled electron transfer pathways might affect kinetics and long-term stability in a practical energy storage system.

4. A polymer backbone is functionalized with multiple azide groups to store chemical energy, where decomposition releases substantial enthalpy. If the polymer's crosslinking density is doubled, qualitatively predict how mechanical rigidity, glass transition temperature, and energy release profile would change. Suggest a computational approach—such as reactive molecular dynamics or kinetic Monte Carlo simulations—to model thermal decomposition kinetics, including the effects of chain mobility and cooperative bond-breaking. Discuss how experimental calorimetric techniques and spectroscopy could validate predicted energy release rates.

5. 5. Consider a thermochemical storage system where 10 moles of magnesium oxide (MgO) react with water to form magnesium hydroxide ($Mg(OH)_2$). Calculate the total enthalpy stored in this reaction. Discuss how reaction kinetics could be enhanced using solid supports, high surface area formulations, or heterogeneous catalysts to enable rapid energy release suitable for small-scale portable applications. Analyze trade-offs between reaction rate, heat management, reversibility, and operational temperature constraints.

6. In a photochemical energy storage experiment, azobenzene derivatives undergo cis-trans isomerization upon irradiation with UV light. If the quantum yield is 0.6 and 1 mmol of azobenzene is irradiated, estimate the moles of molecules converted to the cis isomer. Discuss molecular design strategies—such as electronic substitution, steric protection, or polymer integration—to reduce thermal back-isomerization

and enhance energy retention. Consider the impact of solvent, temperature, and light intensity on quantum efficiency and practical device lifetime.

7. A multi-electron redox system is proposed using polyoxometalates capable of storing six electrons per molecule. If the molecular mass is 500 g/mol and the cell operates at 1.2 V, calculate the theoretical gravimetric energy density in Wh/kg. Discuss challenges associated with maintaining stability over repeated redox cycles, including electron delocalization, solubility limits, aggregation, and counter-ion effects. Propose strategies to optimize both energy density and cycle life using molecular design, electrolyte formulation, and electrode engineering.

8. A polymer is functionalized with high-energy peroxy bonds (O–O) for energy storage. Propose an experimental design to accurately measure the rate of energy release, including calorimetry, gas evolution, and spectroscopic monitoring. Discuss how covalent stabilization, polymer architecture, or encapsulation techniques could enhance safety during storage, transport, and handling. Evaluate the trade-offs between energy density, mechanical properties, and decomposition kinetics.

9. A nitrogen-rich heterocyclic compound decomposes via a first-order reaction with a rate constant of 0.02 s^{-1} at 300 K. Calculate the half-life of the compound and discuss molecular strategies—such as steric shielding, electronic tuning, or substituent effects—to prevent premature decomposition while enabling rapid, triggered energy release. Explore how computational tools (transition state modeling, bond dissociation energy analysis) and experimental kinetics studies could be integrated to optimize both stability and performance.

10. Design a hybrid chemical-electrochemical energy storage device using quinone-functionalized polymers. Explain how tuning redox potential, solubility, polymer architecture (crosslinking, chain length, and porosity), and electrolyte compatibility can maximize energy density while minimizing side reactions such as irreversible polymer oxidation or dimerization. Discuss volumetric versus gravimetric energy trade-offs, the role of electrolyte conductivity, and strategies to integrate such polymers into flow or solid-state devices for practical applications.

11. A ruthenium pincer complex catalyzes hydrogen release from a metal hydride at 80 °C. Propose a detailed catalytic mechanism, including oxidative addition, hydride transfer, and reductive elimination steps. If the reaction rate doubles when

the temperature is increased by 10 °C, calculate the activation energy using the Arrhenius equation. Discuss how electronic properties of the ligands (donating or withdrawing groups) influence both the thermodynamics and reversibility of hydrogen release, and how steric effects could enhance catalyst turnover frequency while minimizing side reactions.

12. A fulvalene molecule is used for solar energy storage via photochemical conversion. It absorbs at 400 nm with an extinction coefficient of 10^4 M^{-1} cm^{-1}. For a 1 cm pathlength solution at 1 mM, estimate the fraction of light absorbed using the Beer-Lambert law. Discuss how molecular engineering—altering conjugation length, substituents, and planarity—affects absorption wavelength, quantum yield, and energy density, and propose strategies to reduce thermal back-conversion to extend storage lifetime.

13. A nitrogen-rich compound is predicted by DFT to have a decomposition enthalpy of 450 kJ/mol. If a practical micropropulsion device requires 1 MJ of energy, calculate the mass of compound required. Discuss computational screening methods to identify analogues with similar energy output but improved safety profiles, considering bond strain, steric protection, and electronic effects. Propose experimental validation strategies, such as calorimetry or controlled ignition tests, and discuss encapsulation for handling safety.

14. A redox flow battery uses an aqueous viologen derivative at 1.5 M concentration, transferring 2 electrons per molecule with a cell voltage of 1.0 V. Calculate the theoretical volumetric energy density in Wh/L. Discuss strategies to enhance solubility, prevent dimerization, and optimize membrane compatibility. Consider molecular modifications (e.g., alkylation, sulfonation) to improve both electrochemical reversibility and long-term cycling stability.

15. A photochemical polymer stores energy through azobenzene cis-trans isomerization. If thermal relaxation has a half-life of 48 hours at room temperature, estimate the fraction of energy lost over one week. Discuss polymer modifications (crosslinking density, polymer backbone rigidity, steric shielding of azobenzene) to increase fatigue resistance and reduce photochemical degradation. Evaluate how matrix effects and aggregation could influence quantum yield and thermal stability.

16. A hydrogen storage material absorbs H_2 with an enthalpy of −25 kJ/mol H_2. Using van't Hoff analysis, estimate the equilibrium hydrogen pressure at 300 K and 350 K. Discuss how tuning the metal coordination environment, ligand electronics, or particle morphology can optimize both energy density and release kinetics. Consider the trade-off between thermodynamic favorability, reversibility, and operating temperature for practical storage applications.

17. A supramolecular host encapsulates a high-energy molecule to prevent accidental decomposition. Discuss factors controlling release kinetics when triggered chemically, thermally, or photochemically, including diffusion rates, host cavity size, binding affinity, and host-guest interactions. Propose computational methods to predict binding energies and barrier heights, and experimental methods to quantify controlled release (NMR, calorimetry, gas evolution).

18. A strained cyclic molecule releases energy upon ring-opening polymerization. If the reaction enthalpy is 120 kJ/mol and the molar mass is 150 g/mol, calculate energy released per gram of monomer. Discuss how molecular strain, substituents, and polymerization kinetics influence both stability and energy content. Propose computational approaches (DFT, molecular mechanics) to optimize strain-energy release while maintaining safe handling.

19. A lithium hydroxide hydration system is proposed for rapid energy release. Compare the thermodynamics (ΔH, ΔG) and kinetics to magnesium oxide hydration. Propose a heterogeneous or nanoparticle catalyst to accelerate the reaction rate without compromising reversibility. Discuss design considerations for heat management, energy density, and scale-up for portable or stationary applications.

20. Design a molecular system for on-demand energy release triggered by visible light. Consider absorption spectra, quantum yield, photostability, and thermal stability. Discuss strategies to integrate the molecule into a polymeric matrix or solid-state support for safe handling and repeatable operation. Analyze trade-offs between light penetration, energy density, and response time for practical devices.

21. A polyoxometalate molecule stores 12 electrons per molecule. Compute the theoretical gravimetric energy density if the molecular mass is 600 g/mol and the average redox potential is 1.0 V. Discuss how multi-electron capacity impacts redox reversibility, stability, and side reactions. Consider electrolyte design,

counter-ion selection, and electrode engineering to optimize performance in a flow or solid-state device.

22. A peroxynitroalkane derivative releases energy upon thermal decomposition. Propose an experimental protocol to measure the energy output safely, including calorimetric methods and gas detection. Discuss hazards (oxygen evolution, sensitivity to impact/heat) and propose polymer encapsulation or stabilization strategies to improve handling safety and controllable release.

23. A hybrid system combines thermochemical energy storage with a phase-change material (PCM). Explain how this combination enhances both energy density and thermal management. Quantitatively discuss heat transfer limitations, thermal conductivity of PCM, reaction enthalpy of the thermochemical component, and energy retention over multiple cycles. Propose modeling appr

24. A computational chemist uses DFT to evaluate a high-energy organometallic compound. Explain how bond order, electron density distribution, frontier molecular orbitals, and HOMO-LUMO gap inform predictions of both stability and energy release kinetics. Discuss how computationally predicted decomposition pathways can guide experimental safety testing and molecular design optimization.

25. A nitrogen-rich compound is synthesized for space propulsion. Discuss how steric hindrance, electronic effects, and substituent patterns influence decomposition rate, energy density, and thermal stability. Propose a computational workflow (screening of analogues, DFT energetics, transition state calculations) for safe candidate selection, considering practical device requirements.

26. A diarylethene derivative absorbs sunlight and undergoes reversible photoisomerization. If 2 mmol absorbs sunlight over 3 hours, assuming photon energy of 3 eV and 80% conversion efficiency, calculate the energy stored. Discuss molecular modifications to improve fatigue resistance, thermal stability, and quantum efficiency for long-term storage applications.

27. A quinone-functionalized redox polymer is integrated into an aqueous flow battery. Explain how polymer architecture, crosslinking density, and redox potential tuning optimize volumetric energy density, solubility, and long-term stability. Discuss potential side reactions, dimerization, or solubility limits, and propose experimental approaches to monitor degradation and performance over multiple cycles.

28. A metal-organic framework catalyzes reversible hydrogenation of an organic carrier. Discuss how pore size, metal identity, and ligand environment influence both thermodynamics and kinetics of H_2 uptake/release. Propose computational approaches (DFT, molecular dynamics, adsorption modeling) to predict optimal MOF structures, and experimental methods to validate storage capacity and reversibility.

29. A bioinspired ATP analogue is proposed as a high-energy chemical carrier. Discuss how proton-coupled electron transfer, multi-step reaction cascades, and phosphate bond energetics can be exploited to design artificial molecules with high energy storage efficiency. Propose strategies to optimize stability, control release kinetics, and integrate into hybrid chemical or electrochemical devices.

30. A norbornadiene derivative stores solar energy via photochemical conversion. Design an experiment to measure quantum yield, thermal relaxation, and energy release kinetics. Discuss molecular modifications to extend storage lifetime, minimize back-conversion, and enhance photostability, including substitution patterns and electronic conjugation strategies.

31. A nitrogen-rich heterocycle releases 600 kJ/mol upon decomposition. Calculate the mass required to deliver 2 MJ of energy. Discuss safety considerations such as sensitivity to heat, shock, or friction, and propose computational screening and encapsulation strategies to identify safer analogues while maintaining energy output.

32. In a hydrogen storage experiment, a cobalt pincer complex catalyzes H_2 release. Using kinetic data at 300 K and 310 K, calculate the activation energy. Discuss ligand design modifications to lower activation barriers, improve reversibility, and optimize hydrogen release kinetics for portable applications.

33. A redox-active organic molecule undergoes reversible three-electron transfer with a standard potential of 0.9 V. Determine the Gibbs free energy change per mole for three electrons, and discuss how solvent polarity, ionic strength, and electrolyte composition influence redox reversibility and energy storage density.

34. A strained cyclic molecule is incorporated into a polymer backbone. Explain how crosslinking density, polymer rigidity, and stimuli-responsive groups affect

mechanical stability, glass transition temperature, and triggered energy release. Discuss computational modeling approaches to optimize structure-property relationships.

35. A hybrid redox-photochemical system stores sunlight chemically. Propose a design maximizing energy density, reversibility, and cycle stability. Discuss photodegradation, side reactions, and strategies such as polymer encapsulation, conjugation tuning, or multi-step redox cascades to improve performance.

36. A high-energy polynitrogen species is considered for grid-scale energy storage. Discuss how molecular strain, bond energy distribution, and decomposition kinetics influence energy density, safety, and scalability. Suggest encapsulation, solid supports, or composite materials to enable safe handling and practical application.

37. A polymer functionalized with azobenzene undergoes cis-trans isomerization upon UV irradiation. Calculate theoretical energy stored per gram if each isomerization stores 50 kJ/mol and the polymer contains 0.01 mol azobenzene per gram. Discuss polymer matrix effects on energy retention, quantum yield, and photochemical fatigue resistance.

38. A thermochemical cycle uses ammonium salt decomposition for solar energy storage. Compare enthalpy and entropy contributions to Gibbs free energy and discuss catalysts that could enhance both reversibility and reaction kinetics. Evaluate design considerations for cycle efficiency and thermal management.

39. A machine-learning algorithm screens 10^5 molecules for high-energy storage potential. Discuss which descriptors (bond strain, HOMO-LUMO gap, redox potential, steric hindrance) are most relevant for predicting energy density and stability. Propose validation strategies combining computational screening and experimental calorimetry.

40. A polymeric redox flow battery uses phenothiazine derivatives. Propose strategies to enhance solubility, prevent side reactions, and improve multi-cycle stability, considering molecular architecture, polymer crosslinking, and electrolyte formulation. Discuss trade-offs between gravimetric and volumetric energy density.

41. A photochromic molecule is embedded in a conductive polymer. Explain how

charge transfer, conjugation length, and polymer rigidity affect energy storage efficiency and photochemical lifetime. Discuss design strategies to balance electronic conductivity with long-term photostability.

42. A hydrogen storage material releases 40 kJ/mol H_2 at 298 K. Using van't Hoff analysis, estimate equilibrium H_2 pressure at 350 K. Discuss how ligand electronics and metal center selection influence both thermodynamics and kinetics, and how these parameters affect practical storage operation.

43. A supramolecular assembly stores a high-energy compound. Suggest methods for selective energy release (thermal, chemical, photochemical) and discuss how host-guest interactions, cavity size, and diffusion barriers control reaction rates. Propose computational and experimental validation strategies.

44. A polynitrogen compound exhibits high energy density but limited stability. Design a computational and experimental workflow to evaluate candidates for safe energy storage, considering encapsulation, host-guest protection, and structural modification to balance energy density and safety.

45. A hybrid system couples lithium hydroxide hydration with a redox-active polymer. Discuss optimization of energy management, heat transfer, reaction kinetics, and cycle efficiency for both high power and high density. Consider integration strategies for portable or stationary applications.

46. A bioinspired molecule mimics NADH for chemical energy storage. Discuss proton-coupled electron transfer pathways and how enzyme-mimetic catalysts could improve reversibility, energy retention, and release kinetics. Propose design modifications to enhance photochemical or electrochemical integration.

47. A norbornadiene derivative captures sunlight. Calculate maximum theoretical energy stored if 0.5 mmol absorbs sunlight over 2 hours at 70% efficiency. Discuss molecular strategies, such as electronic tuning and steric hindrance, to minimize back-conversion and enhance photostability.

48. A high-energy organometallic compound decomposes exothermically ($\Delta G = -400$ kJ/mol). Propose polymer encapsulation strategies and discuss trade-offs between safety, energy density, and release rate. Include considerations of mechanical stability, thermal conductivity, and potential side reactions.

49. A redox-active polymer integrated into a flow battery exhibits 90% coulombic efficiency. Discuss molecular design strategies to improve cycle life, volumetric energy density, and minimize side reactions. Consider polymer architecture, crosslinking, solubility, and electrolyte compatibility.

50. Design a smart, self-healing chemically activated energy storage system. Propose a molecular architecture that responds to environmental triggers (temperature, light, or chemical stimuli), explaining both the energy storage mechanism and safety considerations. Discuss potential polymer matrices, reversible covalent bonds, or supramolecular interactions enabling repeated activation and healing.

Chapter 44

Molecular Origami and 3D Self-Assembly

Molecular origami, a field at the intersection of chemistry, nanotechnology, and molecular engineering, represents a paradigm shift in the design and construction of complex three-dimensional (3D) nanostructures with atomic precision. Inspired by traditional paper origami, molecular origami employs the inherent folding properties of macromolecules, particularly nucleic acids, peptides, and synthetic polymers, to form defined geometries under thermodynamically controlled conditions. Unlike conventional self-assembly, which often results in bulk aggregates or randomly oriented structures, molecular origami exploits programmable interactions such as hydrogen bonding, base stacking, metal-ligand coordination, hydrophobic forces, and van der Waals interactions to drive predictable folding and assembly into precise nanostructures. The foundational principle of molecular origami rests on the encoding of structural information within the primary sequence of a polymeric molecule. In the case of DNA origami, pioneered by Rothemund, long single-stranded DNA scaffolds are folded into target 2D and 3D architectures by hundreds of short staple strands, which selectively hybridize to complementary sequences to define edges, angles, and junctions. This approach not only allows precise control over nanoscale geometry but also enables the incorporation of functional moieties, including fluorophores, nanoparticles, and enzymes, at defined positions within the structure. The concept of sequence-encoded folding can be extended to peptides and synthetic polymers, where amphiphilic block copolymers, peptoids, and β-sheet forming sequences fold into well-defined secondary and tertiary structures that further assemble into higher-order 3D architectures through hierarchical self-assembly. These folding pathways are governed by the delicate interplay of intramolecular interactions and environmental conditions such as ionic strength, pH, temperature, and solvent polarity, which modulate the free energy landscape and kinetic accessibility of target structures.

The thermodynamic basis of molecular origami and 3D self-assembly lies in the minimization of Gibbs free energy, whereby molecules adopt conformations and supramolecular arrangements that maximize favorable enthalpic interactions while minimizing entropic penalties. For nucleic acid-based systems, the specificity of Watson-Crick base pairing provides a predictable and programmable route to hierarchical assembly. Complementary strands selectively hybridize, and cooperative interactions among multiple duplexes stabilize defined motifs such as Holliday junctions, tensegrity triangles, and multi-arm branched junctions. These motifs can further interact to form extended 2D lattices or 3D polyhedra, including cubes, tetrahedra, octahedra, and icosahedra. The stability and folding efficiency of these structures are highly sensitive to staple design, strand stoichiometry, and annealing protocols, with iterative computational optimization enabling precise control over curvature, twist, and overall topology. In peptide-based molecular origami, amphiphilic sequences fold into helices, sheets, and turns, which subsequently engage in directional hydrogen bonding, π-π stacking, and hydrophobic clustering to drive self-assembly into nanofibers, cages, and vesicular structures. Computational tools such as molecular dynamics simulations, coarse-grained modeling, and Monte Carlo methods have become indispensable in predicting folding pathways, identifying metastable intermediates, and guiding the rational design of sequence-defined polymers for target architectures. Moreover, energy landscape analysis enables the identification of kinetic traps and competing assembly pathways, providing insights into error-correction mechanisms and design principles for robust self-assembly.

Hierarchical assembly is a defining feature of molecular origami, where discrete molecular motifs first fold into secondary and tertiary structures before organizing into complex superstructures. In DNA origami, secondary motifs such as double helices and crossover junctions organize into tertiary polyhedral frameworks, which can further self-assemble into quasicrystalline lattices or programmable nanocages. The concept of hierarchical folding is mirrored in peptide and protein-based systems, where secondary structural elements such as α-helices and β-sheets assemble into tertiary folds, which then oligomerize into quaternary assemblies. Supramolecular polymers, block copolymers, and dendrimers also leverage hierarchical self-assembly to form nanoscale architectures with tunable porosity, mechanical rigidity, and surface functionality. The interplay of local folding energetics and long-range intermolecular interactions allows for emergent properties, including stimuli-responsiveness, self-healing, and adaptive behavior, which are crucial for functional applications in nanomedicine, sensing, catalysis, and photonic materials. For instance, stimuli-responsive molecular origami structures can undergo conformational transitions in response to changes in temperature, pH, ionic environment, or light exposure, enabling switchable encapsulation, targeted cargo release, and dynamic reconfiguration.

The design principles for 3D self-assembly extend beyond the choice of molecular building blocks to include topological and geometric considerations. Polyhedral cages and wireframe structures can be generated by designing junction connectivity and edge lengths, guided by Eulerian graph theory, symmetry analysis, and geometric constraints. Topological design allows for the creation of structures with controlled curvature, internal cavities, and external surfaces suitable for functionalization. For example, DNA tensegrity triangles can be assembled into periodic 3D crystals, where the rigidity and curvature of each motif dictate the resulting lattice parameters. In protein-based origami, rational design of coiled-coil motifs and β-sheet tapes can yield cages, cages-on-a-cage assemblies, and filamentous networks with pre-programmed curvature and porosity. The incorporation of modular domains and orthogonal interaction motifs facilitates combinatorial assembly, allowing a single molecular scaffold to generate diverse structures with minimal sequence variation. Moreover, synthetic chemists have extended these principles to entirely non-biological systems, using metal-organic frameworks (MOFs), covalent organic frameworks (COFs), and supramolecular polymers as building blocks for programmable 3D assemblies with precise pore architectures and functional surfaces.

Experimental methods for constructing molecular origami structures encompass a combination of solution-phase folding, templated assembly, and surface-mediated techniques. DNA origami typically involves the thermal annealing of scaffold and staple strands in buffered solutions containing divalent cations, which stabilize duplex formation. Folding kinetics are controlled by the rate of cooling, the stoichiometry of strands, and the concentration of ions, which influence the formation of secondary and tertiary interactions. Peptide and polymer-based assemblies often rely on solvent conditions, concentration, and sequence design to drive co-assembly, with self-assembly monitored using spectroscopic techniques such as circular dichroism, fluorescence resonance energy transfer (FRET), and nuclear magnetic resonance (NMR). High-resolution structural characterization is achieved using atomic force microscopy (AFM), cryogenic electron microscopy (cryo-EM), transmission electron microscopy (TEM), and small-angle X-ray scattering (SAXS), which provide detailed insights into folding fidelity, defect formation, and supramolecular organization. Surface-assisted assembly enables the templating of molecular origami onto solid supports, providing additional control over orientation, packing density, and higher-order patterning. Combined with lithographic and nanoimprinting methods, these techniques allow integration of molecular origami into hybrid devices, sensors, and nanoelectronic platforms.

Computational modeling plays an essential role in guiding the design of molecular origami and 3D self-assembly. Sequence optimization algorithms predict secondary structure formation, minimize misfolding, and optimize staple placement for DNA origami. Coarse-grained simulations capture long-range interactions and assembly dynamics in larger constructs, enabling the study of folding kinetics and cooperative effects. For protein and peptide origami, molecular dynamics simulations provide atomistic resolution of folding pathways, energy landscapes, and solvent effects. Machine learning approaches are increasingly employed to predict folding outcomes, suggest sequence modifications, and identify emergent structural motifs. These computational tools accelerate the rational design of complex 3D architectures, allowing for predictive control over size, shape, rigidity, and functionality.

Applications of molecular origami and 3D self-assembly are rapidly expanding across nanomedicine, materials science, and molecular devices. In nanomedicine, DNA origami cages and peptide-based nanocarriers serve as programmable drug delivery vehicles capable of encapsulating therapeutics, targeting specific cell types, and releasing payloads in response to stimuli. Functionalization with aptamers, antibodies, or small molecules enhances specificity, while the precise nanoscale geometry enables multivalent interactions and cooperative effects. In materials science, self-assembled nanostructures serve as scaffolds for photonic crystals, plasmonic arrays, and enzyme-mimetic catalysts. Hierarchical assemblies allow for tunable porosity, mechanical robustness, and surface chemistry, providing versatile platforms for sensing, separation, and energy conversion. Molecular origami has also been employed in constructing nanoscale devices such as molecular switches, logic gates, and artificial ion channels, demonstrating the potential for bottom-up fabrication of functional nanomachines. Furthermore, the ability to integrate inorganic nanoparticles, quantum dots, and metallic clusters within DNA and peptide frameworks provides avenues for hybrid materials with emergent optical, electronic, and catalytic properties.

Challenges in molecular origami and 3D self-assembly include controlling folding fidelity, minimizing defects, and achieving scalability for practical applications. Misfolded or partially assembled structures can compromise functionality, particularly in applications requiring precise spatial organization. Advances in error-correction strategies, cooperative folding motifs, and hierarchical assembly pathways have mitigated some of these issues, but achieving robust, large-scale fabrication remains an active area of research. Stability under physiological conditions, resistance to nucleases and proteases, and functional integration into complex biological environments are additional considerations for biomedical applications. For synthetic polymer systems, controlling polydispersity, conformational heterogeneity, and solvent-dependent behavior poses ongoing challenges, requiring iterative design, characterization, and modeling.

Future directions in molecular origami and 3D self-assembly involve expanding the diversity of building blocks, enhancing functional complexity, and integrating dynamic and adaptive behaviors. Light-responsive, redox-sensitive, and enzyme-responsive motifs enable stimuli-triggered reconfiguration, allowing molecular assemblies to act as sensors, actuators, and responsive materials. The combination of multiple orthogonal interaction modes, including DNA-peptide hybrids, metal-ligand coordination, and host-guest chemistry, enables modular and programmable architectures with multifunctional capabilities. Emerging computational approaches, including deep learning and AI-driven sequence design, promise to accelerate the discovery of novel folding motifs and optimize assembly efficiency. Moreover, the translation of molecular origami principles to mesoscale and macroscale materials may enable the construction of functional devices with hierarchical organization across length scales, bridging the gap between molecular precision and practical engineering.

Molecular origami and 3D self-assembly represent a transformative approach to designing and constructing nanoscale architectures with atomic-level precision. By leveraging programmable interactions, hierarchical assembly, and computational design, chemists and materials scientists can create complex, functional nanostructures for applications spanning nanomedicine, materials science, and molecular machinery. The field continues to evolve rapidly, driven by advances in sequence design, experimental techniques, computational modeling, and functional integration. As understanding of folding mechanisms, error-correction strategies, and dynamic behavior deepens, molecular origami is poised to revolutionize the bottom-up fabrication of nanoscale materials and devices, offering unprecedented control over structure, function, and complexity in the molecular world.

Building upon the foundational principles of molecular origami, one of the most striking aspects of this field is its ability to bridge the gap between abstract molecular design and tangible functional outcomes. The design of DNA, RNA, and peptide-based origami structures relies on the precise control of complementary interactions and steric constraints, allowing researchers to dictate the three-dimensional conformations of molecules with near-atomic accuracy. Computational tools have become indispensable in this endeavor, facilitating the prediction of folding pathways, the identification of energetically favorable conformations, and the minimization of misfolded intermediates. Algorithms rooted in thermodynamics and kinetics enable the anticipation of potential structural conflicts, permitting the preemptive design of error-correcting motifs that enhance the fidelity of self-assembly. Consequently, the integration of computational and experimental approaches forms a synergistic feedback loop, wherein predictive modeling informs experimental synthesis, and empirical observations refine computational models, creating a robust platform for rational nanoscale engineering.

Beyond mere structural elegance, the functionalization of molecular origami extends its utility into realms traditionally dominated by top-down fabrication techniques. By strategically positioning reactive groups, metal-binding sites, or hydrophobic/hydrophilic domains, chemists can endow these nanoscale constructs with catalytic, photonic, or recognition capabilities. For instance, the incorporation of enzyme-mimetic moieties into DNA origami scaffolds has facilitated the development of nanoscale reaction vessels capable of orchestrating multi-step chemical transformations with unprecedented spatial control. Similarly, the precise spatial arrangement of fluorophores, quenchers, and energy transfer pairs within an origami framework enables the design of highly efficient nanoscale light-harvesting systems, mimicking the energy capture and transfer processes found in natural photosynthetic complexes. These functional assemblies underscore the versatility of molecular origami as a platform not only for structural design but also for the creation of chemically active nanosystems with programmable behaviors.

The dynamic nature of self-assembled molecular architectures further expands their potential applications. Unlike rigid, static nanostructures, many origami systems are capable of conformational flexibility, responsive motion, and environmental adaptability. Stimuli-responsive elements, including pH-sensitive motifs, light-activated switches, and redox-responsive linkages, can be incorporated to produce nanoscale devices that change shape, release cargo, or modulate function in response to external cues. Such dynamic behavior opens avenues for precision drug delivery, where molecular carriers can encapsulate therapeutic agents and release them at targeted sites in response to specific biochemical signals. Furthermore, the integration of dynamic origami structures into nanoscale robotics—often termed "molecular machines"—paves the way for autonomous nanoscale operations, including targeted molecular transport, mechanical work at the molecular level, and programmable assembly/disassembly cycles.

Hierarchical assembly represents another frontier within molecular origami research. While single-layer or single-unit origami structures provide a foundation for nanoscale design, the ability to organize multiple units into larger, ordered superstructures enables the translation of nanoscale precision into mesoscale functionality. Techniques such as tile-based assembly, scaffolded multi-origami networks, and programmable inter-unit interactions facilitate the construction of two- and three-dimensional lattices, frameworks, and metamaterials. These hierarchically assembled systems exhibit emergent properties, including collective mechanical stability, tunable porosity, and cooperative functional behavior, which are unattainable in isolated nanoscale units. Hierarchical design principles thus allow chemists to move beyond the molecule and the single origami motif toward the creation of complex, multi-scale materials with integrated functionalities.

Integration with other chemical and materials domains further enhances the impact of molecular origami. Hybrid systems that combine organic and inorganic components, such as DNA-templated nanoparticle arrays or peptide-origami-inorganic composites, allow the precise organization of metallic, semiconducting, or magnetic elements at the nanoscale. These hybrid structures are particularly promising for applications in nanoelectronics, plasmonics, and sensing, where the spatial arrangement of components directly influences electronic, optical, and magnetic properties. Moreover, the compatibility of origami scaffolds with biomolecules such as proteins, antibodies, and nucleic acids enables their incorporation into biologically relevant systems, facilitating the design of biosensors, molecular diagnostic devices, and programmable therapeutic platforms. By bridging the gap between synthetic chemistry, materials science, and biology, molecular origami offers a versatile platform for interdisciplinary innovation.

Despite these advances, several chemical and technical challenges remain. Achieving long-term stability, scalability, and cost-effective synthesis of origami structures is critical for transitioning laboratory innovations into practical applications. The intrinsic sensitivity of nucleic acid-based systems to enzymatic degradation, chemical hydrolysis, and thermal fluctuations necessitates the development of protective strategies, including chemical modifications, covalent cross-linking, and encapsulation within stabilizing matrices. Additionally, controlling the kinetics of assembly to minimize defects while maximizing throughput remains a central challenge, particularly for large or complex structures. Addressing these issues requires not only chemical ingenuity but also the integration of microfluidic systems, automated synthesis platforms, and real-time monitoring techniques, creating a fully integrated approach to molecular fabrication.

Looking forward, the field of molecular origami is likely to intersect increasingly with emerging areas such as synthetic biology, adaptive materials, and quantum nanotechnology. In synthetic biology, origami-based scaffolds can provide spatially defined platforms for organizing enzymes, regulatory elements, and signaling molecules, enabling the construction of synthetic metabolic pathways and regulatory circuits with nanoscale precision. In adaptive materials, dynamic origami motifs may serve as stimuli-responsive building blocks for materials that self-heal, reconfigure, or adapt their properties in response to environmental changes. In quantum nanotechnology, the precise positioning of quantum dots, spin centers, or superconducting elements within an origami scaffold could facilitate the creation of coherent, scalable quantum architectures with tailored interactions. Each of these directions exemplifies the profound potential of molecular origami to reshape not only nanotechnology but also the broader landscape of chemistry and materials science.

In conclusion, molecular origami represents a paradigm shift in chemical design, offering unprecedented control over the structure, function, and dynamic behavior of nanoscale systems. By integrating computational design, programmable interactions, hierarchical assembly, and functionalization, chemists are creating nanoscale architectures that transcend the limitations of traditional synthetic approaches. The ongoing convergence of molecular origami with materials science, biology, and advanced manufacturing promises to unlock new frontiers in drug delivery, molecular machinery, energy capture, and functional materials. As the field continues to evolve, driven by innovations in chemistry, computation, and engineering, molecular origami is poised to redefine our approach to constructing matter at the nanoscale, transforming abstract chemical concepts into tangible, functional, and responsive nanosystems with the potential to impact science, technology, and medicine profoundly.

Chapter 44: Word Problems

1. In a DNA origami experiment, a 7,000-nucleotide single-stranded scaffold is folded using 200 short staple strands, each designed to be 32 nucleotides long. Assume

that each staple binds to a complementary region on the scaffold to define edges and junctions of a complex 3D architecture. Considering incomplete hybridization, potential strand displacement, and steric hindrance between adjacent staples, discuss how these factors could reduce folding fidelity and overall structural yield. Additionally, calculate the theoretical maximum number of unique binding sites available and explain how computational staple optimization could mitigate misfolding.

2. A peptide sequence composed of alternating hydrophobic and hydrophilic amino acids is designed to form a β-sheet-based nanofiber capable of hierarchical assembly into a 3D scaffold. Explain how environmental factors such as ionic strength, pH, temperature, and solvent polarity modulate the thermodynamic stability of the β-sheet and affect aggregation kinetics. Predict the impact of introducing a single charged residue in a hydrophobic stretch on the folding landscape and hierarchical assembly pathway.

3. A DNA tetrahedron is being constructed from four Holliday junctions, with each edge designed to be 15 base pairs in length. Consider how altering the edge length affects curvature, internal angles, and mechanical strain within the tetrahedron. Discuss the cooperative effects between junctions and the influence of sequence-dependent flexibility on the assembly kinetics. Additionally, explain how temperature-controlled annealing and divalent cation concentration might optimize folding fidelity while preventing kinetic traps.

4. In designing a DNA origami lattice composed of tensegrity triangles, symmetry considerations, junction placement, and edge lengths are critical for defect-free assembly. Using Eulerian graph theory and geometric constraints, propose a design strategy for minimizing lattice defects and misalignments. Discuss how mismatched crossover placements could propagate structural errors in higher-order assemblies and suggest computational approaches to predict and correct these errors before experimental implementation.

5. Molecular dynamics simulations of a peptide-based nanocage reveal the presence of metastable intermediates that persist longer than anticipated. Construct a kinetic model describing the folding pathway, including potential energy barriers, intermediate states, and transition rates. Discuss how the persistence of these

intermediates might impact the overall yield of fully folded cages and propose experimental approaches, such as altering solvent conditions or introducing catalytic folding aids, to overcome kinetic traps.

6. A researcher intends to incorporate gold nanoparticles at specific vertices of a DNA cube for use in plasmonic applications. Considering scaffold flexibility, staple placement, and steric effects, propose a design strategy to achieve precise nanoparticle positioning. Discuss how multivalent attachment, linker chemistry, and nanoparticle size influence structural rigidity and folding efficiency, and explain how computational modeling could predict the optimal configuration to minimize strain and prevent aggregation.

7. You are developing a pH-responsive DNA nanocarrier for targeted drug delivery in tumor microenvironments. Explain how to design pH-sensitive motifs that enable conformational transitions at acidic pH while remaining stable in physiological conditions. Consider thermodynamic principles governing duplex stability, hydrogen bonding, and base stacking, and discuss how staple sequence design, strand stoichiometry, and annealing protocols could be optimized to enhance folding efficiency and functional responsiveness.

8. A peptide-based nanostructure exhibits self-healing behavior after partial mechanical disruption. Analyze the roles of hydrogen bonding, π-π stacking, hydrophobic clustering, and solvent-mediated interactions in facilitating repair. Discuss how hierarchical assembly and secondary structure organization contribute to emergent self-healing properties, and predict the limits of repairability under varying environmental stresses, such as temperature fluctuations or changes in ionic strength.

9. In a coarse-grained simulation of DNA origami folding, increasing the concentration of divalent cations stabilizes duplex formation by screening electrostatic repulsion. Quantitatively discuss how this stabilization alters the Gibbs free energy landscape, affects nucleation rates, and influences the folding pathway. Explain potential trade-offs between enhanced stability and increased kinetic trapping due to overly strong interactions, and propose experimental validation methods using thermal annealing and fluorescence assays.

10. A hierarchical self-assembly experiment involves α-helices folding into tertiary bundles, which then oligomerize into quaternary peptide cages. Evaluate how

variations in secondary structure stability, helix length, and amphiphilic patterning influence the kinetics of quaternary assembly, the likelihood of misfolding, and the error propagation to higher-order structures. Suggest computational and experimental strategies for optimizing folding efficiency while minimizing undesired aggregation.

11. Compare the design considerations and functional advantages of DNA versus peptide origami for constructing nanoscale cages intended to encapsulate enzymes. Discuss factors including stability under physiological conditions, tunability of pore size, responsiveness to environmental stimuli, and the potential for multivalent interactions. Consider how sequence programmability and hierarchical assembly pathways affect folding fidelity and structural robustness.

12. A synthetic amphiphilic block copolymer is designed to assemble into vesicular structures via solvent-mediated self-assembly. Explain how the relative lengths of hydrophilic and hydrophobic blocks, solvent polarity, and concentration influence curvature, bilayer formation, and vesicle size distribution. Predict how slight variations in temperature or solvent composition could shift the system toward lamellar or micellar morphologies, and discuss experimental methods to monitor these changes.

13. Atomic force microscopy (AFM) images of DNA origami sheets reveal regions of partial unfolding and irregular topology. Propose a method for quantifying defect density and correlating it with staple stoichiometry, annealing rates, and scaffold length. Discuss how such quantitative metrics could inform iterative redesign of staple sequences to improve folding fidelity and structural uniformity.

14. In designing a DNA-based photonic crystal lattice, edge lengths, triangle angles, and motif curvature determine both lattice periodicity and optical properties. Discuss how variations in local curvature and crossover placement influence the formation of defects, the uniformity of interparticle spacing, and ultimately the photonic bandgap. Suggest computational and experimental approaches to optimize lattice parameters for maximal light-manipulating efficiency.

15. A protein origami cage is constructed using modular coiled-coil motifs. Evaluate how the strategic design of orthogonal interaction domains can enable combinatorial assembly, producing structural diversity without altering the primary

sequence. Discuss potential challenges such as cross-reactivity, misfolding, and the propagation of errors to higher-order assemblies, and propose experimental validation strategies.

16. Cryo-EM analysis of peptide nanofibers shows partially misaligned β-sheets. Identify possible causes rooted in sequence design, solvent conditions, kinetic folding traps, or intermolecular interactions. Discuss how modifying amphiphilic patterns, adding chaperone-like molecules, or tuning temperature and pH could mitigate misalignment and improve assembly fidelity.
17. In a DNA-peptide hybrid assembly, metal-ligand coordination is employed to stabilize a tetrahedral framework. Analyze the influence of ligand geometry, metal ion selection, coordination number, and kinetic lability on assembly efficiency, thermodynamic stability, and the likelihood of competing structures. Propose a computational approach to predict optimal metal-ligand placement and design parameters.
18. Design an experiment to monitor folding kinetics of a 3D DNA origami nanocage using Förster resonance energy transfer (FRET). Discuss criteria for selecting donor and acceptor fluorophores, ideal labeling positions, and the type of kinetic data that can be obtained. Consider potential complications such as fluorophore quenching, non-specific interactions, and dynamic conformational changes.
19. A DNA origami cube is functionalized with enzymes at each vertex. Discuss how steric constraints, cooperative effects among enzymes, and scaffold rigidity influence catalytic efficiency. Propose strategies for optimizing spatial arrangement to maximize multivalent interactions and overall turnover, and describe experimental methods to measure functional output.
20. Monte Carlo simulations of block copolymer origami reveal competition between metastable lamellar phases and the target vesicular structure. Discuss strategies to bias assembly toward the desired morphology, including sequence modification, solvent choice, annealing temperature, and the introduction of hierarchical assembly motifs. Analyze the thermodynamic and kinetic factors that govern phase selection and folding fidelity.
21. A DNA octahedron is being assembled from 12 edges, each designed with complementary staples. Consider how altering the length of a single edge or

introducing a base mismatch would affect overall curvature, junction angles, and lattice symmetry. Discuss how scaffold flexibility, cation concentration, and annealing protocols could mitigate misfolding and maintain structural integrity. Propose computational modeling strategies to predict the effects of such modifications.

22. A peptide-based nanocarrier is designed to release its payload upon exposure to light. Explain the design principles of incorporating photoresponsive linkages, including chromophores or light-cleavable bonds. Analyze the energetic considerations for reversible folding, potential off-target cleavage, and how the dynamic conformational change can be quantified experimentally using spectroscopic methods.

23. Gold nanoparticles are integrated into a DNA origami framework to create a light-harvesting system. Predict how nanoparticle placement, interparticle distance, and scaffold rigidity influence plasmonic coupling and energy transfer efficiency. Discuss the interplay between structural stability, optical properties, and the potential for cooperative light-harvesting behavior.

24. In a surface-mediated assembly of DNA origami, local orientation and packing density are critical for higher-order patterning. Discuss how substrate functionalization, divalent cation concentration, ionic strength, and annealing rate influence adsorption, orientation, and defect formation. Propose experimental strategies to maximize uniform coverage and minimize folding distortions induced by the substrate.

25. Compare the relative contributions of enthalpy and entropy in the folding of DNA, peptide, and synthetic polymer origami. Provide specific examples of interactions dominating each system (e.g., Watson-Crick base pairing, hydrophobic clustering, π-π stacking) and explain how these energetic contributions influence hierarchical assembly and stability under variable environmental conditions.

26. A DNA origami nanorobot is engineered to undergo a conformational change upon aptamer-target binding. Describe the design of hinge regions, sequence-specific interactions, and staple placement to ensure reliable actuation. Discuss potential sources of kinetic trapping or misfolding and propose computational and experimental methods to validate folding dynamics and functional responsiveness.

27. Hierarchical assembly in peptide origami involves secondary α-helices assembling into tertiary bundles, which then form quaternary cages. Predict how mutations in critical residues within α-helices could propagate errors to higher-order structures, and discuss strategies, such as sequence redesign or stabilizing co-factors, to mitigate misfolding while maintaining target geometry.
28. In a DNA origami lattice, thermal annealing is performed at slow and rapid cooling rates. Analyze the impact of cooling rate on nucleation, defect formation, misfolding, and overall folding efficiency. Suggest experimental readouts (e.g., AFM, FRET, gel electrophoresis) to quantify folding fidelity and correlate it with computationally predicted energy landscapes.
29. Design a DNA nanocage capable of encapsulating a small protein. Discuss how cage rigidity, internal cavity volume, pore size, and internal surface chemistry influence encapsulation efficiency, protein stability, and release kinetics. Consider how scaffold flexibility, staple density, and environmental factors such as ionic strength and temperature affect assembly outcomes.
30. Small-angle X-ray scattering (SAXS) analysis of supramolecular polymers reveals unexpected aggregation into irregular structures. Propose hypotheses related to block copolymer sequence, solvent interactions, temperature, and ionic conditions. Discuss experimental approaches to test these hypotheses and design modifications to restore the desired vesicular or lamellar morphology.
31. A stimuli-responsive peptide vesicle is engineered to release cargo in response to redox changes. Explain the placement, chemical nature, and stability of disulfide linkages that enable controlled disassembly. Discuss potential challenges, including competing reduction pathways, environmental sensitivity, and effects on hierarchical assembly fidelity.
32. A hybrid DNA-metal-organic framework (MOF) structure is designed for catalysis. Analyze how scaffold rigidity, pore architecture, edge lengths, and connectivity influence substrate accessibility, turnover rate, and product selectivity. Propose experimental strategies and computational models to predict and optimize assembly and functional performance.
33. Computational modeling predicts off-target binding of certain staple strands in a DNA origami cube. Propose a redesign strategy considering thermodynamic

parameters (ΔG of hybridization, melting temperature, base stacking energy) to minimize misfolding. Discuss how iterative simulation and experimental validation can improve structural fidelity and assembly yield.

34. Compare the advantages and limitations of AFM, TEM, and cryo-EM in evaluating folding fidelity, defect density, and hierarchical organization of 3D DNA origami structures. Suggest experimental workflows that combine these techniques to obtain complementary structural information at different resolution scales.

35. A multi-layered DNA origami lattice is being constructed. Explain how the choice of interlayer motifs, crossover placement, and scaffold tension influence mechanical rigidity, cooperative folding, and emergent functional properties. Discuss computational and experimental approaches to optimize interlayer alignment and prevent misfolding or layer delamination.

36. A peptide-based nanocarrier is engineered for targeted drug delivery. Analyze how amphiphilic patterning, secondary structure, and hierarchical folding affect circulation time, cellular uptake, and cargo release. Propose strategies to tune folding stability, surface functionality, and stimuli responsiveness to improve therapeutic performance.

37. In a FRET-based assay for DNA origami folding, unexpected energy transfer occurs between non-adjacent sites. Hypothesize structural misalignments, transient dynamics, or conformational flexibility that could account for this observation. Suggest computational simulations and experimental controls to validate and resolve the source of anomalous energy transfer.

38. Design a computational workflow integrating molecular dynamics and machine learning to optimize peptide origami sequences. Discuss how energy landscape exploration, folding kinetics, and sequence variability could be combined with predictive modeling to maximize folding fidelity, minimize kinetic traps, and ensure hierarchical assembly.

39. Assembly of a DNA icosahedron involves 30 edges and multiple junctions. Discuss how junction flexibility, crossover design, and edge lengths influence global curvature, topological constraints, and overall structural stability. Suggest computational modeling and experimental validation strategies to predict and ensure correct assembly.

40. DNA origami photonic arrays incorporate metallic nanoparticles to enhance plasmonic properties. Analyze how interparticle spacing, scaffold rigidity, and edge design influence plasmon coupling, energy transfer, and optical response. Propose methods to experimentally measure these effects and optimize array performance.

41. In peptide origami, increasing hydrophobic cluster size unexpectedly causes aggregation. Propose sequence redesign strategies, solvent modifications, and hierarchical assembly adjustments to restore target folding while maintaining structural integrity and functional sites. Discuss how computational modeling can predict these outcomes.

42. Design a DNA origami scaffold capable of reversible expansion and contraction in response to temperature changes. Explain how staple selection, crossover density, and edge flexibility control thermomechanical behavior, folding reversibility, and structural integrity. Propose experimental methods to measure expansion/contraction cycles and assess fatigue resistance.

43. In hybrid DNA-protein origami, cooperative interactions govern assembly order. Predict consequences of misfolding in a single module, including propagation of errors, loss of function, and structural destabilization. Suggest strategies for hierarchical error correction, sequence optimization, or modular redesign to enhance robustness.

44. Compare the Gibbs free energy landscapes of nucleic acid, peptide, and synthetic polymer origami. Discuss differences in metastable intermediates, kinetic traps, folding barriers, and hierarchical assembly pathways. Explain how these differences affect folding fidelity and the design of error-correction strategies.

45. Design a DNA nanocage with multiple enzyme immobilization sites. Discuss how spatial arrangement, scaffold rigidity, and local steric effects influence multivalent interactions, cooperative catalysis, and overall enzymatic efficiency. Propose computational and experimental approaches to optimize catalytic performance.

46. AFM imaging shows uneven adhesion of DNA origami sheets to the substrate. Propose experimental modifications, including surface treatment, ionic environment, annealing rate, or scaffold functionalization, to improve uniformity,

minimize structural distortion, and enable reproducible higher-order patterning.

47. A DNA origami-based molecular machine rotates upon stimulus binding. Discuss how hinge design, strand tension, and input energy dictate rotational motion, conformational reversibility, and mechanical efficiency. Suggest experimental techniques to quantify motion, torque, and functional output.

48. Compare the effects of annealing temperature profiles on folding fidelity for DNA origami cubes versus peptide cages. Discuss underlying differences in thermodynamics, kinetics, and environmental sensitivity that account for differential folding behavior, and propose experimental approaches to optimize both systems.

49. A light-activated peptide origami nanocarrier is designed for controlled cargo release. Explain how the placement of photoresponsive residues affects folding dynamics, stability, and release kinetics. Discuss potential challenges, including premature activation, photodegradation, or misfolding under illumination, and propose strategies to mitigate these issues.

50. A supramolecular polymer is engineered to assemble into a 3D lattice. Discuss strategies to integrate hierarchical assembly principles and error-correction motifs to maximize yield and structural precision. Consider solvent conditions, temperature, polymer sequence, and cooperative interactions in achieving defect-free, functional architectures.

Chapter 45

Bioluminescence and Biofluorescence

Bioluminescence and biofluorescence represent two remarkable phenomena at the intersection of chemistry and biology, reflecting the intricate molecular strategies that organisms employ to manipulate light. Bioluminescence, the production of visible light by living organisms, is underpinned by enzyme-catalyzed chemiluminescent reactions, typically involving a luciferin substrate and a luciferase enzyme. The precise chemistry varies among taxa: in fireflies, the luciferin-luciferase reaction is ATP-dependent and proceeds via adenylation of luciferin to form luciferyl-adenylate, followed by oxidation in the presence of molecular oxygen, yielding an electronically excited oxyluciferin species that emits a photon as it relaxes to the ground state. In marine organisms, such as the dinoflagellates and certain deep-sea fish, bioluminescence often involves coelenterazine and photoproteins such as aequorin, which form a calcium-dependent complex that produces light through oxidative decarboxylation. These reactions are exquisitely tuned in wavelength to suit ecological niches, with fireflies emitting yellow-green light (~560 nm) and deep-sea organisms often emitting blue light (~470 nm), the latter corresponding to maximal transmission through seawater. The chemical basis of biofluorescence, in contrast, is the absorption of high-energy photons by a fluorophore and their re-emission at longer wavelengths, a process dictated by the fluorophore's electronic structure and environment.

Green fluorescent protein (GFP) and its derivatives exemplify this phenomenon, with an autocatalytically formed chromophore within a beta-barrel scaffold facilitating excitation-dependent fluorescence. The underlying chemistry involves cyclization, oxidation, and resonance stabilization of the chromophoric structure, a process influenced by local pH, polarity, and protein conformation. Fluorescent proteins have diversified through natural evolution and bioengineering, enabling tunable spectral properties ranging from cyan to near-infrared emission, achieved through subtle modifications to the pi-conjugated system and hydrogen-bonding network surrounding the chromophore. Bioluminescent and fluorescent systems both operate under the principles of photochemistry and excited-state dynamics, wherein the interplay of electronic transitions, intersystem crossing, and non-radiative decay pathways determines quantum yield, emission lifetime, and spectral output. The kinetic parameters of luciferase-catalyzed reactions, such as turnover number and substrate affinity, and the photophysical parameters of fluorophores, such as molar absorptivity and Stokes shift, are central to understanding and harnessing these processes. Bioluminescence also exhibits remarkable chemical diversity, encompassing the lumazine derivatives in bacteria, tetrapyrrole-like luciferins in some marine polychaetes, and unique imidazopyrazinone-based luciferins in ostracods, each characterized by distinct oxidative chemistries and emission spectra. The energetic efficiency of these reactions is notable, often exceeding conventional chemiluminescent systems, as biological evolution has optimized photon output relative to metabolic cost. Beyond natural light emission, bioluminescence and biofluorescence have been harnessed for analytical chemistry and molecular imaging. Bioluminescent reporters are widely employed in high-sensitivity assays for ATP, reactive oxygen species, and gene expression, leveraging the direct conversion of chemical energy into detectable photons without the need for external excitation, thereby minimizing background signal and phototoxicity. Fluorescent proteins and synthetic fluorophores, conversely, enable live-cell imaging, Förster resonance energy transfer (FRET) studies, and super-resolution microscopy, relying on precise chemical control of excitation-emission profiles and photostability. Chemically, modifications to luciferins, such as halogenation, alkylation, or incorporation of extended conjugation, can red-shift emission, enhance solubility, and improve stability, illustrating the intimate relationship between molecular structure and photophysical properties. Environmental factors, such as pH, ionic strength, and oxygen availability, further modulate luminescent output, providing insights into biochemical regulation and the adaptive evolution of light-emitting systems. Structural studies, including X-ray crystallography and NMR spectroscopy, have elucidated the active site architecture of luciferases and the chromophore microenvironment of fluorescent proteins, revealing critical hydrogen-bonding networks, electrostatic interactions, and steric constraints that govern catalysis, quantum yield, and spectral tuning.

Computational chemistry and molecular dynamics simulations complement these experimental approaches, enabling the rational design of novel luciferases and fluorophores with optimized emission wavelengths, stability, and reaction kinetics. At the molecular level, the interplay between excited-state chemistry, electron transfer, and vibrational relaxation determines the efficiency of photon emission, with bioluminescent systems often minimizing non-radiative decay through constrained geometries and local environmental optimization. Biofluorescent systems exploit conjugated pi-electron networks within rigid scaffolds to stabilize excited states, modulating fluorescence lifetime, anisotropy, and photobleaching susceptibility. Evolutionary pressures have also shaped spectral tuning, with organisms adapting emission wavelengths for camouflage, mating, predation, and intraspecific communication. For instance, deep-sea organisms have evolved bioluminescent blue emission to maximize transmission in the photic-depleted marine environment, while terrestrial bioluminescent insects produce yellow-green light optimized for visibility against vegetation.

Chemical ecology studies have revealed that luciferin biosynthesis and recycling pathways, often involving S-adenosylmethionine and amino acid precursors, are metabolically integrated with other cellular processes, highlighting the intersection of primary and secondary metabolism in light production.

Fluorescent proteins and small-molecule fluorophores are similarly integrated into cellular architecture, with targeting motifs, pH-sensitivity, and environmental responsiveness allowing precise chemical reporting of intracellular conditions. Synthetic analogs of natural luciferins and fluorophores have expanded the chemical toolkit, enabling bioluminescence resonance energy transfer (BRET) systems, multi-color imaging, and high-throughput screening assays. The design of these molecules often involves systematic chemical modifications to the chromophore or substrate backbone, such as electron-donating or withdrawing substituents, heteroatom incorporation, and ring expansion or contraction, all of which modulate redox potential, emission wavelength, and quantum efficiency. Bioluminescence has further inspired chemiluminescent probes in analytical chemistry, such as luminol and isoluminol derivatives, which generate light upon oxidation with hydrogen peroxide, serving as sensitive reporters for reactive oxygen species, enzymatic activity, and immunoassays. The fundamental chemistry underlying these processes reflects principles of photophysics, enzymology, and reaction thermodynamics, with bioluminescent reactions characterized by exergonic electron transfer, formation of high-energy intermediates, and controlled photon release.

Fluorescence, by contrast, is governed by the absorption of light to promote electrons to singlet excited states, with subsequent radiative relaxation yielding emitted photons. Quantum yield, defined as the ratio of emitted to absorbed photons, is influenced by local environment, chromophore rigidity, and solvent interactions, necessitating careful chemical engineering for optimized fluorescence. Advances in genetic engineering, directed evolution, and chemical synthesis have converged to expand the palette of luminescent and fluorescent tools, producing red-shifted luciferases, near-infrared fluorescent proteins, and long-lived chemiluminescent substrates for in vivo imaging. Recent discoveries include the identification of entirely novel luciferins in previously unexplored taxa, such as polychaete worms, bioluminescent fungi, and deep-sea crustaceans, whose chemistries involve unprecedented oxidative mechanisms and electronic structures. Investigations into these systems reveal sophisticated control of excited-state dynamics, with evolutionary optimization balancing energy efficiency, emission wavelength, reaction kinetics, and ecological function. Chemical studies have elucidated mechanisms of photon emission, including single-electron transfer, radical-mediated decarboxylation, and peroxy intermediate formation, highlighting the diversity of chemical pathways evolved to harness molecular oxygen and metabolic energy for light production. In biofluorescent proteins, fine-tuning of emission spectra relies on chemical modulation of the conjugated chromophore, hydrogen bonding networks, protonation states, and surrounding electrostatics, demonstrating the precise chemical choreography underlying functional fluorescence.

Applications in chemical biology have leveraged these properties, with bioluminescent assays enabling real-time monitoring of cellular metabolism, gene expression, and protein-protein interactions, while fluorescent reporters provide spatial and temporal resolution for live-cell imaging, super-resolution microscopy, and optogenetic control. The chemistry of luminescence extends to materials science, where bio-inspired light-emitting systems inform the design of organic LEDs, fluorescent sensors, and biohybrid devices. Photophysical studies, including time-resolved spectroscopy, single-molecule imaging, and fluorescence lifetime measurements, provide detailed chemical insight into excited-state behavior, electron transfer dynamics, and conformational flexibility. Bioluminescent and fluorescent chemistries are also integral to environmental and ecological studies, enabling sensitive detection of pollutants, tracking of biological populations, and elucidation of interspecies signaling. Metabolic engineering has expanded the accessibility of these systems, with microbial production of luciferins, engineered luciferases, and recombinant fluorescent proteins facilitating scalable applications in medicine, environmental monitoring, and synthetic biology.

Chemical challenges remain, including optimization of photostability, enhancement of red-shifted emission for tissue imaging, and control of reactive intermediate formation in luciferase reactions. Multidisciplinary studies, integrating organic chemistry, enzymology, photophysics, and computational modeling, continue to uncover the principles governing luminescent and fluorescent phenomena, elucidating the interplay between molecular structure, reaction energetics, and ecological function. In conclusion, the chemistry of bioluminescence and biofluorescence represents a rich domain where molecular design, enzymatic catalysis, and photophysical principles converge, enabling organisms to harness chemical energy for light emission and providing chemists with tools for visualization, analysis, and bioinspired innovation. These systems exemplify the intricate interplay between structure, reactivity, and function, illustrating how chemical insight can illuminate biological phenomena and drive technological advances. The ongoing exploration of natural and engineered luminescent systems continues to deepen our understanding of molecular photochemistry, enzyme catalysis, and chemical evolution, highlighting the profound connections between chemistry, biology, and light.

Beyond the foundational principles of luminescent chemistry, contemporary research increasingly emphasizes the fine-tuning of molecular properties to address specific functional demands. One central chemical challenge lies in achieving precise control over the photophysical parameters that dictate emission color, intensity, and duration. In natural systems, bioluminescent proteins exhibit a remarkable diversity of emission wavelengths, ranging from blue and green to yellow and red, with each wavelength optimized for the ecological niche of the organism. Chemists seeking to harness or replicate these phenomena must consider subtle structural modifications of the chromophore, including electron-donating or electron-withdrawing substituents, conjugation length, and steric effects, all of which influence excited-state energies and emission spectra. Such modifications must also preserve the inherent stability of the molecule, ensuring resistance to photobleaching, oxidative degradation, or nonradiative decay, which can drastically reduce functional lifespan in both biological and technological applications.

Parallel to chromophore design, the enzymatic environment in which these molecules operate provides another layer of chemical complexity. Luciferases, for instance, not only catalyze substrate oxidation but also orchestrate the precise orientation and electronic environment of the chromophore to maximize light output. Active site residues modulate protonation states, stabilize transition states, and influence the partitioning of energy between fluorescence and phosphorescence pathways. Computational chemistry and molecular dynamics simulations have emerged as essential tools for predicting how mutations, solvent interactions, and local dielectric effects modulate luminescent efficiency. By integrating theoretical models with high-resolution structural data, chemists can rationally design enzyme-substrate pairs with tailored emission properties, bridging the gap between natural evolution and synthetic optimization.

Another frontier in luminescent chemistry is the engineering of systems for deep-tissue imaging and in vivo applications. Biological tissues exhibit significant light scattering and absorption, particularly at shorter wavelengths, which limits the utility of traditional blue- or green-emitting systems. Red-shifted and near-infrared emitters are therefore highly sought, yet the chemical design of such chromophores presents unique challenges. Extended conjugation often increases molecular flexibility, introducing pathways for nonradiative decay that reduce quantum yield. Moreover, the introduction of electron-rich substituents or heavy atoms to induce bathochromic shifts can inadvertently destabilize the chromophore or alter its reactivity with enzymatic partners. Addressing these competing factors requires a delicate balance of synthetic strategy, including the development of rigidified π-systems, the use of bioorthogonal protective groups, and the exploration of novel scaffolds that combine photostability with emission in the desired spectral range.

The dynamic nature of luminescent reactions also calls attention to transient intermediates, whose fleeting existence can govern both efficiency and specificity. In luciferase-catalyzed reactions, high-energy intermediates, such as dioxetanones or excited-state oxyluciferins, are central to photon generation. However, these intermediates are prone to side reactions, including oxidative cleavage, radical formation, or undesired isomerization, which can reduce overall light output or generate reactive species detrimental to cellular health. Understanding the electronic structure, protonation dynamics, and solvation effects of these intermediates is therefore critical. Techniques such as time-resolved spectroscopy, ultrafast photolysis, and quantum chemical modeling have been instrumental in capturing these transient species and elucidating the mechanistic pathways that convert chemical energy into light with near-perfect efficiency.

In addition to fundamental chemical insights, the study of luminescence offers profound implications for bioinspired innovation. Synthetic analogs of natural luminescent systems are being integrated into diagnostic tools, imaging probes, and environmental sensors. Chemists are designing modular fluorogenic and bioluminescent reporters that respond to specific stimuli—such as pH changes, ion concentrations, or enzymatic activity—allowing real-time visualization of biological processes with high spatial and temporal resolution. These applications illustrate a key principle: by decoding the chemical logic underpinning natural luminescence, one can engineer synthetic systems that mimic, enhance, or even surpass the functional sophistication of their biological counterparts. This iterative interplay between observation, chemical design, and functional testing epitomizes the synergy of chemical and biological research in contemporary science.

The evolutionary context of luminescent chemistry provides additional layers of intrigue. Organisms have harnessed bioluminescence not only for communication, camouflage, and predation but also as an adaptive chemical strategy, where subtle alterations in enzyme kinetics or chromophore structure confer selective advantages. Studying these natural variations can inspire novel synthetic approaches, such as tunable emission scaffolds or enzyme-coupled reaction networks that achieve controlled light output under defined environmental conditions. Such explorations further underscore the centrality of chemical structure-function relationships, illustrating how even minor changes at the molecular level can propagate to macroscopic ecological effects.

Finally, the integration of interdisciplinary approaches is essential for advancing luminescent chemistry. Photophysics provides insight into energy transfer and quenching mechanisms; enzymology elucidates catalytic strategies and substrate specificity; organic synthesis enables the creation of novel chromophores and analogs; and computational modeling allows predictive design and mechanistic exploration. Together, these approaches establish a comprehensive framework for understanding and harnessing luminescence. In this sense, the field serves as a microcosm of chemical research at large: a domain where precise molecular manipulation, energetic considerations, and functional outcomes converge to generate phenomena of both natural wonder and technological utility.

In summary, the study of bioluminescence and biofluorescence exemplifies the rich interplay between chemical principles and biological function. From chromophore design and enzymatic catalysis to photophysical tuning and synthetic replication, these systems illuminate the profound connections between structure, reactivity, and ecological purpose. Ongoing research continues to refine our understanding of transient intermediates, emission optimization, and the translation of natural mechanisms into synthetic and biomedical applications. As chemists probe deeper into these luminous systems, they not only reveal the molecular underpinnings of light emission but also lay the groundwork for innovative technologies that exploit chemistry in service of visualization, diagnostics, and bioinspired engineering.

Chapter 45: Word Problems

1. A firefly luciferase catalyzes the conversion of luciferin to oxyluciferin in an ATP-dependent reaction, producing light in the yellow-green region (~560 nm). Given that the enzyme's turnover number is 40 s^{-1} and the cellular ATP concentration is 2 mM, estimate the rate of photon emission per cell assuming that each oxyluciferin molecule emits a single photon. Discuss the factors, such as substrate availability, enzyme saturation, and non-radiative decay pathways, that could influence the actual photon yield in a biological environment.

2. In marine dinoflagellates, the luminescent protein aequorin forms a calcium-dependent complex with coelenterazine to produce bioluminescence via oxidative decarboxylation. If an intracellular calcium spike increases from 50 nM to 5 μM, model the resulting kinetics of light emission using a Michaelis-Menten-like approach. Predict how the luminescence intensity changes over time and explain the role of calcium in triggering the conformational changes necessary for photon generation.

3. Deep-sea fish luciferases typically emit blue light at ~470 nm to maximize transmission through seawater. Using published absorption spectra of seawater, estimate the fraction of photons transmitted over a distance of 50 meters. Explain how this spectral tuning provides an adaptive advantage for prey detection or communication in the photic-depleted marine environment.

4. A green fluorescent protein (GFP) variant exhibits excitation at 480 nm and emission at 510 nm. If the local pH shifts from 7.4 to 6.5, predict the change in fluorescence intensity based on protonation of the chromophore and alterations in the hydrogen-bonding network within the beta-barrel scaffold. Discuss the implications for live-cell imaging experiments where intracellular pH varies spatially and temporally.

5. A synthetic luciferin derivative is halogenated to induce a red-shift in emission. If the photon energy decreases by 0.1 eV compared to the unmodified luciferin, calculate the resulting emission wavelength and discuss how electronic substituents alter the pi-conjugated system to affect both spectral properties and chemical stability. Consider potential impacts on solubility and enzymatic turnover in luciferase-catalyzed reactions.

6. In designing a high-sensitivity ATP assay using firefly luciferase, background luminescence must be minimized. Construct a quantitative experimental plan that leverages the intrinsic advantage of bioluminescence—namely, the absence of external excitation—and describe why this approach reduces phototoxicity and enhances the signal-to-noise ratio relative to fluorescent assays.

7. Compare the quantum yields of a bacterial luciferase (0.2) and a fluorescent protein (0.6). Calculate the expected number of photons emitted per 10^6 absorbed energy units and discuss how these differences affect sensitivity for cellular imaging applications. Include consideration of photobleaching, non-radiative decay, and environmental factors that might modulate quantum efficiency.

8. A single-site mutation in firefly luciferase replaces an active-site arginine with lysine. Predict how this change could alter transition-state stabilization, protonation dynamics, and overall photon output. Discuss the possible consequences for both the kinetics of luciferin oxidation and the wavelength of emitted light, considering electrostatic interactions in the enzyme's active site.

9. Time-resolved spectroscopy of the firefly luciferase reaction reveals that the excited oxyluciferin species decays with a lifetime of 3 ms. Develop a kinetic model describing the competition between radiative and non-radiative decay pathways, and estimate the photon emission efficiency. Discuss how solvent environment, protein conformational flexibility, and quenching by molecular oxygen may influence the observed lifetime.

10. A cyan fluorescent protein exhibits a Stokes shift of 30 nm and is illuminated with increased excitation intensity. Analyze how increased excitation affects photobleaching rates and discuss strategies, such as chromophore rigidity or beta-barrel stabilization, to mitigate photodamage while preserving fluorescence intensity for long-term imaging experiments.

11. Lumazine-based luciferins in a polychaete worm generate blue-green bioluminescence. Propose a detailed biosynthetic pathway starting from amino acid precursors, including the role of S-adenosylmethionine in methyl group donation, and explain how enzymatic control of oxidation steps ensures high quantum efficiency and minimal formation of deleterious byproducts.

12. A bioluminescent probe intended for deep-tissue imaging requires near-infrared emission (~700 nm). Considering that extending conjugation to red-shift emission also increases molecular flexibility, calculate the expected decrease in quantum yield if non-radiative decay probability increases by 10% per additional conjugated bond. Discuss strategies to mitigate this loss while maintaining emission in the desired spectral range.

13. A luciferase-catalyzed reaction involves formation of a high-energy dioxetanone intermediate. Predict how halving the oxygen concentration would affect the rate of excited-state formation, photon emission efficiency, and potential side reactions. Consider how cellular oxygen availability could influence in vivo bioluminescence assays.

14. A bioluminescence resonance energy transfer (BRET) system couples a blue-emitting luciferase donor with a yellow fluorescent protein acceptor. Given a Förster distance of 5 nm and a donor-acceptor separation of 6 nm, calculate the energy transfer efficiency. Discuss how the orientation factor, spectral overlap, and quantum yield of the donor contribute to the efficiency of the system.

15. A GFP variant is engineered to be highly pH-sensitive. If the chromophore pKa shifts from 6.0 to 6.5, calculate the expected fluorescence intensity at pH 7.0 using the Henderson-Hasselbalch equation, and explain how protonation-deprotonation dynamics influence chromophore resonance stabilization and emission properties.

16. An oxidized coelenterazine derivative produces bioluminescence with a quantum yield of 0.35. Calculate the number of photons emitted per mole of substrate and explain how oxidative decarboxylation contributes to the high energy efficiency of this bioluminescent system compared to conventional chemiluminescent reactions.

17. Fluorescence anisotropy measurements of a GFP-tagged protein reveal a rotational correlation time of 5 ns. Estimate the hydrodynamic radius of the protein and discuss how the structural rigidity of the chromophore within the beta-barrel scaffold influences anisotropy and fluorescence lifetime measurements.

18. A luciferin derivative is chemically modified via alkylation to improve solubility. Predict the impact of this modification on Km and Vmax of luciferase, and discuss

how changes in substrate binding and enzyme kinetics could influence light output in cellular assays.

19. In super-resolution microscopy, a fluorescent protein with a 4 ns excited-state lifetime is used. Design a pulsed excitation scheme that maximizes photon detection while minimizing photobleaching. Discuss how excitation intensity, pulse duration, and chromophore stability influence imaging resolution.

20. Firefly luciferase is compared under ATP-rich and ATP-depleted conditions. Quantify the effect of substrate limitation on photon output using Michaelis-Menten kinetics, and discuss how energy availability and enzyme saturation influence the efficiency of bioluminescence in living cells.

21. A bacterial luciferase can oxidize multiple lumazine derivatives through competing pathways. Propose how active-site architecture could selectively favor one pathway over another, thereby modulating emission spectrum, and explain how enzyme evolution might optimize photon output for ecological function.

22. Site-directed mutagenesis alters hydrogen-bonding residues surrounding a GFP chromophore. Predict effects on absorption and emission maxima, Stokes shift, and quantum yield, explaining the chemical basis for spectral tuning through modulation of chromophore electrostatics and resonance stabilization.

23. A marine organism emits light at 470 nm. Calculate photon energy and discuss why blue emission maximizes visibility in seawater compared to green or red light. Analyze how extinction coefficients of water and the absorption spectrum of surrounding organisms influence the ecological advantage of this emission wavelength.

24. Time-dependent density functional theory (TD-DFT) predicts the excited-state energy of a fluorescent protein to be 20 nm red-shifted from experimental observation. Discuss environmental factors, such as hydrogen bonding, local dielectric, and chromophore flexibility, that could account for the discrepancy and suggest experimental methods to validate the predictions.

25. A luciferase-catalyzed reaction produces reactive oxygen species (ROS) as a side product. Calculate the expected ROS concentration per 10^6 photons emitted, and discuss potential phototoxic effects in live-cell imaging, considering how cellular antioxidants or oxygen scavengers might modulate the bioluminescent assay.

26. A fluorescent protein is engineered to emit in the near-infrared (NIR) region for deep-tissue imaging (~700 nm). Discuss chemical strategies to extend conjugation of the chromophore while maintaining rigidity to prevent non-radiative decay. Analyze how electron-donating and electron-withdrawing substituents, steric hindrance, and hydrogen-bonding networks contribute to both red-shifting the emission and preserving quantum yield.

27. In designing a luminescent assay for reactive oxygen species, researchers must choose between luminol derivatives and luciferin-based probes. Compare the chemical mechanisms of light production, highlighting differences in substrate oxidation, intermediate formation, and photon yield. Discuss how reaction kinetics and environmental sensitivity influence the selection of one system over the other for real-time cellular measurements.

28. A fluorophore exhibits solvent-dependent emission shifts. Using data for solvent dielectric constants, predict changes in emission wavelength when moving from aqueous to nonpolar environments. Explain how solvation, dipole stabilization, and chromophore hydrogen bonding influence excited-state energy levels and the magnitude of the spectral shift.

29. A yellow-green-emitting luciferase is mutated to introduce additional steric hindrance in the active site. Predict how this modification would affect substrate binding, turnover number, and photon emission, and explain the consequences for both reaction kinetics and spectral properties. Discuss how enzyme flexibility and substrate orientation contribute to the overall efficiency of bioluminescence.

30. A bioluminescent reporter is used to monitor intracellular ATP fluctuations in real time. Develop a kinetic model relating luminescence intensity to ATP concentration changes, including factors such as enzyme saturation, substrate diffusion, and non-radiative losses. Explain how temporal resolution and light emission efficiency influence the accuracy of cellular energy monitoring.

31. Compare GFP and mCherry in terms of quantum yield, emission lifetime, and spectral overlap for FRET experiments. Predict the relative energy transfer efficiency when these proteins are used as donor-acceptor pairs, and discuss how chromophore rigidity, orientation, and local protein environment modulate FRET performance.

32. A bioluminescent bacterium is engineered to detect heavy metals. Propose a chemical mechanism by which metal binding to a luciferase active-site residue could modulate enzymatic activity and photon output. Discuss how changes in emission intensity could be quantitatively correlated to metal concentration in environmental samples.

33. Polychaete luciferins contain tetrapyrrole-like moieties with extended conjugation. Predict how substituents, conjugation length, and oxidation state influence absorption and emission spectra. Discuss how evolutionary pressures may have optimized these chemical properties to match the organism's ecological niche and maximize photon transmission in seawater.

34. A fluorescent protein derivative is engineered for high photostability under 488 nm excitation. Suggest chemical modifications, such as rigidification of the chromophore scaffold, hydrogen-bonding optimization, or protective group incorporation, to minimize photobleaching while maintaining spectral characteristics suitable for long-term imaging.

35. The quantum efficiency of a luciferase-catalyzed reaction decreases with temperature. Model this effect using Arrhenius kinetics and discuss contributions from enzyme flexibility, intermediate stability, and non-radiative decay pathways. Predict how changes in temperature affect both photon yield and emission wavelength.

36. A coelenterazine analog is synthesized with extended conjugation to red-shift emission. Predict how this modification affects luciferase binding, reaction kinetics, and quantum yield. Discuss potential steric or electronic effects that could alter substrate orientation and the rate of photon generation.

37. Time-resolved emission studies of a fluorescent protein reveal two decay components: 3 ns and 15 ns. Propose a chemical explanation for multi-exponential decay, considering chromophore microenvironments, conformational heterogeneity, and excited-state proton transfer dynamics.

38. A firefly emits 10^{12} photons per second. If the metabolic cost per photon is 2×10^{-19} J, calculate the total energy expenditure per second. Discuss how natural selection may have optimized the chemical efficiency of light production to balance metabolic cost with ecological advantage in mating or predation.

39. A FRET pair is designed to monitor protein-protein interactions. With donor emission at 500 nm and acceptor absorption at 520 nm, calculate the spectral overlap integral and energy transfer efficiency at a 5 nm separation. Discuss how chromophore orientation, distance, and quantum yield modulate the FRET signal.

40. Bioluminescent probes for in vivo imaging require NIR emission to overcome tissue absorption. Design a chemical modification strategy to shift emission from 560 nm to 700 nm, considering factors such as π-conjugation extension, electron-donating substituents, steric rigidity, and solubility. Predict how these modifications might affect luciferase kinetics and photon output.

41. A luciferase produces excited-state oxyluciferin that decays with a lifetime of 5 ms. If oxygen concentration is doubled, predict the effect on photon emission rate and intermediate stability. Discuss mechanistic considerations, including the formation of peroxy intermediates and quenching pathways.

42. GFP variants exhibit differing Stokes shifts depending on the chromophore environment. Model how chromophore polarity, hydrogen-bonding network, and beta-barrel confinement influence the magnitude of the Stokes shift, and discuss implications for multicolor imaging and FRET applications.

43. Compare bioluminescent and fluorescent reporters in live-cell imaging in terms of phototoxicity, background signal, and sensitivity. Quantify expected photon flux per cell at 10 nM probe concentration and discuss experimental conditions that favor one system over the other.

44. A luciferase active site contains a critical glutamate residue. Predict how changes in protonation state at pH 6 versus pH 8 affect substrate binding, reaction rate, and photon emission efficiency. Discuss the role of electrostatics and active-site microenvironment in modulating enzymatic catalysis.

45. A fluorescent protein is engineered with a rigidified beta-barrel scaffold. Explain how structural rigidity modulates fluorescence lifetime, quantum yield, and resistance to photobleaching. Discuss trade-offs between chromophore mobility and emission efficiency in live-cell imaging applications.

46. A BRET system is applied to monitor intracellular signaling. Develop a quantitative model relating donor emission intensity, acceptor concentration, and energy

transfer efficiency. Explain how factors such as spectral overlap, donor quantum yield, and distance dependence affect signal interpretation.

47. A marine organism emits bioluminescence at 470 nm with a quantum yield of 0.25. Calculate photon flux per mole of luciferin and discuss the energetic efficiency relative to ATP consumed in generating this light. Consider how emission wavelength and quantum yield are tuned for ecological function.

48. A fluorescent protein exhibits pH-dependent fluorescence. Using the Henderson-Hasselbalch equation, calculate the fraction of protonated chromophore at pH 7.2 if the chromophore pKa is 6.8, and predict the corresponding change in emission intensity. Discuss how this property could be exploited for intracellular pH sensing.

49. A fungal luciferase is being studied for novel reaction intermediates. Propose plausible chemical structures of transient species formed during oxidation of tetrapyrrole-like luciferins and suggest spectroscopic techniques, such as ultrafast photolysis or time-resolved fluorescence, to detect them and elucidate reaction pathways.

50. A chemist aims to engineer a multi-color bioluminescent system for simultaneous imaging of multiple cellular targets. Discuss chemical strategies to tune emission wavelengths, including halogenation, alkylation, conjugation extension, and active-site mutations in luciferase. Predict how these modifications could influence substrate affinity, turnover number, and photon yield, and explain considerations for avoiding spectral overlap in multi-color imaging experiments.

Chapter 46

Chemistry in Non-Aqueous Life Possibilities

The possibility of life arising and persisting in non-aqueous environments represents one of the most compelling frontiers in modern chemical and astrobiological research. While terrestrial life is overwhelmingly water-based, driven by the unique properties of H_2O, it is critical to examine the chemical criteria that might allow for life to emerge in solvents other than water. The search for non-aqueous life is motivated not only by astrobiology, in the context of extraterrestrial bodies such as Titan with its hydrocarbon lakes, but also by fundamental chemical inquiry into the universality of life-like self-organization. Water's unparalleled ability to stabilize complex macromolecules through hydrogen bonding, facilitate hydrolysis and condensation reactions, and provide a polar medium for ionic interactions forms a benchmark against which other solvents must be evaluated. The challenge is to identify liquids that exhibit sufficient polarity, chemical stability, and solvation capacity to support analogous biochemical functions, while allowing for the formation of macromolecular structures capable of information storage, catalysis, and compartmentalization.

Ammonia (NH_3) has long been proposed as a candidate solvent for non-aqueous life due to its hydrogen-bonding ability, high dielectric constant relative to non-polar solvents, and capacity to solvate both polar and moderately nonpolar species. Unlike water, ammonia is less cohesive due to weaker hydrogen bonding, but it remains liquid over a temperature range conducive to chemical kinetics sufficiently fast to sustain metabolism-like reactions. The nucleophilicity of ammonia allows for unique reaction pathways, including amination and imine formation, that could support polymeric backbone structures analogous to peptides or nucleic acid analogs. However, the lower hydrogen-bond strength also imposes constraints on macromolecular stability, necessitating either shorter informational polymers or structural stabilization through alternative non-covalent interactions, such as dipole-dipole interactions or van der Waals forces. Furthermore, ammonia's relatively low freezing point raises questions about the thermal regimes in which life could operate, particularly in extraterrestrial environments far colder than Earth, suggesting that life in ammonia might require kinetic stabilization or the use of eutectic mixtures to depress freezing points.

Hydrocarbons, particularly methane (CH_4) and ethane (C_2H_6), present a strikingly different chemical milieu, representing an apolar, cryogenic solvent environment. Titan's lakes of methane and ethane exemplify a potential habitat for extremely cold, non-aqueous life, though the absence of conventional hydrogen bonding challenges the very notion of molecular self-organization. In hydrocarbon solvents, solvation of polar molecules is minimal, and ionic species are poorly stabilized, demanding alternative chemistries for energy transduction and macromolecular assembly. Hypothetical life in such media might rely on entirely nonpolar polymers, stabilized primarily through van der Waals interactions and steric complementarity. Metabolic processes could involve radical-mediated reactions or other low-temperature chemistries, possibly driven by photochemistry or catalytic surfaces such as ice or mineral substrates. The cryogenic nature of these solvents also slows reaction kinetics dramatically, suggesting that non-aqueous life in hydrocarbons would operate on time scales far longer than terrestrial biochemical processes.

Formamide ($HCONH_2$) emerges as another intriguing solvent candidate due to its exceptional hydrogen-bonding capacity, high dielectric constant, and liquid stability over a broad temperature range. Formamide is also chemically versatile, participating in condensation reactions and facilitating the formation of nucleobase analogs and peptides under laboratory conditions. Its reactivity allows for prebiotic chemistry under non-aqueous conditions, including polymerization pathways that are difficult to achieve in water due to competing hydrolysis. Importantly, formamide's ability to dissolve both polar and moderately nonpolar molecules, coupled with its relatively low vapor pressure, makes it an attractive medium for hypothetical life chemistry that requires complex macromolecular interactions in a liquid environment. The theoretical viability of formamide-based life depends on a careful balance between its high reactivity, which can promote macromolecule formation, and the need to avoid uncontrolled degradation of critical informational molecules.

The exploration of non-aqueous life possibilities also requires re-examining the fundamental nature of biological macromolecules. In water-based life, proteins, nucleic acids, and polysaccharides provide the structural and functional framework necessary for catalysis, genetic information storage, and structural organization. Non-aqueous life would likely necessitate analogous macromolecules with backbones and side chains adapted to the solvent chemistry. In polar aprotic solvents such as dimethyl sulfoxide or acetonitrile, carbon-based polymers with hydrogen-bonding or dipolar interactions may serve similar roles to peptides and nucleic acids, though the precise folding energetics and solvation dynamics would differ markedly. In nonpolar solvents, alternative folding strategies, possibly involving macrocycles, cages, or supramolecular assemblies stabilized primarily by hydrophobic and van der Waals interactions, would be essential to achieve defined three-dimensional structures. The study of such hypothetical polymers challenges the traditional dogma that water is uniquely suited to support life, instead suggesting that life may be a chemical phenomenon contingent on the solvent-mediated balance of enthalpy and entropy in macromolecular assembly.

Energy transduction in non-aqueous life represents another critical dimension of chemical plausibility. Terrestrial life relies heavily on redox reactions involving water and oxygen or other electron carriers, coupled to proton gradients across membranes. In alternative solvents, the solubility of ions and redox-active species is often severely restricted, necessitating entirely novel energy transduction mechanisms. For instance, in hydrocarbon solvents, radical or photochemical reactions might serve as the primary drivers of metabolic processes, while in polar aprotic solvents, redox-active organometallic complexes could play roles analogous to nicotinamide adenine dinucleotide or flavin cofactors. Membrane structures, which in terrestrial cells rely on amphiphilic lipids self-assembling in aqueous environments, would require adapted chemistries capable of forming stable compartments in low-polarity media. Amphiphilic analogs suitable for hydrocarbon or ammonia solvents could include long-chain hydrocarbons with polar headgroups compatible with the solvent, or macrocyclic structures with internal cavities that stabilize internal chemistry against solvent intrusion.

Solvent-dependent chemical kinetics also imposes strict limitations on reaction networks. In extremely cold solvents such as liquid methane, diffusion-limited processes dominate, suggesting that life, if it exists, would be characterized by slow but highly efficient catalytic pathways. In contrast, solvents with higher temperatures and moderate polarity, such as formamide, could permit rapid dynamic chemistry, potentially supporting complex regulatory networks similar to terrestrial metabolism. These considerations highlight the interplay between solvent properties, molecular design, and system-level chemical networks, emphasizing that solvent choice fundamentally constrains both the architecture and dynamics of hypothetical life.

Experimental studies on non-aqueous chemical systems provide valuable insights into the plausibility of such life. Laboratory simulations have demonstrated the formation of peptide-like polymers in ammonia and the assembly of nucleotide analogs in formamide, while studies of hydrocarbon chemistry under Titan-like conditions reveal the potential for complex radical chemistry. Computational chemistry and molecular dynamics simulations further enable the prediction of macromolecular folding, solvation energetics, and reaction kinetics under conditions vastly different from Earth's aqueous environment. These studies collectively suggest that while water is uniquely well-suited for life as we know it, chemical principles permit the possibility of life in a diversity of solvent environments, provided the molecular architecture is appropriately adapted.

Theoretical considerations also extend to the question of information storage and replication. In terrestrial life, nucleic acids encode genetic information via sequence-specific hydrogen bonding and base stacking. In non-aqueous environments, alternative polymers must fulfill analogous informational roles. Candidate systems may include oligomers stabilized by non-traditional hydrogen bonding motifs, π–π stacking, or even reversible covalent linkages in highly polar solvents. The replication of such polymers would require catalysis adapted to the solvent's chemical landscape, potentially involving organometallic centers, radical-mediated chemistry, or surface-mediated templating. The chemical robustness of information storage in non-aqueous solvents depends critically on minimizing hydrolysis or solvent-mediated degradation while maintaining sufficient chemical lability to allow replication and evolution.

Beyond ammonia, hydrocarbons, and formamide, other solvents present unique chemical landscapes with potential implications for life. Sulfuric acid, a highly polar, strongly acidic medium, has been proposed as a candidate for life in extremely hostile environments, such as the clouds of Venus. Its extraordinary proton activity and high density enable the dissolution of certain polar organic molecules and provide a medium for acid-catalyzed reactions. In such environments, molecular architectures must be exceptionally resistant to protonation and hydrolysis, favoring molecules with delocalized electron systems, stable heterocycles, or sulfur-based backbones. The energetic landscape in sulfuric acid allows for acid-mediated condensation and polymerization reactions that are otherwise impossible in water. Such chemistry would likely produce dense, highly cross-linked polymers capable of forming structurally robust compartments or protocellular analogs. Additionally, the ability of sulfuric acid to stabilize carbocations opens alternative pathways for energy transduction, allowing proton-coupled reactions to drive molecular motion or synthetic cycles analogous to metabolism. However, the extreme acidity and oxidative potential of such environments impose severe constraints on chemical diversity, necessitating specialized molecular adaptations to survive, replicate, and evolve.

Ionic liquids, composed entirely of ions and often liquid at room temperature, provide another intriguing framework for non-aqueous life. Their negligible vapor pressures, broad liquid ranges, and tunable polarity allow for solvation of a wide variety of organic, inorganic, and organometallic species. In particular, ionic liquids can stabilize charged or polar macromolecules in environments where conventional polar solvents are unavailable, while simultaneously supporting nonpolar interactions through alkylated cations or anions. These solvents enable reaction pathways inaccessible in aqueous media, including electrochemical transformations, organocatalytic cycles, and polymerization reactions mediated by ionic pairing or charge delocalization. Their high chemical stability and ability to facilitate compartmentalization through phase separation suggest that ionic liquids could support self-organized chemical systems capable of information processing, catalysis, and energy transduction. The modular design of ionic liquids allows for tuning the dielectric environment and viscosity, potentially enabling adaptive chemical systems capable of maintaining homeostasis in response to environmental fluctuations.

Supercritical fluids, such as supercritical carbon dioxide, present yet another non-aqueous medium with properties intermediate between gases and liquids. Supercritical CO_2 exhibits low viscosity, high diffusivity, and tunable density, offering unique opportunities for solute transport and reaction kinetics that differ markedly from those in conventional liquids. Despite its nonpolarity, supercritical CO_2 can solvate moderately polar organometallic complexes or fluorinated molecules, supporting catalytic networks that exploit radical chemistry or electron transfer processes. Self-assembly in such environments would likely rely on amphiphilic molecules that form micelle-like or vesicle-like structures stabilized through van der Waals interactions, fluorophilic effects, or π–π stacking. The dynamic nature of supercritical fluids also allows rapid redistribution of reactants, potentially enabling emergent behaviors akin to primitive metabolic cycles, albeit under conditions dramatically different from aqueous biochemistry. These environments highlight the adaptability of chemical systems to unconventional solvents, expanding the conceptual boundaries of life chemistry.

The thermodynamics and kinetics of chemical reactions in non-aqueous solvents impose fundamental constraints on the types of metabolism and macromolecular architecture that can exist. Solvent polarity, viscosity, and dielectric constant influence the rate of bond formation, the stability of intermediates, and the feasibility of complex reaction networks. In polar aprotic solvents, nucleophilic substitution and condensation reactions are favored, potentially supporting polymeric backbones and informational macromolecules analogous to peptides or nucleic acids. In nonpolar or cryogenic solvents, reaction kinetics slow drastically, and radical or photochemical pathways become more dominant, necessitating molecular adaptations that maximize reaction efficiency under kinetic constraints. Entropic contributions also differ significantly: in non-aqueous environments, the balance between solute-solvent interactions and intramolecular interactions dictates folding, self-assembly, and molecular recognition. Computational models of solvation and folding energetics demonstrate that even modest changes in solvent properties can shift the stability of secondary structures, influence catalytic site accessibility, and alter polymerization equilibria, emphasizing the solvent-dependence of life-like chemical systems.

Membrane formation and compartmentalization are central challenges for non-aqueous life. In water-based systems, amphiphilic lipids spontaneously form bilayers, vesicles, or micelles, creating isolated microenvironments that allow the concentration and regulation of biochemical reactions. In non-aqueous solvents, membrane analogs must exploit solvent-specific interactions to achieve compartmentalization. Hydrocarbon-based solvents favor amphiphilic molecules with polar cores and long hydrophobic tails, whereas polar aprotic solvents may require polymers with complementary dipolar or hydrogen-bonding motifs. Supramolecular assemblies stabilized by macrocyclic scaffolds, host-guest interactions, or noncovalent crosslinking provide alternative strategies for creating functional compartments. Such compartmentalization is not merely structural; it modulates diffusion, local concentration, and reaction kinetics, enabling primitive metabolic networks and feedback mechanisms. The design of these systems requires careful consideration of solvent properties, molecular geometry, and noncovalent interaction patterns, highlighting the intimate interplay between solvent chemistry and life-like organization.

Energy storage and chemical transduction in non-aqueous life also diverge from terrestrial paradigms. Terrestrial biochemistry relies heavily on ATP and redox-active cofactors in aqueous media, exploiting proton gradients and electron transport chains. In non-aqueous media, equivalent energy currencies may be entirely different, exploiting solvent-specific chemistry. In hydrocarbon solvents, long-lived radicals or photochemically activated species could serve as transient energy carriers, while in polar aprotic solvents, organometallic complexes, conjugated polymers, or charge-transfer assemblies might fulfill analogous roles. The generation, stabilization, and utilization of chemical energy must be compatible with the solvent environment, implying that catalysis and energy transduction are inseparable from molecular solvation and folding dynamics. The study of these hypothetical pathways provides a conceptual framework for understanding the diversity of energetically feasible life chemistries beyond water.

The evolution of chemical systems in non-aqueous solvents requires consideration of both molecular adaptability and environmental constraints. For life to emerge, chemical networks must be capable of replication, variation, and selection. Solvent-dependent stability of informational polymers, the efficiency of catalytic cycles, and the robustness of compartmentalization collectively determine the potential for evolutionary processes. Laboratory experiments simulating ammonia, hydrocarbon, or formamide chemistries reveal that polymerization, templating, and catalytic processes are feasible under a range of conditions, though reaction rates and error frequencies differ substantially from aqueous systems. Computational studies further illustrate that chemical networks can exhibit autocatalysis, feedback loops, and network-level resilience in non-aqueous media, supporting the hypothesis that life-like evolution may be a general property of suitably organized chemical systems, independent of water as a solvent.

Astrobiological observations provide compelling contexts for these chemical possibilities. Titan's methane-ethane lakes, Venus' sulfuric acid clouds, and icy bodies with ammonia-rich subsurfaces demonstrate the ubiquity of non-aqueous liquid environments in the Solar System. The chemistry of these environments, constrained by temperature, pressure, and solvent properties, suggests that alternative life chemistries may not only be plausible but may represent robust ecological niches. Laboratory analog experiments under these conditions reveal complex organic chemistry, from radical-driven polymerization to the formation of nitrogenous heterocycles, indicating that molecular complexity can arise without water. Such studies challenge the assumption that water is a prerequisite for life, emphasizing that life may emerge wherever the interplay of solvent, molecular architecture, and energy flux supports self-organization, information processing, and replication.

The storage and transmission of information is a defining hallmark of life, and in non-aqueous media, the design of informational polymers poses unique chemical challenges. Traditional nucleic acids rely on Watson-Crick hydrogen bonding, π-stacking interactions, and aqueous solvation to maintain sequence specificity and structural stability. In solvents such as ammonia, formamide, or ionic liquids, alternative hydrogen-bonding motifs, reversible covalent linkages, or dipole-dipole interactions may serve analogous functions. Polyimines, peptide nucleic acids, and other backbone-modified polymers have been proposed as hypothetical informational carriers capable of templated replication in non-aqueous environments. The chemical robustness of these polymers depends critically on their resistance to solvent-mediated degradation while retaining sufficient lability to allow sequence copying and evolutionary adaptation. In nonpolar media, π–π stacking, hydrophobic collapse, or van der Waals-mediated folding could maintain defined tertiary structures, providing an environment for catalytic or regulatory functions. The flexibility in backbone chemistry and base analogs significantly expands the diversity of potential life chemistries, suggesting that information processing is a generalizable chemical principle rather than a water-dependent phenomenon.

Catalysis, the acceleration of chemical transformations with selectivity, is equally crucial for non-aqueous life. Terrestrial enzymes utilize intricate folding, polar side chains, and cofactors to stabilize transition states in aqueous media. In non-aqueous environments, analogous catalysts may adopt radically different chemical strategies. In ammonia or formamide, proton transfer and nucleophilic catalysis remain feasible, allowing the design of small-molecule or polymeric catalysts that mediate polymerization, energy transduction, or redox reactions. In nonpolar solvents, radical or photochemical catalysis could dominate, with supramolecular assemblies providing spatial confinement and orientational control analogous to enzyme active sites. Computational studies reveal that catalysis in these media is highly solvent-dependent, with transition state stabilization, diffusion limitations, and solvation energetics dictating the feasibility and efficiency of reaction pathways. Catalytic networks in non-aqueous media may also display emergent properties, such as cooperative behavior, feedback loops, and network-level robustness, highlighting the potential for complex metabolism without water.

Computational modeling and molecular simulations have become indispensable tools for exploring non-aqueous life chemistries. Molecular dynamics, Monte Carlo simulations, and quantum chemical calculations allow the prediction of solvation energies, folding dynamics, and reaction kinetics under conditions far removed from terrestrial biochemistry. These models provide insight into the stability of hypothetical polymers, the energetics of self-assembly, and the feasibility of autocatalytic networks. In cryogenic hydrocarbons, simulations suggest that radical intermediates can persist for long durations, allowing slow but directional chemical evolution. In polar aprotic solvents, folding simulations of polymeric backbones indicate the potential for stable secondary and tertiary structures capable of catalytic or informational functions. By integrating computational predictions with laboratory analog experiments, researchers can explore the chemical parameter space necessary for non-aqueous life, testing hypotheses about macromolecular stability, reaction network feasibility, and environmental constraints.

Integration of metabolism, information storage, and compartmentalization is a hallmark of life-like chemical systems, and non-aqueous life is no exception. Compartmentalized reaction networks can concentrate reactants, regulate diffusion, and couple energy transduction to macromolecular synthesis, forming the basis for primitive protocells. In ammonia or formamide, vesicle-like structures stabilized by amphiphilic polymers may enclose catalytic networks, while in hydrocarbon solvents, macrocycles or micelle-like assemblies could provide confined chemical environments. The coupling of chemical energy sources—such as radical generation, photochemistry, or proton gradients—to the replication and maintenance of informational polymers enables a feedback-driven, self-organizing system. This integrated approach suggests that life is not a function of a specific solvent but rather an emergent property of chemical networks capable of maintaining organization, reproducing, and evolving under suitable energetic and structural constraints.

Experimental approaches to studying non-aqueous life continue to expand, bridging synthetic chemistry, prebiotic simulation, and astrobiology. Laboratory studies have successfully demonstrated the formation of oligomers, nucleobase analogs, and catalytically active polymeric species in ammonia, formamide, and hydrocarbon media. Photochemical and radical-driven reactions in hydrocarbon solvents under cryogenic conditions produce complex organic molecules resembling primitive metabolic intermediates. These experiments, often coupled with computational modeling, provide proof-of-concept evidence that chemical evolution is feasible outside aqueous environments. In addition, the study of extreme environments on Earth, such as superacidic pools, hydrocarbon seeps, and polar ice matrices, informs our understanding of the chemical flexibility required for non-aqueous life, revealing that molecular complexity can arise in conditions previously considered prohibitive.

Astrobiological implications of non-aqueous life are profound. Titan, with its extensive hydrocarbon lakes and cryogenic temperatures, represents a natural laboratory for hydrocarbon-based chemistries, potentially supporting radical-driven metabolic networks and macrocyclic polymers. Venus' sulfuric acid clouds demonstrate that extreme acid environments may host life-like chemistry using acid-stable polymers and proton-mediated catalysis. Icy moons such as Europa or Enceladus may contain subsurface ammonia-water eutectics capable of supporting polymerization and catalytic reactions at moderate pressures and temperatures. These diverse environments expand the search for life beyond the classical water-centric paradigm, emphasizing that life is a chemical phenomenon contingent on solvent properties, molecular architecture, and energy availability rather than a single universal chemistry.

Finally, the theoretical exploration of non-aqueous life challenges conventional definitions of life and encourages a broader perspective on chemistry's role in biology. Life is often viewed as a product of water's unique chemistry; however, the study of alternative solvents underscores that the principles of self-organization, information storage, catalysis, and energy transduction are generalizable. Solvent chemistry dictates molecular stability, reaction kinetics, and compartmentalization, but the underlying requirements for life—sustained chemical organization, replication, and adaptability—can, in principle, be achieved in a variety of chemical media. This realization has profound implications for synthetic biology, astrobiology, and our understanding of the chemical foundations of life, suggesting that the universe may harbor life in forms and environments far removed from terrestrial expectations.

In conclusion, the chemistry of non-aqueous life possibilities represents a vibrant intersection of physical chemistry, organic chemistry, astrobiology, and theoretical biology. By examining ammonia, hydrocarbons, formamide, sulfuric acid, ionic liquids, and supercritical fluids, we uncover diverse chemical landscapes capable of supporting life-like processes. Informational polymers, catalytic networks, and compartmentalized reaction systems illustrate that life's defining features can emerge in solvent-specific chemical contexts. Experimental and computational studies validate these theoretical possibilities, providing concrete frameworks for exploring life beyond water. Ultimately, the study of non-aqueous life expands the conceptual boundaries of chemistry, offering a vision of life as a general chemical phenomenon, adaptable, emergent, and not limited to a single solvent, thereby redefining our understanding of biological potential in the cosmos.

Chapter 46: Word Problems

1. Considering ammonia's hydrogen-bonding capacity and its lower cohesion relative to water, calculate the maximum chain length of a peptide-like polymer that could remain stable in liquid ammonia at 200 K. Take into account the average hydrogen bond energy of 16 kJ/mol and consider how intramolecular hydrogen bonding competes with solvent interactions. Discuss the implications for informational polymer storage, including the potential for replication fidelity and how these limitations might influence the design of alternative polymer backbones suitable for ammonia-based life.

2. Titan's methane lakes present a cryogenic, nonpolar environment with temperatures near 95 K. Propose a detailed mechanism for radical-mediated polymerization of simple hydrocarbons under these conditions, taking into account the diffusion-limited mobility of reactants. Estimate the approximate rate constant for polymer chain elongation using collision theory modified for cryogenic viscosity, and discuss how extremely slow kinetics might affect the feasibility of sustaining a metabolic-like network over geological timescales.

3. Formamide has a high dielectric constant and low vapor pressure, making it an attractive candidate for non-aqueous prebiotic chemistry. Design a condensation reaction pathway leading to the formation of nucleobase analogs from simple precursors, highlighting solvent-mediated stabilization of transition states. Compare the thermodynamic favorability of this reaction pathway in formamide with the same pathway in water at 298 K, and discuss how solvent reactivity, polarity, and hydrogen-bonding potential influence both the yield and stability of the resulting oligomers.

4. Compute the solvation energy difference for a polar amino acid side chain in ammonia versus formamide using a Born solvation model, considering both dielectric constants and ion pairing effects. Discuss how differences in solvation energy influence the folding stability of short polymer chains, the formation of

secondary and tertiary structures, and the potential for functional catalysis or information storage in each solvent environment. Reflect on the constraints this imposes for designing stable macromolecules suitable for non-aqueous life.

5. Using the dielectric properties of ethane and methane at Titan-like temperatures, estimate the probability of stabilizing an ionic intermediate required for a hypothetical metabolic cycle in these hydrocarbon solvents. Incorporate the effects of extreme cold on molecular diffusion, radical lifetimes, and solvent viscosity, and discuss how these physical constraints may necessitate alternative chemistries, such as radical-mediated transformations, for energy transduction and polymer synthesis in nonpolar environments.

6. Design a hypothetical amphiphilic molecule capable of forming stable vesicles in liquid ammonia, considering hydrogen bonding, dipole interactions, and van der Waals forces. Calculate the free energy of self-assembly and evaluate the role of solvent interactions in stabilizing compartmentalized microenvironments. Discuss how such vesicles could support localized chemical networks, including replication of informational polymers or catalytic cycles, and highlight the challenges in maintaining structural integrity under low-temperature and low-cohesion conditions.

7. Compare the activation energy barriers for imine formation in ammonia with peptide bond formation in water, and predict the relative feasibility of synthesizing informational polymers in each environment. Consider the influence of solvent polarity, hydrogen bonding, and nucleophilicity on the reaction kinetics, and discuss how these chemical differences constrain the maximum polymer length, sequence fidelity, and potential for evolutionary adaptation in ammonia-based life systems.

8. For a nonpolar hydrocarbon solvent, calculate the van der Waals interaction energy between two hypothetical nonpolar polymers, each 10 monomer units long, separated by 0.5 nm. Discuss how this interaction contributes to folding, stability, and self-organization of macromolecular structures in cryogenic nonpolar solvents. Evaluate the role of steric complementarity and low-temperature diffusion in maintaining functional tertiary structures capable of supporting radical-mediated catalysis or information storage.

9. Evaluate the kinetic limitations for enzyme-like catalysis in liquid methane at 95 K. If diffusion limits the encounter rate between substrate and catalytic radical species, estimate the mean reaction time for a polymerization of 50 monomer units. Discuss how slow kinetics might affect the efficiency of metabolic networks, the persistence of transient intermediates, and the potential timescales for evolutionary adaptation in hydrocarbon-based life.

10. Using the concept of eutectic mixtures, calculate the minimum concentration of ammonia required with a co-solvent to depress the freezing point below 180 K, allowing slow chemical reactions to occur. Discuss how eutectic formation could enable dynamic chemical networks in low-temperature environments, the trade-offs between solvent mobility and reaction kinetics, and how these conditions could support the emergence of primitive metabolic cycles and compartmentalized reaction systems in ammonia-based life.

11. In formamide, nucleophilic reactions are particularly favorable due to high dielectric stabilization. Design a polymerization network where monomers undergo selective amide bond formation without significant hydrolysis. Model the relative rates of forward and reverse reactions at 320 K, and discuss how solvent reactivity, hydrogen-bonding, and polarity influence both polymer yield and potential for generating sequence-specific informational polymers.

12. Sulfuric acid, proposed as a solvent on Venus, exhibits extraordinary proton activity. Determine the protonation equilibrium for a heterocyclic molecule with pKa = 4 in concentrated H_2SO_4, considering the high ionic strength and activity coefficients. Discuss the implications of protonation for macromolecular stability, polymer backbone selection, and the feasibility of maintaining informational polymers capable of catalysis in such a highly acidic environment.

13. Compare diffusion coefficients of small radicals in methane at 95 K and in ethane at 90 K, using kinetic theory adjusted for cryogenic viscosity. Predict how these differences influence the efficiency of radical-mediated polymerization and hypothetical metabolic networks, including considerations for reaction timescales, transient intermediate lifetimes, and the potential for long-term chemical evolution in hydrocarbon solvents.

14. Propose a detailed membrane architecture using macrocycles capable of forming

stable compartments in hydrocarbon solvents. Calculate the Gibbs free energy of self-assembly, incorporating van der Waals, π–π stacking, and hydrophobic effects, and discuss how such compartments could spatially organize catalytic networks, regulate diffusion of reactive species, and protect informational polymers under cryogenic conditions.

15. For ammonia-based life, estimate the lifetime of a short informational polymer at 180 K, assuming a rate constant for spontaneous hydrolysis of 10^{-5} s^{-1}. Analyze how temperature, solvent polarity, and hydrogen-bonding limitations influence polymer stability, replication fidelity, and the potential for evolutionary adaptation in a low-cohesion environment.

16. Using molecular dynamics simulations, model the folding of a peptide nucleic acid analog in formamide at 300 K. Predict secondary and tertiary structure stability compared to an aqueous environment, taking into account solvation energetics, hydrogen-bonding patterns, and entropic contributions. Discuss how solvent-specific folding dynamics influence the potential for catalysis and information storage in non-aqueous systems.

17. In a supercritical CO_2 environment, predict the diffusion-limited encounter rate of a small organometallic catalyst with a substrate molecule. Calculate the expected turnover frequency for a simple redox cycle, considering the effects of tunable density, low viscosity, and moderate polarity on reaction kinetics. Discuss the implications for emergent metabolic networks in supercritical fluids.

18. Ionic liquids stabilize both polar and nonpolar species, offering a broad chemical landscape. Propose a network of oligomerization reactions in a 1-ethyl-3-methylimidazolium-based ionic liquid, including possible ion-pair mediated mechanisms. Analyze how charge delocalization, solvent viscosity, and polarity influence reaction selectivity, polymer folding, and potential for protocellular compartmentalization.

19. Consider an ammonia-based metabolic reaction involving amination of a small aldehyde. Determine the equilibrium constant at 190 K, taking into account solvent nucleophilicity, hydrogen-bonding effects, and competing side reactions. Predict the net reaction direction, possible accumulation of intermediates, and implications for the design of persistent catalytic cycles.

20. Design a nonpolar polymer capable of information storage in liquid ethane at 95 K. Estimate the maximum sequence length that can maintain a stable tertiary structure using van der Waals and steric interactions. Discuss how folding energetics, solvent polarity, and cryogenic kinetic limitations affect information retention, potential catalysis, and the plausibility of evolutionary adaptation.

21. Compare viscosities of methane, ethane, and ammonia at 100 K. Predict how the differences in solvent mobility affect diffusion-limited reactions, encounter rates of catalysts and substrates, and the potential timescales for metabolic network activity and evolutionary processes in these non-aqueous solvents.

22. In formamide, small organocatalysts can accelerate condensation reactions. Calculate the expected rate enhancement for peptide bond formation compared to uncatalyzed reactions at 320 K. Consider both solvent stabilization of the transition state and reorganization energy, and discuss how catalysis in non-aqueous environments may enable the formation of longer informational polymers.

23. In concentrated sulfuric acid, proton-mediated polymerization is possible. Compute the Gibbs free energy change for a carbocation-driven condensation of a simple organic monomer at 350 K. Analyze the feasibility of sustained polymer formation, considering both kinetic and thermodynamic limitations, and discuss implications for informational polymer stability under extreme acidity.

24. Evaluate the effect of solvent polarity on π–π stacking interactions between nucleobase analogs in ammonia and formamide. Quantify the interaction energies and discuss how differences in hydrogen-bonding, solvation, and dielectric properties affect replication fidelity, duplex stability, and potential for evolution of information-carrying polymers.

25. Model the temperature-dependent folding dynamics of a macrocyclic polymer in methane at 95 K. Predict how slow kinetics, van der Waals interactions, and steric constraints affect the formation of functional tertiary structures. Discuss implications for catalysis, energy transduction, and the storage of information in cryogenic nonpolar environments.

26. For a hydrocarbon-based protocell, calculate the minimum chain length of a radical-stabilized polymer required to support catalytic activity at 95 K.

Consider contributions from van der Waals stabilization, steric complementarity, and diffusion-limited reaction kinetics, and discuss the potential for energy transduction and self-sustaining reaction networks.

27. Using molecular dynamics, predict the self-assembly propensity of amphiphilic polymers in ionic liquids. Determine the critical concentration for vesicle formation, incorporating both Coulombic interactions and solvent polarity effects, and discuss how these compartments could concentrate reactants and facilitate prebiotic chemistry in non-aqueous systems.

28. Design an energy transduction cycle in formamide-based life using organometallic redox couples. Calculate the standard Gibbs free energy change, estimate the maximum energy yield per reaction, and discuss the role of solvent polarity, hydrogen-bonding, and solvation dynamics in sustaining cyclic chemical networks.

29. Compare reaction rate constants for radical-mediated C–C bond formation in methane versus ethane at Titan-like temperatures. Analyze how solvent density, polarity, and diffusion limitations affect the feasibility of metabolic pathways, and discuss potential strategies for stabilizing reactive intermediates in cryogenic nonpolar environments.

30. Calculate the solubility limit of a charged monomer in liquid ammonia with a dielectric constant of 22. Discuss how solubility constraints affect polymerization rates, informational polymer stability, and the potential for replication in ammonia-based life systems.

31. For supercritical CO_2, estimate the diffusion time of a 1-nm macromolecule across a 10 μm distance. Analyze how high diffusivity and low viscosity influence the coupling of catalytic cycles and the emergence of primitive metabolic behavior under supercritical conditions.

32. Predict the relative stability of a nucleobase analog in formamide versus water. Quantify hydrolytic degradation rates and discuss the implications for solvent choice in prebiotic polymerization and informational polymer replication.

33. Consider a nonpolar polymer designed for information storage in hydrocarbons. Using Lennard-Jones potentials, model the free energy difference between folded and unfolded states and discuss how folding stabilization contributes to structural and functional integrity in cryogenic solvents.

34. Propose a mechanism for proton-coupled energy transduction in concentrated sulfuric acid. Identify the rate-limiting step and estimate the reaction velocity at 350 K. Discuss the compatibility of such proton-driven cycles with polymer stability and potential protocellular structures.

35. Design a hypothetical peptide nucleic acid (PNA)-like polymer capable of templated replication in an ionic liquid. Calculate the Gibbs free energy of duplex formation, discuss the effect of solvent polarity on hybridization, and predict replication fidelity in non-aqueous environments.

36. For ammonia, calculate the nucleophilicity index for NH_3 versus H_2O and predict its effect on amination reactions critical for forming polymeric backbones. Discuss implications for the synthesis and stability of information-carrying polymers.

37. Model diffusion-limited catalysis in a methane-based metabolic network at 95 K. Predict reaction turnover rates for a 100-monomer polymeric substrate, considering cryogenic limitations and radical intermediate lifetimes, and discuss implications for the persistence of chemical evolution over long timescales.

38. Compute the effect of solvent polarity on van der Waals stabilization of a macrocyclic folding motif in ethane versus formamide. Discuss how tertiary structure formation, functional folding, and information retention depend on the chemical environment.

39. Design an autocatalytic network in supercritical CO_2 capable of sustaining energy flux. Quantify the timescale over which steady-state concentrations of intermediates are reached, considering solvent density, diffusivity, and reaction kinetics, and discuss the potential for emergent metabolic cycles.

40. Estimate the equilibrium constant for an acid-catalyzed condensation reaction in concentrated H_2SO_4, and compare it to the same reaction in water at 298 K. Discuss the chemical feasibility of polymerization under extreme acidity and its implications for informational polymer formation.

41. For ionic liquid solvents, calculate the Debye length and predict its effect on diffusion of charged macromolecules within compartmentalized assemblies. Discuss implications for self-organization, information retention, and catalysis.

42. Predict the lifetime of radical intermediates in methane at 95 K using Arrhenius kinetics. Discuss how extended intermediate lifetimes affect slow but directional chemical evolution in cryogenic hydrocarbon solvents.

43. Design a photochemically-driven radical network in ethane capable of energy transduction. Estimate photon flux requirements needed to sustain a cyclic reaction, considering cryogenic diffusion limitations and radical stabilization energies.

44. Compare entropic contributions to folding for a polar polymer in formamide versus a nonpolar polymer in methane. Discuss how solvent-dependent entropy influences macromolecular architecture, folding stability, and potential catalytic functionality.

45. For ammonia-based life, calculate the thermal energy available at 180 K to overcome activation barriers for polymerization reactions. Discuss how temperature, kinetic limitations, and solvent interactions constrain polymer length, fidelity, and chemical evolution.

46. Propose a mechanism for supramolecular compartmentalization using macrocycles in hydrocarbon solvents. Quantify the free energy contribution of host-guest interactions and discuss how compartmentalization enhances localized catalysis and informational polymer replication.

47. In formamide, design a network of condensation reactions leading to oligomer formation. Calculate relative rates, predict equilibrium polymer length distributions, and discuss implications for prebiotic chemical complexity.

48. Evaluate the feasibility of nucleotide analog replication in a low-polarity ionic liquid. Calculate the Gibbs free energy for duplex formation and predict potential error frequency during sequence copying in non-aqueous environments.

49. Consider supercritical CO_2 as a medium for catalytic cycles. Compute the viscosity-dependent diffusion coefficient for a 2-nm catalytic complex and predict the rate-limiting step of the network, analyzing how solvent properties influence reaction timescales and stability.

50. Integrate solvent chemistry, polymer stability, and energy transduction to propose a minimal protocellular system in ammonia or formamide. Estimate the minimal set of molecules and interactions required to sustain replication, primitive metabolism, and self-organized chemical networks, and discuss the constraints imposed by solvent-dependent kinetics, macromolecular folding, and compartmentalization.

Chapter 47

EPIGENETIC CHEMISTRY

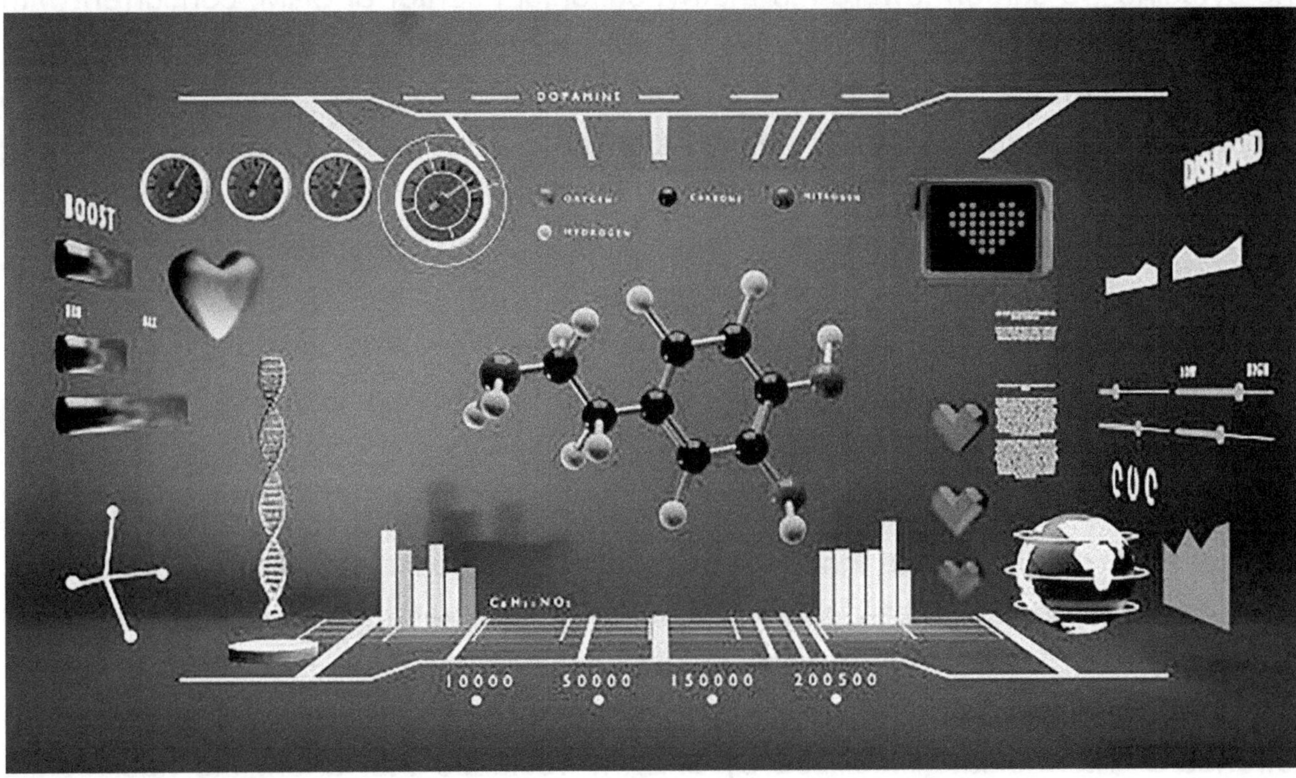

Epigenetic chemistry represents the intricate network of molecular transformations through which heritable changes in gene expression are encoded, maintained, and reversed without alteration of the DNA sequence itself, and it is fundamentally governed by covalent modifications, non-covalent interactions, enzymatic catalysis, and the dynamic chemical environment of chromatin. Central to this chemistry is DNA methylation, predominantly occurring at the C5 position of cytosine residues within CpG dinucleotides. The chemical reaction involves nucleophilic attack by the C5 carbon of cytosine on the electrophilic methyl group of S-adenosylmethionine (SAM), which exists as a sulfonium cation stabilized by resonance in the positively charged sulfur center. DNMTs facilitate this process by positioning the cytosine in an active site pocket, aligning the nucleophilic carbon with the methyl donor and stabilizing the transition state through hydrogen bonding and π-stacking interactions with surrounding aromatic residues. The reaction proceeds via the formation of a covalent intermediate between the C6 position of cytosine and a catalytic cysteine thiol, which temporarily stabilizes the negative charge developed on the cytosine ring during methyl transfer. Electron-pushing involves the donation of a lone pair from the cytosine C5 carbon toward the methyl sulfonium center of SAM, concurrent with the displacement of the sulfur-bound methyl group, followed by proton abstraction from the thiol to restore the cytosine aromaticity. The oxidation of 5-methylcytosine to 5-hydroxymethylcytosine by TET enzymes involves a Fe^{2+}/α-ketoglutarate-dependent dioxygenation mechanism, in which a highly reactive Fe(IV)=O intermediate abstracts a hydrogen atom from the methyl group, generating a radical that recombines with the hydroxyl radical derived from molecular oxygen, ultimately producing 5hmC. Successive oxidation generates 5-formylcytosine and 5-carboxylcytosine, each with distinct electronic properties that influence protein-DNA recognition, and base excision repair completes the demethylation cycle.

Histone proteins, particularly the N-terminal tails, undergo a series of post-translational modifications that exemplify the chemical versatility of epigenetic regulation. Lysine acetylation, catalyzed by HATs, involves nucleophilic attack by the ε-amino group of lysine on the carbonyl carbon of acetyl-CoA. Electron flow begins with the lone pair of nitrogen attacking the electrophilic carbonyl carbon, forming a tetrahedral intermediate stabilized by enzyme residues that orient the nucleophile and stabilize the oxyanion. Collapse of the tetrahedral intermediate transfers the acetyl group to the lysine side chain and releases Coenzyme A. The deacetylation reaction, catalyzed by HDACs, often utilizes a Zn^{2+} ion coordinated by histidine and aspartate residues to polarize a water molecule, which acts as a nucleophile attacking the carbonyl carbon of acetylated lysine, generating a tetrahedral intermediate that collapses to release acetate and restore the lysine's positive charge. The kinetics and thermodynamics of these reactions are finely tuned by active site electrostatics and substrate positioning, highlighting the precision of chemical control in epigenetic regulation.

Histone methylation, occurring on lysine and arginine residues, demonstrates additional chemical subtlety. Lysine can be mono-, di-, or trimethylated, each state modifying steric bulk and hydrogen bonding without altering charge. The reaction is catalyzed by lysine methyltransferases (KMTs), which utilize SAM as the methyl donor. Mechanistically, the ε-amino group of lysine acts as a nucleophile, attacking the methyl sulfonium center of SAM via an S_N2-like displacement. The enzyme stabilizes the developing negative charge in the transition state through hydrogen bonding and electrostatic interactions, and specificity for mono-, di-, or trimethylation is determined by steric constraints within the active site. Demethylation of lysine residues occurs via FAD-dependent amine oxidases or Jumonji C domain-containing dioxygenases. In the FAD-dependent pathway, the ε-amino group forms a Schiff base with FAD, facilitating a hydride transfer to the flavin ring, generating an iminium intermediate that is hydrolyzed to release formaldehyde and regenerate unmodified lysine. In the dioxygenase pathway, α-ketoglutarate coordinates with Fe^{2+} to generate an Fe(IV)=O species, which abstracts a hydrogen atom from the methyl group, forming a radical that reacts with hydroxyl radical to generate hydroxymethyllysine, which spontaneously decomposes to formaldehyde. Arginine methylation, catalyzed by protein arginine methyltransferases (PRMTs), follows similar nucleophilic attack mechanisms but introduces additional symmetry considerations, as the guanidinium side chain can be mono- or dimethylated in symmetric or asymmetric arrangements, affecting hydrogen bonding and cation-π interactions within the chromatin environment.

Cross-talk between histone modifications adds chemical complexity, wherein one modification influences the reactivity or recognition of another. Phosphorylation of serine or threonine residues introduces a negatively charged phosphate, altering electrostatics and local hydrogen bonding networks, thereby affecting nearby lysine acetylation or methylation. Ubiquitination and sumoylation covalently attach small proteins to lysine residues, imposing steric constraints that can allosterically modulate enzymatic activities. The chemical interplay between these modifications exemplifies combinatorial control, in which the sum of individual covalent modifications defines chromatin accessibility, nucleosome stability, and transcription factor recruitment.

RNA epigenetics further extends the chemical dimension of chromatin regulation. N6-methyladenosine (m^6A) modification, catalyzed by METTL3/14, alters the electronic distribution at the N6 position of adenine, affecting hydrogen bonding with complementary nucleotides and altering RNA folding dynamics. Mechanistically, the nucleophilic exocyclic amino group attacks the methyl group of SAM in an S_N2

displacement, facilitated by enzyme active site residues that stabilize the transition state. Demethylation by FTO or ALKBH5 involves oxidative hydroxylation of the methyl group, with subsequent spontaneous decomposition to release formaldehyde, linking redox chemistry to RNA regulation. These chemical transformations demonstrate that RNA, like DNA and histones, operates under a precise chemical code responsive to enzymatic catalysis and cofactor availability.

Chromatin remodeling complexes exemplify chemical energy transduction in epigenetics. ATP-dependent remodelers, including SWI/SNF, ISWI, and CHD complexes, hydrolyze ATP in a reaction that begins with nucleophilic attack on the γ-phosphate by a water molecule coordinated with Mg^{2+} ions. The resultant ADP and inorganic phosphate release chemical energy, which is converted into mechanical work, driving nucleosome sliding, eviction, or histone variant incorporation. These processes are chemically coupled to histone and DNA modifications, integrating energetic and covalent chemical regulation.

Metabolic state directly influences epigenetic chemistry via cofactor availability. SAM levels, derived from one-carbon metabolism, determine methylation potential for both DNA and histones. Acetyl-CoA concentrations modulate histone acetylation, while NAD^+ availability regulates sirtuin-mediated deacetylation. α-Ketoglutarate and Fe^{2+} levels influence TET and Jumonji dioxygenase activity, linking cellular redox state and metabolism to chromatin chemistry. These interactions exemplify a chemical network in which small-molecule metabolites dictate the efficiency, reversibility, and specificity of covalent modifications, establishing a direct chemical bridge between environment, metabolism, and gene expression.

Chemical modifications also regulate phase separation phenomena within the nucleus, as intrinsically disordered regions of chromatin-associated proteins undergo charge modulation, hydrophobic changes, and altered interaction surfaces upon post-translational modification. Phosphorylation introduces negative charges, acetylation neutralizes positive charges, and methylation modifies hydrogen-bonding potential, collectively modulating liquid-liquid phase separation that compartmentalizes transcriptional machinery. These mesoscale chemical effects translate molecular modifications into organizational control over transcription and chromatin architecture.

Oxidative DNA modifications, including 5-hydroxymethylcytosine, 5-formylcytosine, and 5-carboxylcytosine, illustrate the chemical integration of DNA repair and epigenetic regulation. TET-mediated oxidation alters electronic density and hydrogen bonding, providing recognition sites for repair enzymes and influencing chromatin binding. Endogenous oxidative damage products such as 8-oxoguanine also intersect with epigenetic chemistry, requiring specialized recognition and repair to maintain fidelity. These processes highlight the sensitivity of the chemical epigenetic landscape to both endogenous enzymatic and exogenous environmental influences.

Chemical strategies in epigenetic therapeutics exploit the specificity and reactivity of enzyme active sites. DNMT inhibitors, such as 5-azacytidine, incorporate into DNA and trap enzymes via covalent adduct formation, irreversibly halting methylation. HDAC inhibitors chelate Zn^{2+} in the active site, preventing deacetylation. Bromodomain inhibitors mimic acetylated lysine to competitively block reader domain interactions. Rational design of these molecules relies on precise mechanistic understanding of electron flow, transition-state stabilization, substrate positioning, and stereoelectronic complementarity, underscoring the centrality of chemical principles in therapeutic intervention.

The chemical logic of epigenetics orchestrates developmental programming, cellular memory, and disease pathogenesis. Aberrant methylation, misregulated histone modifications, or altered RNA modifications perturb chemical networks, leading to oncogenic transformation, neurodegeneration, and metabolic dysfunction. The reversible, chemically specific nature of epigenetic modifications offers avenues for intervention, wherein targeted manipulation of chemical pathways restores normal chromatin chemistry. Understanding epigenetic chemistry requires integration of molecular reactivity, cofactor dynamics, enzyme catalysis, energy transduction, and structural biology, demonstrating the exquisite precision by which chemistry governs the functional genome. Through covalent modifications, non-covalent interactions, oxidative transformations, and metabolite-dependent regulation, the chemical framework of epigenetics provides a comprehensive mechanistic foundation for life, emphasizing the profound role of chemical principles in controlling gene expression, cellular identity, and biological plasticity.

The chemical architecture of epigenetics extends far beyond static modifications, revealing a dynamic interplay between metabolites, enzymes, and the genomic substrate. At the center of this system are methylation reactions, wherein methyl groups are transferred from donor molecules such as S-adenosylmethionine

(SAM) to DNA cytosines or to specific lysine and arginine residues on histone tails. These methyl transfers, governed by DNA methyltransferases (DNMTs) and histone methyltransferases (HMTs), exemplify classical nucleophilic substitution chemistry under enzymatic catalysis, where precise positioning of reactive groups and stabilization of transition states ensures fidelity in epigenetic programming. The reversibility of these reactions, mediated by demethylases such as TET proteins for DNA or lysine demethylases for histones, highlights a chemical logic analogous to redox cycles, in which electron transfer and cofactor chemistry modulate the reactivity of the methylated substrate. The interplay between methyl donor availability, cofactor redox potential, and enzyme kinetics forms a chemical nexus linking metabolism to chromatin regulation, emphasizing that epigenetic states are inextricably tied to cellular chemical milieu.

Beyond methylation, histone acetylation demonstrates a complementary chemical strategy for modulating chromatin accessibility. Acetyl groups derived from acetyl-CoA are transferred to ε-amino groups of lysine residues by histone acetyltransferases (HATs), neutralizing positive charges and weakening electrostatic interactions with the negatively charged DNA backbone. This seemingly simple nucleophilic acyl transfer reaction is chemically significant because it exemplifies how covalent modification can directly influence macromolecular assembly and dynamics. Deacetylation by histone deacetylases (HDACs) represents the reverse reaction, often requiring metal cofactors such as zinc or NAD^+, which participate in hydrolysis or redox-like chemical transformations. The chemical precision of these reactions underpins chromatin plasticity, allowing cells to rapidly switch transcriptional programs in response to metabolic cues or environmental signals. Importantly, the interconnection between acetyl-CoA availability, central carbon metabolism, and histone acetylation highlights a broader principle: metabolites are not merely intermediates in energy pathways, but active chemical participants in gene regulation.

Expanding the chemical landscape of epigenetics further, post-transcriptional RNA modifications—including N^6-methyladenosine (m6A), pseudouridylation, and 2′-O-methylation—represent a parallel layer of chemical information. Enzymes such as methyltransferases and pseudouridine synthases catalyze reactions that alter the electronic and steric properties of RNA bases, influencing secondary structure, stability, and recognition by RNA-binding proteins. These modifications involve nucleophilic attack, proton abstraction, and stabilization of high-energy intermediates, illustrating classical principles of chemical kinetics within a biological context. Moreover, the reversibility of certain RNA modifications, mediated by demethylases like FTO and ALKBH5, underscores an overarching chemical theme: biological systems employ finely tuned, reversible chemical transformations to encode, read, and erase information at multiple molecular levels. By considering RNA modifications as dynamic chemical signals, one can appreciate how cells integrate chemical logic to orchestrate temporal and spatial control of gene expression, complementing the chromatin-level regulation mediated by DNA and histone modifications.

A crucial dimension of epigenetic chemistry involves oxidative modifications, which serve both regulatory and stress-responsive roles. For instance, the TET family of dioxygenases catalyzes the oxidation of 5-methylcytosine to 5-hydroxymethylcytosine and further oxidized derivatives, employing Fe(II) and α-ketoglutarate cofactors to activate molecular oxygen and insert oxygen atoms into the substrate. These reactions embody sophisticated inorganic chemical principles, including ligand coordination, electron transfer, and radical stabilization. Oxidative modifications not only facilitate demethylation but also generate chemical intermediates that can recruit specific reader proteins or alter nucleosome dynamics. Similarly, reactive oxygen species (ROS) can induce chemical changes in nucleic acids and histones, highlighting the sensitivity of epigenetic chemistry to the cellular redox environment. Such interactions underscore the broader notion that chemical reactivity is not isolated within a single pathway but permeates regulatory networks, linking metabolism, signaling, and epigenetic plasticity.

Another layer of complexity in epigenetic chemistry arises from non-covalent interactions, including hydrogen bonding, π-π stacking, cation-π interactions, and van der Waals forces, which are critical for the recognition and interpretation of chemical marks. Reader proteins, such as bromodomains and chromodomains, discriminate among specific histone modifications through highly selective binding pockets, where subtle electronic differences and steric complementarity govern binding affinity. From a chemical perspective, these interactions exemplify principles of supramolecular chemistry, wherein multivalent interactions and cooperative binding govern molecular recognition. The specificity of these interactions is further modulated by local chromatin architecture and nucleosome positioning, demonstrating how chemical principles at the atomic and molecular scale dictate emergent biological behaviors at the cellular level.

Metabolite-dependent regulation adds yet another chemical dimension. Small molecules such as NAD^+, SAM, acetyl-CoA, α-ketoglutarate, and ATP act not only as cofactors but as chemical signals linking cellular nutritional and energetic status to epigenetic output. For instance, fluctuations in α-ketoglutarate levels directly influence TET-mediated DNA demethylation, while acetyl-CoA availability modulates histone acetylation. This integration of metabolic chemistry with chromatin chemistry illustrates the concept of chemical flux as an informational currency, where the quantitative and qualitative properties of small molecules convey regulatory instructions. Perturbations in these metabolite pools, as observed in cancer or metabolic disorders, can lead to widespread dysregulation of epigenetic chemistry, resulting in altered gene expression, disrupted cellular identity, and disease progression.

Finally, the chemical principles of epigenetics provide a mechanistic foundation for therapeutic intervention. Small molecule inhibitors of DNMTs, HDACs, HMTs, and histone demethylases exploit the inherent chemical specificity of these enzymes, competing with natural substrates or cofactors to modulate epigenetic marks. Rational design of such molecules requires an intimate understanding of transition states, active site geometry, electronic effects, and cofactor chemistry. Beyond inhibition, strategies for chemical epigenetic editing now harness programmable DNA- or RNA-targeted effectors fused to catalytic domains, enabling precise installation or removal of chemical modifications. These approaches exemplify the translation of fundamental chemical knowledge into applied molecular medicine, highlighting the centrality of chemistry in decoding, manipulating, and ultimately restoring the regulatory circuits that govern life.

In conclusion, the chemistry of epigenetics demonstrates an intricate, multilayered network of covalent and non-covalent transformations, metabolite-mediated regulation, and redox-sensitive modifications that collectively encode cellular memory and developmental instructions. By integrating concepts from nucleophilic substitution, acyl transfer, oxidation, supramolecular chemistry, and cofactor dynamics, researchers can unravel the chemical logic underlying gene expression, cellular differentiation, and disease etiology. The study of epigenetic chemistry not only illuminates fundamental principles of molecular biology but also provides a rational framework for the design of targeted therapeutics, emphasizing that life itself is orchestrated by a delicate balance of chemical forces finely tuned across space and time. Ultimately, appreciating epigenetic regulation as a chemical phenomenon enriches our understanding of biology, positioning chemistry at the very heart of cellular identity, adaptability, and the evolution of complex organisms.

Chapter 47: Word Problems

1. DNA methylation at cytosine residues is catalyzed by DNMTs using S-adenosylmethionine (SAM) as the methyl donor. Describe the stepwise nucleophilic mechanism of methyl transfer, including the formation of the covalent intermediate at the C6 position of cytosine, the electron-pushing mechanism, and how the enzyme stabilizes the transition state via hydrogen bonding and π-stacking interactions. Discuss how chromatin context or local DNA accessibility might influence the reaction kinetics.
2. TET enzymes catalyze the oxidation of 5-methylcytosine to 5-hydroxymethylcytosine through a Fe^{2+}/α-ketoglutarate-dependent dioxygenation mechanism. Explain the formation of the highly reactive Fe(IV)=O intermediate, the radical-mediated hydrogen abstraction from the methyl group, and recombination with hydroxyl radicals. Include how enzyme positioning and local electrostatics influence the specificity and efficiency of oxidation.
3. The sequential oxidation of 5-methylcytosine (5mC) by TET enzymes produces 5-hydroxymethylcytosine (5hmC), 5-formylcytosine (5fC), and 5-carboxylcytosine (5caC). Analyze in detail how each successive oxidation step modifies the electronic density on the pyrimidine ring, affects hydrogen-bonding capabilities with complementary bases, and alters the affinity of DNA-binding proteins such as methyl-CpG-binding domain proteins. Discuss the chemical rationale for why these oxidative intermediates are selectively recognized and excised during base excision repair, including how the changing electronic and steric properties of the cytosine derivatives influence glycosylase catalysis and transition state stabilization.
4. Lysine acetylation on histone tails, catalyzed by histone acetyltransferases (HATs), proceeds via nucleophilic attack of the ε-amino group of lysine on the carbonyl carbon of acetyl-CoA. Detail the formation of the tetrahedral intermediate,

the electron flow during nucleophilic acyl substitution, and the subsequent collapse of the intermediate to release CoA and form an acetylated lysine. Explain how the geometry of the enzyme active site, hydrogen bonding networks, and electrostatic interactions enhance reaction specificity and catalytic efficiency. Include discussion of stereoelectronic effects, transition state stabilization, and how modifications to surrounding residues could influence the rate and selectivity of acetyl transfer.

5. Histone deacetylases (HDACs) mediate lysine deacetylation via a Zn^{2+}-dependent mechanism, in which a metal-coordinated water molecule acts as a nucleophile attacking the carbonyl carbon of acetylated lysine. Describe the stepwise electron flow during nucleophilic attack, formation of the tetrahedral intermediate, and its collapse to yield the deacetylated lysine and free acetate. Emphasize how Zn^{2+} polarization, hydrogen bonding from active site residues, and precise substrate positioning stabilize negative charges and orient the substrate, providing chemical reasoning for high specificity and avoidance of off-target hydrolysis.

6. Lysine residues on histone tails can undergo mono-, di-, or tri-methylation by lysine methyltransferases (KMTs) using S-adenosylmethionine (SAM) as the methyl donor. Describe the S_N2-like nucleophilic attack on the methyl sulfonium of SAM, detailing how steric constraints, electronic effects, and active site architecture dictate the methylation state. Explain the chemical rationale for why transition state stabilization differs among mono-, di-, and tri-methylation, including how partial positive charge distribution, enzyme-substrate complementarity, and steric hindrance influence reaction kinetics and methyl group transfer efficiency.

7. FAD-dependent lysine demethylases remove methyl groups through oxidation via a Schiff base intermediate followed by hydride transfer to the flavin ring. Describe the chemical mechanism in detail, including formation of the iminium intermediate, electron flow during hydride transfer, and hydrolytic release of formaldehyde. Discuss how the enzyme ensures selectivity toward specific lysine residues, avoids non-specific oxidation, and stabilizes reactive intermediates, considering flavin redox potential, substrate orientation, and active site hydrogen bonding networks.

8. Jumonji C (JmjC) domain-containing demethylases utilize Fe^{2+}/α-ketoglutarate-dependent oxidation to hydroxylate methylated lysines. Outline the stepwise radical mechanism, including formation of the high-valent Fe(IV)=O species,

hydrogen atom abstraction from the methyl group, hydroxylation, and spontaneous formaldehyde release. Discuss the chemical rationale for cofactor dependency, how redox chemistry governs substrate selectivity, and how electron flow within the Fe-oxo complex is coupled to precise radical-mediated oxidation.

9. Arginine methylation by protein arginine methyltransferases (PRMTs) can result in mono- or symmetric/asymmetric dimethylation. Explain how guanidinium nucleophilicity, resonance delocalization, and steric constraints of the enzyme active site influence whether mono- or dimethylation occurs. Discuss the chemical consequences of methylation on hydrogen bonding, cation-π interactions, and the overall chromatin context, including how these electronic changes influence reader protein recognition and transcriptional regulation.

10. Phosphorylation introduces negatively charged phosphate groups to serine or threonine residues on histone tails. Analyze in chemical detail how this modification alters local electrostatics, hydrogen-bonding networks, and the reactivity of neighboring lysine residues toward acetylation or methylation. Include discussion of how the presence of a phosphate group can modulate nucleophilic attack, influence transition state energetics, and impact enzyme-substrate recognition in multisite histone modifications.

11. Covalent attachment of ubiquitin or SUMO to lysine residues imposes significant steric constraints on histone tails. Examine the chemical consequences of such post-translational modifications on neighboring enzymatic reactions, including steric hindrance, electronic redistribution of nearby residues, and allosteric modulation of histone-modifying enzymes. Discuss how this can influence the thermodynamics and kinetics of adjacent acylation or methylation events at a molecular level.

12. N^6-methyladenosine (m^6A) in RNA is installed via nucleophilic attack of the exocyclic amino group on SAM. Describe the S_N2-like mechanism in detail, including transition state stabilization by active site residues, stereoelectronic effects, and the influence of RNA structure on reaction kinetics. Discuss how methylation affects adenine hydrogen bonding, RNA secondary and tertiary structure, and recognition by m^6A reader proteins, connecting chemical modification to functional outcomes.

13. Demethylation of m^6A by FTO or ALKBH5 involves oxidative hydroxylation of the methyl group. Provide a detailed mechanistic explanation of formaldehyde release, including redox chemistry, cofactor activation, electron flow, and transition state stabilization. Discuss how enzyme active site geometry and electrostatics ensure selective demethylation, avoiding unintended modification of adjacent nucleotides.

14. ATP-dependent chromatin remodelers reposition nucleosomes via ATP hydrolysis. Explain in chemical detail how nucleophilic attack on the γ-phosphate is facilitated by Mg^{2+} coordination, the stepwise bond cleavage mechanism, and the chemical-to-mechanical energy transduction that drives nucleosome sliding. Include analysis of electron flow, phosphate transition state stabilization, and coupling of energy release to histone-DNA contact disruption.

15. SAM levels, regulated by one-carbon metabolism, influence DNA and histone methylation reactions. Discuss how fluctuations in SAM concentration affect the nucleophilicity of cytosine or lysine residues, enzyme kinetics, transition state stabilization, and overall methylation efficiency. Include a chemical rationale connecting metabolic state, methyl donor availability, and epigenetic modification dynamics.

16. Acetyl-CoA availability directly modulates histone acetylation reactions. Analyze how changes in acetyl-CoA concentration influence the thermodynamics and kinetics of lysine acylation, including nucleophile orientation, tetrahedral intermediate stability, active site interactions, and chemical driving forces for acetyl transfer. Discuss how metabolic flux impacts epigenetic regulation at the chemical level.

17. NAD^+ availability regulates sirtuin-mediated histone deacetylation. Explain the chemical mechanism involving formation of ADP-ribose intermediates, electron flow, bond cleavage, and hydrolysis, highlighting how NAD^+ concentration modulates catalytic efficiency, reaction reversibility, and transition state energetics. Include discussion of chemical coupling between metabolism and enzyme-mediated deacetylation.

18. α-Ketoglutarate and Fe^{2+} levels influence the activity of TET and Jumonji dioxygenases. Examine how cofactor concentration affects radical formation, oxidation kinetics, substrate positioning, and efficiency of DNA or histone

demethylation. Discuss the chemical basis of electron transfer, coordination chemistry, and radical-mediated reaction control.

19. Post-translational modifications modulate phase separation of chromatin-associated proteins. Discuss how acetylation, methylation, and phosphorylation chemically alter charge distribution, hydrogen bonding, and hydrophobic interactions, and how these changes drive mesoscale reorganization and compartmentalization within the nucleus. Include mechanistic reasoning at the molecular level for modification-induced phase separation.
20. Oxidative DNA damage, such as 8-oxoguanine formation, intersects with epigenetic regulation. Analyze how alterations in electronic density and hydrogen bonding patterns influence recognition by repair enzymes and chromatin reader proteins. Discuss chemical mechanisms by which oxidative lesions can impact transcriptional outcomes and enzymatic processing of neighboring epigenetic marks.
21. DNMT inhibitors such as 5-azacytidine trap DNA methyltransferases via covalent adduct formation. Explain the stepwise nucleophilic mechanism by which the enzyme attacks the C6 position of the cytosine analog, forms a covalent intermediate, and becomes irreversibly inhibited. Discuss the chemical rationale for how the modified nucleobase blocks methyl transfer, including electronic effects, transition state stabilization, and enzyme conformational consequences.
22. HDAC inhibitors chelate the active site Zn^{2+} ion, preventing hydrolysis of acetylated lysine. Analyze how geometric and electronic factors contribute to tight metal coordination, stabilization of the inhibitor-bound complex, and disruption of the catalytic mechanism. Include discussion of nucleophile orientation, charge polarization, and how metal-based catalysis is chemically impeded.
23. Bromodomain inhibitors mimic acetylated lysine side chains to competitively occupy reader domains. Explain the chemical principles underlying this binding, including hydrogen bonding, π-stacking interactions, steric complementarity, and transition state mimicry. Discuss how these interactions interfere with enzyme-substrate recognition within chromatin at a molecular level.
24. Histone modification cross-talk can influence enzymatic activity. Propose a chemical mechanism by which phosphorylation of a serine residue could enhance

or inhibit acetylation or methylation of nearby lysines. Include discussion of electrostatics, steric effects, local hydrogen bonding networks, and how these changes modulate transition state stabilization for adjacent reactions.

25. Histone ubiquitination impacts methylation on adjacent lysine residues. Explain the chemical basis for steric hindrance, electronic redistribution, and transition state destabilization imposed by the ubiquitin moiety. Discuss mechanistically how this affects nucleophilic attack, intermediate stabilization, and methyltransferase catalysis.

26. DNA cytosine methylation occurs within nucleosome-compacted chromatin. Discuss how chromatin accessibility, local electrostatic environment, and DNA-histone interactions influence nucleophilic attack, methyl transfer efficiency, and transition state stabilization. Include chemical reasoning for how nucleosome positioning and histone tail dynamics modulate methylation kinetics.

27. Aromatic residues in DNMT active sites contribute to π-stacking interactions that stabilize the transition state. Explain how these interactions orient the nucleophilic cytosine, enhance electrophilicity at the methyl donor, and facilitate efficient methyl transfer. Discuss the interplay between sterics, electronics, and enzyme-substrate complementarity.

28. The oxidation state of Fe cofactors in TET enzymes modulates DNA demethylation. Analyze how electron density changes, radical stability, and O_2 activation are chemically linked to the reaction efficiency. Include discussion of Fe(II)/Fe(IV) redox cycling, electron flow, and how cofactor state determines substrate oxidation selectivity.

29. Compare the electronic properties and reactivity of 5-methylcytosine, 5-hydroxymethylcytosine, 5-formylcytosine, and 5-carboxylcytosine. Discuss how changes in electron density, hydrogen bonding potential, and polarity affect recognition by reader proteins and subsequent enzymatic processing. Include a mechanistic rationale for selectivity in DNA demethylation pathways.

30. The tetrahedral intermediate in HAT-catalyzed acetylation is stabilized by active site residues. Discuss stereoelectronic effects, nucleophile orientation, hydrogen bonding, and charge stabilization that ensure efficient acetyl transfer. Include consideration of how minor changes in active site geometry could lead to side reactions or decreased catalytic efficiency.

31. In HDAC catalysis, Zn^{2+} polarizes a coordinated water molecule for nucleophilic attack. Analyze the electronic redistribution during formation of the oxyanion intermediate, the role of active site residues in stabilizing negative charge, and how precise geometry enables selective deacetylation while minimizing hydrolysis of non-target residues.

32. Lysine methylation by KMTs proceeds through an S_N2 mechanism. Explain how the positive charge on the SAM sulfonium enhances electrophilicity, stabilizes the transition state, and how active site constraints dictate the degree of methylation. Discuss chemical principles governing reaction kinetics and specificity for mono-, di-, or tri-methylation.

33. FAD-dependent lysine demethylation involves hydride transfer to the flavin ring. Discuss why the flavin is an effective hydride acceptor, how electron flow regenerates unmodified lysine, and the chemical rationale for selective oxidation of the target methyl group without affecting other residues.

34. Arginine methylation introduces symmetry considerations. Explain how resonance delocalization of the guanidinium affects nucleophilicity, how steric factors within the enzyme active site influence substrate orientation, and the chemical basis for producing symmetric versus asymmetric dimethylarginine.

35. Phosphorylation of histone residues can allosterically regulate nearby methyltransferases. Discuss chemical principles of how negative charges alter local electrostatics, hydrogen bonding networks, and transition state stabilization, ultimately influencing enzyme activity. Include considerations of cooperative or inhibitory effects on adjacent catalytic sites.

36. RNA methylation by METTL3/14 modifies hydrogen bonding at the N6 position of adenine. Explain how this chemical change affects RNA secondary structure, folding energetics, base-pairing interactions, and recognition by RNA-binding proteins. Include stereoelectronic reasoning for how methylation influences reaction kinetics and molecular recognition.

37. FTO-mediated m^6A demethylation involves oxidative cleavage to release formaldehyde. Provide a detailed chemical mechanism, highlighting electron flow, redox transformations, cofactor involvement, and transition state stabilization. Discuss how enzymatic selectivity is achieved despite multiple potential oxidation sites.

38. ATP hydrolysis in chromatin remodelers releases chemical energy. Explain in chemical detail how γ-phosphate bond cleavage is coupled to nucleosome sliding, including Mg^{2+} coordination, electron flow, and the conversion of chemical energy into mechanical work at histone-DNA interfaces.

39. SAM and acetyl-CoA levels connect metabolism to epigenetic chemistry. Discuss how variations in metabolite concentrations influence nucleophilic substitution and acyl transfer reactions, affecting transition state stabilization, reaction kinetics, and enzymatic efficiency. Include chemical rationale for metabolic regulation of epigenetic marks.

40. Post-translational modifications influence phase separation of chromatin-associated proteins. Analyze how acetylation, methylation, and phosphorylation chemically alter local charge, hydrogen bonding, and hydrophobicity. Discuss mechanistic principles by which these changes promote or prevent condensate formation at a molecular and mesoscale level.

41. Reactive oxygen species modify nucleic acids and histones. Discuss how oxidative chemical changes affect electron density, hydrogen bonding, and molecular recognition by reader proteins. Include mechanistic analysis of how these modifications influence downstream enzymatic processing and transcriptional regulation.

42. Rational drug design for epigenetic enzymes often relies on transition state mimicry. Explain the chemical principles behind designing inhibitors that resemble high-energy intermediates, including stereoelectronic complementarity, charge stabilization, and hydrogen bonding, and discuss how these factors translate to potent and selective inhibition.

43. Cross-talk between acetylation and methylation can influence nucleosome dynamics. Discuss the chemical mechanisms by which a modification at one site affects nucleophilic accessibility, enzyme recognition, and modification efficiency at adjacent residues, including stereoelectronic effects and transition state perturbation.

44. RNA pseudouridylation converts uridine to pseudouridine via nucleophilic rearrangement. Describe the detailed chemical mechanism, including intermediate stabilization, electron flow, and energetic considerations that facilitate this post-

transcriptional modification. Discuss how the chemical changes influence RNA folding and protein interactions.

45. Oxidative DNA modifications are recognized differently by repair enzymes. Explain the chemical basis for differential recognition of 5-hydroxymethylcytosine versus 5-methylcytosine, including changes in electron density, hydrogen bonding, and polarity. Discuss how these properties affect enzymatic excision and subsequent epigenetic regulation.

46. Supramolecular interactions govern histone mark recognition. Discuss how π-π stacking, hydrogen bonding, and cation-π interactions collectively determine reader domain specificity. Include mechanistic reasoning for how local chromatin structure and residue environment influence these chemical interactions.

47. NAD^+-dependent sirtuin deacetylation involves ADP-ribose intermediates. Provide a detailed electron-flow mechanism, including nucleophilic attack, bond cleavage, and hydrolysis. Explain the chemical logic behind cofactor-mediated catalysis, including transition state stabilization and selectivity toward acetylated lysines.

48. Chemical modifications of chromatin influence nucleosome dynamics. Analyze how acetylation, methylation, and phosphorylation collectively alter DNA-histone electrostatics, hydrogen bonding networks, and nucleosome stability. Include mechanistic reasoning for how multiple covalent changes synergize or antagonize structural effects.

49. α-Ketoglutarate and Fe^{2+}-dependent dioxygenases act as chemical sensors of cellular metabolic state. Discuss how cofactor availability regulates radical formation, oxidation reactions, and electron flow. Explain the chemical modulation of epigenetic marks in response to metabolic cues, highlighting transition state energetics and substrate selectivity.

50. Integrating multiple layers of chemical epigenetics, propose a mechanistic model explaining how covalent modifications, non-covalent interactions, cofactor levels, and oxidative chemistry collectively encode cellular memory. Include detailed discussion of electron flow, transition state stabilization, stereoelectronic effects, and chemical energetics in shaping chromatin state and transcriptional output.

Chapter 48

Synthetic Chemical Ecosystems

The concept of synthetic chemical ecosystems represents an emerging frontier in chemical research, wherein chemists aim to emulate the dynamic behaviors, interactions, and adaptive capacities of natural ecosystems through controlled chemical networks. Unlike traditional chemical systems, which are designed to achieve a specific reaction outcome or product, synthetic chemical ecosystems are engineered to maintain dynamic equilibrium, support self-regulation, and exhibit emergent behaviors reminiscent of ecological communities. At the core of these systems is the principle of chemical communication, wherein molecules interact not only through conventional reaction pathways but also through feedback loops, cross-catalysis, and competitive interactions, allowing the system to respond to environmental perturbations in a manner analogous to living systems. Central to the design of such ecosystems is the careful selection of reactive components, kinetic parameters, and environmental constraints, which collectively define the system's stability, resilience, and capacity for adaptation.

Synthetic chemical ecosystems are frequently constructed using networks of autocatalytic and cross-catalytic reactions, which provide the foundation for self-amplifying behaviors and the emergence of hierarchical organization. Autocatalytic motifs, such as the formose reaction and peptide-based self-replicating networks, illustrate how simple molecules can give rise to complex patterns of growth and decay, including oscillatory and bistable dynamics. Cross-catalytic networks, in which two or more species mutually accelerate each other's formation, introduce interdependence akin to symbiotic interactions in biological ecosystems. These networks can be further tuned by incorporating inhibitory feedback loops, resource limitations, or compartmentalization strategies, which allow for the stabilization of species distributions and the mitigation of runaway reaction cascades. Computational modeling and kinetic simulations play a pivotal role in designing these systems, enabling the prediction of steady states, oscillatory regimes, and chaotic behaviors that may arise under varying conditions of concentration, temperature, and solvent environment.

The notion of resource limitation in synthetic chemical ecosystems draws inspiration from ecological nutrient cycles, wherein the availability of essential substrates governs population dynamics and interspecies interactions. In chemical analogs, reactant depletion and byproduct accumulation serve as natural constraints that modulate reaction fluxes and drive system-level adaptations. For instance, the controlled introduction of a limiting cofactor or redox mediator can selectively enhance the propagation of certain chemical species while suppressing others, creating a chemical analogue of competitive exclusion or niche differentiation. Similarly, the temporal modulation of resource input, analogous to seasonal variations in natural ecosystems, can induce periodic shifts in chemical composition and promote diversity within the network. These strategies facilitate the emergence of adaptive behaviors, including self-maintenance, recovery from perturbation, and the stabilization of cooperative reaction motifs.

A key structural element in synthetic chemical ecosystems is compartmentalization, which allows for spatial segregation and localized interactions. Lipid vesicles, microfluidic droplets, and polymer-based protocells serve as discrete reaction domains, providing both confinement and selective permeability. Compartmentalization introduces gradients of concentration and energy, enabling differential reaction kinetics and the emergence of spatially organized patterns. For example, chemical waves, Turing patterns, and oscillatory fronts can spontaneously arise in compartmentalized systems, reflecting a rich interplay between reaction kinetics and diffusive transport. Moreover, compartmentalization allows for the co-existence of otherwise incompatible reactions by physically isolating reactive intermediates, thereby expanding the repertoire of achievable behaviors and facilitating the emergence of higher-order chemical organization.

Energy transduction is another critical component of synthetic chemical ecosystems, as it underpins the capacity for sustained dynamics and adaptive responses. Energy sources can be chemical, such as activated monomers, redox couples, or high-energy cofactors, or physical, such as light, heat, or electrical fields. The coupling of energy input to specific reaction pathways establishes a form of chemical metabolism, wherein energy fluxes sustain non-equilibrium states and drive the propagation of selected species. Photochemically driven oscillatory reactions, fuel-dependent autocatalytic cycles, and dissipative self-assembly exemplify mechanisms through which energy input can maintain system dynamics far from equilibrium. Dissipative processes, in particular, highlight the intimate connection between energy dissipation and emergent chemical order, revealing parallels to the thermodynamic principles that govern living ecosystems.

Chemical communication within these ecosystems extends beyond direct reaction pathways to include signaling motifs reminiscent of quorum sensing and hormonal regulation in biology. Small-molecule mediators, redox-active species, or pH-sensitive intermediates can serve as messengers that convey information across compartments or between reaction modules. These signaling molecules modulate reaction rates, trigger feedback loops, and orchestrate synchronized behaviors, enabling the system to implement conditional responses to environmental or internal changes. The design of robust signaling networks often incorporates redundancy, hierarchical regulation, and competitive inhibition, ensuring that communication remains effective under fluctuating conditions. Such principles allow synthetic chemical ecosystems to display emergent phenomena such as collective oscillations, population cycling, and adaptive switching between metabolic modes.

A central challenge in constructing synthetic chemical ecosystems is the management of complexity, as the number of interacting species and reactions can rapidly escalate. Approaches to mitigate this include modular design, in which discrete reaction motifs are developed independently and then coupled, and hierarchical control, wherein higher-order regulatory elements govern the behavior of lower-level reaction modules. This mirrors strategies observed in natural ecosystems, where trophic hierarchies and keystone species stabilize the broader network. Computational and experimental integration is essential to navigate this complexity, enabling systematic exploration of parameter spaces, identification of robust motifs, and iterative refinement of network topology. Network analysis tools, including graph theory, flux balance analysis, and dynamical systems modeling, are increasingly employed to quantify connectivity, resilience, and potential emergent behaviors.

Emergent behaviors in synthetic chemical ecosystems manifest in multiple forms, ranging from oscillatory concentration dynamics to adaptive reconfiguration of network topology. Oscillatory reactions, such as the Belousov–Zhabotinsky reaction and glycolytic analogs, provide canonical examples of temporally structured chemical behavior arising from non-linear feedback. Bistability and multistability enable chemical networks to exhibit memory-like functions, retaining information about prior states or environmental conditions. More complex phenomena, including chemical phase separation, self-sorting, and cooperative catalysis, can give rise to spatial patterning and selective amplification of specific species, analogous to population structuring and niche specialization in ecological systems. These behaviors underscore the capacity of synthetic chemical ecosystems to act as information-processing platforms, capable of sensing, responding, and adapting to perturbations.

The integration of evolution-like processes into synthetic chemical ecosystems represents a frontier in the field, blurring the line between chemistry and protocellular life. By implementing cycles of replication, selection, and mutation, chemical species can undergo directed diversification, leading to the preferential amplification of functional motifs. Laboratory demonstrations of such chemical evolution have employed template-directed polymerization, self-replicating peptides, and network-mediated autocatalytic selection. When combined with compartmentalization and resource limitation, these systems can undergo adaptive dynamics, wherein the chemical composition evolves to optimize survival or replication efficiency. Such approaches open avenues for exploring prebiotic chemistry, the origins of life, and the design of self-optimizing chemical technologies.

Synthetic chemical ecosystems also provide a platform for exploring principles of ecological network theory in a chemical context. Concepts such as keystone species, resilience, redundancy, and modularity can be translated into reaction network design, guiding the selection of catalysts, intermediates, and feedback loops. For instance, the identification of keystone reactions—those whose removal disproportionately destabilizes the network—enables targeted interventions to modulate system behavior. Redundant pathways can confer robustness, allowing the system to maintain function under perturbation or stochastic fluctuations. Modularity facilitates the hierarchical assembly of complex networks, permitting incremental expansion and integration of novel reaction motifs without compromising overall stability.

Experimental realization of synthetic chemical ecosystems requires careful control of environmental parameters, including temperature, solvent composition, pH, and ionic strength. Microfluidic platforms and droplet-based reactors enable precise spatiotemporal control, allowing systematic investigation of emergent behaviors under defined constraints. Analytical techniques, including mass spectrometry, NMR spectroscopy, fluorescence imaging, and microfluidic monitoring, provide quantitative insight into species distributions, reaction fluxes, and temporal dynamics. These experimental frameworks facilitate the iterative design and refinement of chemical ecosystems, bridging theoretical predictions with empirical validation.

Beyond fundamental scientific inquiry, synthetic chemical ecosystems hold promise for technological applications. Potential avenues include adaptive chemical sensing, self-regulating material synthesis, programmable catalysis, and energy conversion systems. Chemical networks capable of environmental sensing and responsive adaptation could serve as intelligent materials, altering their structure or function in response to stimuli. Similarly, fuel-driven autocatalytic systems may be harnessed for energy storage or release, exploiting non-equilibrium chemical dynamics to mediate controlled energy transduction. The exploration of chemical computation, whereby network topology and kinetics encode information processing, further expands the potential of these systems to operate as chemical analogues of neural or logical networks.

Theoretical approaches continue to evolve alongside experimental advances, with a growing emphasis on predictive modeling, design automation, and machine learning integration. Machine learning algorithms can assist in identifying viable network motifs, optimizing kinetic parameters, and predicting emergent behaviors across high-dimensional chemical spaces. Coupled with high-throughput experimentation, these approaches enable accelerated exploration of chemical ecosystem design principles, reducing the reliance on trial-and-error methods. Furthermore, the application of information theory and network science provides a quantitative framework for characterizing complexity, redundancy, and adaptability within chemical systems, facilitating rigorous comparison with biological ecosystems.

The field of synthetic chemical ecosystems also intersects with prebiotic chemistry and the origins-of-life research. By emulating ecological interactions, compartmentalized energy fluxes, and resource-dependent adaptation, these systems provide models for understanding how life-like behavior can emerge from purely chemical substrates. The study of chemical ecosystem dynamics sheds light on the minimal requirements for self-maintenance, reproduction, and evolution, offering experimental platforms for testing hypotheses about the emergence of metabolic networks and early replicators. In this context, synthetic chemical ecosystems serve both as tools for fundamental exploration and as testbeds for engineering novel chemical behaviors.

Challenges remain in scaling synthetic chemical ecosystems to higher levels of complexity while maintaining controllability and reproducibility. Stochastic fluctuations, reaction crosstalk, and environmental noise pose significant hurdles, necessitating robust network design, feedback regulation, and compartmentalization strategies. Advances in microfabrication, high-resolution monitoring, and automated reaction control are essential to overcome these barriers, enabling precise orchestration of chemical interactions across multiple spatial and temporal scales. The integration of evolutionary dynamics, energy transduction, and signaling networks represents a promising pathway toward the realization of fully adaptive, self-organizing chemical systems.

Future directions in the field include the incorporation of hybrid materials, such as enzyme-mimetic catalysts, stimuli-responsive polymers, and nanostructured scaffolds, to expand the functional repertoire of chemical ecosystems. Coupling synthetic networks with bio-inspired modules, including membranous compartments, light-harvesting assemblies, or molecular machines, could enable sophisticated behaviors such as self-repair, signal amplification, and dynamic restructuring. The development of predictive design principles, informed by a combination of chemical kinetics, thermodynamics, and ecological theory, will be critical for translating these concepts into practical applications. Ultimately, synthetic chemical ecosystems may serve as a bridge between chemistry and biology, providing a framework for engineering adaptive, information-processing, and energy-utilizing chemical systems that mirror the complexity and resilience of living ecosystems.

In conclusion, synthetic chemical ecosystems represent a paradigm shift in chemistry, moving beyond static reaction schemes toward dynamic, adaptive, and self-organizing networks. By integrating principles of autocatalysis, cross-catalysis, compartmentalization, energy transduction, signaling, and evolutionary selection, chemists are beginning to emulate ecological behaviors in purely chemical contexts. These systems offer unprecedented opportunities to study emergent phenomena, explore prebiotic chemistry, and engineer novel adaptive materials and technologies. As experimental, computational, and theoretical tools continue to advance, synthetic chemical ecosystems are poised to become a foundational component of next-generation chemical science, providing insights into the principles of life-like behavior and the design of functional, self-organizing chemical networks.

An emerging area within synthetic chemical ecosystems is the incorporation of information processing capabilities, where chemical networks are designed to sense, integrate, and respond to multiple inputs in a programmable manner. By leveraging reaction kinetics and feedback control, chemical systems can implement logic operations, signal amplification, and decision-making processes analogous to biological regulatory networks. For instance, networks of bistable switches and oscillators can serve as chemical memory units or timers, storing information about previous environmental conditions and triggering context-dependent responses. The use of DNA-based strand displacement circuits, peptide assemblies, and redox-responsive chemical motifs exemplifies how molecular components can be engineered to perform computational tasks, bridging chemistry with principles of synthetic biology and molecular programming. These approaches enable the development of chemical systems that do not merely react, but "behave" in a context-sensitive manner, dynamically adjusting their function in response to environmental cues or internal states.

Another frontier is the integration of multi-scale interactions, where chemical ecosystems are designed to operate across molecular, mesoscopic, and macroscopic scales. At the molecular level, individual reaction networks establish local feedback and autocatalytic dynamics, while at larger scales, compartmentalized units or microreactor arrays can exchange chemical information via diffusion, signaling molecules, or active transport. This hierarchical organization allows for emergent collective behaviors, including spatial patterning, synchronization, and oscillatory waves that resemble community-level ecological dynamics. By tuning the connectivity and coupling strength between compartments, researchers can explore phenomena such as chemical "colonization," competitive exclusion, and cooperative catalysis, revealing principles that govern self-organization across scales. The integration of multi-scale dynamics also enables the investigation of robustness and adaptability under fluctuating environmental conditions, providing a chemically tractable model for studying resilience in complex networks.

Stochasticity and noise are additional critical factors in chemical ecosystem design. Unlike classical deterministic reaction systems, synthetic chemical ecosystems often operate under conditions where the probabilistic nature of molecular interactions significantly influences system behavior. Fluctuations in reactant concentrations, random reaction events, and diffusive transport can induce variability in network outcomes, leading to diversity in emergent patterns or bistable state selection. Researchers exploit these stochastic effects to mimic evolutionary processes, explore adaptive landscape dynamics, and enhance the functional diversity of chemical networks. For example, stochastic fluctuations can drive transitions between quasi-stable states, enable exploration of alternative chemical pathways, or facilitate the emergence of rare but functionally important species. Understanding and controlling noise is therefore essential for designing robust yet adaptable chemical ecosystems capable of sustained dynamic behavior.

Finally, the interface between synthetic chemical ecosystems and materials science presents exciting opportunities. Chemical networks embedded within soft matter matrices, polymer gels, or responsive hydrogels can couple reaction dynamics with mechanical or structural changes, yielding materials that self-reconfigure, self-heal, or adapt to environmental stimuli. For example, fuel-driven chemical oscillations can generate rhythmic mechanical motion or modulate material porosity, creating active materials with life-like behaviors. By combining chemical computation, energy transduction, and adaptive pattern formation, such systems pave the way for programmable matter and autonomous materials capable of interacting with their environment in complex ways. These innovations highlight the potential of synthetic chemical ecosystems not only as models for understanding emergent life-like phenomena, but also as platforms for engineering functional, adaptive, and intelligent chemical technologies that operate far beyond the constraints of classical static chemistry.

Chapter 48: Word Problems

1. Design a synthetic chemical ecosystem using interconnected autocatalytic and cross-catalytic reaction networks, and explain how you would systematically tune the kinetic parameters, environmental constraints, and feedback mechanisms to achieve a desired oscillatory behavior. Discuss how inhibitory feedback loops could prevent runaway reactions, and consider how stochastic fluctuations in reactant concentrations might influence the stability of the oscillatory regime.

2. Propose an experimental setup to investigate resource limitation in a chemical network, including the controlled depletion of a key cofactor or redox mediator. Analyze how such limitations could influence species diversity, competitive interactions, and emergent hierarchical organization, drawing analogies to niche differentiation and competitive exclusion in natural ecosystems.

3. Consider a compartmentalized chemical ecosystem confined within lipid vesicles or microfluidic droplets. Describe how gradients of concentration, selective

permeability, and differential diffusion could drive emergent spatial patterns, such as Turing structures or oscillatory fronts. Predict how altering inter-compartment communication or diffusion rates might modulate these patterns and affect the overall network dynamics.

4. A cross-catalytic network exhibits bistable dynamics under specific concentration and temperature regimes. Identify potential strategies to extend this network to multistable behavior, and discuss how such dynamics could be exploited to encode chemical memory. Include considerations of energy input, feedback regulation, and the role of autocatalytic motifs in sustaining long-term memory states.

5. Devise a chemical energy transduction pathway that maintains a non-equilibrium steady state in a synthetic ecosystem. Compare the relative effectiveness of chemical energy sources, such as activated monomers or high-energy cofactors, versus physical sources like light or electrical fields. Discuss the role of dissipative processes in establishing emergent chemical order and maintaining adaptive responses over time.

6. In a multi-species synthetic ecosystem, design a signaling network using redox-active mediators, pH-sensitive intermediates, or small-molecule messengers. Predict how introducing stochastic fluctuations in concentrations or reaction timing could affect the synchronization of collective oscillations and the robustness of conditional behaviors in response to environmental perturbations.

7. Formulate a computational model of a fuel-driven autocatalytic cycle subjected to periodic resource input. Determine how seasonal-like variations in resource availability would affect network composition, diversity, and emergent dynamics. Discuss the implications for self-maintenance, recovery from perturbations, and long-term adaptive chemical behaviors.

8. Design an experimental system that combines compartmentalization with cross-catalytic networks. Explain how microfluidic droplet arrays or polymer-based protocells could control interaction strength between compartments, and predict the types of emergent collective behaviors, such as synchronized oscillations, traveling chemical waves, or spatial patterning, that might arise under varying coupling conditions.

9. Propose a strategy for embedding an evolutionary selection process within a synthetic chemical ecosystem. Identify which autocatalytic or template-directed motifs could undergo replication, mutation, and selection, and explain how compartmentalization and resource limitation would influence the chemical fitness landscape, leading to preferential amplification of functional species.
10. Construct a computational framework to predict oscillatory, bistable, and chaotic behaviors in multi-species chemical ecosystems. Include considerations of concentration gradients, temperature fluctuations, solvent effects, and energy input, and discuss how inhibitory feedback loops or modular network design could stabilize the system and prevent collapse under non-equilibrium conditions.
11. Design a chemical network capable of exhibiting quorum-sensing-like behavior. Specify which small-molecule mediators could coordinate population-level adaptive responses, and outline how experimental readouts such as oscillation frequency, phase synchronization, or concentration thresholds could be used to quantify effective communication between reaction modules.
12. Develop a modular synthetic chemical ecosystem composed of independent reaction motifs. Explain how hierarchical control elements could regulate the behavior of lower-level modules, and discuss the benefits of modularity for maintaining robustness, allowing incremental network expansion, and facilitating emergent behaviors despite environmental perturbations or stochastic fluctuations.
13. Formulate an experimental protocol to investigate bistable switches in autocatalytic networks. Describe how you would demonstrate that the network retains information about prior environmental conditions, and identify critical parameters, including reaction rates, energy input, and feedback inhibition, that govern the reliability and repeatability of memory-like switching behavior.
14. Propose a chemical computation platform that integrates bistable switches, oscillators, and feedback loops to perform logic operations and memory storage. Using DNA-based strand displacement or peptide assembly motifs, explain how reaction kinetics, signal propagation, and redundancy could be engineered to achieve context-dependent computation within a dynamic chemical environment.
15. Design a cross-catalytic network that emulates symbiotic interactions among multiple chemical species. Predict how inhibitory feedback, competitive interactions,

and resource limitations could shape the long-term stability, species distributions, and emergent cooperative dynamics, and discuss how compartmentalization might enhance coexistence of otherwise incompatible reactions.

16. Formulate a strategy to integrate chemical energy fluxes with compartmentalized networks. Describe how energy gradients could drive spatial pattern formation, differential reaction kinetics, and localized oscillations, and specify which analytical or imaging techniques would allow real-time observation of energy-dependent behaviors across multiple compartments.

17. Develop a synthetic ecosystem to explore chemical evolution. Explain how iterative cycles of replication, mutation, and selection could produce functional diversity, and discuss how resource limitation, compartmentalization, and feedback regulation might guide the adaptive dynamics of chemical composition toward optimized survival or replication efficiency.

18. Construct a microfluidic experimental platform to study multi-scale interactions in chemical ecosystems. Explain how molecular-level autocatalysis within individual compartments and coupling between mesoscopic droplet arrays could produce emergent collective behaviors, such as synchronized chemical waves, spatial patterning, or competitive colonization dynamics.

19. Design a stochastic chemical ecosystem that exploits noise to explore quasi-stable states and generate functional diversity. Explain how probabilistic reaction events, diffusion fluctuations, and rare species emergence could contribute to adaptive behaviors, and discuss strategies to control stochasticity while maintaining robustness in overall network dynamics.

20. Create a chemical network that demonstrates dissipative self-assembly far from equilibrium. Describe how energy input, feedback loops, and autocatalytic motifs could be tuned to maintain ordered structures, and discuss metrics, such as oscillation amplitude or spatial correlation length, for quantifying emergent chemical order.

21. Develop a synthetic chemical ecosystem capable of adapting to environmental perturbations. Identify key signaling motifs, energy transduction pathways, and compartmentalization strategies that would enable the network to sense, respond, and recover from transient changes while preserving long-term stability and functional diversity.

22. Propose a chemical network that integrates energy transduction, feedback control, and compartmentalized signaling to sustain oscillatory dynamics. Discuss how energy-limited versus feedback-limited regimes could be distinguished experimentally, and explain how parameter tuning could optimize network resilience under fluctuating environmental conditions.

23. Construct a model to quantify redundancy, robustness, and keystone reactions in a cross-catalytic network. Predict the consequences of removing or inhibiting critical reactions, and discuss strategies, including modular design and compensatory feedback, to preserve overall network functionality and prevent collapse.

24. Design an experiment to induce Turing patterns in a compartmentalized chemical ecosystem. Identify critical reaction and diffusion parameters, including autocatalytic and inhibitory rates, and explain how compartmental boundaries or spatial heterogeneity might enhance or suppress the emergence of these patterns.

25. Propose a chemical computation platform using coupled bistable and oscillatory motifs. Explain how information about prior chemical states could be stored and retrieved, and discuss how reaction kinetics, compartmentalization, and feedback loops could be integrated to implement timing, logic, and memory functions in a dynamic chemical network.

26. Design a fuel-driven chemical ecosystem that exhibits periodic concentration oscillations. Analyze how temporal modulation of resource input would influence species diversity, network resilience, and emergent population dynamics, and draw analogies to seasonal environmental fluctuations in natural ecosystems.

27. Construct a hierarchical chemical network composed of multiple interdependent feedback loops. Discuss how modularity, redundancy, and hierarchical control could influence emergent behaviors, such as population cycling, adaptive switching, and cooperative catalysis, under varying energy and resource constraints.

28. Propose an experiment to couple chemical reaction networks with responsive hydrogel or polymer matrices. Explain how reaction-driven structural changes could produce self-reconfiguring or self-healing materials, and discuss methods to monitor mechanical or structural responses as a function of chemical oscillations or energy dissipation.

29. Develop a computational model to explore chaotic dynamics in cross-catalytic networks. Identify parameters most sensitive to perturbations, such as reaction rates or energy fluxes, and discuss how stochastic fluctuations could influence the emergence of alternative chemical pathways or rare but functionally significant species.

30. Formulate a chemical ecosystem that mimics trophic hierarchies and interspecies dependence. Explain how cross-catalysis, inhibitory feedback, and resource limitations contribute to long-term network stability, and propose experimental strategies for testing these interactions in a controlled compartmentalized system.

31. Design a synthetic chemical ecosystem capable of self-organization across multiple spatial scales. Describe how molecular-level autocatalysis and mesoscopic compartment coupling could generate emergent oscillatory waves, spatial patterning, or collective chemical behaviors, and discuss how tunable inter-compartment communication could influence system-level dynamics.

32. Construct a chemical network to study bistability, multistability, and conditional responses in autocatalytic motifs. Explain how energy input, feedback inhibition, and compartmentalization might be leveraged to implement context-sensitive adaptive responses to environmental fluctuations.

33. Propose a strategy to integrate chemical information processing with energy transduction. Explain how oscillatory motifs, bistable switches, and feedback loops could be coupled to produce adaptive decision-making, timing, or logic operations within a dynamically maintained chemical ecosystem.

34. Design a compartmentalized chemical ecosystem in which stochastic fluctuations drive exploration of alternative chemical pathways. Explain how such probabilistic events could generate rare but functionally important species, and propose experimental methods to monitor diversity, emergent patterns, and network resilience.

35. Develop a chemical ecosystem capable of adaptive switching between metabolic modes. Identify key reaction motifs, energy transduction pathways, and feedback loops, and discuss how compartmentalization, signaling, and resource modulation would enable context-dependent network responses.

36. Formulate an experiment to study prebiotic chemical evolution using synthetic ecosystems. Explain how autocatalysis, compartmentalization, and resource limitation could favor the emergence of self-replicating chemical species and functional motifs that resemble early metabolic networks.

37. Design a chemical network with integrated signaling, energy input, and compartmentalization to achieve synchronized oscillations. Propose quantitative metrics, such as phase coherence or amplitude correlation, to evaluate synchronization, and discuss how network parameters would influence collective behavior.

38. Construct a computational model to investigate stochasticity in chemical information processing. Explain how noise could facilitate transitions between functional states, explore alternative pathways, or enhance adaptability, and discuss design strategies to balance robustness with functional diversity.

39. Propose a synthetic ecosystem for programmable catalysis. Discuss how reaction network topology, kinetic parameter tuning, and feedback regulation could be used to modulate reaction rates, selectively amplify species, and adaptively respond to environmental stimuli.

40. Design an experiment to couple chemical oscillators with adaptive materials. Explain how fuel-driven reactions could modulate mechanical properties, porosity, or morphology in polymer matrices or hydrogels, producing self-organizing, life-like material responses.

41. Develop a computational framework to explore emergent properties in large-scale chemical networks. Discuss how graph theory, flux balance analysis, and dynamical systems modeling could be combined to quantify robustness, oscillatory behavior, multistability, and network adaptability under complex conditions.

42. Construct a chemical ecosystem to investigate cooperative catalysis, self-sorting, and spatial patterning. Explain how compartmentalization, feedback loops, and energy input could enhance selective amplification of specific species and facilitate emergent hierarchical organization.

43. Propose a chemical information-processing network using redox-responsive, pH-sensitive, or small-molecule motifs. Explain how the network could implement

decision-making, conditional responses, or environmental sensing, and discuss strategies to ensure robustness under fluctuating chemical conditions.

44. Design a synthetic chemical ecosystem to emulate ecological succession. Describe how initial network composition, autocatalytic dynamics, resource limitation, and feedback loops could influence temporal progression, species turnover, and emergent adaptive patterns over multiple "generations" of the chemical network.

45. Develop an experiment to explore dissipative phase separation in a synthetic chemical ecosystem. Identify which parameters, such as reaction rates, energy input, and compartmental boundaries, control pattern emergence, and discuss how compartmentalization could stabilize otherwise transient structures.

46. Construct a chemical ecosystem to study the interplay of stochasticity, feedback, and energy flux. Predict how these factors collectively influence emergent oscillatory or chaotic behaviors, population cycling, and adaptive responses, and discuss strategies to enhance robustness in such complex networks.

47. Propose a chemical network capable of self-optimization through iterative adaptation and selection. Explain how cycles of mutation, template-directed replication, and selection could lead to enhanced functional efficiency, and discuss experimental approaches to monitor network evolution over time.

48. Design a multi-scale synthetic ecosystem that integrates molecular-level networks with mesoscopic compartment arrays. Explain how tuning inter-compartment communication, coupling strength, and diffusion rates could allow the exploration of cooperative versus competitive interactions across spatial scales.

49. Formulate a strategy to embed memory and timing functions in a chemical network. Describe how bistable switches, oscillators, and hierarchical feedback loops could be combined to encode environmental history and implement context-sensitive adaptive responses.

50. Propose an adaptive chemical ecosystem that interfaces with responsive materials to produce autonomous behaviors. Explain how integrating energy transduction, signaling, compartmentalization, and material reconfiguration could create self-regulating chemical networks capable of sensing, processing information, and dynamically adapting to environmental stimuli.

Chapter 49

Acoustic and Mechanical Chemistry

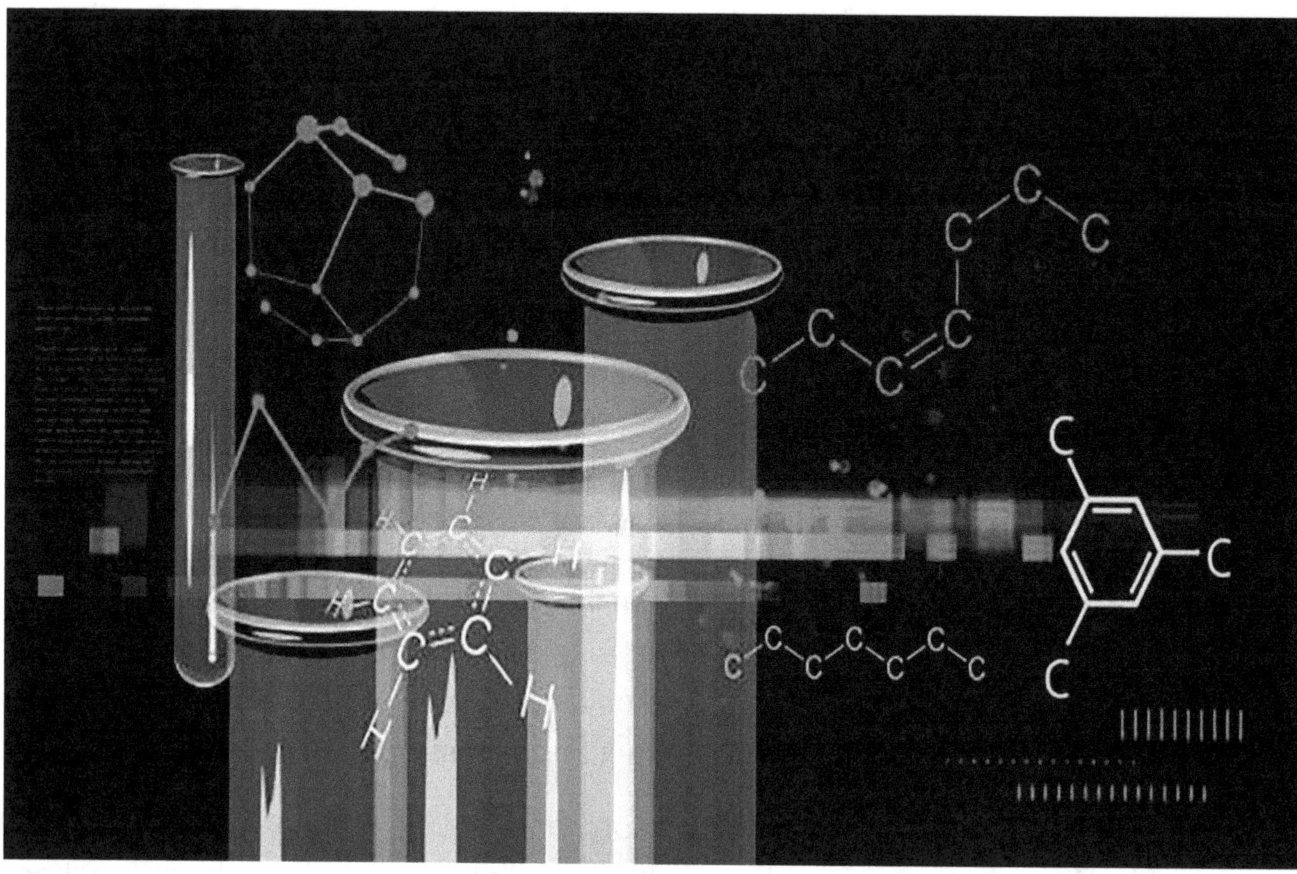

The practical applications of acoustic and mechanical chemistry are both diverse and transformative, spanning the synthesis of complex materials, catalysis, pharmaceuticals, and environmentally sustainable processes. In materials chemistry, ultrasonically assisted reactions have been extensively employed to produce nanoparticles, nanocomposites, and hierarchical structures with controlled morphology and enhanced surface properties. The ability of acoustic energy to induce cavitation and localized high-temperature, high-pressure events facilitates rapid nucleation and growth, allowing for uniform particle size distributions and tailored crystallinity. For example, the synthesis of metal oxide nanoparticles under sonochemical conditions often results in smaller particle sizes and higher surface areas compared to conventional hydrothermal or sol-gel methods. This is particularly significant in catalysis, where surface area and particle morphology directly influence catalytic efficiency. In mechanochemistry, the application of mechanical force to solid-state reactions has enabled the solvent-free synthesis of complex organic and inorganic compounds, reducing environmental impact while providing access to reaction pathways that are otherwise kinetically inaccessible. Mechanochemical methods have been successfully applied in the synthesis of pharmaceuticals, where ball milling can facilitate cocrystallization, polymorph control, and even the formation of covalent bonds under ambient conditions. The convergence of ultrasonic and mechanical approaches, often termed sonomechanochemistry, allows chemists to exploit both energy modes simultaneously, further expanding the scope of achievable reactions and products.

The use of acoustic and mechanical chemistry in catalysis illustrates the profound influence of physical energy on chemical reactivity. In homogeneous catalysis, ultrasonic irradiation can enhance mass transfer, accelerate ligand exchange, and increase the turnover frequency of transition metal catalysts. For heterogeneous systems, cavitation-induced microjets and shockwaves improve the contact between reactants and catalytic surfaces, reducing diffusion limitations and enhancing reaction rates. Mechanochemical activation in heterogeneous catalysis, such as the grinding of solid reactants with a catalyst, can generate fresh active sites, modify surface defects, and even induce phase transitions that alter catalytic behavior. These phenomena highlight the ability of physical energy to modulate both chemical and structural parameters of catalytic systems, demonstrating that reaction efficiency can be controlled not only through chemical composition but also through the intelligent application of acoustic and mechanical stimuli.

In pharmaceutical chemistry, mechanochemical techniques have revolutionized drug formulation and synthesis by providing solvent-free, energy-efficient alternatives to traditional solution-based methods. Ball milling and grinding approaches have been used to form cocrystals, improve solubility, and control polymorphism, which directly affects drug stability and bioavailability. Acoustic energy has also been employed to facilitate crystallization, reduce particle aggregation, and promote uniform distribution of active pharmaceutical ingredients. These techniques not only improve process efficiency but also align with the principles of green chemistry by minimizing solvent use, reducing energy consumption, and avoiding harsh reaction conditions. The combination of acoustic and mechanical approaches in pharmaceutical synthesis exemplifies the ability to manipulate molecular interactions with precision, offering a new level of control over complex chemical systems.

Environmental applications of acoustic and mechanical chemistry have garnered increasing attention due to the potential for sustainable chemical processing. Ultrasonically driven degradation of pollutants, for instance, utilizes cavitation-generated radicals to oxidize organic contaminants in water, achieving high degradation rates without the need for toxic reagents. Similarly, mechanochemical processes enable solvent-free transformations that minimize waste generation and reduce the environmental footprint of chemical manufacturing. The integration of acoustic and mechanical methods provides a toolkit for environmentally conscious chemists, allowing reactions to proceed under mild conditions while maintaining efficiency and selectivity. Case studies in water purification, pollutant degradation, and the synthesis of bio-based polymers demonstrate the practical utility of these approaches, highlighting their role in addressing contemporary environmental challenges.

From a theoretical perspective, the interplay between acoustic and mechanical energy with chemical systems can be understood through the lens of molecular energy landscapes. Mechanical forces alter the potential energy surfaces of reactants, lowering activation barriers along specific reaction coordinates and enabling otherwise forbidden pathways. Acoustic energy, particularly through cavitation, introduces transient high-energy microenvironments that can overcome kinetic limitations and promote radical formation. The combination of these effects allows for precise modulation of reaction kinetics, selectivity, and product distribution. Computational models, including molecular dynamics simulations and quantum chemical calculations, have been employed to elucidate the mechanisms underlying these processes, revealing how energy is transferred from macroscopic stimuli to individual molecules. These studies provide critical insight into the design of reactions that harness physical energy, offering predictive capability and guiding experimental strategies.

Despite their promise, acoustic and mechanical chemistries are not without limitations. Controlling energy distribution remains a central challenge, as the highly localized nature of cavitation events or mechanical stress can lead to heterogeneous reaction environments. Reproducibility is a key concern, particularly when scaling from laboratory experiments to industrial processes, where uniform energy delivery and reaction control are essential. Advances in reactor design, ultrasonic probe technology, and mechanochemical equipment have mitigated some of these challenges, yet precise quantification of energy input and its correlation with chemical outcomes remains an active area of research. Furthermore, the mechanistic complexity of combined acoustic and mechanical processes requires a detailed understanding of both macroscopic and molecular phenomena, necessitating interdisciplinary approaches that integrate chemistry, physics, and materials science.

Looking forward, the future of acoustic and mechanical chemistry lies in the integration of advanced experimental techniques with computational modeling to achieve unprecedented control over chemical reactivity. The development of tunable ultrasonic reactors, high-precision mechanochemical devices, and real-time spectroscopic monitoring will enable chemists to probe reaction dynamics with exceptional detail. Emerging fields such as single-molecule mechanochemistry and sonomechanically assisted catalysis are poised to expand the frontiers of chemical synthesis, allowing for the creation of complex molecules and materials with atomic-level precision. As environmental sustainability becomes an increasingly critical consideration, the solvent-free and energy-efficient nature of these approaches positions acoustic and mechanical chemistry as a cornerstone of green chemistry, offering pathways to reduce chemical waste, minimize energy consumption, and design reactions that are both efficient and environmentally responsible.

Recent advances in acoustic and mechanical chemistry have enabled researchers to explore complex mechanistic phenomena at unprecedented resolution, providing insight into how energy transduction at the molecular level influences chemical reactivity. One particularly illustrative example is the sonochemical synthesis of metal nanoparticles, where the interplay between cavitation-induced hotspots and solvent dynamics governs particle nucleation, growth, and aggregation. In these systems, the implosive collapse of cavitation bubbles generates local temperatures exceeding several thousand Kelvin and pressures of hundreds of atmospheres, which not only promote bond cleavage but also transiently ionize solvent molecules. The resulting reactive radicals act as nucleation centers for nanoparticle formation, while the rapid cooling rates prevent uncontrolled growth, yielding uniform nanoparticles with tunable size and morphology. Mechanistic studies using in situ spectroscopic techniques have revealed that the frequency and amplitude of the applied ultrasound directly influence the distribution of these hotspots, thereby controlling the kinetics of nucleation and the final particle characteristics. Such precise control highlights the ability of acoustic energy to modulate chemical processes at the nanoscale, transforming traditional synthetic strategies in materials chemistry.

Mechanochemistry, in parallel, provides complementary pathways for influencing chemical reactivity through the direct application of mechanical force. In mechanochemical ball milling, for instance, the repeated impact of grinding media generates localized strain within solid reactants, effectively lowering activation barriers for bond-breaking events. The mechanical stress not only disrupts intermolecular interactions but can also induce phase transitions, as observed in the formation of polymorphic pharmaceutical compounds. Studies using single-crystal X-ray diffraction and solid-state NMR spectroscopy have demonstrated that mechanochemical activation can selectively target specific functional groups, allowing chemists to bias reaction pathways toward desired products. Moreover, computational investigations employing density functional theory have shown that applied stress can distort molecular geometries along the reaction coordinate, facilitating bond cleavage while minimizing competing side reactions. This capacity to exert directional control over chemical transformations is particularly valuable in the synthesis of complex organic molecules and coordination compounds, where conventional solution-based methods often require harsh conditions or extensive purification steps.

The synergistic combination of acoustic and mechanical forces, commonly termed sonomechanochemistry, has emerged as a powerful strategy for driving chemical transformations that are challenging under conventional conditions. By coupling ultrasonic irradiation with mechanochemical techniques such as ball milling or high-shear grinding, chemists can simultaneously exploit the high-energy environments generated by cavitation and the localized strain imposed by mechanical force. One compelling case study involves the sonomechanochemical synthesis of covalent organic frameworks (COFs), where ultrasound facilitates nucleation and promotes homogeneous mixing, while mechanical force accelerates bond formation and promotes crystallinity. In this context, the ability to control energy input on both macroscopic and microscopic scales allows for the design of materials with highly ordered architectures, tunable porosity, and enhanced functional properties. Real-time monitoring using Raman spectroscopy and X-ray diffraction has enabled researchers to correlate specific energy inputs with structural evolution, providing mechanistic insight into the interplay between acoustic and mechanical stimuli.

Beyond synthesis, acoustic and mechanical chemistry has found transformative applications in catalysis. Mechanochemical activation can modify the electronic structure of solid catalysts by generating defects, exposing new active sites, or inducing phase transitions, thereby altering catalytic activity and selectivity. Ultrasound, when applied concurrently, enhances mass transport, prevents agglomeration of catalyst particles, and promotes radical generation in liquid-phase reactions. For example, the oxidative coupling of amines or alcohols under sonomechanochemical conditions has been shown to proceed with higher selectivity and faster rates compared to conventional heating, highlighting the potential of these methods for industrially relevant transformations. Mechanistic studies suggest that the observed enhancements arise from a combination of transient high-energy conditions, improved reactant-catalyst contact, and preferential activation of specific bonds, illustrating the multifaceted influence of physical energy on reaction dynamics.

In polymer chemistry, acoustic and mechanical techniques offer precise control over macromolecular architecture. Ultrasound can initiate polymerization by generating radicals in situ, enabling controlled chain growth without the addition of chemical initiators. Mechanochemical approaches, such as twin-screw extrusion or ball milling, can drive polymer–polymer reactions, induce cross-linking, or facilitate depolymerization in a solvent-free environment. Combining these approaches has allowed for the production of advanced polymeric materials with unique mechanical, thermal, and electronic properties, such as conductive polymer composites, shape-memory polymers, and recyclable thermosets. These methods not only improve material performance but also align with sustainable chemistry principles by reducing solvent use and energy consumption.

Mechanistic insights into these processes have been further enriched by the integration of computational modeling with experimental studies. Molecular dynamics simulations, for example, have provided atomistic-level understanding of cavitation phenomena, revealing the formation, growth, and collapse of bubbles in complex solvent environments. Similarly, simulations of mechanochemical reactions have elucidated how stress propagates through crystalline lattices, identifying regions of high strain concentration that correspond to sites of chemical reactivity. Quantum chemical calculations complement these studies by quantifying changes in activation barriers, electronic density distributions, and transition-state geometries under applied mechanical or acoustic forces. Together, these computational tools allow chemists to predict reaction outcomes, optimize energy input, and design experiments that exploit the synergistic effects of acoustic and mechanical stimulation with precision.

The versatility of acoustic and mechanical chemistry extends to emerging areas such as biomolecular engineering and nanomedicine. Ultrasound can induce transient permeability in biological membranes, facilitating the delivery of drugs, genes, or nanoparticles into cells, while mechanochemical strategies enable solvent-free synthesis of bioactive molecules with minimal degradation. Sonomechanochemistry has been explored for the fabrication of nanostructured drug carriers, where combined acoustic and mechanical energy produces uniform particle size distributions and enhances encapsulation efficiency. These applications illustrate how fundamental principles of energy transfer and molecular activation can be harnessed to address complex challenges in biology and medicine, bridging the gap between chemical synthesis and practical utility.

Despite these advances, several challenges remain in the field. Scaling laboratory findings to industrial processes requires precise control over energy distribution, reproducibility, and safety, particularly when dealing with high-intensity ultrasound or mechanochemical equipment. Furthermore, the heterogeneity inherent in cavitation and mechanical stress can complicate mechanistic interpretation and reaction optimization. Ongoing research seeks to develop standardized methodologies, advanced reactor designs, and real-time monitoring techniques to overcome these limitations, ensuring that acoustic and mechanical chemistry can be reliably applied at scale. As understanding of the underlying principles deepens, the potential for these methods to transform chemical synthesis, materials design, and sustainable processing continues to expand.

Emerging technologies in acoustic and mechanical chemistry are expanding the boundaries of what is achievable in both synthesis and analysis. Innovations in ultrasonic reactor design now allow precise control over frequency, amplitude, and pulse duration, enabling chemists to tailor cavitation events for specific reactions. Microfluidic reactors combined with ultrasonic irradiation have been developed to study sonochemical reactions under highly controlled laminar flow conditions, reducing variability and enhancing reproducibility. Similarly, advances in mechanochemical apparatus, including programmable ball mills, twin-screw extruders, and shear reactors, permit precise regulation of mechanical force, energy input, and reaction environment. These technologies facilitate systematic studies of mechanochemical kinetics, enabling the optimization of reaction conditions and the exploration of previously inaccessible chemical transformations. The integration of acoustic and mechanical methods with high-throughput screening and automated reaction platforms promises to accelerate the discovery of new materials, catalysts, and bioactive molecules, providing a convergence of precision engineering and chemical innovation.

Analytical techniques play a central role in elucidating the mechanisms underlying acoustic and mechanical chemical processes. Vibrational spectroscopy, including infrared and Raman spectroscopy, provides direct insight into molecular vibrations and bond activation, allowing researchers to monitor the impact of applied forces on specific functional groups. Complementary techniques, such as X-ray diffraction, nuclear magnetic resonance, and electron microscopy, enable structural characterization of reaction intermediates, products, and nanomaterials synthesized under sonomechanochemical conditions. In situ monitoring, particularly through real-time spectroscopic probes, has revealed dynamic changes in bond order, crystallinity, and molecular conformation during reactions, bridging the gap between applied physical energy and molecular response. Coupled with computational modeling, these analytical approaches offer predictive capability, allowing chemists to design reactions with targeted outcomes and to rationally manipulate energy inputs to achieve desired reactivity.

Industrial applications of acoustic and mechanical chemistry are increasingly prevalent, particularly in sectors where precision, efficiency, and sustainability are paramount. In pharmaceutical manufacturing, solvent-free mechanochemical methods are employed to produce cocrystals, polymorphs, and active pharmaceutical ingredients with improved solubility, stability, and bioavailability. Sonochemical techniques facilitate scale-up of nanoparticle production, catalyst preparation, and polymer synthesis, providing energy-efficient and reproducible methods for high-value chemical products. The use of ultrasound and mechanical activation in catalysis enables higher reaction rates, improved selectivity, and reduced catalyst loading, directly translating to cost savings and reduced environmental impact. In materials science, combined acoustic and mechanical approaches allow the fabrication of advanced composites, functional nanomaterials, and porous frameworks with precise control over structure and performance, supporting applications in energy storage, electronics, and biomedical devices.

Green chemistry benefits substantially from the application of acoustic and mechanical principles, as these methods often obviate the need for hazardous solvents, high temperatures, or excessive energy input. Ultrasonically induced radical generation provides a clean alternative for oxidative transformations, pollutant degradation, and wastewater treatment, minimizing the use of toxic reagents. Mechanochemical synthesis reduces solvent waste and energy consumption while enabling solvent-free reactions that retain high selectivity and yield. The synergistic application of ultrasound and mechanical force, or sonomechanochemistry, exemplifies the potential for environmentally conscious chemical innovation, where precise control over energy delivery achieves chemical transformation efficiently and sustainably. These approaches are increasingly integrated into the design of industrial processes, highlighting the growing relevance of acoustic and mechanical chemistry in addressing contemporary environmental and economic challenges.

Case studies in sonomechanochemistry illustrate both the mechanistic sophistication and practical utility of these approaches. For instance, the mechanochemical synthesis of metal–organic frameworks under ultrasonic irradiation demonstrates how energy input can influence crystal nucleation, growth kinetics, and defect formation, yielding materials with enhanced porosity and adsorption capacity. In polymer chemistry, the combination of ultrasonic initiation and mechanochemical cross-linking produces highly uniform networks with controlled mechanical properties, suitable for applications ranging from biomedical scaffolds to recyclable thermosets. In catalysis, ultrasonic-assisted grinding has enabled the selective oxidation of alcohols and amines with remarkable efficiency and minimal byproduct formation, underscoring the capacity of combined physical forces to direct chemical selectivity. These examples highlight the critical importance of understanding both macroscopic energy input and molecular-level response in designing advanced chemical processes.

Looking toward the future, acoustic and mechanical chemistry is poised to continue transforming chemical synthesis, materials design, and process engineering. Emerging directions include the use of ultrafast spectroscopy to probe transient reaction intermediates generated by cavitation, the integration of machine learning with mechanochemical experimentation to predict optimal reaction conditions, and the development of hybrid systems that combine acoustic, mechanical, photonic, and electrochemical stimuli to achieve unprecedented control over chemical reactivity. The ongoing convergence of experimental innovation, computational modeling, and sustainability considerations ensures that these approaches will play an increasingly central role in next-generation chemical research and industrial application.

In conclusion, acoustic and mechanical chemistry exemplifies the transformative potential of harnessing physical energy to control chemical processes. Through the strategic application of ultrasonic waves, mechanical forces, or their synergistic combination, chemists can modulate reaction pathways, influence molecular interactions, and achieve outcomes that are unattainable under traditional thermal or chemical conditions. The integration of advanced analytical techniques, computational modeling, and precise energy control has deepened mechanistic understanding and facilitated the rational design of reactions. Applications spanning materials science, catalysis, pharmaceuticals, and environmental chemistry underscore the practical significance and versatility of these approaches. As technological innovations continue to refine our ability to deliver and control physical energy at the molecular level, acoustic and mechanical chemistry is set to remain at the forefront of sustainable, efficient, and highly controlled chemical synthesis, offering profound opportunities for both fundamental discovery and industrial application.

Chapter 49: Word Problems

1. In the sonochemical synthesis of titanium dioxide (TiO_2) nanoparticles, the extreme localized temperatures and pressures generated during cavitation events create hotspots that profoundly influence nucleation and crystal growth pathways. How do these extreme conditions affect particle size distribution, specific surface area, phase composition, and overall crystallinity? Propose a detailed experimental plan to systematically vary ultrasonic parameters such as frequency, amplitude, pulse

duration, and power density, and discuss complementary analytical techniques—including X-ray diffraction (XRD), transmission electron microscopy (TEM), and Brunauer–Emmett–Teller (BET) surface area measurements—to monitor structural properties. Additionally, consider potential challenges in scaling up the process for industrial nanoparticle production and strategies to maintain uniformity.

2. Mechanochemical ball milling is increasingly applied for solvent-free cocrystallization of poorly soluble pharmaceutical compounds. How does the input of mechanical energy influence polymorphic selection, particle size reduction, and long-term stability of the resulting cocrystals? Design an experiment comparing ball-milled cocrystals to those grown from solution, detailing critical milling parameters such as rotational speed, ball-to-powder ratio, milling duration, and temperature control. Discuss analytical approaches, including powder XRD, differential scanning calorimetry (DSC), Fourier-transform infrared spectroscopy (FTIR), and solid-state NMR, to assess structural integrity, polymorphism, and reproducibility across multiple batches.

3. In synthesizing covalent organic frameworks (COFs) via sonomechanochemistry, how do ultrasonic irradiation and mechanochemical milling synergistically enhance nucleation rates, crystallinity, and framework ordering? Propose a systematic experimental plan to optimize parameters such as ultrasonic power, frequency, milling speed, and reaction duration. Discuss the influence of solvent selection, if any, on crystallization pathways and reproducibility. Include in situ characterization techniques such as Raman spectroscopy or small-angle X-ray scattering (SAXS) to monitor nucleation and growth kinetics, and evaluate potential challenges when scaling the process for gram- or kilogram-scale production.

4. Ultrasonic cavitation is known to improve heterogeneous catalysis by enhancing reactant–catalyst contact, mass transfer, and surface reactivity. Explain the specific roles of microjets, shockwaves, and localized hotspots on reaction kinetics and catalytic turnover. Design an experiment capable of isolating ultrasonic effects from thermal and bulk mixing contributions, including appropriate control reactions. Discuss analytical methods such as XRD, scanning electron microscopy (SEM), and X-ray photoelectron spectroscopy (XPS) to monitor catalyst surface changes, and evaluate the influence of solvent viscosity, temperature, and scale-up considerations on reaction efficiency.

5. Mechanochemical activation exposes fresh catalytic sites and introduces surface defects in solid-state reactions. In the context of a mechanochemical oxidation reaction, how do milling intensity, duration, ball size, and impact energy distribution affect reaction kinetics, selectivity, and product yield? Propose a series of experiments correlating energy input to structural and electronic modifications using techniques such as TEM, BET surface area analysis, and electron paramagnetic resonance (EPR) spectroscopy. Discuss strategies to ensure reproducibility and scalability, including the effects of batch versus continuous milling operations.

6. Sonochemical degradation of organic pollutants relies on radicals generated from cavitation bubble collapse. How does the extreme environment within collapsing bubbles produce reactive species such as hydroxyl radicals, and how are these influenced by ultrasonic frequency, power, and solvent properties? Propose an experimental setup to monitor radical formation in real time using spin-trapping EPR spectroscopy, coupled with pollutant degradation measurements via high-performance liquid chromatography (HPLC). Consider how solvent choice, dissolved gas concentration, and scale-up affect efficiency and radical lifetime.

7. Mechanochemical techniques enable solvent-free pharmaceutical cocrystallization with improved environmental sustainability. How does mechanical force affect particle size, polymorph formation, and long-term stability? Design an experiment to evaluate reproducibility and scalability by systematically varying milling parameters, temperature, and atmosphere. Include analytical methods such as powder XRD, DSC, FTIR, and solid-state NMR to monitor crystal structure, and discuss potential environmental effects of mechanochemical processing on product integrity.

8. During sonomechanochemical oxidative coupling of amines, ultrasound and milling synergistically accelerate reaction progress. Explain the distinct mechanisms by which cavitation-induced radicals and mechanical strain influence bond activation, nucleation, and reaction rates. Propose a strategy to separate ultrasonic and mechanochemical contributions experimentally, and outline optimization of ultrasonic power, milling speed, reaction time, and solvent choice to maximize yield and selectivity. Include analytical monitoring of intermediates and products using chromatographic and spectroscopic techniques.

9. Ultrasound can initiate polymerization reactions through radical formation without chemical initiators. How do ultrasonic frequency, amplitude, pulse duration, and exposure time affect chain growth, branching, molecular weight distribution, and polydispersity? Design an experiment to systematically study these parameters, and propose analytical techniques such as gel permeation chromatography (GPC), nuclear magnetic resonance (NMR), and rheological measurements to characterize polymer structure. Discuss strategies to minimize side reactions, chain termination, and ensure reproducibility during scale-up.

10. Mechanochemical reactions are known to lower activation barriers and selectively activate bonds in solids. For a nucleophilic substitution occurring in a crystalline solid, explain how applied stress facilitates bond cleavage and enhances reaction kinetics. Propose both computational approaches (e.g., DFT, molecular dynamics simulations) and experimental strategies to predict reactive sites, verify bond activation, and assess product selectivity. Include analytical techniques for monitoring reaction progress, evaluating structural changes, and addressing challenges in reproducibility and scale-up.

11. In sonochemical water treatment, reactive oxygen species (ROS) such as hydroxyl radicals (•OH) and superoxide ions ($O_2^{\bullet-}$) degrade organic pollutants. How are these radicals formed from cavitation events, and how do ultrasonic frequency, power, pulse duration, and solvent properties affect their generation, lifetime, and reactivity? Propose an experimental design to quantify radical formation in real time using spin-trapping EPR or chemiluminescence techniques, coupled with pollutant degradation measurements via HPLC or UV–Vis spectroscopy. Discuss strategies to optimize radical production, ensure reproducibility, and scale the process for industrial water treatment applications.

12. Mechanochemical milling is an effective route for producing metal nanoparticles with controlled size, morphology, and defect density. How do milling parameters—such as ball size, material, milling speed, duration, and atmosphere—influence nucleation, growth, crystallinity, and surface defect formation? Propose a comprehensive study linking energy input to catalytic activity, using TEM, XRD, BET, and XPS to correlate structural and surface properties with performance. Include considerations for reproducibility, batch-to-batch variation, and challenges in scaling up from laboratory to industrial quantities.

13. Ultrasound-assisted crystallization can improve particle uniformity, solubility, and dissolution rates of pharmaceutical solids. How do pulse duration, amplitude, frequency, and solvent viscosity affect nucleation rate, crystal growth, morphology, and particle size distribution? Design an experiment to systematically vary these ultrasonic parameters, and include in situ techniques such as optical microscopy, turbidity measurements, and SAXS to monitor crystallization. Discuss analytical methods for final particle characterization, including dissolution testing, and strategies to ensure reproducibility and scalability.

14. Mechanochemical milling introduces defects in metal-organic frameworks (MOFs), which can significantly alter porosity, surface area, and catalytic properties. How does mechanical stress and energy input influence defect formation, framework integrity, and adsorption properties? Propose experiments to quantify defect density using techniques such as positron annihilation spectroscopy, N_2 adsorption, and X-ray pair distribution function analysis. Include strategies to correlate defects with catalytic activity, optimize milling parameters, and ensure reproducibility across scales.

15. Sonomechanochemistry can enhance slow or kinetically hindered reactions by accelerating nucleation and bond formation. Provide an example of such a reaction where combined ultrasonic irradiation and mechanical milling significantly improve crystallinity, yield, and reaction rate. Propose a monitoring strategy using in situ spectroscopic or diffraction techniques to track nucleation and growth, and outline a systematic experimental plan varying ultrasonic power, milling speed, and duration to optimize both reaction efficiency and reproducibility.

16. Ultrasonic irradiation can accelerate ligand exchange in homogeneous catalysis by enhancing mass transport and generating localized high-energy environments. Explain the roles of cavitation, microjets, and shockwaves in promoting ligand substitution reactions. Design experiments to isolate chemical activation effects from mass transport enhancements, using analytical methods such as NMR, UV–Vis spectroscopy, or cyclic voltammetry. Discuss solvent effects, reproducibility strategies, and scale-up considerations for industrial catalytic applications.

17. Mechanochemistry enables solvent-free covalent bond formation with improved sustainability. How can selectivity, reaction rate, and formation of side products

be controlled by adjusting milling parameters, atmosphere, and energy input? Propose experiments to systematically optimize these variables, including analytical techniques such as NMR, FTIR, and HPLC to confirm product formation and purity. Discuss strategies for ensuring reproducibility and scalability while minimizing environmental impact.

18. Combining acoustic and mechanical energy affects polymer network formation, cross-linking, and filler dispersion. How do ultrasonic cavitation-generated radicals and mechanical stress contribute to bond activation and polymer chain alignment? Propose a comprehensive experimental plan varying ultrasonic power, frequency, milling speed, and duration. Include analytical methods such as GPC, dynamic mechanical analysis (DMA), TEM, and rheometry to monitor cross-link density, molecular weight distribution, and mechanical properties, and discuss scale-up challenges.

19. In sonomechanochemical nanoparticle synthesis, cavitation and mechanical energy affect nucleation, growth, and polydispersity. Explain the underlying physical and chemical mechanisms governing size distribution, defect formation, and crystallinity. Propose combined computational modeling and experimental approaches to optimize ultrasonic and milling parameters. Include strategies to monitor structural evolution in situ and techniques to ensure reproducibility and scalability for larger production volumes.

20. Mechanochemical grinding can trigger phase transitions in crystalline materials. How do mechanical stress, energy distribution, and milling kinetics drive polymorph formation, stability, and phase purity? Propose experiments to selectively synthesize specific polymorphs, including analytical monitoring using XRD, DSC, solid-state NMR, and Raman spectroscopy. Discuss challenges in achieving reproducibility during scale-up and strategies for controlling energy input and environmental conditions.

21. Ultrasound improves nanoparticle dispersion in suspensions by affecting aggregation and colloidal stability. How do cavitation, microjets, and shockwaves influence particle–particle interactions, zeta potential, and uniformity? Propose experiments to quantify ultrasonic effects on dispersion using dynamic light scattering (DLS), TEM, and rheological measurements. Consider solvent effects,

ultrasonic frequency, amplitude, and duration, and address reproducibility and scale-up challenges for industrial formulations.

22. Mechanochemical activation generates surface defects in catalysts that can enhance activity and modify electronic structure. How does the choice of grinding or milling parameters, ball material, and duration affect defect density and distribution? Propose studies correlating structural modifications with catalytic activity using XPS, TEM, and BET surface analysis. Discuss strategies to optimize milling for maximum performance while maintaining reproducibility and feasibility in scaled-up operations.
23. Sonochemical water treatment relies on reactive radical species for pollutant degradation. How do ultrasonic parameters (frequency, power, duty cycle) and dissolved gas concentration affect radical formation, lifetime, and pollutant conversion efficiency? Propose experimental setups to quantify radical formation and degradation kinetics, incorporating in situ monitoring techniques such as EPR spectroscopy and UV–Vis analysis. Discuss solvent effects, energy efficiency, and scale-up considerations for practical water treatment applications.
24. Mechanochemical cocrystallization improves solubility and bioavailability of poorly soluble pharmaceuticals. How does mechanical energy influence polymorphic selection, particle size, and long-term crystal stability? Design experiments comparing milling to solution-based cocrystallization, specifying analytical techniques for polymorph identification (XRD, DSC, FTIR) and environmental effects such as humidity or temperature. Discuss strategies for reproducibility and scalable production.
25. Sonomechanochemical polymer nanocomposites rely on combined acoustic and mechanical energy to improve filler dispersion and polymer chain alignment. How do cavitation and mechanical stress influence polymer network structure, interfacial interactions, and material performance? Propose experiments varying ultrasonic power, milling speed, and duration, with analytical techniques such as TEM, rheology, DMA, and spectroscopy to monitor structure–property relationships, reproducibility, and scale-up feasibility.
26. Solvent properties play a critical role in sonochemical processes by influencing cavitation dynamics and radical formation. How do solvent polarity, viscosity,

surface tension, and vapor pressure affect bubble nucleation, collapse intensity, and reactive species generation? Propose experiments to systematically investigate solvent effects on reaction rates, product distribution, and radical yields, using analytical techniques such as EPR spectroscopy, HPLC, and GC–MS. Discuss implications for reproducibility, scale-up, and the optimization of laboratory versus industrial sonochemical reactions.

27. Combining ultrasound and mechanochemistry can significantly enhance heterogeneous catalysis by increasing surface area, generating defects, and improving mass transport. How do these synergistic effects accelerate reaction kinetics and selectivity? Propose a study to separately quantify the contributions of ultrasonic irradiation and mechanical milling, incorporating analytical methods such as XPS, TEM, BET surface area analysis, and in situ spectroscopy. Discuss strategies for optimizing energy input, ensuring reproducibility, and scaling up to industrially relevant conditions.

28. Mechanochemistry enables multi-step, solvent-free synthesis of complex molecules. Outline a synthetic route for a target molecule that highlights steps particularly benefiting from mechanical activation. Propose strategies for real-time monitoring of reactions, optimization of milling parameters, and verification of product formation using analytical techniques such as NMR, HPLC, and XRD. Discuss considerations for reproducibility, energy efficiency, and scale-up challenges in transitioning from laboratory to industrial production.

29. Ultrasound-assisted oxidation reactions are strongly influenced by cavitation-generated microenvironments, which can accelerate bond activation and radical formation. How do ultrasonic parameters such as power, frequency, and pulse duration affect reaction pathways, selectivity, and yield? Design experiments to systematically vary these parameters, implement temperature control, and employ analytical methods including HPLC, GC–MS, and EPR spectroscopy to monitor both radicals and reaction intermediates. Discuss reproducibility strategies and challenges in scaling up ultrasonic oxidation for industrial processes.

30. Acoustic energy strongly influences crystallization kinetics, including nucleation rates, crystal growth, and final particle size. How do ultrasonic parameters (frequency, amplitude, duty cycle) and solvent properties (viscosity, polarity,

saturation) collectively determine nucleation, growth mechanisms, and polymorph formation? Propose experiments with in situ monitoring using turbidity, optical microscopy, and SAXS to study crystallization dynamics, and discuss strategies for reproducible, scalable production of crystals with controlled morphology and size distribution.

31. Mechanochemistry can selectively cleave bonds in solid-state systems, lowering activation barriers and enabling reactions inaccessible under conventional conditions. How does mechanical stress distribute across crystalline lattices to selectively activate specific bonds? Propose experiments combining milling with computational modeling (DFT, molecular dynamics) to predict reactive sites, and employ analytical techniques such as solid-state NMR, XRD, and FTIR to monitor bond activation and product formation. Discuss challenges in reproducibility and scale-up.

32. Ultrasound enhances catalytic reactions through the formation of reactive radicals and improved mass transport. How do cavitation, microjets, and shockwaves contribute to reaction acceleration and selectivity? Design experiments to quantify the contribution of ultrasonic radicals to reaction rates, including controls with radical scavengers, and use analytical methods such as GC–MS, HPLC, and in situ spectroscopy. Discuss solvent effects, parameter optimization, reproducibility, and industrial scalability.

33. Mechanochemical synthesis of metal oxide nanoparticles can improve surface area, defect density, and catalytic activity. How do milling parameters, such as ball size, rotational speed, and duration, influence particle size, defect formation, and crystallinity? Propose experiments linking energy input to catalytic performance using TEM, XRD, BET, and XPS, and discuss strategies to maintain reproducibility and scale-up feasibility for larger-scale nanoparticle production.

34. Sonomechanochemistry can be used to synthesize covalent organic frameworks (COFs) with controlled porosity and crystallinity. How does the synergy between ultrasonic irradiation and mechanical milling affect nucleation, bond formation, and framework ordering? Propose an experimental plan to monitor structural evolution, optimize energy distribution, and evaluate solvent versus solvent-free conditions. Include analytical methods such as SAXS, PXRD, and solid-state NMR, and discuss reproducibility and scale-up challenges.

35. Ultrasound-assisted polymerization allows radical chain growth without chemical initiators. How do ultrasonic frequency, amplitude, and exposure duration influence chain growth, branching, molecular weight, and dispersity? Design experiments to vary these parameters systematically, and employ analytical techniques such as GPC, NMR, and rheological measurements to characterize polymer structure and properties. Discuss strategies to minimize side reactions and ensure reproducibility.

36. Mechanochemistry can control polymorphism in pharmaceutical solids, affecting solubility and stability. How do stress distribution, milling intensity, and milling duration influence the formation of metastable or stable polymorphs? Propose experiments to selectively synthesize polymorphs, monitor phase purity using PXRD, DSC, and solid-state NMR, and discuss reproducibility and scale-up considerations.

37. Sonomechanochemistry facilitates covalent bond formation by combining ultrasonic and mechanical energies. How do ultrasound and milling individually contribute to bond activation, particle dispersion, and reaction kinetics? Propose an experimental system to separate the effects of each energy input, optimize parameters for yield and selectivity, and monitor structural changes using TEM, XRD, and spectroscopy, while considering reproducibility challenges.

38. Ultrasound influences crystallization kinetics through cavitation-induced heating, microjets, and shockwaves. How do these factors affect nucleation, growth, and polymorph selection in solution or solid-state crystallization? Propose experiments to study ultrasonic effects using in situ monitoring techniques such as optical microscopy, SAXS, and Raman spectroscopy, while considering solvent effects, reproducibility, and scale-up.

39. Mechanochemical activation drives solvent-free reactions by lowering activation barriers and selectively activating bonds. How does energy input, milling intensity, and duration correlate with reaction kinetics and product formation? Propose experiments with systematic variation of milling conditions, supported by analytical monitoring via PXRD, FTIR, and NMR, and discuss computational guidance for selective bond activation, reproducibility, and scale-up.

40. Sonomechanochemistry improves polymer composites by enhancing filler dispersion, polymer chain alignment, and cross-linking. How do combined acoustic

and mechanical energies influence material microstructure and performance? Propose experiments optimizing ultrasonic power, milling speed, and duration, with analytical techniques such as TEM, DMA, rheology, and spectroscopy, while addressing reproducibility and scalability challenges.

41. Mechanochemistry enables solid-state reactions that are inaccessible under conventional conditions. How does mechanical stress induce defects, facilitate bond cleavage, and activate reaction pathways in crystalline solids? Design experiments for condensation or coupling reactions, specifying milling parameters, atmosphere control, and analytical methods. Discuss computational modeling approaches, reproducibility, and scale-up strategies.

42. Sonomechanochemistry enhances heterogeneous catalysis by generating radicals, microjets, and mechanical strain, which can improve reaction rates and selectivity. How do these combined effects influence surface properties and catalytic performance? Propose experiments to separate energy contributions, monitor structural evolution, optimize parameters, and evaluate reproducibility and scale-up.

43. Ultrasound-assisted degradation of pollutants relies on reactive radicals such as hydroxyl and superoxide species. How are these radicals generated from cavitation, and how do ultrasonic parameters and solvent properties affect degradation efficiency? Propose experiments to monitor radicals and pollutant removal using EPR, HPLC, and UV–Vis spectroscopy, and discuss scaling considerations for sustainable water treatment.

44. Mechanochemistry controls polymorphism in pharmaceuticals, which directly affects solubility, bioavailability, and stability. How do stress, milling speed, and duration influence polymorphic transformations and crystal packing? Propose experiments for selective synthesis, monitoring phase purity using PXRD, DSC, and NMR, and discuss reproducibility and scale-up strategies, including environmental effects.

45. Sonochemical and mechanochemical methods can be combined to synthesize COFs with uniform porosity and enhanced crystallinity. How do cavitation and mechanical milling influence nucleation, crystal growth, and defect formation? Propose an experimental study to optimize energy input, monitor structural

evolution using in situ techniques, and evaluate reproducibility and solvent effects for potential industrial applications.

46. Ultrasound can initiate polymerization without chemical initiators. How do frequency, amplitude, and duration affect chain growth, branching, molecular weight distribution, and side reactions? Design experiments to study polymerization kinetics, apply analytical techniques such as GPC, NMR, and rheology, and discuss strategies to ensure reproducibility and scale-up.

47. Mechanochemistry drives solvent-free organic reactions by selectively activating bonds and lowering activation barriers. How does mechanical energy influence reactivity and selectivity in solid-state systems? Propose experiments to optimize milling parameters, monitor reactions with analytical techniques, and discuss reproducibility, sustainability, and scalability benefits compared to conventional methods.

48. Sonomechanochemistry can improve nanoparticle synthesis by combining cavitation and mechanical energy to control nucleation, growth, and uniformity. How do ultrasonic power, milling speed, and duration affect particle size, polydispersity, and surface defects? Propose experiments and analytical methods such as TEM, XRD, and BET to characterize particles, while considering reproducibility and industrial scale-up.

49. Ultrasound-assisted heterogeneous catalysis enhances reaction rates via cavitation-induced high-energy zones and microjets. How do these effects influence active site availability, surface restructuring, and catalytic turnover? Propose experiments to quantify ultrasonic contributions, monitor structural and chemical changes, optimize parameters, and evaluate reproducibility and scalability in industrial contexts.

50. Mechanochemical and sonomechanochemical methods offer sustainable routes for multi-step synthesis. How do mechanical and ultrasonic forces enable solvent-free reactions, lower energy usage, and improve reaction efficiency? Propose a multi-step synthetic route highlighting steps benefiting from each type of energy input, and discuss monitoring strategies, analytical verification, reproducibility, and scale-up for industrial application.

Chapter 50

Chemistry of Consciousness

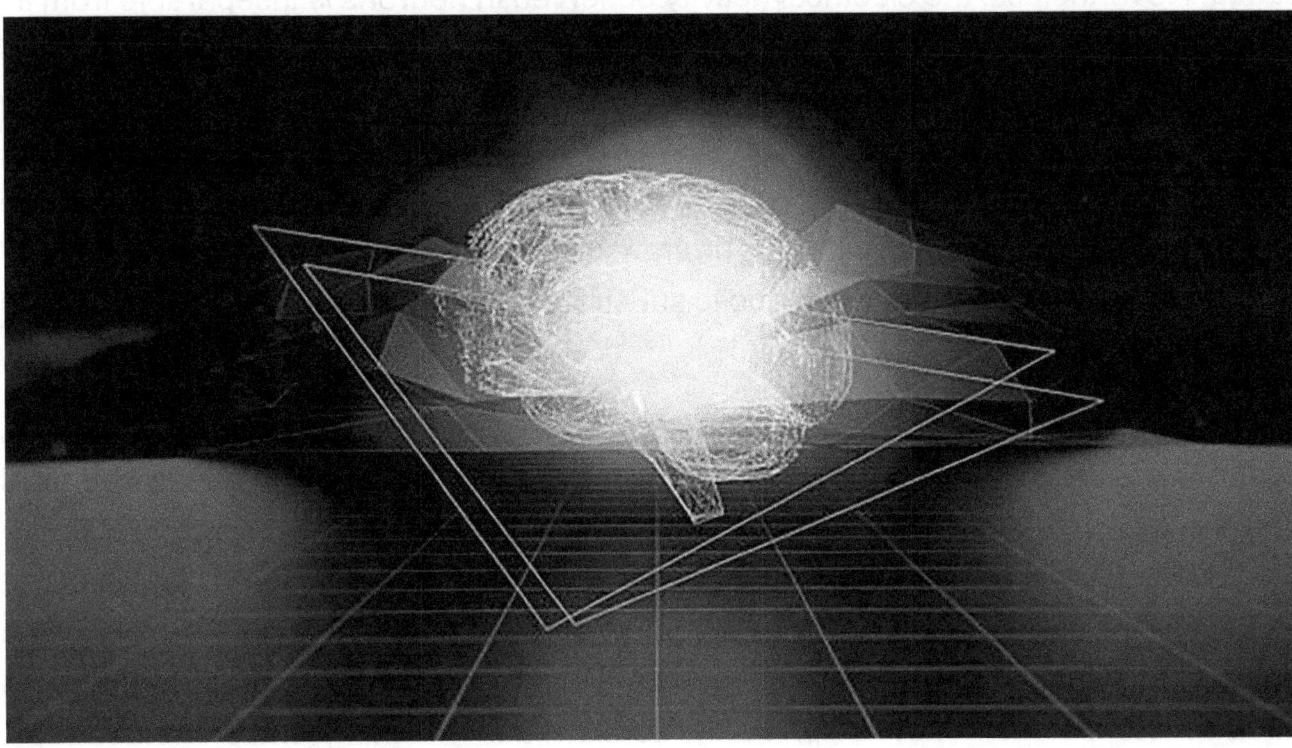

Consciousness, the awareness of self and environment, has long captivated philosophers, psychologists, and scientists alike. Historically, discussions of consciousness centered on philosophical and phenomenological perspectives, pondering the nature of subjective experience and the "hard problem" of how the brain generates awareness. In contemporary science, however, consciousness is increasingly approached through neurobiology and chemistry, emphasizing the molecular underpinnings of neural activity. From a chemical standpoint, consciousness emerges from the orchestrated interactions among neurons, glial cells, and the molecular networks that mediate their communication. At its core, conscious experience is dependent on a vast array of chemical signals that regulate neuronal excitability, synaptic plasticity, and network integration, forming the substrate for perception, attention, and self-awareness.

Historically, early studies of consciousness focused heavily on electrical correlates, such as action potentials, electroencephalographic (EEG) rhythms, and oscillatory synchronization. These approaches provided critical insights into temporal patterns of neural activity, revealing correlations between brain waves and cognitive states. Nevertheless, the electrical activity observed in neurons is inseparable from its chemical basis. Action potentials and oscillatory rhythms are themselves the result of finely tuned ionic fluxes and receptor-mediated signaling, emphasizing the centrality of chemistry in conscious processes. Therefore, a comprehensive understanding of consciousness requires examining not only the electrical signals but also the molecular mechanisms—including neurotransmitters, neuromodulators, ion channels, intracellular messengers, and metabolic substrates—that support and regulate neural activity.

Neurotransmitters constitute the primary chemical agents of synaptic communication, translating presynaptic electrical signals into postsynaptic chemical and electrical responses. Glutamate, the principal excitatory neurotransmitter in the central nervous system, binds to ionotropic receptors such as AMPA, NMDA, and kainate receptors, as well as to metabotropic glutamate receptors (mGluRs). Ionotropic receptor activation results in rapid depolarization of the postsynaptic membrane through the flow of cations, primarily sodium and calcium, while metabotropic receptor activation initiates slower, modulatory signaling cascades involving second messengers such as inositol triphosphate (IP3) and diacylglycerol (DAG). Notably, NMDA receptors possess a voltage-dependent magnesium block, allowing them to function as coincidence detectors. This unique chemical property is critical for synaptic plasticity, enabling associative learning by linking coincident presynaptic and postsynaptic activity to intracellular calcium signaling.

Conversely, gamma-aminobutyric acid (GABA) serves as the dominant inhibitory neurotransmitter, binding primarily to GABA A and GABA B receptors. GABA A receptors are ligand-gated chloride channels that hyperpolarize the postsynaptic membrane, reducing neuronal excitability, whereas GABA B receptors are G-protein coupled and indirectly modulate potassium and calcium channels, contributing to slower inhibitory postsynaptic potentials. The chemical balance between excitatory and inhibitory neurotransmission is fundamental for maintaining network stability and preventing pathological hyperexcitability, which could impair conscious awareness.

Beyond glutamate and GABA, monoamines—dopamine, serotonin, and norepinephrine—act as neuromodulators, exerting diffuse chemical effects across broad neural networks. Dopamine, synthesized from the amino acid tyrosine through the enzymatic actions of tyrosine hydroxylase and aromatic L-amino acid decarboxylase, regulates reward, attention, and working memory. Its signaling involves D1-like and D2-like G-protein coupled receptors (GPCRs), which modulate adenylate cyclase activity and downstream cAMP-dependent pathways, ultimately influencing synaptic plasticity and neuronal excitability. Serotonin, derived from tryptophan, interacts with a variety of receptor subtypes, including ionotropic 5-HT3 and metabotropic 5-HT1 to 5-HT7 receptors, affecting mood regulation, sensory perception, and cognitive flexibility. Norepinephrine, synthesized primarily in the locus coeruleus, fine-tunes arousal and attention by modulating cortical network gain through α- and β-adrenergic receptor signaling. Collectively, these neuromodulators form a dynamic chemical environment that shapes the neural substrate of consciousness, influencing both global brain states and local circuit dynamics.

Neurotransmitter function is mediated by the chemical properties of their receptors. Ionotropic receptors, such as AMPA, NMDA, GABA A, and nicotinic acetylcholine receptors, consist of pore-forming protein complexes that transduce chemical ligand binding into ionic currents. The kinetics of channel opening and closing, as well as the receptor subunit composition, determine the amplitude and duration of postsynaptic potentials, influencing temporal summation, network synchronization, and the integration of sensory input. In contrast, metabotropic receptors function through more complex intracellular signaling cascades. Ligand binding to GPCRs induces conformational changes that promote the exchange of GDP for GTP on G-proteins, liberating α and βγ subunits to interact with downstream effectors such as adenylate cyclase, phospholipase C, or ion channels. These pathways modulate second messenger systems, including cAMP, IP3/DAG, and intracellular calcium levels, amplifying and diversifying chemical signals. Calcium ions, in particular, serve as universal second messengers, regulating kinase activity, transcription factor phosphorylation, and vesicle trafficking, thus enabling nuanced control of synaptic efficacy and neuronal network behavior.

The chemical composition of neuronal membranes also plays a critical role in consciousness. Lipid constituents, cholesterol, and embedded protein complexes influence the function of ion channels and receptors, thereby modulating excitability and synaptic transmission. Ion gradients, maintained by ATP-dependent pumps such as Na^+/K^+-ATPase and Ca^{2+}-ATPase, represent chemical potential energy that drives action potentials and neurotransmitter release. Disruptions in these gradients compromise synaptic fidelity, highlighting the inseparable link between chemical homeostasis and the maintenance of conscious states.

Synaptic plasticity—the brain's ability to modify its connectivity in response to experience—is central to consciousness, and chemical mechanisms underlie these changes. Long-term potentiation (LTP) and long-term depression (LTD) exemplify activity-dependent synaptic modifications mediated by precise molecular events. At glutamatergic synapses, LTP induction involves calcium influx through NMDA receptors, which activates calcium/calmodulin-dependent protein kinase II (CaMKII). CaMKII phosphorylates AMPA receptor subunits, increasing their conductance and promoting their insertion into the postsynaptic membrane, chemically strengthening the synapse. Additionally, retrograde messengers such as nitric oxide modulate presynaptic release, demonstrating a bidirectional chemical dialogue that stabilizes synaptic potentiation. LTD, in contrast, relies on lower levels of calcium influx, activating phosphatases such as calcineurin that dephosphorylate AMPA receptors and promote their internalization, fine-tuning circuit function. These chemically orchestrated processes provide the substrate for learning, memory, and the dynamic experience of consciousness.

Slow-acting neuromodulators further shape global states of consciousness, including attention, arousal, and perceptual awareness. Dopamine, norepinephrine, and serotonin influence large-scale network dynamics through GPCR-mediated chemical signaling. For example, dopaminergic modulation in the prefrontal cortex regulates working memory and executive function, while norepinephrine enhances signal-to-noise ratios in cortical circuits, facilitating selective attention. Endocannabinoids, lipid-derived neuromodulators, act retrogradely to transiently inhibit presynaptic neurotransmitter release, modulating network excitability.

Neuropeptides such as orexin and substance P also affect arousal and alertness through GPCR-mediated kinase signaling pathways. This complex chemical interplay establishes the neurochemical landscape of consciousness, with variations corresponding to different behavioral and experiential states, including sleep, meditation, and pharmacologically induced alterations of awareness.

The brain's energy metabolism is another critical chemical determinant of consciousness. Neurons are metabolically demanding, requiring continuous ATP supply to sustain ion gradients and action potentials. Oxidative phosphorylation within mitochondria generates the bulk of neuronal ATP, fueling Na^+/K^+-ATPase and Ca^{2+} pumps that maintain ionic homeostasis. Glucose metabolism provides both energy and precursors for neurotransmitter synthesis, linking metabolic chemistry directly to synaptic function. Reactive oxygen species (ROS), although potentially damaging at high levels, also serve as chemical signaling molecules at physiological concentrations, influencing kinase activity, gene transcription, and synaptic plasticity. Mitochondrial dysfunction or oxidative stress, however, can disrupt these chemical processes, impairing synaptic transmission and altering conscious experience.

Chemical insights have also informed theoretical frameworks of consciousness. Integrated Information Theory (IIT), while computationally oriented, implies chemical constraints on information processing, since neuronal signaling, synaptic plasticity, and network connectivity are mediated by molecules with specific kinetic properties. Molecular hypotheses, such as those implicating microtubules in conscious processes, emphasize chemical and quantum properties of protein assemblies. Tubulin dimers, for instance, undergo conformational changes and exhibit dipolar interactions, suggesting that intraneuronal molecular states may contribute to coherent conscious processes. Though these theories remain controversial, they underscore that consciousness is not solely an emergent electrical phenomenon but a complex chemical interplay spanning multiple scales of organization.

Immune signaling within the brain adds an additional chemical dimension to consciousness. Microglia, astrocytes, and peripheral immune cells release cytokines and chemokines that interact with neuronal receptors, modulating synaptic plasticity and network function. Proinflammatory cytokines, such as interleukin-1β and tumor necrosis factor-alpha, can reduce synaptic efficacy by altering calcium dynamics and receptor trafficking, producing cognitive deficits and altered states of awareness.

Anti-inflammatory signaling, conversely, maintains homeostasis and supports normal conscious processing. Furthermore, immune-mediated chemical alterations can influence neurotransmitter systems—for instance, elevated cytokines can diminish dopamine synthesis or enhance glutamate excitotoxicity—highlighting the integration of systemic chemical processes with neuronal function.

Emerging chemical tools provide unprecedented opportunities to dissect consciousness. Optogenetics and chemogenetics enable precise molecular control of specific neuronal populations, revealing causal relationships between chemical signaling and behavioral states. Advances in single-cell metabolomics and imaging of neurotransmitter dynamics allow real-time mapping of chemical correlates of consciousness, while pharmacological manipulations combined with computational modeling illuminate the chemical logic underlying awareness. These technologies, integrating molecular, metabolic, and signaling perspectives, promise to refine our understanding of how chemistry generates and modulates conscious experience.

In conclusion, consciousness is a multifaceted phenomenon deeply rooted in chemistry. From neurotransmitter signaling and receptor dynamics to ion channel function, energy metabolism, neuromodulation, and immune interactions, chemical processes underpin every aspect of awareness. Synaptic plasticity, neuromodulatory signaling, and metabolic homeostasis provide the molecular scaffolding that supports learning, perception, and adaptive behavior. Chemical perspectives also inform theoretical approaches, offering mechanistic constraints and testable hypotheses about the emergence of conscious experience. Future research will continue to bridge the molecular and systems levels, elucidating how the dynamic interplay of chemical signals gives rise to the rich tapestry of consciousness. Understanding these processes not only advances neuroscience but also deepens our appreciation for the chemical nature of awareness itself.

Beyond classical neurotransmission and neuromodulation, consciousness emerges from a highly dynamic chemical interplay among diverse molecular actors within neural networks. One important aspect involves astrocytes and glial cells, which were long thought to play only supportive roles but are now recognized as active participants in chemical signaling. Astrocytes regulate extracellular ion concentrations, particularly potassium and calcium, via chemical buffering and transporters, shaping neuronal excitability. They also release gliotransmitters such as glutamate, D-serine, and ATP, which act on nearby neurons to modulate synaptic plasticity. For example, astrocytic release of D-serine acts as a co-agonist at NMDA receptors, enhancing calcium influx during synaptic activity and facilitating LTP. Chemically, this establishes a feedback loop in which glial cells tune the excitatory-inhibitory balance and influence the spatiotemporal dynamics of conscious processing.

Calcium signaling in glial-neuronal interactions represents another critical chemical layer. Intracellular calcium waves within astrocytes can propagate over hundreds of micrometers, coordinating neuronal ensembles and affecting network oscillations linked to attention and perception. These chemical calcium waves interact with second messenger systems, including IP3 and DAG, ultimately influencing synaptic release probability and receptor sensitivity. Thus, consciousness is not solely the product of neuronal chemistry but arises from a distributed molecular network encompassing both neurons and glia, orchestrated through intricate chemical signaling.

Epigenetic and transcriptional regulation adds a further chemical dimension to consciousness. Activity-dependent changes in neuronal gene expression rely on chemically mediated signaling cascades initiated at synapses. Calcium influx and second messenger activation trigger phosphorylation of transcription factors, such as CREB, which binds to DNA response elements to modulate gene transcription. Post-translational modifications, including histone acetylation and methylation, dynamically alter chromatin accessibility, providing a chemical basis for long-term adaptations in neuronal circuits. For example, histone acetyltransferases catalyze the addition of acetyl groups to lysine residues on histones, reducing their positive charge and loosening DNA-histone interactions, chemically enabling transcription. These molecular modifications integrate transient synaptic events with long-term changes in circuit connectivity, illustrating how chemical processes shape the persistent substrate of conscious experience.

Neuromodulator interactions further amplify the chemical complexity underlying consciousness. Dopamine, serotonin, and norepinephrine do not operate in isolation; their chemical interactions modulate network dynamics through synergistic or antagonistic effects. For instance, dopamine receptor activity can influence serotonin signaling by altering receptor phosphorylation or trafficking, while norepinephrine can potentiate or dampen dopaminergic effects depending on receptor subtype activation. These interactions extend to second messenger cross-talk, where cAMP, IP3, DAG, and calcium signaling converge to fine-tune synaptic strength, excitability, and oscillatory activity. The emergent properties of these chemical networks are critical for higher-order cognitive functions, including decision-making, working memory, and selective attention, all central features of conscious awareness.

Lipid signaling and membrane microdomains add yet another molecular layer. Lipid rafts, cholesterol- and sphingolipid-enriched microdomains in the neuronal membrane, chemically organize receptors, ion channels, and signaling proteins, facilitating rapid and coordinated responses to neurotransmitters and neuromodulators. For example, GPCRs localized in lipid rafts may exhibit altered ligand affinity or signaling efficiency due to local lipid chemistry, influencing downstream cAMP or calcium signaling. Phosphoinositides, a class of signaling lipids, are phosphorylated by specific kinases to generate PIP2 and PIP3, which act as docking sites for intracellular effectors and influence actin cytoskeleton dynamics. By chemically scaffolding receptors and signaling molecules, lipid microdomains contribute to the precision, speed, and integration of synaptic signals underlying conscious processes.

Redox chemistry and reactive oxygen species (ROS) also influence consciousness at a molecular level. ROS, including superoxide and hydrogen peroxide, act as chemical modulators of kinase and phosphatase activity, gene expression, and ion channel function. Low levels of ROS can enhance synaptic plasticity and neuronal excitability, while excessive ROS production leads to oxidative stress, damaging membranes, proteins, and nucleic acids. The brain's intricate antioxidant systems—including glutathione, superoxide dismutase, and catalase—maintain chemical homeostasis, ensuring that ROS levels remain within a range that supports adaptive neural function. This chemical balance between pro-oxidants and antioxidants is therefore crucial for sustaining the molecular processes that underlie conscious awareness.

Intracellular organelles, particularly mitochondria, provide additional chemical complexity. Beyond ATP production, mitochondria regulate calcium buffering and release, lipid metabolism, and apoptotic signaling, all of which influence synaptic function. Mitochondrial calcium uptake affects neurotransmitter release probability, while mitochondrial dynamics—fusion, fission, and transport along dendrites—chemically modulate local energy supply and calcium homeostasis. Disruptions in mitochondrial chemistry, such as altered electron transport or impaired oxidative phosphorylation, lead to synaptic dysfunction and altered conscious states, highlighting the chemical underpinnings of energy-dependent awareness.

Emergent chemical network dynamics illustrate how consciousness arises from multiscale interactions. At the microscale, individual molecules such as neurotransmitters, ions, and second messengers interact chemically to influence receptor function and synaptic efficacy. At the mesoscale, chemical interactions across neuronal populations produce oscillations, synchrony, and network motifs. At the macroscale, global chemical states—modulated by neuromodulators, metabolic substrates, and hormonal signals—shape attention, perception, and awareness. This multilevel chemical integration suggests that consciousness emerges not from single molecules or cells but from the dynamic orchestration of chemical processes spanning multiple spatial and temporal scales.

Pharmacological modulation of consciousness provides further evidence for the chemical basis of awareness. Psychoactive substances, from classical psychedelics to anesthetics, act primarily by chemically altering neurotransmitter or receptor function. For example, psychedelics such as LSD and psilocybin bind to serotonin 5-HT2A receptors, altering downstream calcium and IP3 signaling and producing profound changes in perception, cognition, and self-awareness. General anesthetics, in contrast, often potentiate GABAergic inhibition or inhibit excitatory NMDA receptor signaling, chemically suppressing synaptic integration and global network activity, leading to reversible unconsciousness. These pharmacological interventions underscore the fundamental role of chemical signaling in regulating conscious states.

Finally, the gut-brain axis highlights a surprising chemical influence on consciousness. Microbial metabolites, such as short-chain fatty acids, neurotransmitter precursors, and bile acids, can cross the blood-brain barrier or modulate peripheral signaling pathways, influencing central neurotransmitter levels, receptor sensitivity, and neuromodulatory tone. For example, gut-derived tryptophan can enhance central serotonin synthesis, chemically affecting mood, perception, and attentional states. This demonstrates that consciousness is not solely determined by brain chemistry but is shaped by systemic chemical interactions spanning multiple organ systems.

In sum, these additional chemical perspectives reinforce the concept that consciousness is an emergent property of complex, multiscale chemical networks. Astrocytic gliotransmission, epigenetic regulation, lipid signaling, redox dynamics, mitochondrial chemistry, neuromodulator cross-talk, pharmacological modulation, and peripheral chemical influences all converge to orchestrate neural activity in a manner that gives rise to awareness. By integrating these diverse chemical mechanisms, we gain a richer and more nuanced understanding of consciousness, highlighting its dependence on precise molecular interactions and chemical homeostasis.

Chapter 50: Word Problems

1. NMDA receptors act as coincidence detectors in glutamatergic synapses due to their voltage-dependent magnesium block. Imagine a scenario in which a pharmacological agent partially removes this magnesium block, allowing increased calcium influx even at subthreshold presynaptic activity. Describe, at a molecular and chemical level, how this alteration could affect the intracellular calcium signaling cascades, including the activation of CaMKII and downstream phosphorylation of AMPA receptors. Predict how such chemical changes might influence long-term potentiation, synaptic plasticity, and the associative learning processes that underpin conscious perception, paying special attention to the balance of excitatory and inhibitory neurotransmission.

2. The inhibitory neurotransmitter GABA maintains neural network stability through activation of GABA A and GABA B receptors. GABA A receptors function as ligand-gated chloride channels, while GABA B receptors are G-protein coupled receptors that indirectly regulate potassium and calcium channels. Suppose a mutation selectively reduces GABA B receptor efficacy while leaving GABA A signaling intact. Analyze, in terms of chemical signaling, second messenger interactions, and ionic fluxes, how this selective inhibition could alter network oscillations, the timing of inhibitory postsynaptic potentials, and ultimately the capacity for conscious integration of sensory information.

3. Dopaminergic modulation of prefrontal cortex activity plays a critical role in working memory and attention. Dopamine acts through D1-like and D2-like GPCRs, influencing adenylate cyclase activity and cAMP-dependent signaling cascades. Imagine an experimental intervention that either hyperactivates or inhibits adenylate cyclase selectively in prefrontal cortical neurons. Describe the expected chemical and molecular consequences on downstream kinases, AMPA receptor phosphorylation, and neuronal excitability, and explain how these changes could manifest behaviorally in terms of altered working memory performance and conscious decision-making processes.

4. Serotonin influences mood, perception, and cognitive flexibility through a diverse array of receptor subtypes, including ionotropic 5-HT3 and metabotropic 5-HT1–5-HT7 receptors. Consider the administration of a pharmacological agent that selectively activates 5-HT3 receptors while blocking 5-HT1 receptors. Explain, at the chemical signaling level, how such modulation would impact ion flux, second messenger cascades, and neuronal excitability, and predict the resulting effects on network integration, emotional processing, and the phenomenological aspects of conscious awareness.

5. Norepinephrine released from the locus coeruleus modulates cortical network gain through α- and β-adrenergic receptor signaling. Suppose that during heightened attentional demand, norepinephrine concentrations rise in specific cortical microdomains. Discuss, in chemical terms, how the binding of norepinephrine to α- versus β-adrenergic receptors differentially alters second messenger pathways, modulates ion channel conductances, and influences signal-to-noise ratios. Evaluate the implications of these chemical dynamics for selective attention, conscious perception of salient stimuli, and the integration of sensory inputs across cortical networks.

6. Long-term potentiation (LTP) at glutamatergic synapses requires calcium influx through NMDA receptors, leading to CaMKII activation and phosphorylation of AMPA receptor subunits. Consider a pathological condition in which intracellular calcium buffering is impaired, leading to prolonged calcium transients. Describe the chemical consequences for kinase activity, AMPA receptor trafficking, and postsynaptic excitability. Predict how these molecular changes could disrupt LTP induction, alter synaptic strength, and affect the conscious experience of learning and memory.

7. Long-term depression (LTD) involves lower levels of calcium influx activating phosphatases such as calcineurin, which dephosphorylate AMPA receptors and promote receptor internalization. Imagine a scenario in which pharmacological inhibition of calcineurin is applied selectively to a cortical microcircuit. Analyze, at the chemical level, how this inhibition would alter AMPA receptor dynamics, postsynaptic potential amplitude, and network oscillatory behavior. Discuss the potential consequences for synaptic plasticity, memory consolidation, and the fine-tuning of conscious experience.

8. Astrocytes actively contribute to synaptic plasticity by releasing gliotransmitters such as D-serine, which acts as a co-agonist at NMDA receptors. Design a molecular experiment to quantify the influence of astrocytic D-serine on LTP induction, detailing the chemical measurements you would perform, such as intracellular calcium levels, NMDA receptor activation kinetics, and AMPA receptor phosphorylation. Predict how varying D-serine availability would modulate synaptic strength, neuronal network synchrony, and conscious perception of stimuli.

9. Consider a scenario in which the enzymatic degradation of IP_3 is experimentally slowed, prolonging the duration and spatial spread of astrocytic calcium waves. Using chemical reasoning, analyze how this prolonged calcium signaling could influence the kinetics of gliotransmitter release (e.g., glutamate, ATP, D-serine), including vesicular and non-vesicular mechanisms, as well as receptor occupancy at postsynaptic sites. Predict how sustained astrocytic calcium elevations might alter postsynaptic receptor sensitivity, synaptic integration, and the temporal dynamics of network oscillations involved in attention, working memory, and conscious awareness. Include discussion of feedback loops mediated by metabotropic receptors and downstream second messenger cascades.

10. Suppose experimental manipulation reduces cholesterol content, altering raft composition and membrane fluidity. Explain, at the molecular and chemical level, how changes in lipid packing and lateral mobility of proteins could influence receptor localization, ligand-binding affinity, and coupling efficiency to G-proteins or ion channels. Predict how these chemical modifications could impact the integration of excitatory and inhibitory inputs, spike timing-dependent plasticity, network synchronization, and higher-order emergent properties such as conscious perception or attentional focus.

11. Consider a condition in which ROS levels are modestly elevated within physiological limits. Chemically analyze how ROS might enhance kinase activity, transcription factor phosphorylation, and the oxidative modulation of ion channels to facilitate synaptic potentiation. Contrast this with scenarios of excessive ROS, detailing the chemical pathways leading to lipid peroxidation, protein carbonylation, calcium dysregulation, and impaired neurotransmission. Predict the resultant effects on network oscillations, synaptic integration, and the fidelity of conscious processing.

12. Describe, in chemical terms, how reduced ATP availability alters intracellular ion gradients, membrane potential, and action potential propagation. Evaluate the cascading effects on synaptic vesicle cycling, neurotransmitter release, and the temporal coordination of neuronal firing. Predict how these chemical deficits compromise network-level excitability, oscillatory synchronization, and ultimately the capacity for conscious awareness.

13. Explain the chemical mechanism of lysine acetylation and how altered acetylation patterns in cortical neurons could influence gene transcription profiles, synaptic protein synthesis, dendritic spine morphology, circuit connectivity, and long-term substrates of conscious processing. Discuss how feedback between synaptic activity and epigenetic modifications could reinforce or impair network-level integration.

14. Detail the chemical steps from endocannabinoid synthesis (e.g., 2-AG, anandamide) to CB1 receptor binding, G-protein activation, suppression of presynaptic calcium influx, and modulation of vesicle release machinery. Predict how these molecular events influence network excitability, oscillatory dynamics, temporal precision of spike trains, and aspects of conscious experience such as attentional gating or sensory discrimination.

15. Construct a computational model incorporating receptor kinetics for NMDA, AMPA, and GABA A/B channels, including conductance changes, ligand affinity, and desensitization rates. Simulate how shifts in excitatory-inhibitory balance chemically modify postsynaptic potentials, temporal summation, dendritic integration, and emergent oscillatory patterns. Discuss the implications for conscious awareness, including attentional focus, working memory capacity, and sensory binding.

16. Explain the chemical signaling pathways from cytokine receptor activation to intracellular second messenger cascades, post-translational modifications of receptors, and changes in synaptic vesicle release probability. Predict how acute versus chronic elevation of cytokines might disrupt network integration, cognitive function, and conscious processing, considering feedback interactions with astrocytes and neurons.

17. Propose a chemical mechanism by which D1 receptor activation could affect 5-HT receptor phosphorylation, trafficking, or G-protein coupling. Predict how these

molecular interactions might alter network oscillatory patterns, attentional focus, reward processing, and the conscious experience of mood or decision-making.

18. Analyze how acute and chronic administration alters kinase/phosphatase signaling, postsynaptic receptor trafficking, and synaptic plasticity. Predict the downstream network effects on oscillatory synchrony, information integration, learning, memory, and perceptual awareness.

19. Using chemical potential equations, calculate the energy cost required to sustain ion gradients in cortical neurons during sustained firing. Discuss how ATP depletion affects action potential generation, synaptic transmission, network excitability, and conscious processing at both molecular and systems levels.

20. Discuss how receptor kinetics, neurotransmitter turnover, enzymatic degradation, and second messenger lifetimes limit the temporal resolution and integration of information in cortical networks. Predict how chemical factors influence the complexity, richness, and stability of conscious experience.

21. Imagine applying receptor-specific agonists to cortical neurons while recording network activity. Describe the chemical signaling pathways activated by α_1, α_2, and β receptors, including IP_3/DAG production, calcium release, cAMP-dependent PKA activation, and downstream phosphorylation of ion channels or transcription factors. Predict how these pathways differentially influence neuronal excitability, spike timing, oscillatory synchrony, and attentional focus, and discuss the consequent effects on conscious processing of sensory inputs and working memory.

22. Detail the chemical sequence from ligand binding to receptor activation, including Gq-protein coupling, PLC-mediated IP_3 production, calcium mobilization, and activation of kinases such as PKC and MAPK. Discuss how these molecular events alter neuronal firing patterns, network-level excitability, sensory integration, and the subjective conscious experience, emphasizing molecular mechanisms underlying perceptual distortions, altered time perception, and cognitive flexibility.

23. If kinase-mediated phosphorylation of PIP_2 to PIP_3 is enhanced, predict the chemical consequences on the abundance of PIP_3, recruitment of PH-domain-containing proteins, and actin cytoskeleton dynamics. Discuss how these molecular

changes modulate receptor localization, synaptic vesicle trafficking, dendritic spine morphology, and network synchronization, with downstream effects on conscious processing of spatial and temporal information.

24. Describe the chemical sequence from calcium-dependent nitric oxide synthase (nNOS) activation to NO diffusion, interaction with presynaptic soluble guanylate cyclase, cGMP production, and modulation of neurotransmitter release. Predict how variations in NO levels could alter synaptic efficacy, oscillatory synchrony, and conscious perception, and discuss the implications for short-term versus long-term plasticity.

25. Consider a condition where mitochondrial calcium buffering is impaired, causing elevated cytosolic calcium. Describe the chemical consequences for vesicle fusion machinery (e.g., synaptotagmin), second messenger cascades, and kinase activity. Predict the downstream effects on network oscillations, spike timing, dendritic integration, and conscious perception of sensory stimuli.

26. Design a comprehensive molecular experiment to quantify how specific alterations in kinase activity (e.g., CaMKII, PKA, PKC) influence postsynaptic currents during high-frequency synaptic stimulation. Include considerations of experimental variables such as kinase inhibitors/activators, genetically encoded reporters of phosphorylation, and voltage-clamp measurements of receptor currents. Explain in detail the chemical interactions governing receptor opening probability, desensitization kinetics, phosphorylation-dependent conductance changes, and temporal summation of postsynaptic potentials. Predict how these chemically mediated modifications influence the excitability of local neuronal networks, the gating of attentional signals, the timing and amplitude of oscillatory patterns, and the broader substrate for conscious awareness. Discuss potential feedback loops between kinase activity, dendritic spine morphology, and synaptic plasticity that could amplify or dampen these effects.

27. Trace, in a stepwise chemical manner, the pathway from microbial production of tryptophan in the gut to central serotonin synthesis in cortical neurons, including transport across the blood-brain barrier, enzymatic hydroxylation by tryptophan hydroxylase, and subsequent decarboxylation by aromatic L-amino acid decarboxylase. Evaluate how variations in microbial metabolite availability,

enzymatic kinetics, and cofactor concentrations (e.g., BH_4) modulate cortical serotonin levels, receptor occupancy, downstream second messenger cascades, and network excitability. Predict how these chemical fluctuations impact the conscious experience of mood, affective regulation, attentional flexibility, and the coordination of cortical-subcortical networks involved in cognition and emotional processing.

28. Describe the detailed chemical mechanism by which ligand-bound DREADDs (Designer Receptors Exclusively Activated by Designer Drugs) modulate GPCR signaling, from ligand binding and conformational receptor change to G-protein dissociation, modulation of second messengers (cAMP, IP_3/DAG), and regulation of ion channels (e.g., potassium or calcium channels). Predict how targeted chemogenetic activation or inhibition chemically modifies synaptic plasticity, network oscillations, dendritic integration, and emergent cognitive behaviors, including attentional focus, sensory discrimination, and conscious decision-making. Discuss potential cross-talk with endogenous neurotransmitter systems and compensatory biochemical pathways.

29. Discuss how low physiological concentrations of reactive oxygen species (ROS) enhance calcium-dependent kinase activity (e.g., CaMKII), promote phosphorylation of synaptic proteins, and facilitate long-term potentiation, whereas high ROS levels chemically impair calcium channel function, disrupt phosphatase activity, induce oxidative modifications of proteins and lipids, and compromise synaptic efficacy. Predict the consequences of these chemically distinct ROS conditions on network oscillatory dynamics, synchronization of neuronal ensembles, information integration, and conscious processing. Include discussion of redox-sensitive feedback mechanisms and antioxidant buffering capacity.

30. Detail the extracellular enzymatic conversion of ATP to adenosine via ectonucleotidases (CD39, CD73) and describe the chemical kinetics of adenosine binding to A_1 and A_2A receptors. Predict how variations in ATP release, extracellular degradation rates, and receptor sensitivity chemically alter the excitatory-inhibitory balance, synaptic integration, dendritic summation, and the temporal resolution of neural coding, ultimately influencing the conscious perception of temporal sequences, rhythmic stimuli, and attentional focus.

31. Design a detailed chemical experiment to measure the temporal dynamics of PKA phosphorylation in response to phasic dopamine release in cortical neurons. Include the use of fluorescent or FRET-based reporters, phospho-specific antibodies, and kinetic modeling of cAMP accumulation. Explain how these phosphorylation events chemically modulate AMPA receptor trafficking, postsynaptic excitability, dendritic spine remodeling, and network-level plasticity. Predict the influence of these molecular changes on the neural substrates of working memory, decision-making, and conscious choice behavior, incorporating both temporal and spatial network integration.

32. Consider an experimental alteration of phosphoinositide ratios in neuronal membranes. Predict the chemical consequences for GPCR and ion channel localization, second messenger efficiency, and interactions with cytoskeletal elements (actin remodeling). Discuss how these molecular changes affect synaptic integration, dendritic spine morphology, spike timing-dependent plasticity, network synchrony, and conscious awareness, emphasizing feedback between lipid signaling and activity-dependent synaptic modification.

33. Detail the chemical reaction catalyzed by histone methyltransferases (HMTs) and predict how altered methylation patterns (mono-, di-, or tri-methylation) influence gene expression, synaptic protein synthesis, dendritic spine architecture, and structural-functional plasticity necessary for conscious cognition. Discuss potential feedback mechanisms linking synaptic activity, calcium-dependent signaling, and epigenetic remodeling that can stabilize or disrupt conscious information processing.

34. Describe the chemical mechanism of GABA A receptor potentiation by general anesthetics, including ligand binding, conformational changes, chloride influx, and enhanced inhibitory signaling. Predict how these modifications chemically suppress network oscillations, dendritic integration, and information transfer, and how they affect the subjective experience of consciousness. Include considerations of dose-dependence, receptor subtype specificity, and network resilience mechanisms.

35. Discuss the chemical pathways through which retrograde messengers such as NO and endocannabinoids act, including enzymatic synthesis, diffusion, receptor binding, second messenger cascades, and modulation of presynaptic release

machinery. Predict network-level consequences on synaptic strength, oscillatory synchrony, temporal integration, and the continuity of conscious experience, with attention to feedback between pre- and postsynaptic compartments.

36. Analyze the chemical interactions that differentiate moderate versus excessive ROS effects, including protein oxidation, lipid peroxidation, calcium channel modulation, DNA damage, and redox-sensitive signaling pathways. Predict the impact of these chemically mediated changes on network oscillations, synaptic integration, plasticity, and conscious processing across molecular, cellular, and systems scales, highlighting thresholds for functional disruption versus facilitation.

37. Using receptor-specific chemical pathways, predict how simultaneous modulation of dopaminergic (D1/D2) and adrenergic (α/β) signaling alters neuronal firing patterns, spike timing, network synchrony, attentional focus, and working memory. Discuss the implications for chemical regulation of conscious awareness, cognitive flexibility, and the integration of multi-modal sensory and motivational information.

38. Describe the chemical sequence from GPCR activation to liberation of α and βγ G-protein subunits and their modulation of potassium and calcium channels. Predict downstream effects on postsynaptic potentials, temporal summation, oscillatory dynamics, and conscious perception of sensory events, including how channel kinetics and second messenger cross-talk influence network-level integration.

39. Construct a detailed chemical model simulating how NMDA receptor co-agonist (D-serine or glycine) concentration affects calcium influx, LTP induction thresholds, postsynaptic excitability, dendritic spine remodeling, and network synchrony. Predict the behavioral and conscious consequences of altered co-agonist availability, including learning, memory consolidation, and sensory integration.

40. Describe molecular reactions that occur during oxidative stress, including protein carbonylation, lipid peroxidation, calcium channel dysfunction, and redox-mediated enzyme inactivation. Predict how these chemical alterations compromise network oscillations, synaptic plasticity, and conscious processing, and discuss potential compensatory mechanisms such as antioxidant enzyme induction and redox homeostasis.

41. Examine how simultaneous receptor modulation of dopamine, serotonin, and norepinephrine influences kinase activity, second messenger cross-talk, and

synaptic plasticity. Predict the resulting effects on decision-making, working memory, attentional control, and conscious cognition, emphasizing network-level integration of chemical signals.

42. Detail the chemical steps from GABA B receptor activation to G-protein-mediated potassium channel opening, including second messenger involvement, effector interactions, and resulting hyperpolarization. Predict how altered kinetics affect network rhythm, temporal integration, dendritic summation, and conscious processing of ongoing sensory inputs.

43. Analyze the chemical consequences of altered membrane lipid composition on ion channel opening/closing rates, postsynaptic potential amplitudes, and action potential propagation. Predict how these changes influence network synchronization, dendritic integration, and conscious awareness, including potential compensatory mechanisms involving lipid signaling or cytoskeletal remodeling.

44. Describe the chemical pathways through which microglial cytokines and chemokines maintain synaptic homeostasis, including receptor activation, second messenger signaling, and transcriptional feedback loops. Predict how suppression or dysregulation of anti-inflammatory signaling could disrupt synaptic plasticity, network oscillations, and conscious experience.

45. Quantify, using chemical principles, how variations in endocannabinoid levels modulate presynaptic neurotransmitter release probability, postsynaptic potentials, and network excitability. Predict the downstream impact on conscious perception, attentional flexibility, cognitive performance, and the integration of excitatory-inhibitory balance across cortical networks.

46. Detail the chemical cascade linking calcium influx to CREB phosphorylation, histone modification, chromatin remodeling, and gene transcription. Predict how these molecular events sustain synaptic plasticity, circuit connectivity, and the enduring substrate of conscious experience, including potential feedback between neuronal activity and epigenetic regulation.

47. Analyze how enhanced or reduced PLC activity alters intracellular calcium release from stores, PKC activation, DAG signaling, and synaptic vesicle release probability. Discuss consequences for temporal integration, oscillatory dynamics, information coding, and conscious awareness.

48. Model chemically how glucose deprivation reduces ATP production, perturbs Na^+/K^+ and Ca^{2+} gradients, compromises action potential generation, and disrupts postsynaptic signaling. Predict network-level effects, oscillatory disruptions, and consequences for conscious processing under energy-limited conditions.
49. Evaluate chemically how psychedelic compounds acting on 5-HT2A receptors modify calcium signaling, kinase activation, receptor trafficking, and cortical network integration. Predict changes in perception, cognition, sensory integration, and subjective conscious experience, including altered time perception, sensory binding, and thought patterns.
50. Propose a comprehensive framework linking microscale interactions of neurotransmitters, ions, second messengers, and redox molecules to mesoscale oscillatory patterns and macroscale behavioral states. Discuss how chemical perturbations at any level—enzymatic, receptor-mediated, or redox—could alter emergent conscious experience, network integration, cognitive flexibility, and behavioral output. Include consideration of feedback across scales, from molecular to systems-level consciousness.

Science References

Chapter 1

Bernard Jaffe, Crucibles: The Story of Chemistry. Dover Publications, 1976.

Mary Elvira Weeks, The Discovery of the Elements. Journal of Chemical Education, multiple editions.

William H. Brock, *The History of Chemistry*. W.W. Norton, 1992.

Trevor H. Levere, *Transforming Matter: A History of Chemistry from Alchemy to the Buckyball*. Johns Hopkins University Press, 2001.

Lawrence M. Principe, *The Secrets of Alchemy*. University of Chicago Press, 2013.

E.J. Holmyard, *Alchemy*. Dover Publications, 1990.

Jack Lindsay, *The Origins of Alchemy in Graeco-Roman Egypt*. Barnes & Noble, 1970.

Joseph Needham, *Science and Civilisation in China: Volume 5, Part 4 – Chemistry and Chemical Technology*. Cambridge University Press.

David Gordon White, *The Alchemical Body: Siddha Traditions in Medieval India*. University of Chicago Press, 1996.

Syed Nomanul Haq, *Names, Natures, and Things: The Alchemist Jabir ibn Hayyan and His Kitab al-Ahjar*. Kluwer Academic, 1994.

George Sarton, articles on Islamic science in *Isis* (multiple volumes).

Ahmad Y. al-Hassan, *Arab Civilization: Contributions to Science and Technology*. Center for Arab Unity Studies.

Walter Pagel, Paracelsus: An Introduction to Philosophical Medicine in the Era of the Renaissance. Karger, 1958.

Allen G. Debus, *The Chemical Philosophy: Paracelsian Science and Medicine in the Sixteenth and Seventeenth Centuries*. Dover, 2002.

Robert Boyle, *The Sceptical Chymist* (1661).

Michael Hunter, *Boyle: Between God and Science*. Yale University Press, 2009.

Antoine Lavoisier, *Elements of Chemistry* (1789).

Henry Guerlac, *Lavoisier: The Crucial Year*. Cornell University Press, 1961.

Peter Atkins & Julio de Paula, *Physical Chemistry*, 10th ed. Oxford University Press.

Raymond Chang & Kenneth Goldsby, *Chemistry*, McGraw-Hill.

Nivaldo Tro, *Chemistry: A Molecular Approach*. Pearson.

Lehninger Principles of Biochemistry, by Nelson & Cox. W.H. Freeman.

Jeremy M. Berg et al., *Biochemistry*. W.H. Freeman.

Clayden, Greeves, Warren & Wothers, *Organic Chemistry*. Oxford University Press.

Jonathan Clayden, *Organic Chemistry* (several editions).

Gary L. Miessler, Paul J. Fischer, *Inorganic Chemistry*. Pearson.

Catherine Housecroft & Alan G. Sharpe, *Inorganic Chemistry*. Pearson.

Harris, *Quantitative Chemical Analysis*. W.H. Freeman.

Skoog, Holler, & Crouch, *Principles of Instrumental Analysis*. Cengage.

Callister & Rethwisch, *Materials Science and Engineering*. Wiley.

Mark Ratner & Daniel Ratner, *Nanotechnology: A Gentle Introduction*. Prentice Hall.

Manahan, *Environmental Chemistry*. Lewis Publishers/CRC Press.

George Odian, *Principles of Polymerization*. Wiley.

Allen J. Bard, *Chemical Reviews* (for materials and nanotechnology overviews).

Thomas S. Kuhn, *The Structure of Scientific Revolutions*. University of Chicago Press.

James R. Partington, *A History of Chemistry* (4 volumes). Macmillan.

Holmyard, E. J. (1990). *Alchemy*. Dover Publications.

Jaffe, B. (1976). *Crucibles: The story of chemistry*. Dover Publications.

Lindsay, J. (1970). *The origins of alchemy in Graeco-Roman Egypt*. Barnes & Noble.

Needham, J. (1974). *Science and civilisation in China: Vol. 5, Part 4. Chemistry and chemical technology*. Cambridge University Press.

Principe, L. M. (2013). *The secrets of alchemy*. University of Chicago Press.

White, D. G. (1996). *The alchemical body: Siddha traditions in medieval India*. University of Chicago Press.

al-Hassan, A. Y. (1986). *Arab civilization: Contributions to science and technology*. Center for Arab Unity Studies.

Haq, S. N. (1994). *Names, natures, and things: The alchemist Jabir ibn Hayyan and his Kitab al-Ahjar*. Kluwer Academic.

Pagel, W. (1958). *Paracelsus: An introduction to philosophical medicine in the era of the Renaissance*. Karger.

Debus, A. G. (2002). *The chemical philosophy: Paracelsian science and medicine in the sixteenth and seventeenth centuries*. Dover Publications.

Boyle, R. (1661). *The sceptical chymist*.

Hunter, M. (2009). *Boyle: Between God and science*. Yale University Press.

Guerlac, H. (1961). *Lavoisier: The crucial year*. Cornell University Press.

Lavoisier, A. (1789). *Elements of chemistry*.

Levere, T. H. (2001). *Transforming matter: A history of chemistry from alchemy to the buckyball*. Johns Hopkins University Press.

Brock, W. H. (1992). *The history of chemistry*. W. W. Norton.

Weeks, M. E. (1932–1968). *The discovery of the elements*. Journal of Chemical Education.

Atkins, P., & de Paula, J. (2014). *Physical chemistry* (10th ed.). Oxford University Press.

Berg, J. M., Tymoczko, J. L., Gatto, G. J., & Stryer, L. (2015). *Biochemistry* (8th ed.). W. H. Freeman.

Callister, W. D., & Rethwisch, D. G. (2020). *Materials science and engineering: An introduction* (10th ed.). Wiley.

Chang, R., & Goldsby, K. (2016). *Chemistry* (13th ed.). McGraw-Hill.

Clayden, J., Greeves, N., Warren, S., & Wothers, P. (2012). *Organic chemistry* (2nd ed.). Oxford University Press.

Harris, D. C. (2015). *Quantitative chemical analysis* (9th ed.). W. H. Freeman.

Housecroft, C., & Sharpe, A. G. (2012). *Inorganic chemistry* (4th ed.). Pearson.

Manahan, S. E. (2017). *Environmental chemistry* (10th ed.). CRC Press.

Miessler, G. L., Fischer, P. J., & Tarr, D. A. (2014). *Inorganic chemistry* (5th ed.). Pearson.

Nelson, D. L., & Cox, M. M. (2017). *Lehninger principles of biochemistry* (7th ed.). W. H. Freeman.

Odian, G. (2004). *Principles of polymerization* (4th ed.). Wiley.

Ratner, M., & Ratner, D. (2003). *Nanotechnology: A gentle introduction to the next big idea*. Prentice Hall.

Skoog, D. A., Holler, F. J., & Crouch, S. R. (2017). *Principles of instrumental analysis* (7th ed.). Cengage.

Tro, N. J. (2020). *Chemistry: A molecular approach* (5th ed.). Pearson.

Kuhn, T. S. (2012). *The structure of scientific revolutions* (4th ed.). University of Chicago Press.

Partington, J. R. (1961–1970). *A history of chemistry* (Vols. 1–4). Macmillan.

Chapter 2

American Association for the Advancement of Science. (2011). *Vision and change in undergraduate biology education: A call to action*. AAAS.

Berg, J. M., Tymoczko, J. L., Gatto, G. J., & Stryer, L. (2019). *Biochemistry* (9th ed.). W. H. Freeman.

Brown, T. L., LeMay, H. E., Bursten, B. E., Murphy, C., Woodward, P., & Stoltzfus, M. (2021). *Chemistry: The central science* (15th ed.). Pearson.

Chang, R., & Goldsby, K. A. (2016). *Chemistry* (12th ed.). McGraw-Hill Education.

Cohen, E. R., Cvitas, T., Frey, J. G., Holmström, B., Kuchitsu, K., Marquardt, R., Mills, I., Pavese, F., Quack, M., Stohner, J., Straus, J., Takami, M., & Thor, A. J. (2007). *Quantities, units and symbols in physical chemistry* (3rd ed.). Royal Society of Chemistry.

Crowl, D. A., & Louvar, J. F. (2020). *Chemical process safety: Fundamentals with applications* (4th ed.). Pearson.

Giancoli, D. C. (2013). *Physics: Principles with applications* (7th ed.). Pearson.

International Bureau of Weights and Measures. (2019). *The International System of Units (SI)* (9th ed.). Bureau International des Poids et Mesures.

Masters, G. M., & Ela, W. P. (2015). *Introduction to environmental engineering and science* (3rd ed.). Pearson.

Moore, J. W. (Ed.). (2014). *Chemistry: The molecular science* (5th ed.). Cengage Learning.

Ohanian, H., & Markert, J. T. (2019). *Physics* (5th ed.). Pearson.

Pickering, M., Borden, R., & McNally, D. (2017). *Dimensions, units, and conversion factors*. In *Chemical engineering essentials for modular chemical process design*. CRC Press.

Silberberg, M. S., & Amateis, P. G. (2021). *Chemistry: The molecular nature of matter and change* (9th ed.). McGraw-Hill Education.

Taylor, J. R. (1997). *An introduction to error analysis: The study of uncertainties in physical measurements* (2nd ed.). University of Science Books.

Young, H. D., Freedman, R. A., & Ford, A. L. (2019). *University physics with modern physics* (15th ed.). Pearson.

Chapter 3

Atkins, P., & Jones, L. (2016). *Chemical principles: The quest for insight* (7th ed.). W. H. Freeman.

Atkins, P., Overton, T., Rourke, J., Weller, M., & Armstrong, F. (2010). *Shriver & Atkins' inorganic chemistry* (5th ed.). Oxford University Press.

Chandrasekhar, S. (1991). *Liquid crystals*. Cambridge University Press. *(General reference for molecular structure and bonding.)*

Dalton, J. (1808). *A new system of chemical philosophy* (Vol. 1). Bickerstaff.

Emsley, J. (2011). *Nature's building blocks: An A–Z guide to the elements* (Revised ed.). Oxford University Press.

Laidler, K. J., Meiser, J. H., & Sanctuary, B. C. (2008). *Physical chemistry* (4th ed.). Houghton Mifflin.

Moeller, T., Bailar, J. C., Kleinberg, J., Guss, C. O., Castellion, M. E., & Metz, C. (1982). *Chemistry with inorganic qualitative analysis* (5th ed.). Academic Press.

Moore, J. T., & Stanitski, C. L. (2015). *Principles of chemistry: The molecular science* (3rd ed.). Cengage Learning.

Oxtoby, D. W., Gillis, H. P., & Campion, A. (2015). *Principles of modern chemistry* (8th ed.). Cengage Learning.

Pauling, L. (1960). *The nature of the chemical bond* (3rd ed.). Cornell University Press.

Scerri, E. R. (2007). *The periodic table: Its story and its significance*. Oxford University Press.

Silberberg, M. S., & Amateis, P. (2018). *Chemistry: The molecular nature of matter and change* (9th ed.). McGraw-Hill Education.

Zumdahl, S. S., & Zumdahl, S. A. (2020). *Chemistry* (11th ed.). Cengage Learning.

Chapter 4

Brown, T. L., LeMay, H. E., Bursten, B. E., Murphy, C. J., Woodward, P. M., & Stoltzfus, M. W. (2018). *Chemistry: The central science* (14th ed.). Pearson.

Chang, R., & Goldsby, K. A. (2016). *Chemistry* (12th ed.). McGraw-Hill Education.

Hill, J. W., Petrucci, R. H., McCreary, T. W., & Perry, S. S. (2017). *General chemistry* (5th ed.). Pearson.

Moore, J. W., Stanitski, C. L., & Jurs, P. C. (2015). *Principles of chemistry: The molecular science* (3rd ed.). Cengage Learning.

Silberberg, M. S., & Amateis, P. A. (2021). *Chemistry: The molecular nature of matter and change* (9th ed.). McGraw-Hill Education.

Zumdahl, S. S., & Zumdahl, S. A. (2020). *Chemistry* (11th ed.). Cengage Learning.

Atkins, P., & de Paula, J. (2014). *Physical chemistry* (10th ed.). W. H. Freeman.

Housecroft, C. E., & Sharpe, A. G. (2018). *Inorganic chemistry* (5th ed.). Pearson.

American Chemical Society. (2020). *ACS style guide: Effective communication of scientific information* (3rd ed.). Oxford University Press.

Dalton, J. (1808). *A new system of chemical philosophy* (Vol. 1). Manchester University Press. *(Classical source related to atomic theory.)*

Chapter 5

Brown, T. L., LeMay, H. E., Bursten, B. E., Murphy, C., & Woodward, P. (2018). *Chemistry: The central science* (14th ed.). Pearson.

Chang, R., & Goldsby, K. (2016). *Chemistry* (12th ed.). McGraw-Hill Education.

Zumdahl, S. S., & Zumdahl, S. A. (2020). *Chemistry* (11th ed.). Cengage Learning.

Atkins, P., & Jones, L. (2015). *Chemical principles: The quest for insight* (7th ed.). W. H. Freeman.

Silberberg, M. S., & Amateis, P. (2021). *Chemistry: The molecular nature of matter and change* (9th ed.). McGraw-Hill Education.

Masterton, W. L., Hurley, C. N., & Neth, E. A. (2015). *Chemistry: Principles and reactions* (8th ed.). Cengage Learning.

Petrucci, R. H., Herring, F. G., Madura, J. D., & Bissonnette, C. (2017). *General chemistry: Principles and modern applications* (11th ed.). Pearson.

Chang, R. (2010). *Physical chemistry for the biosciences*. University Science Books.

Housecroft, C. E., & Sharpe, A. G. (2018). *Inorganic chemistry* (5th ed.). Pearson.

Atkins, P., & de Paula, J. (2014). *Atkins' physical chemistry* (10th ed.). Oxford University Press.

Choppin, G., Liljenzin, J.-O., Rydberg, J., & Ekberg, C. (2013). *Radiochemistry and nuclear chemistry* (4th ed.). Academic Press.

Krane, K. S. (1987). *Introductory nuclear physics*. Wiley.

Chapter 6

Atkins, P., & de Paula, J. (2018). *Atkins' physical chemistry* (11th ed.). Oxford University Press.

Atkins, P., Jones, L., & Laverman, L. (2020). *Chemical principles: The quest for insight* (8th ed.). W. H. Freeman.

Ball, D. W. (2015). *Physical chemistry* (2nd ed.). Cengage Learning.

Chang, R., & Goldsby, K. A. (2016). *Chemistry* (12th ed.). McGraw-Hill Education.

Dalton, J. (1803). *On the absorption of gases by water and other liquids*. Memoirs of the Literary and Philosophical Society of Manchester, 5, 271–287.

Graham, T. (1846). *On the motion of gases. Philosophical Transactions of the Royal Society of London*, 136, 573–631.

Masterton, W. L., Hurley, C. N., & Neth, E. A. (2015). *Chemistry: Principles and reactions* (8th ed.). Cengage Learning.

Moore, J. W., Stanitski, C. L., & Jurs, P. C. (2014). *Principles of chemistry: The molecular science* (3rd ed.). Cengage Learning.

Petrucci, R. H., Herring, F. G., Madura, J. D., & Bissonnette, C. (2017). *General chemistry: Principles and modern applications* (11th ed.). Pearson.

Seinfeld, J. H., & Pandis, S. N. (2016). *Atmospheric chemistry and physics: From air pollution to climate change* (3rd ed.). Wiley.

Van der Waals, J. D. (1873). *On the continuity of the gaseous and liquid states* (Doctoral dissertation). Leiden University.

Zumdahl, S. S., & Zumdahl, S. A. (2020). *Chemistry* (11th ed.). Cengage Learning.

Chapter 7

Alberts, B., Johnson, A., Lewis, J., Morgan, D., Raff, M., Roberts, K., & Walter, P. (2022). *Molecular biology of the cell* (7th ed.). W. W. Norton & Company.

Atkins, P., & de Paula, J. (2018). *Atkins' physical chemistry* (11th ed.). Oxford University Press.

Berg, J. M., Tymoczko, J. L., Gatto, G. J., & Stryer, L. (2019). *Biochemistry* (9th ed.). W. H. Freeman and Company.

Boyle, G. (Ed.). (2012). *Renewable energy: Power for a sustainable future* (3rd ed.). Oxford University Press.

Brown, T. L., Lemay, H. E., Bursten, B. E., Murphy, C., & Woodward, P. (2021). *Chemistry: The central science* (15th ed.). Pearson.

Demirbas, A. (2011). *Biofuels: Securing the planet's future energy needs*. Springer.

Hocking, M. B. (2020). *Handbook of chemical processes in industry: Energy, fuels, and environmental impact* (2nd ed.). CRC Press.

Intergovernmental Panel on Climate Change. (2021). *Climate change 2021: The physical science basis*. Cambridge University Press.

Momirlan, M., & Veziroglu, T. N. (2005). The properties of hydrogen as fuel tomorrow in sustainable energy system for a cleaner planet. *International Journal of Hydrogen Energy, 30*(7), 795–802.

National Research Council. (2011). *Climate change: Evidence, impacts, and choices*. National Academies Press.

Silberberg, M. S., & Amateis, P. (2021). *Chemistry: The molecular nature of matter and change* (9th ed.). McGraw-Hill Education.

Smil, V. (2017). *Energy and civilization: A history*. MIT Press.

Smil, V. (2019). *Growth: From microorganisms to megacities*. MIT Press.

Speight, J. G. (2019). *The chemistry and technology of petroleum* (5th ed.). CRC Press.

Tester, J. W., Drake, E. M., Driscoll, M. J., Golay, M. W., & Peters, W. A. (2012). *Sustainable energy: Choosing among options* (2nd ed.). MIT Press.

Twidell, J., & Weir, T. (2015). *Renewable energy resources* (3rd ed.). Routledge.

U.S. Environmental Protection Agency. (2023). *Inventory of U.S. greenhouse gas emissions and sinks*. U.S. Government Printing Office.

Chapter 8

Dalton, J. (1808). *A new system of chemical philosophy* (Vol. 1). Manchester University Press.

De Broglie, L. (1924). *Recherches sur la théorie des quanta* [Research on the quantum theory] (Doctoral dissertation). University of Paris.

Einstein, A. (1905). Über einen die Erzeugung und Verwandlung des Lichtes betreffenden heuristischen Gesichtspunkt. *Annalen der Physik, 322*(6), 132–148. https://doi.org/10.1002/andp.19053220607

Heisenberg, W. (1927). Über den anschaulichen Inhalt der quantentheoretischen Kinematik und Mechanik. *Zeitschrift für Physik, 43*(3–4), 172–198. https://doi.org/10.1007/BF01397280

Lavoisier, A.-L. (1789). *Traité élémentaire de chimie* [Elementary treatise of chemistry]. Paris: Chez Cuchet.

Mendeleev, D. (1869). On the relation of the properties of the elements to their atomic weights. *Journal of the Russian Chemical Society, 1*, 60–77.

Pauli, W. (1925). Über den Zusammenhang des Abschlusses der Elektronengruppen im Atom mit der Komplexstruktur der Spektren. *Naturwissenschaften, 13*(12), 28–30. https://doi.org/10.1007/BF01558878

Planck, M. (1901). Ueber das Gesetz der Energieverteilung im Normalspectrum. *Annalen der Physik, 309*(3), 553–563. https://doi.org/10.1002/andp.19013090310

Proust, J. L. (1806). Researches on the law of definite proportions. *Annales de Chimie et de Physique.*

Rutherford, E. (1911). The scattering of α and β particles by matter and the structure of the atom. *Philosophical Magazine, 21*(125), 669–688. https://doi.org/10.1080/14786440508637080

Schrödinger, E. (1926). Quantisierung als Eigenwertproblem. *Annalen der Physik, 385*(13), 437–490. https://doi.org/10.1002/andp.19263851302

Thomson, J. J. (1897). Cathode rays. *Philosophical Magazine, 44*(269), 293–316. https://doi.org/10.1080/14786449708621070

Bohr, N. (1913). On the constitution of atoms and molecules. *Philosophical Magazine, 26*(151), 1–25. https://doi.org/10.1080/14786441308634955

Dirac, P. A. M. (1928). The quantum theory of the electron. *Proceedings of the Royal Society A, 117*(778), 610–624. https://doi.org/10.1098/rspa.1928.0023

Feynman, R. P., Leighton, R. B., & Sands, M. (1963). *The Feynman lectures on physics* (Vols. 1–3). Addison-Wesley.

Griffiths, D. J. (2018). *Introduction to quantum mechanics* (3rd ed.). Cambridge University Press.

Atkins, P., & Friedman, R. (2020). *Molecular quantum mechanics* (6th ed.). Oxford University Press.

Ashcroft, N. W., & Mermin, N. D. (1976). *Solid state physics.* Saunders College Publishing.

Kittel, C. (2005). *Introduction to solid state physics* (8th ed.). Wiley.

Cohen-Tannoudji, C., Diu, B., & Laloë, F. (1977). *Quantum mechanics* (Vols. 1–2). Wiley.

Grosso, G., & Parravicini, G. P. (2014). *Solid state physics* (2nd ed.). Academic Press.

Zangwill, A. (2013). *Modern electrodynamics*. Cambridge University Press.

Sakurai, J. J., & Napolitano, J. (2017). *Modern quantum mechanics* (3rd ed.). Cambridge University Press.

Chapter 9

Anderson, J. D. (2007). *The atomic theory: A very short introduction*. Oxford University Press.

Atkins, P., & Friedman, R. (2020). *Molecular quantum mechanics* (6th ed.). Oxford University Press.

Atkins, P., Overton, T., Rourke, J., Weller, M., & Armstrong, F. (2018). *Shriver & Atkins' inorganic chemistry* (6th ed.). Oxford University Press.

Baggott, J. (2013). *Higgs: The invention and discovery of the "God particle"*. Oxford University Press.

Brush, S. G. (1999). *The history of modern science: A guide to the second scientific revolution, 1800–1950*. Iowa State University Press.

Close, F. (2015). *The infinity puzzle: Quantum field theory and the hunt for an orderly universe*. Basic Books.

Cohen, I. B. (1985). *The birth of a new physics* (Rev. ed.). W. W. Norton.

Dalton, J. (1808). *A new system of chemical philosophy* (Vol. 1). Bickerstaff Press. (Original work published 1808)

Dirac, P. A. M. (1981). *The principles of quantum mechanics* (4th ed.). Oxford University Press.

Feynman, R. P., Leighton, R. B., & Sands, M. (2011). *The Feynman lectures on physics* (New Millennium ed.). Basic Books.

Griffiths, D. J. (2018). *Introduction to quantum mechanics* (3rd ed.). Cambridge University Press.

Griffiths, D. J., & Schroeter, D. F. (2020). *Introduction to elementary particles* (3rd ed.). Cambridge University Press.

Hazen, R. M. (2012). *The story of Earth: The first 4.5 billion years*. Viking.

Kragh, H. (2002). *Quantum generations: A history of physics in the twentieth century*. Princeton University Press.

Lavoisier, A.-L. (1965). *Elements of chemistry* (R. Kerr, Trans.). Dover. (Original work published 1789)

Moore, J. T. (2015). *The history of chemistry*. Greenwood.

Pauling, L. (1960). *The nature of the chemical bond* (3rd ed.). Cornell University Press.

Peebles, P. J. E. (2020). *Cosmology's century: An inside history of our modern understanding of the universe*. Princeton University Press.

Rutherford, E. (1911). The scattering of α and β particles by matter and the structure of the atom. *Philosophical Magazine, 21*(125), 669–688.

Schrödinger, E. (1926). Quantization as an eigenvalue problem. *Annalen der Physik, 79*, 361–376.

Shankar, R. (2014). *Principles of quantum mechanics* (2nd ed.). Springer.

Thomson, J. J. (1897). Cathode rays. *Philosophical Magazine, 44*(269), 293–316.

Weinberg, S. (1994). *Dreams of a final theory*. Vintage.

Zee, A. (2010). *Quantum field theory in a nutshell* (2nd ed.). Princeton University Press.

Chapter 10

Atkins, P. W. (2025, October 30). *Chemical bonding — The quantum mechanics of bonding*. In *Chemical bonding*. Encyclopaedia Britannica.

Ding, L., Matito, E., & Schilling, C. (2025). From entanglement to bonds: Chemical bonding concepts from quantum information theory. *arXiv*. https://doi.org/10.48550/arXiv.2501.15699

Levine, D. S., & Head-Gordon, M. (2020). Clarifying the quantum mechanical origin of the covalent chemical bond. *Nature Communications, 11*, 4893. https://doi.

org/10.1038/s41467-020-18670-8

Szalay, S., Barcza, G., Szilvási, T., Veis, L., & Legeza, Ö. (2017). The correlation theory of the chemical bond. *Scientific Reports, 7*, 2237.

"Models of Chemical Bonding – Quantum-mechanical models." (n.d.). In *Chemistry LibreTexts*.

Pauling, L. (1931). The hybrid orbital model. In discussions of bonding models and orbital hybridization. As described in "Models of Chemical Bonding – Quantum-mechanical models." *Chemistry LibreTexts.*

"The source of chemical bonding." (2014). *Studies in History and Philosophy of Science Part A, 45*, 1–13. https://doi.org/10.1016/j.shpsa.2013.10.011

Heitler, W., & London, F. (1927). Quantum-mechanical treatment of the hydrogen molecule — The first successful wave-mechanical account of the covalent bond. As described in "Chemical bonds: quantum chemical theory of chemical bonds." In *Ch 3 Chemical bonds*, Iwanami (n.d.).

Chapter 11

Atkins, P., & de Paula, J. (2014). *Physical chemistry* (10th ed.). Oxford University Press.

Atkins, P., Jones, L., & Laverman, L. (2022). *Chemical principles: The quest for insight* (8th ed.). W. H. Freeman.

Atkins, P., Overton, T., Rourke, J., Weller, M., & Armstrong, F. (2010). *Shriver & Atkins' inorganic chemistry* (5th ed.). Oxford University Press.

Bransden, B. H., & Joachain, C. J. (2003). *Physics of atoms and molecules* (2nd ed.). Prentice Hall.

Brown, T. L., LeMay, H. E., Bursten, B. E., Murphy, C., & Woodward, P. (2018). *Chemistry: The central science* (14th ed.). Pearson.

Chang, R., & Goldsby, K. (2016). *Chemistry* (12th ed.). McGraw-Hill Education.

Cotton, F. A., & Wilkinson, G. (1999). *Advanced inorganic chemistry* (6th ed.). Wiley.

Diebold, U. (2003). The surface science of titanium dioxide. *Surface Science Reports, 48*(5–8), 53–229. https://doi.org/10.1016/S0167-5729(02)00100-0

Graham, T. (1833). On the law of diffusion of gases. *The London and Edinburgh Philosophical Magazine and Journal of Science, 2*(8), 175–190.

Housecroft, C. E., & Sharpe, A. G. (2018). *Inorganic chemistry* (5th ed.). Pearson.

Laidler, K. J., Meiser, J. H., & Sanctuary, B. C. (1998). *Physical chemistry* (4th ed.). Houghton Mifflin.

Levine, I. N. (2014). *Physical chemistry* (6th ed.). McGraw-Hill Education.

McQuarrie, D. A., & Simon, J. D. (1999). *Molecular thermodynamics*. University Science Books.

McQuarrie, D. A., & Simon, J. D. (1997). *Physical chemistry: A molecular approach*. University Science Books.

Petrucci, R. H., Herring, F. G., Madura, J. D., & Bissonnette, C. (2017). *General chemistry: Principles and modern applications* (11th ed.). Pearson.

Rao, C. N. R. (1997). *Solid state chemistry and its applications*. Wiley.

Silbey, R. J., Alberty, R. A., & Bawendi, M. G. (2005). *Physical chemistry* (4th ed.). Wiley.

Sze, S. M., & Ng, K. K. (2006). *Physics of semiconductor devices* (3rd ed.). Wiley.

Tipler, P. A., & Llewellyn, R. (2008). *Modern physics* (5th ed.). W. H. Freeman.

Van der Waals, J. D. (1873). *Over de continuiteit van den gas- en vloeistoftoestand* [On the continuity of the gaseous and liquid state] (Doctoral dissertation, Leiden University).

Zumdahl, S. S., & Zumdahl, S. A. (2020). *Chemistry* (11th ed.). Cengage Learning.

Chapter 12

Atkins, P., & de Paula, J. (2014). *Physical chemistry* (10th ed.). Oxford University Press.

Atkins, P., Jones, L., & Laverman, L. (2020). *Chemical principles: The quest for insight* (8th ed.). W. H. Freeman.

Bard, A. J., & Faulkner, L. R. (2001). *Electrochemical methods: Fundamentals and applications* (2nd ed.). Wiley.

Berg, J. M., Tymoczko, J. L., Gatto Jr., G. J., & Stryer, L. (2019). *Biochemistry* (9th ed.). W. H. Freeman.

Birdi, K. S. (2013). *Surface and colloid chemistry: Principles and applications* (3rd ed.). CRC Press.

Birdi, K. S. (2015). *Self-assembly monolayer structures of lipids and macromolecules at interfaces*. Springer.

Brady, J. E., Senese, F., & Jespersen, N. D. (2016). *Chemistry: Matter and its changes* (7th ed.). Wiley.

Chang, R., & Thoman, J. W. (2019). *Physical chemistry for the chemical sciences* (2nd ed.). University Science Books.

de Gennes, P.-G., & Prost, J. (1993). *The physics of liquid crystals* (2nd ed.). Oxford University Press.

Einstein, A. (1956). *Investigations on the theory of the Brownian movement* (R. Fürth, Ed.; A. D. Cowper, Trans.). Dover. (Original work published 1905)

Fennell Evans, D., & Wennerström, H. (1999). *The colloidal domain: Where physics, chemistry, biology, and technology meet* (2nd ed.). Wiley-VCH.

Hiemenz, P. C., & Rajagopalan, R. (1997). *Principles of colloid and surface chemistry* (3rd ed.). CRC Press.

Israelachvili, J. N. (2011). *Intermolecular and surface forces* (3rd ed.). Academic Press.

Laidler, K. J., Meiser, J. H., & Sanctuary, B. C. (2018). *Physical chemistry* (5th ed.). Houghton Mifflin.

Masel, R. I. (1996). *Principles of adsorption and reaction on solid surfaces*. Wiley.

Ninham, B. W., & Lo Nostro, P. (2010). *Molecular forces and self-assembly: In colloid, nano sciences and biology*. Cambridge University Press.

Pashley, R. M., & Karaman, M. E. (2023). *Applied colloid and surface chemistry* (2nd ed.). Wiley.

Petrucci, R. H., Herring, F. G., Madura, J. D., & Bissonnette, C. (2017). *General chemistry: Principles and modern applications* (11th ed.). Pearson.

Russel, W. B., Saville, D. A., & Schowalter, W. R. (1989). *Colloidal dispersions*. Cambridge University Press.

Shaw, D. J. (1992). *Introduction to colloid and surface chemistry* (4th ed.). Butterworth-Heinemann.

Silberberg, M. S., & Amateis, P. (2021). *Chemistry: The molecular nature of matter and change* (9th ed.). McGraw Hill.

Somasundaran, P. (Ed.). (2006). *Encyclopedia of surface and colloid science* (2nd ed.). CRC Press.

Tanford, C. (1980). *The hydrophobic effect: Formation of micelles and biological membranes* (2nd ed.). Wiley.

van der Schoot, P., & Zwanikken, J. (2018). *Colloids and polymers: Science and applications*. Springer.

Whitten, K. W., Davis, R. E., Peck, M. L., & Stanley, G. G. (2020). *General chemistry* (11th ed.). Cengage Learning.

Chapter 13

Atkins, P., & de Paula, J. (2014). *Physical chemistry* (10th ed.). Oxford University Press.

Atkins, P., de Paula, J., & Keeler, J. (2018). *Atkins' physical chemistry* (11th ed.). Oxford University Press.

Baird, N. C. (1999). *Quantum chemistry* (2nd ed.). W. H. Freeman.

Bell, R. P. (1973). *The tunnel effect in chemistry*. Chapman & Hall.

Bell, R. P. (1974). *The proton in chemistry* (2nd ed.). Chapman & Hall.

Benson, S. W. (1960). *The foundations of chemical kinetics*. McGraw–Hill.

Benson, S. W. (1968). *Thermochemical kinetics* (2nd ed.). Wiley.

Briers, D. J., & Hill, T. L. (1974). *Statistical thermodynamics and kinetics*. Benjamin.

Chandler, D. (1987). *Introduction to modern statistical mechanics*. Oxford University Press.

Cremer, D. (Ed.). (2001). *Computational chemistry: A primer*. Wiley-VCH.

Eyring, H., Walter, J., & Kimball, G. E. (1944). *Quantum chemistry*. Wiley.

Evans, M. G., & Polanyi, M. (1935). Some applications of the transition state method to the calculation of reaction velocities, especially in solution. *Transactions of the Faraday Society, 31*, 875–894.

Espenson, J. H. (1995). *Chemical kinetics and reaction mechanisms* (2nd ed.). McGraw–Hill.

Hammes-Schiffer, S. (2001). Theoretical perspectives on proton-coupled electron transfer reactions. *Accounts of Chemical Research, 34*(4), 273–281.

House, J. E. (2007). *Principles of chemical kinetics* (2nd ed.). Academic Press.

Houser, R. P. (2020). *Bioinorganic chemistry*. Oxford University Press.

Laidler, K. J. (1987). *Chemical kinetics* (3rd ed.). Harper & Row.

Marcus, R. A. (1956). Theory of oxidation–reduction reactions involving electron transfer. *Journal of Chemical Physics, 24*(5), 966–978.

Marcus, R. A., & Sutin, N. (1985). Electron transfers in chemistry and biology. *Biochimica et Biophysica Acta (BBA) – Reviews on Bioenergetics, 811*(3), 265–322.

Masel, R. I. (1996). *Chemical kinetics and catalysis*. Wiley.

Masel, R. I. (1999). *Principles of adsorption and reaction on solid surfaces*. Wiley.

Michl, J., & Bončuk, V. (Ed.). (2015). *Molecular structure and energetics: Mechanochemistry*. Wiley-Interscience.

Nitzan, A. (2006). *Chemical dynamics in condensed phases: Relaxation, transfer, and reactions in condensed molecular systems*. Oxford University Press.

Nørskov, J. K., Studt, F., Abild-Pedersen, F., & Bligaard, T. (2014). *Fundamentals of heterogeneous catalysis*. Wiley.

Reed-Hill, R. E., Abbaschian, R., & Abbaschian, L. (2009). *Physical metallurgy principles* (4th ed.). Cengage.

Schatz, G. C., & Ratner, M. A. (2002). *Quantum mechanics in chemistry*. Prentice Hall.

Schmid, G. (2010). *Nanoparticles: From theory to application* (2nd ed.). Wiley-VCH.

Segel, I. H. (1993). *Enzyme kinetics: Behavior and analysis of rapid equilibrium and steady-state enzyme systems*. Wiley.

Szabo, A., & Ostlund, N. S. (1989). *Modern quantum chemistry: Introduction to advanced electronic structure theory*. Dover.

Transition State Theory. (2004). In W. Koch & M. C. Holthausen, *A chemist's guide to density functional theory* (2nd ed.). Wiley-VCH.

Turro, N. J., Ramamurthy, V., & Scaiano, J. C. (2010). *Modern molecular photochemistry of organic molecules*. University Science Books.

van 't Hoff, J. H. (1884). *Études de dynamique chimique*. Muller.

Zewail, A. H. (2000). Femtochemistry: Atomic-scale dynamics of the chemical bond. *Journal of Physical Chemistry A, 104*(24), 5660–5694.

Chapter 14

Atkins, P., & de Paula, J. (2023). *Atkins' physical chemistry* (12th ed.). Oxford University Press.

Atkins, P., de Paula, J., & Keeler, J. (2018). *Atkins' physical chemistry for the life sciences* (3rd ed.). Oxford University Press.

Chang, R., & Thoman, J. W. (2022). *Physical chemistry for the chemical sciences* (2nd ed.). University Science Books.

Chang, R., & Goldsby, K. A. (2021). *Chemistry* (14th ed.). McGraw-Hill Education.

DeVoe, H. (2015). *Thermodynamics and chemistry* (2nd ed.). Cambridge University Press.

Fogler, H. S. (2021). *Elements of chemical reaction engineering* (6th ed.). Pearson.

House, J. E. (2019). *Principles of chemical kinetics* (3rd ed.). Academic Press.

Huheey, J. E., Keiter, E. A., & Keiter, R. L. (2013). *Inorganic chemistry: Principles of structure and reactivity* (4th ed.). Pearson.

Laidler, K. J., Meiser, J. H., & Sanctuary, B. C. (2022). *Physical chemistry* (5th ed.). Houghton Mifflin.

Laidler, K. J., & Meiser, J. H. (1999). *Chemical kinetics* (2nd ed.). HarperCollins.

McQuarrie, D. A. (2000). *Statistical mechanics*. University Science Books.

McQuarrie, D. A., & Simon, J. D. (1997). *Physical chemistry: A molecular approach*. University Science Books.

Morrison, R. T., & Boyd, R. N. (2010). *Organic chemistry* (7th ed.). Pearson.

Peck, T., & Gelbart, W. M. (2011). *Statistical thermodynamics.* Oxford University Press.

Petrucci, R. H., Herring, F. G., Madura, J., & Bissonnette, C. (2017). *General chemistry: Principles and modern applications* (11th ed.). Pearson.

Pilling, M. J., & Seakins, P. W. (2023). *Reaction kinetics* (3rd ed.). Oxford University Press.

Silbey, R. J., Alberty, R. A., & Bawendi, M. (2004). *Physical chemistry* (4th ed.). Wiley.

Skoog, D. A., West, D. M., Holler, F. J., & Crouch, S. R. (2013). *Fundamentals of analytical chemistry* (9th ed.). Brooks/Cole.

Snoeyink, V. L., & Jenkins, D. (2020). *Water chemistry* (2nd ed.). Wiley.

Steinfeld, J. I., Francisco, J. S., & Hase, W. L. (1998). *Chemical kinetics and dynamics* (2nd ed.). Prentice Hall.

Tinoco, I., Sauer, K., Wang, J., & Puglisi, J. D. (2021). *Physical chemistry: Principles and applications in biological sciences* (6th ed.). Pearson.

Warren, S. C., & Schaefer, H. F. (2012). *Modern quantum chemistry* (2nd ed.). Dover.

Zumdahl, S. S., & Zumdahl, S. A. (2020). *Chemistry* (11th ed.). Cengage Learning.

Chapter 15

Arrhenius, S. (1887). Ü*ber die Dissociation der in Wasser gelösten Stoffe*. Zeitschrift für Physikalische Chemie, 1, 631–648.

Atkins, P., & de Paula, J. (2014). *Atkins' physical chemistry* (10th ed.). Oxford University Press.

Atkins, P., Overton, T., Rourke, J., Weller, M., & Armstrong, F. (2010). *Shriver & Atkins' inorganic chemistry* (5th ed.). Oxford University Press.

Baeyer, H. C. (1884). *Ueber die Constitution und die Bildung der Weinsäure und Traubensäure*. Annalen der Chemie, 224(2), 256–324.

Brønsted, J. N. (1923). Einige Bemerkungen über den Begriff der Säuren und Basen. *Recueil des Travaux Chimiques des Pays-Bas, 42*, 718–728.

de Paula, J., & Atkins, P. (2017). *Elements of physical chemistry* (7th ed.). Oxford University Press.

Ebbing, D. D., & Gammon, S. D. (2020). *General chemistry* (12th ed.). Cengage Learning.

Housecroft, C. E., & Sharpe, A. G. (2018). *Inorganic chemistry* (5th ed.). Pearson.

Jencks, W. P. (1969). *Catalysis in chemistry and enzymology*. McGraw–Hill.

Lewis, G. N. (1923). *Valence and the structure of atoms and molecules*. The Chemical Catalog Company.

Lide, D. R. (Ed.). (2004). *CRC handbook of chemistry and physics* (85th ed.). CRC Press.

Lowry, T. M. (1923). The uniqueness of hydrogen. *Journal of the Society of Chemical Industry, 42*, 43–47.

McMurry, J. (2021). *Organic chemistry* (10th ed.). Cengage Learning.

Mitchell, P. (1961). Coupling of phosphorylation to electron and hydrogen transfer by a chemiosmotic type of mechanism. *Nature, 191*(4784), 144–148.

Morrison, R. T., & Boyd, R. N. (2010). *Organic chemistry* (7th ed.). Pearson.

Petrucci, R. H., Herring, F. G., Madura, J. D., & Bissonnette, C. (2017). *General chemistry: Principles and modern applications* (11th ed.). Pearson.

Pitzer, K. S. (1995). Thermodynamics of electrolytes. I. Theoretical basis and general equations. *Journal of Physical Chemistry, 77*, 268–277.

Schlesinger, H. I., & Brown, H. C. (1932). Metal amides. *Chemical Reviews, 10*(1), 1–41.

Smith, J. M., Van Ness, H. C., & Abbott, M. M. (2005). *Introduction to chemical engineering thermodynamics* (7th ed.). McGraw–Hill.

Solomon, S., Qin, D., Manning, M., & Chen, Z. (Eds.). (2007). *Climate change 2007: The physical science basis*. Cambridge University Press.

Streitwieser, A., & Heathcock, C. H. (1985). *Introduction to organic chemistry* (3rd ed.). Macmillan.

Stumm, W., & Morgan, J. J. (1996). *Aquatic chemistry* (3rd ed.). Wiley–Interscience.

Zumdahl, S. S., & Zumdahl, S. A. (2020). *Chemistry* (11th ed.). Cengage Learning.

Chapter 16

Atkins, P., & de Paula, J. (2022). *Atkins' physical chemistry* (12th ed.). Oxford University Press.

Brønsted, J. N. (1923). Some remarks on the concept of acids and bases. *Journal of the American Chemical Society, 45*(12), 2898–2908. https://doi.org/10.1021/ja01665a015

Chang, R., & Goldsby, K. A. (2023). *Chemistry* (15th ed.). McGraw-Hill Education.

DeVoe, H. (2015). *Thermodynamics and chemistry* (2nd ed.). Pearson.

Housecroft, C. E., & Sharpe, A. G. (2018). *Inorganic chemistry* (5th ed.). Pearson.

Nernst, W. (1921). The theoretical principles of chemistry. *Science, 54*(1388), 569–573. https://doi.org/10.1126/science.54.1388.569

Petrucci, R. H., Herring, F. G., Madura, J. D., & Bissonnette, C. (2017). *General chemistry: Principles and modern applications* (11th ed.). Pearson.

Skoog, D. A., West, D. M., Holler, F. J., & Crouch, S. R. (2013). *Fundamentals of analytical chemistry* (9th ed.). Brooks/Cole.

Zumdahl, S. S., & Zumdahl, S. A. (2020). *Chemistry* (11th ed.). Cengage Learning.

Harris, D. C. (2015). *Quantitative chemical analysis* (9th ed.). W. H. Freeman.

Poon, T., & Enger, E. (2018). *Exploring biological chemistry* (2nd ed.). McGraw-Hill Education.

Stumm, W., & Morgan, J. J. (2012). *Aquatic chemistry: Chemical equilibria and rates in natural waters* (3rd ed.). Wiley.

Berg, J. M., Tymoczko, J. L., & Stryer, L. (2017). *Biochemistry* (9th ed.). W. H. Freeman.

Chapter 17

Atkins, P., & de Paula, J. (2022). *Atkins' physical chemistry* (12th ed.). Oxford University Press.

Baum, E. B. (2020). *Chemical thermodynamics* (2nd ed.). CRC Press.

Callen, H. B. (1985). *Thermodynamics and an introduction to thermostatistics* (2nd ed.). Wiley.

Chang, R., & Thoman, J. W. (2022). *Physical chemistry for the chemical sciences* (2nd ed.). University Science Books.

Clausius, R. (1867). *The mechanical theory of heat: With its applications to the steam-engine and to the physical properties of bodies.* John van Voorst. (Original work published earlier)

Engel, T., & Reid, P. (2020). *Physical chemistry* (5th ed.). Pearson.

Hill, T. L. (2012). *An introduction to statistical thermodynamics.* Dover Publications. (Original work published 1960)

Joule, J. P. (1884). *The scientific papers of James Prescott Joule*. Physical Society of London. (Original works published 1840s–1850s)

Kondepudi, D., & Prigogine, I. (2015). *Modern thermodynamics: From heat engines to dissipative structures* (2nd ed.). Wiley.

McQuarrie, D. A., & Simon, J. D. (1999). *Molecular thermodynamics.* University Science Books.

McQuarrie, D. A., & Simon, J. D. (1997). *Physical chemistry: A molecular approach.* University Science Books.

Reif, F. (2009). *Fundamentals of statistical and thermal physics.* Waveland Press. (Original work published 1965)

Smith, J. M., Van Ness, H. C., & Abbott, M. M. (2021). *Introduction to chemical engineering thermodynamics* (9th ed.). McGraw Hill.

Tipler, P. A., & Mosca, G. (2019). *Physics for scientists and engineers* (7th ed.). W. H. Freeman.

Zemansky, M. W., & Dittman, R. H. (1997). *Heat and thermodynamics* (7th ed.). McGraw-Hill.

Chapter 18

Atkins, P., & de Paula, J. (2014). *Physical chemistry* (10th ed.). Oxford University Press.

Bard, A. J., & Faulkner, L. R. (2001). *Electrochemical methods: Fundamentals and applications* (2nd ed.). John Wiley & Sons.

Brown, T. L., LeMay, H. E., Bursten, B. E., Murphy, C. J., & Woodward, P. (2018). *Chemistry: The central science* (14th ed.). Pearson.

Faulkner, L. R., & Harris, D. C. (2015). *Quantitative chemical analysis* (9th ed.). W. H. Freeman.

Housecroft, C. E., & Sharpe, A. G. (2018). *Inorganic chemistry* (5th ed.). Pearson.

Oxtoby, D. W., Gillis, H. P., & Campion, A. (2015). *Principles of modern chemistry* (8th ed.). Cengage Learning.

Silbey, R. J., Alberty, R. A., & Bawendi, M. G. (2005). *Physical chemistry* (4th ed.). Wiley.

Skoog, D. A., Holler, F. J., & Crouch, S. R. (2017). *Principles of instrumental analysis* (7th ed.). Cengage Learning.

Winter, M. (2024). *Battery technologies: Principles, materials, and applications*. Springer.

Zoski, C. G. (2007). *Handbook of electrochemistry*. Elsevier.

Chapter 19

Cotton, F. A., Murillo, C. A., & Bochmann, M. (1999). *Advanced inorganic chemistry* (6th ed.). Wiley.

Housecroft, C. E., & Sharpe, A. G. (2018). *Inorganic chemistry* (5th ed.). Pearson.

Miessler, G. L., Fischer, P. J., & Tarr, D. A. (2014). *Inorganic chemistry* (5th ed.). Pearson.

Huheey, J. E., Keiter, E. A., & Keiter, R. L. (1997). *Inorganic chemistry: Principles of structure and reactivity* (4th ed.). Prentice Hall.

Crabtree, R. H. (2014). *The organometallic chemistry of the transition metals* (6th ed.). Wiley.

Hartwig, J. F. (2010). *Organotransition metal chemistry: From bonding to catalysis*. University Science Books.

Shilov, A. E., & Shul'pin, G. B. (2002). Activation of C–H bonds by metal complexes. *Chemical Reviews, 97*(8), 2879–2932. https://doi.org/10.1021/cr9601018

Carlin, R. L. (2010). *Magnetochemistry.* Springer.

Figgis, B. N., & Hitchman, M. A. (2000). *Ligand field theory and its applications.* Wiley-VCH.

Kalyanasundaram, K. (1992). *Photochemistry of polypyridine and porphyrin complexes.* Academic Press.

Prier, C. K., Rankic, D. A., & MacMillan, D. W. C. (2013). Visible light photoredox catalysis with transition metal complexes. *Chemical Reviews, 113*(7), 5322–5363. https://doi.org/10.1021/cr300503r

West, A. R. (2014). *Solid state chemistry and its applications* (2nd ed.). Wiley.

Goodenough, J. B. (2015). *Magnetism and the chemical bond.* Wiley.

Imada, M., Fujimori, A., & Tokura, Y. (1998). Metal-insulator transitions. *Reviews of Modern Physics, 70*(4), 1039–1263. https://doi.org/10.1103/RevModPhys.70.1039

Daniel, M.-C., & Astruc, D. (2004). Gold nanoparticles: Assembly, supramolecular chemistry, quantum-size-related properties, and applications. *Chemical Reviews, 104*(1), 293–346. https://doi.org/10.1021/cr030698+

Burda, C., Chen, X., Narayanan, R., & El-Sayed, M. A. (2005). Chemistry and properties of nanocrystals of different shapes. *Chemical Reviews, 105*(4), 1025–1102. https://doi.org/10.1021/cr030063a

Bertini, I., Gray, H. B., Stiefel, E. I., & Valentine, J. S. (2007). *Biological inorganic chemistry: Structure and reactivity.* University Science Books.

Kelland, L. (2007). The resurgence of platinum-based cancer chemotherapy. *Nature Reviews Cancer, 7*(8), 573–584. https://doi.org/10.1038/nrc2167

Goodenough, J. B., & Park, K.-S. (2013). The Li-ion rechargeable battery: A perspective. *Journal of the American Chemical Society, 135*(4), 1167–1176. https://doi.org/10.1021/ja3091438

Zhang, H., & Chen, Y. (2009). Nanostructured photocatalytic materials for environmental purification. *Chemical Society Reviews, 38*(3), 706–721. https://doi.org/10.1039/B808701F

Chapter 20

Bruice, P. Y. (2021). *Organic chemistry* (9th ed.). Pearson.

McMurry, J. (2021). *Organic chemistry* (9th ed.). Cengage Learning.

Carey, F. A., & Sundberg, R. J. (2013). Advanced organic chemistry: Part A: Structure and mechanisms (5th ed.). Springer.

Vollhardt, K. P. C., & Schore, N. E. (2018). *Organic chemistry: Structure and function* (8th ed.). W. H. Freeman.

Smith, M. B., & March, J. (2007). March's advanced organic chemistry: Reactions, mechanisms, and structure (6th ed.). Wiley.

Geim, A. K., & Novoselov, K. S. (2007). The rise of graphene. *Nature Materials, 6*(3), 183–191. https://doi.org/10.1038/nmat1849

Kroto, H. W., Heath, J. R., O'Brien, S. C., Curl, R. F., & Smalley, R. E. (1985). C_{60}: Buckminsterfullerene. *Nature, 318*(6042), 162–163. https://doi.org/10.1038/318162a0

Robert, C., & Stevens, R. (2010). Heterocyclic chemistry in drug discovery. *Journal of Medicinal Chemistry, 53*(2), 539–556. https://doi.org/10.1021/jm901660p

Royal Society of Chemistry. (n.d.). *Organic chemistry: An introduction*. https://www.rsc.org/learn-chemistry

National Center for Biotechnology Information. (n.d.). *PubChem compound database*. https://pubchem.ncbi.nlm.nih.gov

Chapter 21

Brasseur, G. P., Orlando, J. J., & Tyndall, G. S. (1999). *Atmospheric chemistry and global change*. Oxford University Press.

Seinfeld, J. H., & Pandis, S. N. (2016). *Atmospheric chemistry and physics: From air pollution to climate change* (3rd ed.). John Wiley & Sons.

Jacob, D. J. (1999). *Introduction to atmospheric chemistry*. Princeton University Press.

IPCC. (2021). *Climate change 2021: The physical science basis. Contribution of Working Group I to the Sixth Assessment Report of the Intergovernmental Panel on Climate Change*. Cambridge University Press. https://www.ipcc.ch/report/ar6/wg1/

Finlayson-Pitts, B. J., & Pitts, J. N. (2000). *Chemistry of the upper and lower atmosphere: Theory, experiments, and applications*. Academic Press.

Charlson, R. J., & Heintzenberg, J. (1995). *Aerosol forcing of climate*. John Wiley & Sons.

Schlesinger, W. H., & Bernhardt, E. S. (2013). *Biogeochemistry: An analysis of global change* (3rd ed.). Academic Press.

Liu, S., & Millero, F. J. (2002). The solubility of gases in seawater. In *Treatise on geochemistry* (Vol. 6, pp. 17–47). Elsevier.

Vitousek, P. M., Aber, J. D., Howarth, R. W., Likens, G. E., Matson, P. A., Schindler, D. W., Schlesinger, W. H., & Tilman, D. G. (1997). Human alteration of the global nitrogen cycle: Sources and consequences. *Ecological Applications, 7*(3), 737–750. https://doi.org/10.1890/1051-0761(1997)007[0737:HAOTGN]2.0.CO;2

Mann, M. E., & Kump, L. R. (2015). *Dire predictions: Understanding climate change*. DK Publishing.

Chapter 22

Atkins, P., & de Paula, J. (2014). *Atkins' physical chemistry* (10th ed.). Oxford University Press.

Brown, T. L., LeMay, H. E., Bursten, B. E., Murphy, C., Woodward, P., & Stoltzfus, M. W. (2018). *Chemistry: The central science* (14th ed.). Pearson.

Emsley, J. (2011). *Nature's building blocks: An A-Z guide to the elements* (2nd ed.). Oxford University Press.

Greenwood, N. N., & Earnshaw, A. (2012). *Chemistry of the elements* (2nd ed.). Butterworth-Heinemann.

Lide, D. R. (Ed.). (2005). *CRC handbook of chemistry and physics* (86th ed.). CRC Press.

Scerri, E. (2007). *The periodic table: Its story and its significance*. Oxford University Press.

Zumdahl, S. S., & Zumdahl, S. A. (2014). *Chemistry* (10th ed.). Cengage Learning.

Royal Society of Chemistry. (n.d.). *The periodic table of elements*. https://www.rsc.org/periodic-table

PubChem. (n.d.). *Chemical and physical properties of elements*. https://pubchem.ncbi.nlm.nih.gov

Chapter 23

Greenwood, N. N., & Earnshaw, A. (2012). *Chemistry of the elements* (2nd ed.). Butterworth-Heinemann.

Lide, D. R. (Ed.). (2005). *CRC handbook of chemistry and physics* (86th ed.). CRC Press.

Emsley, J. (2011). *Nature's building blocks: An A–Z guide to the elements* (2nd ed.). Oxford University Press.

Scerri, E. R. (2013). *The periodic table: Its story and its significance* (2nd ed.). Oxford University Press.

Cotton, F. A., Wilkinson, G., Murillo, C. A., & Bochmann, M. (1999). *Advanced inorganic chemistry* (6th ed.). Wiley.

WebElements. (n.d.). *The periodic table on the web.* https://www.webelements.com

Royal Society of Chemistry. (n.d.). *Periodic table of elements.* https://www.rsc.org/periodic-table

U.S. Geological Survey. (2023). *Mineral commodity summaries.* https://www.usgs.gov/centers/nmic/mineral-commodity-summaries

International Union of Pure and Applied Chemistry (IUPAC). (n.d.). *Periodic Table of the Elements.* https://iupac.org/what-we-do/periodic-table-of-elements

Chapter 24

Carey, F. A., & Sundberg, R. J. (2007). Advanced organic chemistry: Part A: Structure and mechanisms (5th ed.). Springer.

Patrick, G. L. (2017). An introduction to medicinal chemistry (6th ed.). Oxford University Press.

Brunton, L. L., Hilal-Dandan, R., & Knollmann, B. C. (2018). *Goodman & Gilman's: The pharmacological basis of therapeutics* (13th ed.). McGraw-Hill Education.

Katzung, B. G., Trevor, A. J., Masters, S. B., & Kruidering-Hall, M. (2018). *Basic and clinical pharmacology* (14th ed.). McGraw-Hill Education.

Silverman, R. B., & Holladay, M. W. (2014). The organic chemistry of drug design and drug action (3rd ed.). Academic Press.

March, J. (1992). Advanced organic chemistry: Reactions, mechanisms, and structure (4th ed.). Wiley.

Nelson, D. L., Cox, M. M., Lehninger, A. L., & Cox, M. M. (2021). *Lehninger principles of biochemistry* (8th ed.). W. H. Freeman.

Green, T. W., & Wuts, P. G. M. (2018). *Protective groups in organic synthesis* (5th ed.). Wiley.

Tropp, B. E. (2012). *Molecular biology: Genes to proteins* (3rd ed.). Jones & Bartlett Learning.

Lodish, H., Berk, A., Kaiser, C. A., Krieger, M., Bretscher, A., Ploegh, H., Amon, A., & Scott, M. P. (2021). *Molecular cell biology* (9th ed.). W. H. Freeman.

Chapter 25

Berg, J. M., Tymoczko, J. L., & Stryer, L. (2019). *Biochemistry* (9th ed.). W.H. Freeman.

Bowen, W. H., & Koo, H. (2011). Biology of Streptococcus mutans–derived glucosyltransferases: Role in extracellular matrix formation of cariogenic biofilms. *Caries Research, 45*(1), 69–86. https://doi.org/10.1159/000324598

Darvell, B. W. (2018). *Materials science for dentistry* (10th ed.). Woodhead Publishing.

Featherstone, J. D. B. (2000). The science and practice of caries prevention. *Journal of the American Dental Association, 131*(7), 887–899. https://doi.org/10.14219/jada.archive.2000.0307

Hannig, M., & Joiner, A. (2006). The structure, function, and properties of the acquired pellicle. *Monographs in Oral Science, 19*, 29–64. https://doi.org/10.1159/000090585

Kidd, E. A. M., & Fejerskov, O. (2016). *Dental caries: The disease and its clinical management* (3rd ed.). Wiley-Blackwell.

Lussi, A., Hellwig, E., & Zero, D. (2019). Erosive tooth wear: Diagnosis, risk factors, prevention, and therapy. *Karger.*

Marsh, P. D., & Zaura, E. (2017). Dental biofilm: Ecological interactions in health and disease. *Journal of Clinical Periodontology, 44*(S18), S12–S22. https://doi.org/10.1111/jcpe.12679

Newbrun, E. (2012). *Cariology* (4th ed.). Quintessence Publishing.

O'Brien, W. J. (2008). *Dental materials and their selection* (4th ed.). Quintessence Publishing.

Pashley, D. H., Tay, F. R., Breschi, L., & Carvalho, R. M. (2011). State of the art etch-and-rinse adhesives. *Dental Materials, 27*(1), 1–16. https://doi.org/10.1016/j.dental.2010.10.016

Phillips, R. W., & Anusavice, K. J. (2013). *Phillips' science of dental materials* (12th ed.). Saunders.

Ten Cate, J. M. (1999). Current concepts on the theories of the mechanism of action of fluoride. *Acta Odontologica Scandinavica, 57*(6), 325–329. https://doi.org/10.1080/000163599428562

van Loveren, C. (2004). Sugar alcohols: What's new? *Caries Research, 38*(3), 286–293. https://doi.org/10.1159/000077763

Wang, Y., & Spencer, P. (2002). Hybridization efficiency of the adhesive/dentin interface with wet bonding. *Journal of Dental Research, 81*(2), 80–85. https://doi.org/10.1177/154405910208100205

Zhao, J., Liu, Y., Sun, W., & Zhang, H. (2018). Artificial enamel via biomimetic mineralization. *Journal of Dental Research, 97*(2), 148–157. https://doi.org/10.1177/0022034517735292

Chapter 26

Baker, M. J., & Bell, S. (2023). *Forensic chemistry* (3rd ed.). Pearson.

Bell, S. (2020). *Forensic chemistry*. CRC Press.

Bertol, E., Vaiano, F., Mari, F., & Favretto, D. (2015). Advances in LC–MS/MS analysis of new psychoactive substances (NPSs). *Current Pharmaceutical Biotechnology, 16*(9), 1010–1020. https://doi.org/10.2174/1389201016666150619095809

Brennan, J. S., & Day, J. (2018). Chemical analysis of explosives and explosive residues. *Forensic Science Review, 30*(2), 119–150.

Casey, E. (2019). *Environmental forensics: Principles & applications* (2nd ed.). Academic Press.

Cody, R. B. (2019). Mass spectrometry in forensic science. *Mass Spectrometry Reviews, 38*(4–5), 297–303. https://doi.org/10.1002/mas.21581

DeHaan, J. D., & Icove, D. J. (2020). *Kirk's fire investigation* (8th ed.). Pearson.

Houck, M. M., & Siegel, J. A. (2015). *Fundamentals of forensic science* (3rd ed.). Academic Press.

Kobilinsky, L., Liotti, T. F., & Oeser-Sweat, J. (Eds.). (2011). *Forensic chemistry handbook*. Wiley.

Moffat, A. C., Osselton, M. D., & Widdop, B. (Eds.). (2016). *Clarke's analysis of drugs and poisons* (4th ed.). Pharmaceutical Press.

Moore, J. C. (2019). Forensic analysis of controlled substances. *Analytical Chemistry, 91*(1), 97–115. https://doi.org/10.1021/acs.analchem.8b05230

Orfila, M. (1848). *Traité de toxicologie* (3rd ed.). Paris: Chez Crochard. (Historical reference)

Pyrak-Nolte, L. J., & Wenk, H. R. (2017). Microscopy and microanalysis in forensic science. *Microscopy Today, 25*(5), 24–31. https://doi.org/10.1017/S1551929517000920

Saferstein, R. (2021). *Criminalistics: An introduction to forensic science* (13th ed.). Pearson.

Skoog, D. A., Holler, F. J., & Crouch, S. R. (2017). *Principles of instrumental analysis* (7th ed.). Cengage Learning.

Smith, F. P. (2018). *Forensic toxicology: Principles and concepts* (2nd ed.). Springer.

Verma, R., & Kumar, R. (2018). Recent advances in portable analytical instrumentation for on-site forensic analysis. *TrAC Trends in Analytical Chemistry, 105*, 88–102. https://doi.org/10.1016/j.trac.2018.05.001

Chapter 27

Drysdale, D. (2011). *An introduction to fire dynamics* (3rd ed.). Wiley.

Glassman, I., Yetter, R. A., & Glumac, N. G. (2014). *Combustion* (5th ed.). Academic Press.

Turns, S. R. (2012). *An introduction to combustion: Concepts and applications* (3rd ed.). McGraw-Hill.

Luca, Ș., & Kozempel, S. (2020). *Fire chemistry*. Springer.

Law, C. K. (2010). *Combustion physics*. Cambridge University Press.

Atkins, P., de Paula, J., & Keeler, J. (2018). *Atkins' physical chemistry* (11th ed.). Oxford University Press.

Laidler, K. J. (1995). *Chemical kinetics* (3rd ed.). HarperCollins.

Kee, R. J., Coltrin, M. E., & Glarborg, P. (2003). *Chemically reacting flow: Theory & practice*. Wiley.

Seinfeld, J. H., & Pandis, S. N. (2016). *Atmospheric chemistry and physics: From air pollution to climate change* (3rd ed.). Wiley.

Bockhorn, H. (Ed.). (1994). *Soot formation in combustion*. Springer.

Drysdale, D., & Macmillan, A. (2019). *Fire behavior and fire protection engineering*. CRC Press.

DeHaan, J. D., & Icove, D. J. (2020). *Kirk's fire investigation* (8th ed.). Pearson.

Quintiere, J. G. (2016). *Principles of fire behavior* (2nd ed.). CRC Press.

NFPA. (2021). *NFPA 921: Guide for fire and explosion investigations*. National Fire Protection Association.

Hull, T. R., & Witkowski, A. (2020). *Fire retardancy of polymers: New strategies and mechanisms*. Royal Society of Chemistry.

Alongi, J., Carosio, F., & Malucelli, G. (2015). *Current emerging techniques to impart flame retardancy to fabrics: A review. Polymer Degradation and Stability, 113*, 10–35

Chapter 28

Berg, J. M., Tymoczko, J. L., Gatto, G. J., Jr., & Stryer, L. (2019). *Biochemistry* (9th ed.). W. H. Freeman.

Chaplin, M. (2021). *Water structure and science*. London South Bank University. https://water.lsbu.ac.uk

Czernuszewicz, R. S., & Pemberton, J. E. (Eds.). (2020). *Aqueous chemistry and geochemistry of oxides, oxyhydroxides, and related materials*. American Chemical Society.

Debenedetti, P. G., & Stanley, H. E. (2003). Supercooled and glassy water. *Physics Today, 56*(6), 40–46. https://doi.org/10.1063/1.1580053

Eisenberg, D., & Kauzmann, W. (2005). *The structure and properties of water*. Oxford University Press. (Original work published 1969)

Grotthuss, C. J. T. (2006). Memoir on the decomposition of water and of the bodies that it holds in solution by means of galvanic electricity. *Biochimica et Biophysica Acta, 1757*(8), 871–875. (Original work published 1806)

Huang, C., & Chandler, D. (2000). Temperature and length scale dependence of hydrophobic effects and their possible implications for protein folding. *Proceedings of the National Academy of Sciences, 97*(15), 8324–8327. https://doi.org/10.1073/pnas.120176397

Israelachvili, J. N. (2011). *Intermolecular and surface forces* (3rd ed.). Academic Press.

Klamt, A. (2011). *COSMO-RS: From quantum chemistry to fluid thermodynamics and drug design*. Elsevier.

Levin, Y., dos Santos, A. P., & Diehl, A. (2019). Ions at the air–water interface: An end to a hundred-year-old mystery? *Physical Review Letters, 103*(25), 257802. https://doi.org/10.1103/PhysRevLett.103.257802

Mulder, M. (2020). *Basic principles of membrane technology* (3rd ed.). Springer.

Natelson, D. (2022). *Nanostructures and nanotechnology* (2nd ed.). Cambridge University Press.

Nie, X., & Hu, W. (2017). Proton transport in water and water-filled materials: Insights from molecular simulations. *Accounts of Chemical Research, 50*(12), 2915–2923. https://doi.org/10.1021/acs.accounts.7b00133

Qiang, Z., Liu, W., & Wilson, C. (2020). *Advanced oxidation processes for water purification*. Elsevier.

Sposito, G. (2017). *The chemistry of soils* (3rd ed.). Oxford University Press.

Ullmann's Encyclopedia of Industrial Chemistry. (2021). *Water treatment*. Wiley-VCH.

Wang, Z., Liu, G., & Payne, G. F. (2019). Water-enabled self-assembly and its biological implications. *Chemical Reviews, 119*(24), 12495–12540. https://doi.org/10.1021/acs.chemrev.9b00374

Zhang, H., Liu, D., & Li, J. (2018). Semiconductor photocatalysis for water purification: Fundamentals and applications. *Chemical Society Reviews, 47*(7), 1711–1737. https://doi.org/10.1039/C7CS00571F

Zhao, S., Zou, L., Tang, C. Y., & Mulcahy, D. (2012). Recent developments in forward osmosis: Opportunities and challenges. *Journal of Membrane Science, 396*, 1–21. https://doi.org/10.1016/j.memsci.2011.12.023

Chapter 29

Ball, P. (2001). *Bright Earth: Art and the invention of color.* University of Chicago Press.

Berrie, B. H. (Ed.). (2007). *Artists' pigments: A handbook of their history and characteristics* (Vol. 4). National Gallery of Art.

Brunetti, B. G., Sgamellotti, A., & Miliani, C. (2016). *Science and art: The painted surface*. Royal Society of Chemistry.

Carlyle, L. (2001). *The artist's assistant: Oil painting instruction manuals and handbooks in Britain, 1800–1900*. Archetype Publications.

Coddington, J., van den Berg, K. J., Osmond, G., Centeno, S. A., & Learner, T. (Eds.). (2020). *Modern paints: Uncovered*. Springer.

de la Rie, E. R. (1988). Aging and yellowing of resin varnishes. *Studies in Conservation, 33*(1), 1–13.

Doménech-Carbó, M. T. (2016). *Electrochemistry of pigments and conservation of artworks*. Springer.

Feller, R. L. (Ed.). (1986). *Artist's pigments: A handbook of their history and characteristics* (Vol. 1). National Gallery of Art.

Hermens, E. (Ed.). (1998). *Looking through paintings: The study of painting techniques and materials in support of art historical research*. Archetype Publications.

Lluveras-Tenorio, A., Mazurek, J., van den Berg, K. J., & Bonaduce, I. (2021). Aging of oil paint: A chemical review. *Chemical Reviews, 121*(19), 13004–13043.

Mayer, R. (1991). *The artist's handbook of materials and techniques* (5th ed.). Viking.

Mills, J. S., & White, R. (1994). *The organic chemistry of museum objects* (2nd ed.). Butterworth-Heinemann.

Ploeger, R., & Scalarone, D. (2015). The chemistry of modern acrylic paints. *Progress in Organic Coatings, 78*, 35–47.

Roy, A. (Ed.). (1993). *Artists' pigments: A handbook of their history and characteristics* (Vol. 2). National Gallery of Art.

Salvadori, O. (2019). Degradation processes in pigments and binder media. In A. Casini (Ed.), *Science and art: The painted surface* (pp. 45–78). Royal Society of Chemistry.

Schreiner, M., Melcher, M., & Uhlir, K. (2007). Scanning electron microscopy and energy-dispersive X-ray analysis in materials science in conservation. *Analytical and Bioanalytical Chemistry, 387*, 737–747.

Stols-Witlox, M. (2017). Ground layers in European easel paintings. In S. Centeno et al. (Eds.), *The science of a great painting* (pp. 95–124). Conservation Institute Press.

van den Berg, J. D. J., et al. (2019). Metal soap formation in oil paintings: History, chemistry, and prevention. *Studies in Conservation, 64*(sup1), 1–15.

Zhang, X., & Schilling, M. (2015). Light-induced degradation of organic colorants. *Journal of Cultural Heritage, 16*(6), 882–889.

Chapter 30

Barel, A. O., Paye, M., & Maibach, H. I. (Eds.). (2014). *Handbook of cosmetic science and technology* (4th ed.). CRC Press.

Butler, H. (Ed.). (2016). *Poucher's perfumes, cosmetics and soaps* (11th ed.). Springer.

de Groot, A. C. (2019). *Cosmetics ingredients: A complete guide* (4th ed.). CRC Press.

Draelos, Z. D. (2016). *Cosmetic dermatology: Products and procedures* (3rd ed.). Wiley-Blackwell.

Elsner, P., & Maibach, H. I. (Eds.). (2018). *Cosmetic formulation of skin-care products* (2nd ed.). CRC Press.

Garg, A., & Maibach, H. I. (2005). Cosmetic moisturizers. *Skin Pharmacology and Physiology, 18*(6), 297–302. https://doi.org/10.1159/000088068

Laba, D. (Ed.). (1993). *Rheological properties of cosmetics and toiletries*. CRC Press.

Lodzki, M., & Barenholz, Y. (2020). Liposomes and other lipid-based nanocarriers in cosmetic and dermatologic delivery. *Journal of Controlled Release, 325*, 129–146. https://doi.org/10.1016/j.jconrel.2020.06.041

Loretz, L. J., Api, A. M., Barraj, L. M., Burdick, J., & Gennings, C. (2008). Exposure data for cosmetic products: Lipstick, lip gloss, and lip balm. *Food and Chemical Toxicology, 46*(5), 1516–1524. https://doi.org/10.1016/j.fct.2007.12.011

Neto, R. M. B., & Grimaldi, R. (2021). Advances in cosmetic surfactant science: Self-assembly, structure, and function. *Colloids and Surfaces A: Physicochemical and Engineering Aspects, 623*, 126757. https://doi.org/10.1016/j.colsurfa.2021.126757

Pardeike, J., Schwabe, K., & Müller, R. H. (2010). Nanoemulsions in cosmetic and pharmaceutical applications. *International Journal of Pharmaceutics, 393*(1–2), 6–15. https://doi.org/10.1016/j.ijpharm.2010.03.012

Rieger, M. M., & Rhein, L. D. (Eds.). (2020). *Surfactants in cosmetics* (3rd ed.). CRC Press.

Roberts, M. S., & Cross, S. E. (2002). Skin transport. In Z. D. Draelos & L. A. Thaman (Eds.), *Cosmetic formulation of skin care products* (pp. 89–100). Taylor & Francis.

SCCS (Scientific Committee on Consumer Safety). (2021). *Notes of guidance for the testing of cosmetic ingredients and their safety evaluation* (11th ed.). European Commission.

Schueller, R., & Romanowski, P. (2021). *Conditioning agents for hair and skin* (2nd ed.). Allured Publishing.

Stevenson, D. E., & Haskell-Ramsey, C. (2016). Flavonoids and their role as antioxidants in cosmetics. *Phytotherapy Research, 30*(8), 1230–1240. https://doi.org/10.1002/ptr.5643

Tadros, T. F. (2018). *Emulsion formation and stability* (2nd ed.). Wiley-VCH.

U.S. Food and Drug Administration. (2022). *Cosmetics labeling guide*. https://www.fda.gov/cosmetics

Zhang, Z., & Michniak-Kohn, B. (2012). Nanoparticles and skin penetration. *Journal of Controlled Release, 153*(1), 40–48. https://doi.org/10.1016/j.jconrel.2011.05.026

Chapter 31

Adleman, L. M. (1994). Molecular computation of solutions to combinatorial problems. *Science, 266*(5187), 1021–1024. https://doi.org/10.1126/science.7973651

Balzani, V., Credi, A., & Venturi, M. (2008). *Molecular devices and machines: Concepts and perspectives for the nanoworld* (2nd ed.). Wiley-VCH.

Benenson, Y. (2012). Biomolecular computing systems: Principles, progress and potential. *Nature Reviews Genetics, 13*(7), 455–468. https://doi.org/10.1038/nrg3197

De Silva, A. P., Gunaratne, H. Q. N., Gunnlaugsson, T., Huxley, A. J. M., McCoy, C. P., Rademacher, J. T., & Rice, T. E. (1997). Signaling recognition events with fluorescent sensors and switches. *Chemical Reviews, 97*(5), 1515–1566. https://doi.org/10.1021/cr960386p
(Seminal work in molecular logic gates using supramolecular chemistry.)

De Silva, A. P., Uchiyama, S. (2007). Molecular logic and computing. *Nature Nanotechnology, 2*(7), 399–410. https://doi.org/10.1038/nnano.2007.170

Ding, Y., & Seelig, G. (2010). Analysis of digital logic circuits in biomolecular computation. *Nature Physics, 6*(6), 408–412. https://doi.org/10.1038/nphys1617

Katz, E., & Privman, V. (2010). Enzyme-based logic systems for information processing. *Chemical Society Reviews, 39*(5), 1835–1857. https://doi.org/10.1039/b906708c

Katz, E. (Ed.). (2012). *Molecular and supramolecular information processing: From molecular switches to logic systems.* Wiley-VCH.

Kubiak, C. P., & Stoddart, J. F. (2012). MIMICs: Mechanically interlocked molecular information carriers. *Chemical Society Reviews, 41*(16), 5645–5664. https://doi.org/10.1039/c2cs35177a

Ladder, J., & Winfree, E. (2000). Turing-universal chemical reaction networks. *Journal of Theoretical Biology, 205*(4), 543–557. https://doi.org/10.1006/jtbi.2000.2078
(Foundational work showing reaction networks can perform universal computation.)

Ornes, S. (2017). Computing with chemistry. *Proceedings of the National Academy of Sciences, 114*(21), 5285–5288. https://doi.org/10.1073/pnas.1705076114

Qian, L., & Winfree, E. (2011). Scaling up digital circuit computation with DNA strand-displacement cascades. *Science, 332*(6034), 1196–1201. https://doi.org/10.1126/science.1200520

Reif, J. H. (2002). *DNA computing: Models and designs.* Springer.
(A foundational monograph for sequence-based computation.)

Stojanovic, M. N., Mitchell, T. E., & Stefanovic, D. (2002). Deoxyribozyme-based molecular automaton. *Journal of the American Chemical Society, 124*(14), 3555–3561. https://doi.org/10.1021/ja017376x

Szaciłowski, K. (2008). Digital information processing in molecular systems.

Chemical Reviews, 108(9), 3481–3548. https://doi.org/10.1021/cr068405x *(Comprehensive review of molecular logic and computing.)*

Tolley, M. T., & Olson, E. J. (2019). Chemical reaction networks: Toward biological and synthetic computation. *Accounts of Chemical Research, 52*(5), 1260–1268. https://doi.org/10.1021/acs.accounts.8b00644

Toth, R., & Showalter, K. (2007). Logic gates in excitable chemical media. *Journal of Chemical Physics, 127*(5), 054707. https://doi.org/10.1063/1.2752172 *(Demonstrates logic using BZ reaction–diffusion dynamics.)*

Chapter 32

Becker, H., & Wang, Z. (2019). *High-pressure chemistry: Effects on structure and bonding of materials.* Chemical Reviews, 119(3), 907–932. https://doi.org/10.1021/acs.chemrev.8b00430

Bibring, J.-P., & Lunine, J. (Eds.). (2018). *Planetary exploration and astrobiology.* Oxford University Press.

Buseck, P. R., Tschauner, O., & Huang, S. (2020). *Mineralogy and chemistry at extreme conditions.* American Mineralogist, 105(4), 449–465. https://doi.org/10.2138/am-2020-7051

Cavicchioli, R. (2016). *Extremophiles and the limits of life.* Astrobiology, 16(3), 221–223. https://doi.org/10.1089/ast.2016.29002.rca

Cavicchioli, R., Baker, P. W., & Charlesworth, J. (2019). *Extremophiles: Diversity, adaptation, and applications.* F1000Research, 8, 1910. https://doi.org/10.12688/f1000research.20569.1

Chopra, A., & Lineweaver, C. H. (2016). *The habitability of icy moons.* Annual Review of Earth and Planetary Sciences, 44, 21–45. https://doi.org/10.1146/annurev-earth-060115-012647

Glein, C. R., Waite, J. H., & Postberg, F. (2018). *The geochemistry of Enceladus: Composition and structure of the plume and ocean.* In O. Grasset et al. (Eds.), *Enceladus and the icy moons of Saturn* (pp. 297–336). University of Arizona Press.

Gibbs, J. W., & Chidsey, C. E. (2017). *Supercritical water chemistry: Principles and applications.* Accounts of Chemical Research, 50(7), 1701–1709. https://doi.org/10.1021/acs.accounts.7b00122

Gross, J. (2018). *Hydrothermal vent chemistry and the origin of life*. Nature Reviews Chemistry, 2, 0101. https://doi.org/10.1038/s41570-018-0101

Holt, K. A., & Stevenson, K. (2020). *Radiation chemistry in nuclear systems: Mechanisms and materials responses*. Progress in Nuclear Energy, 123, 103317. https://doi.org/10.1016/j.pnucene.2020.103317

Kawamura, K., & Saito, A. (2020). *Chemistry of icy interstellar grains and prebiotic molecules*. Chemical Society Reviews, 49(24), 9020–9047. https://doi.org/10.1039/D0CS00433C

McCollom, T. M. (2016). *Serpentinization, hydrogen generation, and early Earth environments*. Geochimica et Cosmochimica Acta, 179, 240–262. https://doi.org/10.1016/j.gca.2016.01.036

Merino, N., Aronson, H. S., Bojanova, D. P., Feyhl-Buska, J., Wong, M. L., Zhang, S., & Giovannelli, D. (2019). *Living at the extremes: Extremophiles and the limits of life in a planetary context*. Frontiers in Microbiology, 10, 780. https://doi.org/10.3389/fmicb.2019.00780

Ohtani, E. (2020). *Deep water cycle in Earth: Role of high-pressure minerals and subduction*. Elements, 16(6), 413–418. https://doi.org/10.2138/gselements.16.6.413

Parnell, J. (2018). *Radiolysis-powered life in the deep subsurface*. Astrobiology, 18(9), 1291–1302. https://doi.org/10.1089/ast.2017.1716

Porter, S. M., & Stern, J. C. (2022). *Perchlorate chemistry on Mars: Implications for habitability*. Planetary and Space Science, 214, 105530. https://doi.org/10.1016/j.pss.2022.105530

Price, P. B. (2007). *Microbial life in glacial ice and implications for astrobiology*. Proceedings of the National Academy of Sciences, 104(3), 827–828. https://doi.org/10.1073/pnas.0610735104

Sanchez-Lavega, A., & Lebonnois, S. (2018). *Venus atmospheric chemistry and dynamics*. In L. F. Wilson (Ed.), *Venus: The atmosphere, geology, and climate* (pp. 55–86). Cambridge University Press.

Shock, E. L., & Canovas, P. (2010). *Geochemical constraints on the origin of life*. Elements, 6(1), 21–26. https://doi.org/10.2113/gselements.6.1.21

Span, R., & Wagner, W. (1996). *A new equation of state for carbon dioxide covering the fluid region from the triple-point temperature to 1100 K at pressures up to 800*

MPa. Journal of Physical and Chemical Reference Data, 25(6), 1509–1596. https://doi.org/10.1063/1.555991

Tian, F., & Kasting, J. F. (2019). *Upper atmospheric chemistry under extreme radiation*. Journal of Geophysical Research: Planets, 124(1), 77–94. https://doi.org/10.1029/2018JE005780

Totten, G. E., & MacKenzie, D. S. (Eds.). (2018). *Handbook of high-temperature materials science*. CRC Press.

Chapter 33

Al-Khalili, J., & McFadden, J. (2014). *Life on the edge: The coming of age of quantum biology*. Crown Publishers.

Arndt, M., Juffmann, T., & Vedral, V. (2009). Quantum physics meets biology. *HFSP Journal, 3*(6), 386–400. https://doi.org/10.2976/1.3244985

Ball, P. (2011). Physics of life: The dawn of quantum biology. *Nature, 474*(7351), 272–274. https://doi.org/10.1038/474272a

Benderskii, V. A., Makarov, D. E., & Wight, C. A. (1994). *Chemical dynamics at low temperatures*. Wiley.

Berrisford, J. M., Blackburn, N. J., & Richards, J. H. (2013). Hydrogen tunneling in biology. *Current Opinion in Chemical Biology, 17*(5), 761–767. https://doi.org/10.1016/j.cbpa.2013.07.007

Boxer, S. G. (2009). Mechanisms of charge transfer in biological systems. *Annual Review of Biochemistry, 78*, 35–60. https://doi.org/10.1146/annurev.biochem.78.070907.103350

Cao, J., & Silbey, R. J. (2009). Optimization of exciton trapping in energy transfer processes. *The Journal of Physical Chemistry A, 113*(50), 13825–13838. https://doi.org/10.1021/jp908408s

Engel, G. S., Calhoun, T. R., Read, E. L., Ahn, T. K., Mančal, T., Cheng, Y. C., Blankenship, R. E., & Fleming, G. R. (2007). Evidence for wavelike energy transfer through quantum coherence in photosynthetic systems. *Nature, 446*(7137), 782–786. https://doi.org/10.1038/nature05678

Fischer, G., & Pulay, P. (1993). Proton tunneling and tautomerization in DNA base pairs. *Chemical Physics Letters, 205*(1–4), 29–35. https://doi.org/10.1016/0009-2614(93)85004-H

Fleming, G. R., & Scholes, G. D. (2004). Physical chemistry: Quantum mechanics for plants. *Nature, 431*(7009), 256–257. https://doi.org/10.1038/431256a

Gallagher, R., & Buckman, B. (2018). *Quantum biology: The hidden nature of life*. Firefly Books.

Gauger, E. M., Rieper, E., Morton, J. J. L., Benjamin, S. C., & Vedral, V. (2011). Sustained quantum coherence and entanglement in the avian compass. *Physical Review Letters, 106*(4), 040503. https://doi.org/10.1103/PhysRevLett.106.040503

Hammes-Schiffer, S. (2015). Proton-coupled electron transfer: Moving together and charging forward. *The Journal of the American Chemical Society, 137*(28), 8860–8871. https://doi.org/10.1021/jacs.5b04034

Huelga, S. F., & Plenio, M. B. (2013). Vibrations, quanta and biology. *Contemporary Physics, 54*(4), 181–207. https://doi.org/10.1080/00107514.2013.803516

Lambert, N., Chen, Y.-N., Cheng, Y.-C., Li, C.-M., Chen, G.-Y., & Nori, F. (2013). Quantum biology. *Nature Physics, 9*(1), 10–18. https://doi.org/10.1038/nphys2474

McFadden, J., & Al-Khalili, J. (1999). A quantum mechanical model of adaptive mutation. *BioSystems, 50*(3), 203–211. https://doi.org/10.1016/S0303-2647(99)00013-6

Mohseni, M., Rebentrost, P., Lloyd, S., & Aspuru-Guzik, A. (2008). Environment-assisted quantum walks in photosynthetic energy transfer. *The Journal of Chemical Physics, 129*(17), 174106. https://doi.org/10.1063/1.3002335

Plenio, M. B., & Huelga, S. F. (2008). Dephasing-assisted transport: Quantum networks and biomolecules. *New Journal of Physics, 10*(11), 113019. https://doi.org/10.1088/1367-2630/10/11/113019

Ritz, T., Adem, S., & Schulten, K. (2000). A model for photoreceptor-based magnetoreception in birds. *Biophysical Journal, 78*(2), 707–718. https://doi.org/10.1016/S0006-3495(00)76629-X

Turin, L. (1996). A spectroscopic mechanism for primary olfactory reception. *Chemical Senses, 21*(6), 773–791. https://doi.org/10.1093/chemse/21.6.773

van der Kamp, M. W., & Mulholland, A. J. (2013). Combined quantum mechanics/molecular mechanics (QM/MM) methods in computational enzymology. *Biochemistry, 52*(16), 2708–2728. https://doi.org/10.1021/bi400215w

Whaley, K. B. (2014). Quantum biology: Better living through quantum mechanics? *Nature, 512*(7515), 375–376. https://doi.org/10.1038/512375a

Woodward, J. R. (2015). Radical pairs in magnetic fields. *Progress in Reaction Kinetics and Mechanism, 40*(1), 1–40. https://doi.org/10.3184/146867815X14270566611078

Chapter 34

Abdelrahman, Y., & Rohe, B. (2023). *Self-healing polymers: Chemistry, design strategies, and applications.* Progress in Polymer Science, 140, 101678. https://doi.org/10.1016/j.progpolymsci.2023.101678

Bergman, S. D., & Wudl, F. (2008). Mendable polymers. *Journal of Materials Chemistry, 18*(1), 41–62. https://doi.org/10.1039/B713953P

Chen, X., Dam, M. A., Ono, K., Mal, A., Shen, H., Nutt, S. R., Sheran, K., & Wudl, F. (2002). A thermally re-mendable cross-linked polymeric material. *Science, 295*(5560), 1698–1702. https://doi.org/10.1126/science.1065879

Cordier, P., Tournilhac, F., Soulié-Ziakovic, C., & Leibler, L. (2008). Self-healing and thermoreversible rubber from supramolecular assembly. *Nature, 451*(7181), 977–980. https://doi.org/10.1038/nature06669

Denissen, W., Winne, J. M., & Du Prez, F. E. (2016). Vitrimers: Permanent organic networks with glass-like fluidity. *Chemical Science, 7*(1), 30–38. https://doi.org/10.1039/C5SC02223A

Ducrot, E., & Creton, C. (2016). Self-healing in supramolecular polymers. *Smart Materials and Structures, 25*(10), 103001. https://doi.org/10.1088/0964-1726/25/10/103001

Garcia, S. J. (2014). Self-healing polymer systems: Properties, synthesis and applications. *Journal of Materials Chemistry A, 2*(10), 19344–19361. https://doi.org/10.1039/C4TA02841J

Hager, M. D., Greil, P., Leyens, C., van der Zwaag, S., & Schubert, U. S. (2010). Self-healing materials. *Advanced Materials, 22*(47), 5424–5430. https://doi.org/10.1002/adma.201003036

Li, C., Ma, S., & Xu, Y. (2021). Nanocellulose-reinforced dynamic covalent polymer networks: Toward strong, healable, and sustainable composites. *ACS Sustainable Chemistry & Engineering, 9*(27), 9150–9166. https://doi.org/10.1021/acssuschemeng.1c02516

Liu, Y., & Zhang, H. (2019). Dynamic covalent chemistry in polymer networks: A molecular toolkit for self-healing materials. *Accounts of Chemical Research, 52*(4), 1156–1166. https://doi.org/10.1021/acs.accounts.9b00024

Montarnal, D., Capelot, M., Tournilhac, F., & Leibler, L. (2011). Silica-like malleable materials from permanent organic networks. *Science, 334*(6058), 965–968. https://doi.org/10.1126/science.1212648

Omonov, T. S., Koo, J. M., Ilyin, S. O., & Park, J. W. (2020). Supramolecular polymers with UPy hydrogen bonding for self-healing elastomers. *Polymer, 208*, 122949. https://doi.org/10.1016/j.polymer.2020.122949

Toohey, K. S., Sottos, N. R., Lewis, J. A., Moore, J. S., & White, S. R. (2007). Self-healing materials with microvascular networks. *Nature Materials, 6*(8), 581–585. https://doi.org/10.1038/nmat1934

Van Gemert, G. M. L., Peeters, J. W., Sijbesma, R. P., & Meijer, E. W. (2012). Biological inspiration for autonomous self-healing polymers: Supramolecular UPy motifs. *Macromolecular Chemistry and Physics, 213*(2), 234–242. https://doi.org/10.1002/macp.201100447

Wang, S., Urban, M. W. (2020). Self-healing polymers. *Chemical Society Reviews, 49*(20), 7253–7278. https://doi.org/10.1039/D0CS00342E

Wang, Z., Zhang, Y., & Yang, J. (2019). Dual-dynamic crosslinked polymers toward self-healing and reprocessable materials. *Advanced Functional Materials, 29*(37), 1902392. https://doi.org/10.1002/adfm.201902392

White, S. R., Sottos, N. R., Geubelle, P. H., Moore, J. S., Kessler, M. R., Sriram, S. R., Brown, E. N., & Viswanathan, S. (2001). Autonomic healing of polymer composites. *Nature, 409*(6822), 794–797. https://doi.org/10.1038/35057232

Yang, Y., Davydovich, D., Kelly, C. N., White, S. R., Sottos, N. R., & Moore, J. S. (2020). Strain-induced mechanochemical activation enables self-healing polymers. *Nature Chemistry, 12*(5), 482–490. https://doi.org/10.1038/s41557-020-0457-4

Zhao, Q., Zou, W., Luo, Y., & Xie, T. (2016). Covalent adaptive networks: A unique paradigm of dynamic covalent chemistry. *Macromolecules, 49*(18), 7079–7086. https://doi.org/10.1021/acs.macromol.6b01429

Chapter 35

Brereton, R. G. (2018). *Chemometrics: Data analysis for the laboratory and chemical plant* (3rd ed.). John Wiley & Sons.

Kowalski, B. R., & Bender, C. F. (2017). Chemometrics in process analytical technology. *Analytical Chemistry, 89*(1), 34–45. https://doi.org/10.1021/acs.analchem.6b04567

Gemperline, P. (2021). *Practical guide to chemometrics* (2nd ed.). CRC Press.

Lavine, B. K., & Workman, J. (2018). Chemometrics in spectroscopy. *Annual Review of Analytical Chemistry, 11*, 13–34. https://doi.org/10.1146/annurev-anchem-061417-125150

Massart, D. L., Vandeginste, B. G. M., Buydens, L. M. C., De Jong, S., Lewi, P. J., & Smeyers-Verbeke, J. (1997). *Handbook of chemometrics and qualimetrics: Part A.* Elsevier.

Eriksson, L., Johansson, E., Kettaneh-Wold, N., Trygg, J., Wikström, C., & Wold, S. (2006). *Multi- and megavariate data analysis: Part I: Basic principles and applications* (2nd ed.). Umetrics AB.

Wold, S., Sjöström, M., & Eriksson, L. (2001). PLS-regression: A basic tool of chemometrics. *Chemometrics and Intelligent Laboratory Systems, 58*(2), 109–130. https://doi.org/10.1016/S0169-7439(01)00155-1

Chapter 36

Simons, M. A. J., Lamberts, T., & Cuppen, H. M. (2020). *Formation of COMs through CO hydrogenation on interstellar grains. Astronomy & Astrophysics, 629*, A28. A&A+1

Cuppen, H. M., & collaborators. (2017). *Grain Surface Models and Data for Astrochemistry. Space Science Reviews.* SpringerLink

Jin, M., & Garrod, R. T. (2020). *Formation of complex organic molecules in cold interstellar environments through non-diffusive grain-surface and ice-mantle chemistry. The Astrophysical Journal (preprint).* arXiv

Mathew, T., & collaborators. (2022). *Methanol in the RNA world: An astrochemical perspective. Frontiers in Astronomy and Space Sciences, 9*, 809928. Frontiers+1

Li, J., Shen, Z., Wang, J., Chen, X., Li, D., & others. (2017). *Widespread presence of glycolaldehyde and ethylene glycol around Sagittarius B2. arXiv:1709.10247 [astro-ph.GA].* arXiv+1

Rivilla, V. M., et al. (2023). *First glycine isomer detected in the interstellar medium: glycolamide ($NH_2C(O)CH_2OH$). The Astrophysical Journal Letters, 953*(L20). AAS Nova+1

Furuya, K. (2017). *Isotopic fractionation in interstellar molecules. Proceedings of IAU Symposium 332: Astrochemistry VII: Through the Cosmos from Galaxies to Planets.* arXiv:1711.04406 [astro-ph.GA]. arXiv

Sandford, S. A., Allamandola, L. J., & collaborators. (2002). *Interstellar processes leading to molecular deuterium enrichment. Planetary and Space Science, 50*, 1145–1154. Astrochem+1

Belloche, A., & collaborators. (2016). *Deuterated complex organic molecules in Sagittarius B2(N2). Astronomy & Astrophysics, 59 (2016).* A&A+1

Ceccarelli, C. (2023). *Spiers Memorial Lecture: Astrochemistry at high resolution. Chemical Reviews (Perspective).* RSC Publishing+1

Balucani, N., Ceccarelli, C., & Taquet, V. (2015). *Formation of complex organic molecules in cold objects: the role of gas-phase reactions. The Astrophysical Journal, 813*(1). arXiv:1501.03668. arXiv+1

Chapter 37

Mukhopadhya, I., & Louis, P. (2025). Gut microbiota-derived short-chain fatty acids and their role in human health and disease. *Nature Reviews Microbiology, 23*(5), 635–651. https://doi.org/10.1038/s41579-025-01183-w Nature

Messengers from the gut: Gut microbiota-derived metabolites on host regulation. (2022). *Cellular & Molecular Life Sciences, 79*(6), 1–17. https://doi.org/10.1007/s00018-022-04123-8 PMC

Quinn, R. A. (2021). Microbial transformations of human bile acids. *Microbiome, 9*, 140. https://doi.org/10.1186/s40168-021-01101-1 SpringerLink

Human gut microbiota plays a role in the metabolism of drugs. (2016). *Proceedings of the National Academy of Sciences, 113*(25), 707–715. https://doi.org/10.1073/pnas.1524645113 PubMed+1

Zeng, M., et al. (2025). Gut microbial metabolites modulate host immune responses via receptor-mediated and epigenetic pathways. *Frontiers in Microbiology, 16*, 1675065. https://doi.org/10.3389/fmicb.2025.1675065

Chapter 38

BHC Company. (n.d.). Presidential Green Chemistry Challenge: 1997 Greener Synthetic Pathways Award — BHC Company Ibuprofen Process. U.S. EPA. US EPA

Göller, A., & Löwe, H. (2023). Green-chemistry routes enable sustainable production — The BHC Company ibuprofen synthesis (3-step process) vs. the six-step Boots process. *Green Chemistry*, *25*, 1–14. RSC Publishing

Drumright, R. E., Gruber, P. R., & Henton, D. E. (2000). Poly(lactic acid) technology. In *Advanced Materials and Technologies for Plastics Recycling* (pp. 92–105). Elsevier. (as cited in foundational reviews of PLA's sustainability)

Dar, E., & Kim, Y. (2021). The diffusion of bioplastics: What can we learn from poly(lactic acid)? *Sustainability*, 15(6). https://doi.org/10.3390/su15064699 MDPI

Zeng, L., & Chen, H. (2021). Poly(lactic acid): A versatile biobased polymer for the future with multifunctional properties — From monomer synthesis, polymerization techniques and molecular-weight increase to PLA applications. *Polymers*, 13, 1–29. https://doi.org/10.3390/polym13213784 PMC

Yao, X.-R., Cui, J.-W., Jia, M.-Z., & Zhang, J. (2025). Photoredox catalysis with pyridinium and structurally related onium salts: Sustainable synthesis via reactive oxygen species and electron transfer. *Green Chemistry*. https://doi.org/10.1039/D5GC03738G RSC Publishing

Báez, R. R., & Smith, P. T. (2024). Efficient catalytic oxidation of biomass to formic acid coupled with low-energy formaldehyde production from methanol. *Green Chemistry*, 26, 2170–2182. https://doi.org/10.1039/D3GC03892K RSC Publishing

Neyerlin, K. C., Hu, L., & colleagues. (2024). A scalable membrane electrode assembly architecture for efficient electrochemical conversion of CO_2 to formic acid. *Nature Communications*. https://doi.org/10.1038/s41467-024-00000-w NREL+1

Chapter 39

Holliman Jr., J. E., Schaef, H. T., McGrail, B. P., & Miller, Q. R. S. (2022). Review of foundational concepts and emerging directions in metamaterial research: Design, phenomena, and applications. *Materials Advances, 3*, 8390-8406. https://doi.org/10.1039/D2MA00497F RSC Publishing

Liu, Y., & Zhang, X. (2011). Metamaterials: A new frontier of science and technology.

Chemical Society Reviews, 40(5), 2494–2507. https://doi.org/10.1039/C0CS00184H RSC Publishing+1

Liang, T.-W., Chen, C., Kusaka, S., Siddique, S. K., Chang, C.-Y., Matsuda, R., & Ho, R.-M. (2025). Mesoporous metal–organic framework from templated synthesis as mechanical metamaterials. *Journal of the American Chemical Society.* Advance online publication. https://doi.org/10.1021/jacs.5c04214 PMC+2PubMed+2

(2023, December 7). Metamaterials and origamic metal–organic frameworks. *Phys.org.* https://phys.org/news/2023-12-metamaterials-origamic-metal-organic-frameworks.html Phys.org

Authors. (2025). The potential of covalent organic frameworks to act as mechanical metamaterials. *Computational Materials Science, 248*, Article 113555. https://doi.org/10.1016/j.commatsci.2024.113555 ScienceDirect

Authors. (Year). Biomolecule-based optical metamaterials: Design and applications. *[Journal].* (As of access, publication details as in source.)

Chapter 40

Bear, M. F., Connors, B. W., & Paradiso, M. A. (2020). *Neuroscience: Exploring the brain* (4th ed.). Wolters Kluwer.

Berridge, M. J. (2016). The inositol trisphosphate/calcium signaling pathway in health and disease. *Physiological Reviews, 96*(4), 1261–1296. https://doi.org/10.1152/physrev.00006.2016

Bliss, T. V. P., & Collingridge, G. L. (2013). Expression of NMDA receptor-dependent LTP in the hippocampus: Bridging the divide. *Molecular Brain, 6*(5). https://doi.org/10.1186/1756-6606-6-5

Fuxe, K., Agnati, L. F., & Borroto-Escuela, D. O. (2010). Volume transmission in central dopamine and serotonin neurons and its relevance to neuropsychopharmacology. *CNS & Neurological Disorders - Drug Targets, 9*(2), 153–165. https://doi.org/10.2174/187152710791012631

Gage, F. H., & Temple, S. (2013). Neural stem cells: Generating and regenerating the brain. *Neuron, 80*(3), 588–601. https://doi.org/10.1016/j.neuron.2013.10.037

Huang, E. J., & Reichardt, L. F. (2001). Neurotrophins: Roles in neuronal development and function. *Annual Review of Neuroscience, 24*, 677–736. https://doi.org/10.1146/annurev.neuro.24.1.677

Kandel, E. R., Dudai, Y., & Mayford, M. R. (2014). The molecular and systems biology of memory. *Cell, 157*(1), 163–186. https://doi.org/10.1016/j.cell.2014.03.001

Purves, D., Augustine, G. J., Fitzpatrick, D., Hall, W. C., LaMantia, A.-S., Mooney, R. D., ... White, L. E. (2018). *Neuroscience* (6th ed.). Oxford University Press.

Semyanov, A., & Kullmann, D. M. (2000). Modulation of GABAergic signaling among interneurons by GABA transporters. *Nature Neuroscience, 3*(3), 608–614. https://doi.org/10.1038/75705

Südhof, T. C. (2018). Towards an understanding of synapse formation. *Neuron, 100*(2), 276–293. https://doi.org/10.1016/j.neuron.2018.09.040

Zhou, Y., Danbolt, N. C., & Kanner, B. I. (2019). Glutamate and GABA transporters: Structure, function, and regulation. *Progress in Neurobiology, 179*, 101614. https://doi.org/10.1016/j.pneurobio.2019.101614

Chapter 41

Belitz, H.-D., Grosch, W., & Schieberle, P. (2009). *Food chemistry* (4th ed.). Springer.

Brennan, C. S., & Tudorica, C. M. (2008). Food hydrocolloids: Functional and technological properties. *International Journal of Food Science & Technology, 43*(6), 1021–1031. https://doi.org/10.1111/j.1365-2621.2008.01603.x

Ciecierski, W. (2012). *Molecular gastronomy: Exploring the science of flavor*. Wiley-Blackwell.

McGee, H. (2004). *On food and cooking: The science and lore of the kitchen* (2nd ed.). Scribner.

Reinhart, P. (2006). *The bread baker's apprentice: Mastering the art of extraordinary bread*. Ten Speed Press.

Shahidi, F., & Zhong, Y. (2010). Lipid oxidation: Measurement methods and control. *Food Science & Nutrition, 50*(1), 1–21. https://doi.org/10.1111/j.1745-4557.2010.00320.x

Velikov, K. P., & Pelan, E. G. (2008). Colloidal delivery systems for functional foods. *Current Opinion in Colloid & Interface Science, 13*(3), 213–227. https://doi.org/10.1016/j.cocis.2008.06.005

Wikström, M. (2015). *The molecular gastronomy guide: Chemistry and physics in the kitchen*. Royal Society of Chemistry.

Zhang, Y., & Zhang, L. (2018). Enzyme applications in modern gastronomy: Texture modification and flavor development. *Trends in Food Science & Technology, 79*, 151–160. https://doi.org/10.1016/j.tifs.2018.07.003

Chapter 42

Hartmann, J., et al. (2025). Review and syntheses: Ocean alkalinity enhancement and carbon dioxide removal through marine enhanced rock weathering using olivine. *Biogeosciences*, 22, 355–??. Biogeosciences+1

Meng, H., Xie, Z., & others. (2022). Mineralization and CO_2 capturing methods over the last two decades. *International Journal of Environmental Science and Technology*. SpringerLink

Stoffel, B., Meng, H., & Xie, Z. (2022). Mineralization as a carbon dioxide sequestration strategy: chemical pathways for silicate and oxide minerals. *International Journal of Environmental Science and Technology*. SpringerLink

Duan, H., & others. (2025). A Review of Materials for Carbon Dioxide Capture. *Catalysts*, 15(3), 273. MDPI

Guo, M., & others. (2025). Biochar: A carbon dioxide removal strategy — current knowledge, mechanisms, and future perspectives. *Biochar*, 9(3), 67. MDPI+1

Woolf, D., Lehmann, J., Lee, D., et al. (2010). Sustainable biochar to mitigate global climate change. *Nature Communications*, 1, 56. (Estimates net CO_2-C mitigation potential over century scale.) Nature
Jones, A. C., Haywood, J. M., & Jones, A. (2016). Climatic impacts of stratospheric geoengineering with sulfate, black carbon and titania injection. *Atmospheric Chemistry and Physics*, 16, 2843–2862. ACP

Visioni, D., Quaglia, I., Pitari, G., & Kravitz, B. (2022). An approach to sulfate geoengineering with surface emissions of carbonyl sulfide. *Atmospheric Chemistry and Physics*, 22, 5757–5773. ACP

Author(s). (2025). Shortwave absorption by alumina aerosol amplifies stratospheric warming. arXiv. arXiv

Author(s). (2025). Stratospheric aerosol injection can weaken the carbon dioxide greenhouse effect. *Communications Earth & Environment*. Nature+1

Chapter 43

Yuan, Y., Li, Y., & Zhao, J. (2018). *Development on thermochemical energy storage based on CaO-based materials: A review. Sustainability, 10*(8), 2660. https://doi.org/10.3390/su10082660 MDPI

Bhojani, S., et al. (2025). Metal-oxide thermochemical energy storage systems: trends, materials and applications. *Frontiers in Energy Research.* https://doi.org/10.3389/fenrg.2025.1651471 Frontiers

Wang, F., Li, Y., Qiu, T., et al. (2022). *Computational design of quinone electrolytes for redox flow batteries using high-throughput machine learning and theoretical calculations. Frontiers in Chemical Engineering.* https://doi.org/10.3389/fceng.2022.1086412 Frontiers

Duan, J., et al. (2025). *Multi-redox organic species for high-energy-density aqueous redox flow batteries: Mechanism, characteristics, and applications. Energy Storage Materials, 83*, 104684. https://doi.org/10.1016/j.ensm.2025.104684 ScienceDirect

Lin, K., et al. (Year unspecified). *Organic electroactive materials for aqueous redox flow batteries. Journal of Energy Storage.* (See review summary in PubMed.) PubMed+1

AuYeung, N., & Kreider, P. (2017). Solar thermochemical energy storage. *CEP Magazine.* (Special Section) AIChE

National Renewable Energy Laboratory (NREL). (2025). *Industrial energy storage review.* (NREL Publication)

Chapter 44

Tørring, T., Voigt, N. V., Nangreave, J., Yan, H., & Gothelf, K. V. (2011). DNA origami: a quantum leap for self-assembly of complex structures. *Chemical Society Reviews, 40*(12), 5636–5646. https://doi.org/10.1039/C1CS15057J RSC Publishing

Rothemund, P. W. K. (2006). Design of DNA origami. In Proceedings of the IEEE/ACM International Conference on Computer-Aided Design – ICCAD '05 (pp. …). [Original method for scaffolded DNA origami]. Caltech DNA+1

Douglas, S. M., Dietz, H., Liedl, T., Högberg, B., Graf, F., & Shih, W. M. (2009). Self-assembly of DNA into nanoscale three-dimensional shapes. *Nature*, 459, 414–418. (Often cited as the first demonstration of 3D DNA origami.) — as reviewed in overview articles. PubMed+2PMC+2

Andersen, E. S., Dong, M., Nielsen, M. M., Jahn, K., Subramani, R., Mamdouh, W., Golas, M. M., Sander, B., Stark, H., Oliveira, C. L., Leslie, A. G., Birkedal, V.,

Besenbacher, F., Gothelf, K. V., & Kjems, J. (2009). Self-assembly of a nanoscale DNA box with a controllable lid. *Nature*, 459, 73–76. (An early example of functional 3D DNA nanostructures enabling encapsulation and controlled access.) — as part of the broader "DNA origami → nanodevice / nanomedicine" narrative. PubMed+1

Sabouri, S., Sheng, J., Vilar, R., White, J., & Hart, S. (2025). Molecular Origami: Designing Functional Molecules of the Future. *Molecules, 30*(2), 242. https://doi.org/10.3390/molecules30020242 MDPI+1

Endo, M., & Sugiyama, H. (2018). DNA origami nanomachines. *Molecules, 23*(7), 1766. https://doi.org/10.3390/molecules23071766 MDPI

Junge, F., & Dietz, H. (2024). Economical and versatile subunit design principles for self-assembled DNA origami structures. *preprint arXiv:2411.09801.* (Demonstrating modular subunit design enabling a variety of self-assembled architectures.)

Chapter 45

Haddock, S. H. D., Moline, M. A., & Case, J. F. (2010). Bioluminescence in the sea. *Annual Review of Marine Science, 2*(1), 443–493. https://doi.org/10.1146/annurev-marine-120308-081028

Shimomura, O. (2006). *Bioluminescence: Chemical principles and methods*. World Scientific.

Branchini, B. R., Southworth, T. L., & Ruggiero, M. T. (2015). Firefly luciferase: A tool for molecular imaging. *Methods in Molecular Biology, 1316*, 1–23. https://doi.org/10.1007/978-1-4939-2742-5_1

Chudakov, D. M., Matz, M. V., Lukyanov, S., & Lukyanov, K. A. (2010). Fluorescent proteins and their applications in imaging living cells and tissues. *Physiological Reviews, 90*(3), 1103–1163. https://doi.org/10.1152/physrev.00038.2009

Hastings, J. W. (1983). Biological diversity, chemical mechanisms, and the evolutionary origins of bioluminescent systems. *Journal of Molecular Evolution, 19*(5), 309–321. https://doi.org/10.1007/BF01734363

Mezzanotte, L., van 't Root, M., Karatas, H., Goun, E. A., & Löwik, C. W. G. M. (2017). In vivo molecular bioluminescence imaging: New tools and applications. *Trends in*

Biotechnology, 35(7), 640–652. https://doi.org/10.1016/j.tibtech.2017.03.002

Tsien, R. Y. (1998). The green fluorescent protein. *Annual Review of Biochemistry, 67*, 509–544. https://doi.org/10.1146/annurev.biochem.67.1.509

Wilson, T., & Hastings, J. W. (1998). Bioluminescence: Living lights, lights for living. *Harvard University Press.*

Viviani, V. R. (2002). The origin, diversity, and structure-function relationships of insect luciferases. *Cellular and Molecular Life Sciences, 59*(11), 1833–1850. https://doi.org/10.1007/s00018-002-8582-0

Martini, S., & Strassert, J. F. (2021). Marine biofluorescence: Chemistry and applications in research and technology. *Marine Drugs, 19*(12), 665. https://doi.org/10.3390/md19120665

Chapter 46

Bains, W., Petkowski, J., & Seager, S. (2024). *Alternative solvents for life: framework for evaluation, current status and future research*. arXiv. arXiv

Chaabouni, H., Diana, S., Nguyen, T., & Dulieu, F. (2018). Thermal desorption of formamide and methylamine from graphite and amorphous water ice surfaces. *Astronomy & Astrophysics, 612*, A47. https://doi.org/10.1051/0004-6361/201731006 A&A

Kaspin-Powell, L. (2018, May 3). Why formamide may have been early life's alternative to water. Astrobiology News. NASA Astrobiology

Sánchez, I. E., Galpern, E. A., & Ferreiro, D. U. (2023). Solvent constraints for biopolymer folding and evolution in extraterrestrial environments. arXiv. arXiv

Schulze-Makuch, D., & Irwin, L. (n.d.). *Life in the Universe*. (Excerpts discussing non-aqueous solvents and hydrocarbon environments.) Old Mis+1

Vandevere, L. (n.d.). Non-aqueous life: The search for life beyond the comfort of water. *Journal of Student Research*. https://doi.org/10.47611/jsrhs.v12i3.4493 Journal of Student Research+1

Cowing, K. (2024, January 22). Alternative solvents for life: framework for evaluation, current status and future research. *Astrobiology.com*. (Summary of Bains et al., 2024) Astrobiology

Prebiotic organic chemistry of formamide and the origin of life in planetary conditions: what we know and what is the future. *Origins of Life and Evolution of Biospheres.* (Review article summarizing formamide-based prebiotic chemistry)

Chapter 47

Allis, C. D., Jenuwein, T., & Reinberg, D. (2023). *Epigenetics* (4th ed.). Cold Spring Harbor Laboratory Press.

Bannister, A. J., & Kouzarides, T. (2011). Regulation of chromatin by histone modifications. *Cell Research, 21*(3), 381–395. https://doi.org/10.1038/cr.2011.22

Chedin, F. (2011). The DNMT3 family of mammalian de novo DNA methyltransferases. *Progress in Molecular Biology and Translational Science, 101*, 255–285. https://doi.org/10.1016/B978-0-12-387685-0.00010-9

Dawson, M. A., & Kouzarides, T. (2012). Cancer epigenetics: From mechanism to therapy. *Cell, 150*(1), 12–27. https://doi.org/10.1016/j.cell.2012.06.013

Jones, P. A., & Takai, D. (2001). The role of DNA methylation in mammalian epigenetics. *Science, 293*(5532), 1068–1070. https://doi.org/10.1126/science.1063852

Kouzarides, T. (2007). Chromatin modifications and their function. *Cell, 128*(4), 693–705. https://doi.org/10.1016/j.cell.2007.02.005

Li, X., & Zhao, X. (2020). Epigenetic regulation of chromatin structure and function. *Nature Reviews Molecular Cell Biology, 21*, 179–196. https://doi.org/10.1038/s41580-019-0202-8

Liu, N., & Pan, T. (2016). N6-methyladenosine–encoded epitranscriptomics. *Nature Structural & Molecular Biology, 23*, 98–102. https://doi.org/10.1038/nsmb.3144

Smith, Z. D., & Meissner, A. (2013). DNA methylation: Roles in mammalian development. *Nature Reviews Genetics, 14*(3), 204–220. https://doi.org/10.1038/nrg3354

Wu, X., & Zhang, Y. (2017). TET-mediated active DNA demethylation: Mechanism, function, and beyond. *Nature Reviews Genetics, 18*, 517–534. https://doi.org/10.1038/nrg.2017.33

Yang, X., & Seto, E. (2008). Lysine acetylation: Codified crosstalk with other posttranslational modifications. *Molecular Cell, 31*(4), 449–461. https://doi.org/10.1016/j.molcel.2008.07.002

Zhao, B. S., Roundtree, I. A., & He, C. (2017). Post-transcriptional gene regulation by mRNA modifications. *Nature Reviews Molecular Cell Biology, 18*, 31–42. https://doi.org/10.1038/nrm.2016.132

Chapter 48

Peng, Z., Plum, A. M., Gagrani, P., & Baum, D. A. (2020). An ecological framework for the analysis of prebiotic chemical reaction networks and their dynamical behavior. *Journal of Theoretical Biology, 507*, 110451. https://doi.org/10.1016/j.jtbi.2020.110451 NSF Public Access Repository+2PubMed+2

Ashkenasy, G. (2023). Collectively autocatalytic sets. *Cell Reports Physical Science, 4*(10), 101594. https://doi.org/10.1016/j.xcrp.2023.101594 ScienceDirect

Joshi, B., & Craciun, G. (2020). Autocatalytic networks: An intimate relation between network topology and dynamics. *arXiv*. https://arxiv.org/abs/2006.01384 arXiv

Kriukov, D. V., Huskens, J., & Wong, A. S. Y. (2024). Exploring the programmability of autocatalytic chemical reaction networks. *Nature Communications, 15*, 8289. https://doi.org/10.1038/s41467-024-52649-z Nature+1

Loutchko, D. (2023). An algebraic characterization of self-generating chemical reaction networks using semigroup models. *Journal of Mathematical Biology, 86*, 76. https://doi.org/10.1007/s00285-023-01899-4 SpringerLink+1

Plum, A. M., Kempes, C. P., Peng, Z., & Baum, D. A. (2022). Spatial structure supports diversity in prebiotic autocatalytic chemical ecosystems. *npj Complexity*, (Article). https://doi.org/10.1038/s44260-025-00045-z Nature+1

Forrest, S., & … (2025). Maintaining an autocatalytic reaction system … *Physical Review E, 111*(014424). (2025).

Chapter 49

Dheyab, M. A., Aziz, A. A., & Jameel, M. S. (2021). Recent advances in inorganic nanomaterials synthesis using sonochemistry: A comprehensive review on iron oxide, gold and iron-oxide-coated gold nanoparticles. *Molecules, 26*(9), 2453. MDPI

Boldyreva, E. (2017). Chalcogenide mechanochemistry in materials science: insight into synthesis and applications. *Journal of Materials Science*. SpringerLink

Crawford, D. E. (2017). Solvent-free sonochemistry: Sonochemical organic synthesis in the absence of a liquid medium. *Beilstein Journal of Organic Chemistry, 13*, 1850–1856. https://doi.org/10.3762/bjoc.13.179 Beilstein Journals

Cravotto, G., et al. (2015). Advances in green organic sonochemistry. *Ultrasonics Sonochemistry, review* (2006–2016). PubMed

Fabritius, M., & others. (2025). Sonomechanochemical interplays: localized cavitation-induced stress and its overlap with mechanochemical force. *RSC Mechanochem*, 2(3), 325–490. RSC Publishing

Henglein, A. (1987). Sonochemistry: Basic effects and synthesis of nanomaterials and their application to environmental remediation and catalysis. *Ultrasonics Sonochemistry* (as discussed in reviews of sonochemical synthesis). PubMed+1

Ultrasonics Sonochemistry journal editors. (2023). Ultrasonics, sonochemistry and sonoprocessing: Editors' perspective. *Ultrasonics Sonochemistry*. PubMed

Xu, H., Zeiger, B. W., & Suslick, K. S. (2013). Sonochemical synthesis of nanomaterials: a tutorial review. *Chemical Society Reviews, 42*, 2555. suslick.scs.illinois.edu

Chapter 50

Larner, A. J. (2002). *Neurochemistry of consciousness: Neurotransmitters in mind. Brain, 125*(11), 2581–2582. https://doi.org/10.1093/brain/awf244 OUP Academic

Xiao-Jing, W. (2010). Neurophysiological and computational principles of cortical rhythms in cognition. *Physiological Reviews, 90*(3), 1195–1268. https://doi.org/10.1152/physrev.00035.2008 PMC

Pretel, J., Torres, J. J., & Marro, J. (2021). EEGs disclose significant brain activity correlated with synaptic fickleness. *Biology, 10*(7), Article 647. https://doi.org/10.3390/biology10070647 MDPI

Modolo, J., Hassan, M., Wendling, F., & Benquet, P. (2019). Decoding the circuitry of consciousness: From local microcircuits to brain-scale networks. *arXiv*. https://arxiv.org/abs/1907.11570 arXiv

StatPearls. (2023). Biochemistry, Glutamate. In *StatPearls [Internet]*. https://www.ncbi.nlm.nih.gov/books/NBK556062/ (see "Glutamate is the principal excitatory neurotransmitter … and acts on NMDA, AMPA, kainite, and metabotropic receptors"). NCBI+1

López-González, I., et al. (2024). Restoring consciousness with pharmacologic therapy: Mechanisms, targets, and future directions. *Neuroscience & Biobehavioral Reviews*. (Note: this article discusses how disruptions in excitatory/inhibitory neurotransmission — e.g., via glutamate and GABA — affect consciousness.) ScienceDirect

Ross, C., & Redpath, S. (2009). Physical foundations of consciousness: Brain organisation: the role of synapses. *arXiv*. https://arxiv.org/abs/0907.2192 arXiv

About the Author

Harminder Gill is a distinguished chemistry educator, academic mentor, and accomplished author, currently teaching advanced chemistry courses at the University of California, Los Angeles Extension Center. He holds a Bachelor of Science in Biological Chemistry with an Emphasis in Chemistry from the University of California, Riverside, and a Master's degree in Organic Chemistry from the University of California, Davis—where he is a life member of both alumni associations. His academic training combines rigorous scientific inquiry with a profound understanding of pedagogical methods, enabling him to cultivate excellence in students across a wide range of disciplines.

Over the course of his extensive career, Harminder has taught at nine different community colleges across Southern California, led comprehensive homeschooling programs, and provided online tutoring to students worldwide, reaching learners in over 30 U.S. states and Canada. He has designed and instructed courses spanning the sciences, humanities, social sciences, and standardized test preparation, demonstrating an exceptional ability to adapt complex material to diverse learning styles. Under his mentorship, countless students have achieved extraordinary outcomes, including top-tier scores on national examinations, acceptance into highly competitive universities, graduate programs, and specialized training academies, and the successful pursuit of advanced academic and professional goals. His students consistently cite his guidance as transformative, not only in their academic development but also in shaping their critical thinking, scientific curiosity, and lifelong learning habits.

Beyond his unparalleled instructional expertise, Harminder is an accomplished author in both science-fiction and cosmetics chemistry, merging creativity with scientific rigor. His science-fiction works explore speculative concepts rooted in real scientific principles, inspiring readers to engage with science in imaginative and meaningful ways. In cosmetics chemistry, he distills complex formulations and chemical interactions into accessible insights, advancing both scientific understanding and practical application for a broad readership.

Harminder is also deeply engaged in civic, cultural, and professional leadership. He has served on the Board of Directors for an art association in Corona, California, and on the Board of Library Trustees for Ward 4 in Riverside. A Distinguished Toastmaster (DTM), he exemplifies excellence in public speaking, leadership development, and organizational mentorship, further enhancing the personal and professional growth of his students and colleagues alike.

Committed to lifelong learning and scientific innovation, Harminder maintains a cutting-edge understanding of developments in chemistry, mathematics, physics, and technology. He is particularly noted for his ability to integrate interdisciplinary knowledge into teaching, writing, and research, inspiring students and readers to pursue ambitious intellectual and creative endeavors. His rare combination of global reach, pedagogical mastery, literary contribution, and unwavering dedication to student success has earned him recognition as an inspiring and transformative educator, shaping the next generation of leaders, scientists, and thinkers both nationally and internationally. Harminder Gill's career represents a unique synthesis of rigorous academic training, expansive teaching experience, literary creativity, and civic engagement. His relentless commitment to excellence in education, mentorship, and scientific communication continues to leave a lasting impact on students, colleagues, and readers worldwide, establishing him as one of the most respected and visionary figures in contemporary chemistry education and science communication.

www.ingramcontent.com/pod-product-compliance
Lightning Source LLC
LaVergne TN
LVHW062131200926
848484LV00007B/1800